Lecture Notes in Computer Science 12262

More information about this series at http://www.springer.com/series/7412

Anne L. Martel · Purang Abolmaesumi ·
Danail Stoyanov · Diana Mateus ·
Maria A. Zuluaga · S. Kevin Zhou ·
Daniel Racoceanu · Leo Joskowicz (Eds.)

Medical Image Computing and Computer Assisted Intervention – MICCAI 2020

23rd International Conference
Lima, Peru, October 4–8, 2020
Proceedings, Part II

 Springer

Editors
Anne L. Martel ⓘD
University of Toronto
Toronto, ON, Canada

Purang Abolmaesumi ⓘD
The University of British Columbia
Vancouver, BC, Canada

Danail Stoyanov ⓘD
University College London
London, UK

Diana Mateus ⓘD
École Centrale de Nantes
Nantes, France

Maria A. Zuluaga ⓘD
EURECOM
Biot, France

S. Kevin Zhou ⓘD
Chinese Academy of Sciences
Beijing, China

Daniel Racoceanu ⓘD
Sorbonne University
Paris, France

Leo Joskowicz ⓘD
The Hebrew University of Jerusalem
Jerusalem, Israel

ISSN 0302-9743　　　　　ISSN 1611-3349　(electronic)
Lecture Notes in Computer Science
ISBN 978-3-030-59712-2　　　ISBN 978-3-030-59713-9　(eBook)
https://doi.org/10.1007/978-3-030-59713-9

LNCS Sublibrary: SL6 – Image Processing, Computer Vision, Pattern Recognition, and Graphics

This Springer imprint is published by the registered company Springer Nature Switzerland AG
The registered company address is: Gewerbestrasse 11, 6330 Cham, Switzerland

Preface

The 23rd International Conference on Medical Image Computing and Computer-Assisted Intervention (MICCAI 2020) was held this year under the most unusual circumstances, due to the COVID-19 pandemic disrupting our lives in ways that were unimaginable at the start of the new decade. MICCAI 2020 was scheduled to be held in Lima, Peru, and would have been the first MICCAI meeting in Latin America. However, with the pandemic, the conference and its program had to be redesigned to deal with realities of the "new normal", where virtual presence rather than physical interactions among attendees, was necessary to comply with global transmission control measures. The conference was held through a virtual conference management platform, consisting of the main scientific program in addition to featuring 25 workshops, 8 tutorials, and 24 challenges during October 4–8, 2020. In order to keep a part of the original spirit of MICCAI 2020, SIPAIM 2020 was held as an adjacent LatAm conference dedicated to medical information management and imaging, held during October 3–4, 2020.

The proceedings of MICCAI 2020 showcase papers contributed by the authors to the main conference, which are organized in seven volumes of *Lecture Notes in Computer Science* (LNCS) books. These papers were selected after a thorough double-blind peer-review process. We followed the example set by past MICCAI meetings, using Microsoft's Conference Managing Toolkit (CMT) for paper submission and peer reviews, with support from the Toronto Paper Matching System (TPMS) to partially automate paper assignment to area chairs and reviewers.

The conference submission deadline had to be extended by two weeks to account for the disruption COVID-19 caused on the worldwide scientific community. From 2,953 original intentions to submit, 1,876 full submissions were received, which were reduced to 1,809 submissions following an initial quality check by the program chairs. Of those, 61% were self-declared by authors as Medical Image Computing (MIC), 6% as Computer Assisted Intervention (CAI), and 32% as both MIC and CAI. Following a broad call to the community for self-nomination of volunteers and a thorough review by the program chairs, considering criteria such as balance across research areas, geographical distribution, and gender, the MICCAI 2020 Program Committee comprised 82 area chairs, with 46% from North America, 28% from Europe, 19% from Asia/Pacific/Middle East, 4% from Latin America, and 1% from Australia. We invested significant effort in recruiting more women to the Program Committee, following the conference's emphasis on equity, inclusion, and diversity. This resulted in 26% female area chairs. Each area chair was assigned about 23 manuscripts, with suggested potential reviewers using TPMS scoring and self-declared research areas, while domain conflicts were automatically considered by CMT. Following a final revision and prioritization of reviewers by area chairs in terms of their expertise related to each paper, over 1,426 invited reviewers were asked to bid for the papers for which they had been suggested. Final reviewer allocations via CMT took account of reviewer bidding, prioritization of area chairs, and TPMS scores, leading to allocating about 4 papers per reviewer. Following an initial double-blind review phase by reviewers, area chairs provided a meta-review summarizing key points of reviews and a recommendation for each paper. The program chairs then evaluated the reviews and their scores, along with the recommendation from the area chairs, to directly accept 241 papers (13%) and reject 828 papers (46%); the remainder of the papers were sent for rebuttal by the authors. During the rebuttal phase, two additional area chairs were assigned to each paper using the CMT and TPMS scores while accounting for domain conflicts. The three area chairs then independently scored each paper to accept or reject, based on the reviews, rebuttal, and manuscript, resulting in clear paper decisions using majority voting. This process resulted in the acceptance of a further 301 papers for an overall acceptance rate of 30%. A virtual Program Committee meeting was held on July 10, 2020, to confirm the final results and collect feedback of the peer-review process.

For the MICCAI 2020 proceedings, 542 accepted papers have been organized into seven volumes as follows:

- Part I, LNCS Volume 12261: Machine Learning Methodologies
- Part II, LNCS Volume 12262: Image Reconstruction and Machine Learning
- Part III, LNCS Volume 12263: Computer Aided Intervention, Ultrasound and Image Registration
- Part IV, LNCS Volume 12264: Segmentation and Shape Analysis
- Part V, LNCS Volume 12265: Biological, Optical and Microscopic Image Analysis
- Part VI, LNCS Volume 12266: Clinical Applications
- Part VII, LNCS Volume 12267: Neurological Imaging and PET

For the main conference, the traditional emphasis on poster presentations was maintained; each author uploaded a brief pre-recorded presentation and a graphical abstract onto a web platform and was allocated a personal virtual live session in which they talked directly to the attendees. It was also possible to post questions online allowing asynchronous conversations – essential to overcome the challenges of a global conference spanning many time zones. The traditional oral sessions, which typically included a small proportion of the papers, were replaced with 90 "mini" sessions where all of the authors were clustered into groups of 5 to 7 related papers; a live virtual session allowed the authors and attendees to discuss the papers in a panel format.

We would like to sincerely thank everyone who contributed to the success of MICCAI 2020 and the quality of its proceedings under the most unusual circumstances of a global pandemic. First and foremost, we thank all authors for submitting and presenting their high-quality work that made MICCAI 2020 a greatly enjoyable and successful scientific meeting. We are also especially grateful to all members of the Program Committee and reviewers for their dedicated effort and insightful feedback throughout the entire paper selection process. We would like to particularly thank the MICCAI society for support, insightful comments, and continuous engagement with organizing the conference. Special thanks go to Kitty Wong, who oversaw the entire process of paper submission, reviews, and preparation of conference proceedings. Without her, we would have not functioned effectively. Given the "new normal", none of the workshops, tutorials, and challenges would have been feasible without the true leadership of the satellite events organizing team led by Mauricio Reyes: Erik Meijering (workshops), Carlos Alberola-López (tutorials), and Lena Maier-Hein (challenges). Behind the scenes, MICCAI secretarial personnel, Janette Wallace and Johanne Langford, kept a close eye on logistics and budgets, while Mehmet Eldegez and his team at Dekon Congress and Tourism led the professional conference organization, working tightly with the virtual platform team. We also thank our sponsors for financial support and engagement with conference attendees through the virtual platform. Special thanks goes to Veronika Cheplygina for continuous engagement with various social media platforms before and throughout the conference to publicize the conference. We would also like to express our gratitude to Shelley Wallace for helping us in Marketing MICCAI 2020, especially during the last phase of the virtual conference organization.

The selection process for Young Investigator Awards was managed by a team of senior MICCAI investigators, led by Julia Schnabel. In addition, MICCAI 2020 offered free registration to the top 50 ranked papers at the conference whose primary authors were students. Priority was given to low-income regions and Latin American students. Further support was provided by the National Institutes of Health (support granted for MICCAI 2020) and the National Science Foundation (support granted to MICCAI 2019 and continued for MICCAI 2020) which sponsored another 52 awards for USA-based students to attend the conference. We would like to thank Marius Linguraru and Antonion Porras, for their leadership in regards to the NIH sponsorship for 2020, and Dinggang Shen and Tianming Liu, MICCAI 2019 general chairs, for keeping an active bridge and engagement with MICCAI 2020.

Marius Linguraru and Antonion Porras were also leading the young investigators early career development program, including a very active mentorship which we do hope, will significantly catalyze young and briliant careers of future leaders of our scientific community. In link with SIPAIM (thanks to Jorge Brieva, Marius Linguraru, and Natasha Lepore for their support), we also initiated a Startup Village initiative, which, we hope, will be able to bring in promising private initiatives in the areas of MICCAI. As a part of SIPAIM 2020, we note also the presence of a workshop for Peruvian clinicians. We would like to thank Benjaming Castañeda and Renato Gandolfi for this initiative.

MICCAI 2020 invested significant efforts to tightly engage the industry stakeholders in our field throughout its planning and organization. These efforts were led by Parvin Mousavi, and ensured that all sponsoring industry partners could connect with the conference attendees through the conference's virtual platform before and during the meeting. We would like to thank the sponsorship team and the contributions of Gustavo Carneiro, Benjamín Castañeda, Ignacio Larrabide, Marius Linguraru, Yanwu Xu, and Kevin Zhou.

We look forward to seeing you at MICCAI 2021.

October 2020

Anne L. Martel
Purang Abolmaesumi
Danail Stoyanov
Diana Mateus
Maria A. Zuluaga
S. Kevin Zhou
Daniel Racoceanu
Leo Joskowicz

Organization

General Chairs

Daniel Racoceanu   Sorbonne Université, Brain Institute, France
Leo Joskowicz   The Hebrew University of Jerusalem, Israel

Program Committee Chairs

Anne L. Martel   University of Toronto, Canada
Purang Abolmaesumi   The University of British Columbia, Canada
Danail Stoyanov   University College London, UK
Diana Mateus   Ecole Centrale de Nantes, LS2N, France
Maria A. Zuluaga   Eurecom, France
S. Kevin Zhou   Chinese Academy of Sciences, China

Keynote Speaker Chair

Rene Vidal   The John Hopkins University, USA

Satellite Events Chair

Mauricio Reyes   University of Bern, Switzerland

Workshop Team

Erik Meijering (Chair)   The University of New South Wales, Australia
Li Cheng   University of Alberta, Canada
Pamela Guevara   University of Concepción, Chile
Bennett Landman   Vanderbilt University, USA
Tammy Riklin Raviv   Ben-Gurion University of the Negev, Israel
Virginie Uhlmann   EMBL, European Bioinformatics Institute, UK

Tutorial Team

Carlos Alberola-López   Universidad de Valladolid, Spain
    (Chair)
Clarisa Sánchez   Radboud University Medical Center, The Netherlands
Demian Wassermann   Inria Saclay Île-de-France, France

Challenges Team

Lena Maier-Hein (Chair)   German Cancer Research Center, Germany
Annette Kopp-Schneider   German Cancer Research Center, Germany
Michal Kozubek   Masaryk University, Czech Republic
Annika Reinke   German Cancer Research Center, Germany

Sponsorship Team

Parvin Mousavi (Chair)   Queen's University, Canada
Marius Linguraru   Children's National Institute, USA
Gustavo Carneiro   The University of Adelaide, Australia
Yanwu Xu   Baidu Inc., China
Ignacio Larrabide   National Scientific and Technical Research Council,
    Argentina
S. Kevin Zhou   Chinese Academy of Sciences, China
Benjamín Castañeda   Pontifical Catholic University of Peru, Peru

Local and Regional Chairs

Benjamín Castañeda   Pontifical Catholic University of Peru, Peru
Natasha Lepore   University of Southern California, USA

Social Media Chair

Veronika Cheplygina   Eindhoven University of Technology, The Netherlands

Young Investigators Early Career Development Program Chairs

Marius Linguraru   Children's National Institute, USA
Antonio Porras   Children's National Institute, USA

Student Board Liaison Chair

Gabriel Jimenez   Pontifical Catholic University of Peru, Peru

Submission Platform Manager

Kitty Wong   The MICCAI Society, Canada

Conference Management

DEKON Group
Pathable Inc.

Program Committee

Ehsan Adeli	Stanford University, USA
Shadi Albarqouni	ETH Zurich, Switzerland
Pablo Arbelaez	Universidad de los Andes, Colombia
Ulas Bagci	University of Central Florida, USA
Adrien Bartoli	Université Clermont Auvergne, France
Hrvoje Bogunovic	Medical University of Vienna, Austria
Weidong Cai	The University of Sydney, Australia
Chao Chen	Stony Brook University, USA
Elvis Chen	Robarts Research Institute, Canada
Stanley Durrleman	Inria, France
Boris Escalante-Ramírez	National Autonomous University of Mexico, Mexico
Pascal Fallavollita	University of Ottawa, Canada
Enzo Ferrante	CONICET, Universidad Nacional del Litoral, Argentina
Stamatia Giannarou	Imperial College London, UK
Orcun Goksel	ETH Zurich, Switzerland
Alberto Gomez	King's College London, UK
Miguel Angel González Ballester	Universitat Pompeu Fabra, Spain
Ilker Hacihaliloglu	Rutgers University, USA
Yi Hong	University of Georgia, USA
Yipeng Hu	University College London, UK
Heng Huang	University of Pittsburgh and JD Finance America Corporation, USA
Juan Eugenio Iglesias	University College London, UK
Madhura Ingalhalikar	Symbiosis Center for Medical Image Analysis, India
Pierre Jannin	Université de Rennes, France
Samuel Kadoury	Ecole Polytechnique de Montreal, Canada
Bernhard Kainz	Imperial College London, UK
Marta Kersten-Oertel	Concordia University, Canada
Andrew King	King's College London, UK
Ignacio Larrabide	CONICET, Argentina
Gang Li	University of North Carolina at Chapel Hill, USA
Jianming Liang	Arizona State University, USA
Hongen Liao	Tsinghua University, China
Rui Liao	Siemens Healthineers, USA
Feng Lin	Nanyang Technological University, China
Mingxia Liu	University of North Carolina at Chapel Hill, USA
Jiebo Luo	University of Rochester, USA
Xiongbiao Luo	Xiamen University, China
Andreas Maier	FAU Erlangen-Nuremberg, Germany
Stephen McKenna	University of Dundee, UK
Bjoern Menze	Technische Universität München, Germany
Mehdi Moradi	IBM Research, USA

Dong Ni	Shenzhen University, China
Marc Niethammer	University of North Carolina at Chapel Hill, USA
Jack Noble	Vanderbilt University, USA
Ipek Oguz	Vanderbilt University, USA
Gemma Piella	Pompeu Fabra University, Spain
Hedyeh Rafii-Tari	Auris Health Inc., USA
Islem Rekik	Istanbul Technical University, Turkey
Nicola Rieke	NVIDIA Corporation, USA
Tammy Riklin Raviv	Ben-Gurion University of the Negev, Israel
Hassan Rivaz	Concordia University, Canada
Holger Roth	NVIDIA Corporation, USA
Sharmishtaa Seshamani	Allen Institute, USA
Li Shen	University of Pennsylvania, USA
Feng Shi	Shanghai United Imaging Intelligence Co., China
Yonggang Shi	University of Southern California, USA
Michal Sofka	Hyperfine Research, USA
Stefanie Speidel	National Center for Tumor Diseases (NCT), Germany
Marius Staring	Leiden University Medical Center, The Netherlands
Heung-Il Suk	Korea University, South Korea
Kenji Suzuki	Tokyo Institute of Technology, Japan
Tanveer Syeda-Mahmood	IBM Research, USA
Amir Tahmasebi	CodaMetrix, USA
Xiaoying Tang	Southern University of Science and Technology, China
Tolga Tasdizen	The University of Utah, USA
Pallavi Tiwari	Case Western Reserve University, USA
Sotirios Tsaftaris	The University of Edinburgh, UK
Archana Venkataraman	Johns Hopkins University, USA
Satish Viswanath	Case Western Reserve University, USA
Hongzhi Wang	IBM Almaden Research Center, USA
Linwei Wang	Rochester Institute of Technology, USA
Qian Wang	Shanghai Jiao Tong University, China
Guorong Wu	University of North Carolina at Chapel Hill, USA
Daguang Xu	NVIDIA Corporation, USA
Ziyue Xu	NVIDIA Corporation, USA
Pingkun Yan	Rensselaer Polytechnic Institute, USA
Xin Yang	Huazhong University of Science and Technology, China
Zhaozheng Yin	Stony Brook University, USA
Tuo Zhang	Northwestern Polytechnical University, China
Guoyan Zheng	Shanghai Jiao Tong University, China
Yefeng Zheng	Tencent, China
Luping Zhou	The University of Sydney, Australia

Mentorship Program (Mentors)

Ehsan Adeli	Stanford University, USA
Stephen Aylward	Kitware, USA
Hrvoje Bogunovic	Medical University of Vienna, Austria
Li Cheng	University of Alberta, Canada
Marleen de Bruijne	University of Copenhagen, Denmark
Caroline Essert	University of Strasbourg, France
Gabor Fichtinger	Queen's University, Canada
Stamatia Giannarou	Imperial College London, UK
Juan Eugenio Iglesias Gonzalez	University College London, UK
Bernhard Kainz	Imperial College London, UK
Shuo Li	Western University, Canada
Jianming Liang	Arizona State University, USA
Rui Liao	Siemens Healthineers, USA
Feng Lin	Nanyang Technological University, China
Marius George Linguraru	Children's National Hospital, George Washington University, USA
Tianming Liu	University of Georgia, USA
Xiongbiao Luo	Xiamen University, China
Dong Ni	Shenzhen University, China
Wiro Niessen	Erasmus MC - University Medical Center Rotterdam, The Netherlands
Terry Peters	Western University, Canada
Antonio R. Porras	University of Colorado, USA
Daniel Racoceanu	Sorbonne University, France
Islem Rekik	Istanbul Technical University, Turkey
Nicola Rieke	NVIDIA, USA
Julia Schnabel	King's College London, UK
Ruby Shamir	Novocure, Switzerland
Stefanie Speidel	National Center for Tumor Diseases Dresden, Germany
Martin Styner	University of North Carolina at Chapel Hill, USA
Xiaoying Tang	Southern University of Science and Technology, China
Pallavi Tiwari	Case Western Reserve University, USA
Jocelyne Troccaz	CNRS, Grenoble Alpes University, France
Pierre Jannin	INSERM, Université de Rennes, France
Archana Venkataraman	Johns Hopkins University, USA
Linwei Wang	Rochester Institute of Technology, USA
Guorong Wu	University of North Carolina at Chapel Hill, USA
Li Xiao	Chinese Academy of Science, China
Ziyue Xu	NVIDIA, USA
Bochuan Zheng	China West Normal University, China
Guoyan Zheng	Shanghai Jiao Tong University, China
S. Kevin Zhou	Chinese Academy of Sciences, China
Maria A. Zuluaga	EURECOM, France

Additional Reviewers

Alaa Eldin Abdelaal
Ahmed Abdulkadir
Clement Abi Nader
Mazdak Abulnaga
Ganesh Adluru
Iman Aganj
Priya Aggarwal
Sahar Ahmad
Seyed-Ahmad Ahmadi
Euijoon Ahn
Alireza Akhondi-asl
Mohamed Akrout
Dawood Al Chanti
Ibraheem Al-Dhamari
Navid Alemi Koohbanani
Hanan Alghamdi
Hassan Alhajj
Hazrat Ali
Sharib Ali
Omar Al-Kadi
Maximilian Allan
Felix Ambellan
Mina Amiri
Sameer Antani
Luigi Antelmi
Michela Antonelli
Jacob Antunes
Saeed Anwar
Fernando Arambula
Ignacio Arganda-Carreras
Mohammad Ali Armin
John Ashburner
Md Ashikuzzaman
Shahab Aslani
Mehdi Astaraki
Angélica Atehortúa
Gowtham Atluri
Kamran Avanaki
Angelica Aviles-Rivero
Suyash Awate
Dogu Baran Aydogan
Qinle Ba
Morteza Babaie

Hyeon-Min Bae
Woong Bae
Wenjia Bai
Ujjwal Baid
Spyridon Bakas
Yaël Balbastre
Marcin Balicki
Fabian Balsiger
Abhirup Banerjee
Sreya Banerjee
Sophia Bano
Shunxing Bao
Adrian Barbu
Cher Bass
John S. H. Baxter
Amirhossein Bayat
Sharareh Bayat
Neslihan Bayramoglu
Bahareh Behboodi
Delaram Behnami
Mikhail Belyaev
Oualid Benkarim
Aicha BenTaieb
Camilo Bermudez
Giulia Bertò
Hadrien Bertrand
Julián Betancur
Michael Beyeler
Parmeet Bhatia
Chetan Bhole
Suvrat Bhooshan
Chitresh Bhushan
Lei Bi
Cheng Bian
Gui-Bin Bian
Sangeeta Biswas
Stefano B. Blumberg
Janusz Bobulski
Sebastian Bodenstedt
Ester Bonmati
Bhushan Borotikar
Jiri Borovec
Ilaria Boscolo Galazzo

Alexandre Bousse
Nicolas Boutry
Behzad Bozorgtabar
Nadia Brancati
Christopher Bridge
Esther Bron
Rupert Brooks
Qirong Bu
Tim-Oliver Buchholz
Duc Toan Bui
Qasim Bukhari
Ninon Burgos
Nikolay Burlutskiy
Russell Butler
Michał Byra
Hongmin Cai
Yunliang Cai
Sema Candemir
Bing Cao
Qing Cao
Shilei Cao
Tian Cao
Weiguo Cao
Yankun Cao
Aaron Carass
Heike Carolus
Adrià Casamitjana
Suheyla Cetin Karayumak
Ahmad Chaddad
Krishna Chaitanya
Jayasree Chakraborty
Tapabrata Chakraborty
Sylvie Chambon
Ming-Ching Chang
Violeta Chang
Simon Chatelin
Sudhanya Chatterjee
Christos Chatzichristos
Rizwan Chaudhry
Antong Chen
Cameron Po-Hsuan Chen
Chang Chen
Chao Chen
Chen Chen
Cheng Chen
Dongdong Chen

Fang Chen
Geng Chen
Hao Chen
Jianan Chen
Jianxu Chen
Jia-Wei Chen
Jie Chen
Junxiang Chen
Li Chen
Liang Chen
Pingjun Chen
Qiang Chen
Shuai Chen
Tianhua Chen
Tingting Chen
Xi Chen
Xiaoran Chen
Xin Chen
Yuanyuan Chen
Yuhua Chen
Yukun Chen
Zhineng Chen
Zhixiang Chen
Erkang Cheng
Jun Cheng
Li Cheng
Xuelian Cheng
Yuan Cheng
Veronika Cheplygina
Hyungjoo Cho
Jaegul Choo
Aritra Chowdhury
Stergios Christodoulidis
Ai Wern Chung
Pietro Antonio Cicalese
Özgün Çiçek
Robert Cierniak
Matthew Clarkson
Dana Cobzas
Jaume Coll-Font
Alessia Colonna
Marc Combalia
Olivier Commowick
Sonia Contreras Ortiz
Pierre-Henri Conze
Timothy Cootes

Luca Corinzia
Teresa Correia
Pierrick Coupé
Jeffrey Craley
Arun C. S. Kumar
Hui Cui
Jianan Cui
Zhiming Cui
Kathleen Curran
Haixing Dai
Xiaoliang Dai
Ker Dai Fei Elmer
Adrian Dalca
Abhijit Das
Neda Davoudi
Laura Daza
Sandro De Zanet
Charles Delahunt
Herve Delingette
Beatrice Demiray
Yang Deng
Hrishikesh Deshpande
Christian Desrosiers
Neel Dey
Xinghao Ding
Zhipeng Ding
Konstantin Dmitriev
Jose Dolz
Ines Domingues
Juan Pedro Dominguez-Morales
Hao Dong
Mengjin Dong
Nanqing Dong
Qinglin Dong
Suyu Dong
Sven Dorkenwald
Qi Dou
P. K. Douglas
Simon Drouin
Karen Drukker
Niharika D'Souza
Lei Du
Shaoyi Du
Xuefeng Du
Dingna Duan
Nicolas Duchateau

James Duncan
Jared Dunnmon
Luc Duong
Nicha Dvornek
Dmitry V. Dylov
Oleh Dzyubachyk
Mehran Ebrahimi
Philip Edwards
Alexander Effland
Jan Egger
Alma Eguizabal
Gudmundur Einarsson
Ahmed Elazab
Mohammed S. M. Elbaz
Shireen Elhabian
Ahmed Eltanboly
Sandy Engelhardt
Ertunc Erdil
Marius Erdt
Floris Ernst
Mohammad Eslami
Nazila Esmaeili
Marco Esposito
Oscar Esteban
Jingfan Fan
Xin Fan
Yonghui Fan
Chaowei Fang
Xi Fang
Mohsen Farzi
Johannes Fauser
Andrey Fedorov
Hamid Fehri
Lina Felsner
Jun Feng
Ruibin Feng
Xinyang Feng
Yifan Feng
Yuan Feng
Henrique Fernandes
Ricardo Ferrari
Jean Feydy
Lucas Fidon
Lukas Fischer
Antonio Foncubierta-Rodríguez
Germain Forestier

Reza Forghani
Nils Daniel Forkert
Jean-Rassaire Fouefack
Tatiana Fountoukidou
Aina Frau-Pascual
Moti Freiman
Sarah Frisken
Huazhu Fu
Xueyang Fu
Wolfgang Fuhl
Isabel Funke
Philipp Fürnstahl
Pedro Furtado
Ryo Furukawa
Elies Fuster-Garcia
Youssef Gahi
Jin Kyu Gahm
Laurent Gajny
Rohan Gala
Harshala Gammulle
Yu Gan
Cong Gao
Dongxu Gao
Fei Gao
Feng Gao
Linlin Gao
Mingchen Gao
Siyuan Gao
Xin Gao
Xinpei Gao
Yixin Gao
Yue Gao
Zhifan Gao
Sara Garbarino
Alfonso Gastelum-Strozzi
Romane Gauriau
Srishti Gautam
Bao Ge
Rongjun Ge
Zongyuan Ge
Sairam Geethanath
Yasmeen George
Samuel Gerber
Guido Gerig
Nils Gessert
Olivier Gevaert

Muhammad Usman Ghani
Sandesh Ghimire
Sayan Ghosal
Gabriel Girard
Ben Glocker
Evgin Goceri
Michael Goetz
Arnold Gomez
Kuang Gong
Mingming Gong
Yuanhao Gong
German Gonzalez
Sharath Gopal
Karthik Gopinath
Pietro Gori
Maged Goubran
Sobhan Goudarzi
Baran Gözcü
Benedikt Graf
Mark Graham
Bertrand Granado
Alejandro Granados
Robert Grupp
Christina Gsaxner
Lin Gu
Shi Gu
Yun Gu
Ricardo Guerrero
Houssem-Eddine Gueziri
Dazhou Guo
Hengtao Guo
Jixiang Guo
Pengfei Guo
Yanrong Guo
Yi Guo
Yong Guo
Yulan Guo
Yuyu Guo
Krati Gupta
Vikash Gupta
Praveen Gurunath Bharathi
Prashnna Gyawali
Stathis Hadjidemetriou
Omid Haji Maghsoudi
Justin Haldar
Mohammad Hamghalam

Bing Han
Hu Han
Liang Han
Xiaoguang Han
Xu Han
Zhi Han
Zhongyi Han
Jonny Hancox
Christian Hansen
Xiaoke Hao
Rabia Haq
Michael Hardisty
Stefan Harrer
Adam Harrison
S. M. Kamrul Hasan
Hoda Sadat Hashemi
Nobuhiko Hata
Andreas Hauptmann
Mohammad Havaei
Huiguang He
Junjun He
Kelei He
Tiancheng He
Xuming He
Yuting He
Mattias Heinrich
Stefan Heldmann
Nicholas Heller
Alessa Hering
Monica Hernandez
Estefania Hernandez-Martin
Carlos Hernandez-Matas
Javier Herrera-Vega
Kilian Hett
Tsung-Ying Ho
Nico Hoffmann
Matthew Holden
Song Hong
Sungmin Hong
Yoonmi Hong
Corné Hoogendoorn
Antal Horváth
Belayat Hossain
Le Hou
Ai-Ling Hsu
Po-Ya Hsu

Tai-Chiu Hsung
Pengwei Hu
Shunbo Hu
Xiaoling Hu
Xiaowei Hu
Yan Hu
Zhenhong Hu
Jia-Hong Huang
Junzhou Huang
Kevin Huang
Qiaoying Huang
Weilin Huang
Xiaolei Huang
Yawen Huang
Yongxiang Huang
Yue Huang
Yufang Huang
Zhi Huang
Arnaud Huaulmé
Henkjan Huisman
Xing Huo
Yuankai Huo
Sarfaraz Hussein
Jana Hutter
Khoi Huynh
Seong Jae Hwang
Emmanuel Iarussi
Ilknur Icke
Kay Igwe
Alfredo Illanes
Abdullah-Al-Zubaer Imran
Ismail Irmakci
Samra Irshad
Benjamin Irving
Mobarakol Islam
Mohammad Shafkat Islam
Vamsi Ithapu
Koichi Ito
Hayato Itoh
Oleksandra Ivashchenko
Yuji Iwahori
Shruti Jadon
Mohammad Jafari
Mostafa Jahanifar
Andras Jakab
Amir Jamaludin

Won-Dong Jang
Vincent Jaouen
Uditha Jarayathne
Ronnachai Jaroensri
Golara Javadi
Rohit Jena
Todd Jensen
Won-Ki Jeong
Zexuan Ji
Haozhe Jia
Jue Jiang
Tingting Jiang
Weixiong Jiang
Xi Jiang
Xiang Jiang
Jianbo Jiao
Zhicheng Jiao
Amelia Jiménez-Sánchez
Dakai Jin
Taisong Jin
Yueming Jin
Ze Jin
Bin Jing
Yaqub Jonmohamadi
Anand Joshi
Shantanu Joshi
Christoph Jud
Florian Jug
Yohan Jun
Alain Jungo
Abdolrahim Kadkhodamohammadi
Ali Kafaei Zad Tehrani
Dagmar Kainmueller
Siva Teja Kakileti
John Kalafut
Konstantinos Kamnitsas
Michael C. Kampffmeyer
Qingbo Kang
Neerav Karani
Davood Karimi
Satyananda Kashyap
Alexander Katzmann
Prabhjot Kaur
Anees Kazi
Erwan Kerrien
Hoel Kervadec

Ashkan Khakzar
Fahmi Khalifa
Nadieh Khalili
Siavash Khallaghi
Farzad Khalvati
Hassan Khan
Bishesh Khanal
Pulkit Khandelwal
Maksym Kholiavchenko
Meenakshi Khosla
Naji Khosravan
Seyed Mostafa Kia
Ron Kikinis
Daeseung Kim
Geena Kim
Hak Gu Kim
Heejong Kim
Hosung Kim
Hyo-Eun Kim
Jinman Kim
Jinyoung Kim
Mansu Kim
Minjeong Kim
Seong Tae Kim
Won Hwa Kim
Young-Ho Kim
Atilla Kiraly
Yoshiro Kitamura
Takayuki Kitasaka
Sabrina Kletz
Tobias Klinder
Kranthi Kolli
Satoshi Kondo
Bin Kong
Jun Kong
Tomasz Konopczynski
Ender Konukoglu
Bongjin Koo
Kivanc Kose
Anna Kreshuk
AnithaPriya Krishnan
Pavitra Krishnaswamy
Frithjof Kruggel
Alexander Krull
Elizabeth Krupinski
Hulin Kuang

Serife Kucur
David Kügler
Arjan Kuijper
Jan Kukacka
Nilima Kulkarni
Abhay Kumar
Ashnil Kumar
Kuldeep Kumar
Neeraj Kumar
Nitin Kumar
Manuela Kunz
Holger Kunze
Tahsin Kurc
Thomas Kurmann
Yoshihiro Kuroda
Jin Tae Kwak
Yongchan Kwon
Aymen Laadhari
Dmitrii Lachinov
Alexander Ladikos
Alain Lalande
Rodney Lalonde
Tryphon Lambrou
Hengrong Lan
Catherine Laporte
Carole Lartizien
Bianca Lassen-Schmidt
Andras Lasso
Ngan Le
Leo Lebrat
Changhwan Lee
Eung-Joo Lee
Hyekyoung Lee
Jong-Hwan Lee
Jungbeom Lee
Matthew Lee
Sangmin Lee
Soochahn Lee
Stefan Leger
Étienne Léger
Baiying Lei
Andreas Leibetseder
Rogers Jeffrey Leo John
Juan Leon
Wee Kheng Leow
Annan Li

Bo Li
Chongyi Li
Haohan Li
Hongming Li
Hongwei Li
Huiqi Li
Jian Li
Jianning Li
Jiayun Li
Junhua Li
Lincan Li
Mengzhang Li
Ming Li
Qing Li
Quanzheng Li
Shulong Li
Shuyu Li
Weikai Li
Wenyuan Li
Xiang Li
Xiaomeng Li
Xiaoxiao Li
Xin Li
Xiuli Li
Yang Li (Beihang University)
Yang Li (Northeast Electric Power
   University)
Yi Li
Yuexiang Li
Zeju Li
Zhang Li
Zhen Li
Zhiyuan Li
Zhjin Li
Zhongyu Li
Chunfeng Lian
Gongbo Liang
Libin Liang
Shanshan Liang
Yudong Liang
Haofu Liao
Ruizhi Liao
Gilbert Lim
Baihan Lin
Hongxiang Lin
Huei-Yung Lin

Jianyu Lin
C. Lindner
Geert Litjens
Bin Liu
Chang Liu
Dongnan Liu
Feng Liu
Hangfan Liu
Jianfei Liu
Jin Liu
Jingya Liu
Jingyu Liu
Kai Liu
Kefei Liu
Lihao Liu
Luyan Liu
Mengting Liu
Na Liu
Peng Liu
Ping Liu
Quande Liu
Qun Liu
Shengfeng Liu
Shuangjun Liu
Sidong Liu
Siqi Liu
Siyuan Liu
Tianrui Liu
Xianglong Liu
Xinyang Liu
Yan Liu
Yuan Liu
Yuhang Liu
Andrea Loddo
Herve Lombaert
Marco Lorenzi
Jian Lou
Nicolas Loy Rodas
Allen Lu
Donghuan Lu
Huanxiang Lu
Jiwen Lu
Le Lu
Weijia Lu
Xiankai Lu
Yao Lu

Yongyi Lu
Yueh-Hsun Lu
Christian Lucas
Oeslle Lucena
Imanol Luengo
Ronald Lui
Gongning Luo
Jie Luo
Ma Luo
Marcel Luthi
Khoa Luu
Bin Lv
Jinglei Lv
Ilwoo Lyu
Qing Lyu
Sharath M. S.
Andy J. Ma
Chunwei Ma
Da Ma
Hua Ma
Jingting Ma
Kai Ma
Lei Ma
Wenao Ma
Yuexin Ma
Amirreza Mahbod
Sara Mahdavi
Mohammed Mahmoud
Gabriel Maicas
Klaus H. Maier-Hein
Sokratis Makrogiannis
Bilal Malik
Anand Malpani
Ilja Manakov
Matteo Mancini
Efthymios Maneas
Tommaso Mansi
Brett Marinelli
Razvan Marinescu
Pablo Márquez Neila
Carsten Marr
Yassine Marrakchi
Fabio Martinez
Antonio Martinez-Torteya
Andre Mastmeyer
Dimitrios Mavroeidis

Jamie McClelland
Verónica Medina Bañuelos
Raghav Mehta
Sachin Mehta
Liye Mei
Raphael Meier
Qier Meng
Qingjie Meng
Yu Meng
Martin Menten
Odyssée Merveille
Pablo Mesejo
Liang Mi
Shun Miao
Stijn Michielse
Mikhail Milchenko
Hyun-Seok Min
Zhe Min
Tadashi Miyamoto
Aryan Mobiny
Irina Mocanu
Sara Moccia
Omid Mohareri
Hassan Mohy-ud-Din
Muthu Rama Krishnan Mookiah
Rodrigo Moreno
Lia Morra
Agata Mosinska
Saman Motamed
Mohammad Hamed Mozaffari
Anirban Mukhopadhyay
Henning Müller
Balamurali Murugesan
Cosmas Mwikirize
Andriy Myronenko
Saad Nadeem
Ahmed Naglah
Vivek Natarajan
Vishwesh Nath
Rodrigo Nava
Fernando Navarro
Lydia Neary-Zajiczek
Peter Neher
Dominik Neumann
Gia Ngo
Hannes Nickisch

Dong Nie
Jingxin Nie
Weizhi Nie
Aditya Nigam
Xia Ning
Zhenyuan Ning
Sijie Niu
Tianye Niu
Alexey Novikov
Jorge Novo
Chinedu Nwoye
Mohammad Obeid
Masahiro Oda
Thomas O'Donnell
Benjamin Odry
Steffen Oeltze-Jafra
Ayşe Oktay
Hugo Oliveira
Marcelo Oliveira
Sara Oliveira
Arnau Oliver
Sahin Olut
Jimena Olveres
John Onofrey
Eliza Orasanu
Felipe Orihuela-Espina
José Orlando
Marcos Ortega
Sarah Ostadabbas
Yoshito Otake
Sebastian Otalora
Cheng Ouyang
Jiahong Ouyang
Cristina Oyarzun Laura
Michal Ozery-Flato
Krittin Pachtrachai
Johannes Paetzold
Jin Pan
Yongsheng Pan
Prashant Pandey
Joao Papa
Giorgos Papanastasiou
Constantin Pape
Nripesh Parajuli
Hyunjin Park
Sanghyun Park

Seyoun Park
Angshuman Paul
Christian Payer
Chengtao Peng
Jialin Peng
Liying Peng
Tingying Peng
Yifan Peng
Tobias Penzkofer
Antonio Pepe
Oscar Perdomo
Jose-Antonio Pérez-Carrasco
Fernando Pérez-García
Jorge Perez-Gonzalez
Skand Peri
Loic Peter
Jorg Peters
Jens Petersen
Caroline Petitjean
Micha Pfeiffer
Dzung Pham
Renzo Phellan
Ashish Phophalia
Mark Pickering
Kilian Pohl
Iulia Popescu
Karteek Popuri
Tiziano Portenier
Alison Pouch
Arash Pourtaherian
Prateek Prasanna
Alexander Preuhs
Raphael Prevost
Juan Prieto
Viswanath P. S.
Sergi Pujades
Kumaradevan Punithakumar
Elodie Puybareau
Haikun Qi
Huan Qi
Xin Qi
Buyue Qian
Zhen Qian
Yan Qiang
Yuchuan Qiao
Zhi Qiao

Chen Qin
Wenjian Qin
Yanguo Qin
Wu Qiu
Hui Qu
Kha Gia Quach
Prashanth R.
Pradeep Reddy Raamana
Jagath Rajapakse
Kashif Rajpoot
Jhonata Ramos
Andrik Rampun
Parnesh Raniga
Nagulan Ratnarajah
Richard Rau
Mehul Raval
Keerthi Sravan Ravi
Daniele Ravì
Harish RaviPrakash
Rohith Reddy
Markus Rempfler
Xuhua Ren
Yinhao Ren
Yudan Ren
Anne-Marie Rickmann
Brandalyn Riedel
Leticia Rittner
Robert Robinson
Jessica Rodgers
Robert Rohling
Lukasz Roszkowiak
Karsten Roth
José Rouco
Su Ruan
Daniel Rueckert
Mirabela Rusu
Erica Rutter
Jaime S. Cardoso
Mohammad Sabokrou
Monjoy Saha
Pramit Saha
Dushyant Sahoo
Pranjal Sahu
Wojciech Samek
Juan A. Sánchez-Margallo
Robin Sandkuehler

Rodrigo Santa Cruz
Gianmarco Santini
Anil Kumar Sao
Mhd Hasan Sarhan
Duygu Sarikaya
Imari Sato
Olivier Saut
Mattia Savardi
Ramasamy Savitha
Fabien Scalzo
Nico Scherf
Alexander Schlaefer
Philipp Schleer
Leopold Schmetterer
Julia Schnabel
Klaus Schoeffmann
Peter Schueffler
Andreas Schuh
Thomas Schultz
Michael Schwier
Michael Sdika
Suman Sedai
Raghavendra Selvan
Sourya Sengupta
Youngho Seo
Lama Seoud
Ana Sequeira
Saeed Seyyedi
Giorgos Sfikas
Sobhan Shafiei
Reuben Shamir
Shayan Shams
Hongming Shan
Yeqin Shao
Harshita Sharma
Gregory Sharp
Mohamed Shehata
Haocheng Shen
Mali Shen
Yiqiu Shen
Zhengyang Shen
Luyao Shi
Xiaoshuang Shi
Yemin Shi
Yonghong Shi
Saurabh Shigwan

Hoo-Chang Shin
Suprosanna Shit
Yucheng Shu
Nadya Shusharina
Alberto Signoroni
Carlos A. Silva
Wilson Silva
Praveer Singh
Ramandeep Singh
Rohit Singla
Sumedha Singla
Ayushi Sinha
Rajath Soans
Hessam Sokooti
Jaemin Son
Ming Song
Tianyu Song
Yang Song
Youyi Song
Aristeidis Sotiras
Arcot Sowmya
Rachel Sparks
Bella Specktor
William Speier
Ziga Spiclin
Dominik Spinczyk
Chetan Srinidhi
Vinkle Srivastav
Lawrence Staib
Peter Steinbach
Darko Stern
Joshua Stough
Justin Strait
Robin Strand
Martin Styner
Hai Su
Pan Su
Yun-Hsuan Su
Vaishnavi Subramanian
Gérard Subsol
Carole Sudre
Yao Sui
Avan Suinesiaputra
Jeremias Sulam
Shipra Suman
Jian Sun

Liang Sun
Tao Sun
Kyung Sung
Chiranjib Sur
Yannick Suter
Raphael Sznitman
Solale Tabarestani
Fatemeh Taheri Dezaki
Roger Tam
José Tamez-Peña
Chaowei Tan
Jiaxing Tan
Hao Tang
Sheng Tang
Thomas Tang
Xiongfeng Tang
Zhenyu Tang
Mickael Tardy
Eu Wern Teh
Antonio Tejero-de-Pablos
Paul Thienphrapa
Stephen Thompson
Felix Thomsen
Jiang Tian
Yun Tian
Aleksei Tiulpin
Hamid Tizhoosh
Matthew Toews
Oguzhan Topsakal
Jordina Torrents
Sylvie Treuillet
Jocelyne Troccaz
Emanuele Trucco
Vinh Truong Hoang
Chialing Tsai
Andru Putra Twinanda
Norimichi Ukita
Eranga Ukwatta
Mathias Unberath
Tamas Ungi
Martin Urschler
Verena Uslar
Fatmatulzehra Uslu
Régis Vaillant
Jeya Maria Jose Valanarasu
Marta Vallejo

Fons van der Sommen
Gijs van Tulder
Kimberlin van Wijnen
Yogatheesan Varatharajah
Marta Varela
Thomas Varsavsky
Francisco Vasconcelos
S. Swaroop Vedula
Sanketh Vedula
Harini Veeraraghavan
Gonzalo Vegas Sanchez-Ferrero
Anant Vemuri
Gopalkrishna Veni
Ruchika Verma
Ujjwal Verma
Pedro Vieira
Juan Pedro Vigueras Guillen
Pierre-Frederic Villard
Athanasios Vlontzos
Wolf-Dieter Vogl
Ingmar Voigt
Eugene Vorontsov
Bo Wang
Cheng Wang
Chengjia Wang
Chunliang Wang
Dadong Wang
Guotai Wang
Haifeng Wang
Hongkai Wang
Hongyu Wang
Hua Wang
Huan Wang
Jun Wang
Kuanquan Wang
Kun Wang
Lei Wang
Li Wang
Liansheng Wang
Manning Wang
Ruixuan Wang
Shanshan Wang
Shujun Wang
Shuo Wang
Tianchen Wang
Tongxin Wang

Wenzhe Wang
Xi Wang
Xiangxue Wang
Yalin Wang
Yan Wang (Sichuan University)
Yan Wang (Johns Hopkins University)
Yaping Wang
Yi Wang
Yirui Wang
Yuanjun Wang
Yun Wang
Zeyi Wang
Zhangyang Wang
Simon Warfield
Jonathan Weber
Jürgen Weese
Donglai Wei
Dongming Wei
Zhen Wei
Martin Weigert
Michael Wels
Junhao Wen
Matthias Wilms
Stefan Winzeck
Adam Wittek
Marek Wodzinski
Jelmer Wolterink
Ken C. L. Wong
Jonghye Woo
Chongruo Wu
Dijia Wu
Ji Wu
Jian Wu (Tsinghua University)
Jian Wu (Zhejiang University)
Jie Ying Wu
Junyan Wu
Minjie Wu
Pengxiang Wu
Xi Wu
Xia Wu
Xiyin Wu
Ye Wu
Yicheng Wu
Yifan Wu
Zhengwang Wu
Tobias Wuerfl

Pengcheng Xi
James Xia
Siyu Xia
Yingda Xia
Yong Xia
Lei Xiang
Deqiang Xiao
Li Xiao (Tulane University)
Li Xiao (Chinese Academy of Science)
Yuting Xiao
Hongtao Xie
Jianyang Xie
Lingxi Xie
Long Xie
Xueqian Xie
Yiting Xie
Yuan Xie
Yutong Xie
Fangxu Xing
Fuyong Xing
Tao Xiong
Chenchu Xu
Hongming Xu
Jiaofeng Xu
Kele Xu
Lisheng Xu
Min Xu
Rui Xu
Xiaowei Xu
Yanwu Xu
Yongchao Xu
Zhenghua Xu
Cheng Xue
Jie Xue
Wufeng Xue
Yuan Xue
Faridah Yahya
Chenggang Yan
Ke Yan
Weizheng Yan
Yu Yan
Yuguang Yan
Zhennan Yan
Changchun Yang
Chao-Han Huck Yang
Dong Yang

Fan Yang (IIAI)
Fan Yang (Temple University)
Feng Yang
Ge Yang
Guang Yang
Heran Yang
Hongxu Yang
Huijuan Yang
Jiancheng Yang
Jie Yang
Junlin Yang
Lin Yang
Xiao Yang
Xiaohui Yang
Xin Yang
Yan Yang
Yujiu Yang
Dongren Yao
Jianhua Yao
Jiawen Yao
Li Yao
Chuyang Ye
Huihui Ye
Menglong Ye
Xujiong Ye
Andy W. K. Yeung
Jingru Yi
Jirong Yi
Xin Yi
Yi Yin
Shihui Ying
Youngjin Yoo
Chenyu You
Sahar Yousefi
Hanchao Yu
Jinhua Yu
Kai Yu
Lequan Yu
Qi Yu
Yang Yu
Zhen Yu
Pengyu Yuan
Yixuan Yuan
Paul Yushkevich
Ghada Zamzmi
Dong Zeng

Guodong Zeng
Oliver Zettinig
Zhiwei Zhai
Kun Zhan
Baochang Zhang
Chaoyi Zhang
Daoqiang Zhang
Dongqing Zhang
Fan Zhang (Yale University)
Fan Zhang (Harvard Medical School)
Guangming Zhang
Han Zhang
Hang Zhang
Haopeng Zhang
Heye Zhang
Huahong Zhang
Jianpeng Zhang
Jinao Zhang
Jingqing Zhang
Jinwei Zhang
Jiong Zhang
Jun Zhang
Le Zhang
Lei Zhang
Lichi Zhang
Lin Zhang
Ling Zhang
Lu Zhang
Miaomiao Zhang
Ning Zhang
Pengfei Zhang
Pengyue Zhang
Qiang Zhang
Rongzhao Zhang
Ru-Yuan Zhang
Shanzhuo Zhang
Shu Zhang
Tong Zhang
Wei Zhang
Weiwei Zhang
Wenlu Zhang
Xiaoyun Zhang
Xin Zhang
Ya Zhang
Yanbo Zhang
Yanfu Zhang

Yi Zhang
Yifan Zhang
Yizhe Zhang
Yongqin Zhang
You Zhang
Youshan Zhang
Yu Zhang
Yue Zhang
Yulun Zhang
Yunyan Zhang
Yuyao Zhang
Zijing Zhang
Can Zhao
Changchen Zhao
Fenqiang Zhao
Gangming Zhao
Haifeng Zhao
He Zhao
Jun Zhao
Li Zhao
Qingyu Zhao
Rongchang Zhao
Shen Zhao
Tengda Zhao
Tianyi Zhao
Wei Zhao
Xuandong Zhao
Yitian Zhao
Yiyuan Zhao
Yu Zhao
Yuan-Xing Zhao
Yue Zhao
Zixu Zhao
Ziyuan Zhao
Xingjian Zhen
Hao Zheng
Jiannan Zheng
Kang Zheng

Yalin Zheng
Yushan Zheng
Jia-Xing Zhong
Zichun Zhong
Haoyin Zhou
Kang Zhou
Sanping Zhou
Tao Zhou
Wenjin Zhou
Xiao-Hu Zhou
Xiao-Yun Zhou
Yanning Zhou
Yi Zhou (IIAI)
Yi Zhou (University of Utah)
Yuyin Zhou
Zhen Zhou
Zongwei Zhou
Dajiang Zhu
Dongxiao Zhu
Hancan Zhu
Lei Zhu
Qikui Zhu
Weifang Zhu
Wentao Zhu
Xiaofeng Zhu
Xinliang Zhu
Yingying Zhu
Yuemin Zhu
Zhe Zhu
Zhuotun Zhu
Xiahai Zhuang
Aneeq Zia
Veronika Zimmer
David Zimmerer
Lilla Zöllei
Yukai Zou
Gerald Zwettler
Reyer Zwiggelaa

Contents – Part II

Cross-Domain Methods and Reconstruction

Domain Adaptation

Machine Learning Applications

Generative Adversarial Networks

Image Reconstruction

Improving Amide Proton Transfer-Weighted MRI Reconstruction Using T2-Weighted Images

Puyang Wang[1]($\boxtimes$), Pengfei Guo[2], Jianhua Lu[3], Jinyuan Zhou[3], Shanshan Jiang[3], and Vishal M. Patel[1,2]

[1] Department of Electrical and Computer Engineering, Johns Hopkins University, Baltimore, MD, USA
pwang47@jhu.edu
[2] Department of Computer Science, Johns Hopkins University, Baltimore, MD, USA
[3] Department of Radiology, Johns Hopkins University, Baltimore, MD, USA

Abstract. Current protocol of Amide Proton Transfer-weighted (APTw) imaging commonly starts with the acquisition of high-resolution T2-weighted (T_2w) images followed by APTw imaging at particular geometry and locations (i.e. slice) determined by the acquired T_2w images. Although many advanced MRI reconstruction methods have been proposed to accelerate MRI, existing methods for APTw MRI lacks the capability of taking advantage of structural information in the acquired T_2w images for reconstruction. In this paper, we present a novel APTw image reconstruction framework that can accelerate APTw imaging by reconstructing APTw images directly from highly undersampled k-space data and corresponding T_2w image at the same location. The proposed framework starts with a novel sparse representation-based slice matching algorithm that aims to find the matched T_2w slice given only the undersampled APTw image. A Recurrent Feature Sharing Reconstruction network (RFS-Rec) is designed to utilize intermediate features extracted from the matched T_2w image by a Convolutional Recurrent Neural Network (CRNN), so that the missing structural information can be incorporated into the undersampled APT raw image thus effectively improving the image quality of the reconstructed APTw image. We evaluate the proposed method on two real datasets consisting of brain data from rats and humans. Extensive experiments demonstrate that the proposed RFS-Rec approach can outperform the state-of-the-art methods.

Keywords: Magnetic resonance imaging · Image reconstruction · Amide proton transfer imaging

Electronic supplementary material The online version of this chapter (https://doi.org/10.1007/978-3-030-59713-9_1) contains supplementary material, which is available to authorized users.

A. L. Martel et al. (Eds.): MICCAI 2020, LNCS 12262, pp. 3–12, 2020.
https://doi.org/10.1007/978-3-030-59713-9_1

1   Introduction

Amide Proton Transfer-weighted (APTw) imaging is an emerging molecular Magnetic Resonance Imaging (MRI) method that can generate image contrast unique from the conventional MRI. As a type of chemical exchange saturation transfer (CEST) MRI, APTw signal intensity is based on concentrations of endogenous mobile proteins and peptides or tissue pH. Moreover, APTw MRI does not require any contrast agent administration. Previous studies in animals and humans have demonstrated that APT imaging is capable of detecting brain tumors [18] and ischemic stroke [10]. In a recent preclinical study [11], APT imaging was shown to accurately detect intracerebral hemorrhage and distinctly differentiate hyperacute hemorrhage from cerebral ischemia. Notably, the capability and uniqueness of APT imaging for the detection of primary and secondary brain injuries in experimental Controlled Cortical Impact (CCI) Traumatic Brain Injury (TBI) models have recently been explored with promising results [14].

However, relatively long acquisition times due to the use of multiple RF saturation frequencies and multiple acquisitions to increase the signal-to-noise ratio (SNR) hinders the wide spread clinical use of APTw imaging. A typical CEST MRI acquisition currently requires long scan times in the range of 5 to 10 min. Recently, several methods have been developed to accelerate CEST/APT acquisitions. These can be classified into conventional fast imaging techniques (e.g. turbospin-echo [17]) and reduced k-space acquisition techniques (including spectroscopy with linear algebraic modeling [15] and compressed sensing (CS) [3]) that require more advanced data processing. Due to recent advances in deep learning, deep learning-based methods have shown to provide much better generic MRI image reconstruction results from undersampled k-space data than conventional CS-based methods. The combination of convolutional autoencoder and generative adversarial networks can perform faster and more accurate reconstruction [7]. In [12], a pyramid convolutional Recurrent Neural Network (RNN) was designed to iteratively refine reconstructed image in three different feature scales.

Despite the success of deep learning-based MR image reconstruction methods for single contrast/modality imaging, multi-contrast reconstruction still remains a challenge. In multiple-contrast MR imaging it is beneficial to utilize fully sampled images acquired at one contrast for the reconstruction of undersampled MR images in another contrast [4]. For instance, information pertaining to undersampled T_1w images and undersampled T_2w images can be mutually beneficial when reconstructing both images. A joint reconstruction network of T_1, T_2 and PD images was proposed in [9] and was shown to outperform single-contrast models. Furthermore, undersampled T_2w image scan be reconstructed more accurately using the information from fully sampled high-resolution T_1w images [2]. To this end, Y-net was proposed in [2] by modifying U-net which takes two inputs and produces a single output. Features extracted from two independent encoders are concatenated together to generate the final output reconstruction. However, these methods are only evaluated on structural MR images and can be affected by slice mismatch between different scans. To deal with this issue, additional registration process between the images might be required.

Current 2D APTw imaging protocol starts with a high-resolution 3D T_2w scan that is used to locate the slice of interest (usually contains lesion region) by examination. After setting the interested slice, to reduce the effect of B_0 field inhomogeneity on APT imaging, high-order localized slab shimming is performed around the lesion. The final APTw image is defined as the difference of ± 3.5 ppm image normalized by unsaturated image. While one can accelerate APTw imaging by reducing the raw k-space measurement data and apply reconstruction using off-the-shelf algorithms, no existing methods take 3D T_2w scan into reconstruction process as the idea of multi-contrast MR reconstruction suggests.

In this paper, in order to leverage the structural information of T_2w images, we present a Recurrent Feature Sharing Reconstruction network (RFS-Rec) that has two convolutional RNNs (CRNN). These two CRNNs are connected by the proposed recurrent feature sharing approach to encourage bi-directional flow of information. In addition, we propose a sparse representation-based slice matching algorithm to find the corresponding slice in T_2w volume given the undersampled APT k-space data. As a result, input T_2w and APT raw images are aligned and mutual information can be maximized.

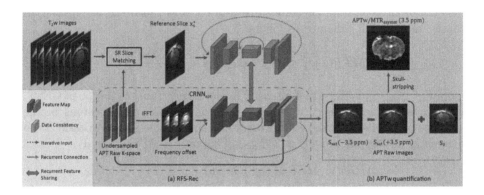

Fig. 1. An overview of the proposed framework.

2 Methodology

In this section, we first give a brief introduction of APTw imaging. Then we describe our recurrent feature sharing reconstruction network and sparse representation (SR) based slice matching algorithm. As shown in Fig. 1, the slice matching step in the proposed framework takes T_2w images and undersampled k-space as input and selects out a reference T_2w slice. The APT raw images are reconstructed by RFS-Rec using both reference T_2w slice and undersampled APT k-space data.

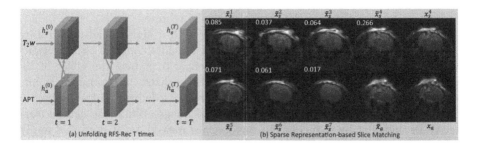

Fig. 2. (a) The proposed recurrent neural network, RFS-Rec, can be unfolded T times. Hidden states of T_2w and APT RNN are connected by two-way feature sharing. (b) Absolute weights in sparse vector w_a are shown at the top left corner of the corresponding T_2w slices. x_a is the average of fully sampled APT raw images.

2.1 APTw Imaging

CEST effects are usually analyzed using Z-spectrum, in which the intensity of the water signal during saturation at a frequency offset from water, $S_{\text{sat}}(\Delta\omega)$, normalized by the signal without saturation S_0, is displayed as a function of irradiation frequency using the water frequency as a zero-frequency reference. The sum of all saturation effects at a certain offset is called the magnetization transfer ratio (MTR), defined as follows

$$MTR(\Delta\omega) = 1 - Z(\Delta\omega) = 1 - \frac{S_{\text{sat}}(\Delta\omega)}{S_0}, \tag{1}$$

where $Z = S_{\text{sat}}/S_0$ is the signal intensity in the Z-spectrum. As a type of CEST, APTw imaging is designed to detect the exchangeable amide protons in the backbone of mobile proteins and are assessed using magnetization transfer ratio asymmetry at 3.5ppm, namely $MTR_{\text{asymm}}(3.5\text{ppm})$ as APTw signal

$$\begin{aligned} APTw &= MTR_{\text{asymm}}(3.5\text{ppm}) \\ &= MTR(+3.5\text{ppm}) - MTR(-3.5\text{ppm}) \\ &= \frac{S_{\text{sat}}(-3.5\text{ppm}) - S_{\text{sat}}(+3.5\text{ppm})}{S_0}. \end{aligned} \tag{2}$$

Hence, the quality of APTw image solely depends on the above three images at different frequency offsets. An example of APTw quantification is shown in the right part of Fig. 1. For visualization purpose, skull-stripping procedure is usually performed on APTw image. In the rest of paper, we refer $S_{\text{sat}}(\pm3.5\text{ppm})$ and S_0 as APT raw images and $MTR_{\text{asymm}}(3.5\text{ppm})$ as an APTw image.

2.2 Recurrent Feature Sharing Reconstruction

The data acquisition process of accelerated MRI can be formulated as follows

$$y = F_D x + \epsilon, \tag{3}$$

where $x \in \mathbb{C}^M$ is the fully sampled image, $y \in \mathbb{C}^N$ is the observed k-space, and ϵ is the noise. Both x and y are image data represented in vector forms. F_D is the undersampling Fourier encoding matrix which is defined as the multiplication of the Fourier transform matrix F and the binary undersampling matrix D. We define the acceleration factor R as the ratio of the amount of k-space data required for a fully sampled image to the amount collected in an accelerated acquisition. The goal of MRI image reconstruction is to estimate image x from the observed k-space y. MRI reconstruction problem is an ill-posed problem due to the information loss in the forward process as $N \ll M$.

We solve the MRI image reconstruction problem in an iterative manner using CRNN as the base reconstruction network in RFS-Rec. A single contrast CRNN can be divided into four parts: 1) encoder f_{enc}, 2) decoder f_{dec}, 3) hidden state transition f_{res} consisting of two residual convolution blocks (ResBlock), and 4) data consistency (DC) layer. f_{enc} and f_{dec} are constructed using strided and transposed convolutions. Input images to CRNN are zero-filled undersampled complex APT raw images $x^{(0)} = F_D^H y$ with real and imaginary values as two channels. The output of the $(t+1)^{th}$ iteration of a single CRNN_{apt} model can be described as follows:

$$
\begin{aligned}
x^{(t+1)} &= \text{DC}(f(x^{(t)}, h^{(t)}, y, D)), \\
&= F^{-1}[Dy + (1-D)F f_{dec}(f_{res}(h^{(t)}) + f_{enc}(x^{(t)}))],
\end{aligned}
\tag{4}
$$

where $h^{(t)} = f_{res}(h^{(t-1)}) + f_{enc}(x^{(t-1)})$ is the hidden state from the previous iteration and $h^{(0)} = 0$.

As discussed above, by using the information from other contrast, one can more accurately reconstruct an image of another contrast. This approach is known as multi-contrast MRI reconstruction [1]. Information or feature sharing has been shown to be the key for multi-contrast MR image reconstruction [2,9].

Note that CRNNs have been proposed for MRI reconstruction [6] and it has been demonstrated that they can outperform cascaded models and U-net [8]. However, feature sharing in CRNN has not been studied in the literature for MRI reconstruction. In this paper, we present a novel recurrent feature sharing method that exchanges intermediate hidden state features of two CRNNs (see Fig. 2(a)). This allows us to use CRNNs for multi-contrast MR image reconstruction in a more efficient way.

The proposed RFS-Rec consists of two CRNNs, CRNN_{apt} and CRNN_{t2w}. CRNN_{t2w} for T2w images are constructed similar to CRNN_{apt} which is defined in Eq. 4 but without the DC layer. CRNN_{t2w} takes the reference slice x_s^* which is assumed to be aligned with underlying full sampled x.

To enable two-way information flow between APT features h_a and structural T2w features h_s, we add bi-directional skip connection links (Fig. 2(a)) in each iteration, which is inspired by the one-time feature concatenation in Y-net [2].

Thus, the overall dynamics of our proposed RFS-Rec is given as follows

$$
\begin{aligned}
h_a^{(t)} &= f_{res}(h_a^{(t-1)} \oplus h_s^{(t-1)}) + f_{enc}(x^{(t-1)}), \text{ and} \\
h_s^{(t)} &= f_{res}(h_s^{(t-1)} \oplus h_a^{(t-1)}) + f_{enc}(x_s^{(t-1)}),
\end{aligned}
\tag{5}
$$

where $\oplus$ stands for channel-wise concatenation. We refer to this hidden state exchange design as recurrent feature sharing.

In terms of the loss function, we use a combination of the Normalised Mean Square Error (NMSE) loss and the Structural Similarity Index (SSIM) loss as our training loss. The overall loss function we use to train the network is defined as follows

$$
\begin{aligned}
\mathcal{L}(\hat{x}, x) &= \mathcal{L}_{\text{NMSE}} + \beta \mathcal{L}_{\text{SSIM}}, \\
&= \frac{\|\hat{x} - x\|_2^2}{\|x\|_2^2} + \beta \frac{(2\mu_{\hat{x}}\mu_x + c_1)(2\sigma_{\hat{x}x} + c_2)}{(\mu_{\hat{x}}^2 + \mu_x^2 + c_1)(\sigma_{\hat{x}}^2 + \sigma_x^2 + c_2)},
\end{aligned}
\tag{6}
$$

where $\mu_{\hat{x}}$ and μ_x are the average pixel intensities in $\hat{x}$ and x, respectively, $\sigma_{\hat{x}}^2$ and σ_x^2 are their variances, $\sigma_{\hat{x}x}$ is the covariance between $\hat{x}$ and x, and $c_1 = (k_1 L)^2$, $c_2 = (k_2 L)^2$. In this paper, we choose a window size of 7×7, and set $k_1 = 0.01$, $k_2 = 0.03$, and define L as the maximum magnitude value of the target image x, i.e. $L = \max(|x|)$. We use $\beta = 0.5$ to balance the two loss functions.

2.3   Sparse Representation-Based Slice Matching

As mentioned earlier, a 3D T$_2$w scan is normally acquired prior to 2D APTw imaging. In order to fully leverage the T$_2$w scan, it is important to identify the matching slices between T$_2$w and the undersampled APT raw image. We propose a simple yet effective sparse representation-based slice matching algorithm that can find the closest slice in T$_2$w scan in terms of location given undersampled APT raw images.

Sparse representation-based approach, first described in [13], exploits the discriminative nature of sparsity. The average undersampled APT raw image can be represented by a set of T$_2$w images as a linear combination of all elements. This representation is naturally sparse and can be recovered efficiently via ℓ_1-minimization, seeking the sparsest representation of the APT raw image. Let $X_s = [\tilde{x}_s^1, \tilde{x}_s^2, \dots, \tilde{x}_s^n]$ be the matrix that contains all n undersampled structural T$_2$w slices, $\tilde{x}_s = F_D^H F_D x_s$ and $\tilde{x}_a = \sum_{i=1}^3 F_D^H y_i/3$ be the average of the undersampled APT raw images. The sparsest vector w_a that represents $\tilde{x}_a$ in X_s and gives small reconstruction error $\|\tilde{x}_a - X_s w_a\|_2$ can be found by solving the following l_1-minimization problem

$$
w_a = \underset{w}{\operatorname{argmin}} \|w\|_1 \quad \text{s.t.} \quad \|\tilde{x}_a - X_s w\|_2 \leq \sigma.
\tag{7}
$$

After solving the optimization problem, the matching T$_2$w slice x_s^* is determined by the slice index $i = \operatorname{argmax}|w_a^i|$ (i.e. the slice with the largest absolute weight). From an example of SR slice matching ($\sigma = 0.1$) shown in Fig. 2(b), x_s^4 which has the largest absolute weight $w_a^4 = 0.266$ is the one matched to the APT raw image $\tilde{x}_a$ and will be used as the reference T_2w image in the reconstruction phase.

3 Experiments

Datasets. We evaluate the proposed image reconstruction framework on two datasets.

Rat TBI Data: 300 MRI scans are performed on 65 open-skull rats with controlled cortical impact model of TBI at different time point after TBI. Each MRI scan includes high-resolution T_2w imaging with a fast spin echo sequence in coronal plane (number of slices= 7; matrix= 256×256; field of view (FOV) = 32×32 mm^2; slice thickness = 1.5 mm) and 2D APT (frequency labeling offsets of ± 3.5 ppm; matrix= 64×64; FOV = 32×32 mm^2; single slice; slice thickness = 1.5 mm). An unsaturated image S_0 in the absence of radio-frequency saturation was also acquired for APT imaging signal intensity normalization.

Human Brain Tumor Data: 144 3D T_2w and APTw MRI volumes were collected from 90 patients with pathologically proven primary malignant glioma. Imaging parameters for APTw can be summarized as follows: FOV = $212 \times 212 \times 66$ mm3; resolution = $0.82 \times 0.82 \times 4.4$ mm^3; size in voxel = $256 \times 256 \times 15$. T_2w sequences were acquired with the following imaging parameters: FOV = $212 \times 212 \times 165$ mm3; resolution, $0.41 \times 0.41 \times 1.1$ mm3; size in voxel, $512 \times 512 \times 150$. Co-registration between APTw and T_2w sequences [16], and MRI standardization [5] were performed. After preprocessing, the final volume size of each sequence is $256 \times 256 \times 15$. Data collection and processing are approved by the Institutional Review Board.

Training Details: We simulated undersampled k-space measurements of APT raw images using the Cartesian sampling method with a fixed 0.08% center frequency sampled and random sampling in other frequencies uniformly. Training and testing subsets are randomly selected with 80/20% split. We conducted model training under the acceleration factors R=4 and 8. All models are implemented in Pytorch and trained on NVIDIA GPUs. Hyperparameters are set as follows: learning rate of 10^{-3} with decreasing rate of 0.9 for every 5 epochs, 60 maximum epochs, batch size of 6. Adam optimizer is used in training all the networks. For CRNN and RFS-Rec, the number of iterations T is set equal to 7.

We compare our proposed RFS-Rec against U-net [8], Y-net [2], single contrast CRNN $_{apt}$, CRNN with concatenation of center T_2w slice and undersampled APT raw images as input and CRNN using the proposed SR slice matching to select the reference slice. Regarding the U-net implementation, a DC layer was added at the end of the network. The quantitative metrics, including NMSE, PSNR and SSIM, are computed between fully sampled APT raw

Table 1. Quantitative results of APT raw image reconstruction under the acceleration factors $R = 4$ and $R = 8$. T_2w indicates whether T_2w image is used during reconstruction. SM denotes the use of the proposed SR slice matching instead of always using the center T_2w slice. Note that, for Human brain dataset, the slice matching does not apply because T_2w and APT volume are already well co-registered.

Dataset	Method	T_2w	SM	$R = 4$			$R = 8$		
				NMSE	PSNR	SSIM	NMSE	PSNR	SSIM
Rat	U-net [8]			0.144	34.29	0.920	0.242	31.96	0.878
	Y-net [2]	✓		0.111	35.35	0.932	0.218	32.29	0.889
	$CRNN_{apt}$			0.087	36.41	0.939	0.217	32.31	0.889
	CRNN	✓		0.085	36.43	0.940	0.219	32.28	0.893
	CRNN	✓	✓	0.084	36.56	0.941	0.212	32.37	0.893
	RFS-Rec	✓	✓	**0.076**	**36.94**	**0.950**	**0.187**	**33.11**	**0.906**
Human	U-net [8]		N/A	0.022	37.19	0.910	0.045	33.76	0.872
	Y-net [2]	✓	N/A	0.014	39.27	0.938	0.037	34.65	0.889
	$CRNN_{apt}$		N/A	0.014	39.64	0.943	0.041	34.30	0.887
	CRNN	✓	N/A	0.012	40.35	0.950	0.038	34.84	0.896
	RFS-Rec	✓	N/A	**0.010**	**40.99**	**0.956**	**0.034**	**35.27**	**0.903**

images ($S_{\mathrm{sat}}(\pm 3.5\mathrm{ppm})$ and S_0) and their reconstructions. Detailed quantitative experimental results are shown in Table 1. It can be seen from the table that the proposed RFS-Rec approach outperforms all the other compared methods on both datasets. Furthermore, the individual contribution of the modules in the proposed method (SR-based slice matching and RFS), are demonstrated by

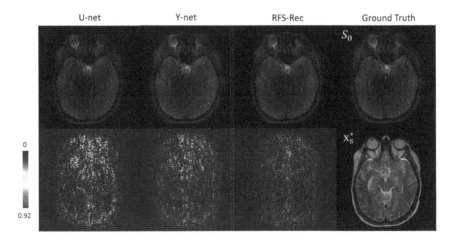

Fig. 3. S_0 reconstructions at $R = 4$ and the corresponding error maps.

an ablation study (i.e. CRNN with/without SM and RFS-Rec). One interesting observation from Table 1 is that the difference between $CRNN_{apt}$ and Y-net, when $R = 8$, on the human dataset is inverse of what we observe on the rat dataset. This may be caused by the good registration of T_2w and APT in the human dataset. The issue of shape inconsistency of the APT raw image and T_2w image in the rat dataset can also be observed by comparing $CRNN_{apt}$ and CRNN with T_2w.

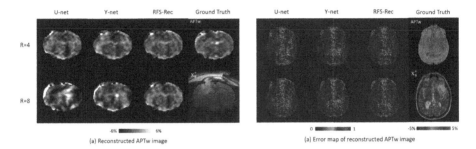

Fig. 4. Results of APTw images derived from the reconstructed APT raw images using Eq. 2. Skull-stripping is performed for better visualization. Reference T_2w slice x_s^* used for reconstruction are also shown.

Results of reconstructed S_0 and APTw images compared to the ground truth in Figs. 3 and 4 show consistent findings as quantitative results suggest. Our method yields not only better $S_{sat}(\pm 3.5ppm)$ and S_0 reconstruction but also more accurate APTw images.

4   Conclusion

We proposed an APTw image reconstruction network RFS-Rec for accelerating APTw imaging, which can more accurately reconstruct APT raw images by using the information of fully sampled T_2w images. We achieved this goal by incorporating a novel recurrent feature sharing mechanism into two CRNNs which enable two-way information flow between APT and T_2w features. In addition, to maximize the effectiveness of RFS-Rec, we use a sparse representation-based slice matching algorithm to locate reference T_2w slice. Extensive experiments on two real datasets consisting of brain data from rats and humans showed the significance of the proposed work.

Acknowledgment. This work was supported in part by grant UG3NS106937 from the National Institutes of Health.

References

1. Bilgic, B., Goyal, V.K., Adalsteinsson, E.: Multi-contrast reconstruction with Bayesian compressed sensing. Magn. Reson. Med. **66**(6), 1601–1615 (2011)

2. Do, W.J., Seo, S., Han, Y., Ye, J.C., Hong Choi, S., Park, S.H.: Reconstruction of multi-contrast MR images through deep learning. Med. Phys. **47**(3), 983–997 (2019)

3. Heo, H.Y., Zhang, Y., Lee, D.H., Jiang, S., Zhao, X., Zhou, J.: Accelerating chemical exchange saturation transfer (CEST) MRI by combining compressed sensing and sensitivity encoding techniques. Magn. Reson. Med. **77**(2), 779–786 (2017)

4. Huang, J., Chen, C., Axel, L.: Fast multi-contrast MRI reconstruction. Magn. Reson. Imaging **32**(10), 1344–1352 (2014)

5. Nyúl, L.G., Udupa, J.K., Zhang, X.: New variants of a method of MRI scale standardization. IEEE Trans. Med. Imaging **19**(2), 143–150 (2000)

6. Qin, C., Schlemper, J., Caballero, J., Price, A.N., Hajnal, J.V., Rueckert, D.: Convolutional recurrent neural networks for dynamic MR image reconstruction. IEEE Trans. Med. Imaging **38**(1), 280–290 (2018)

7. Quan, T.M., Nguyen-Duc, T., Jeong, W.K.: Compressed sensing MRI reconstruction using a generative adversarial network with a cyclic loss. IEEE Trans. Med. Imaging **37**(6), 1488–1497 (2018)

8. Ronneberger, O., Fischer, P., Brox, T.: U-Net: convolutional networks for biomedical image segmentation. In: Navab, N., Hornegger, J., Wells, W.M., Frangi, A.F. (eds.) MICCAI 2015. LNCS, vol. 9351, pp. 234–241. Springer, Cham (2015). https://doi.org/10.1007/978-3-319-24574-4_28

9. Sun, L., Fan, Z., Fu, X., Huang, Y., Ding, X., Paisley, J.: A deep information sharing network for multi-contrast compressed sensing mri reconstruction. IEEE Trans. Image Process. **28**(12), 6141–6153 (2019)

10. Sun, P.Z., Cheung, J.S., Wang, E., Lo, E.H.: Association between PH-weighted endogenous amide proton chemical exchange saturation transfer MRI and tissue lactic acidosis during acute ischemic stroke. J. Cereb. Blood Flow Metab. **31**(8), 1743–1750 (2011)

11. Wang, M., et al.: Simultaneous detection and separation of hyperacute intracerebral hemorrhage and cerebral ischemia using amide proton transfer MRI. Magn. Reson. Med. **74**(1), 42–50 (2015)

12. Wang, P., Chen, E.Z., Chen, T., Patel, V.M., Sun, S.: Pyramid convolutional RNN for MRI reconstruction. arXiv preprint arXiv:1912.00543 (2019)

13. Wright, J., Yang, A.Y., Ganesh, A., Sastry, S.S., Ma, Y.: Robust face recognition via sparse representation. IEEE Trans. Pattern Anal. Mach. Intell. **31**(2), 210–227 (2008)

14. Zhang, H., et al.: Amide proton transfer-weighted MRI detection of traumatic brain injury in rats. J. Cereb. Blood Flow Metab. **37**(10), 3422–3432 (2017)

15. Zhang, Y., Heo, H.Y., Jiang, S., Lee, D.H., Bottomley, P.A., Zhou, J.: Highly accelerated chemical exchange saturation transfer (CEST) measurements with linear algebraic modeling. Magn. Reson. Med. **76**(1), 136–144 (2016)

16. Zhang, Y., et al.: Selecting the reference image for registration of cest series. J. Magn. Reson. Imaging **43**(3), 756–761 (2016)

17. Zhao, X., et al.: Three-dimensional turbo-spin-echo amide proton transfer MR imaging at 3-tesla and its application to high-grade human brain tumors. Mol. Imag. Biol. **15**(1), 114–122 (2013)

18. Zhou, J., et al.: Differentiation between glioma and radiation necrosis using molecular magnetic resonance imaging of endogenous proteins and peptides. Nat. Med. **17**(1), 130–134 (2011)

Compressive MR Fingerprinting Reconstruction with Neural Proximal Gradient Iterations

Dongdong Chen[1], Mike E. Davies[1], and Mohammad Golbabaee[2(✉)]

[1] School of Engineering, University of Edinburgh, Edinburgh, UK
{d.chen,mike.davies}@ed.ac.uk
[2] Computer Science department, University of Bath, Bath, UK
mg2105@bath.ac.uk

Abstract. Consistency of the predictions with respect to the physical forward model is pivotal for reliably solving inverse problems. This consistency is mostly un-controlled in the current end-to-end deep learning methodologies proposed for the Magnetic Resonance Fingerprinting (MRF) problem. To address this, we propose PGD-NET, a learned proximal gradient descent framework that directly incorporates the forward acquisition and Bloch dynamic models within a recurrent learning mechanism. The PGD-NET adopts a compact neural proximal model for de-aliasing and quantitative inference, that can be flexibly trained on scarce MRF training datasets. Our numerical experiments show that the PGD-NET can achieve a superior quantitative inference accuracy, much smaller storage requirement, and a comparable runtime to the recent deep learning MRF baselines, while being much faster than the dictionary matching schemes. Code has been released at https://github.com/edongdongchen/PGD-Net.

Keywords: Magnetic resonance fingerprinting · Compressed Sensing · Deep learning · Learned proximal gradient descent

1 Introduction

Magnetic resonance fingerprinting (MRF) is an emerging technology that enables simultaneous quantification of multitudes of tissues' physical properties in short and clinically feasible scan times [20]. Iterative reconstruction methods based on Compressed Sensing (CS) have proven efficient to help MRF overcome the challenge of computing accurate quantitative images from the undersampled k-space measurements taken in aggressively short scan times [3, 10, 11, 30]. However, these methods require dictionary matching (DM) that is non-scalable and can create enormous storage and computational overhead. Further, such approaches often do not fully account for the joint spatiotemporal structures of the MRF data which can lead to poor reconstructions [14].

© Springer Nature Switzerland AG 2020
A. L. Martel et al. (Eds.): MICCAI 2020, LNCS 12262, pp. 13–22, 2020.
https://doi.org/10.1007/978-3-030-59713-9_2

Deep learning methodologies have emerged to address DM's computational bottleneck [9,15,24,27], and in some cases to perform joint spatiotemporal MRF processing through using convolutional layers [4,8,12,13,17,18,26]. These models are trained in an end-to-end fashion *without* an explicit account for the known physical acquisition model (i.e. the forward operator) and a mechanism for explicitly enforcing measurement consistency according to this sampling model which can be crucial in the safety-first medical applications. Further, ignoring the structure of the forward model could lead to building unnecessary large inference models and possible overfitted predictions, especially for the extremely *scarce* labelled anatomical quantitative MRI datasets that are available for training.

Our Contributions: we propose PGD-NET a deep convolutional model that is able to learn and perform robust spatiotemporal MRF processing, and work with limited access to the ground-truth (i.e. labelled) quantitative maps. Inspired by iterative proximal gradient descent (PGD) methods for CS reconstruction [22], we adopt learnable, compact and shared convolutional layers within a data-driven proximal step, meanwhile explicitly incorporating the acquisition model as a non-trainable gradient step in all iterations. The proximal operator is an auto-encoder network whose decoder embeds the Bloch magnetic responses and its convolutional encoder embeds a de-aliasing projector to the tissue maps' quantitative properties. Our work is inspired by recent general CS methodologies [1,2,7,22,25] that replace traditional hand-crafted image priors by deep data-driven models. To the best of our knowledge, this is the first work to adopt and investigate the feasibility of such an approach for solving the MRF inverse problem.

2   Methodology

MRF adopts a linear spatiotemporal compressive acquisition model:

$$\mathbf{y} = \overline{\mathbf{H}}(\overline{\mathbf{x}}) + \xi \tag{1}$$

where $\mathbf{y} \in \mathbb{C}^{m \times L}$ are the k-space measurements collected at L temporal frames and corrupted by some noise ξ, and $\overline{\mathbf{x}} \in \mathbb{C}^{n \times L}$ is the Time-Series of Magnetisation Images (TSMI) with n voxels across L timeframes. The forward operator $\overline{\mathbf{H}} : \mathbb{C}^{n \times L} \to \mathbb{C}^{m \times L}$ models Fourier transformations subsampled according to a set of temporally-varying k-space locations in each timeframe. Accelerated MRF acquisition implies working with heavily under-sampled data $m \ll n$, which makes $\overline{\mathbf{H}}$ becomes ill-posed for the inversion.

Bloch Response Model: Per-voxel TSMI temporal signal evolution is related to the quantitative NMR parameters/properties such as $\{T1_v, T2_v\}$ relaxation times, through the solutions of the *Bloch differential equations* $\overline{\mathbf{x}}_v \approx \rho_v \overline{\mathcal{B}}(T1_v, T2_v)$, scaled by the ρ_v proton density (PD) in each voxel v [19,20].

The Subspace Dimension-Reducing Model: In many MRF applications (including ours) a low $s \ll L$ dimensional subspace $V \in \mathbb{C}^{L \times s}$ embeds the Bloch solutions $\mathcal{B}(.) \approx V V^H \mathcal{B}(.)$. This subspace can be computed through

PCA decomposition of the MRF dictionary [23], and enables re-writing (1) in a compact form that is beneficial to the storage, runtime and accuracy of the reconstruction [3, 15]:

$$\mathbf{y} = \mathbf{H}(\mathbf{x}) + \xi \quad \text{where, for each voxel} \quad \mathbf{x}_v \approx \rho_v \mathcal{B}(T1_v, T2_v), \tag{2}$$

and $\mathbf{x} \in \mathbb{C}^{n \times s}$ is the dimension-reduced TSMI, $\mathbf{H} := \overline{\mathbf{H}} \circ V$ and $\mathcal{B} = V^H \overline{\mathcal{B}}$ denotes the subspace-compressed Bloch solutions (for more details see [14]).

Tissue Quantification: Given the compressed measurements $\mathbf{y}$, the goal of MRF is to solve the inverse problem (2) and to compute the underlying multi-parametric maps $\mathbf{m} = \{T1, T2, \rho\}$ (and $\mathbf{x}$ as a bi-product). Such problems are typically casted as an optimisation problem of the form:

$$\arg\min_{\mathbf{x}, \mathbf{m}} \|\mathbf{y} - \mathbf{H}\mathbf{x}\|_2^2 + \phi(\mathbf{x}, \mathbf{m}), \tag{3}$$

and solved iteratively by the proximal gradient descent (PGD):

$$\text{PGD} : \begin{cases} \mathbf{g}^{(t+1)} = \mathbf{x}^{(t)} + \alpha^{(t)} \mathbf{H}^H (\mathbf{y} - \mathbf{H}\mathbf{x}^{(t)}) \to \text{Gradient with step size } \alpha \\ \{\mathbf{x}^{(t+1)}, \mathbf{m}^{(t+1)}\} = \text{Prox}_\phi(\mathbf{g}^{(t+1)}) \quad \to \text{Proximal update} \end{cases} \tag{4}$$

where the gradient updates encourage k-space fidelity (the first term of (3)), and the *proximal operator* $\text{Prox}_\phi(\cdot)$ enforces image structure priors through a *regularisation* term $\phi(\cdot)$ that makes the inverse problem well-posed. The Bloch dynamics in (2) place an important temporal constraint (prior) for per-voxel trajectories of $\mathbf{x}$. Projecting onto this model (i.e. a temporal Prox model) has been suggested via iterative dictionary search schemes [3, 10]. This approach boost MRF reconstruction accuracy compared to the non-iterative DM [20], however, DM is non-scalable and can create enormous storage and computational overhead. Further, such approach processes data independently per voxel and neglects important spatial domain regularities in the TSMIs and quantitative maps.

3   PGD-Net for MRF Quantification

We propose to learn a data-driven proximal operator within the PGD mechanism for solving the MRF problem. Implemented by compact networks with convolutional layers, the *neural* Prox improves the storage overhead and the sluggish runtime of the DM-based PGD by orders of magnitudes. Further, trained on quantitative MR images, the neural Prox network learns to simultaneously enforce spatial- and temporal-domain data structures within PGD iterations.

Prox Auto-encoder: We implement Prox : $\mathbf{g} \to \{\mathbf{x}, \mathbf{m}\}$ through a deep convolutional auto-encoder network:

$$\text{Prox} := \text{BLOCH} \circ \mathcal{G}, \tag{5}$$

consisting of an encoder $\mathcal{G} : \mathbf{g} \to \mathbf{m}$ and a decoder BLOCH: $\mathbf{m} \to \mathbf{x}$ subnetworks. The *information bottleneck* in the (neural) Prox auto-encoder corresponds to projecting multichannel TSMIs to the low-dimensional manifold of the tissues' intrinsic (quantitative) property maps [14].

Decoder Network: Creates a differentiable model for *generating* the Bloch magnetic responses. This network uses 1×1 filters to process image time-series in a voxel-wise manner. Given quantitative properties $\mathbf{m}_v = \{T1_v, T2_v, \gamma_v\}$, the decoder *approximates* (dimension-reduced) Bloch responses in voxel v i.e. $\mathrm{BLOCH}(\mathbf{m}_v) \approx \rho_v \mathcal{B}(T1_v, T2_v)$. This network is trained separately from the encoder. Training uses physical (Bloch) simulations for many combinations of the T1, T2 and PD values which can flexibly produce a rich training dataset [14].

Encoder Network: Projects $\mathbf{g}$ the gradient-updated TSMIs in each iteration (i.e. the first line of (4)) to the quantitative property maps $\mathbf{m}$. Thus, $\mathcal{G}$ must simultaneously (i) learn to incorporate spatial-domain regularities to de-alias TSMIs from the undersampling artefacts, and (ii) resolve the temporal-domain *inverse* mapping from the (noisy) TSMIs to the quantitative property maps. For this, and unlike BLOCH which applies pixel-wise temporal-only processing, $\mathcal{G}$ uses multichannel convolution filters with wider receptive fields to learn/enable spatiotemporal processing of the TSMIs.

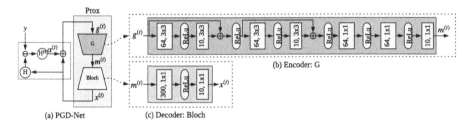

Fig. 1. Overview of the proposed proximal gradient descent network (PGD-Net) for tissue quantification in the compressive MR fingerprinting.

PGD-Net: Fig. 1a shows the *recurrent* architecture of the proposed *learned* PGD algorithm, coined as the PGD-NET. The trainable parameters within the PGD-NET are those of the encoder network $\mathcal{G}$ (Fig. 1b) and the step sizes α_t. Other operators such as $\mathbf{H}, \mathbf{H}^H$ and BLOCH (pre-trained separately, Fig. 1c) are kept frozen during training. Further, $\mathcal{G}$'s parameters are *shared* through all iterations. In practice, a truncated $T \geq 1$ recurrent iterations is used for training. Supervised training requires the MRF measurements, TSMIs, and the ground truth property maps to form the training input $\mathbf{y}$ and target $\mathbf{x}, \mathbf{m}$ samples.

Note there are many arts of engineering to determine the optimal network architecture, including different ways to encode temporal [17] or spatial-temporal information [5], these aspects are somewhat orthogonal to the model consistency question. Indeed, such mechanisms could also be incorporated in PGD-NET.

Training Loss: Given a training set $\{\mathbf{x}_i, \mathbf{m}_i, \mathbf{y}_i\}_{i=1}^N$, and $T \geq 1$ recurrent iterations of the PGD-NET (i.e. iterations used in PGD), the loss is defined as

$$\mathcal{L} = \gamma \sum_{i=1}^N \ell\left(\mathbf{x}_i, \mathbf{x}_i^{(T)}\right) + \sum_{j \in \{T_1, T_2, \rho\}} \beta_j \sum_{i=1}^N \ell\left(\mathbf{m}_{ij}, \mathbf{m}_{ij}^{(T)}\right) + \lambda \sum_{t=1}^T \sum_{i=1}^N \ell\left(\mathbf{y}_i, \mathbf{H}(\mathbf{x}_i^{(t)})\right),$$

(6)

where ℓ is the MSE loss defined with appropriate weights γ, β_j, λ on the reconstructed TSMIs $\mathbf{x}$ (which measures the Bloch dynamic consistency) and tissue property maps $\mathbf{m}$, as well as on $\mathbf{y}$ to maximise k-space data consistency with respect to the (physical) forward acquisition model. In this paper, the scaling between parameters γ, β_j and λ were initialized based on the physics (see 4.3).

4   Numerical Experiments

4.1   Anatomical Dataset

We construct a dataset of brain scans acquired using the 1.5T GE HDxT scanner with 8-channel receive-only head RF coil. For setting ground-truth (GT) values for the T1, T2 and PD parameters, gold standard anatomical maps were acquired using MAGIC quantification protocol [21]. Ground-truth quantitative maps were acquired from 8 healthy volunteers (16 axial brain slices each, at the spatial resolution of 128×128 pixels). From these parametric maps, we then construct the TSMIs and MRF measurements using the MRF acquisition protocol mentioned below to form the training/testing tuples $(\mathbf{m}_i, \mathbf{x}_i, \mathbf{y}_i)$. Data from 7 subjects were used for training our models, and one subject was kept for performance testing. We augmented training data into total 224 samples using random rotations (uniform angles in $[-8°, 8°]$), and left-right flipping of the GT maps. Training batches at each learning epoch were corrupted by i.i.d Gaussian noises of $30\,\mathrm{dB}$ SNR added to $\mathbf{y}$ (we similarly add noise to the k-space test data).

4.2   MRF Acquisition

Our experiments use an excitation sequence of $L = 200$ repetitions which jointly encodes T1 and T2 values using an inversion pulse followed by a flip angle schedule that linearly ramps up from $1°$ to $40°$, i.e. $\times 4$ truncated sequence than [14,16]. Following [16], we set acquisition parameters Tinv $= 18\,\mathrm{ms}$ (inversion time), fixed TR $= 10\,\mathrm{ms}$ (repetition time), and TE $= 0.46\,\mathrm{ms}$ (echo time). Spiral readouts subsample the k-space frequencies (the 128×128 Cartesian FFT grid) across 200 repetition times. We sample spatial frequencies $k(t) \propto (1.05)^{\frac{16\pi\tau}{1000}} e^{j\frac{16\pi\tau}{1000}}$ for $\tau = 1, 2, ..., 1000$, which after quantisation to the nearest FFT grid, results in $m = 654$ samples per timeframe. In every repetition, similar to [20], this spiral pattern rotates by $7.5°$ in order to sub-sample new k-space frequencies. Given the anatomical T1, T2 and PD maps, we simulate magnetic responses using the Extended Phase Graph (EPG) formalism [29] and construct TMSIs and k-space measurements datasets, and use them for training and retrospective validations.

4.3 Reconstruction Algorithms

Two DM baselines namely, the non-iterative Fast Group Matching (FGM) [6] and the model-based iterative algorithm BLIP empowered by the FGM's fast searches, were used for comparisons. For this, a MRF dictionary of 113'781 fingerprints was simulated over a dense grid of (T1, T2) = [100:10:4000] × [20:2:600] ms values. We implemented FGM searches on GPU using 100 groups for clustering this dictionary. The BLIP algorithm uses backtracking step size search and runs for maximum 20 iterations if is not convergent earlier. Further, we compared against related deep learning MRF baselines MRFCNN [8] and SCQ [12]. In particular, MRFCNN is a fully convolutional network and SCQ mainly uses 3 U-nets to separately infer T1, T2 and PD maps. The input to these networks is the dimension-reduced back-projected TSMIs $\mathbf{H}^H(\mathbf{y})$, and their training losses only consider quantitative maps consistency i.e. the second term in (6).

We trained PGD-NET with recurrent iterations $T = 2$ and 5 to learn appropriate proximal encoder $\mathcal{G}$ and the step sizes $\alpha^{(t)}$. The architectures of $\mathcal{G}$ and BLOCH networks are illustrated in Fig. 1. Similar to [14], the MRF dictionary was used for pre-training the BLOCH decoder that embeds a differentiable model for generating Bloch magnetic responses. A compact shallow network with one hidden layer and 1×1 filters (for pixel-wise processing) implements our BLOCH model [14]. On the other hand, our encoder $\mathcal{G}$ has two residual blocks with 3×3 filters (for de-aliasing) followed by three convolutional layers with 1×1 filters for quantitative inference. The final hyper-parameters were $\beta = [1, 20, 2.5]$, $\gamma = 10^{-3}$ and $\lambda = 10^{-2}$ selected via a multiscale grid search to minimize error w.r.t. the ground truth. The inputs were normalized such that PD ranged in $[0, 1]$; smaller weights were used for $\mathbf{x}$ and $\mathbf{y}$ since they have higher energy than PD; we set $\lambda > \gamma$ since $\mathbf{x}$'s norm is larger than $\mathbf{y}$; $T1/T2$ values typically exhibit different ranges with $T1 \gg T2$, justifying their relative weightings in β to balance these terms. We used ADAM optimiser with 2000 epochs, mini-batch size 4 and learning rate 10^{-4}. We pre-trained our encoder $\mathcal{G}$ using back-projected TSMIs to initialise the recurrent training, and also to compare the *encoder alone* predictions to the PGD-NET. All algorithms use a $s = 10$ dimensional MRF subspace representation for temporal-domain dimensionality reduction. The input and output channels are respectively 10 and 3 for MRFCNN, SCQ and $\mathcal{G}$. All networks were implemented in PyTorch, and trained and tested on NVIDIA 2080Ti GPUs.

4.4 Results and Discussions

Table 1 and Fig. 2 compare the performances of the MRF baselines against our proposed PGD-NET using $T = 2$ and 5 recurrent iterations. We also include inference results using the proposed *encoder alone* $\mathcal{G}$, without proximal iterations. Reconstruction performances were measured by the Normalised RMSE $= \frac{\|T1 - T1^{GT}\|}{\|T1^{GT}\|}$, MAE $= |T1 - T1^{GT}|$, Structural Similarity Index Metric (SSIM) [28], the required storage for the MRF dictionary (in DM methods) or the networks, and the algorithm runtimes averaged over the test image slices.

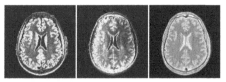

Ground truth (T1, T2, PD) anatomical maps acquired by the MAGIC gold standard [21]

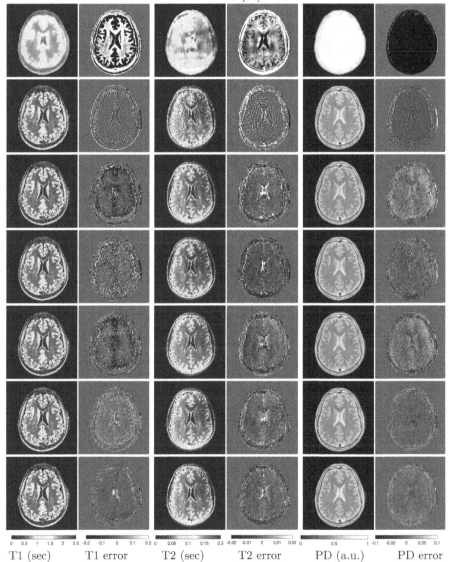

| T1 (sec) | T1 error | T2 (sec) | T2 error | PD (a.u.) | PD error |

Fig. 2. A slice of the true T1, T2 and PD maps acquired by the gold standard MAGIC (top), and the corresponding MRF reconstructions using (from top to bottom) FGM, BLIP+FGM, MRFCNN, SCQ, encoder alone $\mathcal{G}$, and the PROXNET with $T = 2$ and $T = 5$ algorithms.

Table 1. Average errors (NRMSE, SSIM, MAE), memory (for storing a dictionary or a network) and runtimes (per image sclice) required for computing T1, T2 and PD maps using the MRF baselines and our PGD-NET algorithm.

	NRMSE			SSIM			MAE (msec)		Time (sec)	Memory (MB)
	T1	T2	PD	T1	T2	PD	T1	T2		
FGM	0.475	0.354	1.12	0.614	0.652	0.687	350.0	14.6	1.29	8.81
BLIP+FGM	0.230	0.545	0.073	0.886	0.880	0.984	91.7	8.0	79.28	8.81
MRFCNN	0.155	0.158	0.063	0.943	0.972	0.987	80.3	5.4	0.083	4.72
SCQ	0.172	0.177	0.064	0.929	0.967	0.984	91.7	6.1	0.132	464.51
$\mathcal{G}$ (encoder alone)	0.142	0.155	0.065	0.948	0.973	0.987	77.1	5.6	**0.067**	**0.55**
PGD-NET ($T = 2$)	0.104	0.138	0.050	0.973	0.979	0.991	59.9	5.0	0.078	0.57
PGD-NET ($T = 5$)	**0.100**	**0.132**	**0.045**	**0.975**	**0.981**	**0.992**	**50.8**	**4.6**	0.103	0.57

The non-iterative FGM results in incorrect maps due to the severe under-sampling artefacts. The model-based BLIP iterations improve this, however, due to lacking spatial regularisation, BLIP has limited accuracy and cannot fully remove aliasing artefacts (e.g. see T2 maps in Fig. 2) despite 20 iterations and very long runtime. In contrast, all deep learning methods outperform BLIP not only in accuracy but also in having 2 to 3 orders of magnitude faster reconstruction times—an important advantage of the learning-based methods. The proposed PGD-NET consistently outperforms all baselines, including DM and learning-based methods, over all defined accuracy metrics. This is achieved due to learning an effective spatiotemporal model (only) for the proximal operator i.e. the $\mathcal{G}$ and BLOCH networks, directly incorporating the physical acquisition model **H** into the recurrent iterations to avoid over-parameterisation of the overall inference model, as well as enforcing reconstructions to be consistent with the Bloch dynamics and the k-space data through the multi-term training loss (6). The MRFCNN and SCQ over-parametrise the inference by 1 and 3 orders of magnitude larger model sizes (the SCQ requires larger memory than DM) and are unable to achieve PGD-NET's accuracy e.g. see the corresponding over-smoothed T2 maps in Fig. 2. Finally, we observe that despite having roughly the same model size (storage), the *encoder alone* $\mathcal{G}$ predictions are not as accurate as the results of the PGD-NET's recurrent iterations. By increasing the number of iterations T we observe that the PGD-NET's accuracy consistently improves despite having an acceptable longer inference time.

5   Conclusions

In this work we showed that the consistency of the computed quantitative maps with respect to the physical forward acquisition model and the Bloch dynamics is important for reliably solving the MRF inverse problem using compact deep neural networks. For this, we proposed PGD-NET, a learned model-based iterative reconstruction framework that directly incorporates the forward acquisition and Bloch dynamic models within a recurrent learning mechanism with a multi-term training loss. The PGD-NET adopts a data-driven neural proximal model for spatiotemporal processing of the MRF data, TSMI de-aliasing and quantitative inference. A chief advantage of this model is its compactness (a small number of weights/biases to tune), which might makes it particularly suitable for supervised training using scarce quantitative MRI datasets. Through our numerical validations we showed that the proposed PGD-NET achieves a superior quantitative inference accuracy, much smaller storage requirement, and a comparable runtime to the recent deep learning MRF baselines, while being much faster than the MRF fast dictionary matching schemes. In future work, we plan to evaluate the non-simulated scanner datasets with higher diversities and possible pathologies to further validate the method's potential for clinical usage.

Acknowledgements. We thank Pedro Gómez, Carolin Prikl and Marion Menzel from GE Healthcare for the quantitative anatomical maps dataset. DC and MD are supported by the ERC C-SENSE project (ERCADG-2015-694888).

References

1. Adler, J., Öktem, O.: Solving ill-posed inverse problems using iterative deep neural networks. Inverse Prob. **33**(12), 124007 (2017)
2. Aggarwal, H.K., Mani, M.P., Jacob, M.: MoDL: model-based deep learning architecture for inverse problems. IEEE Trans. Med. Imaging **38**(2), 394–405 (2018)
3. Assländer, J., Cloos, M.A., Knoll, F., Sodickson, D.K., Hennig, J., Lattanzi, R.: Low rank alternating direction method of multipliers reconstruction for MR fingerprinting. Magn. Reson. Med. **79**(1), 83–96 (2018)
4. Balsiger, F., et al.: Magnetic resonance fingerprinting reconstruction via spatiotemporal convolutional neural networks. In: Knoll, F., Maier, A., Rueckert, D. (eds.) MLMIR 2018. LNCS, vol. 11074, pp. 39–46. Springer, Cham (2018). https://doi.org/10.1007/978-3-030-00129-2_5
5. Balsiger, F., Scheidegger, O., Carlier, P.G., Marty, B., Reyes, M.: On the spatial and temporal influence for the reconstruction of magnetic resonance fingerprinting. In: International Conference on Medical Imaging with Deep Learning, pp. 27–38 (2019)
6. Cauley, S.F., et al.: Fast group matching for MR fingerprinting reconstruction. Magn. Reson. Med. **74**(2), 523–528 (2015)
7. Chen, D., Davies, M.E.: Deep decomposition learning for inverse imaging problems. In: Proceedings of the European Conference on Computer Vision (ECCV) (2020)
8. Chen, D., Golbabaee, M., Gómez, P.A., Menzel, M.I., Davies, M.E.: Deep fully convolutional network for MR fingerprinting. In: International Conference on Medical Imaging with Deep Learning (MIDL), London, United Kingdom, 08–10 July 2019
9. Cohen, O., Zhu, B., Rosen, M.S.: MR fingerprinting deep reconstruction network (drone). Magn. Reson. Med. **80**(3), 885–894 (2018)
10. Davies, M., Puy, G., Vandergheynst, P., Wiaux, Y.: A compressed sensing framework for magnetic resonance fingerprinting. SIAM J. Imaging Sci. **7**(4), 2623–2656 (2014)
11. Doneva, M., Amthor, T., Koken, P., Sommer, K., Börnert, P.: Matrix completion-based reconstruction for undersampled magnetic resonance fingerprinting data. Magn. Reson. Imaging **41**, 41–52 (2017)

12. Fang, Z., et al.: Deep learning for fast and spatially-constrained tissue quantification from highly-accelerated data in magnetic resonance fingerprinting. IEEE Trans. Med. Imaging **38**(10), 2364–2374 (2019)
13. Fang, Z., Chen, Y., Nie, D., Lin, W., Shen, D.: RCA-U-Net: residual channel attention U-Net for fast tissue quantification in magnetic resonance fingerprinting. In: Shen, D., et al. (eds.) MICCAI 2019. LNCS, vol. 11766, pp. 101–109. Springer, Cham (2019). https://doi.org/10.1007/978-3-030-32248-9_12
14. Golbabaee, M., et al.: Compressive MRI quantification using convex spatiotemporal priors and deep auto-encoders. arXiv preprint arXiv:2001.08746 (2020)
15. Golbabaee, M., Chen, D., Gómez, P.A., Menzel, M.I., Davies, M.E.: Geometry of deep learning for magnetic resonance fingerprinting. In: IEEE International Conference on Acoustics, Speech and Signal Processing (ICASSP), pp. 7825–7829 (2019)
16. Gómez, P.A., et al.: Rapid three-dimensional multiparametric MRI with quantitative transient-state imaging. arXiv preprint arXiv:2001.07173 (2020)
17. Hoppe, E., et al.: Deep learning for magnetic resonance fingerprinting: a new approach for predicting quantitative parameter values from time series. Stud. Health Technol. Inf. **243**, 202 (2017)
18. Hoppe, E., et al.: RinQ fingerprinting: recurrence-informed quantile networks for magnetic resonance fingerprinting. In: Shen, D., et al. (eds.) MICCAI 2019. LNCS, vol. 11766, pp. 92–100. Springer, Cham (2019). https://doi.org/10.1007/978-3-030-32248-9_11
19. Jiang, Y., Ma, D., Seiberlich, N., Gulani, V., Griswold, M.A.: MR fingerprinting using fast imaging with steady state precession (FISP) with spiral readout. Magn. Reson. Med. **74**(6), 1621–1631 (2015)
20. Ma, D., et al.: Magnetic resonance fingerprinting. Nature **495**(7440), 187 (2013)
21. Marcel, W., AB, S.: New technology allows multiple image contrasts in a single scan. SPRING, 6–10 (2015). GESIGNAPULSE.COM/MR
22. Mardani, M., et al.: Neural proximal gradient descent for compressive imaging. In: Advances in Neural Information Processing Systems, pp. 9573–9583 (2018)
23. McGivney, D.F., et al.: SVD compression for magnetic resonance fingerprinting in the time domain. IEEE Trans. Med. Imaging **33**(12), 2311–2322 (2014)
24. Oksuz, I., et al.: Magnetic resonance fingerprinting using recurrent neural networks. In: IEEE International Symposium on Biomedical Imaging (ISBI), pp. 1537–1540 (2019)
25. Rick Chang, J., Li, C.L., Poczos, B., Vijaya Kumar, B., Sankaranarayanan, A.C.: One network to solve them all-solving linear inverse problems using deep projection models. In: Proceedings of the IEEE International Conference on Computer Vision, pp. 5888–5897 (2017)
26. Song, P., Eldar, Y.C., Mazor, G., Rodrigues, M.R.: HYDRA: hybrid deep magnetic resonance fingerprinting. Med. Phys. **46**(11), 4951–4969 (2019)
27. Virtue, P., Yu, S.X., Lustig, M.: Better than real: complex-valued neural nets for MRI fingerprinting. arXiv preprint arXiv:1707.00070 (2017)
28. Wang, Z., Bovik, A.C., Sheikh, H.R., Simoncelli, E.P.: Image quality assessment: from error visibility to structural similarity. IEEE Trans. Image Process. **13**(4), 600–612 (2004)
29. Weigel, M.: Extended phase graphs: dephasing, RF pulses, and echoes-pure and simple. J. Magn. Reson. Imaging **41**(2), 266–295 (2015)
30. Zhao, B., et al.: Improved magnetic resonance fingerprinting reconstruction with low-rank and subspace modeling. Magn. Reson. Med. **79**(2), 933–942 (2018)

Active MR k-space Sampling with Reinforcement Learning

Luis Pineda[1(✉)], Sumana Basu[2], Adriana Romero[1], Roberto Calandra[1], and Michal Drozdzal[1]

[1] Facebook AI Research, Montreal, Canada
lep@fb.com
[2] McGill University, Montreal, Canada

Abstract. Deep learning approaches have recently shown great promise in accelerating magnetic resonance image (MRI) acquisition. The majority of existing work have focused on designing better reconstruction models given a pre-determined acquisition trajectory, ignoring the question of trajectory optimization. In this paper, we focus on learning acquisition trajectories given a fixed image reconstruction model. We formulate the problem as a sequential decision process and propose the use of reinforcement learning to solve it. Experiments on a large scale public MRI dataset of knees show that our proposed models significantly outperform the state-of-the-art in active MRI acquisition, over a large range of acceleration factors.

Keywords: Active MRI acquisition · Reinforcement learning

1 Introduction

Magnetic resonance imaging (MRI) is a powerful imaging technique used for medical diagnosis.

Its advantages over other imaging modalities, such as computational tomography, are its superior image quality and zero radiation exposure. Unfortunately, MRI acquisition is slow (taking up to an hour) resulting in patient discomfort and in artifacts due to patient motion.

MRI scanners sequentially acquire k-space measurements, collecting raw data from which an image is reconstructed (e.g., through an inverse Fourier transform). A common way to accelerate MRI acquisition is to collect less measurements and reconstruct the image using a partially observed k-space. Since this results in image blur or aliasing artifacts, traditional techniques enhance the image reconstruction using regularized iterative optimization techniques such as compressed sensing [6]. More recently, the inception of large scale MRI reconstruction datasets, such as [22], have enabled the successful use of deep learning

Electronic supplementary material The online version of this chapter (https://doi.org/10.1007/978-3-030-59713-9_3) contains supplementary material, which is available to authorized users.

A. L. Martel et al. (Eds.): MICCAI 2020, LNCS 12262, pp. 23–33, 2020.
https://doi.org/10.1007/978-3-030-59713-9_3

approaches to MRI reconstruction [1,3,7,12,18,19,23,26]. However, these methods focus on designing models that improve image reconstruction quality for a *fixed* acceleration factor and set of measurements.

In this paper, we consider the problem of optimizing the sequence of k-space measurements (i.e., trajectories) to reduce the number of measurements taken. Previous research on optimizing k-space measurement trajectories is extensive and include Compressed Sensing-based techniques [2,11,13,24], SVD basis techniques [9,27,28], and region-of-interest techniques [20]. However, all these solutions work with *fixed trajectories* at inference time. Only recently, on-the-fly acquisition trajectory optimization methods for deep-learning-based MRI reconstruction have emerged in the literature [4,25]. On the one hand, [4] showed that jointly optimizing acquisition trajectory and reconstruction model can lead to slight increase in image quality for a *fixed* acceleration factor with subject-specific acquisition trajectories. On the other hand, [25] introduced *active* MRI acquisition, where both acceleration factor and acquisition trajectory are subject-specific. Their proposed approach performs trajectory optimization over a *full range* of possible accelerations but makes a myopic approximation during training that ignores the sequentiality of k-space acquisition.

In contrast, we focus on Cartesian sampling trajectories and expand the formulation of [25]. More precisely, we specify the active MRI acquisition problem as a Partially Observable Markov Decision Process (POMDP) [5,14], and propose the use of deep reinforcement learning [8] to solve it.[1] Our approach, by formulation, optimizes the reconstruction over the whole range of acceleration factors while considering the sequential nature of the acquisition process – future scans and reconstructions are used to determine the next measurement to take. We evaluate our approach on a large scale single-coil knee dataset [22][2] and show that it outperforms common acquisition heuristics as well as the myopic approach of [25]. Our contributions are: 1) formulating active MRI acquisition as a POMDP; 2) showing state-of-the-art results in active MRI acquisition on a large scale single-coil knee dataset; 3) performing an in-depth analysis of the learned trajectories. The code to reproduce our experiments is available at: https://github.com/facebookresearch/active-mri-acquisition.

2   Background

Partially Observable Markov Decision Processes (POMDP). A POMDP is a highly expressive model for formulating problems that involve sequential

---

[1] For a comprehensive overview of reinforcement learning applied to healthcare data and medical imaging please refer to [21].

[2] Data used in the preparation of this article were obtained from the NYU fastMRI Initiative database (fastmri.med.nyu.edu) [22]. As such, NYU fastMRI investigators provided data but did not participate in analysis or writing of this report. A listing of NYU fastMRI investigators, subject to updates, can be found at: fastmri.med.nyu.edu. The primary goal of fastMRI is to test whether machine learning can aid in the reconstruction of medical images.

decisions [5,10,14]. A solution to a POMDP is a *policy* that maps history of observations (in our case, the k-space measurements) to actions (in our case, an index of k-space measurement to acquire). A POMDP is formally defined by: (1) a set of *states*, summarizing all the relevant information to determine what happens next; (2) a set of *observations*, containing indirect information about the underlying states, which are themselves not directly observable; (3) a set of *actions*, that cause the transition between states; (4) a *transition function*, specifying a probability distribution for moving from one state to another, after taking an action; (5) an *emission function*, defining the probability of obtaining each possible observation after taking an action in some state; (6) a *reward function*, specifying the numeric feedback obtained when moving from one state to another; and (7) and a *discount factor*, defining the impact of immediate versus more distant rewards.

Double Deep Q-Networks (DDQN). (DDQN) [17] is a state-of-the-art deep reinforcement learning method for solving high-dimensional POMDPs. Since active MRI acquisition involves discrete actions (e.g., indexes of k-space measurements to acquire), we focus on the Double DQN (DDQN) [17] algorithm, given its training stability and recent success. In DDQN, policies are not explicitly modelled. Instead, a *value network* is used to predict the *value* of each possible action—defined as the expectation of the future cumulative reward. A policy is then recovered by greedily choosing actions with maximal estimated value. The value prediction problem is posed as a supervised regression problem, trained to minimize the temporal difference error [15] over data sampled from a *replay memory buffer*, which contains tuples of states, actions and rewards obtained by running the learned policy in an exploratory way. The target values used to update the value network are computed as bootstrapped estimates of the estimated value. Further details can be found in [17].

3   Learning Active MR k-space Sampling

Let $\mathbf{y} \in \mathbb{C}^{M \times N}$ be a complex matrix representing a fully sampled k-space, and $\mathbf{x}$ be a reconstruction of $\mathbf{y}$ obtained via Inverse Fourier Transform, $\mathcal{F}^{-1}$; i.e., $\mathbf{x} = \mathcal{F}^{-1}(\mathbf{y})$. We simulate partially observed k-space by masking $\mathbf{y}$ with a Cartesian binary mask $\mathbf{M}$, resulting in $\tilde{\mathbf{y}} = \mathbf{M} \odot \mathbf{y}$. We denote the corresponding *zero-filled* reconstruction as $\tilde{\mathbf{x}} = \mathcal{F}^{-1}(\tilde{\mathbf{y}})$. In this paper, we use a deep-learning-based reconstruction model $\mathbf{r}(\tilde{\mathbf{x}}; \phi) \rightarrow \hat{\mathbf{x}}$ that takes $\tilde{\mathbf{x}}$ as input, and outputs a de-aliased image reconstruction $\hat{\mathbf{x}}$. The reconstruction model is a convolutional neural network (CNN) parametrized by ϕ; in particular, we use the network proposed in [25]. Finally, we use subindices to denote time, e.g., $\hat{\mathbf{x}}_t = \mathbf{r}(\mathcal{F}^{-1}(\mathbf{M}_t \odot \mathbf{y}))$ represents the reconstruction obtained at time step t of the acquisition process.

3.1   Active MRI Acquisition as POMDP

In our active acquisition formulation, the goal is to learn a *policy* function $\pi(\hat{\mathbf{x}}_t, \mathbf{M}_t; \theta) \rightarrow a_t$ that, at time t, maps the tuple of de-aliased image reconstruction $\hat{\mathbf{x}}_t$, and mask representing observed frequencies $\mathbf{M}_t$, to the *action* a_t.

In Cartesian active MRI acquisition, actions are represented by the unobserved columns of the fully sampled k-space, $\mathbf{y}$ (the ones for which the mask $\mathbf{M}_t$ has value of 0). Once an action is observed, both the mask and the de-aliased image reconstruction are updated to $\mathbf{M}_{t+1} = \mathbf{M}_t + \mathbf{M}^{a_t}$ and $\hat{\mathbf{x}}_{t+1} = \mathbf{r}(\mathcal{F}^{-1}(\mathbf{M}_{t+1} \odot \mathbf{y}))$, where $\mathbf{M}^{a_t}$ is a binary matrix with all zeros except of the column indicated by the action a_t. The parameters θ of the policy are optimized to select a sequence of actions (k-space measurements) $[a_t, a_{t+1}, ..., a_T]$ that minimize the acquisition cost function (i.e., maximize the reward function) over all future reconstructions up to time step T. In this work, we consider acquisition costs of the form $f(\mathcal{C}(\hat{\mathbf{x}}_1, \mathbf{x}), \mathcal{C}(\hat{\mathbf{x}}_2, \mathbf{x}), ..., \mathcal{C}(\hat{\mathbf{x}}_T, \mathbf{x}))$, where $\mathcal{C}$ represents a pre-defined cost of interest (e.g., Mean Squared Error or negative SSIM), and f is a function that aggregates the costs observed throughout an acquisition trajectory (e.g., sum, area under the metric curve, final metric value at time T). Thus, the objective is to minimize the aggregated cost over the *whole range* of MRI acquisition speed-ups. Further, we assume that f can be expressed as a discounted sum of T rewards, one per time step. Under this assumption, we formally introduce the active acquisition problem as a POMDP defined by:

- **State set:** The set of all possible tuples $\mathbf{s}_t \triangleq \langle \mathbf{x}, \mathbf{M}_t \rangle$. Note that the ground truth image, $\mathbf{x}$, is hidden and the current mask, $\mathbf{M}_t$, is fully visible.
- **Observation set:** The set of all possible tuples $\mathbf{o}_t \triangleq \langle \hat{\mathbf{x}}, \mathbf{M}_t \rangle$.
- **Action set:** The set of all possible k-space column indices that can be acquired, i.e., $\mathcal{A} \triangleq \{1, 2, ..., W\}$, where W is the image width. Since sampling an already observed k-space column does not improve reconstruction, we specify that previously observed columns are invalid actions.
- **Transition function:** Given current mask $\mathbf{M}_t$ and a valid action $a_t \in \mathcal{A}$, the mask component of the state transitions deterministically to $\mathbf{M}_{t+1} = \mathbf{M}_t + \mathbf{M}^{a_t}$, and $\mathbf{x}$ remains unchanged. After T steps the system reaches the final time step.
- **Emission function:** In our formulation, we assume that the reconstruction model returns a deterministic reconstruction $\hat{\mathbf{x}}_t$ at each time step t; thus, the observation after taking action a_t at state $\mathbf{s}_t$ is defined as $\mathbf{o}_t \triangleq \hat{\mathbf{x}}_{t+1}$ (i.e., the reconstruction after adding the new observed column to the mask).
- **Reward function:** We define the reward as the decrease in reconstruction metric with respect to the previous reconstruction: $R(\mathbf{s}_t, a_t) = \mathcal{C}(\hat{\mathbf{x}}_{t+1}, \mathbf{x}) - \mathcal{C}(\hat{\mathbf{x}}_t, \mathbf{x})$. This assumes that $f(\mathcal{C}(\hat{\mathbf{x}}_1, \mathbf{x}), \mathcal{C}(\hat{\mathbf{x}}_2, \mathbf{x}), ..., \mathcal{C}(\hat{\mathbf{x}}_T, \mathbf{x})) \triangleq \mathcal{C}(\hat{\mathbf{x}}_T, \mathbf{x})$. We found this reward to be easier to optimize than rewards corresponding to more complex aggregations.
- **Discount factor:** We treat the discount, $\gamma \in [0, 1]$, as a hyperparameter.

Note that the above-mentioned POMDP is *episodic*, since the acquisition process has a finite number of steps T. At the beginning of each episode, the acquisition system is presented with an initial reconstruction, $\hat{\mathbf{x}}_0$, of an unobserved ground truth image $\mathbf{x}$, as well with an initial subsampling mask $\mathbf{M}_0$. The system then proceeds to iteratively suggest k-space columns to sample, and receives updated reconstructions from $\mathbf{r}$.

3.2   Solving the Active MRI Aquisition POMDP with DDQN

We start by defining a subject-specific DDQN value network. As mentioned in Sect. 2, POMDP policies are functions of observation histories. However, in active Cartesian MRI acquisition, the whole history of observations is captured by the current observation $\mathbf{o}_t$. Thus, we use $\mathbf{o}_t$ as single input to our value network. We design the value network architecture following [25]'s evaluator network, which receives as input a reconstructed image, $\hat{\mathbf{x}}_t$, and a mask $\mathbf{M}_t$. To obtain the reconstructed image $\mathbf{x}_t$ at each time step, we use a pre-trained reconstruction network. Additionally, we also consider a dataset-specific DDQN variant, which only takes time step information as input (which is equivalent to considering the number of non-zero elements in the mask $\mathbf{M}_t$), and thus, uses the same acquisition trajectory for all subjects in the dataset.

In both cases, we restrict the value network to select among valid actions by setting the value of all previously observed k-space columns to $-\infty$. Additionally, we use a modified ϵ-greedy policy [16] as exploration policy to fill the replay memory buffer. This policy chooses the best action with probability $1 - \epsilon$, and chooses an action from the set of valid actions with probability ϵ.

4   Experimental Results

Datasets and Baselines. To train and evaluate our models, we use the single-coil knee RAW acquisitions from the fastMRI dataset [22], composed of 536 public training set volumes and 97 public validation set volumes. We create a held-out test set by randomly splitting the public validation set into a new validation set with 48 volumes, and a test set with 49. This results in 19,878 2D images for training, 1785 images for validation, and 1851 for test. All data points are composed of a 640×368 complex valued k-space matrix with the 36 highest frequencies zero-padded. In all experiments, we use vertical Cartesian masks, such that one action represent acquisition of one column in the k-space matrix. Hence, the total number of possible actions for this dataset is 332.

We compare our approach to the following heuristics and baselines: (1) Random policy (RANDOM): Randomly select an unobserved k-space column; (2) Random with low frequency bias (RANDOM-LB): Randomly select an unobserved k-space column, favoring low frequency columns; (3) Low-to-high policy (LOWTOHIGH): Select an unobserved k-space column following a low to high frequency order; and (4) Evaluator policy (EVALUATOR): Select an unobserved k-space column following the observation scoring function introduced by [25]. To ensure fair comparisons among all methods, we fix same number of low-frequency observations (for details see Sect. 4). Moreover, for reference, we also include results for a one-step oracle policy that, having access to ground truth at test time, chooses the frequency that will reduce $\mathcal{C}$ the most (denoted as ORACLE).

Training Details. In our experiments, we use the reconstruction architecture from [25] and train it using negative log-likelihood on the training set. Following [25], the reconstructor is trained on the whole range of acceleration

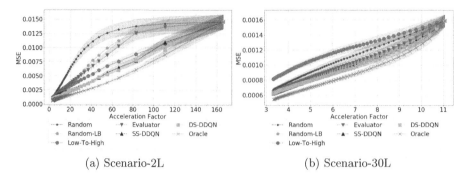

(a) Scenario-2L   (b) Scenario-30L

Fig. 1. Reconstruction MSE vs. acceleration factor for different strategies. Both DDQN models outperform all baselines and heuristics for the vast majority of acceleration factors considered.

factors, by randomly sampling masks with different patterns and acceleration factors each time an image is loaded. Similarly to [25], we also force a fixed number of low frequencies to be always observed, and train two versions of the reconstruction model: one that always observes 30 low frequencies—referred to as Scenario-30L—, corresponding to $\sim 3\times -11\times$ accelerations, and another that always observes 2 low frequencies—referred to as Scenario-2L—, corresponding to extreme acceleration factors of up to $166\times$ acceleration. In contrast to [25], we train the reconstructor and the policy networks in stages, as we noticed that it leads to better results. Note that none of the architectures used in the experiments uses complex convolutions. For both DDQN approaches, we experimented with the reward defined in Sect. 3.1 and four choices of $\mathcal{C}$: Mean Squared Error (MSE), Normalized MSE, negative Peak Signal to Noise Ratio (PSNR) and negative Structural Similarity (SSIM). For each metric, we trained a value network using a budget of $T = 100 - L$ actions, where $L = 2$ or $L = 30$ is the number of initial fixed low frequencies, a discount factor $\gamma = 0.5$, and a replay buffer of 20,000 tuples. Each training episode starts with a mask with a fixed number of low frequencies L. The DDQN models are trained for 5,000,000 state transition steps, roughly equivalent to 3.6 iterations over the full training set for $L = 30$, or 2.6 for $L = 2$. We periodically evaluate the current greedy policy on the validation set, and use the best validation result to select the final model for testing. To reduce computation, we used a random subset of 200 validation images when training DDQN.

4.1   Results

Figure 1(a–b) depicts the average test set MSE as a function of acceleration factor, with Scenario-2L on Fig. 1(a) and Scenario-30L on Fig. 1(b).[3] Results show that both DDQN models outperform all considered heuristics and baselines for the

---

[3] Plots for NMSE, SSIM and PSNR are available in the supplementary material.

Table 1. Average test set AUC with 95% confidence intervals (one AUC value/image) for 6 different active acquisition policies (Scenario-30L).

Metric	RANDOM	RANDOM-LB	LOWTOHIGH	EVALUATOR	SS-DDQN	DS-DDQN
MSE ($\times 10^{-3}$) ↓	8.90 (0.41)	8.24 (0.37)	9.64 (0.45)	8.33 (0.38)	8.00 (0.35)	**7.94 (0.35)**
NMSE ($\times 10^{-1}$) ↓	3.02 (0.16)	2.93 (0.16)	3.13 (0.17)	3.06 (0.17)	2.88 (0.17)	**2.87 (0.16)**
PSNR ($\times 10^2$) ↑	2.23 (1.28)	2.25 (1.75)	2.21 (1.23)	2.24 (1.33)	**2.27 (1.34)**	2.26 (1.35)
SSIM ↑	4.77 (0.06)	4.82 (0.07)	4.71 (0.06)	4.78 (0.07)	**4.86 (0.07)**	**4.86 (0.07)**

Table 2. Average test set AUC with 95% confidence intervals (one AUC value/image) for 6 different active acquisition policies (Scenario-2L).

Metric	RANDOM	RANDOM-LB	LOWTOHIGH	EVALUATOR	SS-DDQN	DS-DDQN
MSE ↓	1.97 (0.17)	1.73 (0.15)	1.39 (0.10)	1.70 (0.15)	1.31 (0.11)	**1.24 (0.10)**
NMSE ↓	20.6 (0.43)	18.7 (0.41)	17.5 (0.37)	17.7 (0.42)	15.5 (0.35)	**15.1 (0.34)**
PSNR ($\times 10^3$) ↑	3.63 (0.02)	3.73 (0.02)	3.76 (0.02)	3.79 (0.02)	3.89 (0.02)	**3.90 (0.02)**
SSIM ↑	73.2 (1.19)	75.3 (1.22)	73.3 (1.19)	76.0 (1.23)	**78.0 (1.25)**	**78.0 (1.25)**

vast majority of acceleration factors. In Scenario-30L, the mean MSE obtained with our models is between 3–7% lower than the best baseline (EVALUATOR), for all accelerations between 4× and 10×. For the case of extreme accelerations (Scenario-2L), our best model (dataset-specific DDQN) outperforms the best heuristic (LOWTOHIGH) by at least 10% (and up to 35%) on all accelerations below 100×. Note that for this scenario and the MSE metric, the performance of RANDOM and EVALUATOR deteriorated significantly compared to Scenario-30L. To further facilitate comparison among all methods, we summarize the overall performance of an acquisition policy into a single number, by estimating, for each image, the area under curve (AUC), and averaging the resulting values over the test set. Results are summarized in Table 1 for Scenario-30L and Table 2 for Scenario-2L. In all cases, the DDQN policies outperform all other models. In Scenario-30L, the improvements of our models over the best baselines range from 0.55% to 2.9%, depending on the metric. In Scenario-2L, the improvements range from 2.68% to 11.6%. Further, paired t-tests (pairing over images) between the AUCs obtained with our models and those of the best baseline, for each metric, indicate highly significant differences, with p-values generally lower than 10^{-8}. Interestingly, we found that the data-specific DDQN slightly outperforms the subject-specific one for most metrics. While this seems to suggest that subject-specific trajectories are not necessary, we point out that the gap between the performance of ORACLE and the models considered indicates the opposite. In particular, policy visualizations for ORACLE (see Sect. 4.2) show wide subject-specific variety. Therefore, we hypothesize that the better performance of DS-DDQN is due to optimization and learning stability issues.

Qualitative Results. Figure 2 shows examples of reconstructions together with error maps for four different acquisition policies (RANDOM, LOWTOHIGH, EVALUATOR, subject-specific DDQN). We display the 10× and 8× acceleration for

Scenario-2L and Scenario-30L, respectively. Looking at the error maps (bottom row), the differences in reconstruction quality between the subject-specific DDQN and the rest of the policies are substantial, and the reconstruction is visibly sharper and more detailed than the ones obtained with the baselines.

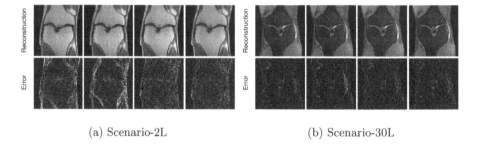

(a) Scenario-2L   (b) Scenario-30L

Fig. 2. Reconstructions and error maps under 4 acquisition policies (from left: RANDOM, LOWTOHIGH, EVALUATOR, SS-DDQN) at 10× acceleration for Scenario-2L and 8× for Scenario-30L. The images depict magnitude information. Note that the subfigures (a) and (b) depict different knee images. Additional images are shown in the supplementary material.

4.2   Policy Visualization

Figure 3 illustrates DDQN policies on the test set for DDQN trained with MSE and negative SSIM acquisition costs. Each row in the heat maps represents a cumulative distribution function of the time (x-axis) at which the corresponding k-space frequency is chosen (y-axis); the darker color the higher the values. Note that low frequencies are closer to the center of the heat map, while high frequencies are closer to the edges. In the dataset-specific DDQN heat maps, each row instantly transitions from light to dark intensities at the time where the frequency is chosen by the policy (recall that this policy is a deterministic function of time). In the subject-specific DDQN, smoother transitions indicate that frequencies are not always chosen at the same time step. Furthermore, one can notice that some frequencies are more likely to be chosen earlier, indicated by a dark intensity appearing closer to the left side of the plot. Overall, for both models and costs, we observe a tendency to start the acquisition process by choosing low and middle frequencies, while incorporating high frequencies relatively early in the process. However, when comparing MSE-based to SSIM-based policies, we observe that the SSIM policy is more biased towards low frequencies and it seems to take advantage of k-space Hermitian symmetry – only few actions selected in the upper-center part of the SSIM heat maps.

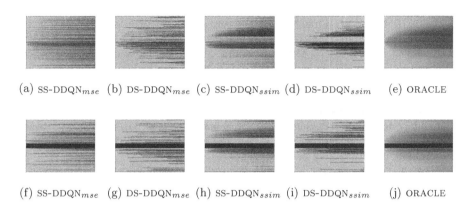

(a) SS-DDQN$_{mse}$ (b) DS-DDQN$_{mse}$ (c) SS-DDQN$_{ssim}$ (d) DS-DDQN$_{ssim}$ (e) ORACLE

(f) SS-DDQN$_{mse}$ (g) DS-DDQN$_{mse}$ (h) SS-DDQN$_{ssim}$ (i) DS-DDQN$_{ssim}$ (j) ORACLE

Fig. 3. Policy visualizations for DDQN models: Scenario-2L (a–e) and Scenario-30L (e–j). Subfigures (e) and (j) shows oracle policy obtained with SSIM criteria. See main text for details. See supplementary for additional results.

5   Conclusion

In this paper, we formulated the active MRI acquisition problem as a Partially Observable Markov Decision Process and solved it using the Double Deep Q-Network algorithm. On a large scale single-coil knee dataset, we learned policies that outperform, in terms of four metrics (MSE, NMSE, SSIM and PSNR), simple acquisition heuristics and the scoring function introduced in [25]. We also observed that the dataset-specific DDQN slightly outperforms the subject-specific DDQN, and that a gap still exists between our models and the best possible performance (illustrated by an oracle policy). This performance gap encourages further research to improve models and algorithms to address the active MRI acquisition problem. Finally, it is important to note that our experimental setup is simplified for the purpose of model exploration (e.g. we do not consider all the practical MRI phase-encoding sampling issues).

Acknowledgements. The authors would like to thank the fastMRI team at FAIR and at NYU for engaging discussions. We would like to express our gratitude to Amy Zhang and Joelle Pineau for helpful pointers and to Matthew Muckley for providing feedback on an early draft of this work.

References

1. Chen, F., et al.: Variable-density single-shot fast spin-echo MRI with deep learning reconstruction by using variational networks. Radiology **289**, 180445 (2018)
2. Gözcü, B., et al.: Learning-based compressive MRI. IEEE Trans. Med. Imaging **37**(6), 1394–1406 (2018)
3. Hammernik, K., et al.: Learning a variational network for reconstruction of accelerated MRI data. arXiv preprint arXiv:1704.00447 (2017)

4. Jin, K.H., Unser, M., Yi, K.M.: Self-supervised deep active accelerated MRI. arXiv preprint arXiv:1901.04547 (2019)
5. Kaelbling, L.P., Littman, M.L., Cassandra, A.R.: Planning and acting in partially observable stochastic domains. Artif. Intell. **101**(1–2), 99–134 (1998)
6. Lustig, M., Donoho, D., Pauly, J.M.: Sparse MRI: the application of compressed sensing for rapid MR imaging. Magn. Reson. Med. Off. J. Soc. Magn. Reson. Med./Soc. Magn. Reson. Med. **58**, 1182–1195 (2007)
7. Lønning, K., Putzky, P., Sonke, J.-J., Reneman, L., Caan, M., Welling, M.: Recurrent inference machines for reconstructing heterogeneous MRI data. Med. Image Anal. **53**, 04 (2019)
8. Mnih, V., et al.: Playing atari with deep reinforcement learning. arXiv preprint arXiv:1312.5602 (2013)
9. Panych, L.P., Oesterle, C., Zientara, G.P., Hennig, J.: Implementation of a fast gradient-echo SVD encoding technique for dynamic imaging. Magn. Reson. Med. **35**(4), 554–562 (1996)
10. Puterman, M.L.: Markov decision processes. In: Heyman, D.P., Sobel, M.J. (eds.) Handbooks in Operations Research and Management Science, vol. 2, pp. 331–434. Elsevier, Amsterdam (1990)
11. Ravishankar, S., Bresler, Y.: Adaptive sampling design for compressed sensing MRI. In: Engineering in Medicine and Biology Society (EMBC), pp. 3751–3755 (2011)
12. Schlemper, J., Caballero, J., Hajnal, J.V., Price, A.N., Rueckert, D.: A deep cascade of convolutional neural networks for dynamic MR image reconstruction. IEEE Trans. Med. Imaging **37**(2), 491–503 (2018)
13. Seeger, M., Nickisch, H., Pohmann, R., Schölkopf, B.: Optimization of k-space trajectories for compressed sensing by Bayesian experimental design. Magn. Reson. Med. Off. J. Int. Soc. Magn. Reson. Med. **63**(1), 116–126 (2010)
14. Sondik., E.J.: The optimal control of partially observable Markov processes. Technical report, Stanford Univ Calif Stanford Electronics Labs (1971)
15. Sutton, R.S.: Learning to predict by the methods of temporal differences. Mach. Learn. **3**(1), 9–44 (1988). https://doi.org/10.1007/BF00115009
16. Sutton, R.S., Barto, A.G.: Reinforcement Learning: An Introduction. MIT Press, Cambridge (2018)
17. Van Hasselt, H., Guez, A., Silver, D.: Deep reinforcement learning with double q-learning. In Thirtieth AAAI Conference on Artificial Intelligence (2016)
18. Wang, S., et al.: DeepcomplexMRI: exploiting deep residual network for fast parallel MR imaging with complex convolution. Magn. Reson. Imaging **68**, 136–147 (2020)
19. Wang, S., et al.: Accelerating magnetic resonance imaging via deep learning. In: 2016 IEEE 13th International Symposium on Biomedical Imaging (ISBI), pp. 514–517, April 2016
20. Yoo, S.-S., Guttmann, C.R.G., Zhao, L., Panych, L.P.: Real-time adaptive functional MRI. Neuroimage **10**(5), 596–606 (1999)
21. Yu, C., Liu, J., Nemati, S.: Reinforcement learning in healthcare: a survey. arXiv preprint arXiv:1908.08796 (2019)
22. Zbontar, J., Knoll, F., Sriram, A., Muckley, M.J., Bruno, M., et al.: fastMRI: an open dataset and benchmarks for accelerated MRI. arXiv preprint arXiv:1811.08839 (2018)

23. Zhang, P., Wang, F., Xu, W., Li, Yu.: Multi-channel generative adversarial network for parallel magnetic resonance image reconstruction in k-space. In: Frangi, A.F., Schnabel, J.A., Davatzikos, C., Alberola-López, C., Fichtinger, G. (eds.) MICCAI 2018. LNCS, vol. 11070, pp. 180–188. Springer, Cham (2018). https://doi.org/10.1007/978-3-030-00928-1_21

24. Zhang, Y., Peterson, B.S., Ji, G., Dong, Z.: Energy preserved sampling for compressed sensing MRI. Comput. Math. Methods Med. **2014**, 12 (2014)

25. Zhang, Z., Romero, A., Muckley, M.J., Vincent, P., Yang, L., Drozdzal, M.: Reducing uncertainty in undersampled MRI reconstruction with active acquisition. In: The IEEE Conference on CVPR, June 2019

26. Zhu, B., Liu, J.Z., Rosen, B., Rosen, M.: Image reconstruction by domain transform manifold learning. Nature **555**, 487–492 (2018)

27. Zientara, G.P., Panych, L.P., Jolesz, F.A.: Dynamically adaptive MRI with encoding by singular value decomposition. Magn. Reson. Med. **32**(2), 268–274 (1994)

28. Zientara, G.P., Panych, L.P., Jolesz, F.A.: Applicability and efficiency of near-optimal spatial encoding for dynamically adaptive MRI. Magn. Reson. Med. **39**(2), 204–213 (1998)

Fast Correction of Eddy-Current and Susceptibility-Induced Distortions Using Rotation-Invariant Contrasts

Sahar Ahmad, Ye Wu, Khoi Minh Huynh, Kim-Han Thung, Weili Lin, Dinggang Shen, Pew-Thian Yap$^{(\boxtimes)}$, and the UNC/UMN Baby Connectome Project Consortium

Department of Radiology and Biomedical Research Imaging Center (BRIC), University of North Carolina, Chapel Hill, NC, USA
ptyap@med.unc.edu

Abstract. Diffusion MRI (dMRI) is typically time consuming as it involves acquiring a series of 3D volumes, each associated with a wave-vector in q-space that determines the diffusion direction and strength. The acquisition time is further increased when "blip-up blip-down" scans are acquired with opposite phase encoding directions (PEDs) to facilitate distortion correction. In this work, we show that geometric distortions can be corrected without acquiring with opposite PEDs for each wave-vector, and hence the acquisition time can be halved. Our method uses complimentary rotation-invariant contrasts across shells of different diffusion weightings. Distortion-free structural T1-/T2-weighted MRI is used as reference for nonlinear registration in correcting the distortions. Signal dropout and pileup are corrected with the help of spherical harmonics. To demonstrate that our method is robust to changes in image appearance, we show that distortion correction with good structural alignment can be achieved within minutes for dMRI data of infants between 1 to 24 months of age.

Keywords: Eddy-current distortion · Susceptibility-induced distortion · Rotation-invariant contrasts

1 Introduction

Diffusion MRI (dMRI) is widely used in neuroimaging studies to investigate brain micro-architecture [11,14] and structural connectivity [6]. Diffusion-weighted (DW) images acquired using spin-echo echo-planar imaging (EPI) suffers from geometric distortions caused by gradient-switching-induced eddy currents and susceptibility-induced off-resonance fields associated with the low bandwidth in the phase-encode direction.

This work was supported in part by NIH grants (NS093842, AG053867, EB006733, MH104324, and MH110274) and the efforts of the UNC/UMN Baby Connectome Project Consortium.

A. L. Martel et al. (Eds.): MICCAI 2020, LNCS 12262, pp. 34–43, 2020.
https://doi.org/10.1007/978-3-030-59713-9_4

Existing distortion correction techniques are based on (i) B_0 field mapping [10,17], (ii) nonlinear spatial normalization of distorted images to undistorted structural T1-/T2-weighted (T1w/T2w) images [12,16], or (iii) unwarping of blip-up blip-down acquisitions to a common undistorted mid-point [1,5,13]. The first approach requires an accurate estimation of the field map, which can be challenging due to motion artifacts and phase errors. The second approach often results in poor alignment due to inter-modality contrast differences. The third approach improves the intensity uniformity of the distortion-corrected images, but has the following limitations: (i) The geometric distortions are corrected by warping opposing-PED non-DW ($b = 0\,\mathrm{s/mm^2}$) images to a hypothetical mid-point space that does not necessarily correspond with the undistorted space according to structural MRI (sMRI). This is problematic particularly when information from sMRI, such as cortical surface geometry, is needed for subsequent analyses. Additional registration and interpolation might cause errors and image degradation. (ii) The correction is dependent on PED-reversed scans, prolonging acquisition time. This can be prohibitive when imaging pediatric, elderly, or claustrophobic individuals. (iii) Typically only non-DW images are used for estimating the distortion-induced displacements. Irfanoglu et al. [9] proposed to include DW images to improve registration in homogeneous regions with large distortions. However, signal attenuation at high b-values results in low signal-to-noise ratio (SNR) and poor tissue contrast, impeding displacement field estimation. (iv) Correcting the distortions of a large number of DW images typically requires a long computation time.

To overcome these limitations, we show in this paper that eddy-current and susceptibility-induced geometric distortions can be corrected without acquiring with opposite PEDs for each wave-vector, and hence the acquisition time can be halved. Our method utilizes complimentary rotation-invariant contrasts (RICs) across shells of different diffusion weightings. Distortion-free structural T1-/T2-weighted MRI is used as reference for nonlinear registration in correcting the distortions. Signal dropout and pileup are corrected with the help of spherical harmonics. To demonstrate that our method is robust to changes in image appearance, we show that distortion correction with good structural alignment can be achieved within minutes for data of infants between 1 to 24 months of age.

2   Methods

The proposed post-acquisition distortion correction method consists of three main steps: (i) Correction of eddy-current-induced distortions; (ii) Multi-contrast correction of susceptibility-induced distortion; and (iii) Correction for signal intensity due to dropout and pileup.

2.1   Interleaved PEDs

Approaches based on "blip-up blip-down" scans require scanning for each wavevector in q-space using opposing PEDs, essentially doubling the acquisition time. We propose, in line with [4], to acquire each q-space sample with only one PED, but interlace q-space samples with different PEDs. In this paper, we will demonstrate with two PEDs, i.e., anterior-posterior (AP) or posterior-anterior (PA), but our method can be easily generalized to more than two PEDs. Unlike the constrained image reconstruction method proposed in [4], we will introduce a fast method for distortion correction using RICs, and hence avoiding the time-consuming process of explicitly correcting for each individual DW image. The appearances of DW images are direction dependent. The RICs remove directional dependency and allow DW images acquired with different PEDs to be registered efficiently. RICs, by agglomerating information from all directions, also improve SNR and facilitate registration of DW images with high b-values (above $1500\,\mathrm{s/mm^2}$), a problem that was reported in [3].

2.2 Distortion Correction

The eddy-current-induced distortions of images $I_p = \{I_p^n\}_{n=1}^N$, acquired with PED $p \in \{\mathrm{AP}, \mathrm{PA}\}$, can be corrected by affine registration, followed by coarse nonlinear registration, of all DW images to their corresponding non-DW image. This is implemented using ANTs registration toolkit [2], resulting in affine transformation parameters $\{\mathcal{A}_p^n\}_{n=1}^N$ and nonlinear displacement fields $\{\phi_p^n\}_{n=1}^N$.

DW images at high b-values exhibit low SNR, low tissues contrast, and therefore present insufficient structural details for accurate registration [3]. Existing methods that register only non-DW images for distortion correction do not necessarily guarantee alignment of high b-value DW images, which are necessary for sensitivity and specificity to tissue microstructure. Here, we propose to use the spherical mean images (SMIs) of DW images corrected for eddy-current distortions in a multi-contrast registration framework for correction of susceptibility-induced distortions. The SMIs—computed across gradient directions for each shell—exhibit higher SNR and are rotation invariant, facilitating the registration of DW images acquired with interleaved PEDs (see Fig. 4). For PED p, we denote the SMIs for M shells, in addition to $b = 0\,\mathrm{s/mm^2}$, as $J_p = \{J_p^i\}_{i=0}^M$. These SMIs are first jointly affine registered to aligned structural T1w and/or T2w images $S_{\mathrm{T_1}}$ and $S_{\mathrm{T_2}}$, producing $\tilde{J}_p = \{\tilde{J}_p^i\}_{i=0}^M$ with affine transformation parameters $\mathcal{A}_p$. Multi-contrast nonlinear registration is then performed using symmetric normalization (SyN) [2] with a mutual information (MI) data matching term:

$$\zeta_p = \int \sum_{i=0}^M \Big(\mathrm{MI}\big(\tilde{J}_p^i(\phi_{p\to \mathrm{S_{T1}}}), S_{\mathrm{T1}}\big) + \mathrm{MI}\big(\tilde{J}_p^i(\phi_{p\to \mathrm{S_{T2}}}), S_{\mathrm{T2}}\big)\Big) d\boldsymbol{x}, \qquad (1)$$

where $\phi_{p\to \mathrm{S}}$ is the displacement field that maps the SMIs to the structural reference space S. SyN multiplied each component of a displacement vector with a weight before updating the displacement field in the next iteration. The displacement field was weighted more towards the PED.

DW images collected with two opposing PEDs typically exhibit distortions in opposite directions. To promote the inverse relationship between $\phi_{\mathrm{AP}\to \mathrm{S}}$ and

$\phi_{\text{PA}\rightarrow\text{S}}$, we first estimate the displacement fields between the SMIs for opposite PEDs as follows:

$$\phi_{\text{AP}\rightarrow\text{PA}} = \phi_{\text{AP}\rightarrow\text{S}} \circ \phi_{\text{PA}\rightarrow\text{S}}^{-1}, \tag{2}$$

$$\phi_{\text{PA}\rightarrow\text{AP}} = \phi_{\text{PA}\rightarrow\text{S}} \circ \phi_{\text{AP}\rightarrow\text{S}}^{-1}. \tag{3}$$

The inverse displacement fields to the undistorted reference space are then given by

$$\phi_{\text{AP}\rightarrow\text{S}} \leftarrow 0.5 \times \phi_{\text{AP}\rightarrow\text{PA}}, \tag{4}$$

$$\phi_{\text{PA}\rightarrow\text{S}} \leftarrow 0.5 \times \phi_{\text{PA}\rightarrow\text{AP}}. \tag{5}$$

This is illustrated in Fig. 1. In principle, $\phi_{\text{AP}\rightarrow\text{S}}$ and $\phi_{\text{PA}\rightarrow\text{S}}$ can be refined by using them to initialize the registration algorithm, and then recomputing Eqs. (2), (3), (4), and (5), but we found that the improvement is negligible for our data. The distortion-corrected DW images are then obtained by warping with the corresponding displacement field

$$\phi_p^n = \mathcal{A}_p^n \circ \phi_p^n \circ \mathcal{A}_p \circ \phi_{\text{p}\rightarrow\text{S}}. \tag{6}$$

The undistorted images are given as $\tilde{I}_p^n = I_p^n(\phi_p^n)$. The signal pileup due to signal compression and signal dropout due to expansion are corrected by modulating the undistorted images $\tilde{I}_p^n$ with the Jacobian determinant of the displacement field ϕ_p^n [7]:

$$\tilde{I}_p^n \leftarrow \tilde{I}_p^n \times \mathfrak{J}(\phi_p^n). \tag{7}$$

Signal pileup might not always be completely rectifiable using images acquired with a single PED. Correction can be improved by using information from images collected with the opposing PED. Unlike blip-up blip-down approaches, our interleaved-PED acquisition scheme did not acquire each q-space sample with opposing PEDs. Therefore, we fitted the spherical harmonics (SHs) to the DW images acquired for a PED, and then estimated the DW images of the opposing PED. The signal was reoriented via the local Jacobian of the deformation field prior to SH fitting.

The final corrected DW images are obtained by combining the DW images associated with the two PEDs. This is achieved by computing the harmonic mean [9] of the signals associated with the two PEDs.

3   Results and Discussion

The dataset consisted of dMRI and sMRI data of 54 infant subjects between 1 and 24 months of age, enrolled as part of the UNC/UMN Baby Connectome Project (BCP) [8]. The subjects were divided into three cohorts according to their first scheduled visits and were scanned every three months. Since not all subjects can be scanned every three months, each subject has a different number of scans, resulting in 68 scans in total. The sMRI data had 208 sagittal slices with 320×320 matrix size and $(0.8\,\text{mm})^3$ voxel resolution. The dMRI data

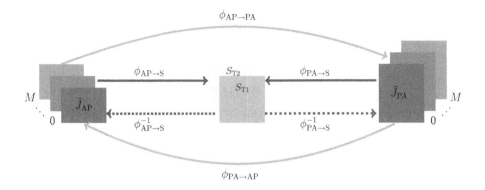

Fig. 1. Displacement fields associated with two opposite PEDs.

had 95 axial slices with 140×140 matrix size and $(1.5\,\text{mm})^3$ voxel resolution. The diffusion data were acquired with b-values 500, 1000, 1500, 2000, 2500, and $3000\,\text{s/mm}^2$. The DW images were acquired with AP and PA PEDs, with a total of 144 DW images and six non-DW images for each PED. To demonstrate the effectiveness of our method, we removed the images associated with half of the gradient directions, resulting in 72 directions per PED. For comparison, the full number of gradients were used for FSL TOPUP+Eddy [15]. The parameters of TOPUP+Eddy were adjusted to obtain optimal results for infant dMRI.

We show the fractional anisotropy (FA) maps computed using the two methods in Fig. 2. The FA map given by TOPUP+Eddy shows structural discrepancies with respect to sMRI. Further evaluation was carried out by overlaying the white matter surface generated from the sMRI data onto the FA maps computed by the two methods. It is clear from Fig. 3 that the FA map generated by our method is well-aligned with sMRI at the interface between white matter and gray matter. Figure 4 shows the spherical mean images of the corrected dMRI data of 12-month-old infant, indicating that our method yields good correction outcomes comparable with TOPUP+Eddy.

We also qualitatively evaluated the Jacobian determinant maps of the deformation fields estimated by the two methods. Figure 5(a) shows that the deformation field generated by TOPUP+Eddy is not diffeomorphic as evident from the negative Jacobian determinant values. In contrast, our method gives positive Jacobian determinant values for all voxels. We validated the inverse relationship of the displacement fields given by our method. The displacement fields for the two opposing PEDs are shown in Fig. 5(b). The inverse consistency error (ICE) [18] map in Fig. 5(c) shows that the two displacement fields are the inverses of each other.

Figure 6 shows the spherical mean images after geometric and intensity correction, indicating that signal inhomogeneity is reduced.

In addition, we compared the FA values given by the two methods. The mean $\pm$ std FA values are 0.19 ± 0.14 and 0.24 ± 0.16 for TOPUP+Eddy and the

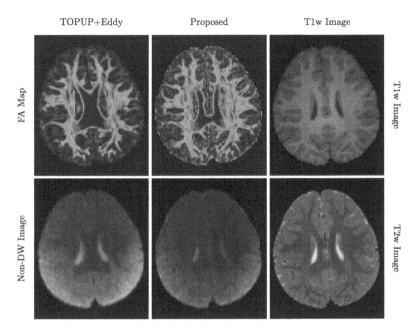

Fig. 2. FA maps and non-DW images of the corrected dMRI data of an 18-month-old infant.

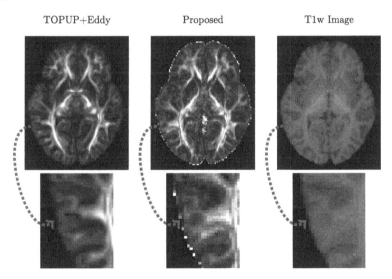

Fig. 3. White matter surface overlaid on the FA maps and T1w image of a 23-month-old infant.

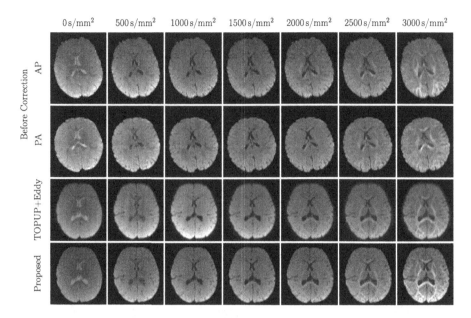

Fig. 4. Spherical mean images for different b-values for a 12-month-old infant subject. Image brightness is adjusted for each shell for visibility.

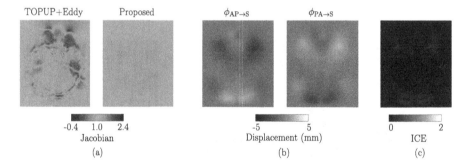

Fig. 5. (a) Jacobian maps of the deformation fields given by TOPUP+Eddy and the proposed method. (b) Displacement fields for the two opposing PEDs estimated by the proposed method, and (c) the corresponding inverse consistency error (ICE) map.

proposed method (Fig. 7(a)), respectively. We quantified the similarity between the FA maps and the T1w images using MI. The boxplot for MI is given in Fig. 7(b), where the mean ± std MI values are 0.44 ± 0.07 and 0.47 ± 0.06 for TOPUP+Eddy and the proposed method, respectively. We also compared the two methods by evaluating the structural similarity (SSIM) between the reference T2w image and the corrected non-DW images for each subject. The mean ± std of SSIM is $(28 \pm 17)\%$ for TOPUP+Eddy and $(61 \pm 7)\%$ for the proposed method (Fig. 7(c)). We further validated the results by computing

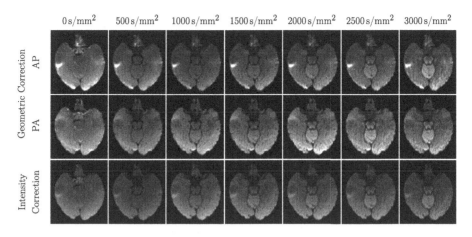

Fig. 6. Spherical mean images after geometric and intensity correction of the dMRI data of a 12-months-old infant. Image brightness is adjusted for each shell for visibility.

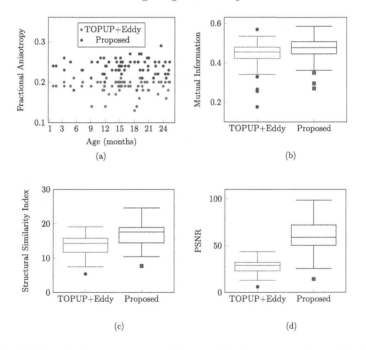

Fig. 7. Quantitative comparison between distortion correction methods.

the PSNR of the spherical mean images for all the b-values obtained with TOPUP+Eddy and the proposed method, using the corresponding T1w as reference. Our method gives an average PSNR of 17.80, whereas TOPUP+Eddy gives an average of 13.27 (Fig. 7(d)). The results for all metrics are statistically significant ($p < 0.01$).

Our method is very time efficient and took 25 min per dMRI dataset, whereas TOPUP+Eddy took a few hours to complete.

4   Conclusion

In this paper, we proposed a distortion correction method for both eddy-current and susceptibility-induced distortions and demonstrated its effectiveness in infant brain dMRI. Our method does not require images of multiple phase-encoding directions to be acquired for each q-space sample, therefore reducing the acquisition time. Our method relies on high SNR spherical mean images for guiding registration. We have also shown that the commonly used TOPUP+Eddy does not fully remove distortions as compared to undistorted T1- and T2-weighted images, potentially causing problems when the different modalities need to be analyzed in a common space. The results show that our method outperforms TOPUP+Eddy both in terms of accuracy and speed.

References

1. Andersson, J.L., Skare, S., Ashburner, J.: How to correct susceptibility distortions in spin-echo echo-planar images: application to diffusion tensor imaging. NeuroImage **20**(2), 870–888 (2003). https://doi.org/10.1016/S1053-8119(03)00336-7
2. Avants, B., Epstein, C., Grossman, M., Gee, J.: Symmetric diffeomorphic image registration with cross-correlation: evaluating automated labeling of elderly and neurodegenerative brain. Med. Image Anal. **12**(1), 26–41 (2008). https://doi.org/10.1016/j.media.2007.06.004
3. Ben-Amitay, S., Jones, D.K., Assaf, Y.: Motion correction and registration of high b-value diffusion weighted images. Magn. Reson. Med. **67**, 1694–1702 (2012). https://doi.org/10.1002/mrm.23186
4. Bhushan, C., Joshi, A.A., Leahy, R.M., Haldar, J.P.: Improved b_0-distortion correction in diffusion MRI using interlaced q-space sampling and constrained reconstruction. Magn. Reson. Med. **72**, 1218–1232 (2014). https://doi.org/10.1002/mrm.25026
5. Embleton, K.V., Haroon, H.A., Morris, D.M., Ralph, M.A.L., Parker, G.J.: Distortion correction for diffusion-weighted MRI tractography and fMRI in the temporal lobes. Hum. Brain Mapp. **31**(10), 1570–1587 (2010). https://doi.org/10.1002/hbm.20959
6. Frau-Pascual, A., Fogarty, M., Fischl, B., Yendiki, A., Aganj, I.: Quantification of structural brain connectivity via a conductance model. NeuroImage **189**, 485–496 (2019). https://doi.org/10.1016/j.neuroimage.2019.01.033
7. Holland, D., Kuperman, J.M., Dale, A.M.: Efficient correction of inhomogeneous static magnetic field-induced distortion in echo planar imaging. NeuroImage **50**(1), 175–183 (2010). https://doi.org/10.1016/j.neuroimage.2009.11.044
8. Howell, B.R., et al.: The UNC/UMN baby connectome project (BCP): an overview of the study design and protocol development. NeuroImage **185**, 891–905 (2019). https://doi.org/10.1016/j.neuroimage.2018.03.049

9. Irfanoglu, M.O., Modi, P., Nayak, A., Hutchinson, E.B., Sarlls, J., Pierpaoli, C.: DR-BUDDI (Diffeomorphic Registration for Blip-Up blip-Down Diffusion Imaging) method for correcting echo planar imaging distortions. NeuroImage **106**, 284–299 (2015). https://doi.org/10.1016/j.neuroimage.2014.11.042

10. Jezzard, P., Balaban, R.S.: Correction for geometric distortion in echo planar images from B0 field variations. Magn. Reson. Med. **34**(1), 65–73 (1995). https://doi.org/10.1002/mrm.1910340111

11. Kunz, N., et al.: Assessing white matter microstructure of the newborn with multi-shell diffusion MRI and biophysical compartment models. NeuroImage **96**, 288–299 (2014). https://doi.org/10.1016/j.neuroimage.2014.03.057

12. Kybic, J., Thevenaz, P., Nirkko, A., Unser, M.: Unwarping of unidirectionally distorted EPI images. IEEE Trans. Med. Imaging **19**(2), 80–93 (2000). https://doi.org/10.1109/42.836368

13. Morgan, P.S., Bowtell, R.W., McIntyre, D.J., Worthington, B.S.: Correction of spatial distortion in EPI due to inhomogeneous static magnetic fields using the reversed gradient method. J. Magn. Reson. Imaging **19**(4), 499–507 (2004). https://doi.org/10.1002/jmri.20032

14. Ouyang, M., Dubois, J., Yu, Q., Mukherjee, P., Huang, H.: Delineation of early brain development from fetuses to infants with diffusion MRI and beyond. NeuroImage **185**, 836–850 (2019). https://doi.org/10.1016/j.neuroimage.2018.04.017

15. Smith, S.M., et al.: Advances in functional and structural MR image analysis and implementation as FSL. NeuroImage **23**, S208–S219 (2004). https://doi.org/10.1016/j.neuroimage.2004.07.051

16. Tao, R., Fletcher, P.T., Gerber, S., Whitaker, R.T.: A variational image-based approach to the correction of susceptibility artifacts in the alignment of diffusion weighted and structural MRI. In: Prince, J.L., Pham, D.L., Myers, K.J. (eds.) IPMI 2009. LNCS, vol. 5636, pp. 664–675. Springer, Heidelberg (2009). https://doi.org/10.1007/978-3-642-02498-6_55

17. Techavipoo, U., et al.: Toward a practical protocol for human optic nerve DTI with EPI geometric distortion correction. J. Magn. Reson. Imaging **30**(4), 699–707 (2009). https://doi.org/10.1002/jmri.21836

18. Yang, D., Li, H., Low, D., Deasy, J., Naqa, I.E.: A fast inverse consistent deformable image registration method based on symmetric optical flow computation. Phys. Med. Biol. **53**(21), 6143–6165 (2008). https://doi.org/10.1088/0031-9155/53/21/017

Joint Reconstruction and Bias Field Correction for Undersampled MR Imaging

Mélanie Gaillochet, Kerem Can Tezcan$^{(\boxtimes)}$, and Ender Konukoglu

Computer Vision Lab, ETH Zürich, Zürich, Switzerland
gamelani@student.ethz.ch, {tezcan,kender}@vision.ee.ethz.ch

Abstract. Undersampling the k-space in MRI allows saving precious acquisition time, yet results in an ill-posed inversion problem. Recently, many deep learning techniques have been developed, addressing this issue of recovering the fully sampled MR image from the undersampled data. However, these learning based schemes are susceptible to differences between the training data and the image to be reconstructed at test time. One such difference can be attributed to the bias field present in MR images, caused by field inhomogeneities and coil sensitivities. In this work, we address the sensitivity of the reconstruction problem to the bias field and propose to model it explicitly in the reconstruction, in order to decrease this sensitivity. To this end, we use an unsupervised learning based reconstruction algorithm as our basis and combine it with a N4-based bias field estimation method, in a joint optimization scheme. We use the HCP dataset as well as in-house measured images for the evaluations. We show that the proposed method improves the reconstruction quality, both visually and in terms of RMSE.

Keywords: MRI reconstruction · Deep learning · Bias field.

1 Introduction

Magnetic resonance imaging (MRI) is a non-invasive imaging technique that allows studying anatomy and tissue properties without ionizing radiation. However, acquiring a detailed high-quality image is time-consuming.

Reducing scan time is therefore essential in order to increase patient comfort and throughput, and to open up more possibilities for optimized examinations. Since acquisition time in MRI is directly related to the number of samples collected in k-space, shortening it usually means reducing the number of samples collected - for instance, by skipping phase-encoding lines in k-space in the Cartesian case. Yet doing so violates the Nyquist criterion, causing aliasing artifacts

---

M. Gaillochet and K. C. Tezcan – Equal contribution.

This work was partially funded by grant 205321_173016 Schweizerischer Nationalfonds zur Förderung der Wissenschaftlichen Forschung.

A. L. Martel et al. (Eds.): MICCAI 2020, LNCS 12262, pp. 44–52, 2020.
https://doi.org/10.1007/978-3-030-59713-9_5

and requiring reconstruction techniques that can recover the fully sampled MR image from the undersampled data.

The problem of designing such techniques has received considerable attention in the clinical and signal processing research communities. Conventional approaches involve compressed sensing [3,4] and parallel imaging [5]. More recently, the research efforts have focused on using deep learning to tackle the problem and achieved state-of-the-art results. However, despite their multiple advantages, these learning based algorithms are susceptible to discrepancies between the training set and the image to reconstruct at test time. One such difference is due the different bias fields present in the images. Such discrepancies, sometimes referred to as 'domain shift', typically lead to a loss in performance. Various methods have been developed to tackle the domain shift problem in the deep learning community. One approach is to augment the training data to hopefully resemble the test data better [17,18]. Although this approach increases the generalization capabilities of the learned model, it does not optimize the solution specifically for the test image at hand. Another approach is to modify the network parameters to reduce the domain gap [19], which indirectly optimizes the solution by modifying the network rather than the solution.

Yet, while the effect of the bias field on the image encoding is easy to model and there are readily available methods that can estimate it well, none of the aforementioned approaches takes these into account.

In this work we propose a joint reconstruction algorithm, which estimates and explicitly models the bias field throughout the reconstruction process. By doing so, we remove one degree of variation between the training set and the image to be reconstructed. The training is done on images without bias field, and the bias field itself is modeled as a multiplicative term in the image encoding process, linking the training and test domains. In order to be able to do this we use a reconstruction algorithm which decouples the learned prior information from the image generation process, namely the DDP algorithm [2]. Using the N4 algorithm, we iteratively estimate the bias field in the test image throughout the reconstruction, and this estimation improves as the reconstructed image become better. We compare the proposed method to reconstruction without bias field estimation, using a publicly available dataset as well as in-house measured images, and show improvement in performance.

2   Methods

2.1   Problem Formulation

The measured k-space data, $\mathbf{y} \in \mathcal{C}^M$ and the underlying true MR image, $\mathbf{x} \in \mathcal{C}^N$ are related through the encoding operation $\mathbf{E} : \mathcal{C}^N \to \mathcal{C}^M$ (which incorporates the coil sensitivites, Fourier transformation and the undersampling operation) as $\mathbf{y} = \mathbf{E}\mathbf{x} + \eta$, where η is complex Gaussian noise.

2.2 DDP Reconstruction

The deep density prior reconstruction is based on maximum-a-posteriori estimation, i.e. $\max_{\mathbf{x}} \log p(\mathbf{x}|\mathbf{y}) = \max_{\mathbf{x}} \log p(\mathbf{y}|\mathbf{x}) + \log p(\mathbf{x})$, where $\log p(y|x)$ is the data consistency and $\log p(x)$ is the prior term. The data consistency can be written exactly as the log Gaussian and the method approximates the prior term using the evidence lower bound (ELBO) of a variational auto-encoder (VAE) [23,24]. With these, the reconstruction problem becomes

$$\min_{x} ||\mathbf{Ex} - \mathbf{y}|| - ELBO(\mathbf{x}). \tag{1}$$

The VAE is trained on patches from fully sampled images and the ELBO term operates on a set of overlapping patches that cover the whole image. As evident in the equation above, the ELBO term is independent of the image encoding. The DDP method solves the problem using the projection onto convex sets (POCS) algorithm. In this scheme the optimization is implemented as successive applications of projection operations for prior $\mathcal{P}_{prior}$, data consistency $\mathcal{P}_{DC}$ and the phase of the image $\mathcal{P}_{phase}$, i.e. $\mathbf{x}^{t+1} = \mathcal{P}_{DC}\mathcal{P}_{phase}\mathcal{P}_{prior}\mathbf{x}^t$. The prior projection is defined as a gradient ascent for a fixed number of steps for the ELBO term w.r.t. the image magnitude, i.e. $\mathcal{P}_{prior}\mathbf{x} = \mathbf{x}^N$, where $\mathbf{x}^{n+1} = \mathbf{x}^n + \alpha\frac{d}{d\mathbf{x}}ELBO(|\mathbf{x}|)|_{\mathbf{x}=\mathbf{x}^n}$ for $n = 0...N$. The data consistency projection is given as $\mathcal{P}_{DC}\mathbf{x} = \mathbf{x} - \mathbf{E}^H(\mathbf{Ex} - \mathbf{y})$. For the phase projection, we use the one defined by [2], to which we add the minimization of the data consistency with respect to the image phase. For further details, we refer the reader to the original paper [2].

2.3 Modeling and Estimating the Bias Field

Signal intensities are often not constant across the MR images, even inside the same tissue. Instead, they usually vary smoothly, with fluctuations of 10%–20%, across the measured 3D volume [6]. These variations, collectively called the bias field, can be attributed to factors like patient anatomy, differences in coil sensitivity or standing wave effects [6,9], which are difficult or impossible to control during an acquisition. Hence they can introduce a degree of variation between images used for training reconstruction models and an image to be reconstructed at test time. In order to prevent a loss of performance, this variation has to be taken into account.

To this end, we model the bias field explicitly and incorporate it into Eq. 1 as a multiplicative term $\mathbf{B}$ before the encoding operation, modifying the image intensities pixelwise. The bias field is an additional unknown in the reconstruction process that is estimated alongside x as

$$\min_{\mathbf{x},\mathbf{B}} ||\mathbf{EBx} - \mathbf{y}|| - ELBO(\mathbf{x}). \tag{2}$$

Here, the measured k-space $\mathbf{y}$ carries the effect of a bias field, while $\mathbf{x}$ is bias field free as the bias field in the image is explicitly modeled using the $\mathbf{B}$ term. This setting allows us to learn the VAE model on images without bias field. The advantage of this idea is two fold. Firstly, since training can be performed on bias field free images, it is easier for the VAE to learn the distribution as there is less spurious variation in the data. Secondly, we make the reconstruction problem easier by explicitly providing the bias field information, which otherwise would have to be reconstructed from the undersampled k-space as well. We solve Eq. 2 as a joint iterative reconstruction problem by minimizing alternatively two subproblems:

$$1. \quad \mathbf{x}^t = \min_x ||\mathbf{EB}^{t-1}\mathbf{x} - \mathbf{y}|| - ELBO(\mathbf{x}) \tag{3}$$

$$2. \quad \mathbf{B}^t = \text{N4}(\mathbf{B}^{t-1}\mathbf{x}^t), \tag{4}$$

where N4 denotes the bias field estimation algorithm, which we will explain below. To account for the bias field, the data consistency projection $\mathcal{P}_{DC}$ needs to be adapted and becomes $\mathcal{P}_{DC}^B\mathbf{x} = \mathbf{B}^{-1}[\mathbf{Bx} - \mathbf{E}^H(\mathbf{EBx} - \mathbf{y})]$. In this case the reconstructed image corresponding to $\mathbf{y}$ is given as $\mathbf{Bx}$. This modification can be interpreted as doing a forward-backward projection with $\mathcal{P}_{bias} = \mathbf{B}$ before and after the data consistency projection to move the image between the "normalized" bias field free domain and the bias field corrupted acquisition domain, i.e. $\mathbf{x}^{t+1} = \mathcal{P}_{bias}^{-1}\mathcal{P}_{DC}\mathcal{P}_{bias}\mathcal{P}_{phase}\mathcal{P}_{prior}\mathbf{x}^t$. The pseudocode for the described joint optimization scheme is presented in Algorithm 1.

N4 Bias Field Estimation. N4 is a variant of the widely used N3 algorithm [7], a non-parametric iterative method that approximates intensity non-uniformity fields in 3-dimensional images. Given an image, its objective is to find a smooth, slowly varying, multiplicative field [6]. N4 improves upon the N3 algorithm by modifying the B-spline smoothing strategy and the iterative optimization scheme used in the original framework. We use the implementation available as N4ITK [8]. We denote this method as $N4(\cdot)$ in our formulations.

Algorithm 1. Joint reconstruction

1: $\mathbf{y} \leftarrow$ undersampled k-space data
2: $\mathbf{E} \leftarrow$ undersampling encoding operator
3: VAE $\leftarrow$ trained VAE
4: NumIter, BiasEstimFreq, DCProjFreq
5: **procedure** JOINTRECON(y, E, VAE)
6:     $\mathbf{B} \leftarrow \text{N4}(\mathbf{E}^H\mathbf{y})$
7:     $\mathbf{x}^0 \leftarrow (\mathbf{B})^{-1}\mathbf{E}^H\mathbf{y}$
8:     **for** t: 0 to NumIter - 1 **do**
9:         $\mathbf{x}^{t+1} = \mathcal{P}_{prior}\mathbf{x}^t$
10:         $\mathbf{x}^{t+1} \leftarrow \mathcal{P}_{phase}\mathbf{x}^{t+1}$ ▷ Optional
11:         **if** $t\%$ DCProjFreq $== 0$ and $t \neq 0$ **then**
12:             $\mathbf{x}^{t+1} \leftarrow \mathbf{B}^{-1}\left[\mathbf{Bx}^{t+1} - \mathbf{E}^H(\mathbf{EBx}^{t+1} - \mathbf{y})\right]$ ▷ $\mathcal{P}_{DC}^B$
13:         **if** $t\%$ BiasEstimFreq $== 0$ and $t \neq 0$ **then**
14:             $\mathbf{B} \leftarrow \text{N4}(\mathbf{Bx}^{t+1})$
        return $\mathbf{x}^{t+1}$, $\mathbf{B}$

2.4   Datasets Used

To train the VAE we used 5 central, non-adjacent slices with 0.7 mm isotropic resolution from T1 weighted images of 360 subjects from the Human Connectome Project (HCP) preprocessed dataset [15,16], which by default have a bias field. We used N4 on the images to also create a bias field free training set.

For test images, we took a central slice from 20 different test subjects from the HCP data. As the HCP images tend to have a similar bias field, we additionally created a modified test set where we estimated the bias fields with N4, took their inverse and multiplied them with the bias field free images. In addition to HCP data, we also tested the proposed method with central slices from 9 in-house measured subjects. These images were acquired using a 16 element head coil and have similar acquisition parameters as the HCP dataset with a 1 mm isotropic resolution. We used ESPIRiT [21] to estimate the coil sensitivity maps for these images.

2.5   Training the VAE

We trained four patch-wise VAEs - for two different resolution levels to match the datasets, each with and without bias field. We used patches of size 28×28 with a batch size of 50 and ran the training for 500,000 iterations. The patches were extracted randomly from the training images with replacement. All the VAEs were trained with the same training images extracted from the HCP dataset, as described above.

2.6   Experimental Setup

We used random Cartesian undersampling patterns with 15 fully sampled central profiles. We generated a different pattern for each subject, and applied the same pattern for a given subject throughout all experiments for comparability of results. When reconstructing test images from the HCP dataset, we used 302 iterations (NumIter) for $R = 2$, 602 for $R = 3$, 1002 for $R = 4$ and 1502 for $R = 5$, to allow for convergence of reconstructed images. Since the in-house measured images have multiple coils, the successive applications of data consistency projections coincide with a POCS-SENSE reconstruction [20], speeding up convergence and requiring less iterations. Hence, when performing reconstruction on images from the in-house measured dataset, we ran the first 10 iterations without prior projection, applying only data consistency projections. Additionally, the discrepancy between the actual coil sensitivities and the ESPIRiT [21] estimations may lead to divergence after too many iterations. Hence, a lower number of iterations was used for reconstruction experiments on the in-house measure dataset: 32 iters for $R = 2$, 102 for $R = 3$ and $R = 4$, and 202 for $R = 5$. For all test datasets, the parameters were set as $\alpha = $ 1e-4, BiasEstimFreq = 10, DCProjFreq = 10. As for the N4 bias field estimation algorithm, the default parameters were used.

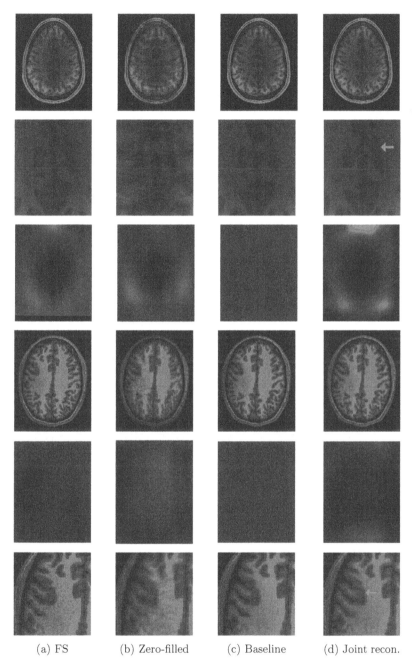

(a) FS   (b) Zero-filled   (c) Baseline   (d) Joint recon.

Fig. 1. Reconstruction results for R = 3, with (a) the fully sampled image (b) the zero-filled image (c) the reconstruction with no bias field estimation (d) the joint reconstruction with bias field estimation using N4. The first three rows show reconstruction results for an HCP image, its zoomed-in version and the corresponding bias field. The next three rows show results for an in-house measured image. For visualization purposes, MR images are clipped to [0, 1.2] and bias fields, to [0.5, 1.8].

To evaluate the reconstruction quality of the different models, we computed, for each image, the percentage RMSE. More specifically, the evaluation metric was given as $\text{RMSE} = 100 \times \sqrt{\frac{\sum_i (|\mathbf{Bx}| - |\hat{\mathbf{x}}|)_i^2}{\sum_i |\hat{\mathbf{x}}|_i^2}}$ where $\hat{\mathbf{x}}$ is the original, fully sampled test image, and where the summation was applied pixel-wise. The error was computed on the skull-stripped images only in the brain area.

To evaluate the statistical significance of our results, we performed a permutation test [22] with 10,000 permutations to assess the null hypothesis that the RMSE's for the reconstruction with and without bias field estimation are from the same distribution. From these tests, we reported the p-values.

3   Experiments and Results

To assess the hypothesis that correcting the bias field improves the overall reconstruction quality, we performed reconstructions on the original and modified HCP test sets as well as on the in-house measured images.

When applying our proposed method, we reconstructed undersampled images with a bias field, and used a VAE trained on images without bias field as well as the N4 bias field estimation algorithm. To evaluate our approach, we also ran baseline experiments on the same images, where we used a VAE trained on images with a bias field and did not apply bias field estimation during reconstruction. Given that the ground truth (fully sampled) images naturally have a bias field in them, we utilized the last bias field estimate multiplied with the reconstructed image, i.e. $\mathbf{Bx}$, for visualisations and RMSE calculations.

Table 1. Table of RMSE values. R is the undersampling factor. Numbers indicate the mean (std). The * indicates a p-value of less than 0.05. The baseline method is the DDP algorithm described in [2] that does not explicitly model the bias field. The proposed joint reconstruction method estimates the bias field using N4 and explicitly models it during reconstruction.

Method	HCP dataset				Modified HCP dataset				In-house measured dataset			
	R = 2	R = 3	R = 4	R = 5	R = 2	R = 3	R = 4	R = 5	R = 2	R = 3	R = 4	R = 5
Baseline	2.24	3.39	4.42	5.72	2.24	3.45	4.26	5.58	4.64	6.852	8.593	11.046
	(0.31)	(0.45)	(0.51)	(1.05)	(0.38)	(0.60)	(0.46)	(1.20)	(0.391)	(0.726)	(1.344)	(1.727)
Joint recon.	2.27	3.34	4.35*	5.52	2.20*	3.33*	4.16	5.12*	4.62	6.714*	8.218*	10.567*
	(0.34)	(0.40)	(0.47)	(0.66)	(0.39)	(0.53)	(0.54)	(0.57)	(0.418)	(0.821)	(1.266)	(1.632)

The results in Table 1 indicate that the proposed joint reconstruction method with bias field estimation improves the reconstruction quality in terms of RMSE when the bias field of the test image is different from those in the training set. In these cases, namely the experiments with the modified HCP and in-house measured images, the improvement is statistically significant with a p-value of less than 0.05 for nearly all undersampling factors. For the unmodified HCP dataset, where the bias field in the test images matches those in the training set, we do not expect a big difference in the performance, which is reflected in the results.

The quantitative improvement is also supported by the visual inspection of the images given in Fig. 1. From the HCP image, one can observe that the level of artifacts is reduced with the proposed method. This becomes more evident in the zoomed-in images. The red arrow points to a part of the image where the proposed method can reconstruct the structures faithfully, whereas the baseline method struggles. Aliasing artifacts are globally suppressed better with the joint reconstruction method. Similarly, for the in-house measured image, the grey matter structure that the red arrow points to is not reconstructed in the baseline method, whereas it again appears with the proposed method.

In this work, we demonstrate the performance loss due to the bias field for a specific algorithm, into which we also integrate our proposed solution. However, the problem may not be specific to the algorithm used as it arises from the domain gap, which is a fundamental problem affecting machine learning based methods in general. Furthermore, the proposed method of estimating and explicitly modeling the bias field in reconstruction is also a generic approach, which can be integrated into different algorithms.

4   Conclusion

In this paper we proposed and evaluated a method for joint reconstruction and bias field estimation to address variations due to the bias field when reconstructing undersampled MR images. The results indicate that the proposed method improves the baseline method (unsupervised learning based reconstruction algorithm), both in RMSE and visually. The improvements can be attributed to two factors. First, the proposed method allows the VAE prior to learn a simpler distribution, and second, providing the bias field explicitly makes the reconstruction problem easier. In essence, estimating the bias field during reconstruction makes the model less sensitive to differences between the data used to train the model and the test data used during reconstruction.

Author contributions. Equal contribution.

References

1. Litjens, G., Kooi, T., Bejnordi, B.E., Setio, A.A.A., Ciompi, F., Ghafoorian, M., van der Laak, J.A.W.M., van Ginneken, B., Sanchez, C.I.: A survey on deep learning in medical image analysis. Med. Image Anal. **42**, 60–88 (2017)
2. Tezcan, K.C., Baumgartner, C.F., Luechinger, R., Pruessmann, K.P., Konukoglu, E.: MR image reconstruction using deep density priors. IEEE Trans. Med. Imaging **38**(7), 1633–1642 (2019)
3. Donoho, D.L.: Compressed sensing. IEEE Trans. Inform. Theor. **52**(4), 1289–1306 (2006)
4. Lustig, M., Donoho, D., Pauly, J.M.: Sparse MRI: the application of compressed sensing for rapid MR imaging. Magn. Reson. Med. **58**(6), 1182–1195 (2007)
5. Deshmane, A., Gulani, V., Griswold, M.A., Seiberlich, N.: Parallel MR imaging. J. Magn. Reson. Imaging **36**(1), 55–72 (2012)

6. Sled, J.G., Zijdenbos, A.P., Evans, A.C.: A nonparametric method for automatic correction of intensity nonuniformity in MRI data. IEEE Trans. Med. Imaging **17**(1), 87–97 (1998)
7. Tustison, N.J., Avants, B.B., Cook, P.A., Zheng, Y., Egan, A., Yushkevich, P.A., Gee, J.C.: N4ITK: improved N3 bias correction. IEEE Trans. Med. Imaging **29**(6), 1310–1320 (2010)
8. SimpleITK: N4 Bias Field Correction, https://simpleitk.readthedocs.io/en/master/link_N4BiasFieldCorrection_docs.html. Accessed 12 March 2020
9. Sled, J.G., Pike, G.B.: Understanding intensity non-uniformity in MRI. In: Wells, W.M., Colchester, A., Delp, S. (eds.) MICCAI 1998. LNCS, vol. 1496, pp. 614–622. Springer, Heidelberg (1998). https://doi.org/10.1007/BFb0056247
10. Meyer, C.R., Bland, P.H., Pipe, J.: Retrospective correction of intensity inhomogeneities in MRI. IEEE Trans. Med. Imaging **14**(1), 36–41 (1995)
11. Van Leemput, K., Maes, F., Vandermeulen, D., Suetens, P.: Automated model-based bias field correction of MR images of the brain. IEEE Trans. Med. Imaging **18**(10), 885–896 (1999)
12. Leemput, K.V., Maes, F., Vandermeulen, D., Suetens, P.: Automated model-based tissue classification of MR images of the brain. IEEE Trans. Med. Imaging **18**(10), 897–908 (1999)
13. Ashburner, J., Friston, K.J.: Unified segmentation. NeuroImage **26**(3), 839–851 (2005)
14. Glasser, M.F., et al.: The minimal preprocessing pipelines for the human connectome project. Neuroimage **80**, 105–124 (2013)
15. Van Essen, D.C., Smith, S.M., Barch, D.M., Behrens, T.E., Yacoub, E., Ugurbil, K.: The wu-minn human connectome project: an overview. Neuroimage **80**, 62–79 (2013)
16. Human Connectome Reference Manual, https://www.humanconnectome.org/storage/app/media/documentation/s1200/HCP_S1200_Release_Reference_Manual.pdf. Accessed 4 March 2020
17. Chaitanya, K., Karani, N., Baumgartner, C.F., Becker, A., Donati, Olivio, Konukoglu, Ender: Semi-supervised and task-driven data augmentation. In: Chung, A.C.S., Gee, J.C., Yushkevich, P.A., Bao, S. (eds.) IPMI 2019. LNCS, vol. 11492, pp. 29–41. Springer, Cham (2019). https://doi.org/10.1007/978-3-030-20351-1_3
18. Volpi, R., Namkoong, H., Sener, O., Duchi, J., Murino, V., Savarese, S.: Generalizing to unseen domains via adversarial data augmentation. NeurIPS **31**, 5334–5344 (2018)
19. Sun, B., Saenko, K.: Deep CORAL: correlation alignment for deep domain adaptation. In: Hua, G., Jégou, H. (eds.) ECCV 2016. LNCS, vol. 9915, pp. 443–450. Springer, Cham (2016). https://doi.org/10.1007/978-3-319-49409-8_35
20. Samsonov, A.A., Kholmovski, E.G., Parker, D.L., Johnson, C.R.: POCSENSE: POCS-based reconstruction for sensitivity encoded magnetic resonance imaging. MRM **52**(6), 1397–1406 (2004)
21. Uecker, M., Lai, P., Murphy, M.J., Virtue, P., Elad, M., Pauly, J.M., Vasanawala, S.S., Lustig, M.: ESPIRiT-an eigenvalue approach to autocalibrating parallel MRI: where SENSE meets GRAPPA. MRM **71**(3), 990–1001 (2014)
22. Collingridge, D.S.: A primer on quantitized data analysis and permutation testing. J. Mixed Meth. Res. **7**(1), 81–97 (2013)
23. Kingma D.P., Welling, M.: Auto-encoding variational bayes. arXiv:1312.6114v10 (2014)
24. Rezende, D.J., Mohamed, S., Wierstra, D.: Stochastic backpropagation and approximate inference in deep generative models. arXiv:1401.4082v3 (2014)

Joint Total Variation ESTATICS
for Robust Multi-parameter Mapping

Yaël Balbastre[1]([envelope]), Mikael Brudfors[1], Michela Azzarito[2], Christian Lambert[1],
Martina F. Callaghan[1], and John Ashburner[1]

[1] Wellcome Centre for Human Neuroimaging, Queen Square Institute of Neurology,
University College London, London, UK
y.balbastre@ucl.ac.uk
[2] Spinal Cord Injury Center Balgrist, University Hospital Zurich,
University of Zurich, Zürich, Switzerland

Abstract. Quantitative magnetic resonance imaging (qMRI) derives tissue-specific parameters – such as the apparent transverse relaxation rate $R_2^\star$, the longitudinal relaxation rate R_1 and the magnetisation transfer saturation – that can be compared across sites and scanners and carry important information about the underlying microstructure. The multi-parameter mapping (MPM) protocol takes advantage of multi-echo acquisitions with variable flip angles to extract these parameters in a clinically acceptable scan time. In this context, ESTATICS performs a joint loglinear fit of multiple echo series to extract $R_2^\star$ and multiple extrapolated intercepts, thereby improving robustness to motion and decreasing the variance of the estimators. In this paper, we extend this model in two ways: (1) by introducing a joint total variation (JTV) prior on the intercepts and decay, and (2) by deriving a nonlinear maximum *a posteriori* estimate. We evaluated the proposed algorithm by predicting left-out echoes in a rich single-subject dataset. In this validation, we outperformed other state-of-the-art methods and additionally showed that the proposed approach greatly reduces the variance of the estimated maps, without introducing bias.

1 Introduction

The magnetic resonance imaging (MRI) signal is governed by a number of tissue-specific parameters. While many common MR sequences only aim to maximise the contrast between tissues of interest, the field of quantitative MRI (qMRI) is concerned with the extraction of the original parameters [30]. This interest stems from the fundamental relationship that exists between the magnetic parameters and the tissue microstructure: the longitudinal relaxation rate $R_1 = 1/T_1$ is sensitive to myelin content [10,27,28]; the apparent transverse relaxation rate $R_2^\star = 1/T_2^\star$ can be used to probe iron content [12,21,22]; the magnetization-transfer saturation (MT_{sat}) indicates the proportion of protons bound to macro-molecules (in contrast to free water) and offers another metric to investigate

© Springer Nature Switzerland AG 2020
A. L. Martel et al. (Eds.): MICCAI 2020, LNCS 12262, pp. 53–63, 2020.
https://doi.org/10.1007/978-3-030-59713-9_6

myelin loss [14,31]. Furthermore, qMRI allows many of the scanner- and centre-specific effects to be factored out, making measures more comparable across sites [3,9,29,33]. In this context, the multi-parameter mapping (MPM) protocol was developed at 3 T to allow the quantification of R_1, $R_2^\star$, MT_{sat} and the proton density (PD) at high resolutions (0.8 or 1 mm) and in a clinically acceptable scan time of 25 min [14,33]. However, to reach these values, compromises must be made so that the signal-to-noise ratio (SNR) suffers, making the parameter maps noisy; Papp et al. [23] found a scan-rescan root mean squared error of about 7.5% for R_1 at 1 mm, in the absence of inter-scan movement. Smoothing can be used to improve SNR, but at the cost of lower spatial specificity.

Denoising methods aim to separate signal from noise. They take advantage of the fact that signal and noise have intrinsically different spatial profiles: the noise is spatially independent and often has a characteristic distribution while the signal is highly structured. Denoising methods originate from partial differential equations, adaptive filtering, variational optimisation or Markov random fields, and many connections exist between them. Two main families emerge:

1. Optimisation of an energy: $\hat{Y} = \mathrm{argmin}_Y\, \mathcal{E}_1\left(X - \mathcal{A}(Y)\right) + \mathcal{E}_2\left(\mathcal{G}(Y)\right)$,
   where X is the observed data, Y is the unknown noise-free data, $\mathcal{A}$ is an arbitrary *forward* transformation (*e.g.*, spatial transformation, downsampling, smoothing) mapping from the reconstructed to the observed data and $\mathcal{G}$ is a linear transformation (*e.g.*, spatial gradients, Fourier transform, wavelet transform) that extracts features of interest from the reconstruction.
2. Application of an adaptive nonlocal filter: $\hat{Y}_i = \sum_{j \in \mathcal{N}_i} w\left(\mathcal{P}_i(X), \mathcal{P}_j(X)\right) X_j$,
   where the reconstruction of a given voxel i is a weighted average all observed voxels j in a given (possibly infinite) neighbourhood $\mathcal{N}_i$, with weights reflecting similarity between patches centred about these voxels.

For the first family of methods, it was found that the denoising effect is stronger when $\mathcal{E}_2$ is an absolute norm (or sum of), rather than a squared norm, because the solution is implicitly sparse in the feature domain [2]. This family of methods include total variation (TV) regularisation [25] and wavelet soft-thresholding [11]. The second family also leverages sparsity in the form of redundancy in the spatial domain; that is, the dictionary of patches necessary to reconstruct the noise-free images is smaller than the actual number of patches in the image. Several such methods have been developed specifically for MRI, with the aim of finding an optimal, voxel-wise weighting based on the noise distribution [6,7,19,20].

Optimisation methods can naturally be interpreted as a maximum *a posteriori* (MAP) solution in a generative model, which eases its interpretation and extension. This feature is especially important for MPMs, where we possess a well-defined (nonlinear) forward function and wish to regularise a small number of maps. In this paper, we use the ESTATICS forward model [32], which assumes a shared $R_2^\star$ decay across contrasts, with a joint total variation (JTV) prior. JTV [26] is an extension of TV to multi-channel images, where the absolute norm is defined across channels, introducing an implicit correlation between them. TV and JTV have been used before in MR reconstruction (*e.g.*, in compressed-sensing [15], quantitative susceptibility mapping [17], super-resolution[4]). JTV is perfectly suited for modelling the multiple contrasts in MPMs and increases the power of the implicit edge-detection problem. However, a challenge stems from the nonlinear forward model that makes the optimisation problem nonconvex.

Our implementation uses a quadratic upper bound of the JTV functional and the surrogate problem is solved using second-order optimisation. Positive-definiteness of the Hessian is enforced by the use of Fisher's scoring, and the quadratic problem is efficiently solved using a mixture of multi-grid relaxation and conjugate gradient. We used a unique dataset – five repeats of the MPM protocol acquired, within a single session, on a healthy subject – to validate the proposed method. Our method was compared to two variants of ESTATICS: loglinear [32] and Tikhonov-regularised. We also compared it with the adaptive optimized nonlocal means (AONLM) method [20], which is recommended for accelerated MR images (as is the case in our validation data). In that case, individual echoes were denoised using AONLM, and maps were reconstructed with the loglinear variant of ESTATICS. In our validation, JTV performed consistently better than all other methods.

2   Methods

Spoiled Gradient Echo. The MPM protocol uses a multi-echo spoiled gradient-echo (SGE) sequence with variable flip angles to generate weighted images. The signal follows the equation:

$$S(\alpha, T_R, T_E) = S_0(\alpha, T_R) \exp(-T_E R_2^\star), \tag{1}$$

where α is the nominal flip angle, T_R is the repetition time and T_E is the echo time. PD and T1 weighting are obtained by using two different flip angles, while MT weighting is obtained by playing a specific off-resonance pulse beforehand. If all three intercepts S_0 are known, rational approximations can be used to compute R_1 and MT_{sat} maps [13,14].

ESTATICS. ESTATICS aims to recover the decay rate $R_2^\star$ and the different intercepts from (1). We therefore write each weighted signal (indexed by c) as:

$$S(c, T_E) = \exp(\theta_c - T_E R_2^\star), \quad \text{with} \quad \theta_c = \ln S_{0c}. \tag{2}$$

At the SNR levels obtained in practice (>3), the noise of the log-transformed data is approximately Gaussian (although with a variance that scales with signal amplitude). Therefore, in each voxel, a least-squares fit can be used to estimate $R_2^\star$ and the log-intercepts S_c from the log-transformed acquired images.

Regularised ESTATICS. Regularisation cannot be easily introduced in logarithmic space because, there, the noise variance depends on the signal amplitude, which is unknown. Instead, we derive a full generative model. Let us assume that all weighted volumes are aligned and acquired on the same grid. Let us define the image acquired at a given echo time t with contrast c as $s_{c,t} \in \mathbb{R}^I$ (where I is the number of voxels). Let $\boldsymbol{\theta}_c \in \mathbb{R}^I$ be the log-intercept with contrast c and let $\boldsymbol{r} \in \mathbb{R}^I$ be the $R_2^\star$ map. Assuming stationary Gaussian noise, we get the conditional probability:

$$p(s_{c,t} \mid \boldsymbol{\theta}_c, \boldsymbol{r}) = \mathcal{N}\left(s_{c,t} \mid \tilde{s}_{c,t}, \sigma_c^2 I\right), \quad \tilde{s}_{c,t} = \exp(\boldsymbol{\theta}_c - t\boldsymbol{r}). \tag{3}$$

The regularisation takes the form of a joint prior probability distribution over $\boldsymbol{\Theta} = [\boldsymbol{\theta}_1, \cdots, \boldsymbol{\theta}_C, \boldsymbol{r}]$. For JTV, we get:

$$p(\boldsymbol{\Theta}) \propto \prod_i \exp\left(-\sqrt{\sum_{c=1}^{C+1} \lambda_c \boldsymbol{\theta}_c^T \boldsymbol{G}_i^T \boldsymbol{G}_i \boldsymbol{\theta}_c}\right), \tag{4}$$

where $\boldsymbol{G}_i$ extracts all forward and backward finite-differences at the i-th voxel and λ_c is a contrast-specific regularisation factor. The MAP solution can be found by maximising the joint loglikelihood with respect to the parameter maps.

Quadratic Bound. The exponent in the prior term can be written as the minimum of a quadratic function [2,8]:

$$\sqrt{\sum_c \lambda_c \boldsymbol{\theta}_c^T \boldsymbol{G}_i^T \boldsymbol{G}_i \boldsymbol{\theta}_c} = \min_{w_i > 0} \left\{ \frac{w_i}{2} + \frac{1}{2w_i} \sum_c \lambda_c \boldsymbol{\theta}_c^T \boldsymbol{G}_i^T \boldsymbol{G}_i \boldsymbol{\theta}_c \right\}. \tag{5}$$

When the weight map $\boldsymbol{w}$ is fixed, the bound can be seen as a Tikhonov prior with nonstationary regularisation, which is a quadratic prior that factorises across channels. Therefore, the between-channel correlations induces by the JTV prior are entirely captured by the weights. Conversely, when the parameter maps are fixed, the weights can be updated in closed-form:

$$w_i = \sqrt{\sum_c \lambda_c \boldsymbol{\theta}_c^T \boldsymbol{G}_i^T \boldsymbol{G}_i \boldsymbol{\theta}_c}. \tag{6}$$

The quadratic term in (5) can be written as $\lambda_c \boldsymbol{\theta}_c^T \boldsymbol{L} \boldsymbol{\theta}_c$, with $\boldsymbol{L} = \sum_i \frac{1}{w_i} \boldsymbol{G}_i^T \boldsymbol{G}_i$.

In the following sections, we will write the full (bounded) model negative loglikelihood as $\mathcal{L}$ and keep only terms that depend on $\boldsymbol{\Theta}$, so that:

$$\mathcal{L} = \sum_{c,t} \mathcal{L}_{c,t}^d + \mathcal{L}^p, \quad \mathcal{L}_{c,t}^d \overset{c}{=} \frac{1}{2\sigma_c^2} \|s_{c,t} - \tilde{s}_{c,t}\|^2, \quad \mathcal{L}^p \overset{c}{=} \frac{1}{2} \sum_c \boldsymbol{\theta}_c^T \boldsymbol{L}_c \boldsymbol{\theta}_c. \tag{7}$$

Fisher's Scoring. The data term (3) does not always have a positive semi-definite Hessian (it is not convex). There is, however, a unique optimum. Here, to ensure that the conditioning matrix that is used in the Newton-Raphson iteration has the correct curvature, we take the expectation of the true Hessian, which is equivalent to setting the residuals to zero – a method known as Fisher's scoring. The Hessian of $\mathcal{L}_{c,t}^d$ with respect to the c-th intercept and $R_2^\star$ map then becomes:

$$\boldsymbol{H}_{c,t}^d = \frac{1}{\sigma^2} \text{diag}(\tilde{s}_{c,t}) \otimes \begin{bmatrix} 1 & -t \\ -t & t^2 \end{bmatrix}. \tag{8}$$

Misaligned Volumes. Motion can occur between the acquisitions of the different weighted volumes. Here, volumes are systematically co-registered using a skull-stripped and bias-corrected version of the first echo of each volume. However, rather than reslicing the volumes onto the same space, which modifies the original intensities, misalignment is handled within the model. To this end, equation (3) is modified to include the projection of each parameter map onto native space, such that $\tilde{s}_{c,t} = \exp(\boldsymbol{\Psi}_c\boldsymbol{\theta}_c - t\boldsymbol{\Psi}_c\boldsymbol{r})$, where $\boldsymbol{\Psi}_c$ encodes trilinear interpolation and sampling with respect to the pre-estimated rigid transformation. The Hessian of the data term becomes $\boldsymbol{\Psi}_c^{\mathrm{T}}\boldsymbol{H}_{c,t}^{\mathrm{d}}\boldsymbol{\Psi}_c$, which is nonsparse. However, an approximate Hessian can be derived [1], so that:

$$\boldsymbol{H}_{c,t}^{\mathrm{d}} \approx \frac{1}{\sigma^2}\mathrm{diag}\left(\boldsymbol{\Psi}_c^{\mathrm{T}}\tilde{s}_{c,t}\right) \otimes \begin{bmatrix} 1 & -t \\ -t & t^2 \end{bmatrix}. \tag{9}$$

Since all elements of $\tilde{s}_{c,t}$ are strictly positive, this Hessian is ensured to be more positive-definite than the true Hessian in the Löwner ordering sense.

Newton-Raphson. The Hessian of the joint negative log-likelihood becomes:

$$\boldsymbol{H} = \boldsymbol{H}^{\mathrm{d}} + \boldsymbol{L} \otimes \mathrm{diag}\left(\boldsymbol{\lambda}\right). \tag{10}$$

Each Newton-Raphson iteration involves solving for $\boldsymbol{H}^{-1}\boldsymbol{g}$, where $\boldsymbol{g}$ is the gradient. Since the Hessian is positive-definite, the method of conjugate gradients (CG) can be used to solve the linear system. CG, however, converges quite slowly. Instead, we first approximate the regularisation Hessian $\boldsymbol{L}$ as $\tilde{\boldsymbol{L}} = \frac{1}{\min(\boldsymbol{w})}\sum_i \boldsymbol{G}_i^{\mathrm{T}}\boldsymbol{G}_i$, which is more positive-definite than $\boldsymbol{L}$. Solving this substitute system therefore ensures that the objective function improves. Since $\boldsymbol{H}^{\mathrm{d}}$ is an easily invertible block-diagonal matrix, the system can be solved efficiently using a multi-grid approach [24]. This result is then used as a warm start for CG. Note that preconditioners have been shown to improve CG convergence rates [5,34], at the cost of slowing down each iteration. Here, we have made the choice of performing numerous cheap CG iterations rather than using an expensive preconditioner.

3   Validation

Dataset. A single participant was scanned five times in a single session with the 0.8 mm MPM protocol, whose parameters are provided in Table 1. Furthermore, in order to correct for flip angles nonhomogeneity, a map of the B_1^+ field was reconstructed from stimulated and spin echo 3D EPI images [18].

Evaluated Methods. Three ESTATICS methods were evaluated: a simple loglinear fit (LOG) [32], a nonlinear fit with Tikhonov regularisation (TKH) and a nonlinear fit with joint total variation regularisation (JTV). Additionally, all echoes were denoised using the adaptive nonlocal means method (AONLM)

Table 1. Sequence parameters of the MPM protocol. The MTw sequence has an off-resonance prepulse (PP): 220°, 4 ms duration, 2 kHz off-resonance.

	FA	TR	TE	Matrix	FOV	PP
T1w	21°	25 ms	[1..8] × 2.3 ms	320 × 280 × 224	256 × 224 × 179.2 mm^3	
PDw	6°	25 ms	[1..8] × 2.3 ms	320 × 280 × 224	256 × 224 × 179.2 mm^3	
MTw	6°	25 ms	[1..6]× 2.3 ms	320 × 280 × 224	256 × 224 × 179.2 mm^3	✓

[20] before performing a loglinear fit. The loglinear and nonlinear ESTATICS fit were all implemented in the same framework, allowing for misalignment between volumes. Regularised ESTATICS uses estimates of the noise variance within each volume, obtained by fitting a two-class Rice mixture to the first echo of each series. Regularised ESTATICS possesses two regularisation factors, one for each intercept and one for the $R_2^\star$ decay, while AONLM has one regularisation factor. These hyper-parameters were optimised by cross-validation (CV) on the first repeat of the MPM protocol.

Leave-One-Echo-Out. Validating denoising methods is challenging in the absence of a ground truth. Classically, one would compute similarity metrics, such as the root mean squared error, the peak signal-to-noise ratio, or the structural similarity index between the denoised images and noise-free references. However, in MR, such references are not artefact free: they are still relatively noisy and, as they require longer sequences, more prone to motion artefacts. A better solution is to use cross-validation, as the forward model can be exploited to predict echoes that were left out when inferring the unknown parameters. We fitted each method to each MPM repeat, while leaving one of the acquired echoes out. The fitted model was then used to predict the missing echo. The quality of these predictions was scored by computing the Rice loglikelihood of the true echo conditioned on the predicted echo within the grey matter (GM), white matter (WM) and cerebro-spinal fluid (CSF). An aggregate score was also computed in the parenchyma (GM+WM). As different echoes or contrasts are not similarly difficult to predict, Z-scores were computed by normalising across repeats, contrasts and left-out echoes. This CV was applied to the first repeat to determine optimal regularisation parameters. We found $\beta = 0.4$ without Rice-specific noise estimation to work better for AONLM, while for JTV we found $\lambda_1 = 5 \times 10^3$ for the intercepts and $\lambda_2 = 10$ for the decay (in s^{-1}) to be optimal.

Quantitative Maps. Rational approximations of the signal equations [13,14] were used to compute R_1 and MT_{sat} maps from the fitted intercepts. The distribution of these quantitative parameters was computed within the GM and WM. Furthermore, standard deviation (S.D.) maps across runs were computed for each method.

4   Results

Leave-One-Echo-Out. The distribution of Rice loglikelihoods and Z-scores for each methods are depicted in Fig. 1 in the form of Tukey's boxplots. In the parenchyma, JTV obtained the best score (mean log-likelihood: -9.15×10^6, mean Z-score: 1.19) followed by TKH (-9.26×10^6 and -0.05), AONLM (-9.34×10^6 and -0.41) and LOG (-9.35×10^6 and -0.72). As some echoes are harder to predict than others (typically, early echoes because their absence impacts the estimator of the intercept the most) the log-pdf has quite a high variance. However, Z-scores show that, for each echo, JTV does consistently better than all other methods. As can be seen in Fig. 1, JTV is particularly good at preserving vessels.

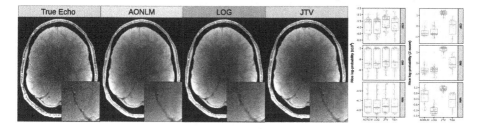

Fig. 1. Leave-one-echo out prediction. Left: the true PDw echo at $T_E = 9.7$ ms from the 5th repeat and three predicted images. Right: boxplots of the Rice log-pdf and corresponding Z-score computed for each method within GM, WM and CSF masks.

Quantitative Maps. R_1, MT_{sat} and $R_2^\star$ maps reconstructed with each method are shown in Fig. 2, along with mean intensity histograms within GM and WM. Note that these maps are displayed for qualitative purposes; low standard deviations are biased toward over-regularised methods and do not necessarily indicate a better predictive performance. It is evident from the histograms that all denoising methods sharpen the peaks without introducing apparent bias. It can be seen that JTV has lower variance than AONLM in the centre of the brain and higher in the periphery. This is because in our probabilistic setting, there is a natural balance between the prior and the quality of the data. In the centre of the brain, the SNR is lower than in the periphery, which gives more weight to the prior and induces a smoother estimate. The mean standard deviation of AONLM, LOG, JTV and TKH is respectively 9.5, 11.5, 11.5, 9.9 $\times 10^{-3}$ in the GM and 8.6, 12, 9.6, 10 $\times 10^{-3}$ in the WM for R_1, 15, 2, 17, 20 in the GM and 11, 20, 10, 13 in the WM for $R_2^\star$, and 4.6, 5.8, 5.1, 4.5 $\times 10^{-2}$ in the GM and 4.9, 8.2, 4.3, 4.7 $\times 10^{-2}$ in the WM for MT_{sat}. Once again, variance is reduced by all denoising methods compared to the nonregularised loglinear fit. Again, a lower variance does not necessarily indicate a better (predictive) fit, which can only be assessed by the CV approach proposed above.

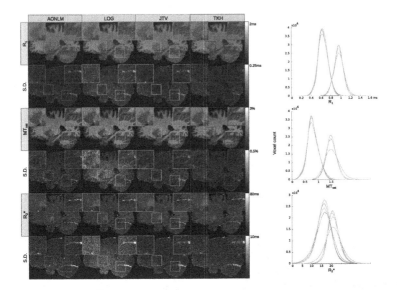

Fig. 2. Quantitative maps. Left: example R_1, $\mathrm{MT_{sat}}$ and $R_2^\star$ maps obtained with each method, and standard deviation (S.D.) maps computed across runs. Right: mean intensity histograms computed within the GM (plain) and WM (dotted) masks.

5   Discussion and Conclusion

In this paper, we introduce a robust, regularisation-based reconstruction method for quantitative MR mapping. The joint total variation prior takes advantage of the multiple MPM contrasts to increase its edge-detection power. Our approach was validated using an unbiased CV scheme, where it compared favourably over other methods, including a state-of-the-art MR denoising technique. It was shown to reduce the variance of the estimated parametric maps over non-regularised approaches, which should translate into increased power in subsequent cross-sectional or longitudinal voxel-wise studies. The use of a well-defined forward model opens the door to multiple extensions: the projection operator could be modified to include other components of the imaging process such as non-homogeneous receive fields or gridding, which would allow for joint reconstruction and super-resolution; parameters that are currently fixed *a priori*, such as the rigid matrices, could be given prior distribution and be optimised in an interleaved fashion; non-linear deformations could be included to account for changes in the neck position between scans; finally, the forward model could be unfolded further so that parameter maps are directly fitted, rather than weighted intercepts. An integrated approach like this one could furthermore include and optimise for other components of the imaging process, such as non-homogeneous transmit fields. In terms of optimisation, our approach should benefit from advances in conjugate gradient preconditioning or other solvers for large linear systems. Alternatively, JTV could be replaced with a patch-based prior.

Nonlocal filters are extremely efficient at denoising tasks and could be cast in a generative probabilistic framework, where images are built using a dictionary of patches [16]. Variational Bayes can then be used to alternatively estimate the dictionary (shared across a neighbourhood, a whole image, or even across subjects) and the reconstruction weights.

Acknowledgements. YB, MFC and JA were funded by the MRC and Spinal Research Charity through the ERA-NET Neuron joint call (MR/R000050/1). MB and JA were funded by the EU Human Brain Project's Grant Agreement No 785907 (SGA2). MB was funded by the EPSRC-funded UCL Centre for Doctoral Training in Medical Imaging (EP/L016478/1) and the Department of Health NIHR-funded Biomedical Research Centre at University College London Hospitals. CL is supported by an MRC Clinician Scientist award (MR/R006504/1). The Wellcome Centre for Human Neuroimaging is supported by core funding from the Wellcome [203147/Z/16/Z].

References

1. Ashburner, J., Brudfors, M., Bronik, K., Balbastre, Y.: An algorithm for learning shape and appearance models without annotations. NeuroImage (2018)
2. Bach, F.: Optimization with sparsity-inducing penalties. FNT Mach. Learn. **4**(1), 1–106 (2011)
3. Bauer, C.M., Jara, H., Killiany, R.: Whole brain quantitative T2 MRI across multiple scanners with dual echo FSE: applications to AD, MCI, and normal aging. NeuroImage **52**(2), 508–514 (2010)
4. Brudfors, M., Balbastre, Y., Nachev, P., Ashburner, J.: MRI super-resolution using multi-channel total variation. In: 22nd Conference on Medical Image Understanding and Analysis, Southampton, UK (2018)
5. Chen, C., He, L., Li, H., Huang, J.: Fast iteratively reweighted least squares algorithms for analysis-based sparse reconstruction. Med. Image Anal. **49**, 141–152 (2018)
6. Coupé, P., Manjón, J., Robles, M., Collins, D.: Adaptive multiresolution non-local means filter for three-dimensional magnetic resonance image denoising. IET Image Process. **6**(5), 558–568 (2012)
7. Coupe, P., Yger, P., Prima, S., Hellier, P., Kervrann, C., Barillot, C.: An optimized blockwise nonlocal means denoising filter for 3-D magnetic resonance images. IEEE Trans. Med. Imaging **27**(4), 425–441 (2008)
8. Daubechies, I., DeVore, R., Fornasier, M., Güntürk, C.S.: Iteratively reweighted least squares minimization for sparse recovery. Commun. Pure Appl. Math. **63**(1), 1–38 (2010)
9. Deoni, S.C.L., Williams, S.C.R., Jezzard, P., Suckling, J., Murphy, D.G.M., Jones, D.K.: Standardized structural magnetic resonance imaging in multicentre studies using quantitative T1 and T2 imaging at 1.5 T. NeuroImage **40**(2), 662–671 (2008)
10. Dick, F., Tierney, A.T., Lutti, A., Josephs, O., Sereno, M.I., Weiskopf, N.: In vivo functional and myeloarchitectonic mapping of human primary auditory areas. J. Neurosci. **32**(46), 16095–16105 (2012)
11. Donoho, D.: De-noising by soft-thresholding. IEEE Trans. Inf. Theory **41**(3), 613–627 (1995)

12. Hasan, K.M., Walimuni, I.S., Kramer, L.A., Narayana, P.A.: Human brain iron mapping using atlas-based T2 relaxometry. Magn. Reson. Med. **67**(3), 731–739 (2012)
13. Helms, G., Dathe, H., Dechent, P.: Quantitative FLASH MRI at 3T using a rational approximation of the Ernst equation. Magn. Reson. Med. **59**(3), 667–672 (2008)
14. Helms, G., Dathe, H., Kallenberg, K., Dechent, P.: High-resolution maps of magnetization transfer with inherent correction for RF inhomogeneity and T1 relaxation obtained from 3D FLASH MRI. Magn. Reson. Med. **60**(6), 1396–1407 (2008)
15. Huang, J., Chen, C., Axel, L.: Fast multi-contrast MRI reconstruction. In: Ayache, N., Delingette, H., Golland, P., Mori, K. (eds.) MICCAI 2012. LNCS, vol. 7510, pp. 281–288. Springer, Heidelberg (2012). https://doi.org/10.1007/978-3-642-33415-3_35
16. Lebrun, M., Buades, A., Morel, J.M.: A nonlocal Bayesian image denoising algorithm. SIAM J. Imaging Sci. **6**(3), 1665–1688 (2013)
17. Liu, T., et al.: Morphology enabled dipole inversion (MEDI) from a single-angle acquisition: comparison with COSMOS in human brain imaging. Magn. Reson. Med. **66**(3), 777–783 (2011)
18. Lutti, A., Hutton, C., Finsterbusch, J., Helms, G., Weiskopf, N.: Optimization and validation of methods for mapping of the radiofrequency transmit field at 3T. Magn. Reson. Med. **64**(1), 229–238 (2010)
19. Manjón, J.V., Coupé, P., Buades, A., Louis Collins, D., Robles, M.: New methods for MRI denoising based on sparseness and self-similarity. Med. Image Anal. **16**(1), 18–27 (2012)
20. Manjón, J.V., Coupé, P., Martí-Bonmatí, L., Collins, D.L., Robles, M.: Adaptive non-local means denoising of MR images with spatially varying noise levels. J. Magn. Reson. Imaging **31**(1), 192–203 (2010)
21. Ogg, R.J., Langston, J.W., Haacke, E.M., Steen, R.G., Taylor, J.S.: The correlation between phase shifts in gradient-echo MR images and regional brain iron concentration. Magn. Reson. Imaging **17**(8), 1141–1148 (1999)
22. Ordidge, R.J., Gorell, J.M., Deniau, J.C., Knight, R.A., Helpern, J.A.: Assessment of relative brain iron concentrations using T2-weighted and T2*-weighted MRI at 3 Tesla. Magn. Reson. Med. **32**(3), 335–341 (1994)
23. Papp, D., Callaghan, M.F., Meyer, H., Buckley, C., Weiskopf, N.: Correction of inter-scan motion artifacts in quantitative R1 mapping by accounting for receive coil sensitivity effects. Magn. Reson. Med. **76**(5), 1478–1485 (2016)
24. Press, W.H., Teukolsky, S.A., Vetterling, W.T., Flannery, B.P.: Numerical Recipes 3rd Edition: The Art of Scientific Computing, 3rd edn. Cambridge University Press, Cambridge, New York (2007)
25. Rudin, L.I., Osher, S., Fatemi, E.: Nonlinear total variation based noise removal algorithms. Physica D **60**(1), 259–268 (1992)
26. Sapiro, G., Ringach, D.L.: Anisotropic diffusion of multivalued images with applications to color filtering. IEEE Trans. Image Process. **5**(11), 1582–1586 (1996)
27. Sereno, M.I., Lutti, A., Weiskopf, N., Dick, F.: Mapping the human cortical surface by combining quantitative T1 with retinotopy. Cereb. Cortex **23**(9), 2261–2268 (2013)
28. Sigalovsky, I.S., Fischl, B., Melcher, J.R.: Mapping an intrinsic MR property of gray matter in auditory cortex of living humans: a possible marker for primary cortex and hemispheric differences. NeuroImage **32**(4), 1524–1537 (2006)
29. Tofts, P.S., et al.: Sources of variation in multi-centre brain MTR histogram studies: Body-coil transmission eliminates inter-centre differences. Magn. Reson. Mater. Phys. **19**(4), 209–222 (2006). https://doi.org/10.1007/s10334-006-0049-8

30. Tofts, P.S.: Quantitative MRI of the Brain, 1st edn. Wiley, Hoboken (2003)
31. Tofts, P.S., Steens, S.C.A., van Buchem, M.A.: MT: magnetization transfer. In: Quantitative MRI of the Brain, pp. 257–298. Wiley, Hoboken (2003)
32. Weiskopf, N., Callaghan, M.F., Josephs, O., Lutti, A., Mohammadi, S.: Estimating the apparent transverse relaxation time (R2*) from images with different contrasts (ESTATICS) reduces motion artifacts. Front. Neurosci. **8**, 278 (2014)
33. Weiskopf, N., et al.: Quantitative multi-parameter mapping of R1, PD*, MT, and R2* at 3T: a multi-center validation. Front. Neurosci. **7**, 95 (2013)
34. Xu, Z., Li, Y., Axel, L., Huang, J.: Efficient preconditioning in joint total variation regularized parallel MRI reconstruction. In: Navab, N., Hornegger, J., Wells, W.M., Frangi, A.F. (eds.) MICCAI 2015. LNCS, vol. 9350, pp. 563–570. Springer, Cham (2015). https://doi.org/10.1007/978-3-319-24571-3_67

End-to-End Variational Networks for Accelerated MRI Reconstruction

Anuroop Sriram[1(✉)], Jure Zbontar[1], Tullie Murrell[1], Aaron Defazio[1],
C. Lawrence Zitnick[1], Nafissa Yakubova[1], Florian Knoll[2],
and Patricia Johnson[2]

[1] Facebook AI Research (FAIR), Menlo Park, CA, USA
{anuroops,jzb,tullie,adefazio,zitnick,nafissay}@fb.com
[2] NYU School of Medicine, New York, NY, USA
{florian.knoll,patricia.johnson3@}@nyulangone.org

Abstract. The slow acquisition speed of magnetic resonance imaging (MRI) has led to the development of two complementary methods: acquiring multiple views of the anatomy simultaneously (parallel imaging) and acquiring fewer samples than necessary for traditional signal processing methods (compressed sensing). While the combination of these methods has the potential to allow much faster scan times, reconstruction from such undersampled multi-coil data has remained an open problem. In this paper, we present a new approach to this problem that extends previously proposed variational methods by learning fully end-to-end. Our method obtains new state-of-the-art results on the fastMRI dataset [16] for both brain and knee MRIs.

Keywords: MRI acceleration · End-to-end learning · Deep learning

1 Introduction

Magnetic Resonance Imaging (MRI) is a powerful diagnostic tool for a variety of disorders, but its utility is often limited by its slow speed compared to competing modalities like CT or X-Rays. Reducing the time required for a scan would decrease the cost of MR imaging, and allow for acquiring scans in situations where a patient cannot stay still for the current minimum scan duration.

One approach to accelerating MRI acquisition, called Parallel Imaging (PI) [3,8,13], utilizes multiple receiver coils to simultaneously acquire multiple views of the underlying anatomy, which are then combined in software. Multi-coil imaging is widely used in current clinical practice. A complementary approach

---

A. Sriram and J. Zbontar—Equal contributions.

Electronic supplementary material The online version of this chapter (https://doi.org/10.1007/978-3-030-59713-9_7) contains supplementary material, which is available to authorized users.

A. L. Martel et al. (Eds.): MICCAI 2020, LNCS 12262, pp. 64–73, 2020.
https://doi.org/10.1007/978-3-030-59713-9_7

to accelerating MRIs acquires only a subset of measurements and utilizes Compressed Sensing (CS) [1,7] methods to reconstruct the final image from these undersampled measurements. The combination of PI and CS, which involves acquiring undersampled measurements from multiple views of the anatomy, has the potential to allow faster scans than is possible by either method alone. Reconstructing MRIs from such undersampled multi-coil measurements has remained an active area of research.

MRI reconstruction can be viewed as an inverse problem and previous research has proposed neural networks whose design is inspired by the optimization procedure to solve such a problem [4,6,9,10]. A limitation of such an approach is that it assumes the forward process is completely known, which is an unrealistic assumption for the multi-coil reconstruction problem. In this paper, we present a novel technique for reconstructing MRI images from undersampled multi-coil data that does not make this assumption. We extend previously proposed variational methods by learning the forward process in conjunction with reconstruction, alleviating this limitation. We show through experiments on the fastMRI dataset that such an approach yields higher fidelity reconstructions.

Our contributions are as follows: 1) we extend the previously proposed variational network model by learning completely end-to-end; 2) we explore the design space for the variational networks to determine the optimal intermediate representations and neural network architectures for better reconstruction quality; and 3) we perform extensive experiments using our model on the fastMRI dataset and obtain new state-of-the-art results for both the knee and the brain MRIs.

2   Background and Related Work

2.1   Accelerated MRI Acquisition

An MR scanner images a patient's anatomy by acquiring measurements in the frequency domain, called *k-space*, using a measuring instrument called a receiver coil. The image can then be obtained by applying an inverse multidimensional Fourier transform $\mathcal{F}^{-1}$ to the measured k-space samples. The underlying image $\mathbf{x} \in \mathbb{C}^M$ is related to the measured k-space samples $\mathbf{k} \in \mathbb{C}^M$ as $\mathbf{k} = \mathcal{F}(\mathbf{x}) + \epsilon$, where ϵ is the measurement noise and $\mathcal{F}$ is the fourier transform operator. A multi-coil scanner acquires several views of the anatomy, which are modulated by the sensitivities of the various coils. The i-th coil measures:

$$\mathbf{k}_i = \mathcal{F}(S_i \mathbf{x}) + \epsilon_i, i = 1, 2, \ldots, N, \tag{1}$$

where S_i is a complex-valued diagonal matrix encoding the position dependent sensitivity map of the i-th coil and N is the number of coils. The sensitivity maps are normalized to satisfy [14]:

$$\sum_{i=1}^{N} S_i^* S_i = 1 \tag{2}$$

The speed of MRI acquisition is limited by the number of k-space samples obtained. This acquisition process can be accelerated by obtaining undersampled k-space data, $\tilde{\mathbf{k}}_i = M\mathbf{k}_i$, where M is a binary mask operator that selects a subset of k-space points and $\mathbf{k}_i$ is the measured k-space. Applying an inverse Fourier transform naively to this under-sampled k-space data results in aliasing artifacts.

Parallel Imaging can be used to accelerate imaging by exploiting redundancies in k-space samples measured by different coils. The sensitivity maps S_i can be estimated using the central region of k-space corresponding to low frequencies, called the *Auto-Calibration Signal (ACS)*, which is typically fully sampled. To accurately estimate these sensitivity maps, the ACS must be sufficiently large, which limits the maximum possible acceleration.

2.2 Compressed Sensing for Parallel MRI Reconstruction

Compressed Sensing [2] enables reconstruction of images by using fewer k-space measurements than is possible with classical signal processing methods by enforcing suitable priors, and solving the following optimization problem:

$$\hat{\mathbf{x}} = \text{argmin}_{\mathbf{x}} \frac{1}{2} \sum_i \left\| M\mathcal{F}(S_i\mathbf{x}) - \tilde{\mathbf{k}}_i \right\|^2 + \lambda\Psi(\mathbf{x}) \tag{3}$$

$$= \text{argmin}_{\mathbf{x}} \frac{1}{2} \left\| A(\mathbf{x}) - \tilde{\mathbf{k}} \right\|^2 + \lambda\Psi(\mathbf{x}), \tag{4}$$

where Ψ is a regularization function that enforces a sparsity constraint, A is the linear forward operator that multiplies by the sensitivity maps, applies 2D fourier transform and then under-samples the data, and $\tilde{\mathbf{k}}$ is the vector of masked k-space data from all coils. This problem can be solved by iterative gradient descent methods. In the t-th step the image is updated from $\mathbf{x}^t$ to $\mathbf{x}^{t+1}$ using:

$$\mathbf{x}^{t+1} = \mathbf{x}^t - \eta^t \left(A^*(A(\mathbf{x}) - \tilde{\mathbf{k}}) + \lambda\Phi(\mathbf{x}^t) \right), \tag{5}$$

where η^t is the learning rate, $\Phi(\mathbf{x})$ is the gradient (or proximal operator) of Ψ with respect to $\mathbf{x}$, and A^* is the hermitian of the forward operator A.

2.3 Deep Learning for Parallel MRI Reconstruction

In the past few years, there has been rapid development of deep learning based approaches to MRI reconstruction [4–6,9,10,12]. A comprehensive survey of recent developments in using deep learning for parallel MRI reconstruction can be found in [5]. Our work builds upon the Variational Network (VarNet) [4], which consists of multiple layers, each modeled after a single gradient update step in Eq. 5. Thus, the t-th layer of the VarNet takes $\mathbf{x}^t$ as input and computes $\mathbf{x}^{t+1}$ using:

$$\mathbf{x}^{t+1} = \mathbf{x}^t - \eta^t A^*(A(\mathbf{x}^t) - \tilde{\mathbf{k}}) + \text{CNN}(\mathbf{x}^t), \tag{6}$$

where CNN is a small convolutional neural network that maps complex-valued images to complex-valued images of the same shape. The η^t values as well as the parameters of the CNNs are learned from data.

The A and A^* operators involve the use of sensitivity maps which are computed using a traditional PI method and fed in as additional inputs. As noted in Sect. 2.1, these sensitivity maps cannot be estimated accurately when the number of auto-calibration lines is small, which is necessary to achieve higher acceleration factors. As a result, the performance of the VarNet degrades significantly at higher accelerations. We alleviate this problem in our model by learning to predict the sensitivity maps from data as part of the network.

3   End-to-End Variational Network

Let $\mathbf{k}_0 = \tilde{\mathbf{k}}$ be the vector of masked multi-coil k-space data. Similar to the Var-Net, our model takes this masked k-space data $\mathbf{k}_0$ as input and applies a number of refinement steps of the same form. We refer to each of these steps as a *cascade* (following [12]), to avoid overloading the term "layer" which is already heavily used. Unlike the VN, however, our model uses k-space intermediate quantities rather than image-space quantities. We call the resulting method the End-to-End Variational Network or E2E-VarNet.

3.1   Preliminaries

To simplify notation, we first define two operators: the expand operator $(\mathcal{E})$ and the reduce operator $(\mathcal{R})$. The *expand* operator $(\mathcal{E})$ takes the image $\mathbf{x}$ and sensitivity maps as input and computes the corresponding image seen by each coil in the idealized noise-free case:

$$\mathcal{E}(\mathbf{x}) = (\mathbf{x}_1, ..., \mathbf{x}_N) = (S_1\mathbf{x}, ..., S_N\mathbf{x}). \tag{7}$$

where S_i is the sensitivity map of coil i. We do not explicitly represent the sensitivity maps as inputs for the sake of readability. The inverse operator, called the *reduce* operator $(\mathcal{R})$ combines the individual coil images:

$$\mathcal{R}(\mathbf{x}_1, ...\mathbf{x}_N) = \sum_{i=1}^{N} S_i^*\mathbf{x}_i. \tag{8}$$

Using the expand and reduce operators, A and A^* can be written succinctly as $A = M \circ \mathcal{F} \circ \mathcal{E}$ and $A^* = \mathcal{R} \circ \mathcal{F}^{-1} \circ M$.

3.2   Cascades

Each cascade in our model applies a refinement step similar to the gradient descent step in Eq. 6, except that the intermediate quantities are in k-space. Applying $\mathcal{F} \circ \mathcal{E}$ to both sides of 6 gives the corresponding update equation in k-space:

$$\mathbf{k}^{t+1} = \mathbf{k}^t - \eta^t M(\mathbf{k}^t - \tilde{\mathbf{k}}) + G(\mathbf{k}^t) \tag{9}$$

where G is the *refinement module* given by:

$$G(\mathbf{k}^t) = \mathcal{F} \circ \mathcal{E} \circ \text{CNN}(\mathcal{R} \circ \mathcal{F}^{-1}(\mathbf{k}^t)). \tag{10}$$

Here, we use the fact that $\mathbf{x}^t = \mathcal{R} \circ \mathcal{F}^{-1}(\mathbf{k}^t)$. CNN can be any parametric function that takes a complex image as input and maps it to another complex image. Since the CNN is applied after combining all coils into a single complex image, the same network can be used for MRIs with different number of coils.

Each cascade applies the function represented by Eq. 9 to refine the k-space. In our experiments, we use a U-Net [11] for the CNN.

3.3   Learned Sensitivity Maps

The expand and reduce operators in Eq. 10 take sensitivity maps $(S_1, ..., S_N)$ as inputs. In the original VarNet model, these sensitivity maps are computed using the ESPIRiT algorithm [14] and fed in to the model as additional inputs. In our model, however, we estimate the sensitivity maps as part of the reconstruction network using a *Sensitivity Map Estimation (SME)* module:

$$H = \text{dSS} \circ \text{CNN} \circ \mathcal{F}^{-1} \circ M_{\text{center}}. \tag{11}$$

The M_{center} operator zeros out all lines except for the autocalibration or ACS lines (described in Sect. 2.1). This is similar to classical parallel imaging approaches which estimate sensitivity maps from the ACS lines. The CNN follows the same architecture as the CNN in the cascades, except with fewer channels and thus fewer parameters in intermediate layers. This CNN is applied to each coil image independently. Finally, the dSS operator normalizes the estimated sensitivity maps to ensure that the property in Eq. 2 is satisfied.

3.4   E2E-VarNet Model Architecture

As previously described, our model takes the masked multi-coil k-space $\mathbf{k}_0$ as input. First, we apply the SME module to $\mathbf{k}_0$ to compute the sensitivity maps. Next we apply a series of cascades, each of which applies the function in Eq. 9, to the input k-space to obtain the final k-space representation K^T. This final k-space representation is converted to image space by applying an inverse Fourier transform followed by a root-sum-squares (RSS) reduction for each pixel:

$$\hat{\mathbf{x}} = RSS(\mathbf{x}_1, ..., \mathbf{x}_N) = \sqrt{\sum_{i=1}^{N} |\mathbf{x}_i|^2} \tag{12}$$

where $\mathbf{x}_i = \mathcal{F}^{-1}(K_i^T)$ and K_i^T is the k-space representation for coil i. The model is illustrated in Fig. 1.

All of the parameters of the network, including the parameters of the CNN model in SME, the parameters of the CNN in each cascade along with the η^ts, are estimated from the training data by minimizing the structural similarity loss, $J(\hat{\mathbf{x}}, \mathbf{x}^*) = -\text{SSIM}(\hat{\mathbf{x}}, \mathbf{x}^*)$, where SSIM is the Structural Similarity index [15] and $\hat{\mathbf{x}}$, $\mathbf{x}^*$ are the reconstruction and ground truth images, respectively.

4   Experiments

4.1   Experimental Setup

We designed and validated our method using the multicoil track of the fastMRI dataset [16] which is a large and open dataset of knee and brain MRIs. To validate the various design choices we made, we evaluated the following models on the knee dataset:

1. Variational network [4] (VN)
2. Variational network with the shallow CNNs replaced with U-Nets (VNU)
3. Similar to VNU, but with k-space intermediate quantities (VNU-K)
4. Our proposed end-to-end variational network model (E2E-VN).

The VN model employs shallow convolutional networks with RBF kernels that have about 150 K parameters in total. VNU replaces these shallow networks with U-Nets to ensure a fair comparison with our model. VNU-K is similar to our proposed model but uses fixed sensitivity maps computed using classical parallel imaging methods. The difference in reconstruction quality between VNU and VNU-K shows the value of using k-space intermediate quantities for reconstruction, while the difference between VNU-K and E2E-VN shows the importance of learning sensitivity maps as part of the network.

We used the same model architecture and training procedure for the VN model as in the original VarNet [4] paper. For each of the other models, we used $T = 12$ cascades, containing a total of about 29.5M parameters. The E2E-VN model contained an additional 0.5M parameters in the SME module, taking the total number of parameters to 30M. We trained these models using the Adam optimizer with a learning rate of 0.0003 for 50 epochs, without using any regularization or data augmentation techniques. Each model was trained jointly on all sequence types.

We used two types of under-sampling masks: *equispaced masks* $M_e(r, l)$, which sample l low-frequency lines from the center of k-space and every r-th line from the remaining k-space; and *random masks* $M_r(a, f)$, which sample a fraction f of the full width of k-space for the ACS lines in addition to a subset of higher frequency lines, selected uniformly at random, to make the overall acceleration equal to a. These random masks are identical to those used in [16]. We also use equispaced masks as they are easier to implement in MRI machines.

4.2   Results

Tables 1 and 2 show the results of our experiments for equispaced and random masks respectively, over a range of down-sampling mask parameters. The VNU model outperforms the baseline VN model by a large margin due to its larger capacity and the multi-scale modeling ability of the U-Nets. VNU-K outperforms VNU demonstrating the value of using k-space intermediate quantities. E2E-VN outperforms VNU-K showing the importance of learning sensitivity maps as part of the network. It is worth noting that the relative performance does not depend on the type of mask or the mask parameters. Some example reconstructions are shown in Fig. 2.

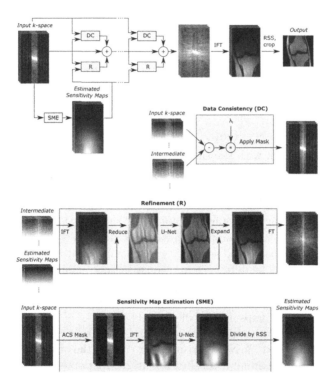

Fig. 1. Top: The E2E-VN model takes under-sampled k-space as input and applies several cascades, followed by IFT and RSS. Bottom: The DC module brings intermediate k-space closer to measured values, the Refinement module maps multi-coil k-space to one image, applies a U-Net and maps back to k-space, and the SME estimates sensitivity maps used in the refinement step.

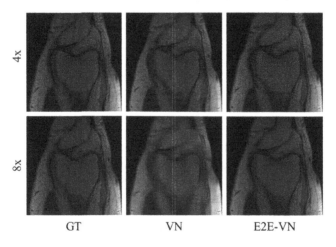

Fig. 2. Examples comparing the ground truth (GT) to VarNet (VN) and E2E-VarNet (E2E-VN). At 8× acceleration, the VN images contain severe artifacts.

Table 1. Experimental results using equispaced masks on the knee MRIs

Accel(r)	Num ACS(l)	Model	SSIM
4	30	VN	0.818
		VNU	0.919
		VNU-K	0.922
		E2E-VN	**0.923**
6	22	VN	0.795
		VNU	0.904
		VNU-K	0.907
		E2E-VN	**0.907**
8	16	VN	0.782
		VNU	0.889
		VNU-K	0.891
		E2E-VN	**0.893**

Table 2. Experimental results using random masks on the knee MRIs

Accel(a)	Frac ACS(f)	Model	SSIM
4	0.08	VN	0.813
		VNU	0.906
		VNU-K	0.907
		E2E-VN	**0.910**
6	0.06	VN	0.767
		VNU	0.886
		VNU-K	0.887
		E2E-VN	**0.892**
8	0.04	VN	0.762
		VNU	0.870
		VNU-K	0.871
		E2E-VN	**0.878**

Significance of Learning Sensitivity Maps. Fig. 3 shows the SSIM values for each model with various equispaced mask parameters. In all cases, learning the sensitivity maps improves the SSIM score. Notably, this improvement in SSIM is larger when the number of low frequency lines is smaller. As previously stated, the quality of the estimated sensitivity maps tends to be poor when there are few ACS lines, which leads to a degradation in the final reconstruction quality. The E2E-VN model is able to overcome this limitation and generate good reconstructions even with a small number of ACS lines.

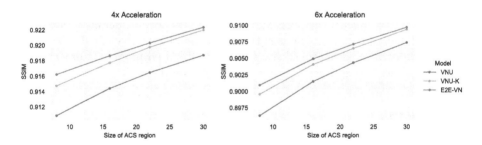

Fig. 3. Effect of different equispaced mask parameters on reconstruction quality at $4\times$ and $6\times$ acceleration.

Experiments on Test Data. Table 3 shows our results on the test datasets for both the brain and knee MRIs compared with the best models on the fastMRI leaderboard[1]. To obtain these results, we used the same training procedure as our

---

[1] http://fastmri.org/leaderboards.

Table 3. Results on the test data for knee and brain MRIs.

Dataset	Model	4× Acceleration			8× Acceleration		
		SSIM	NMSE	PSNR	SSIM	NMSE	PSNR
Knee	E2E-VN	**0.930**	0.005	40	**0.890**	0.009	37
	SubtleMR	0.929	0.005	40	0.888	0.009	37
Brain	E2E-VN	**0.959**	0.004	41	**0.943**	0.008	38
	U-Net	0.945	0.011	38	0.915	0.023	35

previous experiments, except that we trained on both the training and validation sets for 100 epochs. We used the same type of masks that are used for the fastMRI paper [4]. Our model outperforms all other models published on the fastMRI leaderboard for both anatomies.

5 Conclusion

In this paper, we introduced End-to-End Variational Networks for multi-coil MRI reconstruction. While MRI reconstruction can be posed as an inverse problem, multi-coil MRI reconstruction is particularly challenging because the forward process (which is determined by the sensitivity maps) is not completely known. We alleviate this problem by estimating the sensitivity maps within the network, and learning fully end-to-end. Further, we explored the architecture space to identify the best neural network layers and intermediate representation for this problem, which allowed our model to obtain new state-of-the art results on both brain and knee MRIs.

The quantitative measures we have used only provide a rough estimate for the quality of the reconstructions. Many clinically important details tend to be subtle and limited to small regions of the MR image. Rigorous clinical validation needs to be performed before such methods can be used in clinical practice to ensure that there is no degradation in the quality of diagnosis.

References

1. Candès, E.J., et al.: Compressive sampling. In: Proceedings of the International Congress of Mathematicians, Madrid, Spain, vol. 3, pp. 1433–1452 (2006)
2. Donoho, D.: Compressed sensing. IEEE Trans. Inf. Theory **52**(4), 1289–1306 (2006)
3. Griswold, M.A., et al.: Generalized autocalibrating partially parallel acquisitions (GRAPPA). Magn. Reson. Med. **47**(6), 1202–1210 (2002)
4. Hammernik, K., et al.: Learning a variational network for reconstruction of accelerated MRI data. Magn. Reson. Med. **79**(6), 3055–3071 (2018)
5. Knoll, F., et al.: Deep learning methods for parallel magnetic resonance image reconstruction. arXiv preprint arXiv:1904.01112 (2019)
6. Liang, D., Cheng, J., Ke, Z., Ying, L.: Deep MRI reconstruction: unrolled optimization algorithms meet neural networks. arXiv preprint arXiv:1907.11711 (2019)

7. Lustig, M., Donoho, D., Pauly, J.M.: Sparse MRI: the application of compressed sensing for rapid MR imaging. Magn. Reson. Med. **58**(6), 1182–1195 (2007)
8. Pruessmann, K.P., Weiger, M., Scheidegger, M.B., Boesiger, P.: SENSE: sensitivity encoding for fast MRI. Magn. Reson. Med. **42**(5), 952–962 (1999)
9. Putzky, P., Welling, M.: Invert to learn to invert. In: Advances in Neural Information Processing Systems, pp. 444–454 (2019)
10. Putzky, P., et al.: i-RIM applied to the fastMRI challenge. arXiv preprint arXiv:1910.08952 (2019)
11. Ronneberger, O., Fischer, P., Brox, T.: U-Net: convolutional networks for biomedical image segmentation. In: Navab, N., Hornegger, J., Wells, W.M., Frangi, A.F. (eds.) MICCAI 2015. LNCS, vol. 9351, pp. 234–241. Springer, Cham (2015). https://doi.org/10.1007/978-3-319-24574-4_28
12. Schlemper, J., Caballero, J., Hajnal, J.V., Price, A.N., Rueckert, D.: A deep cascade of convolutional neural networks for dynamic MR image reconstruction. IEEE Trans. Med. Imaging **37**(2), 491–503 (2018)
13. Sodickson, D.K., Manning, W.J.: Simultaneous acquisition of spatial harmonics (SMASH): fast imaging with radiofrequency coil arrays. Magn. Reson. Med. **38**(4), 591–603 (1997)
14. Uecker, M., et al.: ESPIRiT -an eigenvalue approach to autocalibrating parallel MRI: where SENSE meets GRAPPA. Magn. Reson. Med. **71**(3), 990–1001 (2014)
15. Wang, Z., Simoncelli, E.P., Bovik, A.C.: Multiscale structural similarity for image quality assessment. In: Asilomar Conference on Signals, Systems & Computers (2003)
16. Zbontar, J., et al.: fastMRI: an open dataset and benchmarks for accelerated MRI. arXiv preprint arXiv:1811.08839 (2018)

3d-SMRnet: Achieving a New Quality of MPI System Matrix Recovery by Deep Learning

Ivo M. Baltruschat[1,2,3(✉)], Patryk Szwargulski[1,2], Florian Griese[1,2], Mirco Grosser[1,2], Rene Werner[3,4], and Tobias Knopp[1,2]

[1] Section for Biomedical Imaging, University Medical Center Hamburg -Eppendorf, Hamburg, Germany
i.baltruschat@uke.de
[2] Institute for Biomedical Imaging, Hamburg University of Technology, Hamburg, Germany
[3] DAISYLabs, Forschungszentrum Medizintechnik Hamburg, Hamburg, Germany
[4] Department of Computational Neuroscience, University Medical Center Hamburg -Eppendorf, Hamburg, Germany
https://www.tuhh.de/ibi

Abstract. Magnetic particle imaging (MPI) data is commonly reconstructed using a system matrix acquired in a time-consuming calibration measurement. Compared to model-based reconstruction, the calibration approach has the important advantage that it takes into account both complex particle physics and system imperfections. However, this has the disadvantage that the system matrix has to be re-calibrated each time the scan parameters, the particle types or even the particle environment (e.g. viscosity or temperature) changes. One way to shorten the calibration time is to scan the system matrix at a subset of the spatial positions of the intended field-of-view and use the system matrix recovery. Recent approaches used compressed sensing (CS) and achieved subsampling factors up to 28, which still allowed the reconstruction of MPI images with sufficient quality. In this work we propose a novel framework with a 3d system matrix recovery network and show that it recovers a 3d system matrix with a subsampling factor of 64 in less than a minute and outperforms CS in terms of system matrix quality, reconstructed image quality, and processing time. The advantage of our method is demonstrated by reconstructing open access MPI datasets. Furthermore, it is also shown that the model is capable of recovering system matrices for different particle types.

Keywords: Magnetic particle imaging · System matrix recovering · Deep learning · Single image super-resolution

© Springer Nature Switzerland AG 2020
A. L. Martel et al. (Eds.): MICCAI 2020, LNCS 12262, pp. 74–82, 2020.
https://doi.org/10.1007/978-3-030-59713-9_8

1 Introduction

Magnetic particle imaging (MPI) is a young tomographic imaging technique that quantitatively images magnetic nanoparticles with a high spatio-temporal resolution and is ideally suited for vascular and targeted imaging [6]. One common way to reconstruct MPI data is the system matrix (SM)-based reconstruction [7]. It requires a complex-valued SM, which is currently determined in a time-consuming calibration measurement. A delta sample is moved through the field-of-view (FOV) using a robot and the system response is measured in a calibration process. The number of voxels encoded in the SM directly determines the image size but also the scan time. The acquisition of a $37 \times 37 \times 37$ voxel SM takes about 32 hours, compared to an $9 \times 9 \times 9$ SM, which takes about 37 min. Therefore, a compromise between image size and scan time is usually made. The resulting SM is only valid for a certain set of scan parameters and for a specific particle type and its settings. If scan parameters (e.g. the size or position of the FOV) or particle settings (e.g. binding state, viscosity or temperature) are changed, SM calibration must be performed again, making it nearly impossible to record high resolution system matrices for each combination of scan parameters and particle settings.

Earlier approaches to reduce calibration time for MPI employed compressed sensing (CS) to recover a subsampled 2d SM [8]. For up to 28-fold subsampling, sufficient image quality after reconstruction was achieved [14]. While the CS approach for SM recovery is promising, it still leaves room for improvement since the sparsification using the DCT is not perfect and row specific. Furthermore, the CS approach currently cannot take prior knowledge from existing high resolution (HR) system matrix measurements into account. In the present work, we will, for the first time, investigate if deep learning (DL) techniques can be used to improve the SM recovery problem in MPI. DL-based super-resolution techniques have been demonstrated to be superior in the up-scaling of images in computer vision [4,11] and recently also for 3d medical image up-scaling [2,3,12]. While super-resolution convolutional networks (SRCNNs) and super-resolution generative adversarial networks (SRGANs) are mostly used directly in the image domain, it is inherently difficult to restore texture and structural details. SRGANs have proven to be successful in computer vision and medical image processing in modeling visually more appealing images than SRCNNs; SRCNNs typically tend to blur the SR image as a whole. Yet, this property may potentially be beneficial if SR is applied *prior* to image reconstruction – like in the current case for MPI SM recovery.

To evaluate the potential of SRCNN-based MPI SM recovery, we present a novel framework that comprises three central steps (see the proposed method branch in Fig. 1). First, we acquire a low resolution (LR) SM on a specific sampling grid. Secondly, we used a ComplexRGB-embedding to transform each complex number of the 3d-SM to RGB vectors, allowing us to use a MSE-loss for optimization. Thirdly, we employ a SRCNN, which we call 3d system matrix recovery network (3d-SMRnet), to recover a high resolution SM by adapting the model to work on 3d data and employing it to each frequency component of the

SM. Finally, we decode each RGB vector of the high resolution SM back to a complex number and use this newly recovered SM to reconstruct an HR image.

We evaluate our method in Sect. 3 on the Open MPI data and will show that our framework reaches superior performance in image quality, SM quality, and processing time compared to the current state-of-the-art. All afore-mentioned aspects (introduction of 3d-SMRnet; conversion of MPI raw data to RGB format; comparison of 3d-SMRnet to compressed sensing) are novel contributions.

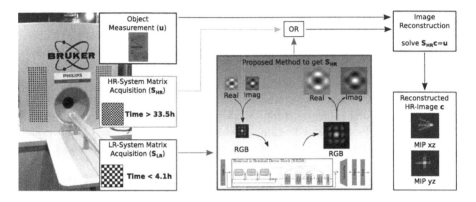

Fig. 1. Overview of the data flow when employing our proposed 3d-SMRnet (blue path). Instead of measuring a high resolution system matrix (SM) and using it for reconstruction (orange path), only a low resolution SM (lower left) is measured and the high resolution SM is retrieved by applying our proposed method to each frequency component of the low resolution SM. The recovered HR SM can be used for reconstruction (upper right).(Color figure online)

2   Methods

In MPI, the relation between the particle concentration $c(\cdot)$ and the Fourier coefficients $\hat{u}_k$ of the induced voltage $u(t)$ at frequency f_k can be described by the linear integral equation $\hat{u}_k = \int_\Omega \hat{s}_k(\mathbf{r})c(\mathbf{r})\mathrm{d}^3 r$ where $\mathbf{r}$ is the spatial position, k is the frequency index, and $\hat{s}_k(\cdot)$ denotes the system function. By sampling the FOV at N positions $\mathbf{r}_n, n = 0, \ldots, N - 1$, one obtains a linear system of equations written in matrix-vector form $\mathbf{Sc} = \hat{\mathbf{u}}$ where $\hat{\mathbf{u}} = (\hat{u}_k)_{k=0}^{K-1} \in \mathbb{C}^K$ and $\mathbf{c} = (c(\mathbf{r}_n))_{n=0}^{N-1} \in \mathbb{R}^N$ are the measurement vector and the particle concentration vector, respectively. K is the total number of frequency components. The goal of this work is to recover a high resolution system matrix

$$\mathbf{S} = (\hat{s}_k(\mathbf{r}_n))_{k=0,\ldots,K-1;n=0,\ldots,N-1} \in \mathbb{C}^{K \times N} \qquad (1)$$

using a subset of the total number of sampling positions N. In our work, we treat the SM rows as independent images and train a network based on the entire set of rows of a measured HR SM. Then for SM recovery one measures an LR SM and infers the HR SM from the trained network. The method consists of three main steps, which are outlined in Fig. 1 and explained in detail below.

System Matrix Sampling: Usually, the system matrix is acquired using a 3-axis linear robot to scan the FOV equidistantly on a predefined grid. In this way, both HR and LR system matrix can be acquired. The points are sampled along a meandering trajectory to minimize the robot's travel time. This is comparable to the generation of an LR SM by sampling every n^{th} voxel from an HR SM. We employ the latter to get an LR and HR SM pair as training data for our 3d-SMRnet.

ComplexRGB-Embedding: Loss functions for regression problems are not intended to deal with imaginary and real parts of complex numbers. When employing standard regression loss functions (i.e. the MSE-loss) for optimization, both parts (i.e. imaginary and real part as channels) are optimized independently. To constrain the loss function, we propose an embedding for the complex numbers $\hat{s}_k(\mathbf{r}_n)$ of the SM before calculating the loss. The embedding transforms the complex number to an RGB color vector $s_k^{\text{RGB}}(\mathbf{r}_n) \in \mathbb{R}^3$. First, we use the hue-saturation-value (HSV) color model to represent the phase $\arg \hat{s}_k(\mathbf{r}_n)$ with hue following the color wheel. Therefore, we employ the transformation $T_{\text{HSV}} : \mathbb{C} \to \mathbb{R}^3$ with $T_{\text{HSV}}(\hat{s}_k(\mathbf{r}_n)) = (H, S, V) = (\arg \hat{s}_k(\mathbf{r}_n), 1, 1)$, where we omit S and V by setting them to 1. Secondly, we convert the HSV color vector $s_k^{\text{HSV}}(\mathbf{r}_n) = T_{\text{HSV}}(\hat{s}_k(\mathbf{r}_n))$ to $s_k^{\text{RGB}}(\mathbf{r}_n)$ with a HSV to RGB transformation [5]. Finally, the RGB color vector is linearly scaled by the amplitude $|\hat{s}_k(\mathbf{r}_n)|$.

For decoding, we recover the complex numbers by extracting the scaling factor of the RGB color vector for the amplitude. Afterwards, we normalize the RGB color vector by the amplitude and convert it to a HSV color vector. The phase is now the hue value.

3d-System Matrix Recovery Network: Following [13] and [9], we extend the SRCNN with residual-in-residual-dense-blocks (RRDBs) from 2d-RGB to 3d-RGB image processing. Our model contains two branches: image reconstruction and feature extraction. The feature extraction branch consists of R stacked RRDBs (here $R = 9$). Each RRDB combines three dense connected blocks and four residual connections as illustrated in Fig. 1. The dense connected blocks are built upon five convolutional layers. The image reconstruction branch generates the final up-scaled image by U up-convolution blocks, followed by two convolutional layers. The up-convolution block contains a nearest-neighbor interpolation upsampling and a convolutional layer as proposed by [10] to reduce checkerboard artifacts from deconvolution. In our model, all 2d convolutions are replaced by 3d convolutions. Hence, the 3d-SMRnet recovers an HR SM by employing it to each frequency component K of an LR SM.

3   Materials and Experiments

We apply our framework to the Open MPI dataset[1]; It contains two HR system matrices, one for the particles Perimag $\mathbf{S}_{\text{HR}}^{\text{Peri}}$ and another for the particles

---

[1] https://magneticparticleimaging.github.io/OpenMPIData.jl/latest/.

Synomag-D $\mathbf{S}_{\mathrm{HR}}^{\mathrm{Syno}}$. Both are acquired using a 4 μL delta sample with a concentration of 100 mmol/L and a grid size of $37 \times 37 \times 37$. Hence, $\mathbf{S}_{\mathrm{HR}}^{\mathrm{Peri}}$ and $\mathbf{S}_{\mathrm{HR}}^{\mathrm{Syno}}$ have the dimensions $37 \times 37 \times 37 \times K$. Furthermore, three different phantom measurements with Perimag are provided in the Open MPI data: shape-, resolution-, and concentration-phantom. In our experiment, we train our 3d-SMRnet on Synomag-D with frequency components of the subsampled $\mathbf{S}_{\mathrm{LR}}^{\mathrm{Syno}}$ as input against $\mathbf{S}_{\mathrm{HR}}^{\mathrm{Syno}}$ and test it on the subsampled $\mathbf{S}_{\mathrm{LR}}^{\mathrm{Peri}}$. This represents the interesting case where the SM for new particles is inferred from a network trained on an established particle system. In addition, this approach prevents overfitting of the data.

We evaluate the recovered SM results in two steps. First, we compare all recovered system matrices with the ground truth $\mathbf{S}_{\mathrm{HR}}^{\mathrm{Peri}}$ by calculating the normalized root mean squared error (NRMSE) for each frequency component. Secondly, we reconstruct the measurements of the Open MPI shape-, resolution- and concentration-phantom with all recovered system matrices using the same standard regularization parameter ($\lambda = 0.01, \mathrm{iter} = 3$).

Implementation Details and Training: We implement three versions 3d-SMRnet$_{8\times}$, 3d-SMRnet$_{27\times}$, and 3d-SMRnet$_{64\times}$. For 3d-SMRnet$_{8\times}$ and 3d-SMRnet$_{27\times}$, we set $U = 1$ and 2-times and 3-times upsampling, respectively. Furthermore, 3d-SMRnet$_{64\times}$ uses $U = 2$ and 2-times upsampling to finally upsample 4-times. To generate an LR and HR SM pair for training, we zero-pad $\mathbf{S}_{\mathrm{HR}}^{\mathrm{Syno}}$ to $40 \times 40 \times 40$ with two rows and one row at the beginning and the end, respectively. Afterwards, we apply 8-fold, 27-fold, and 64-fold subsampling, resulting in $20 \times 20 \times 20$, $13 \times 13 \times 13$ and $10 \times 10 \times 10$ spatial dimensions for the input volume, respectively. After applying a threshold with a signal-to-noise ratio (SNR) of 3, $\mathbf{S}_{\mathrm{HR}}^{\mathrm{Peri}}$ and $\mathbf{S}_{\mathrm{HR}}^{\mathrm{Syno}}$ have $K = 3175$ and $K = 3929$ frequency components, respectively. We split K of $\mathbf{S}_{\mathrm{HR}}^{\mathrm{Syno}}$ into 90% training and 10% validation data. We use random 90° rotations and random flipping as data augmentation. In total, we train for $2 \cdot 10^5$ iterations. Each iteration has a minibatch size of 20 for 3d-SMRnet$_{8\times}$ and 64 for 3d-SMRnet$_{27\times}$ and 3d-SMRnet$_{64\times}$. For optimization, we use ADAM with $\beta_1 = 0.9$, $\beta_2 = 0.999$, and without regularization. We trained with a learning rate of 10^{-5} for 3d-SMRnet$_{8\times}$ and 10^{-4} for 3d-SMRnet$_{27\times}$ and 3d-SMRnet$_{64\times}$. The learning rate is reduced by two every $4 \cdot 10^3$ iteration. As loss function, we employ the *mean squared error* (MSE). Our models are implemented in PyTorch and trained on two Nvidia GTX 1080Ti GPUs. To support the reproduction of our results and further research, our framework and code are publicly available at https://github.com/Ivo-B/3dSMRnet.

Comparison to State-of-the-Art: Compressed sensing exploits the fact that the SM becomes sparse when a discrete cosine transform (DCT) is applied to its rows. As shown in [1,8], such signals can be recovered from an subsampled measurement by solving regularized least squares problems of the form

$$\min_{\mathbf{s}_k} \|\boldsymbol{\Phi}\mathbf{s}_k\|_1 \text{ subject to } \mathbf{P}\mathbf{s}_k = \mathbf{y}_k. \tag{2}$$

Here, $\mathbf{y}_k \in \mathbb{C}^M$ contains the values of the kth SM row at the measured points and $\mathbf{P} \in \mathbb{C}^{M \times N}$ is the corresponding sampling operator. Moreover, $\mathbf{s}_k \in \mathbb{C}^N$ is the SM row to be recovered and $\boldsymbol{\Phi} \in \mathbb{C}^{N \times N}$ denotes the DCT-II.

Since CS requires an incoherent sampling, it cannot be applied directly to the regular sampled LR SMs used for our 3d-SMRnet. Instead, we use 3d Poisson disc patterns to subsample $\mathbf{S}_{HR}^{\text{Peri}}$ and obtain incoherent measurements with the same number of samples as used by the 3d-SMRnet. For every frequency component, we then normalize the measurement $\mathbf{y}_k$ and solve (2) using the Split Bregman method. The solver parameters are chosen manually such that the average NRMSE for all frequency components is minimized.

4   Results and Discussion

Figure 2 (left) shows the NRMSE plot and Fig. 2 (right) shows visualizations of recovery for three frequencies. The results show that all our 3d-SMRnets can correct the noisy characteristics in $\mathbf{S}_{\text{HR}}^{\text{Peri}}$ due to smoothing characteristics of training with a MSE loss function, whereas CS cannot. For all three reduction factors, the 3d-SMRnet has a lower mean NRMSE than CS: 0.040 vs 0.044, 0.048 vs 0.051, and 0.048 vs 0.077 for 8-fold, 27-fold, and 64-fold subsampling, respectively. While CS cannot sufficiently recover the SM for 64-fold subsampling, our 3d-SMRnet$_{64\times}$ still recovers the SM with a 37.66% lower NRMSE. This low NRMSE is comparable to the results of CS with a 8-fold subsampling. Furthermore, our model 3d-SMRnet$_{64\times}$ is over 42 times faster and takes $\approx 23.3\,$s compared to CS$_{64\times}$ with $\approx 17\,$min for SM recovery. Still, some frequency components tend to have a high NRMSE (NRMSE > 0.11) for our 3d-SMRnet$_{64\times}$, whereas for the 3d-SMRnet$_{8\times}$ they do not. This problem can occur because of the equidistant subsampling and the symmetric patterns in the SM.

In Fig. 3, we show one representative slice ($Z = 19$) of the reconstructed shape and resolution phantoms. For CS, all reconstructed phantoms show a "checkerboard" noise and an overestimation of the voxel intensity, which increases for 27- and 64-fold subsampling. Our proposed 3d-SMRnet results produce smoother reconstructed images, with voxel intensities better resembling the ground truth data. Yet, the shape phantom reconstructed with the $\mathbf{S}_{\text{3d-SMRnet}}^{27\times}$ shows some artifacts, while for $\mathbf{S}_{\text{3d-SMRnet}}^{64\times}$ those are not present. Still, the results for the resolution phantom with $\mathbf{S}_{\text{3d-SMRnet}}^{27\times}$ are visually better than $\mathbf{S}_{\text{3d-SMRnet}}^{64\times}$ (see Fig. 3 the second row). Table 1 lists the subject-wise average structural similarity index (SSIM), peak signal to noise ratio (PSNR), and NRMSE. For all three phantoms, the overall best results achieved our 3d-SMRnet$_{8\times}$ with 0.0113, 0.9985, and 64.74 for $\overline{\text{NRMSE}}$, $\overline{\text{SSIM}}$, and $\overline{\text{PSNR}}$. Compared to the second best CS$_{8\times}$, this is an improvement by 31.1% and 5.2% for $\overline{\text{NRMSE}}$ and $\overline{\text{PSNR}}$. Furthermore, our 3d-SMRnet$_{64\times}$ is on par with CS$_{27\times}$ for the resolution phantom and considerably better for the shape and the concentration phantom.

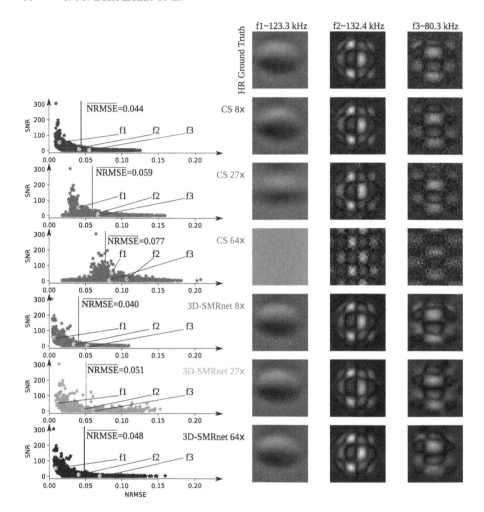

Fig. 2. (left) SNR for each frequency component plotted against their NRMSE for recovered system matrices $\mathbf{S}_{CS}^{8x}$, $\mathbf{S}_{CS}^{27x}$, $\mathbf{S}_{CS}^{64x}$, $\mathbf{S}_{3d\text{-}SMRnet}^{8x}$, $\mathbf{S}_{3d\text{-}SMRnet}^{27x}$ and $\mathbf{S}_{3d\text{-}SMRnet}^{64x}$. (right) Representation of system matrix patterns of exemplary frequency components f_1, f_2 and f_3 for all recovered system matrices and ground truth. The frequency components are presented as RGB converted data

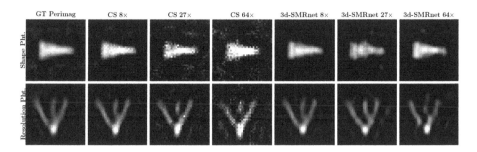

Fig. 3. Exemplary reconstruction of the measurements for the shape and resolution phantom with all recovered system matrices. We selected the slice $Z = 19$.

Table 1. Numerical results for all reconstructed phantoms demonstrates that our 3d-SMRnets clearly outperform CS when comparing the same subsampling factors.

	Shape phantom			Resolution phantom			Concentration pht.			Avg.		
	NRMSE	SSIM	PSNR	NRMSE	SSIM	PSNR	NRMSE	SSIM	PSNR	NRMSE	SSIM	PSNR
CS$_{8\times}$	0.0275	0.9908	51.64	0.0120	0.9996	67.00	0.0098	0.9995	65.93	0.0164	0.9966	61.52
CS$_{27\times}$	0.0628	0.9394	44.47	0.0267	0.9975	60.06	0.0206	0.9972	59.42	0.0367	0.9780	54.60
CS$_{64\times}$	0.0915	0.8804	41.19	0.0452	0.9922	55.50	0.0349	0.9916	54.86	0.0572	0.9547	50.52
3d-SMRnet$_{8\times}$	0.0186	0.9959	55.03	0.0087	0.9998	69.79	0.0066	0.9998	69.39	0.0113	0.9985	64.74
3d-SMRnet$_{27\times}$	0.0320	0.9866	50.31	0.0208	0.9985	62.25	0.0135	0.9988	63.09	0.0221	0.9946	58.55
3d-SMRnet$_{64\times}$	0.0284	0.9874	51.36	0.0249	0.9978	60.68	0.0150	0.9986	62.22	0.0228	0.9946	58.09

5   Conclusion

We presented a novel method based on a 3D system matrix recovery network to significantly reduce the calibration time at the MPI. Our method can recover a heavily subsampled system matrix with 64 times fewer samples compared to the original system matrix and allows for sufficient reconstruction of phantom images. We have also shown that our method exceeds the current state of the art not only in terms of the quality of system matrix recovery, but also in terms of reconstructed image quality and processing time. Furthermore, our 3d-SMRnet can be applied to different types of particles after training. For the future it is of interest to evaluate different types of sampling methods and multi-color MPI where two particles are imaged simultaneously.

References

1. Candès, E., Romberg, J., Tao, T.: Robust uncertainty principles: exact signal reconstruction from highly incomplete frequency information. IEEE Trans. Inform. Theor. **52**, 489–509 (2006)
2. Chen, Y., Shi, F., Christodoulou, A.G., Xie, Y., Zhou, Z., Li, D.: Efficient and accurate mri super-resolution using a generative adversarial network and 3d multi-level densely connected network. In: Medical Image Computing and Computer-Assisted Intervention, pp. 91–99. Springer International Publishing (2018)

3. Chen, Y., Xie, Y., Zhou, Z., Shi, F., Christodoulou, A.G., Li, D.: Brain mri super resolution using 3d deep densely connected neural networks. International Symposium on Biomedical Imaging, pp. 739–742 (2018)

4. Dong, C., Loy, C.C., He, K., Tang, X.: Learning a deep convolutional network for image super-resolution. In: European Conference on Computer Vision, pp. 184–199. Springer International Publishing (2014)

5. Ford, A., Roberts, A.: Colour Space Conversions, vol. 1998, pp. 1–31. Westminster University, London (1998)

6. Gleich, B., Weizenecker, J.: Tomographic imaging using the nonlinear response of magnetic particles. Nature **435**, 1214–1217 (2005)

7. Knopp, T., et al.: Weighted iterative reconstruction for magnetic particle imaging. Phys. Med. Biol. **55**(6), 1577–1589 (2010)

8. Knopp, T., Weber, A.: Sparse reconstruction of the magnetic particle imaging system matrix. IEEE Trans. Med. Imaging **32**(8), 1473–1480 (2013)

9. Ledig, C., et al.: Photo-realistic single image super-resolution using a generative adversarial network. In: Conference on Computer Vision and Pattern Recognition, pp. 4681–4690 (2017)

10. Odena, A., Dumoulin, V., Olah, C.: Deconvolution and checkerboard artifacts. Distill **1**(10), e3 (2016)

11. Tai, Y., Yang, J., Liu, X.: Image super-resolution via deep recursive residual network. In: Conference on Computer Vision and Pattern Recognition (2017)

12. Wang, S., et al.: Accelerating magnetic resonance imaging via deep learning. In: International Symposium on Biomedical Imaging, pp. 514–517 (2016)

13. Wang, X., et al.: Esrgan: enhanced super-resolution generative adversarial networks. In: European Conference on Computer Vision Workshops, pp. 63–79. Springer International Publishing (2019)

14. Weber, A., Knopp, T.: Reconstruction of the magnetic particle imaging system matrix using symmetries and compressed sensing. Advances in Mathematical Physics (2015)

MRI Image Reconstruction via Learning Optimization Using Neural ODEs

Eric Z. Chen, Terrence Chen, and Shanhui Sun[(✉)]

United Imaging Intelligence, Cambridge, MA 02140, USA
{zhang.chen,terrence.chen,shanhui.sun}@united-imaging.com

Abstract. We propose to formulate MRI image reconstruction as an optimization problem and model the optimization trajectory as a dynamic process using ordinary differential equations (ODEs). We model the dynamics in ODE with a neural network and solve the desired ODE with the off-the-shelf (fixed) solver to obtain reconstructed images. We extend this model and incorporate the knowledge of off-the-shelf ODE solvers into the network design (learned solvers). We investigate several models based on three ODE solvers and compare models with fixed solvers and learned solvers. Our models achieve better reconstruction results and are more parameter efficient than other popular methods such as UNet and cascaded CNN. We introduce a new way of tackling the MRI reconstruction problem by modeling the continuous optimization dynamics using neural ODEs.

Keywords: Neural ODE · MRI image reconstruction · Deep learning

1 Introduction

One major limitation of Magnetic Resonance Imaging (MRI) is the slow data acquisition process due to the hardware constraint. To accelerate data acquisition, the undersampled k-space is often acquired, which causes aliasing artifacts in the image domain. Reconstruction of high-quality images from the undersampled k-space data is crucial in the clinical application of MRI.

MRI image reconstruction is an inverse problem, in which the undersampling leads to information loss in the forward model and directly recovering the fully sampled image from undersampled data is intractable. Compressed sensing (CS) provides the theoretical foundation for solving inverse problems by assuming the reconstructed image is sparse itself or in certain transformed domains. With the ability to learn complex distributions from data, deep learning has been applied to MRI reconstruction to learn optimal sparse transformations in an adaptive way [17,19,27]. Methods such as SToRM [24] GANCS [21], and DAGAN [36] follow such strategy and learn the prior distribution of the image from training data. On the other hand, several studies propose to tackle the inverse problem by learning the direct mapping from undersampled data to fully sampled

ⓒ Springer Nature Switzerland AG 2020
A. L. Martel et al. (Eds.): MICCAI 2020, LNCS 12262, pp. 83–93, 2020.
https://doi.org/10.1007/978-3-030-59713-9_9

data in the image domain [18], k-space domain [2,13], or cross domains [42]. Recent works extend this idea by learning such mapping in an iterative way using cascaded networks [1,3,8,10,16,29–31,35], convolutional RNN [26,34] or invertible recurrent models [25]. Many studies design the networks based on the iterative optimization algorithms used in CS [6,7,11,33,39].

Ordinary differential equations (ODEs) are usually used to describe how a system change over time. Solving ODEs involves integration, most of which often have no analytic solutions. Therefore numerical methods are commonly utilized to solve ODEs. For example, the Euler method, which is a first-order method, is one of the basic numerical solvers for ODEs with a given initial value. Runge–Kutta (RK) methods, which is a family of higher-order ODE solvers, are more accurate and routinely used in practice.

Neural ODE [4] was introduced to model the continuous dynamics of hidden states by neural networks and optimize such models as solving ODEs. With the adjoint sensitivity method, the neural ODE models can be optimized without backpropagating through the operations of ODE solvers and thus the memory consumption does not depend on the network depth [4]. ANODE [9,41] further improves the training stability of neural ODEs by using Discretize-Then-Optimize (DTO) differentiation methods. [37] applies ANODE to MRI reconstruction, where the residual blocks in ResNet [14] are replaced with ODE layers. Neural ODE based methods are also applied to image classification [22] and image super-resolution [15].

In this paper, we propose to formulate MRI image reconstruction as an optimization problem and model the optimization trajectory as a dynamic process using ODEs. We model the dynamics in the ODE with a neural network. The reconstructed image can be obtained by solving the ODE with off-the-shelf solvers (fixed solvers). Furthermore, borrowing the ideas from the currently available ODE solvers, we design network structures by incorporating the knowledge of ODE solvers. The network implicitly learns the coefficients and step sizes in the original solver formulation (learned solvers). We investigated several models based on three ODE solvers and compare neural ODE models with fixed solvers and learned solvers. We present a new direction for MRI reconstruction by modeling the continuous optimization dynamics with neural ODEs.

2   Method

MRI reconstruction is an inverse problem and the forward model is

$$y = Ex + \epsilon, \tag{1}$$

where $x \in \mathbb{C}^M$ is the fully sampled image to be reconstructed, $y \in \mathbb{C}^N$ is the observed undersampled k-space and ϵ is the noise. E is the measurement operator that transforms the image into k-space with Fourier transform and undersampling. Since the inverse process is ill-posed, the following regularized objective function is often used:

$$\underset{x}{\operatorname{argmin}} \frac{1}{2}||y - Ex||_2^2 + R(x), \tag{2}$$

where $||y - Ex||_2^2$ is data fidelity term and $R(x)$ is the regularization term . Equation 2 can be optimized with gradient descent based algorithms,

$$x^{(n+1)} = x^n - \eta[E^T(Ex^n - y) + \nabla R(x^n)], \quad \text{for } n = 1, \dots, N, \tag{3}$$

where η is the learning rate.

2.1   ReconODE: Neural ODE for MRI Reconstruction

The iterative optimization algorithm in Eq. 3 can be rewritten as

$$x^{(n+1)} - x^n = f(x^n, y, \theta), \quad \text{for } n = 1, \dots, N. \tag{4}$$

The left hand side of Eq. 4 is the change of the reconstructed image between two adjacent optimization iterations. This equation essentially describes how the reconstructed image changes during the N optimization iterations. The right hand side of Eq. 4 specifies this change by the function f. However, this change is described in discrete states defined by the number of iterations N. If we consider the optimization process as a continuous flow in time, it can be formulated as

$$\frac{dx(t)}{dt} = f(x(t), t, y, \theta). \tag{5}$$

Eq. 5 is an ordinary differential equation, which describes the dynamic optimization trajectory (Fig. 1A). MRI reconstruction can then be regarded as an initial value problem in ODEs, where the dynamics f can be represented by a neural network. The initial condition is the undersampled image and the final condition is the fully sampled image. During model training, given the undersampled image and fully sampled image, the function f is learned from data (Fig. 1B). During inference, given the undersampled image as the initial condition at t_0 and the estimated function f, the fully sampled image can be predicted by evaluating the ODE at the last time point t_N,

$$x(t_N) = x(t_0) + \int_{t_0}^{t_N} f(x(t), t)dt, \tag{6}$$

where y and θ are omitted for brevity. An arbitrary time interval $[0,1]$ is set as in [4]. Evaluating Eq. 6 involves solving the integral, which has no analytic solution due to the complex form of f. The integral needs to be solved by numerical ODE solvers.

Next, we will introduce two models that either use the off-the-shelf solvers (fixed solvers) or learn the solvers by neural networks (learned solvers).

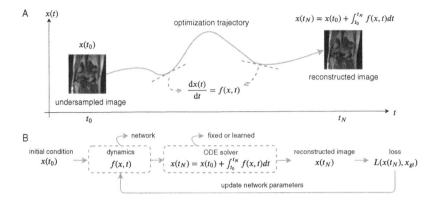

Fig. 1. An illustration of MRI reconstruction via modeling the optimization dynamics using neural ordinary differential equations. (A) MRI reconstruction is formulated as an optimization problem. The dynamic optimization trajectory is described by an ordinary differential equation. (B) We can model the dynamics with a neural network and perform image reconstruction as solving the ODE with the off-the-shelf solver. Alternatively, we can replace both the dynamics and the ODE solver with neural networks to implicitly learn the solver and perform image reconstruction.

2.2 ReconODE with Fixed Solvers

The Euler solver is a first-order ODE solver, which employs the discretization method to approximate the integral with a step size h in an iterative way,

$$x_{n+1} = x_n + hf(x_n, t_n), \tag{7}$$

which is often called the step function of a ODE solver. More complicated methods such as RK solvers with higher orders can provide better accuracy. The step function for a general version of RK methods with stage s is,

$$x_{n+1} = x_n + \sum_{i=1}^{s} a_i F_i \tag{8}$$

$$F_1 = hf(x_n, t_n) \tag{9}$$

$$F_i = hf(x_n + \sum_{j=1}^{i-1} b_{ij} F_j, t_n + c_i h), \quad \text{for } i = 2, \dots, s, \tag{10}$$

where a_i, b_{ij} and c_i are pre-specified coefficients. As an example, for RK2 and RK4 solvers, the coefficients are specified as $s = 2, a_1 = a_2 = \frac{1}{2}, b_{21} = c_2 = 1$ and $s = 4, a_1 = a_4 = \frac{1}{6}, a_2 = a_3 = \frac{2}{6}, b_{21} = b_{32} = c_2 = c_3 = \frac{1}{2}, b_{43} = c_4 = 1$, respectively (all other coefficients are zeros if not specified).

By using one of off-the-shelf ODE solvers to evaluate Eq. 6 numerically, the MRI image can be reconstructed as solving a neural ODE. In our experiment, we model the dynamics f as a CNN (Fig. 2B) with time-dependent convolution (Fig. 2A), in which the time information was incorporated into each convolutional layer by concatenating the scaled time with the input feature maps [4]. To train such models, we can either backpropagate through the operations of solvers (ReconODE-FT) or avoid it with the adjoint sensitivity method [4,9] to compute the gradient (ReconODE-FA). "F" stands for fixed solvers, "T" and "A" indicate backpropagating through solvers and using the adjoint method, respectively. A data consistency layer [29] is added after the output from the ODE solver.

2.3 ReconODE with Learned Solvers

We now extend the above idea further: instead of using the known ODE solvers directly, we incorporate the knowledge of the solvers into the network and let the network learn the coefficients and step size. We can write the step function in Eqs. 8–10 as

$$x_{n+1} = x_n + G(F_1, ..., F_s; \omega) \tag{11}$$

$$F_1 = G_1(x_n, t_n, y; \theta_1) \tag{12}$$

$$F_i = G_i(x_n, F_1, ..., F_{i-1}, t_n, y; \theta_i), \quad \text{for } i = 2, ..., s, \tag{13}$$

where G and G_i are neural networks with parameters ω and θ_i. The basic building block G_i is a CNN with five time dependent convolutional layers, which not only learns the dynamics f but also the coefficients and step size in the original solver. Furthermore, since we observe that during optimization more high-frequency details are recovered and in the original solver, F_{i+1} is evaluated ahead of F_i in time, we expect F_{i+1} to recover more details. Thus to mimic such behavior, the network G_{i+1} has a smaller dilation factor than G_i to enforce the network to learn more detailed information. Based on the knowledge that $a_i \in (0,1)$ in $\sum_{i=1}^{s} a_i F_i$, we replace the weighted sum in Eq. 8 with an attention module G.

We design three networks based on the step functions of Euler, RK2 and RK4, respectively (Fig. 2C-E), named as ReconODE-LT ("L" indicates learned solvers and "T" is backpropagation through solvers). The final network is the cascade of solver step functions and the parameters are shared across iterations.

One potential drawback of incorporating the ODE solver into the network is the increased GPU memory usage during training, especially for complicated ODE solvers. To alleviate this problem, we adopt the gradient checkpoint technique [5] to dramatically reduce the memory consumption while only adding little computation time during training, where the intermediate activations are not saved in the forward pass but re-computed during the backward pass.

2.4 Model Training and Evaluation

We used the single-coil fastMRI knee data [38]. There are 34,742 2D slices in the training data and 7,135 slices in the validation data. The fully sampled k-space

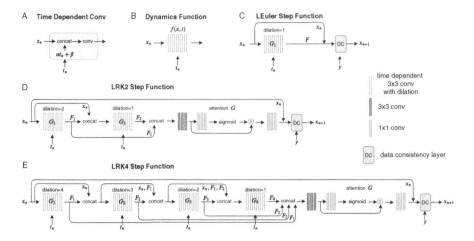

Fig. 2. The building blocks and overall structure of proposed ReconODE networks. (A) In the time-dependent convolutional layer, the time after a linear transformation and input feature maps are concatenated and then fed into the convolution operation. (B) The dynamics in ODE is represented by a CNN with five time-dependent convolutional layers. Using off-the-shelf solvers, the ODE can be solved to obtain reconstructed images. (C-E) We extend this idea and propose three neural networks to learn step functions of Euler, RK2 and RK4 solvers based on the original solver formulations. Different dilation factors are utilized to learn multi-scale features. The ReconODE-LT network is a cascade of the corresponding solver step functions. The parameters are shared across cascade iterations.

data were undersampled with acceleration factors (AF) 4 and 8, respectively. We compared our models with UNet [28,38] and modified it for data with real and imaginary channels and added a data consistency layer [29]. The cascade CNN model (D5C5) [29] and KIKI-Net [8] were also included for comparison. For ReconODE-FA models, we adapted the code from [9]. For a fair comparison, we applied the default settings of channel=32 from fastMRI UNet and N=5 from cascade CNN to all models if applicable. All models were trained using $loss = L1 + 0.5 \times SSIM$ and RAdam [20] with Lookahead [40]. Results were evaluated on the fastMRI validation data using PSNR and SSIM [38].

3   Result

Table 1 shows the reconstruction results on the fastMRI validation dataset. UNet, which learns the direct mapping between the undersampled image to the fully sampled image, has the largest number of parameters and has about 20 to 100 times more parameters than ReconODE models. D5C5 that learns the mapping in an iterative way has slightly better SSIM in 4X but worse SSIM in 8X than UNet. ReconODE-LT-Euler with only 0.9% parameters of UNet achieves similar results as UNet at 4X. ReconODE-LT-RK2 with 64% parameters of D5C5 has similar results as D5C5 at 4X and 8X. The ReconODE-LT-

Table 1. Quantitative results on the fastMRI validation dataset.

		PSNR		SSIM	
Model	Param	4X	8X	4X	8X
UNet	3.35M	31.84	29.87	0.7177	0.6514
KIKI-Net	0.30M	31.83	29.38	0.7178	0.6417
D5C5	0.14M	32.11	29.45	0.7252	0.6425
ReconODE-FA-Euler	0.03M	31.21	28.48	0.7035	0.6172
ReconODE-FA-RK2	0.03M	31.25	28.50	0.7045	0.6175
ReconODE-FA-RK4	0.03M	31.22	28.57	0.7027	0.6183
ReconODE-FT-Euler	0.03M	31.83	29.09	0.7188	0.6355
ReconODE-FT-RK2	0.03M	31.77	29.04	0.7193	0.6345
ReconODE-FT-RK4	0.03M	31.82	29.16	0.7204	0.6388
ReconODE-LT-Euler	0.03M	31.86	29.11	0.7194	0.6363
ReconODE-LT-RK2	0.09M	32.19	29.77	0.7292	0.6498
ReconODE-LT-RK4	0.15M	**32.39**	**30.27**	**0.7333**	**0.6617**

RK4 achieves the best PSNR and SSIM at both 4X and 8X among all models. Figure 3 shows examples of reconstructed images. ReconODE-LT-RK4 achieves the smallest error among all models, which is consistent with the quantitative results.

Using more sophisticated but fixed ODE solvers in ReconODE-FT as well as ReconODE-FA models does not seem to significantly improve the reconstruction results. This is in line with previous results [9]. However, with more complicated but learned ODE solvers, the performance of ReconODE-LT models is improved. Also, ReconODE-LT models outperform ReconODE-FT models with the same ODE solver. These results indicate that learning the ODE solver by the network is beneficial. To demonstrate the effectiveness of the attention in our network, we also trained a ReconODE-FT-RK4 model without attention. We observe that the SSIM at 4X drops from 0.733 (with attention) to 0.726 (without attention). ReconODE-FT models have overall better performance than ReconODE-FA models, which suggests that backpropagation through solvers may be more accurate than the adjoint method. All the improvements described above are statistically significant ($p < 10^{-5}$).

With the help of the gradient checkpoint technique, the GPU memory usage for ReconODE-LT-RK4 can be significantly reduced (73%) with only 0.13 seconds time increase during training. We did not observe any difference in testing when applying the gradient checkpoint (Table 2).

We initially tested the original neural ODE model [4] but the performance was poor (4X SSIM 0.687) and the training was very slow, which may be due to the stability issue [9,41].

Table 2. Benchmark the ReconODE-LT-RK4 model with or without gradient checkpoint technique.

		Train		Test	
Gradient Checkpoint	Param	GPU	Time	GPU	Time
Y	0.15M	3.7G	1.01s	1.4G	0.15s
N	0.15M	13.6G	0.88s	1.4G	0.15s

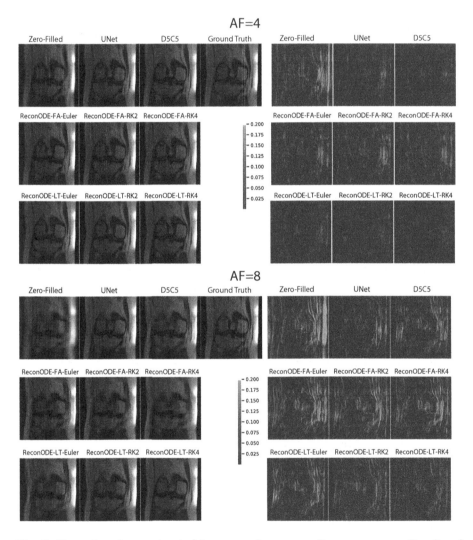

Fig. 3. Examples of reconstructed images and corresponding error maps. Results of other models are omitted due to the space limit.

4   Discussion and Conclusion

In this paper, we propose an innovative idea for MRI reconstruction. We model the continuous optimization process in MRI reconstruction via neural ODEs. Moreover, our proposed ReconODE-LT models integrate the knowledge of the ODE solvers into the network design.

Since the fastMRI leaderboard[1] was closed for submission, we only evaluated the results on the validation dataset. Our results on validation data are not directly comparable to the top leaderboard results on test data, which train on validation data with more epochs [32] and model ensembles [12]. Moreover, compared to models such as E2E-VN (30M) [32], PC-RNN (24M) [34], and Adaptive-CS-Net (33M) [23], our ReconODE models are much smaller ($\leq 0.15M$ parameters). The proposed methods achieve comparable performance with much smaller model size [25]. We expect further studies will lead to a better trade-off between performance and model size. As we intend to propose a new framework rather than a restricted model implementation, further boost of performance can be expected by using larger networks for ODE dynamics and more complicated solvers. This paper provides potential guidance for further research and extension using neural ODE for MRI reconstruction.

References

1. Aggarwal, H.K., Mani, M.P., Jacob, M.: Modl: model-based deep learning architecture for inverse problems. IEEE TMI **38**(2), 394–405 (2018)
2. Akçakaya, M., Moeller, S., Weingärtner, S., Uğurbil, K.: Scan-specific robust artificial-neural-networks for k-space interpolation (raki) reconstruction: database-free deep learning for fast imaging. MRM **81**(1), 439–453 (2019)
3. Bao, L., et al.: Undersampled MR image reconstruction using an enhanced recursive residual network. J. Magn. Reson. **305**, 232–246 (2019)
4. Chen, T.Q., Rubanova, Y., Bettencourt, J., Duvenaud, D.K.: Neural ordinary differential equations. In: NIPS, pp. 6571–6583 (2018)
5. Chen, T., Xu, B., Zhang, C., Guestrin, C.: Training deep nets with sublinear memory cost. arXiv:1604.06174 (2016)
6. Chen, Y., Xiao, T., Li, C., Liu, Q., Wang, S.: Model-based convolutional de-aliasing network learning for parallel MR imaging. In: MICCAI, pp. 30–38. Springer (2019)
7. Duan, J., et al.: VS-Net: variable splitting network for accelerated parallel MRI reconstruction. In: MICCAI, pp. 713–722. Springer (2019)
8. Eo, T., Jun, Y., Kim, T., Jang, J., Lee, H.J., Hwang, D.: KIKI-net: cross-domain convolutional neural networks for reconstructing undersampled magnetic resonance images. MRM **80**(5), 2188–2201 (2018)
9. Gholami, A., Keutzer, K., Biros, G.: ANODE: unconditionally accurate memory-efficient gradients for neural odes. arXiv:1902.10298 (2019)
10. Gilton, D., Ongie, G., Willett, R.: Neumann networks for inverse problems in imaging. arXiv:1901.03707 (2019)
11. Hammernik, K., et al.: Learning a variational network for reconstruction of accelerated MRI data. MRM **79**(6), 3055–3071 (2018)

---

[1] https://fastmri.org/leaderboards.

12. Hammernik, K., Schlemper, J., Qin, C., Duan, J., Summers, R.M., Rueckert, D.: $\sum$-net: systematic evaluation of iterative deep neural networks for fast parallel MR image reconstruction. arXiv:1912.09278 (2019)

13. Han, Y., Sunwoo, L., Ye, J.C.: k-space deep learning for accelerated MRI. IEEE TMI **39**(2), 377–386 (2019)

14. He, K., Zhang, X., Ren, S., Sun, J.: Deep residual learning for image recognition. In: CVPR, pp. 770–778 (2016)

15. He, X., Mo, Z., Wang, P., Liu, Y., Yang, M., Cheng, J.: ODE-inspired network design for single image super-resolution. In: CVPR, pp. 1732–1741 (2019)

16. Huang, Q., Yang, D., Wu, P., Qu, H., Yi, J., Metaxas, D.: MRI reconstruction via cascaded channel-wise attention network. In: ISBI, pp. 1622–1626. IEEE (2019)

17. Knoll, F., et al.: Deep learning methods for parallel magnetic resonance image reconstruction. arXiv:1904.01112 (2019)

18. Lee, D., Yoo, J., Tak, S., Ye, J.C.: Deep residual learning for accelerated MRI using magnitude and phase networks. IEEE T. Biomed. Eng. **65**(9), 1985–1995 (2018)

19. Liang, D., Cheng, J., Ke, Z., Ying, L.: Deep MRI reconstruction: unrolled optimization algorithms meet neural networks. arXiv:1907.11711 (2019)

20. Liu, L., et al.: On the variance of the adaptive learning rate and beyond. arXiv:1908.03265 (2019)

21. Mardani, M., et al.: Deep generative adversarial networks for compressed sensing automates MRI. arXiv:1706.00051 (2017)

22. Paoletti, M.E., Haut, J.M., Plaza, J., Plaza, A.: Neural ordinary differential equations for hyperspectral image classification. IEEE TGRS **58**(3), 1718–1734 (2019)

23. Pezzotti, N., et al.: Adaptive-CS-net: eastMRI with adaptive intelligence. arXiv:1912.12259 (2019)

24. Poddar, S., Jacob, M.: Dynamic MRI using smoothness regularization on manifolds (storm). IEEE TMI **35**(4), 1106–1115 (2015)

25. Putzky, P., Welling, M.: Invert to learn to invert. In: NIPS, pp. 444–454 (2019)

26. Qin, C., Schlemper, J., Caballero, J., Price, A.N., Hajnal, J.V., Rueckert, D.: Convolutional recurrent neural networks for dynamic MR image reconstruction. IEEE TMI **38**(1), 280–290 (2018)

27. Ravishankar, S., Ye, J.C., Fessler, J.A.: Image reconstruction: from sparsity to data-adaptive methods and machine learning. arXiv:1904.02816 (2019)

28. Ronneberger, O., Fischer, P., Brox, T.: U-net: convolutional networks for biomedical image segmentation. In: MICCAI, pp. 234–241. Springer (2015)

29. Schlemper, J., Caballero, J., Hajnal, J.V., Price, A.N., Rueckert, D.: A deep cascade of convolutional neural networks for dynamic MR image reconstruction. IEEE TMI **37**(2), 491–503 (2017)

30. Souza, R., Bento, M., Nogovitsyn, N., Chung, K.J., Lebel, R.M., Frayne, R.: Dual-domain cascade of u-nets for multi-channel magnetic resonance image reconstruction. arXiv:1911.01458 (2019)

31. Souza, R., Frayne, R.: A hybrid frequency-domain/image-domain deep network for magnetic resonance image reconstruction. In: SIBGRAPI, pp. 257–264. IEEE (2019)

32. Sriram, A., et al.: End-to-end variational networks for accelerated MRI reconstruction. arXiv:2004.06688 (2020)

33. Sun, J., Li, H., Xu, Z., et al.: Deep ADMM-Net for compressive sensing MRI. In: NIPS, pp. 10–18 (2016)

34. Wang, P., Chen, E.Z., Chen, T., Patel, V.M., Sun, S.: Pyramid convolutional RNN for MRI reconstruction. arXiv:1912.00543 (2019)

35. Wang, S., et al.: Dimension: dynamic MR imaging with both k-space and spatial prior knowledge obtained via multi-supervised network training. arXiv:1810.00302 (2018)
36. Yang, G., et al.: DAGAN: deep de-aliasing generative adversarial networks for fast compressed sensing MRI reconstruction. IEEE TMI **37**(6), 1310–1321 (2017)
37. Yazdanpanah, A.P., Afacan, O., Warfield, S.K.: ODE-based deep network for MRI reconstruction. arXiv:1912.12325 (2019)
38. Zbontar, J., et al.: FastMRI: an open dataset and benchmarks for accelerated MRI. arXiv:1811.08839 (2018)
39. Zhang, J., Ghanem, B.: ISTA-net: interpretable optimization-inspired deep network for image compressive sensing. In: CVPR, pp. 1828–1837 (2018)
40. Zhang, M., Lucas, J., Ba, J., Hinton, G.E.: Lookahead optimizer: k steps forward, 1 step back. In: NIPS, pp. 9593–9604 (2019)
41. Zhang, T., et al.: ANODEV2: a coupled neural ODE evolution framework. arXiv:1906.04596 (2019)
42. Zhu, B., Liu, J.Z., Cauley, S.F., Rosen, B.R., Rosen, M.S.: Image reconstruction by domain-transform manifold learning. Nature **555**(7697), 487 (2018)

An Evolutionary Framework for Microstructure-Sensitive Generalized Diffusion Gradient Waveforms

Raphaël Truffet[1]([⊠]), Jonathan Rafael-Patino[2], Gabriel Girard[2,3,4],
Marco Pizzolato[2,5,6], Christian Barillot[1], Jean-Philippe Thiran[2,3,4],
and Emmanuel Caruyer[1]

[1] Univ Rennes, Inria, CNRS, Inserm, IRISA UMR 6074, Empenn ERL U-1228,
35000 Rennes, France
`raphael.truffet@irisa.fr`
[2] Signal Processing Lab (LTS5), École Polytechnique Fédérale de Lausanne (EPFL),
Lausanne, Switzerland
[3] Radiology Department, Centre Hospitalier Universitaire Vaudois (CHUV) and
University of Lausanne (UNIL), Lausanne, Switzerland
[4] Center for BioMedical Imaging (CIBM), Lausanne, Switzerland
[5] Department of Applied Mathematics and Computer Science,
Technical University of Denmark, Kongens Lyngby, Denmark
[6] Danish Research Centre for Magnetic Resonance, Centre for Functional and
Diagnostic Imaging and Research, Copenhagen University, Hospital Hvidovre,
Hvidovre, Denmark

Abstract. In diffusion-weighted MRI, general gradient waveforms became of interest for their sensitivity to microstructure features of the brain white matter. However, the design of such waveforms remains an open problem. In this work, we propose a framework for generalized gradient waveform design with optimized sensitivity to selected microstructure features. In particular, we present a rotation-invariant method based on a genetic algorithm to maximize the sensitivity of the signal to the intra-axonal volume fraction. The sensitivity is evaluated by computing a score based on the Fisher information matrix from Monte-Carlo simulations, which offer greater flexibility and realism than conventional analytical models. As proof of concept, we show that the optimized waveforms have higher scores than the conventional pulsed-field gradients experiments. Finally, the proposed framework can be generalized to optimize the waveforms for to any microstructure feature of interest.

Keywords: Diffusion MRI · Acquisition design · Generalized gradient waveforms · Fisher information · Monte-Carlo simulations

1 Introduction

By measuring the displacement of molecules at the micrometer scale, diffusion-weighted MRI (DW-MRI) is sensitive to the fine structure of biological tissues.

R. Truffet and J. Rafael-Patino—These two authors contributed equally.

© Springer Nature Switzerland AG 2020
A. L. Martel et al. (Eds.): MICCAI 2020, LNCS 12262, pp. 94–103, 2020.
https://doi.org/10.1007/978-3-030-59713-9_10

Biophysical models have been proposed to extract useful information about tissue microstructure *in vivo*. In particular in the brain white matter, axons have a coherent organization. Besides estimating the orientation of fibers [21], one can quantify the apparent intra-axonal volume fraction (IAVF), parameters of the distribution of axon radii [5,6] or orientation coherence of the fiber orientations [22].

Complementary to analytical modeling of the DW-MRI signal, Monte-Carlo simulation offers increased flexibility for the description of biological substrates and higher accuracy of the DW-MRI signal for selected microstructure configurations [14,15,18]. The signals computed using Monte-Carlo simulations were shown to be a good predictor of those measured *in vivo* [19].

Another key factor, besides modeling, to improve the accuracy of microstructure features estimation is the experimental design. In [4], the authors present a general framework for experimental design, applied to the optimization of pulsed gradient spin echo (PGSE) sequence parameters for the estimation of the composite hindered and restricted model of diffusion (CHARMED). The same framework was later extended to oscillating gradient spin echo (OGSE) [12], generalized gradient waveforms (linear-encoding) [11] and gradient trajectories (spherical-encoding) [10].

In this work, we focus on the choice of gradient waveforms to optimize the sensitivity of the acquisition to the intra-axonal volume fraction (IAVF) in white matter. Using a three-dimensional substrate model of white matter and Monte-Carlo simulations [18], we optimize the gradient waveforms for an increased rotation-averaged Fisher information. The search for optimized waveforms follows a particle swarm heuristic. Our optimized waveforms are compared with the PGSE sequence.

2   Methods

In this section, we present our novel framework for the generation of waveforms sensitive to IAVF, beginning by designing a score measuring this sensitivity. We then describe the numerical substrate, and how they can be used to compute the scores. Finally, we describe the evolution approach that leads to optimized waveforms.

2.1   A Score Based on the Fisher Information

The Fisher information is a measure of how sensitive a gradient trajectory, $\mathbf{g}$, is to a microstructure parameter of interest. Indeed, the Cramér-Rao bound, defined as the lower bound on the variance of any unbiased estimator, is computed as the inverse of the Fisher information [13]. In this study, we focus on the IAVF, f, while the other microstructure parameters are fixed (see Sect. 2.2). We focused on the IAVF for two main reasons: i- the IAVF is a biologically relevant index of axonal loss, and therefore a biomarker specific to several neurodegenerative diseases; ii- measuring the IAVF from clinically plausible datasets is more accessible and less sensitive to noise than other parameters of interest, such as e.g. axon diameter index, for which a precise estimation remains challenging for diameters below $5\,\mu$m. We derive in the sequel the Fisher information for f, and provide implementation details for its computation using Monte-Carlo simulations.

Signal and Noise Model. The signal attenuation model $A(f, \mathbf{g})$ depends on a the IAVF, f, and an effective DW-MRI gradient trajectory, $\mathbf{g}(t), t \in [0, TE]$ (TE is the echo time). Besides, the magnitude signal in DW-MRI is corrupted by noise. As mentioned in [3], it is important to consider Rician noise rather than Gaussian noise, since the latter leads to an unrealistic choice of higher b-value. In some situations, the Rician model does not adequately describe noise properties. However, in most cases, it can be considered a valid model [2]. The probability density function of the measured signal, $\tilde{A}$, in noise is, with a spread parameter σ:

$$p(\tilde{A}; f, \mathbf{g}, \sigma^2) = \frac{\tilde{A}}{\sigma^2} I_0 \left(\frac{A(f, \mathbf{g})\tilde{A}}{\sigma^2} \right) \exp \left(-\frac{A^2(f, \mathbf{g}) + \tilde{A}^2}{2\sigma^2} \right). \tag{1}$$

Computation of the Fisher Information. The Fisher information for the parameter f is defined as

$$F_0(\mathbf{g}, f, \sigma^2) = E \left[\frac{\partial^2 \log p}{\partial f} (\tilde{A}; f, \mathbf{g}, \sigma^2). \right] \tag{2}$$

The full derivation for the Fisher information F_0 (not reported here) can be found in the Appendix of [3]; it can be computed from the estimation of the partial derivatives $\partial A / \partial f$. Note that in our case, there is no analytical expression for the signal attenuation; instead, A is computed using the Monte-Carlo simulator. Because of the intrinsic uncertainty in the signal, we cannot estimate these partial derivatives using classical finite-difference. Therefore, we empirically propose to perform linear regression from a set of signals generated for 10 values of f around the value of interest.

We note that the Fisher information depends on the value of the IAVF f. Without prior knowledge on this parameter, and to avoid introducing any bias in the acquisition design, we compute the average Fisher information over a set $\Omega = \{0.25, 0.50, 0.75\}$:

$$F(\mathbf{g}, \sigma^2) = \frac{1}{|\Omega|} \sum_{f \in \Omega} F_0(\mathbf{g}, f, \sigma^2). \tag{3}$$

Gradient Waveforms and Rotation Invariance. In this work, we focus on gradient trajectories with a fixed orientation, $\mathbf{u} \in \mathcal{S}^2$:

$$\mathbf{g}(t) = g(t) \, \mathbf{u}.$$

We want to separate the search for optimal waveforms from the search for optimal sampling directions, the latter having already received prior attention in the community [7–9,16]. Besides, to have a rotation-invariant measure, we define the sampling score, U, as the average over directions on the sphere:

$$U(g, \sigma^2) = \int_{\mathcal{S}^2} F(g\,\mathbf{u}, \sigma^2) \mathrm{d}^2 \mathbf{u}. \tag{4}$$

In practice, the integral in Eq. 4 is computed by taking advantage of the cylindrical symmetry of the substrates, and considering a set of nine gradient directions making an angle θ with the cylinders uniformly spread in the range $[0, \pi/2]$. As a result, we needed to run 270 simulations (3 IAVF values, each with 10 values for the linear regression and each with 9 angles θ) to compute the score $U(g, \sigma^2)$ for one waveform. Overall, this corresponds to 30 Monte Carlo particle trajectories files to be generated.

2.2 Numerical Substrate Design

All substrates were generated assuming a two-compartment model composed of intra- and extra-axonal spaces. The intra-axonal space compartment was represented using a collection of parallel cylinders with radius sampled from a Gamma distribution $\Gamma(2.0, 0.35)$ (in μm), which is in the range of values reported from histology samples [1,17]. The extra-axonal space compartment corresponds to space outside the cylindrical axons, which comprise the extra-axonal matrix, glial cells, and cerebrospinal fluid. The substrates were then generated by randomly placing a total of 10^4 sampled cylinders into an isotropic voxel, without intersection between them, and ensuring periodicity at the voxel boundaries [18]. Figure 1, panel a), shows a toy example of a numerical substrate.

A total of 40 numerical substrates were generated for four IAVF of 0.25, 0.5, 0.6, and 0.75. For each of the four nominal IAVF, we generated 10 samples around the target value to compute the numerical derivative of the Fisher information. To generate samples with the same intra-axonal configuration, but slightly different IAVF, we first generated one randomly packed phantom for each of the four selected IAVF, and then scaled each one of them by a small factor $1.0 + \epsilon$, with $\epsilon = \{0.00, 0.001, 0.002, \ldots, 0.010\}$, without scaling the cylinders radii distribution, as is shown in Fig. 1, panel b). This way, the same intra-axonal space configuration is kept, but different IAVF can be computed. Doing so, we avoid any bias arising from radii re-sampling and positioning of the cylinders, which would effectively change the considered geometry, hence the signal.

Principle. To build waveforms with the highest score defined in Eq. 4, we perform a stochastic optimization based on a genetic algorithm. To narrow down the search space, the admissible range of b-values is pre-determined. Intuitively, if the b-value is too low, there is almost no diffusion-weighting; conversely, a too high b-value leads to a highly attenuated signal which has an amplitude comparable to that of the noise floor. We start with an initial set of 100 random waveforms, and we perform cross-overs to build the next generation.

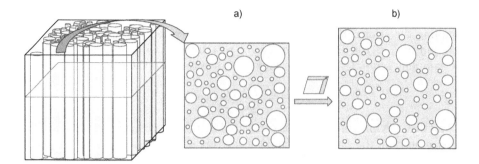

Fig. 1. Example of substrate design composed of intra- and extra- axonal spaces. The intra-axonal space compartment is represented using a collection of parallel cylinders with various radii packed inside a containing voxel, which results in the top view displayed in panel a). Panel b) shows how the base configuration is then transformed by scaling the voxel and the position of the cylinders by a constant factor, but without scaling the cylinder's radii; this transformation reduces the volume ratio between the voxel and the cylinders total volume, thus reducing the represented IAVF.

Initialization. The process to build random gradient waveforms is based on Markov Chains: the value of the waveforms at time $t + dt$ only depends on the value of the waveform at time t. This method allows us to build waveforms that respect the properties of maximum gradient strength $(\text{T} \cdot \text{m}^{-1})$ and slew rate $(\text{T} \cdot \text{m}^{-1} \cdot \text{s}^{-1})$ that are well documented. Other limitations such as the duty cycle could have been added [20], but are not documented enough to add numeric constraints. We referred to the limitations of a SIEMENS Prisma 3T, for possible future experimentations:

$$|g(t)| \leq 80 \, \text{mT} \cdot \text{m}^{-1} \quad ; \quad \left| \frac{dg}{dt}(t) \right| \leq 200 \, \text{T} \cdot \text{m}^{-1} \cdot \text{s}^{-1}. \tag{5}$$

We generate the waveform between $t = 0$ and $t = TE/2$, and we perform a symmetry to obtain the full gradient trajectory. Examples of randomly generated gradients are shown on Fig. 2(a). We generated several waveforms and selected only those that have a b-value in the targeted range.

Evolution Process. We defined the cross-over as the combination of two waveforms. We keep the beginning of a waveform, and the end of another, as shown in Fig. 2. The position of the cut is sampled randomly from a Gaussian centered on $TE/4$, with a standard deviation of $TE/12$ to avoid that the new waveform inherits too much from only one of the two previous waveforms.

2.3 Genetic Evolution

The generation $i + 1$ is built on generation i. First, we randomly select two waveforms. Then, the probability to select a waveform is a linear scaling of the

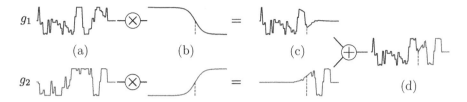

Fig. 2. Illustration of the cross-over between two waveforms (a), g_1 and g_2. We prepare two amplitude modulations (b) which will keep respectively the beginning and the end of waveforms g_1 and g_2. After applying this modulation, we sum the results to obtain a new waveform (d). NB: we only represent the chronogram for $t \in [0, TE/2]$, *i.e.* before the 180° pulse.

score that leads to have a probability to select the waveform with the highest score 10 times higher than selecting the waveform with the lowest score. If the b-value of the resulting waveform is outside the targeted range, we drop this cross-over and try another one. We repeat this process until having 95 waveforms that fall within the target range. The last 5 waveforms needed to complete generation $i + 1$ are randomly generated such as for generation 0 to promote novelty.

2.4 Signal Simulation

The simulated DW-MRI signals were computed using DW-MRI Monte-Carlo simulations as described in [18]. However, since in each iteration of our proposed framework it is required to recompute the DW-MRI signal of all the numerical substrates at each generation of waveforms, this requires generating a high number of new signals for the same substrate. Because of this, we proposed to first generate and store all the Monte-Carlo particles dynamics required to compute the DW-MRI signal using the MC/DC open-source simulator [18], and then use a tailored version of this simulator written in C++ CUDA, which is able to compute the DW-MRI signal in a fraction of the original computation time. Using the latter approach, we were able to produce one generation of waveforms in about 30 min, in contrast to 20 h for the former.

For each numerical substrate, we generated and stored a total of 2×10^5 particles, with a maximum diffusion time of 100 ms, and in time-steps intervals of 20 µs. The diffusion coefficient was set to $D = 1.7 \times 10^{-9}\,\mathrm{m}^2 \cdot \mathrm{s}^{-1}$ [3].

3 Results and Discussion

3.1 Pre-optimization of the b-value

In Fig. 3, we report the Fisher information for several b-values ranging from 0 to $5,000\,\mathrm{s} \cdot \mathrm{mm}^{-2}$, with 100 different waveforms for each b-value. We first notice a high variability within one b-value. This shows that the b-value is not enough to characterize the efficiency of one waveform for the estimation of the IAVF. Then,

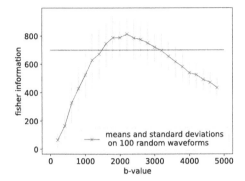

Fig. 3. Fisher information computed for a family of randomly generated gradient wave-forms, spanning a range of b-values from 0 to $5,000\,\text{s}\cdot\text{mm}^{-2}$. The Fisher information shows a dependence on the b-values, but within a given b-value, there is a remaining variability which is explained by waveform. We restricted our waveform search in the b-value range of $[1500, 3100]\,\text{s}\cdot\text{mm}^{-2}$ (value above the red line) to further investigate their sensitivity to the IAVF.

we also notice that the b-value still represents an important parameter since we observe, as expected, a low score for low b-values and too high b-values. This result leads us to restrict our search for waveforms to those within the target range $[1,500\,\text{s}\cdot\text{mm}^{-2}; 3,100\,\text{s}\cdot\text{mm}^{-2}]$.

3.2 Waveforms Optimization

As we can see in Fig. 5a, after 30 generations, the score of the waveforms has increased compared to generation 0. Since most of the waveforms of generation 30 have a higher score than the random waveforms of generation 0, one can hope that this increase is due to the optimization process and not only to new random waveforms incorporated at each generation. The waveforms with the highest score of generation 30 are shown in Fig. 4.

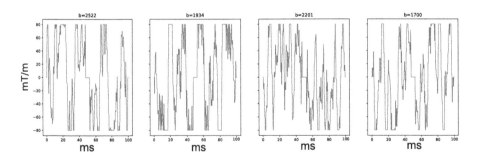

Fig. 4. Waveforms genetically generated to optimize the Fisher Information. From left to right, the figure shows the 4 waveforms of generation 30 with the highest score.

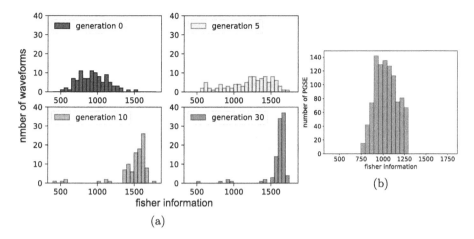

Fig. 5. Histogram of the Fisher information of DW-MRI sequences. (a) Histogram associated to gradient waveforms across four generations of the genetic algorithm. (b) Histogram computed for the family of 1,000 PGSE.

3.3 Comparison of Score Between PGSE and Genetically Optimized Waveforms

As a proof-of-concept, we compared our optimized waveforms with PGSE. We computed the score for a family PGSE covering a wide range of timing and gradient amplitude parameters. We generated all the PGSE sequences with $G_{\max} = 80\,\mathrm{mT} \cdot \mathrm{m}^{-1}$; δ between 0.5 ms and 40 ms with a step of 0.5 ms; and Δ between 10 ms and 80 ms with a step of 0.5 ms. After filtering only feasible sequences, and only selecting those with a b-value between 1500 and 3100, we obtained approximately 1,000 PGSE. Figure 5b shows the distribution of scores among these PGSE sequences.

The maximum score with these b-values is lower than the score obtained with generalized waveforms in generation 30. We can also notice that, even in generation 0, some randomly generated waveforms had a higher score than PGSE sequences. This shows that generalized waveforms have a greater potential than PGSE sequences at identifying the substrate's IAVF.

4 Conclusion

The optimization of gradient waveforms is an important step in designing new acquisition sequences in DW-MRI. The generalized waveforms protocol obtained in this work is optimized independently from the sampling directions. We designed a genetic algorithm to make the waveforms evolve. After a few generations, our algorithm created a family of waveforms that have a higher sensitivity to the IAVF than PGSE.

We have presented a novel framework with several advantages. First, it is based on Monte-Carlo simulations, and thus can be adapted to a variety of microstructure configurations, such as substrate with various compartment shapes and sizes. Second, generalized waveform optimization can be performed on one or several microstructure parameters of interest. In this work, we optimized for the IAVF parameter in two-compartment substrates composed of parallel cylinders with various radii and packing densities, but other biomarkers can be used instead. The only requirement is our ability to compute partial derivatives of the signal with respect to these parameters. Future work will target angular dispersion and spherical pore size parameter estimation. In vivo experiments will be a necessary step for the validation of the generated waveforms. Designing an experiment that shows the efficiency of the waveforms without having the ground truth microstructure is a challenging issue.

Acknowledgements. This research project is part of the MMINCARAV Inria associate team program between Empenn (Inria Rennes Bretagne Atlantique) and LTS5 (École Polytechnique Fédérale de Lausanne—EPFL) started in 2019. Raphaël Truffet's PhD is partly funded by ENS Rennes. We gratefully acknowledge the support of NVIDIA Corporation with the donation of the Titan XP GPU used for this research. This project has received funding from the European Union's Horizon 2020 research and innovation programme under the Marie Skłodowska-Curie grant agreement No. 754462.

References

1. Aboitiz, F., Scheibel, A.B., Fisher, R.S., Zaidel, E.: Fiber composition of the human corpus callosum. Brain Res. **598**(1), 143–153 (1992). https://doi.org/10.1016/0006-8993(92)90178-C, http://www.sciencedirect.com/science/article/pii/000689939290178C
2. Aja-Fernndez, S., Vegas-Snchez-Ferrero, G.: Statistical Analysis of Noise in MRI: Modeling, Filtering and Estimation, 1st edn. Springer Publishing Company, Incorporated (2016)
3. Alexander, D.C.: A general framework for experiment design in diffusion MRI and its application in measuring direct tissue-microstructure features. Magn. Reson. Med. **60**(2), 439–448 (2008). https://doi.org/10.1002/mrm.21646
4. Alexander, D.C.: A general framework for experiment design in diffusion MRI and its application in measuring direct tissue-microstructure features. Mag. Reson. Med. Official J. Int. Soc. Magn. Reson. Med. **60**(2), 439–448 (2008)
5. Assaf, Y., Basser, P.J.: Composite hindered and restricted model of diffusion (charmed) MR imaging of the human brain. Neuroimage **27**(1), 48–58 (2005)
6. Assaf, Y., Blumenfeld-Katzir, T., Yovel, Y., Basser, P.J.: Axcaliber: a method for measuring axon diameter distribution from diffusion MRI. Magn. Reson. Med. Official J. Int. Soc. Magn. Reson. Med. **59**(6), 1347–1354 (2008)
7. Bates, A.P., Khalid, Z., Kennedy, R.A.: An optimal dimensionality sampling scheme on the sphere with accurate and efficient spherical harmonic transform for diffusion MRI. IEEE Signal Process. Lett. **23**(1), 15–19 (2015)
8. Caruyer, E., Lenglet, C., Sapiro, G., Deriche, R.: Design of multishell sampling schemes with uniform coverage in diffusion MRI. Magn. Reson. Med. **69**(6), 1534–1540 (2013)

9. Cheng, J., Shen, D., Yap, P.-T.: Designing single- and multiple-shell sampling schemes for diffusion MRI using spherical code. In: Golland, P., Hata, N., Barillot, C., Hornegger, J., Howe, R. (eds.) MICCAI 2014. LNCS, vol. 8675, pp. 281–288. Springer, Cham (2014). https://doi.org/10.1007/978-3-319-10443-0_36

10. Drobnjak, I., Alexander, D.C.: Optimising time-varying gradient orientation for microstructure sensitivity in diffusion-weighted MR. J. Magn. Reson. **212**(2), 344–354 (2011)

11. Drobnjak, I., Siow, B., Alexander, D.C.: Optimizing gradient waveforms for microstructure sensitivity in diffusion-weighted MR. J. Magn. Reson. **206**(1), 41–51 (2010)

12. Drobnjak, I., Zhang, H., Ianuş, A., Kaden, E., Alexander, D.C.: Pgse, ogse, and sensitivity to axon diameter in diffusion MRI: insight from a simulation study. Magn. Reson. Med. **75**(2), 688–700 (2016)

13. Galdos, J.I.: A cramer-rao bound for multidimensional discrete-time dynamical systems. IEEE Trans. Autom. Control **25**(1), 117–119 (1980). https://doi.org/10.1109/TAC.1980.1102211

14. Ginsburger, K., Matuschke, F., Poupon, F., Mangin, J.F., Axer, M., Poupon, C.: Medusa: a GPU-based tool to create realistic phantoms of the brain microstructure using tiny spheres. NeuroImage **193**, 10–24 (2019)

15. Hall, M.G., Alexander, D.C.: Convergence and parameter choice for Monte-Carlo simulations of diffusion MRI. IEEE Trans. Med. Imaging **28**(9), 1354–1364 (2009)

16. Jones, D.K., Horsfield, M.A., Simmons, A.: Optimal strategies for measuring diffusion in anisotropic systems by magnetic resonance imaging. Magn. Reson. Med. Official J. Int. Soc. Magn. Reson. Med. **42**(3), 515–525 (1999)

17. Lamantia, A.S., Rakic, P.: Cytological and quantitative characteristics of four cerebral commissures in the rhesus monkey. J. Comp. Neurol. **291**(4), 520–537 (1990). https://doi.org/10.1002/cne.902910404, https://onlinelibrary.wiley.com/doi/abs/10.1002/cne.902910404

18. Rafael-Patino, J., Romascano, D., Ramirez-Manzanares, A., Canales-Rodríguez, E.J., Girard, G., Thiran, J.P.: Robust Monte-Carlo simulations in diffusion-MRI: effect of the substrate complexity and parameter choice on the reproducibility of results. Front. Neuroinform. **14**, 8 (2020). https://doi.org/10.3389/fninf.2020.00008

19. Rensonnet, G., Scherrer, B., Girard, G., Jankovski, A., Warfield, S.K., Macq, B., Thiran, J.P., Taquet, M.: Towards microstructure fingerprinting: estimation of tissue properties from a dictionary of Monte Carlo diffusion MRI simulations. NeuroImage **184**, 964–980 (2019)

20. Sjölund, J., Szczepankiewicz, F., Nilsson, M., Topgaard, D., Westin, C.F., Knutsson, H.: Constrained optimization of gradient waveforms for generalized diffusion encoding. J. Magn. Reson. **261**, 157–168 (2015). https://doi.org/10.1016/j.jmr.2015.10.012

21. Tuch, D.S.: Q-ball imaging. Magn. Reson. Med. Official J. Int. Soc. Magn. Reson. Med. **52**(6), 1358–1372 (2004)

22. Zhang, H., Schneider, T., Wheeler-Kingshott, C.A., Alexander, D.C.: Noddi: practical in vivo neurite orientation dispersion and density imaging of the human brain. Neuroimage **61**(4), 1000–1016 (2012)

Lesion Mask-Based Simultaneous Synthesis of Anatomic and Molecular MR Images Using a GAN

Pengfei Guo[1], Puyang Wang[2], Jinyuan Zhou[3], Vishal M. Patel[1,2], and Shanshan Jiang[3(✉)]

[1] Department of Computer Science, Johns Hopkins University, Baltimore, MD, USA
[2] Department of Electrical and Computer Engineering, Johns Hopkins University, Baltimore, MD, USA
[3] Department of Radiology, Johns Hopkins University, Baltimore, MD, USA
sjiang21@jhmi.edu

Abstract. Data-driven automatic approaches have demonstrated their great potential in resolving various clinical diagnostic dilemmas for patients with malignant gliomas in neuro-oncology with the help of conventional and advanced molecular MR images. However, the lack of sufficient annotated MRI data has vastly impeded the development of such automatic methods. Conventional data augmentation approaches, including flipping, scaling, rotation, and distortion are not capable of generating data with diverse image content. In this paper, we propose a method, called synthesis of anatomic and molecular MR images network (SAMR), which can simultaneously synthesize data from arbitrary manipulated lesion information on multiple anatomic and molecular MRI sequences, including T1-weighted (T_1w), gadolinium enhanced T_1w (Gd-T_1w), T2-weighted (T_2w), fluid-attenuated inversion recovery ($FLAIR$), and amide proton transfer-weighted (APTw). The proposed framework consists of a stretch-out up-sampling module, a brain atlas encoder, a segmentation consistency module, and multi-scale label-wise discriminators. Extensive experiments on real clinical data demonstrate that the proposed model can perform significantly better than the state-of-the-art synthesis methods.

Keywords: MRI · Multi-modality synthesis · GAN

1 Introduction

Malignant gliomas, such as glioblastoma (GBM), remain one of the most aggressive forms of primary brain tumor in adults. The median survival of patients with

Electronic supplementary material The online version of this chapter (https://doi.org/10.1007/978-3-030-59713-9_11) contains supplementary material, which is available to authorized users.

© Springer Nature Switzerland AG 2020
A. L. Martel et al. (Eds.): MICCAI 2020, LNCS 12262, pp. 104–113, 2020.
https://doi.org/10.1007/978-3-030-59713-9_11

glioblastomas is only 12 to 15 months with aggressive treatment [15]. For the clinical management in patients who finish surgery and chemoradiation, the treatment responsiveness assessment is relied on the pathological evaluations [16]. In recent years, deep convolutional neural network (CNN) based medical image analysis methods have shown to produce significant improvements over the conventional methods [2,6]. However, a large amount of data with rich diversity is required for training effective CNNs models, which is usually unavailable for medical image analysis. Furthermore, lesion annotations and image prepossessing (e.g. co-registration) are labor-intensive, time-consuming and expensive, since expert radiologists are required to label and verify the data. While deploying conventional data augmentations, such as rotation, flipping, random cropping, and distortion, during training partly mitigates such issues, the performance of CNN models still suffer from the limited diversity of the dataset [18]. In this paper, we propose a generative network which can simultaneously synthesize meaningful high quality T_1w, Gd-T_1w, T_2w, $FLAIR$, and APTw MRI sequences from input lesion mask. In particular, APTw is a novel molecular MRI technique, which yields a reliable marker for treatment responsiveness assessment for patients with post-treatment malignant gliomas [8,17].

Recently Goodfellow et al. [5] proposed generative adversarial network (GAN) which has been shown to synthesize photo-realistic images. Isola et al. [7] and Wang et al. [14] applied GAN under the conditional settings and achieved impressive results on image-to-image translation tasks. When considering the generative models for MRI synthesis alone, several methods have been proposed in the literature. Nguyen et al. [13] and Chartsias et al. [3] proposed CNN-based architectures integrating intensity features from images to synthesize cross-modality MR images. However, their inputs are existing MRI modalities and the diversity of the synthesized images is limited by the training images. Cordier et al. [4] used a generative model for multi-modal MR images with brain tumors from a single label map. However, the input label map contains detailed brain anatomy and the method is not capable of producing manipulated outputs. Shin et al. [11] adopted pix2pix [7] to transfer brain anatomy and lesion segmentation maps to multi-modal MR images with brain tumors. However, it requires to train an extra segmentation network that provides white matter, gray matter, and cerebrospinal fluid (CSF) masks as partial input of synthesis network. Moreover, it is only demonstrated to synthesize anatomical MRI sequences. In this paper, a novel generative model is proposed that can take arbitrarily manipulated lesion mask as input facilitated by brain atlas generated from training data to simultaneously synthesize a diverse set of anatomical and molecular MR images.

To summarize, the following are our key contributions: **1.** A novel conditional GAN-based model is proposed to synthesize meaningful high quality multimodal anatomic and molecular MR images with controllable lesion information. **2.** Multi-scale label-wise discriminators are developed to provide specific supervision on the region of interest (ROI). **3.** Extensive experiments are conducted and comparisons are performed against several recent state-of-the-art image synthesis approaches. Furthermore, an ablation study is conducted to demonstrate the improvements obtained by various components of the proposed method.

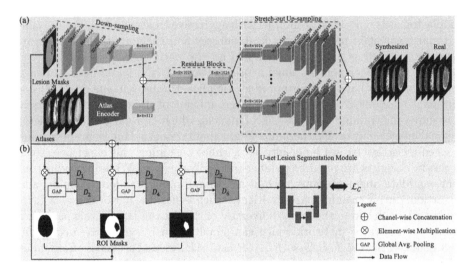

Fig. 1. An overview of the proposed network. (a) Generator network. (b) Multi-scale Label-wise discriminators. Global averaging pooling is used to create the factor of 2 down-sampling input. (c) U-net based lesion segmentation module. We denote lesion shape consistency loss as $\mathcal{L}_C$.

2 Methodology

Figure 1 gives an overview of the proposed framework. Incorporating multi-scale label-wise discriminators and shape consistency-based optimization, the generator aims to produce meaningful high-quality anatomical and molecular MR images with diverse and controllable lesion information. In what follows, we describe different parts of the network in detail.

Multi-modal MRI Sequence Generation. Our generator architecture is inspired by the models proposed by Johnson et al. [9] and Wang et al. [14]. The generator network, consists of four components (see Fig. 1(a): a down-sampling module, an atlas encoder, a set of residual blocks, and a stretch-out up-sampling module. A lesion segmentation map of size $256 \times 256 \times 5$, containing 5 labels: background, normal brain, edema, cavity caused by surgery, and tumor, is passed through the down-sampling module to get a latent feature map. The corresponding multi-model atlas of size $256 \times 256 \times 15$ (details of atlas generation are provided in Sect. 3) is passed through an atlas encoder to get another latent feature map. Then, the two latent feature maps are concatenated and are passed through residual blocks and stretch-out up-sampling module to synthesize multi-model MRI slices of size $256 \times 256 \times 5$.

The down-sampling module consists of a fully-convolutional module with 6 layers. We set the kernel size and stride equal to 7 and 1, respectively for the first layer. For down-sampling, instead of using maximum-pooling, the stride of other 5 layers is set equal to 2. Rectified Linear Unit (ReLu) activation and batch normalization are sequentially added after each layer. The atlas encoder has the same network architecture but the number of channels in the first convolutional layer is modified to match the input size of the multi-model atlas input. The depth of the network is increased by a set of residual blocks, which is proposed to learn better transformation functions and representations through a deeper perception [18]. The stretch-out up-sampling module contains 5 similar sub-modules designed to utilize the same latent representations from residual blocks and perform customized synthesis for each sequence. Each sub-module contains one residual learning block and a symmetric architecture with a down-sampling module. All convolutional layers are replaced by transposed convolutional layers for up-sampling. The synthesized multi-model MR images are produced from each sub-model.

Multi-scale Label-wise Discriminators. In order to efficiently achieve large receptive field in discriminators, we adopt multi-scale PatchGAN discriminators [7], which have identical network architectures but take multi-scale inputs [14]. Conventional discriminators operate on the combination of images and conditional information to distinguish between real and synthesized images. However, optimizing generator to produce realistic images in each ROI cannot be guaranteed by discriminating on holistic images. To address this issue, we propose label-wise discriminators. Based on the radiographic features, original lesion segmentation masks are reorganized into 3 ROIs, including background, normal brain, and lesion. Specifically, the input of each discriminator is the ROI-masked combination of lesion segmentation maps and images. Since proposed discriminators are in a multi-scale setting, for each ROI there are 2 discriminators that operate on original and a down-sampled scales (factor of 2). Thus, there are in total 6 discriminators for 3 ROIs and we refer to these set of discriminators as $\mathbb{D} = \{D_1, D_2, D_3, D_4, D_5, D_6\}$. In particular, $\{D_1, D_2\}, \{D_3, D_4\}$, and $\{D_5, D_6\}$ operate on original and down-sampled versions of background, normal brain, and lesion, respectively. An overview of the proposed discriminators is shown in Fig. 1(b). The objective function for a specific discriminator $\mathcal{L}_{GAN}(G, D_k)$ is as follows:

$$\mathcal{L}_{GAN}(G, D_k) = \mathbb{E}_{(\hat{x}, \hat{y})}[\log D_k(\hat{x}, \hat{y})] + \mathbb{E}_{\hat{x}}[\log(1 - D_k(\hat{x}, \hat{G}(x)))], \qquad (1)$$

where x and y are paired original lesion segmentation masks and real multi-model MR images, respectively. Here, $\hat{x} \triangleq c_k \odot x$, $\hat{y} \triangleq c_k \odot y$, and $\hat{G}(x) \triangleq c_k \odot G(x)$, where $\odot$ denotes element-wise multiplication and c_k corresponds to the ROI mask. For simplicity, we omit the down-sampling operation in this equation.

Multi-task Optimization. A multi-task loss is designed to train the generator and discriminators in an adversarial setting. Instead of only using the conventional adversarial loss $\mathcal{L}_{GAN}$, we also adopt a feature matching loss $\mathcal{L}_{FM}$ [14] to stabilize training, which optimizes generator to match these intermediate representations from the real and the synthesized images in multiple layers of the discriminators. For a specific discriminator, $\mathcal{L}_{FM}(G, D_k)$ is defined as follows:

$$\mathcal{L}_{FM}(G, D_k) = \sum_i^T \frac{1}{N_i} \left[\| D_k^{(i)}(\hat{x}, \hat{y}) - D_k^{(i)}(\hat{x}, \hat{G}(x)) \|_2^2 \right], \quad (2)$$

where $D_k^{(i)}$ denotes the ith layer of the discriminator D_k, T is the total number of layers in D_k and N_i is the number of elements in the ith layer. If we perform lesion segmentation on images, it is worth to note that there is a consistent relation between the prediction and the real one serving as input for the generator. Lesion labels are usually occluded with each other and brain anatomic structure, which causes ambiguity for synthesizing realistic MR images. To tackle this problem, we propose a lesion shape consistency loss $\mathcal{L}_C$ by adding a U-net [10] segmentation module (Fig. 1(c)), which regularizes the generator to obey this consistency relation. We adopt Generalized Dice Loss (GDL) [12] to measure the difference between predicted and real segmentation maps and is defined as follows:

$$GDL(R, S) = 1 - \frac{2 \sum_i^N r_i s_i}{\sum_i^N r_i + \sum_i^N s_i}, \quad (3)$$

where R denotes the ground truth and S is the segmentation result. r_i and s_i represent the ground truth and predicted probabilistic maps at each pixel i, respectively. N is the total number of pixels. The lesion shape consistency loss $\mathcal{L}_C$ is then defined as follows:

$$\mathcal{L}_C(x, U(G(x)), U(y)) = GDL(x, U(y)) + GDL(x, U(G(x))), \quad (4)$$

where $U(y)$ and $U(G(x))$ represent the predicted probabilistic maps by taking y and $G(x)$ as inputs in the segmentation module, respectively. The proposed final multi-task loss function for the generator is defined as:

$$\mathcal{L} = \sum_{k=1}^6 \mathcal{L}_{GAN}(G, D_k) + \lambda_1 \sum_{k=1}^6 \mathcal{L}_{FM}(G, D_k) + \lambda_2 \mathcal{L}_C(x, U(G(x)), U(y)), \quad (5)$$

where λ_1 and λ_2 two parameters that control the importance of each loss.

3   Experiments and Evaluations

Data Acquisition and Implementation Details. 90 postsurgical patients were involved in this study. MRI scans were acquired by a 3T human MRI scanner (Achieva; Philips Medical Systems). Anatomic sequences of size $512 \times 512 \times 150$ voxels and Molecular APTw sequence of size $256 \times 256 \times 15$ voxels were collected. Detailed imaging parameters and preprocessing pipeline can be found in supplementary material. After preprocessing, the final volume size of each sequence is $256 \times 256 \times 15$. Expert neuroradiologist manually annotated five labels (i.e. background, normal brain, edema, cavity and tumor) for each patient. Then, a multivariate template construction tool [1] was used to create the group average for each sequence (atlas). 1350 instances with the size of

$256 \times 256 \times 5$ were extracted from the volumetric data, where the 5 corresponds to five MR sequences. For every instance, the one corresponding atlas slice and two adjacent (in axial direction) atlas slices were extracted to provide human brain anatomy prior. We split these instances randomly into 1080 (80%) for training and 270 (20%) for testing on the patient level. Data collection and processing are approved by the Institutional Review Board.

The synthesis model was trained based on the final objective function Eq. (5) using the Adam optimizer [1]. λ_1 and λ_2 in Eq. (5) were set equal to 5 and 1, respectively. Hyperparameters are set as follows: constant learning rate of 2×10^{-4} for the first 250 epochs then linearly decaying to 0; 500 maximum epochs; batch size of 8. For evaluating the effectiveness of the synthesized MRI sequences on data augmentation, we leveraged U-net [10] to train lesion segmentation models. U-net [10] was trained by the Adam optimizer [1]. Hyperparameters are set as follows: constant learning rate of 2×10^{-4} for the first 100 epochs then linearly decaying to 0; 200 maximum epochs; batch size of 16. In the segmentation training, all the synthesized data was produced by randomly manipulated lesion masks. For evaluation, we always keep 20% of data unseen for both of the synthesis and segmentation models.

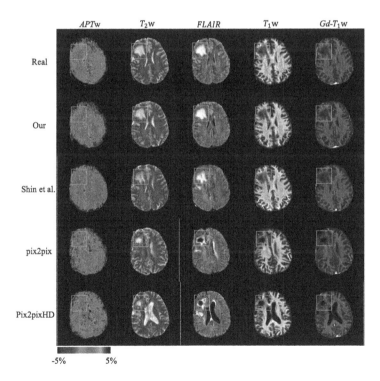

Fig. 2. Qualitative comparison of different methods. The same lesion mask is used to synthesize images from different methods. Red boxes indicate the lesion region. (Color figure online)

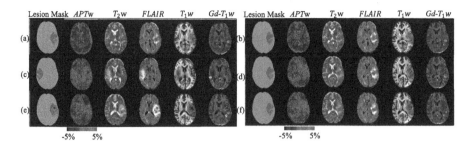

Fig. 3. Examples of lesion mask manipulations. (a) Real images. (b) Synthesized images from the original mask. (c) Synthesized images by mirroring lesion. (d) Synthesized images by increasing tumor size to 100%. (e) Synthesized images by replacing lesion from another patient. (f) Synthesized images by shrinking tumor size to 50%. In lesion masks, gray, green, yellow, and blue represent normal brain, edema, tumor, and cavity caused by surgery, respectively.

MRI Synthesis Results. We evaluate the performance of different synthesis methods by qualitative comparison and human perceptual study. We compare the performance of our method with the following recent state-of-the-art synthesis methods: pix2pix [7], pix2pixHD [14], and Shin et al. [11]. Figure 2 presents the qualitative comparison of the synthesized multi-model MRI sequences from four different methods. It can be observed that pix2pix [7] and pix2pixHD [14] fail to synthesize realistic looking human brain MR images. There is either an unreasonable brain ventricle (see the last row of Fig. 2) or wrong radiographic features in the lesion region (see the fourth row of Fig. 2). With the help of probability maps of white matter, gray matter and CSF, Shin et al. [11] can produce realistic brain anatomic structures for anatomic MRI sequences. However, there is an obvious disparity between the synthesized and real APTw sequence in both normal brain and lesion region. The boundary of the synthezied lesion is also blurred and uncertain (see red boxes in the third row of Fig. 2). The proposed method produces more accurate radiographic features of lesions and more diverse anatomic structure based on the human anatomy prior provided by atlas.

Human Perceptual Study. To verify the pathological information of the synthesized images, we conduct the human perceptual study by an expert neuroradiologist and the corresponding preference rates are reported in Table 1. It is clear that the images generated by our method are more preferred by an expert neuroradiologist than others showing the practicality of our synthesis method.

Table 1. Preference rates corresponding to the human perceptual study.

Real	Our	Shin et al. [11]	pix2pixHD [14]	pix2pix [7]
100%	**72.1%**	65.6%	39.3%	16.4%

Data Augmentation Results. To further evaluate the quality of the synthezied images, we perform data augmentation by using the synthesized images in training and then perform lesion segmentation. Evaluation metrics in BraTS [2] challenge (i.e. Dice score, Hausdorff distance (95%), Sensitivity, and Specificity) are used to measure the performance of different methods. The data augmentation by synthesis is evaluated by the improvement for lesion segmentation models. We arbitrarily control lesion information to synthesize different number of data for augmentation. The detail of mechanism for manipulating lesion mask can be found in supplementary material. To simulate the piratical usage of data augmentation, we conduct experiments in the manner of utilizing all real data. In each experiment, we vary the percentage of synthezied data to observe the contribution for data augmentation. Table 2 shows the calculated segmentation performance. Comparing with the baseline experiment that only uses real data, the synthesized data from pix2pix [7] and pix2pixHD [14] degrade the segmentation performance. The performance of Shin et al. [11] improves when synthesized data is used for segmentation, but the proposed method outperforms other methods by a large margin. Figure 3 demonstrates the robustness of the proposed model under different lesion mask manipulations (e.g. changing the size of tumor, moving lesion location, and even reassembling lesion information between patients). As can be seen from this figure, our method is robust to various lesion mask manipulations.

Table 2. Quantitative results corresponding to image segmentation when the synthesized data is used for data augmentation. For each experiments, the first row reports the percentage of synthesized/real data for training and the number of instances of synthesized/real data in parentheses. Exp.4 reports the results of ablation study.

	Dice score			Hausdorff95 distance			Sensitivity			Specificity		
	Edema	Cavity	Tumor	Edema	Cavity	Tumor	Edema	Cavity	Tumor	Edema	Cavity	Tumor
Exp.1: 50% Synthesized+ 50% Real (1080 + 1080)												
pix2pix [7]	0.589	0.459	0.562	13.180	21.003	10.139	0.626	0.419	0.567	0.995	0.998	0.999
pix2pix HD [14]	0.599	0.527	0.571	17.406	8.606	10.369	0.630	0.494	0.570	0.996	0.998	0.999
Shin et al. [11]	0.731	0.688	0.772	7.306	6.290	6.294	0.701	0.662	0.785	0.997	0.999	0.999
Our	**0.794**	**0.813**	**0.821**	**6.049**	**1.568**	**2.293**	**0.789**	**0.807**	**0.841**	**0.997**	**0.999**	**0.999**
Exp.2: 25% Synthesized+ 75% Real (540 + 1080)												
pix2pix [7]	0.602	0.502	0.569	10.706	9.431	10.147	0.640	0.463	0.579	0.995	0.999	0.998
pix2pix HD [14]	0.634	0.514	0.663	17.754	9.512	9.061	0.670	0.472	0.671	0.996	0.999	0.999
Shin et al. [11]	0.673	0.643	0.708	14.835	7.798	6.688	0.664	0.602	0.733	0.997	0.999	0.998
Our	**0.745**	**0.780**	**0.772**	**8.779**	**6.757**	**4.735**	**0.760**	**0.788**	**0.805**	**0.997**	**0.999**	**0.999**
Exp.3: 0% Synthesized + 100% Real (0 + 1080)												
Baseline	**0.646**	**0.613**	**0.673**	**8.816**	**7.856**	**7.078**	**0.661**	**0.576**	**0.687**	**0.996**	**0.999**	**0.998**
Exp.4: **Ablation study**												
w/o Stretch-out	0.684	0.713	0.705	6.592	5.059	4.002	0.708	0.699	0.719	0.997	0.999	0.999
w/o Label-wise D	0.753	0.797	0.785	7.844	2.570	2.719	0.735	0.780	0.783	0.998	0.999	0.999
w/o Atlas	0.677	0.697	0.679	13.909	11.481	7.123	0.691	0.689	0.723	0.997	0.999	0.998
w/o $\mathcal{L}_C$	0.728	0.795	0771	8.604	3.024	3.233	0.738	0.777	0.777	0.997	0.999	0.999
Our	**0.794**	**0.813**	**0.821**	**6.049**	**1.568**	**2.293**	**0.789**	**0.807**	**0.841**	**0.998**	**0.999**	**0.999**

Ablation Study. We conduct extensive ablation study to separately evaluate the effectiveness of using stretch-out up-sampling module, label-wise discriminators, atlas, and lesion shape consistency loss $\mathcal{L}_C$ in our method using the same experimental setting as exp.1 in Table 2. The contribution of modules for data augmentation by synthesis is reported in Table 2 exp.4. We find that when atlas is not used in our method, it significantly affects the synthesis quality due to the lack of human brain anatomy prior. Losing the customized reconstruction for each sequence (stretch-out up-sampling module) can also degrade the synthesis quality. Moreover, dropping either $\mathcal{L}_C$ or label-wise discriminators in the training reduces the performance, since the shape consistency loss and the specific supervision on ROIs are not used to optimize the generator to produce more realistic images.

4   Conclusion

We proposed an effective generation model for multi-model anatomic and molecular *APT*w MRI sequences. It was shown that the proposed multi-task optimization under adversarial training further improves the synthesis quality in each ROI. The synthesized data can be used for data augmentation, particularly for those images with pathological information of malignant gliomas, to improve the performance of segmentation. Moreover, the proposed approach is an automatic, low-cost solution to produce high quality data with diverse content that can be used for training of data-driven methods.

In our future work, we will generalize data augmentation by synthesis to other tasks, such as classification. Furthermore, the proposed method will be extended to 3D synthesis once better quality molecular MRI data is available for training the models.

Acknowledgment. This work was supported in part by grants from National Institutes of Health (R01CA228188) and National Science Foundation (1910141).

References

1. Avants, B.B., Tustison, N.J., Song, G., Cook, P.A., Klein, A., Gee, J.C.: A reproducible evaluation of ants similarity metric performance in brain image registration. Neuroimage **54**(3), 2033–2044 (2011)
2. Bakas, S., et al.: Identifying the best machine learning algorithms for brain tumor segmentation, progression assessment, and overall survival prediction in the brats challenge. arXiv preprint arXiv:1811.02629 (2018)
3. Chartsias, A., Joyce, T., Giuffrida, M.V., Tsaftaris, S.A.: Multimodal MR synthesis via modality-invariant latent representation. IEEE Trans. Med. Imaging **37**(3), 803–814 (2017)
4. Cordier, N., Delingette, H., Lê, M., Ayache, N.: Extended modality propagation: image synthesis of pathological cases. IEEE Trans. Med. Imaging **35**(12), 2598–2608 (2016)

5. Goodfellow, I., et al.: Generative adversarial nets. In: Advances in Neural Information Processing Systems, pp. 2672–2680 (2014)
6. Guo, P., Li, D., Li, X.: Deep OCT image compression with convolutional neural networks. Biomed. Opt. Express **11**(7), 3543–3554 (2020)
7. Isola, P., Zhu, J.Y., Zhou, T., Efros, A.A.: Image-to-image translation with conditional adversarial networks. In: Proceedings of the IEEE Conference on Computer Vision and Pattern Recognition, pp. 1125–1134 (2017)
8. Jiang, S., et al.: Identifying recurrent malignant glioma after treatment using amide proton transfer-weighted MR imaging: a validation study with image-guided stereotactic biopsy. Clin. Cancer Res. **25**(2), 552–561 (2019)
9. Johnson, J., Alahi, A., Fei-Fei, L.: Perceptual losses for real-time style transfer and super-resolution. In: Leibe, B., Matas, J., Sebe, N., Welling, M. (eds.) ECCV 2016. LNCS, vol. 9906, pp. 694–711. Springer, Cham (2016). https://doi.org/10.1007/978-3-319-46475-6_43
10. Ronneberger, O., Fischer, P., Brox, T.: U-Net: convolutional networks for biomedical image segmentation. In: Navab, N., Hornegger, J., Wells, W.M., Frangi, A.F. (eds.) MICCAI 2015. LNCS, vol. 9351, pp. 234–241. Springer, Cham (2015). https://doi.org/10.1007/978-3-319-24574-4_28
11. Shin, H.-C., et al.: Medical image synthesis for data augmentation and anonymization using generative adversarial networks. In: Gooya, A., Goksel, O., Oguz, I., Burgos, N. (eds.) SASHIMI 2018. LNCS, vol. 11037, pp. 1–11. Springer, Cham (2018). https://doi.org/10.1007/978-3-030-00536-8_1
12. Sudre, C.H., Li, W., Vercauteren, T., Ourselin, S., Jorge Cardoso, M.: Generalised dice overlap as a deep learning loss function for highly unbalanced segmentations. In: Cardoso, M.J., et al. (eds.) DLMIA/ML-CDS 2017. LNCS, vol. 10553, pp. 240–248. Springer, Cham (2017). https://doi.org/10.1007/978-3-319-67558-9_28
13. Van Nguyen, H., Zhou, K., Vemulapalli, R.: Cross-domain synthesis of medical images using efficient location-sensitive deep network. In: Navab, N., Hornegger, J., Wells, W.M., Frangi, A.F. (eds.) MICCAI 2015. LNCS, vol. 9349, pp. 677–684. Springer, Cham (2015). https://doi.org/10.1007/978-3-319-24553-9_83
14. Wang, T.C., Liu, M.Y., Zhu, J.Y., Tao, A., Kautz, J., Catanzaro, B.: High-resolution image synthesis and semantic manipulation with conditional GANs. In: Proceedings of the IEEE Conference on Computer Vision and Pattern Recognition (2018)
15. Wen, P.Y., Kesari, S.: Malignant gliomas in adults. N. Engl. J. Med. **359**(5), 492–507 (2008)
16. Woodworth, G.F., Garzon-Muvdi, T., Ye, X., Blakeley, J.O., Weingart, J.D., Burger, P.C.: Histopathological correlates with survival in reoperated glioblastomas. J. Neurooncol. **113**(3), 485–493 (2013). https://doi.org/10.1007/s11060-013-1141-3
17. Zhou, J., Payen, J.F., Wilson, D.A., Traystman, R.J., van Zijl, P.C.: Using the amide proton signals of intracellular proteins and peptides to detect pH effects in MRI. Nat. Med. **9**(8), 1085–1090 (2003)
18. Zhou, Y., He, X., Cui, S., Zhu, F., Liu, L., Shao, L.: High-resolution diabetic retinopathy image synthesis manipulated by grading and lesions. In: Shen, D., et al. (eds.) MICCAI 2019. LNCS, vol. 11764, pp. 505–513. Springer, Cham (2019). https://doi.org/10.1007/978-3-030-32239-7_56

T2 Mapping from Super-Resolution-Reconstructed Clinical Fast Spin Echo Magnetic Resonance Acquisitions

Hélène Lajous[1,2(✉)], Tom Hilbert[1,3,4], Christopher W. Roy[1],
Sébastien Tourbier[1], Priscille de Dumast[1,2], Thomas Yu[4],
Jean-Philippe Thiran[1,4], Jean-Baptiste Ledoux[1,2], Davide Piccini[1,3],
Patric Hagmann[1], Reto Meuli[1], Tobias Kober[1,3,4], Matthias Stuber[1,2],
Ruud B. van Heeswijk[1], and Meritxell Bach Cuadra[1,2,4]

[1] Department of Radiology, Lausanne University Hospital (CHUV) and University
of Lausanne (UNIL), Lausanne, Switzerland
helene.lajous@unil.ch
[2] Center for Biomedical Imaging (CIBM), Lausanne, Switzerland
[3] Advanced Clinical Imaging Technology (ACIT), Siemens Healthcare,
Lausanne, Switzerland
[4] Signal Processing Laboratory 5 (LTS5), Ecole Polytechnique Fédérale de Lausanne
(EPFL), Lausanne, Switzerland

Abstract. Relaxometry studies in preterm and at-term newborns have
provided insight into brain microstructure, thus opening new avenues for
studying normal brain development and supporting diagnosis in equivocal neurological situations. However, such quantitative techniques require
long acquisition times and therefore cannot be straightforwardly translated to *in utero* brain developmental studies. In clinical fetal brain magnetic resonance imaging routine, 2D low-resolution T2-weighted fast spin
echo sequences are used to minimize the effects of unpredictable fetal
motion during acquisition. As super-resolution techniques make it possible to reconstruct a 3D high-resolution volume of the fetal brain from
clinical low-resolution images, their combination with quantitative acquisition schemes could provide fast and accurate T2 measurements. In this
context, the present work demonstrates the feasibility of using super-
resolution reconstruction from conventional T2-weighted fast spin echo
sequences for 3D isotropic T2 mapping. A quantitative magnetic resonance phantom was imaged using a clinical T2-weighted fast spin echo
sequence at variable echo time to allow for super-resolution reconstruction at every echo time and subsequent T2 mapping of samples whose
relaxometric properties are close to those of fetal brain tissue. We demonstrate that this approach is highly repeatable, accurate and robust when

Electronic supplementary material The online version of this chapter (https://
doi.org/10.1007/978-3-030-59713-9_12) contains supplementary material, which is
available to authorized users.

using six echo times (total acquisition time under 9 minutes) as compared to gold-standard single-echo spin echo sequences (several hours for one single 2D slice).

Keywords: Super-Resolution (SR) reconstruction · T2 mapping · T2-weighted images · Fast spin echo sequences · Fetal brain magnetic resonance imaging (MRI)

1   Introduction

Early brain development encompasses many crucial structural and physiological modifications that have an influence on health later in life. Changes in T1 and T2 relaxation times may provide valuable clinical information about ongoing biological processes, as well as a better insight into the early stages of normal maturation [5]. Indeed, quantitative MRI (qMRI) has revealed biomarkers sensitive to subtle changes in brain microstructure that are characteristic of abnormal patterns and developmental schemes in newborns [6,20]. T1 and T2 mapping of the developing fetal brain would afford physicians new resources for pregnancy monitoring, including quantitative diagnostic support in equivocal situations and prenatal counselling, as well as postnatal management. Unfortunately, current relaxometry strategies require long scanning times that are not feasible in the context of *in utero* fetal brain MRI due to unpredictable fetal motion in the womb [4,8,15,23]. As such, very little work has explored *in vivo* qMRI of the developing fetal brain. Myelination was characterized *in utero* using a mono-point T1 mapping based on fast spoiled gradient echo acquisitions [1], and more recently by fast macromolecular proton fraction mapping [25]. T2* relaxometry of the fetal brain has been explored through fast single-shot multi-echo gradient echo-type echo-planar imaging (GRE-EPI) [24] and, recently, based on a slice-to-volume registration of 2D dual-echo multi-slice EPI with multiple time points reconstructed into a motion-free isotropic high-resolution (HR) volume [3]. To our knowledge, similar strategies have not been investigated for *in utero* T2 mapping yet. Today, super-resolution (SR) techniques have been adopted to take advantage of the redundancy between multiple T2-weighted (T2w) low-resolution (LR) series acquired in orthogonal orientations and thereby reconstruct a single isotropic HR volume of the fetal brain with reduced motion sensitivity for thorough anatomical exploration [7,9,11,19,22]. In clinical routine, 2D thick slices are typically acquired in a few seconds using T2w multi-slice single-shot fast spin echo sequences [8]. We hypothesize that the combination of SR fetal brain MRI with the sensitivity of qMRI would enable reliable and robust 3D HR T2 relaxometry of the fetal brain [2,3]. In this context, we have explored the feasibility of repeatable, accurate and robust 3D HR T2 mapping from SR-reconstructed clinical fast T2w Half-Fourier Acquisition Single-shot Turbo spin Echo (HASTE) with variable echo time (TE) on a quantitative MR phantom [12].

2   Methodology

2.1   Model Fitting for T2 Mapping

The T2w contrast of an MR image is governed by an exponential signal decay characterized by the tissue-specific relaxation time, T2. Since any voxel within brain tissue may contain multiple components, a multi-exponential model is the closest to reality. However, it requires long acquisition times that are not acceptable in a fetal imaging context. The common simplification of a single-compartment model [6,15] allows for fitting the signal according to the following equation:

$$\hat{X}_{\text{TE}} = \mathcal{M}_0 e^{\frac{-TE}{T2}}, \tag{1}$$

where $\mathcal{M}_0$ is the equilibrium magnetization and $\hat{X}_{TE}$ is the signal intensity at a given echo time TE at which the image is acquired. As illustrated in Fig. 1, the time constant T2 can be estimated in every voxel by fitting the signal decay over TE with this mono-exponential analytical model [17].

We aim at estimating a HR 3D T2 map of the fetal brain with a prototype algorithm. Our strategy is based on SR reconstruction from orthogonal 2D multi-slice T2w clinical series acquired at variable TE (see complete framework in Fig. 1). For every TE_i, a motion-free 3D image $\hat{X}_{TE_i}$ is reconstructed using a Total-Variation (TV) SR reconstruction algorithm [21,22] which solves:

$$\hat{\mathbf{X}}_{TE_i} = \arg\min_{\mathbf{X}} \frac{\lambda}{2} \sum_{kl} \| \underbrace{\mathbf{D}_{kl}\mathbf{B}_{kl}\mathbf{M}_{kl}}_{\mathbf{H}_{kl}} \mathbf{X} - \mathbf{X}_{kl,TE_i}^{LR} \|^2 + \|\mathbf{X}\|_{TV}, \tag{2}$$

where the first term relates to data fidelity, k being the k-th LR series $\mathbf{X}_{TE_i}^{LR}$ and l the l-th slice. $\|\mathbf{X}\|_{TV}$ is a TV prior introduced to regularize the solution while λ balances the trade-off between both data and regularization ($\lambda = 0.75$). $\mathbf{D}$ and $\mathbf{B}$ are linear downsampling and Gaussian blurring operators given by the acquisition characteristics. $\mathbf{M}$, which encodes the rigid motion of slices, is set to the identity transform in the absence of motion.

The model fitting described in Eq. 1 is computed in every voxel of a SR 3D volume estimated at time TE. T2 maps are computed using a non-linear least-squares optimization (MATLAB, MathWorks, R2019a). As shown in Fig. 1B, the T2 signal decay may reveal an offset between the first echoes and the rest of the curve that can be explained by stimulated echoes [16] and the sampling order of the HASTE sequence. It is common practice to exclude from the fitting the first points that exhibit the pure spin echo without the stimulated echo contributions [17].

2.2   Validation Study

Quantitative Phantom. Our validation is based on the system standard model 130 that was established by the National Institute for Standards and Technology (NIST) of the United States in collaboration with the International Society

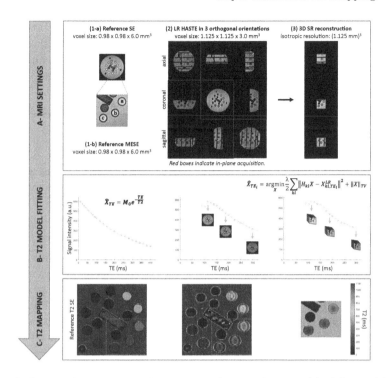

Fig. 1. Evaluation framework. Reference T2 values of elements (a), (b) and (c) (blue dashed area) of the NIST phantom are measured by (A-1-a) single-echo spin echo (SE) and (A-1-b) multi-echo spin echo (MESE) sequences. (A-2) Low-resolution orthogonal HASTE images acquired at variable TE are SR-reconstructed into (A-3) an isotropic volume for every TE. (B) The signal decay as a function of TE is fitted in each voxel by a mono-exponential model. (C) Resulting voxel-wise T2 maps. (Color figure online)

for Magnetic Resonance in Medicine (ISMRM). It is produced by QalibreMD (Boulder, CO, USA) and is hereafter referred to as the NIST phantom [12]. This quantitative phantom was originally developed to assess the repeatability and reproducibility of MRI protocols across vendors and sites. Our study focuses on a region-of-interest (ROI) represented by a blue square in Fig. 1A. It is centered on elements of the NIST phantom that have relaxometry properties close to those reported in the literature for *in vivo* brain tissue of fetuses and preterm newborns at 1.5 T [3,10,18,24,25], namely T2 values higher than 170 ms and 230 ms in grey matter and white matter respectively, and high T1 values. Accordingly, we focus on the following areas: (a) T2 = 428.3 ms, (b) T2 = 258.4 ms and (c) T2 = 186.1 ms, with a relatively high T1/T2 ratio (4.5–6.9), and which fall within a field-of-view similar to that of fetal MRI.

MR Imaging. Acquisitions are performed at 1.5 T (MAGNETOM Sola, Siemens Healthcare, Erlangen, Germany), with an 18-channel body coil and a 32-

channel spine coil (12 elements used). Three clinical T2w series of 2D thick slices are acquired in orthogonal orientations using an ultra-fast multi-slice HASTE sequence (TE = 90 ms, TR = 1200 ms, excitation/refocusing pulse flip angles of $90°/180°$, interslice gap of 10%, voxel size of $1.13 \times 1.13 \times 3.00$ mm^3). For consistency with the clinical fetal protocol, a limited field-of-view (360×360 mm^2) centered on the above-referenced ROI is imaged. Each series contains 23 slices and is acquired in 28 seconds.

We extend the TE of this clinical protocol in order to acquire additional sets of three orthogonal series, leading to six configurations with 4, 5, 6, 8, 10, or 18 TEs uniformly sampled over the range of 90 ms to 298 ms. The acquisition time is about 90 seconds per TE, thus the total acquisition time ranges from 6 minutes (4 TEs) to 27 minutes (18 TEs). Binary masks are drawn on each LR series for reconstruction of a SR volume at every TE, as illustrated in Fig. 1C.

Gold-Standard Sequences for T2 Mapping. A conventional single-echo spin echo (SE) sequence with variable TE is used as a reference for validation (TR = 5000 ms, 25 TEs sampled from 10 to 400 ms, voxel size of $0.98 \times 0.98 \times 6.00$ mm^3). One single 2D slice is imaged in 17.47 minutes for a given TE, which corresponds to a total acquisition time of more than 7 hours. As recommended by the phantom manufacturer, an alternative multi-echo spin echo (MESE) sequence is used for comparison purposes (TR = 5000 ms, 32 TEs equally sampled from 13 to 416 ms, voxel size of $0.98 \times 0.98 \times 6.00$ mm^3). The total acquisition time to image the same 2D slice is of 16.05 minutes.

Gold-standard and HASTE acquisitions are made publicly available in our repository [14] for further reproducibility and validation studies.

Evaluation Procedure. We evaluate the accuracy of the proposed 3D SR T2 mapping framework with regard to T2 maps obtained from HASTE, MESE and SE acquisitions. Since only one single 2D coronal slice is imaged by SE and MESE sequences, quantitative measures are computed on the corresponding slice of the coronal 2D HASTE series and 3D SR images. At a voxel-wise level, T2 standard deviation (SD) and R^2 are computed to evaluate the fitting quality. A region-wise analysis is conducted over the three ROIs previously denoted as (a), (b) and (c). An automated segmentation of these areas in the HASTE and SR images is performed by Hough transform followed by a one-pixel dilation. Mean T2 values $\pm$ SD are estimated within each ROI. The relative error in T2 estimation is computed using SE measurements as reference values. It is defined as the difference in T2 measures between either HASTE or SR and the corresponding SE reference value normalized by the SE reference value. This metric is used to evaluate the accuracy of the studied T2 mapping technique as well as its robustness to noise (see also Supplementary Material - Table 1).

We run the same MRI protocol (SE, MESE and HASTE) on three different days in order to study the repeatability of T2 measurements. The relative error in T2 estimation between two independent experiments (i.e., on two different days) is calculated as described above, using every measure in turn as a reference. Thus, we are able to evaluate the mean absolute percentage error $|\Delta\varepsilon|$ as the average of relative absolute errors in T2 estimation over all possible reference days. The coefficient of variation (CV) for T2 quantification represents the variability (SD) relative to the mean fitted T2 value.

3   Results

3.1   3D Super-Resolution T2 Mapping

Voxel-wise T2 maps as derived from one coronal HASTE series and from the 3D SR reconstruction of three orthogonal HASTE series are shown in Fig. 2 together with associated standard deviation maps. HASTE series show Gibbs ringing in the phase-encoding direction at the interface of the different elements. Since SE and MESE images are corrupted in a similar way across all TEs, a homogeneous T2 map is recovered for every element of the phantom (Fig. 2). Instead, as HASTE acquisitions rely on a variable k-space sampling for every TE, resulting T2 maps are subject to uncompensated Gibbs artifacts that cannot easily be corrected due to reconstruction of HASTE images by partial Fourier techniques [13]. Interestingly though, Gibbs ringing is much less pronounced in

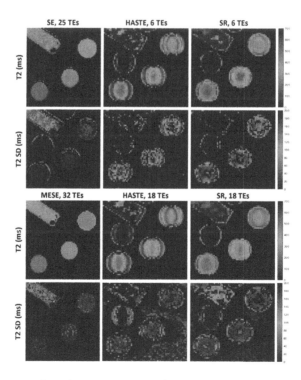

Fig. 2. Comparison of voxel-wise T2 maps and T2 SD maps estimated from SE, MESE, HASTE and corresponding SR reconstruction at variable TE

the SR reconstructions where it is probably attenuated by the combination of orthogonal series. Of note, *in vivo* data are much less prone to this artifact.

3.2   Repeatability Study

As highlighted in Table 1, T2 estimation is highly repeatable over independent measurements with a mean CV of less than 4%, respectively 8%, for T2 quantification from HASTE acquisitions, respectively SR with 5 TEs. The mean absolute percentage error is less than 5% in HASTE images, respectively 10% in SR.

Table 1. Repeatability of T2 mapping strategies between three independent experiments. Mean fitted T2 value ± SD, CV, mean absolute difference and mean absolute percentage error in T2 estimation are presented. The lowest difference for each (ROI, method) pair is shown in bold.

		SE	HASTE/SR					
TEs		25	4	5	6	8	10	18
T2 (ms)	(a)	380±7	422±2/288±7	427±12/312±21	425±2/352±13	451±4/409±18	454±0/403±20	451±2/407±12
	(b)	256±5	314±6/188±8	304±12/223±11	315±4/258±5	337±6/287±8	335±9/288±4	333±10/297±4
	(c)	187±2	252±9/159±10	247±2/180±14	249±4/207±6	267±1/225±5	267±6/223±4	263±3/233±4
CV (%)	(a)	1.8	0.4/**2.3**	2.8/6.7	0.5/3.8	1.0/4.4	**0.0**/5.0	0.4/3.0
	(b)	1.8	1.8/4.4	3.9/4.8	**1.3**/2.0	1.8/2.9	2.8/1.6	2.9/**1.2**
	(c)	1.0	3.6/6.2	0.8/7.7	1.5/2.9	**0.3**/2.4	2.2/**1.7**	1.1/1.9
|ΔT2| (ms)	(a)	9.1	1.9/**8.9**	15.8/24.4	2.7/17.8	5.2/23.3	**0.2**/26.8	2.6/15.4
	(b)	6.1	7.7/11.1	14.2/13.8	**4.9**/6.0	7.7/10.8	11.4/5.6	12.8/**4.2**
	(c)	2.1	12.1/11.8	2.6/16.4	4.5/7.7	**1.1**/7.2	6.9/**4.9**	3.8/5.3
|Δε| (%)	(a)	2.4	0.4/**3.1**	3.7/7.7	0.6/5.1	1.2/5.7	**0.0**/6.7	0.6/3.8
	(b)	2.4	2.4/5.9	4.6/6.3	**1.6**/2.3	2.3/3.8	3.4/1.9	3.9/**1.4**
	(c)	1.1	4.8/7.3	1.1/9.4	1.8/3.7	**0.4**/3.2	2.6/**2.2**	1.4/2.3

3.3   Impact of the Number of Echo Times on T2 Measurements

In an effort to optimize the acquisition scheme, especially regarding energy deposition and reasonable acquisition time in a context of fetal examination, we investigate the influence of the number of TEs on the T2 estimation accuracy. As T2 quantification is highly repeatable throughout independent measurements, the following results are derived from an arbitrarily-selected experiment.

T2 estimation by both clinical HASTE acquisitions and corresponding SR reconstructions demonstrates a high correlation with reference SE values over the 180–400 ms range of interest (Supplementary Material - Fig. 1).

Bland-Altman plots presented in Fig. 3 report the agreement between HASTE-/SR-based T2 quantification and SE reference values in the three ROIs. The average error in T2 estimation from HASTE series is almost the same across

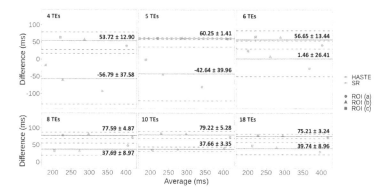

Fig. 3. Bland-Altman plots of differences in T2 quantification between HASTE/ corresponding SR and reference SE in three ROIs for various numbers of TEs

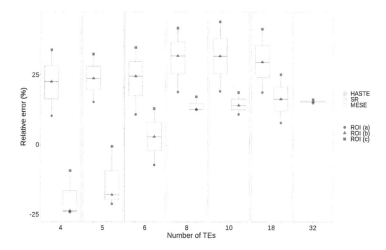

Fig. 4. Relative error in T2 quantification according to the method and number of TEs as compared to reference SE measurements

all configurations. The difference in T2 measurements is independent of the studied ROI. Conversely, the average error in SR T2 quantification varies with the number of TEs, the smallest average difference being for 6 TEs. In a given configuration, the difference in T2 measurements depends on the targeted value.

Figure 4 displays the relative error in T2 quantification from MESE, HASTE and SR as compared to SE measurements according to the number of TEs acquired. MESE provides T2 quantification with a small dispersion but with low out-of-plane resolution and prohibitive scanning times in a context of fetal MRI. In the following, its 16%-average relative error is considered an acceptable reference error level. HASTE-based T2 quantification overestimates T2 values in the range of interest by around 25%. As for MESE-based T2 mapping, such an overestimation can be attributed to stimulated echo contamination [16]. In the case of SR, quantification errors vary with the number of TEs acquired. Below six echoes, SR underestimates T2 values on average, and enables a dramatic improvement in T2 quantification over the HASTE-based technique, but only for T2 values less than 200 ms. Above 6 TEs, SR exhibits approximately the same average error as MESE. For six echoes, SR outperforms HASTE and MESE over the whole range of T2 values studied with a relative error less than 11%. Furthermore, preliminary results on T2 quantification from HASTE images corrupted by higher levels of noise and their corresponding SR reconstructions in this optimized set-up using six echoes demonstrate the robustness of the proposed SR T2 mapping technique (Supplementary Material - Table 1). Overall, SR substantially outperforms HASTE for T2 quantification.

4   Conclusion

This work demonstrates the feasibility of repeatable, accurate and robust 3D isotropic HR T2 mapping in a reasonable acquisition time based on SR-reconstructed clinical fast spin echo MR acquisitions. We show that SR-based T2 quantification performs accurately in the range of interest for fetal brain studies (180–430 ms) as compared to gold-standard methods based on SE or MESE. Moreover, it could be straightforwardly translated to the clinic since only the TE of the HASTE sequence routinely used in fetal exams needs to be adapted. A pilot study will be conducted on an adult brain to replicate these results *in vivo*. Although our study focuses on static data, the robustness of SR techniques to motion makes us hypothesize that 3D HR T2 mapping of the fetal brain is feasible. The number of TEs required for an accurate T2 quantification in this context has to be explored.

Acknowledgements. This work was supported by the Swiss National Science Foundation through grants 141283 and 182602, the Centre d'Imagerie Biomédicale (CIBM) of the UNIL, UNIGE, HUG, CHUV and EPFL, the Leenaards and Jeantet Foundations, and the Swiss Heart Foundation. The authors would like to thank Yasser Alemán-Gómez for his help in handling nifti images.

References

1. Abd Almajeed, A., Adamsbaum, C., Langevin, F.: Myelin characterization of fetal brain with mono-point estimated T1-maps. Magn. Reson. Imaging **22**(4), 565–572 (2004). https://doi.org/10/frdp45
2. Bano, W., et al.: Model-based super-resolution reconstruction of T2 maps. Magn. Reson. Med. **83**(3), 906–919 (2020). https://doi.org/10/gf85n4
3. Blazejewska, A.I., et al.: 3D in utero quantification of T2* relaxation times in human fetal brain tissues for age optimized structural and functional MRI. Magn. Reson. Med. **78**(3), 909–916 (2017). https://doi.org/10/gf2n9z
4. Chen, L.W., Wang, S.T., Huang, C.C., Tu, Y.F., Tsai, Y.S.: T2 relaxometry MRI predicts cerebral palsy in preterm infants. Am. J. Neuroradiol. **39**(3), 563–568 (2018). https://doi.org/10/gdcz66

5. Deoni, S.C.: Quantitative relaxometry of the brain. Top. Magn. Reson. Imaging **21**(2), 101–113 (2010). https://doi.org/10/fj3m42

6. Dingwall, N., et al.: T2 relaxometry in the extremely-preterm brain at adolescence. Magn. Reson. Imaging **34**(4), 508–514 (2016). https://doi.org/10/ggb9qn

7. Ebner, M., et al.: An automated framework for localization, segmentation and super-resolution reconstruction of fetal brain MRI. NeuroImage **206**, 116324 (2020). https://doi.org/10/ggdnsm

8. Gholipour, A., et al.: Fetal MRI: a technical update with educational aspirations. Concepts Magn. Reson. Part A Bridg. Educ. Res. **43**(6), 237–266 (2014). https://doi.org/10/gf4bc6

9. Gholipour, A., Estroff, J.A., Warfield, S.K.: Robust super-resolution volume reconstruction from slice acquisitions: application to fetal brain MRI. IEEE Trans. Med. Imaging **29**(10), 1739–1758 (2010). https://doi.org/10/b2xmdp

10. Hagmann, C.F., et al.: T2 at MR imaging is an objective quantitative measure of cerebral white matter signal intensity abnormality in preterm infants at term-equivalent age. Radiology **252**(1), 209–217 (2009). https://doi.org/10/bqkd9r

11. Kainz, B., et al.: Fast volume reconstruction from motion corrupted stacks of 2D slices. IEEE Trans. Med. Imaging **34**(9), 1901–1913 (2015). https://doi.org/10/f3svr5

12. Keenan, K.E., et al.: Multi-site, multi-vendor comparison of T1 measurement using ISMRM/NIST system phantom. In: Proceedings of the 24th Annual Meeting of ISMRM, Singapore (2016). Program number 3290

13. Kellner, E., Dhital, B., Kiselev, V.G., Reisert, M.: Gibbs-ringing artifact removal based on local subvoxel-shifts. Magn. Reson. Med. **76**(5), 1574–1581 (2016). https://doi.org/10/f9f64r

14. Lajous, H., Ledoux, J.B., Hilbert, T., van Heeswijk, R.B., Meritxell, B.C.: Dataset T2 mapping from super-resolution-reconstructed clinical fast spin echo magnetic resonance acquisitions (2020). https://doi.org/10.5281/zenodo.3931812

15. Leppert, I.R., et al.: T2 relaxometry of normal pediatric brain development. J. Magn. Reson. Imaging **29**(2), 258–267 (2009). https://doi.org/10/c77mvm

16. McPhee, K.C., Wilman, A.H.: Limitations of skipping echoes for exponential T2 fitting. J. Magn. Reson. Imaging **48**(5), 1432–1440 (2018). https://doi.org/10/ggdj43

17. Milford, D., Rosbach, N., Bendszus, M., Heiland, S.: Mono-exponential fitting in T2-relaxometry: relevance of offset and first echo. PLoS ONE **10**, e0145255 (2015). https://doi.org/10/gfc68d

18. Nossin-Manor, R., et al.: Quantitative MRI in the very preterm brain: assessing tissue organization and myelination using magnetization transfer, diffusion tensor and T1 imaging. NeuroImage **64**, 505–516 (2013). https://doi.org/10/f4jgtg

19. Rousseau, F., Kim, K., Studholme, C., Koob, M., Dietemann, J.-L.: On super-resolution for fetal brain MRI. In: Jiang, T., Navab, N., Pluim, J.P.W., Viergever, M.A. (eds.) MICCAI 2010. LNCS, vol. 6362. Springer, Berlin, Heidelberg (2010). https://doi.org/10.1007/978-3-642-15745-5_44

20. Schneider, J., et al.: Evolution of T1 relaxation, ADC, and fractional anisotropy during early brain maturation: a serial imaging study on preterm infants. Am. J. Neuroradiol. **37**(1), 155–162 (2016). https://doi.org/10/f7489d

21. Tourbier, S., Bresson, X., Hagmann, P., Meuli, R., Bach Cuadra, M.: sebastien-tourbier/mialsuperresolutiontoolkit: MIAL Super-Resolution Toolkit v1.0 (2019). https://doi.org/10.5281/zenodo.2598448

22. Tourbier, S., Bresson, X., Hagmann, P., Thiran, J.P., Meuli, R., Bach Cuadra, M.: An efficient total variation algorithm for super-resolution in fetal brain MRI with adaptive regularization. NeuroImage **118** (2015). https://doi.org/10/f7p5zx

23. Travis, K.E., et al.: More than myelin: probing white matter differences in prematurity with quantitative T1 and diffusion MRI. NeuroImage Clin. **22**, 101756 (2019). https://doi.org/10/ggnr3d
24. Vasylechko, S., et al.: T2* relaxometry of fetal brain at 1.5 Tesla using a motion tolerant method. Magn. Reson. Med. **73**(5), 1795–1802 (2015). https://doi.org/10/gf2pbh
25. Yarnykh, V.L., Prihod'ko, I.Y., Savelov, A.A., Korostyshevskaya, A.M.: Quantitative assessment of normal fetal brain myelination using fast macromolecular proton fraction mapping. Am. J. Neuroradiol. **39**(7), 1341–1348 (2018). https://doi.org/10/gdv9nf

Learned Proximal Networks
for Quantitative Susceptibility Mapping

Kuo-Wei Lai[1,2], Manisha Aggarwal[3], Peter van Zijl[2,3], Xu Li[2,3],
and Jeremias Sulam[1(✉)]

[1] Department of Biomedical Engineering, Johns Hopkins University,
Baltimore, MD 21218, USA
{klai10,jsulam1}@jhu.edu
[2] F.M. Kirby Research Center for Functional Brain Imaging,
Kennedy Krieger Institute, Baltimore, MD 21205, USA
[3] Department of Radiology and Radiological Sciences, Johns Hopkins University,
Baltimore, MD 21205, USA

Abstract. Quantitative Susceptibility Mapping (QSM) estimates tissue magnetic susceptibility distributions from Magnetic Resonance (MR) phase measurements by solving an ill-posed dipole inversion problem. Conventional single orientation QSM methods usually employ regularization strategies to stabilize such inversion, but may suffer from streaking artifacts or over-smoothing. Multiple orientation QSM such as calculation of susceptibility through multiple orientation sampling (COSMOS) can give well-conditioned inversion and an artifact free solution but has expensive acquisition costs. On the other hand, Convolutional Neural Networks (CNN) show great potential for medical image reconstruction, albeit often with limited interpretability. Here, we present a Learned Proximal Convolutional Neural Network (LP-CNN) for solving the ill-posed QSM dipole inversion problem in an iterative proximal gradient descent fashion. This approach combines the strengths of data-driven restoration priors and the clear interpretability of iterative solvers that can take into account the physical model of dipole convolution. During training, our LP-CNN learns an implicit regularizer via its proximal, enabling the decoupling between the forward operator and the data-driven parameters in the reconstruction algorithm. More importantly, this framework is believed to be the first deep learning QSM approach that can naturally handle an arbitrary number of phase input measurements without the need for any ad-hoc rotation or re-training. We demonstrate that the LP-CNN provides state-of-the-art reconstruction results compared to both traditional and deep learning methods while allowing for more flexibility in the reconstruction process.

Keywords: Quantitative Susceptibility Mapping · Proximal learning · Deep learning

Electronic supplementary material The online version of this chapter (https://doi.org/10.1007/978-3-030-59713-9_13) contains supplementary material, which is available to authorized users.

A. L. Martel et al. (Eds.): MICCAI 2020, LNCS 12262, pp. 125–135, 2020.
https://doi.org/10.1007/978-3-030-59713-9_13

1   Introduction

Quantitative Susceptibility Mapping (QSM) is a Magnetic Resonance Imaging (MRI) technique that aims at mapping tissue magnetic susceptibility from gradient echo imaging phase [9,39]. QSM has important clinical relevance since bulk tissue magnetic susceptibility provides important information about tissue composition and microstructure [25,39] such as myelin content in white matter and iron deposition in gray matter. Pathological changes in these tissue susceptibility sources are closely related to a series of neurodegenerative diseases such as Multiple Sclerosis [7,18,22] and Alzheimer's Disease [2,4,38].

The QSM processing typically includes two main steps [10]: (i) a phase preprocessing step comprising phase unwrapping and background field removal, which provides the local phase image, and (ii) the more challenging phase-to-susceptibility dipole inversion, which reconstructs the susceptibility map from the local phase. Due to the singularity in the dipole kernel and the limited number of phase measurements in the field of view, inverting the dipole kernel is an ill-posed inverse problem [10]. Different strategies have been used in conventional QSM methods to stabilize this ill-posed inverse. One strategy is through data over-sampling, i.e. by utilizing multiple phase measurements acquired at different head-orientations to eliminate the singularity in the dipole kernel. Such calculation of susceptibility through multiple orientation sampling (COSMOS) is usually used as a gold standard for in vivo QSM [26]. Even though COSMOS QSM exhibits excellent image quality, it is often prohibitively expensive in terms of data acquisition, as phase images at three or more head orientations are required, leading to long scan times and patient discomfort. For single orientation QSM, direct manipulation of the dipole kernel as in thresholded k-space division (TKD) often leads to residual streaking artifacts and susceptibility underestimation [35]. To suppress such streaking artifacts [10,19], different regularization strategies are often used, e.g.. as in morphology enabled dipole inversion (MEDI) [27,28] and in the two-level susceptibility estimation method, STAR-QSM [40]. However, such methods usually need careful parameter tuning or may introduce extra smoothing due to the regularization enforced, leading to sub-optimal reconstruction.

Upon the advent of deep convolutional neural networks in a myriad of computer vision problems, some recent deep learning based QSM algorithms have shown that these data-driven methods can approximate the dipole inversion and generate high-quality QSM reconstructions from single phase measurements [14]. QSMnet [41] used a 3D Unet [34] to learn a single orientation dipole inversion and its successor, QSMnet$^+$ [15], explored model generalization on wider susceptibility value ranges with data augmentation. DeepQSM [5] showed that a deep network trained with synthetic phantom data can be deployed for QSM reconstruction with real phase measurement. More recently, Chen et al. [8] utilized generative adversarial networks [11] to perform dipole inversion, while Polak et al. [33] proposed a new nonlinear dipole inversion and combined it with a variational network (VaNDI) [12] to further improve accuracy. Closest to our work is that from Kames et al. [16], who employed a proximal gradient algorithm with a variational network. However their approach is limited to downsampled data due to GPU memory constraints, limiting its applicability.

While these previous works illustrate the potential benefit of deep learning for QSM, they have known limitations [14]: most of them [5,8,15,41] only learn an unique dipole inversion for a predefined direction (often aligned with the main field), thus having to rotate the input image when it comes to different oblique acquisitions before applying their reconstruction. More importantly, these methods can only take one phase measurement at deployment, preventing further improvements when another phase image is available. They are also quite limited in combining multiple phase measurements for estimation of more complicated models such as used in susceptibility tensor imaging (STI) [21,23]. Inspired by [3,29], we present a Learned Proximal Convolutional Neural Network (LP-CNN) QSM method that provides far more flexibility for solving the ill-posed dipole inversion problem and high quality reconstruction. We devised an unrolled iterative model combining a proximal gradient descent algorithm and a convolutional neural network. In this way, we naturally decouple the utilization of the forward model and the data-driven parameters within the reconstruction algorithm. As a result, our LP-CNN can handle a single or an arbitrary number of phase input measurements without any ad-hoc rotation or re-training, while achieving state-of-the-art reconstruction results.

2 Methods

Our LP-CNN architecture is shown in Fig. 1. We first preprocess the dataset and then conduct the training of our LP-CNN for dipole inversion. Training data pairs include the local phase data as input data, the corresponding dipole kernel as forward operator, and the COSMOS susceptibility maps as ground-truth data. In order to learn the prior information between local phase measurements and susceptibility maps, we train our LP-CNN in a supervised manner. In the rest of this section, we provide further details for each step.

Dataset Acquisition and Preprocessing. A total of 36 MR phase measurements were acquired at 7T (Philips Achieva, 32 channel head coil) using three slightly different 3D gradient echo (GRE) sequences with voxel size of $1 \times 1 \times 1$ mm^3 on 8 healthy subjects (5 orientations for 4 subjects with TR = 28 ms, TE1/δTE = 5/5 ms, 5 echoes, FOV = 224 $\times$ 224 $\times$ 126 mm^3, 4 orientations for 3 subjects with TR=45 ms, TE1/δTE = 2/2 ms, 9 echoes, FOV = 224 $\times$ 224 $\times$ 110 mm^3, 4 orientations for 1 subject with TR = 45 ms, TE1/δTE = 2/2 ms, 16 echoes, FOV = 224 $\times$ 224 $\times$ 110 mm^3) at F.M. Kirby Research Center for Functional Brain Imaging, Kennedy Krieger Institute. The study was IRB approved and signed informed consent was obtained from all participants. GRE magnitude data acquired on each subject at different orientations were first coregistered to the natural supine position using FSL FLIRT [13], after which the dipole kernel corresponding to each head orientation was calculated based on the acquisition rotation parameters and coregistration transformation matrix. Multiple steps of

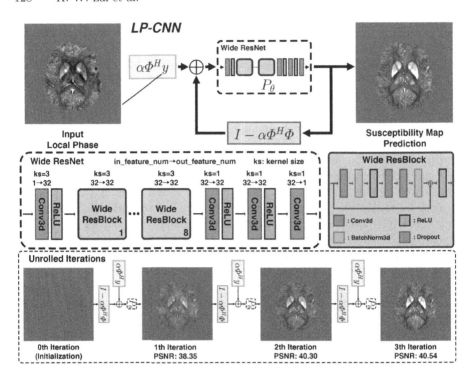

Fig. 1. A scheme of LP-CNN model (top), Wide ResNet architecture (middle) and unrolled iterations (bottom).

phase preprocessing were conducted in the native space of each head orientation including best-path based phase unwrapping [1], brain masking with FSL BET [36], V-SHARP [31] for removing the background field and echo averaging for echoes with TEs between 10 ms and 30 ms. The co-registration transformation matrices were then applied on the corresponding local field maps to transform them to the natural supine position. The COSMOS reconstructions were then generated using all 4 or 5 orientations for each subject. Following [41], we also augmented our dataset by generating, in total, 36 simulated local phase maps at random head orientations (within $\pm 45°$ of the z-axis) using the corresponding COSMOS map as the susceptibility source.

Proximal Gradient Descent QSM. The dipole inversion problem in QSM can be regarded as an ill-posed linear inverse problem. We model the measurement process by $y = \Phi x + v$, where $y \in \mathcal{Y} \subseteq \mathbb{R}^n$ is the local phase measurement, $x \in \mathcal{X} \subseteq \mathbb{R}^n$ is the underlying susceptibility map, and $\mathcal{Y}$ and $\mathcal{X}$ are the spaces of phase measurements and susceptibility maps, respectively. Furthermore, $\Phi : \mathcal{X} \to \mathcal{Y}$ is the corresponding forward operator, and v accounts for the measurement noise (as well as accounting for model mismatch). According to the physical model between local phase and susceptibility [24], the forward operator is determined by the (diagonal) dipole kernel D as $\Phi = F^{-1}DF$, where F is the (discrete) Fourier transform. Since D contains zeros in its diagonal [24], the operator Φ cannot be directly inverted. In other words, there exist an infinite number of feasible reconstructions $\hat{x}$ that explain the measurements y. We can nonetheless formulate this problem as a regularized inverse problem:

$$\min_x \mathcal{L}(x) = \underbrace{\frac{1}{2}\|y - \Phi x\|_2^2}_{f(x)} + \psi(x), \tag{1}$$

where f is a data-likelihood or data-fidelity term and ψ is a regularizer that restricts the set of possible solutions by promoting desirable properties according to prior knowledge[1]. Our goal is to find a minimizer for $\mathcal{L}(x)$ as the estimator for the susceptibility map, $\hat{x}$. Because of its computational efficiency (requiring only matrix-vector multiplications and scalar functions), and ability to handle possibly non-differentiable regularizers ψ, we apply The Proximal Gradient Descent algorithm [32] to minimize (1) iteratively by performing the updates:

$$\hat{x}_{i+1} = P_\psi \left(\hat{x}_i - \alpha\nabla f(\hat{x}_i)\right) \tag{2}$$

where[2] $\nabla f(\hat{x}) = \Phi^H(\Phi\hat{x} - y)$, α is a suitable gradient step size and $P_\psi(x)$ is the proximal operator of ψ, namely $P_\psi(x) = \arg\min_u \psi(u) + \frac{1}{2}\|x - u\|_2^2$. Such a proximal gradient descent algorithm is very useful when a regularizer is non-differentiable but still *proximable* in closed form. Clearly, the quality of the estimate $\hat{x}$ depends on the choice of regularizer, ψ, with traditional options including sparsity [6] or total variation [30] priors. In this work, we take advantage of supervised learning strategies to parameterize these regularizers [3,29,37] by combining proximal gradient descent strategies with a convolutional neural network, as we explain next.

Learned Proximal CNN. In order to exploit the benefits of deep learning in a supervised regime, we parameterize the proximal operator P_ψ by a 3D Wide ResNet [42] which learns an implicit data-driven regularizer by finding its corresponding proximal operator. This way, Eq. (2) becomes:

$$\hat{x}_{i+1} = P_\theta \left(\hat{x}_i - \alpha\Phi^H(\Phi\hat{x}_i - y)\right) = P_\theta \left(\alpha\Phi^H y + (I - \alpha\Phi^H\Phi)\hat{x}_i\right), \tag{3}$$

where θ represents all learnable parameters in the 3D Wide ResNet. The iterates are initialized from $\hat{x}_0 = 0$ for simplicity[3], and the final estimate is then given by the kth iteration, $\hat{x}_k$. We succinctly denote the overall function that produces the estimate as $\hat{x}_k = \phi_\theta(y)$. The training of the network is done through empirical risk minimization of the expected loss. In particular, consider training samples

---

[1] In fact, the expression in (1) can also be interpreted as a Maximum a Posteriori (MAP) estimator.

[2] A^H denotes the conjugate transpose (or Hermitian transpose) of the matrix A.

[3] Any other susceptibility estimation choices can serve as an initialization as well.

$\{y^j, x_c^j\}_{j=1}^N$ of phase measurements y^j and target susceptibility reconstructions x_c^j. Note that one does not have access to ground-truth susceptibility maps in general, and tissue susceptibility is treated as isotropic in QSM [10]. For these reasons, and following [5,8,15,41], we use a high-quality COSMOS reconstruction from multiple phase as ground-truth. We minimize the ℓ_2 loss function[4] over the training samples via the following optimization formulation:

$$\phi_{\hat{\theta}} = \arg\min_{\theta} \frac{1}{N} \sum_{j=1}^N \|x_c^j - \phi_\theta(y^j)\|_2^2, \tag{4}$$

which is minimized with a stochastic first order method. We expand on the experimental details below.

Deployment with Multiple Input Phase Measurements. Our proposed learned proximal network benefits from a clear separation between the utilization of the forward physical model and the learned proximal network parameters. This has two important advantages: (i) no ad-hoc rotation of the input image is required to align the main field (B_0), as this is instead managed by employing the operator Φ in the restoration; and (ii) the model learns a proximal function as a prior to susceptibility maps which is separated from the measurement model. More precisely, our learned proximal function has the same domain and co-domain, namely the space of susceptibility-map vectors, $P_{\hat{\theta}} : \mathcal{X} \to \mathcal{X}$. As a result, if at deployment (test) time the input measurements change, these can naturally be absorbed by modifying the function f in Eq. (1). In the important case where $L \geq 1$ measurements are available at arbitrary orientations, say $\{y_l = \Phi_l x\}_{l=1}^L$, a restored susceptibility map can still be computed with the obtained algorithm in Eq. (3) by simply constructing the operator $\bar{\Phi} = [\Phi_1; \ldots; \Phi_L] \in \mathbb{R}^{nL \times n}$ and concatenating the respective measurements, $\bar{y} = [y_1; \ldots; y_L] \in \mathbb{R}^{nL}$, and performing the updates $\hat{x}_{i+1} = P_{\hat{\theta}}\left(\frac{\alpha}{L}\bar{\Phi}^H \bar{y} + (I - \frac{\alpha}{L}\bar{\Phi}^H \bar{\Phi})\hat{x}_i\right)$. As we will demonstrate in the next section, the same LP-CNN model (trained with a single phase image) benefits from additional local phase inputs at test time, providing improved reconstructions without any re-training or tuning.

3 Experiments and Results

We trained and tested our LP-CNN method with our 8 subject dataset in a 4-fold cross validation manner, with each split containing 6 subjects for training and 2 subjects for testing. Our LP-CNN is implemented[5] in Pytorch and trained on a single NVIDIA GeForce RTX 2080 Ti GPU with mini-batch of size 2. We train our LP-CNN model with $k = 3$ iterates for 100 epochs (96 h) with dropout rate 0.5, using Adam optimizer [17] with an initial learning rate 10^{-4} with weight

---

[4] We chose an ℓ_2 norm for simplicity, and other choices are certainly possible, potentially providing further reconstruction improvements.
[5] Our code is released at https://github.com/Sulam-Group.

decay of 5×10^{-4}, and we decrease the learning rate with 0.8 ratio every 25 epochs. We compare our results with traditional methods (TKD [35], MEDI [27] and STAR-QSM [40] in STI Suite [20]) as well as with the deep learning-based QSMnet [41] across all splits.

Table 1. Quantitative performance metrics of QSM reconstruction estimated using the 4-fold cross validation set with different QSM methods. Upper part is single phase reconstruction and lower part is multiple phase reconstruction. LP-CNN shows comparable performance with QSMnet and outperforms TKD, MEDI and STAR-QSM.

Methods	NRMSE (%)	PSNR(dB)	HFEN (%)	SSIM
TKD	87.79 ± 8.15	35.54 ± 0.799	83.59 ± 10.19	0.9710 ± 0.0054
MEDI	67.40 ± 7.91	37.89 ± 0.996	60.01 ± 9.63	0.9823 ± 0.0041
STAR-QSM	63.70 ± 5.68	38.32 ± 0.758	62.06 ± 6.19	0.9835 ± 0.0032
QSMnet	55.10 ± 4.93	39.56 ± 0.788	58.14 ± 6.21	0.9857 ± 0.0026
LP-CNN	55.00 ± 5.01	39.58 ± 0.792	55.54 ± 5.73	0.9865 ± 0.0026
LP-CNN^2	50.10 ± 4.15	40.43 ± 0.733	48.96 ± 3.58	0.9889 ± 0.0021
LP-CNN^3	47.38 ± 4.17	40.90 ± 0.780	46.14 ± 3.58	0.9903 ± 0.0021

Patch Training. Training convolutional neural networks with 3D medical images often requires a prohibitively large amount of GPU memory. By taking advantage of the shift-invariance properties of CNN, a common workaround is to train these models in small patched data [5,8,15,41], we partition our data into $64 \times 64 \times 64$ 3D patches for training. However, such a modification must be done with caution, since naively employing a patched dipole kernel degrades performance by losing most of the high frequency information. In order to utilize the full dipole kernel during training with patched data, we modify the forward operator Φ by zero-padding, i.e. $\Phi' = CF^{-1}DFP = C\Phi P$, where P is a padding operation from 64^3 to the original image size and C is its (adjoint) cropping operator, $C = P^T$. See supplementary material for visualizing the difference between employing a patched dipole kernel and a full dipole kernel.

Reconstruction Performance. Table 1 shows the results of our LP-CNN and other methods evaluated by quantitative performance metrics of Normalized Root Mean Square Error (NRMSE), Peak Signal-to-Noise Ratio (PSNR), High Frequency Error Norm (HFEN), and Structural Similarity Index (SSIM). We denote LP-CNN^L as training with single input and testing with L number multiple inputs. Figure 2 shows the visualization results. The results of our LP-CNN showcases more high-frequency details and is the closest to the gold standard, COSMOS.

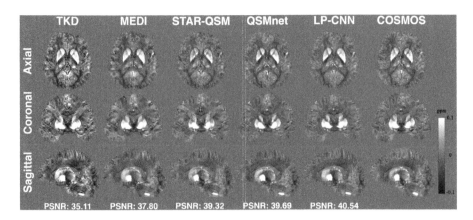

Fig. 2. Three plane views of susceptibility maps reconstructed using different methods. LP-CNN result shows high-quality COSMOS-like reconstruction performance. The TKD reconstruction suffers from streaking artifacts, while MEDI is over smoothing and STAR-QSM shows slight streaking effect.

Multiple Input LP-CNN. We finally demonstrate the flexibility of our LP-CNN by comparing the performance of our obtained model (trained with single local phase measurements) when deployed with increasing number of input measurements, and we compare with NDI [33] and COSMOS using a different number of inputs in Fig. 3. Though our LP-CNN was only trained with single local phase input, it allows to incorporate phase measurements at multiple orientations that improve QSM reconstruction, while none of the previous deep learning-based QSM methods [14] can handle a different number of input phase measurements.

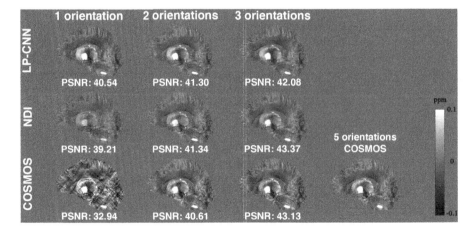

Fig. 3. Comparison between LP-CNN, NDI and COSMOS for multiple numbers of input phase measurement reconstructions. LP-CNN shows high-quality reconstruction on single input and refine the reconstruction with multiple inputs.

4   Discussion and Conclusion

In this work, we proposed a LP-CNN method to resolve the ill-posed dipole inversion problem in QSM. Our approach naturally incorporates the forward operator with arbitrarily oriented dipole and enhances reconstruction accuracy for single orientation QSM giving COSMOS like solution. More importantly, the learned parameters are separated from the measurement operator, providing greater flexibility. In particular, we demonstrated how reconstruction performance increases with the number of input local phase images at test time; even while the LP-CNN network was trained with a single phase measurement. Furthermore, in order to address tissue magnetic susceptibility anisotropy [10], similar tools could be also extended to more complex, higher order, susceptibility models such as in Susceptibility Tensor Imaging (STI) [21,23].

Acknowledgement. The authors would like to thank Mr. Joseph Gillen, Ms. Terri Brawner, Ms. Kathleen Kahl and Ms. Ivana Kusevic for their assistance with data acquisition. This work was partly supported by NCRR and NIBIB (P41 EB015909).

References

1. Abdul-Rahman, H., et al.: Fast three-dimensional phase-unwrapping algorithm based on sorting by reliability following a non-continuous path. In: Optical Measurement Systems for Industrial Inspection IV, vol. 5856, pp. 32–40. International Society for Optics and Photonics (2005)
2. Acosta-Cabronero, J., et al.: In vivo quantitative susceptibility mapping (QSM) in Alzheimer's disease. PloS One **8**(11), e81093 (2013)
3. Adler, J., Öktem, O.: Learned primal-dual reconstruction. IEEE Trans. Med. Imaging **37**(6), 1322–1332 (2018)
4. Ayton, S., et al.: Cerebral quantitative susceptibility mapping predicts amyloid-β-related cognitive decline. Brain **140**(8), 2112–2119 (2017)
5. Bollmann, S., et al.: DeepQSM-using deep learning to solve the dipole inversion for quantitative susceptibility mapping. NeuroImage **195**, 373–383 (2019)
6. Bruckstein, A.M., et al.: From sparse solutions of systems of equations to sparse modeling of signals and images. SIAM Rev. **51**(1), 34–81 (2009)
7. Chen, W., et al.: Quantitative susceptibility mapping of multiple sclerosis lesions at various ages. Radiology **271**(1), 183–192 (2014)
8. Chen, Y., et al.: QSMGAN: improved quantitative susceptibility mapping using 3D generative adversarial networks with increased receptive field. NeuroImage **207**, 116389 (2020)
9. De Rochefort, L., et al.: Quantitative MR susceptibility mapping using piece-wise constant regularized inversion of the magnetic field. Magn. Reson. Med. Off. J. Int. Soc. Magn. Reson. Med. **60**(4), 1003–1009 (2008)
10. Deistung, A., et al.: Overview of quantitative susceptibility mapping. NMR Biomed. **30**(4), e3569 (2017)
11. Goodfellow, I., et al.: Generative adversarial nets. In: Advances in Neural Information Processing Systems, pp. 2672–2680 (2014)
12. Hammernik, K., et al.: Learning a variational network for reconstruction of accelerated MRI data. Magn. Reson. Med. **79**(6), 3055–3071 (2018)

13. Jenkinson, M., Smith, S.: A global optimisation method for robust affine registration of brain images. Med. Image Anal. **5**(2), 143–156 (2001)
14. Jung, W., Bollmann, S., Lee, J.: Overview of quantitative susceptibility mapping using deep learning: current status, challenges and opportunities. NMR Biomed. e4292 (2020). https://onlinelibrary.wiley.com/doi/abs/10.1002/nbm.4292
15. Jung, W., et al.: Exploring linearity of deep neural network trained QSM: QSM-net+. NeuroImage **211**, 116619 (2020)
16. Kames, C., et al.: Proximal variational networks: generalizable deep networks for solving the dipole-inversion problem. In: 5th International QSM Workshop (2019)
17. Kingma, D.P., Ba, J.: Adam: a method for stochastic optimization. arXiv preprint arXiv:1412.6980 (2014)
18. Langkammer, C., et al.: Quantitative susceptibility mapping in multiple sclerosis. Radiology **267**(2), 551–559 (2013)
19. Langkammer, C., et al.: Quantitative susceptibility mapping: report from the 2016 reconstruction challenge. Magn. Reson. Med. **79**(3), 1661–1673 (2018)
20. Li, W., Wu, B., Liu, C.: 5223 STI Suite: a software package for quantitative susceptibility imaging (2013)
21. Li, W., et al.: Susceptibility tensor imaging (STI) of the brain. NMR Biomed. **30**(4), e3540 (2017)
22. Li, X., et al.: Magnetic susceptibility contrast variations in multiple sclerosis lesions. J. Magn. Reson. Imaging **43**(2), 463–473 (2016)
23. Liu, C.: Susceptibility tensor imaging. Magn. Reson. Med. Off. J. Int. Soc. Magn. Reson. Med. **63**(6), 1471–1477 (2010)
24. Liu, C., et al.: Quantitative susceptibility mapping: contrast mechanisms and clinical applications. Tomography **1**(1), 3 (2015)
25. Liu, C., et al.: Susceptibility-weighted imaging and quantitative susceptibility mapping in the brain. J. Magn. Reson. Imaging **42**(1), 23–41 (2015)
26. Liu, T., et al.: Calculation of susceptibility through multiple orientation sampling (COSMOS): a method for conditioning the inverse problem from measured magnetic field map to susceptibility source image in MRI. Magn. Reson. Med. Off. J. Int. Soc. Magn. Reson. Med. **61**(1), 196–204 (2009)
27. Liu, T., et al.: Morphology enabled dipole inversion (MEDI) from a single-angle acquisition: comparison with cosmos in human brain imaging. Magn. Reson. Med. **66**(3), 777–783 (2011)
28. Liu, T., et al.: Nonlinear formulation of the magnetic field to source relationship for robust quantitative susceptibility mapping. Magn. Reson. Med. **69**(2), 467–476 (2013)
29. Mardani, M., et al.: Deep generative adversarial neural networks for compressive sensing MRI. IEEE Trans. Med. Imaging **38**(1), 167–179 (2018)
30. Osher, S., et al.: An iterative regularization method for total variation-based image restoration. Multiscale Model. Simul. **4**(2), 460–489 (2005)
31. Özbay, P.S., et al.: A comprehensive numerical analysis of background phase correction with v-sharp. NMR Biomed. **30**(4), e3550 (2017)
32. Parikh, N., et al.: Proximal algorithms. Found. Trends® Optim. **1**(3), 127–239 (2014)
33. Polak, D., et al.: Nonlinear dipole inversion (NDI) enables robust quantitative susceptibility mapping (QSM). NMR Biomed. (2020). https://doi.org/10.1002/nbm.4271

34. Ronneberger, O., Fischer, P., Brox, T.: U-Net: convolutional networks for biomedical image segmentation. In: Navab, N., Hornegger, J., Wells, W.M., Frangi, A.F. (eds.) MICCAI 2015. LNCS, vol. 9351, pp. 234–241. Springer, Cham (2015). https://doi.org/10.1007/978-3-319-24574-4_28

35. Shmueli, K., et al.: Magnetic susceptibility mapping of brain tissue in vivo using MRI phase data. Magn. Reson. Med. Off. J. Int. Soc. Magn. Reson. Med. **62**(6), 1510–1522 (2009)

36. Smith, S.M.: Fast robust automated brain extraction. Hum. Brain Mapp. **17**(3), 143–155 (2002)

37. Sulam, J., et al.: On multi-layer basis pursuit, efficient algorithms and convolutional neural networks. IEEE Trans. Pattern Anal. Mach. Intell. **42**, 1968–1980 (2019)

38. Van Bergen, J., et al.: Colocalization of cerebral iron with amyloid beta in mild cognitive impairment. Sci. Rep. **6**(1), 1–9 (2016)

39. Wang, Y., Liu, T.: Quantitative susceptibility mapping (QSM): decoding MRI data for a tissue magnetic biomarker. Magn. Reson. Med. **73**(1), 82–101 (2015)

40. Wei, H., et al.: Streaking artifact reduction for quantitative susceptibility mapping of sources with large dynamic range. NMR Biomed. **28**(10), 1294–1303 (2015)

41. Yoon, J., et al.: Quantitative susceptibility mapping using deep neural network: QSMnet. NeuroImage **179**, 199–206 (2018)

42. Zagoruyko, S., Komodakis, N.: Wide residual networks. arXiv preprint arXiv:1605.07146 (2016)

Learning a Gradient Guidance
for Spatially Isotropic MRI
Super-Resolution Reconstruction

Yao Sui[1,2]($\boxtimes$), Onur Afacan[1,2], Ali Gholipour[1,2], and Simon K. Warfield[1,2]

[1] Harvard Medical School, Boston, MA, USA
{yao.sui,onur.afacan,ali.gholipour,simon.warfield}@childrens.harvard.edu
[2] Boston Children's Hospital, Boston, MA, USA

Abstract. In MRI practice, it is inevitable to appropriately balance between image resolution, signal-to-noise ratio (SNR), and scan time. It has been shown that super-resolution reconstruction (SRR) is effective to achieve such a balance, and has obtained better results than direct high-resolution (HR) acquisition, for certain contrasts and sequences. The focus of this work was on constructing images with spatial resolution higher than can be practically obtained by direct Fourier encoding. A novel learning approach was developed, which was able to provide an estimate of the spatial gradient prior from the low-resolution (LR) inputs for the HR reconstruction. By incorporating the anisotropic acquisition schemes, the learning model was trained over the LR images themselves only. The learned gradients were integrated as prior knowledge into a gradient-guided SRR model. A closed-form solution to the SRR model was developed to obtain the HR reconstruction. Our approach was assessed on the simulated data as well as the data acquired on a Siemens 3T MRI scanner containing 45 MRI scans from 15 subjects. The experimental results demonstrated that our approach led to superior SRR over state-of-the-art methods, and obtained better images at lower or the same cost in scan time than direct HR acquisition.

Keywords: MRI · Super-resolution · Deep neural networks

This work was supported in part by the National Institutes of Health (NIH) under grants R01 NS079788, R01 EB019483, R01 EB018988, R01 NS106030, IDDRC U54 HD090255; by a research grant from the Boston Children's Hospital Translational Research Program; by a Technological Innovations in Neuroscience Award from the McKnight Foundation; by a research grant from the Thrasher Research Fund; and by a pilot grant from National Multiple Sclerosis Society under Award Number PP-1905-34002.

© Springer Nature Switzerland AG 2020
A. L. Martel et al. (Eds.): MICCAI 2020, LNCS 12262, pp. 136–146, 2020.
https://doi.org/10.1007/978-3-030-59713-9_14

1   Introduction

It is inevitable to deal with the trade-off between image resolution, signal-to-noise ratio (SNR), and scan time, in magnetic resonance imaging (MRI) practice [1]. Images of higher resolution provide more details, and correspondingly increase the scan time. Higher SNR renders the signals of interest better from noise contamination. According to MRI physics, SNR is proportional to slice thickness and scan time. In pursuit of higher SNR, however, thick slices lower image resolution, while long scans discomfort subjects and potentially lead to subject motion during the acquisition, which adversely affects image quality. Literature has shown many effective methods to acquire images of high resolution, high SNR, and low scan time, such as parallel imaging [15] and robust K-space sampling [12]. Among these methods, super-resolution reconstruction (SRR) has recently been demonstrated to be capable of obtaining better images than direct high-resolution (HR) acquisition, for certain image contrasts and sequences, such as T_2 weighted images [13]. As a result, SRR has become one of the most widely used methods that acquire MR images of high quality.

The idea of SRR originated in [19] for natural images where multiple low-resolution (LR) images were combined into an HR reconstruction. Extensive methods expanded this idea and achieved various SRR schemes in MRI [14,16]. Recently, deep learning-based methods have achieved impressive results for 2D natural images [5]. These 2D learning models were directly applied to 3D MRI volumes by means of slice-by-slice processing [8,23]. However, by the nature that MRI volumes reveal anatomical structures in 3D space, it is straightforward to learn 3D models for capturing richer knowledge. It has been shown that 3D models outperformed their 2D counterparts in MRI [3,11]. To this end, extensive 3D learning models were developed [2–4,11,20]. Unfortunately, these 3D models required large-scale training datasets, that contain HR labels that are practically difficult to obtain at excellent SNR due to subject motion, to learn the patterns mapping from LR to HR images. Although few datasets are publicly available, there is no known theory that indicates a satisfactory amount of training data has been incorporated in the training process. Furthermore, since the LR inputs were manually generated as blurred images, the learned function may be brittle when faced with data from a different scanner or with different intensity properties.

In this work, we aimed at developing the learning model that is not subject to the above limitations. Instead of learning end-to-end mappings from LR to HR images, we turned to learn the inherent structure of the HR reconstruction, and then as prior knowledge the learned structure was incorporated in a forward model-based SRR framework. To this end, we targeted the spatial image gradients in the learning, since they can be equivalently decomposed from 3D space onto 2D space, and correspondingly the required amount of training data was substantially reduced. More importantly, by incorporating the anisotropic acquisition schemes, our learning model was trained over the LR images themselves only, as the training data consisted of the pairs of an in-plane HR slice of an LR image and a through-plane LR slice of another LR image. A gradient-guided

SRR framework [17] was leveraged to incorporate the learned gradients for the HR reconstruction.

The SRR method [17] used the gradients of LR images as HR gradient estimates. An LR image containing n_s slices can provide n_s accurate HR gradients for the HR reconstruction containing N_s slices, as the n_s slices are in-plane HR. Since the upscale factor N_s/n_s is often 4 or 5 in practice, 75–80% gradients in [17] had to be estimated from interpolated slices. The interpolation led to blurry or displaced image edges, and thus resulted in less accurate gradient localization. In this work, we aimed at improving the localization accuracy of those 75–80% gradients. We used a CNN to learn over the n_s in-plane HR slices the patterns mapping from LR slices to HR gradients, enabled by a strategy that is able to decompose a 3D gradient onto 2D space. The learned patterns were used to infer the HR gradients over the rest $N_s - n_s$ slices. As high frequencies were gained in the inference, improved gradient localization was achieved, compared to [17], and thus increased the quality of deblurring done in the SRR. In addition, a closed-form solution to the SRR model was developed, instead of an iterative optimization algorithm used in [17] that may be stuck in local optima.

2   Methods

2.1   Gradient-Guided Framework

Given n LR images $\{\mathcal{Y}_k\}_{k=1}^{n}$, acquired from n scans with arbitrary orientations and displacements, the forward model over the HR image $\mathcal{X}$, which describes the MRI acquisition process, is formulated as

$$\mathbf{y}_k = \mathbf{D}_k \mathbf{H}_k \mathbf{T}_k \mathbf{x} + \boldsymbol{\mu}_k, \tag{1}$$

where $\mathbf{x}$ and $\mathbf{y}_k$ are the column vectors of their volumetric forms $\mathcal{X}$ and $\mathcal{Y}_k$ in a lexicographical order of voxels; $\mathbf{T}_k$ denotes a rigid transform that characterizes subject motion between scans; $\mathbf{H}_k$ denotes a point spread function (PSF); $\mathbf{D}_k$ denotes a downsampling operator; $\boldsymbol{\mu}_k$ denotes the noise in the acquisition. According to [7], $\boldsymbol{\mu}_k$ can be assumed to be additive and Gaussian when SNR > 3. As the problem defined by the forward model is ill-posed, prior knowledge is commonly used, known as the regularization, to restrict the HR estimate. By incorporating the gradient guidance regularization [17], the SRR is achieved by

$$\min_{\mathbf{x}} \sum_{k=1}^{n} \|\mathbf{D}_k \mathbf{H}_k \mathbf{T}_k \mathbf{x} - \mathbf{y}_k\|_2^2 + \lambda \sigma \left(\nabla \mathbf{x} - \mathbf{g} \right), \tag{2}$$

where $\mathbf{g}$ denotes the gradient guidance for the HR estimate $\mathbf{x}$, ∇ computes the spatial gradient, $\sigma \left(\cdot \right)$ penalizes the difference between the actual gradient and the guidance, and $\lambda > 0$ balances between the data fidelity and the regularization.

2.2   Gradient Guidance Learning

As shown in Eq. (2), the accuracy of the gradient guidance estimate $\mathbf{g}$ is critically important to the quality of the HR reconstruction $\mathbf{x}$. Thanks to the fact that a 3D gradient can be decomposed onto 2D space, we are able to compute the gradient of a volume slice by slice. Specifically, the 3D gradient $\nabla_d^s \mathbf{x}$ at a scale s in direction $d \in \{x, y, z, -x, -y, -z\}$ is computed from

$$\nabla_d^s \mathbf{x} = (\mathbf{I} - \mathbf{\Psi}_d^s)\, \mathbf{x} = \prod_{k=1}^{N} \left(\mathbf{I} - \mathbf{S}_d^{s,k}\right) \mathbf{x}, \tag{3}$$

where $\mathbf{I}$ denotes an identity matrix, $\mathbf{\Psi}_d^s$ shifts a volume circularly by s voxels in direction d, and $\mathbf{S}_d^{s,k}$ shifts the k-th slice circularly by s pixels in direction d.

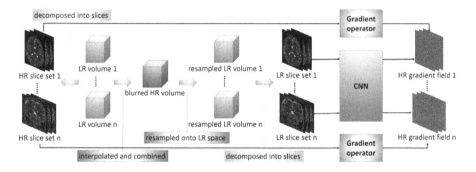

Fig. 1. Overview of our proposed approach to learning the gradient guidance.

Architecture of the Learning Model. By the above decomposition, the gradient guidance is learned over slices by a convolutional neural network (CNN). The overview of our proposed learning model is shown in Fig. 1. The CNN takes as input 2D LR image patches, and outputs the 2D gradients of HR image patches. ℓ_2-loss is used to minimize the difference between the gradient of k-th HR patch $\mathbf{p}_H^k$ and the CNN's output for the k-th inputted LR patch $\mathbf{p}_L^k$:

$$\min_{\theta} \sum_{k} \left\| \nabla \mathbf{p}_H^k - f_\theta\left(\mathbf{p}_L^k\right) \right\|_2^2, \tag{4}$$

where $f_\theta\left(\cdot\right)$ is the function defined by the CNN, and θ denotes the hyperparameters. The CNN has 3 layers: a layer with 64 filters of size $9 \times 9 \times 1$; a layer with 32 filters of size $1 \times 1 \times 64$; and a layer with 2 filters of size $5 \times 5 \times 32$. A ReLU is imposed on the outputs of the first 2 layers. Since a slice contains 2D information, the CNN outputs two gradients in two in-plane perpendicular directions.

Pairing LR Inputs and HR Labels for Training Data. The LR images are of in-plane HR, through-plane LR, and different orientations. Therefore, the gradients of the in-plane HR slices are directly used as the HR labels. The corresponding LR inputs are found in the following steps. First, all the LR images are aligned and interpolated by a third-order B-spline method to the size and resolution the same as the HR reconstruction. These interpolated LR images are combined into a mean image by averaging them out over their voxels. This mean image is then resampled according to the geometric properties of each original LR image, respectively. The slices of the resampled images are HR (small voxel size) but blurry due to the averaging, and are used as the LR inputs. The training dataset is thus constructed from the pairs of the LR inputs and the HR labels.

Training. The LR-HR pairs in the training are of size 33×33 pixels. By adjusting the strides, about 20,000 data pairs are sampled uniformly with overlap from the slices to form the training dataset. The CNN is trained for 2000 epochs on an NVIDIA Quadro P5000 GPU with TensorFlow. Stochastic gradient descent with a learning rate of $1e^{-4}$ is used to optimize the hyperparameters $\boldsymbol{\theta}$ in Eq. (4).

2.3 Super-Resolution Reconstruction

In the SRR, the CNN is first trained over the LR-HR pairs. The mean image is calculated following the steps mentioned above. The slices from each plane of the mean image are inputted to the trained CNN. The outputted gradients that are in the same direction are averaged into a single gradient. All the gradients form a gradient guidance $\mathbf{g}$ for the SRR.

As the gradient guidance is highly accurate due to the learning, a squared error is incorporated in the regularization. Therefore, the SRR is achieved by

$$\min_{\mathbf{x}} \sum_{k=1}^{n} \|\mathbf{D}_k \mathbf{H}_k \mathbf{T}_k \mathbf{x} - \mathbf{y}_k\|_2^2 + \lambda \sum_{\mathbf{g} \in \mathcal{G}} \|\nabla_{\mathbf{g}} \mathbf{x} - \mathbf{g}\|_2^2, \tag{5}$$

where all the gradients at different scales in different directions are organized in a set $\mathcal{G}$, and $\nabla_{\mathbf{g}}$ computes the gradient at the same scale and in the same direction as $\mathbf{g}$. $\mathbf{H}_k$ is designed as a sinc function in the slice selection direction of $\mathbf{y}_k$ (corresponding a boxcar in Fourier domain) convolved with a Gaussian kernel. In this work, the gradient guidance is computed at a scale of 1 in 3 directions of x, y, and z. λ is fixed at 0.5 according to the empirical results.

A closed-form solution to Eq. (5) is developed to obtain the HR reconstruction $\mathbf{x}$ from

$$\widehat{\mathbf{x}} = \frac{\sum_{k=1}^{n} \widehat{\mathbf{H}}_k^* \circ \widehat{\widetilde{\mathbf{y}}}_k + \lambda \sum_{\mathbf{g} \in \mathcal{G}} \widehat{\nabla}_{\mathbf{g}}^* \circ \widehat{\mathbf{g}}}{\sum_{k=1}^{n} \widehat{\mathbf{H}}_k^* \circ \widehat{\mathbf{H}}_k + \lambda \sum_{\mathbf{g} \in \mathcal{G}} \widehat{\nabla}_{\mathbf{g}}^* \circ \widehat{\nabla}_{\mathbf{g}}}, \tag{6}$$

where $\widetilde{\mathbf{y}}_k = \mathbf{T}_k^T \mathbf{D}_k^T \mathbf{y}_k$ denotes the interpolated and aligned form of the LR image $\mathbf{y}_k$, $\widehat{\cdot}$ denotes 3D discrete Fourier transform, $\cdot^*$ computes complex conjugate, $\circ$ denotes Hardamard product, and the fraction denotes element-wise division.

2.4   Experimental Setup

Simulated Data. We simulated a dataset by using the MPRAGE data from the Dryad package [10]. We downsampled the original image of isotropic 0.25 mm to the image of isotropic 0.5 mm, and used it as the ground truth. We then downsampled the ground truth by factors of $\{2, 3, 4, 5, 6, 8\}$ in the directions of x, y, and z, respectively, to form an axial, a coronal, and a sagittal LR image at each factor. Simulated motion with maximum magnitudes of 10 mm for translation and $10°$ for rotation was randomly generated and applied to each LR image. Gaussian noise was also randomly generated with a power of 10% of mean voxel intensities and added to each LR image. We assessed the quality of the HR reconstruction on this dataset in terms of peak signal-to-noise ratio (PSNR), structural similarity (SSIM) [21], and root square mean error (RMSE).

Real Data. We acquired 45 MRI scans from 15 subjects with various acquisition schemes on a Siemens 3T MRI scanner. All scans were performed in accordance with the local institutional review board protocol.

Origin-Shifted Axial LR Acquisitions (OSA). We acquired two axial LR T_2 weighted images with their origins shifted by a distance of half slice thickness. The in-plane resolution is 1 mm $\times$ 1 mm and the slice thickness is 2 mm. We also acquired an HR image of isotropic 1 mm as the ground truth. The HR reconstruction was evaluated in terms of PSNR, SSIM, and sharpness [6].

Axial and Coronal LR Acquisitions (AC). We acquired 42 images from 14 subjects. With each subject, an axial and a coronal LR T_2-TSE image were acquired with in-plane resolution of 0.4 mm $\times$ 0.4 mm and slice thickness of 2 mm. It took about 2 min in acquiring a T_2-TSE image. A 3D T_2-SPACE image of isotropic 0.9 mm was acquired as the reference. The HR reconstruction was of isotropic 0.4 mm. Since the T_2-TSE has different contrast from the 3D T_2-SPACE image, we evaluated the HR reconstruction in terms of the blind assessment metrics that focus on image contrast and sharpness, including average edge strength (AES) [22], gradient entropy (GRENT) [9], and diagonal XSML (LAPD) [18]. As the HR reconstructions were assessed blindly on this dataset, we reported the normalized metrics that were computed from $\rho(\mathbf{x}) / \rho(\mathbf{x}')$, where $\rho(\cdot)$ denotes a metric function, $\mathbf{x}$ denotes a reconstruction, and $\mathbf{x}'$ denotes a reference image. Higher metric values produce better image quality for all metrics. We also investigated the accuracy of the gradient guidance estimate in terms of edge localization error that was computed from $\|\mathbf{g} - \mathbf{g}'\|_1 / \|\mathbf{g}'\|_1$, where $\mathbf{g}$ is the gradient guidance, and $\mathbf{g}'$ is the gradients computed from the reference image.

Baseline Methods. We employed a CNN-based learning model as a baseline, which is similar to [8], trained on the same data, with the same network architecture as our model except that the output is the HR slice. The HR reconstruction was obtained from the fusion over the HR slice stacks estimated in different planes. We denoted this baseline by deepSlice. We compared our app-roach to deepSlice to demonstrate that learning a gradient guidance is superior to learning an end-to-end mapping for HR slices in volumetric SRR. We also compared our approach to the plain gradient guidance regularization method [17], denoted by GGR, to demonstrate that our approach led to a more accurate gradient guidance estimate, and in turn resulted in better HR reconstruction. We denoted our approach by deepGG, that stands for deep gradient guidance.

3   Results

We compared the accuracy of the gradient guidance estimates obtained from GGR and deepGG on the AC dataset in terms of edge localization errors. As shown in Fig. 2(a), deepGG yielded much smaller errors than GGR on all 14 HR reconstructions. The representative slices of the learned gradient guidance obtained from deepGG are shown in Fig. 2(b).

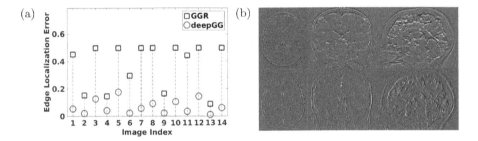

Fig. 2. Evaluation results of the gradient guidance estimation on the AC dataset. (a) Comparisons between GGR and deepGG in the edge localization errors. (b) Representative slices of the learned gradient guidance obtained from deepGG.

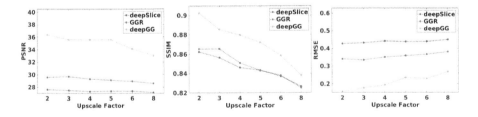

Fig. 3. Evaluation results obtained from the three methods on the simulated dataset.

Figure 3 shows the evaluation results of deepSlice, GGR, and deepGG on the simulated dataset. It is evident that deepGG consistently outperformed the two baselines by large margins at all upscale factors in terms of all metrics.

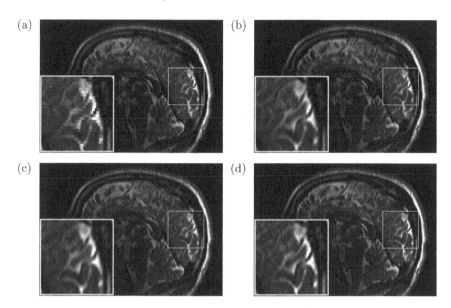

Fig. 4. Representative slices of (a) an LR image and the reconstructed images obtained from (b) deepSlice, (c) GGR, and (d) deepGG (ours), respectively, on the OSA dataset.

On the OSA dataset, the quantitative evaluation results were in terms of 1) PSNR: deepSlice = 35.01, GGR = 34.45, deepGG = 35.71; 2) SSIM: deepSlice = 0.912, GGR = 0.910, deepGG = 0.921; and 3) sharpness: deepSlice = 0.75, GGR = 0.78, deepGG = 0.86. Figure 4 shows the representative slices of an LR image and the reconstructed images obtained from deepSlice, GGR, and deepGG. The deepGG method achieved the sharpest images and the best contrast, which was consistent with the quantitative assessments, in particular with the sharpness metric.

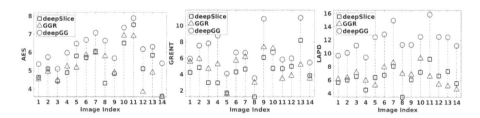

Fig. 5. Evaluation results on the AC datasets in terms of average edge strength (AES), gradient entropy (GRENT), and diagonal XSML (LAPD), respectively.

Figure 5 plots the evaluation results of the three methods on the AC dataset. It is evident that deepGG considerably outperformed the two baselines. Figure 6

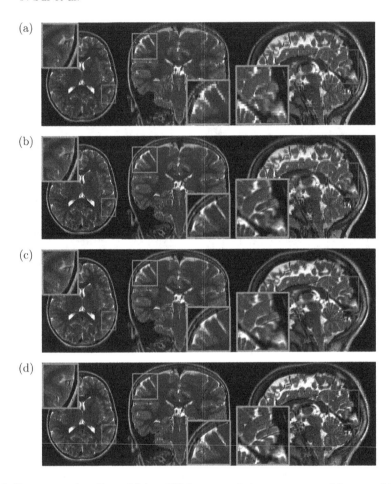

Fig. 6. Representative slices of (a) an LR image, and the reconstructed images obtained from (b) deepSlice, (c) GGR, and (d) deepGG (ours), respectively, on the AC dataset.

shows the representative slices of an LR image, and the HR images reconstructed by the three methods. It is shown that deepGG achieved more image details than the two baselines, and obtained the best contrast and the sharpest images.

4   Conclusions

We have developed a learning-based method to estimate the gradient guidance prior for MRI SRR. The learning tasks have been decomposed from 3D volumes onto 2D slices. The training dataset was constructed by exploiting the correspondences of in-plane HR and through-plane LR slices when incorporating the anisotropic acquisition schemes. Therefore, the learning model has been trained over the LR images themselves only. The learned gradient guidance, as prior knowledge, has been integrated into a forward model-based SRR framework. A closed-form solution to the SRR model, which is globally optimal, has been developed for obtaining the HR reconstruction. Extensive experimental results on both simulated and real data have demonstrated that our approach led to superior SRR over state-of-the-art methods, and obtained better images at lower or the same cost in scan time than direct HR acquisition.

References

1. Brown, R.W., Cheng, Y.C.N., Haacke, E.M., Thompson, M.R., Venkatesan, R.: Magnetic Resonance Imaging: Physical Principles and Sequence Design, 2nd edn. Wiley, Hoboken (2014)
2. Chaudhari, A., et al.: Super-resolution musculoskeletal MRI using deep learning. Magn. Reson. Med. **80**(5), 2139–2154 (2018)
3. Chen, Y., Xie, Y., Zhou, Z., Shi, F., Christodoulou, A.G., Li, D.: Brain MRI super resolution using 3D deep densely connected neural networks. In: International Symposium on Biomedical Imaging, pp. 739–742 (2018)
4. Chen, Y., Shi, F., Christodoulou, A.G., Xie, Y., Zhou, Z., Li, D.: Efficient and accurate MRI super-resolution using a generative adversarial network and 3D multi-level densely connected network. In: Frangi, A.F., Schnabel, J.A., Davatzikos, C., Alberola-López, C., Fichtinger, G. (eds.) MICCAI 2018. LNCS, vol. 11070, pp. 91–99. Springer, Cham (2018). https://doi.org/10.1007/978-3-030-00928-1_11
5. Dong, C., Loy, C.C., He, K., Tang, X.: Image super-resolution using deep convolutional networks. IEEE Trans. Pattern Anal. Mach. Intell. **38**(2), 295–307 (2016)
6. Gholipour, A., Estroff, J.A., Warfield, S.K.: Robust super-resolution volume reconstruction from slice acquisitions: application to fetal brain MRI. IEEE Trans. Med. Imaging **29**(10), 1739–1758 (2010)
7. Gudbjartsson, H., Patz, S.: The Rician distribution of noisy MRI data. Magn. Reson. Med. **34**, 910–914 (1995)
8. Jurek, J., Kocinski, M., Materka, A., Elgalal, M., Majos, A.: CNN-based superresolution reconstruction of 3D MR images using thick-slice scans. Biocybern. Biomed. Eng. **40**(1), 111–125 (2020)
9. Loktyushin, A., Nickisch, H., Pohmann, R., Schölkopf, B.: Blind multirigid retrospective motion correction of MR images. Magn. Reson. Med. **73**(4), 1457–1468 (2015)
10. Lusebrink, F., Sciarra, A., Mattern, H., Yakupov, R., Speck, O.: T1-weighted in vivo human whole brain MRI dataset with an ultrahigh isotropic resolution of 250 μm. Sci. Data **4**, 170032 (2017)
11. Pham, C., Ducournau, A., Fablet, R., Rousseau, F.: Brain MRI super-resolution using deep 3D convolutional networks. In: International Symposium on Biomedical Imaging, pp. 197–200 (2017)
12. Pipe, J.: Motion correction with PROPELLER MRI: application to head motion and free-breathing cardiac imaging. Magn. Reson. Med. **42**, 963–969 (1999)
13. Plenge, E., et al.: Super-resolution methods in MRI: can they improve the trade-off between resolution, signal-to-noise ratio, and acquisition time? Magn. Reson. Med. **68**, 1983–1993 (2012)
14. Poot, D.H.J., Van Meir, V., Sijbers, J.: General and efficient super-resolution method for multi-slice MRI. In: Jiang, T., Navab, N., Pluim, J.P.W., Viergever, M.A. (eds.) MICCAI 2010. LNCS, vol. 6361, pp. 615–622. Springer, Heidelberg (2010). https://doi.org/10.1007/978-3-642-15705-9_75

15. Pruessmann, K.P., Weiger, M., Scheidegger, M.B., Boesiger, P.: SENSE: sensitivity encoding for fast MRI. Magn. Reson. Med. **42**, 952–962 (1999)
16. Shilling, R.Z., Robbie, T.Q., Bailloeul, T., Mewes, K., Mersereau, R.M., Brummer, M.E.: A super-resolution framework for 3-D high-resolution and high-contrast imaging using 2-D multislice MRI. IEEE Trans. Med. Imaging **28**(5), 633–644 (2009)
17. Sui, Y., Afacan, O., Gholipour, A., Warfield, S.K.: Isotropic MRI super-resolution reconstruction with multi-scale gradient field prior. In: Shen, D., et al. (eds.) MICCAI 2019. LNCS, vol. 11766, pp. 3–11. Springer, Cham (2019). https://doi.org/10.1007/978-3-030-32248-9_1
18. Thelen, A., Frey, S., Hirsch, S., Hering, P.: Improvements in shape-from-focus for holographic reconstructions with regard to focus operators, neighborhood-size, and height value interpolation. IEEE Trans. Image Process. **18**(1), 151–157 (2009)
19. Tsai, R.Y., Huang, T.: Multi-frame image restoration and registration. In: Advances in Computer Vision and Image Processing (1984)
20. Wang, J., Chen, Y., Wu, Y., Shi, J., Gee, J.: Enhanced generative adversarial network for 3D brain MRI super-resolution. In: IEEE Winter Conference on Applications of Computer Vision, pp. 3627–3636 (2020)
21. Wang, Z., Bovik, A.C., Sheikh, H.R., Simoncelli, E.P.: Image quality assessment: from error measurement to structural similarity. IEEE Trans. Image Process. **13**(1), 600–612 (2004)
22. Zaca, D., Hasson, U., Minati, L., Jovicich, J.: A method for retrospective estimation of natural head movement during structural MRI. J. Magn. Reson. Imaging **48**(4), 927–937 (2018)
23. Zhao, C., Carass, A., Dewey, B.E., Prince, J.L.: Self super-resolution for magnetic resonance images using deep networks. In: International Symposium on Biomedical Imaging, pp. 365–368 (2018)

Encoding Metal Mask Projection for Metal Artifact Reduction in Computed Tomography

Yuanyuan Lyu[1], Wei-An Lin[2], Haofu Liao[3], Jingjing Lu[4,5], and S. Kevin Zhou[6,7(✉)]

[1] Z²Sky Technologies Inc., Suzhou, China
[2] Adobe, San Jose, CA, USA
[3] Department of Computer Science, University of Rochester, Rochester, NY, USA
[4] Department of Radiology, Beijing United Family Hospital, Beijing, China
[5] Peking Union Medical College Hospital, Beijing, China
[6] Institute of Computing Technology, Chinese Academy of Sciences, Beijing, China
s.kevin.zhou@gmail.com
[7] Peng Cheng Laboratory, Shenzhen, China

Abstract. Metal artifact reduction (MAR) in computed tomography (CT) is a notoriously challenging task because the artifacts are structured and non-local in the image domain. However, they are inherently local in the sinogram domain. Thus, one possible approach to MAR is to exploit the latter characteristic by learning to reduce artifacts in the sinogram. However, if we directly treat the metal-affected regions in sinogram as missing and replace them with the surrogate data generated by a neural network, the artifact-reduced CT images tend to be over-smoothed and distorted since fine-grained details within the metal-affected regions are completely ignored. In this work, we provide analytical investigation to the issue and propose to address the problem by (1) retaining the metal-affected regions in sinogram and (2) replacing the binarized metal trace with the metal mask projection such that the geometry information of metal implants is encoded. Extensive experiments on simulated datasets and expert evaluations on clinical images demonstrate that our novel network yields anatomically more precise artifact-reduced images than the state-of-the-art approaches, especially when metallic objects are large.

Keywords: Artifact reduction · Sinogram inpainting · Image enhancement

1 Introduction

Modern computed tomography (CT) systems are able to provide accurate images for diagnosis [9,15,16]. However, highly dense objects such as metallic implants

---

Electronic supplementary material The online version of this chapter (https://doi.org/10.1007/978-3-030-59713-9_15) contains supplementary material, which is available to authorized users.

© Springer Nature Switzerland AG 2020
A. L. Martel et al. (Eds.): MICCAI 2020, LNCS 12262, pp. 147–157, 2020.
https://doi.org/10.1007/978-3-030-59713-9_15

cause inaccurate sinogram data in projection domain, which leads to non-local streaking artifacts in image domain after reconstruction. The artifacts degrade the image quality of CT and its diagnostic value. The challenge of metal artifact reduction (MAR) aggravates *when metallic objects are large*.

Conventional MAR algorithms can be grouped into three categories: iterative reconstruction, image domain MAR and sinogram domain MAR. Iterative approaches are often time-consuming and require hand-crafted regularizers, which limit their practical impacts [1,4]. Image domain methods aim to directly estimate and then remove the streak artifacts from the original contaminated image by image processing techniques [6,11], but they achieve limited success in suppressing artifacts. Sinogram domain methods treat metal-affected regions in sinogram as missing and replace them by interpolation [5] or forward projection [9] but they would introduce streak artifacts tangent to the metallic objects, as the discontinuity in sinogram is hard to avoid.

Recently, convolutional neural networks (CNNs) has been applied to solve MAR based on sinogram completion [2,14] or image processing [12]. DuDoNet [8] been recently proposed to reduce the artifacts jointly in sinogram and image domains, which offers advantages over the single domain methods. Specifically, DuDoNet consists of two separate networks, one for sinogram enhancement (SE) and the other for image enhancement (IE). These two networks are connected by a Radon inversion layer (RIL) to allow gradient propagation during training.

However, there are still some limitations in DuDoNet [8]. First, in the SE network, a binarized metal trace map is used to indicate the presence of metal in the sinogram. We will theoretically show that such a binarized map is a rather crude representation that *totally discards* the details inside the metal mask projection. Second, in DuDoNet, the dual-domain enhancement is applied to linearly interpolated sinograms and the correspondingly reconstructed CTs. As linear interpolation only provides a rough estimate to the corrupted sinogram data, the artifact reduced images tend to be over-smoothed and severely distorted around regions with high-density materials, e.g. bones. Finally, the training data in DuDoNet are simulated by a limited number of projection angles and rays and consequently, metal artifact is compounded by strong under-sampling effect.

To address these problems of DuDoNet [8], we present a novel approach utilizing the realistic information in the original sinogram and image while clearly specifying the *metal mask projection*, whose importance is justified via our theoretical derivation. Furthermore, we introduce a padding scheme that is designed for sinogram and increase the number of projection angles and rays to mitigate the under-sampling effect. We boost the MAR performance of DuDoNet by a large margin (over 4dB) on a large-scale database of simulated images. The improvement is more evident when metallic objects are large. Expert evaluations confirm the efficacy of our model on clinical images too.

2   Problem Formulation

CT images represent spatial distribution of linear attenuation coefficients, which indicate the underlying anatomical structure within the human body. Let $X(E)$

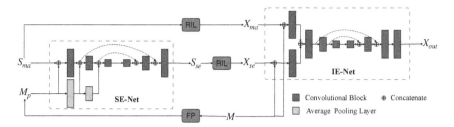

Fig. 1. The proposed network architecture.

denote the linear attenuation image at energy level E. According to Lambert-Beer's Law, the ideal projection data (sinogram) S detected by the CT scanner can be expressed as:

$$S = -ln \int \eta(E)e^{-\mathcal{P}(X(E))} dE, \tag{1}$$

where $\eta(E)$ represents fractional energy at E and $\mathcal{P}$ denotes a forward projection (FP) operator.

When metallic implants are present, $X(E)$ has large variations with respect to E because mass attenuation coefficient of metal $\lambda_m(E)$ varies rapidly against E:

$$X(E) = X_r + X_m(E) = X_r + \lambda_m(E)\rho_m M, \tag{2}$$

where $X_m(E)$ denotes the linear attenuation image of the metallic implants, X_r denotes the residual image without the implants and is almost constant with respect to E, ρ_m is the density of metal, and M denotes a metal mask. According the linearity of $\mathcal{P}$, the forward projection of $X_m(E)$ can be written as:

$$\mathcal{P}(X_m(E)) = \lambda_m(E)\rho_m\mathcal{P}(M) = \lambda_m(E)\rho_m M_p, \tag{3}$$

where $M_p = \mathcal{P}(M)$ is the *metal mask projection*. Substituting (3) into (1) yields

$$S_{ma} = \mathcal{P}(X_r) - ln \int \eta(E)e^{-\lambda_m(E)\rho_m M_p} dE. \tag{4}$$

Here, the first term $\mathcal{P}(X_r)$ is the projection data originated from X_r. The second term brings metal artifacts. Sinogram domain MAR algorithms aim to restore a clean counterpart S^* (ideally $S^* = \mathcal{P}(X_r)$) from the contaminated sinogram S_{ma}. Then, an artifact-reduced image X^* could be inferred by the filtered back projection (FBP) algorithm, that is, $X^* = \mathcal{P}^{-1}(S^*)$.

3　Network Architecture

Following [8], we use a sinogram enhancement network (SE-Net) and an image enhancement network (IE-Net) to jointly restore a clean image. Figure 1 shows the architecture of our proposed network.

SE-Net. To restore a clean sinogram from S_{ma}, conventional methods remove the second term in (4) through inpainting. Following this concept, DuDoNet takes linearly interpolated sinogram S_{LI} and binarized metal trace M_t as inputs for sinogram domain enhancement, where $M_t = \delta[M_p > 0]$ ($\delta[true] = 1$, $\delta[false] = 0$). Here, we observe that the second term in (4) is actually a function of M_p. Therefore, we propose to directly *utilize the knowledge of metal mask projection* M_p. As shown in Fig. 1, our SE-Net uses a pyramid U-Net architecture ϕ_{SE} [7], which takes both X_{ma} and M_p as inputs. To retain the projection information, M_p goes through average pooling layers and then fuse with multi-scale feature maps. As metals only affect part of the sinogram data of the corresponding projection pathway, SE-Net learns to correct sinogram data within the metal trace and outputs the enhanced sinogram S_{se}. Sinogram enhanced image X_{se} is reconstructed by the differentiable RIL first introduced in [8], that is, $X_{se} = \mathcal{P}^{-1}(S_{se})$.

Sinogram data is inherently periodic along the projection direction, while DuDoNet uses zero padding for convolutions in SE-Net which ignores the periodic information. Here, to offer more useful information for convolution, we propose a new padding strategy for sinogram data using periodic padding along the direction of projection angles and zero padding along the direction of detectors, as shown in Fig. 6.

IE-Net. To suppress the secondary artifacts in X_{se}, we apply an image enhancement net, which refines X_{se} with M and X_{ma}. The network contains two initial convolutional layers, a U-net [10] and a final convolutional layer. To pay attention to the strongly distorted regions, we concatenate an image (X_{se} or X_{ma}) with metal mask M separately and obtain mask-aware feature maps by an initial convolutional layer with 64 3×3 kernels. The two sets of mask-aware feature maps are concatenated as the input for the subsequent U-Net. A U-Net of depth 4 is used which outputs a feature map with 64 channels. Finally, a convolutional layer is used as the output layer which generates the enhanced image X_{out}.

Learning. The total loss of our model consists of sinogam enhancement loss, image enhancement loss and Radon consistency loss [8]:

$$\mathcal{L}_{total} = \alpha_{se}||S_{se} - S_{gt}||_1 + (\alpha_{rc}||X_{se} - X_{gt}||_1 + \alpha_{ie}||X_{out} - X_{gt}||_1) \odot (1 - M), \quad (5)$$

where α_{se}, α_{rc}, and α_{ie} are blending weights. We empirically set them to 1.

4　Experiment

4.1　Dataset and Experimental Setup

Simulation Data. We generate 360,000 cases for training and 2,000 cases for testing based on clean CT images. We first resize CT images to a size of

416×416 and use 640 projection angles and 641 rays for imaging geometry to simulate realistic metal artifacts (details are presented in Fig. 2).

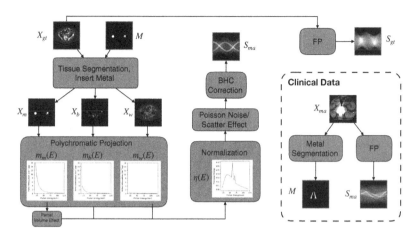

Fig. 2. Flowchart of metal artifact simulation and data generation of clinical images. Images with red borders are the inputs of our model and images with green borders are ground truth. (Color figure online)

Clinical Data. We evaluate the proposed method using two clinical datasets. We refer them to DL and CL. DL represents the DeepLesion dataset [13] and CL is a clinical CT scan for a patient with metal rods and screws after spinal fusion. We randomly select 30 slices from DL and 10 slices from CL with *more than 100 pixels above 3,000 HU and moderate or severe metal artifacts*. The clinical images are resized and processed with the same geometry as the simulation data (see Fig. 2).

Implementation and Training Details. Our model is implemented using the PyTorch framework. We use the Adam optimizer with $(\beta_1, \beta_2) = (0.5, 0.999)$ to train the model. The learning rate starts from 0.0002 and is halved for every 30 epochs. The model is trained on an Nvidia 2080Ti GPU card for 201 epochs with a batch size of 2.

Metrics. We use peak signal-to-noise ratio (PSNR) and structural similarity index (SSIM) to evaluate the corrected image with a soft tissue window in the range of $[-175, +275]$ HU. To evaluate the sinogram restoration performance, we use mean square error (MSE) to compare the enhanced S_{se} with S_{gt}. We group results according to the size of metal implants to investigate the MAR performance.

Rating. A proficient radiologist with about 20 years of reading experience is invited to rate the image quality for each group of the corrected images by paying close attention to ameliorating beam hardening, removing primary streaky

Table 1. Quantitative evaluation (PSNR(dB)/SSIM%/MSE) for different models.

	Large metal		$\rightarrow$	Small metal		Average
X_{ma}	19.42/81.1/1.1e+1	23.07/85.4/7.3e+0	26.12/88.7/2.2e+0	26.60/89.3/1.7e+0	27.69/89.9/3.8e−1	24.58/86.9/4.5e+0
IE-Net	31.19/94.8/ n.a.	30.33/95.9/ n.a.	34.48/96.8/ n.a.	35.52/96.8/ n.a.	36.37/97.0/ n.a.	33.58/96.3/ n.a.
SE$_0$-Net	20.28/86.5/3.0e+0	21.65/89.6/1.6e+0	26.39/91.7/3.0e−2	25.65/91.3/6.4e−2	24.93/91.1/8.4e−2	23.78/90.0/9.5e−1
SE-Net	26.71/91.0/2.7e−3	27.93/92.6/4.3e−4	28.20/93.2/2.4e−4	28.31/93.2/1.8e−4	28.34/93.3/**1.4e−4**	27.90/92.7/7.4e−4
SE$_p$-Net	26.86/91.0/2.2e−3	27.94/92.5/4.4e−4	28.20/93.1/2.4e−4	28.31/93.2/1.9e−4	28.34/93.3/1.7e−4	27.93/92.6/6.5e−4
SE$_p$-IE-Net	34.35/96.1/**1.7e−3**	36.03/96.8/4.4e−4	37.02/97.1/2.4e−4	37.53/97.2/1.9e−4	37.64/97.3/1.5e−4	36.52/96.9/**5.5e−4**
Ours	**34.60/96.2**/3.4e−3	**36.84/97.0**/4.2e−4	**37.84/97.4**/2.2e−4	**38.34/97.4**/1.7e−4	**38.38/97.5**/1.5e−4	**37.20/97.1**/8.8e−4

artifact, reducing secondary streaky artifacts and overall image quality. The radiologist is asked to rate all the images from each group in a random order, with a rating from 1, indicating very good MAR performance, to 4, not effective at all. We use paired T-test to compare the ratings between our model and every state-of-the-art method.

4.2   Ablation Study

In this section, we investigate the effectiveness of different modules of the proposed architecture. We use the following configurations for this ablation study:

a) IE-Net: the IE network with X_{ma} and M,
b) SE$_0$-Net: the SE network with S_{ma} and M_t,
c) SE-Net: the SE network with S_{ma} and M_p,
d) SE$_p$-Net: the SE-Net with sinogram padding,
e) SE$_p$-IE-Net: the SE$_p$-Net with an IE-Net to refine X_{se} with M,
f) Ours: our full model, SE$_p$-IE-Net refined with X_{ma}.

Effect of Metal Mask Projection (SE$_0$-Net vs SE-Net). From Table 1, we can observe the use of M_p instead of M_t improves the performance for at least 4.1 dB in PNSR and reduces MSE from 0.95219 to 0.00074 for all metal sizes. The groups with large metal implants benefit more than groups with small metal implants. As shown in Fig. 3, the artifacts in metal trace of SE$_0$-Net are over-removed or under-removed, which introduces bright and dark bands in the reconstructed CT image. With the help of M_p, SE-Net can suppress the artifacts even when the metallic implants are large and the surrogate data are more consistent with the correct data outside the metal trace.

Effect of Sinogram Padding (SE-Net vs SE$_p$-Net). Sinogram padding mainly improves the performance in the group with the largest metal objects, with a PSNR gain of 0.15 dB and an MSE reduction of 0.00048. As shown in Fig. 3, the model with sinogram padding restores finer details of soft tissue between large metallic objects because more correct information is retained by periodic padding than zero-padding.

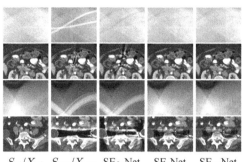

S_{gt}/X_{gt} S_{ma}/X_{ma} SE$_0$-Net   SE-Net   SE$_p$-Net

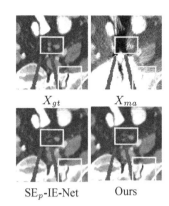

X_{gt} X_{ma}

SE$_p$-IE-Net   Ours

Fig. 3. Comparison of different sinogram enhancement networks. The enhanced sinograms and paired CT images are presented. The red pixels stand for metal implants. (Color figure online)

Fig. 4. Comparison of refinement with and without X_{ma}.

Effect of Learning with X_{ma} (SE$_p$-IE-Net vs Ours). When X_{se} is jointly restored with the corrupted X_{ma}, the sinogram correction performance is affected with an increment of 0.00033 in MSE and of 0.7 dB in PSNR. More details of soft tissue around metal are retained and the image becomes sharper, as shown in Fig. 4.

4.3   Comparison on Simulation Data

We compare our model with multiple state-of-the-art MAR methods. LI [5] and NMAR [9] are traditional algorithms, in which we use the simulated S_{ma} as inputs. Wang et al. [12] propose conditional GAN for MAR purely in image domain. Here we refer their method as cGan-CT and retrain the model using pix2pix [3] on our simulation data. For CNNMAR, we use the trained model provided by [14]. Note that DuDoNet reported here is trained on new simulation data with larger sinogram resolution (641×640), which is different from the sinogram resolution (321×320) used in [8].

Table 2. Quantitative evaluation for proposed network and the state-of-the-arts methods.

Matrics		Large metal	$\rightarrow$	Small metal		Average
X_{ma}	19.42/81.1/1.1e+1	23.07/85.4/7.3e+0	26.12/88.7/2.2e+0	26.60/89.3/1.7e+0	27.69/89.9/3.8e−1	24.58/86.9/4.5e+0
cGAN-CT [12]	16.89/80.7/ n.a.	18.35/83.7/ n.a.	19.94/86.6/ n.a.	21.43/87.6/ n.a.	24.53/89.0/ n.a.	20.23/85.5/ n.a.
LI [5]	20.10/86.7/1.4e−1	22.04/88.7/9.4e−2	25.50/90.2/2.1e−2	26.54/90.7/1.9e−2	27.25/91.2/9.7e−3	24.28/89.5/5.7e−2
NMAR [9]	20.89/86.6/2.3e−1	23.73/89.7/1.3e−1	26.80/91.4/2.7e−2	27.25/91.8/3.6e−2	28.08/92.1/2.2e−2	25.35/90.3/9.0e−2
CNNMAR [14]	23.72/90.1/4.4e−2	25.78/91.6/2.4e−2	28.25/92.6/4.7e−3	28.87/92.9/3.3e−3	29.16/93.1/2.0e−3	27.16/92.0/1.6e−2
DuDoNet [8]	28.98/94.5/5.1e−2	31.00/95.6/3.9e−2	33.80/96.5/5.9e−3	35.61/96.8/3.6e−3	35.67/96.9/2.0e−3	33.01/96.0/2.0e−2
Ours	**34.60/96.2/3.4e−3**	**36.84/97.0/4.2e−4**	**37.84/97.4/2.2e−4**	**38.34/97.4/1.7e−4**	**38.38/97.5/1.5e−4**	**37.20/97.1/8.8e−4**

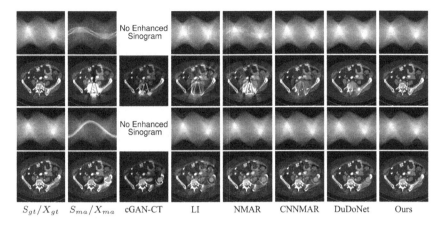

S_{gt}/X_{gt} S_{ma}/X_{ma} cGAN-CT   LI   NMAR   CNNMAR   DuDoNet   Ours

Fig. 5. Comparison with the state-of-the-art methods on simulation data.

Quantitative Comparison. As shown in Table 2, we can see all the sinogram domain MAR algorithms outperform image enhancement approach cGAN-CT in PSNR and SSIM. It is because the sinogram restoration only happens inside the metal trace and the correct sinogram data outside the metal trace help to retain the anatomical structure. CNN-based methods (CNNMAR, DuDoNet, Ours) achieve much better performance than traditional methods, with higher PSNRs and SSIMs in image domain and lower MSEs in sinogram domain. Among all the state-of-the-art methods, CNNMAR achieves the best performance in sinogram enhancement and DuDoNet achieves the best performance in reconstructed images. The proposed method attains the best performance in all metal sizes, with an overall improvement of 4.2 dB in PSNR compared with DuDoNet and 99.4% reduction in MSE compared with CNNMAR.

Visual Comparison. As shown in Fig. 5, metallic implants such as spinal rods and hip prosthesis cause severe streaky artifacts and metal shadows, which obscure bone structures around them. cGan-CT cannot recover image intensity correctly for both cases. Sinogram domain or dual-domain methods perform much better than cGan-CT. LI, NMAR, and CNNMAR introduce strong secondary artifacts and distort the whole images. In NMAR images, there are fake bone structures around the metals, which is related to segmentation error in the prior image from strong metal artifacts. The segmentation error is also visible in NMAR sinogram. CNNMAR cannot restore the correct bone structures between rods in case 1. The tissues around the metals are over-smoothed in DuDoNet because LI sinogram and image are used as inputs, and the missing information cannot be inferred later. Our model retains more structural information than DuDoNet and generates anatomically more faithful artifact-reduced images.

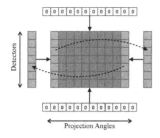

Fig. 6. Sinogram padding.

Table 3. Ratings of clinical CT images.

	DL		CL	
	Rating	P Value	Rating	P Value
cGAN-CT [12]	2.50 ± 0.17	<0.001	4.00 ± 0.00	<0.001
LI [5]	3.80 ± 0.09	<0.001	3.70 ± 0.15	<0.001
NMAR [9]	2.73 ± 0.13	<0.001	2.70 ± 0.15	<0.001
CNNMAR [14]	2.40 ± 0.12	<0.001	2.20 ± 0.20	0.003
DuDoNet [8]	1.46 ± 0.11	0.312	1.70 ± 0.21	0.278
Ours	**1.27 ± 0.13**	n.a	**1.40 ± 0.16**	n.a

4.4 Clinical Study

Rating. Table 3 summarizes the ratings and P values for comparison between our model and the other methods. The performance of our model is significantly better than cGan-CT, LI, NMAR, CNNMAR on both datasets (all P values $\leq$ 0.03). Our model also achieves better ratings than DuDoNet.

Visual Comparison. Figure 7 shows two clinical CT images with metal artifacts. Case 1 is with moderate metal artifacts. cGan-CT does not suppress the artifacts completely and generates some fake details. LI, NMAR, CNN-MAR remove all the artifacts but introduce new streak artifacts, which is caused by the discontinuity in the corrected sinogram. DuDoNet outputs over-smoothed sinogram, which leads to blurred tissues close to the metal implants, such as muscle and bone. Only our model can provide realistic enhanced sinogram and remove the artifacts while retaining the structure of nearby tissues. Case 2 is very challenging as the rods bring strong metal shadows and bright artifacts around the vertebra. cGan-CT recovers the shape of vertebra but changes the overall image intensity. Other sinogram inpainting methods fail as the soft tissue and bone near the rods are heavily distorted. Our model removes part of the dark bands and reproduces correct anatomical structures around the rods.

The results show that our model generalizes well for clinical images with unknown metal materials and geometries. We generate simulate training data using titanium and will retrain the model with multiple metal materials to make it more robust. Meanwhile, images with unknown geometry would be processed in the same simulation space. But it is worth noting that our model is limited to 2D geometry and the metal artifacts in 3D projection (e.g. cone-beam CT) are much more challenging.

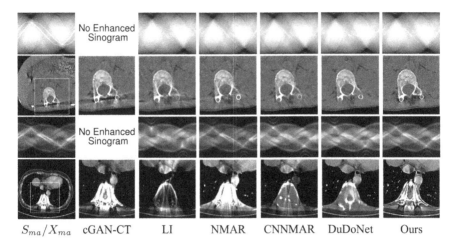

Fig. 7. Comparison with the state-of-the-art methods on clinical CT images with metal artifacts.

5   Conclusion

We present a novel model to better solve the metal artifact reduction problem. We propose encoding mask projection for the sinogram restoration while utilizing the metal-affected real image and sinogram to retain the rich information in dual-domain learning. With the fine details recovered in metal trace, our model can correctly restore the underlying anatomical structure even with large metallic objects present. Visual comparisons and qualitative evaluations demonstrate that our model yields better image quality than competing methods and exhibits a great potential of reducing CT metal artifacts even when applied to clinical images. In the future, we plan to conduct a large scale clinical study to thoroughly evaluate our approach in real clinical practices.

References

1. Chang, Z., Ye, D.H., Srivastava, S., Thibault, J.B., Sauer, K., Bouman, C.: Prior-guided metal artifact reduction for iterative x-ray computed tomography. IEEE Trans. Med. Imaging **38**(6), 1532–1542 (2018)
2. Ghani, M.U., Karl, W.C.: Fast enhanced CT metal artifact reduction using data domain deep learning. IEEE Trans. Comput. Imaging **6**, 181–193 (2019)
3. Isola, P., Zhu, J.Y., Zhou, T., Efros, A.A.: Image-to-image translation with conditional adversarial networks. In: Proceedings of the IEEE Conference on Computer Vision and Pattern Recognition, pp. 1125–1134 (2017)
4. Jin, P., Bouman, C.A., Sauer, K.D.: A model-based image reconstruction algorithm with simultaneous beam hardening correction for x-ray CT. IEEE Trans. Comput. Imaging **1**(3), 200–216 (2015)
5. Kalender, W.A., Hebel, R., Ebersberger, J.: Reduction of CT artifacts caused by metallic implants. Radiology **164**(2), 576–577 (1987)

6. Karimi, S., Martz, H., Cosman, P.: Metal artifact reduction for CT-based luggage screening. J. X-ray Sci. Technol. **23**(4), 435–451 (2015)
7. Liao, H., et al.: Generative mask pyramid network for CT/CBCT metal artifact reduction with joint projection-sinogram correction. In: Shen, D., et al. (eds.) MICCAI 2019. LNCS, vol. 11769, pp. 77–85. Springer, Cham (2019). https://doi.org/10.1007/978-3-030-32226-7_9
8. Lin, W.A., et al.: DudoNet: dual domain network for CT metal artifact reduction. In: Proceedings of the IEEE Conference on Computer Vision and Pattern Recognition, pp. 10512–10521 (2019)
9. Meyer, E., Raupach, R., Lell, M., Schmidt, B., Kachelrieß, M.: Normalized metal artifact reduction (NMAR) in computed tomography. Med. phys. **37**(10), 5482–5493 (2010)
10. Ronneberger, O., Fischer, P., Brox, T.: U-Net: convolutional networks for biomedical image segmentation. In: Navab, N., Hornegger, J., Wells, W.M., Frangi, A.F. (eds.) MICCAI 2015. LNCS, vol. 9351, pp. 234–241. Springer, Cham (2015). https://doi.org/10.1007/978-3-319-24574-4_28
11. Soltanian-Zadeh, H., Windham, J.P., Soltanianzadeh, J.: CT artifact correction: an image-processing approach. In: Medical Imaging 1996: Image Processing, vol. 2710, pp. 477–485. International Society for Optics and Photonics (1996)
12. Wang, J., Zhao, Y., Noble, J.H., Dawant, B.M.: Conditional generative adversarial networks for metal artifact reduction in CT images of the ear. In: Frangi, A.F., Schnabel, J.A., Davatzikos, C., Alberola-López, C., Fichtinger, G. (eds.) MICCAI 2018. LNCS, vol. 11070, pp. 3–11. Springer, Cham (2018). https://doi.org/10.1007/978-3-030-00928-1_1
13. Yan, K., et al.: Deep lesion graphs in the wild: relationship learning and organization of significant radiology image findings in a diverse large-scale lesion database. In: Proceedings of the IEEE Conference on Computer Vision and Pattern Recognition, pp. 9261–9270 (2018)
14. Zhang, Y., Yu, H.: Convolutional neural network based metal artifact reduction in x-ray computed tomography. IEEE Trans. Med. Imaging **37**(6), 1370–1381 (2018)
15. Zhou, S.K.: Medical Image Recognition, Segmentation and Parsing: Machine Learning and Multiple Object Approaches. Academic Press (2015)
16. Zhou, S.K., Greenspan, H., Shen, D.: Deep Learning for Medical Image Analysis. Academic Press (2017)

Acceleration of High-Resolution 3D MR Fingerprinting via a Graph Convolutional Network

Feng Cheng[1], Yong Chen[3], Xiaopeng Zong[2], Weili Lin[2], Dinggang Shen[1,2], and Pew-Thian Yap[1,2(✉)]

[1] Department of Computer Science,
University of North Carolina, Chapel Hill, NC, USA
`fengchan@cs.unc.edu`, {`dgshen,ptyap`}`@med.unc.edu`
[2] Department of Radiology and Biomedical Research Imaging Center,
University of North Carolina, Chapel Hill, NC, USA
[3] Case Western Reserve University, Cleveland, OH, USA
`yxc235@case.edu`

Abstract. Magnetic resonance fingerprinting (MRF) is a novel imaging framework for fast and simultaneous quantification of multiple tissue properties. Recently, 3D MRF methods have been developed, but the acquisition speed needs to be improved before they can be adopted for clinical use. The purpose of this study is to develop a novel deep learning approach to accelerate 3D MRF acquisition along the slice-encoding direction in k-space. We introduce a graph-based convolutional neural network that caters to non-Cartesian spiral trajectories commonly used for MRF acquisition. We improve tissue quantification accuracy compared with the state of the art. Our method enables fast 3D MRF with high spatial resolution, allowing whole-brain coverage within 5 min, making MRF more feasible in clinical settings.

Keywords: 3D MR fingerprinting · K-space interpolation · Graph convolution · GRAPPA

1 Introduction

Quantitative MR imaging is experiencing rapid expansion in the field of medical imaging. Compared to qualitative imaging approaches, quantitative imaging methods have been shown to yield improved consistency across scanners, allowing more objective tissue characterization, disease diagnosis, and treatment assessment [3]. However, most quantitative MRI methods [12,14] are relatively slow and generally provide only a single tissue property at a time, limiting adoption in routine clinical settings. Magnetic resonance fingerprinting (MRF) is a novel quantitative imaging method that is rapid and efficient for simultaneous

This work was supported in part by NIH grant EB006733.

The original version of this chapter was revised: the NIH grant number and typographical errors have been corrected. The correction to this chapter is available at https://doi.org/10.1007/978-3-030-59713-9_75

quantification of multiple tissue properties in a single acquisition [10]. Due to its superior performance over other quantitative imaging approaches, MRF has attracted a lot of interest and has been successfully applied to quantitative imaging of various human organs, including the brain, abdomen, heart, and breast [2,5,15].

Significant efforts have been recently devoted to extending the original 2D MRF to 3D using stack-of-spirals acquisition [5,11]. 3D MRF potentially improves signal-to-noise ratio, spatial resolution and coverage. However, 3D MRF with high spatial resolution and volumetric coverage prolongs the acquisition time. Various methods have been developed to accelerate 3D MRF. Chen et al. [4] proposed a method to combine non-Cartesian parallel imaging with machine learning to accelerate 3D MRF. While an acceleration of factor 4 was achieved along the temporal domain using a deep learning network, only an acceleration of factor 2 was achieved along the slice-encoding direction using non-Cartesian parallel imaging. The total acquisition time with 1 mm isotropic resolution and whole brain coverage still takes 7 min. In addition, image reconstruction for non-Cartesian parallel imaging is time-consuming, hindering the wider adoption of 3D MRF in the clinics.

Recently, graph theory has been integrated with deep learning [8,9] for tackling tasks in non-Euclidean space. In this work, we propose to use the graph-based deep learning to replace GRAPPA in non-Cartesian parallel imaging to accelerate 3D MRF along the slice-encoding direction. This is done in combination with a U-Net to accelerate MRF acquisition along the temporal domain for efficient tissue quantification. We develop a graph neural network to accelerate spiral-based 3D MRF acquisition, resulting in higher acceleration than conventional methods and reduced post-acquisition processing time. Our method, although applied to 3D MRF in the current work, holds great potential for the acceleration of general non-Cartesian acquisition for both qualitative and quantitative imaging purposes.

2   Problem Formulation

Similar to conventional 2D MRF, a typical 3D MRF dataset consists of hundreds of time frames for tissue characterization and each time frame is acquired with the stack-of-spirals trajectory. Only one spiral arm is acquired for each slice, so the acquisition is highly accelerated in-plane. For further speed improvement, we explore the possibility of acceleration along the slice-encoding direction in addition to the temporal dimension. For 3D MRF, the acquired data in k-space can be expressed as $Y \in \mathbb{C}^{T \times S \times Q \times C}$, where T is the number of temporal frames, S is the number of slices, Q is the length of spiral arm, and C is the number of coils.

2.1   Acceleration Along Slice-Encoding Direction in k-space

To accelerate along the slice-encoding direction, an interleaved undersampling pattern as described in [11] was adopted. Specifically, given an acceleration factor

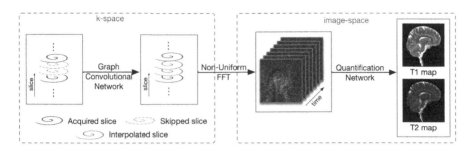

Fig. 1. Method overview.

of r_slice, each slice with index s satisfying $s \equiv f \pmod{r_\text{slice}}$ is acquired and other slices are skipped, where f is the time frame index. Therefore, the acquired k-space data is $X_\text{k-space} \in \mathbb{C}^{T \times (\frac{S-p}{r_\text{slice}}+p) \times Q \times C}$, where p is the number of auto calibration signal (ACS) slices for parallel imaging reconstruction. Conventional methods, such as GRAPPA, aim to obtain Y by interpolating the accelerated acquired data $X_\text{k-space}$. A GRAPPA kernel of size 3×2 (spiral readout direction $\times$ slice-encoding direction) is typically used. Due to the nature of non-Cartesian acquisition with spiral trajectories, Q GRAPPA kernels need to be calculated from the ACS data and applied to fill the missing data along the slice-encoding direction.

2.2   Acceleration Along Temporal Dimension in Image Space

Acceleration along the temporal domain using deep learning has been described in the literature [6,7]. In brief, the acceleration in image-space is achieved by acquiring fewer time frames, i.e., acquiring only the first $T_a = \frac{T}{r_\text{frame}}$ time frames. The images after non-uniform Fourier transform (NUFFT) can be expressed as $X_\text{image-space} \in \mathbb{C}^{T_a \times S \times H \times W}$, where r_frame is the acceleration factor and $H \times W$ is the in-plane image matrix size. The total acceleration factor $r = r_\text{slice} \times r_\text{frame}$.

3   Approach

In this study, we propose to use a graph convolutional network (GCN) to interpolate the undersampled data along the slice-encoding direction. NUFFT was then applied to transform the k-space data to the image space. A network was further applied to generate the tissue property maps. Figure 1 shows the workflow of the proposed method.

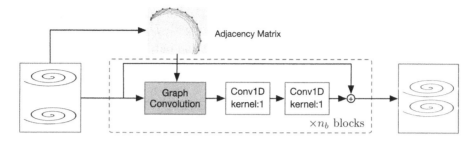

Fig. 2. The proposed graph convolutional network.

3.1 Graph Convolutional Network

We aim to reconstruct the skipped slices from two adjacent slices, as shown in Fig. 1. The data points of an in-plane spiral trajectory are non-uniformly distributed in k-space, with more data points at the center of k-space and less data points at the edge. This non-uniform property requires different kernels to be applied to different data points, which poses challenges for conventional convolution. We adopt graph convolution to solve this problem. The spiral arm is first represented as a graph and then convolution is performed on the graph with the help of the graph adjacency matrix. This is in essence applying different convolutional kernels on different data points.

Figure 2 shows the structure of the proposed network. It is constructed with a cascade of n_b blocks. Each block is constructed by one graph convolution layer and two kernel-1 1D conventional convolution. Residual connections are added to all blocks except the first block. The activation functions of all the layers except the last layer are ReLU. n_b was set to 3 as an inception field that is too small or too large will degrade performance.

To build the graph convolutional layer, we first represent the in-plane non-Cartesian trajectory as a graph $\mathcal{G} = (\mathcal{V}, \mathcal{E}, A)$, where $\mathcal{V}$ is the vertices representing the data points on a trajectory, $\mathcal{E}$ is the edges and $A \in \mathbb{R}^{Q \times Q}$ is the weighted adjacency matrix. The weights of the graph adjacency matrix A are defined as the negative exponential of the square of normalized Euclidean distances between points in the spiral-readout direction:

$$A_{i,j} = e^{-\frac{|v_i - v_j|^2}{\bar{d}^2}}, \tag{1}$$

where v_i, v_j are the spatial coordinates of the spiral trajectory and $\bar{d}$ is the mean distance of all vertex pairs. To maintain locality, we only retain the q nearest neighbors for each vertex by keeping the q largest values in each row of A and setting the others to 0. The kernel size k for graph convolution is set to $q + 1$, i.e., the vertex and its q neighbors.

We follow [9] to construct the graph convolutional layer:

$$H^{l+1} = \sigma(\hat{D}^{-1/2} \hat{A} \hat{D}^{1/2} H^{(l)} W^{(l)}), \tag{2}$$

where $\hat{A} = A + I$ is the adjacency matrix of $\mathcal{G}$ with self connections, I is the identity matrix, D is a diagonal degree matrix with $\hat{D}_{ii} = \sum_j A_{ij}$, $W^{(l)}$ is a learnable kernel, $H^{(l)}$ and $H^{(l+1)}$ are the input and output of the layer, l is the layer's index and $\sigma(\cdot)$ is the ReLU activation function.

Table 1. Evaluation of k-space interpolation, evaluated via NMSE. Reconstruction time per subject reported in minutes.

Methods	NMSE			Time			Params (million)
	r_{slice}						
	2	3	4	2	3	4	
GRAPPA [4]	0.93	1.42	1.56	97.6	132.0	301.5	1.97
RAKI [1]	0.401	0.568	0.731	7.2	5.3	4.8	1.36
GCN-1b	0.384	0.531	0.705	**3.6**	**3.3**	**3.1**	**1.31**
GCN-3b	**0.377**	**0.500**	**0.663**	6.1	5.1	4.7	3.41

3.2   Implementation

We trained the GCN on the ACS slices with stride 1 using a training dataset acquired from 5 subjects and evaluated the network on all the acquired slices of a testing subject. A total of 12 ACS slices from each subject were used. Each sample consists of two slices separated by an interval of r_{slice} as input ($\mathbb{C}^{Q \times (C \times 2)}$) and $r_{\text{slice}} - 1$ slices in between as target ($\mathbb{C}^{Q \times (C \times (r_{\text{slice}} - 1))}$). A total of $n \times T_a \times (12 - r_{\text{slice}})$ training samples were used, where n is the number of training subjects. The complex numbers are converted to real numbers by stacking the real and imaginary parts. In each block the filters of GCN and the following two conv layers are 1024, 512, 512 respectively. Mean square error (MSE) is used as the objective function. The network is trained with a batch size of 1 and optimized using ADAM with an initial learning rate of 0.0005 and was reduced by 99% after each epoch. For tissue property quantification, we use the same network structure and settings as described in [4].

4   Experiments

MRI experiments were performed on a Siemens 3T Prisma scanner. A 3D MRF dataset from 6 subjects (M/F: 3/3; age: 34 ± 10 years) was acquired. Each MRF time frame was highly undersampled in-plane with only one spiral arm (reduction factor: 48). The slice direction was linearly encoded and fully sampled. With a constant TR of 9.2 ms and a waiting time of 2 sec between partitions, the total scan time was ~40 min for each subject. FOV: $25 \times 25 \, \text{cm}^2$; matrix size: 256×256; effective in-plane resolution: 1 mm; slice thickness: 1 mm; number of slices: 144; variable flip angles: $5°–12°$; number of MRF time frames: 768.

Various acceleration factors (2–4) along the slice-encoding direction were evaluated. We considered acceleration of factor 4 along the temporal direction, similar to [4]. Quantitative T1 and T2 maps generated using all 768 time frames with no acceleration along the slice-encoding direction were used as the ground truth. Leave-one-out cross validation was employed.

4.1   Comparison with State-of-the-Art Methods

We evaluated the performance of the proposed method wit respect to two state-of-the-art (SOTA) methods: 1) Non-Cartesian GRAPPA [4] and 2) RAKI [1]. RAKI was originally proposed for Cartesian MRI, and we extended it to non-Cartesian 3D MRF. Performance in both k-space interpolation and tissue quantification was assessed.

k-Space Interpolation: Following [1], normalized mean square error (NMSE) was used to evaluate the accuracy of k-space interpolation. As shown in Table 1, the proposed GCN-1b ($n_b = 1$) with only one block shows improved accuracy over either GRAPPA or RAKI for all three acceleration factors. GCN-3b ($n_b = 3$), with a cascade of multple blocks, achieves the best NMSE over all methods. Performance comparison between GCN-1b and GCN-3b will be illustrated in Sect. 4.2. The reconstruction time is reduced by 15–60 times with deep learning methods (RAKI and GCN) compared to non-Cartesian GRAPPA.

Tissue Quantification Maps: Following [7,13], relative-L1, SNR and PSNR were applied for evaluation of tissue quantification accuracy (Table 2). Representative quantitative T1 and T2 maps are shown in Fig. 3. Compared to GRAPPA and RAKI, the proposed GCN yields the lowest Relative-L1 and highest SNR and PSNR for all three acceleration factors. Improved accuracy in tissue quantification and sharper edges can be observed in both T1 and T2 maps obtained with the proposed method (Fig. 3). All these results suggest that higher acceleration factors along the slice-encoding direction can be achieved using the proposed GCN. This will reduce the total acquisition time to 5 min ($r_{\text{slice}} = 3$) or 4 min ($r_{\text{slice}} = 4$) for 3D MRF with whole-brain coverage, comparable to the time needed to acquire either T1-weighted or T2-weighted images with similar resolution and coverage.

4.2   Ablation Study

We investigated the effects of two key network parameters: 1) the kernel size for graph convolution; 2) the number of cascaded blocks.

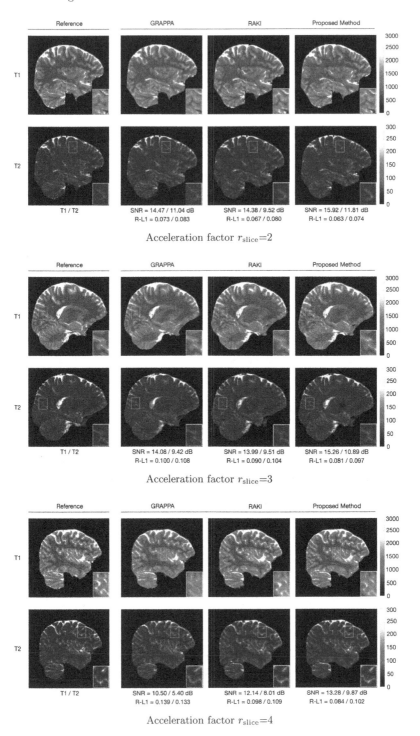

Fig. 3. Representative T1 and T2 maps generated from MRF data acquired with various acceleration factors.

Table 2. Comparison of tissue quantification errors.

Quan. Map	Method	Relative-L1			SNR			**PSNR**		
		r_{slice}								
		2	3	4	2	3	4	2	3	4
T1	GRAPPA	0.077	0.107	0.152	15.96	13.72	10.53	26.26	24.09	21.14
	RAKI	0.069	0.097	0.118	16.03	13.77	12.83	26.16	24.01	23.15
	GCN	**0.061**	**0.085**	**0.103**	**17.48**	**15.23**	**13.50**	**27.70**	**25.51**	**23.95**
T2	GRAPPA	0.101	0.132	0.173	11.32	8.45	5.17	26.59	24.09	21.02
	RAKI	0.093	0.118	0.166	9.98	8.70	6.72	26.72	24.32	21.98
	GCN	**0.088**	**0.110**	**0.132**	**12.16**	**10.42**	**8.75**	**27.66**	**25.90**	**24.26**

Table 3. Effects of kernel size and number of cascaded blocks on the accuracy of k-space interpolation.

Method	**NMSE**		
	r_{slice}		
	2	3	4
GCN-k1	0.397	0.530	0.686
GCN-k3	0.379	0.514	0.671
GCN-k5	**0.377**	**0.500**	**0.663**
GCN-k7	0.395	0.529	0.680
GCN-k9	0.432	0.558	0.699

(a) Different kernel sizes with 3 blocks

Method	**NMSE**		
	r_{slice}		
	2	3	4
GCN-1b	0.384	0.531	0.705
GCN-3b	**0.377**	**0.500**	0.663
GCN-5b	**0.377**	0.505	**0.660**
GCN-7b	0.383	0.513	0.682

(b) Different n_b's with $k = 5$

Kernel Size: The kernel size k for graph convolution determines the neighborhood extent each point on a spiral can draw information from. No neighborhood information is considered when $k = 1$. As shown in Table 3a, kernel size 5 results in the best performace in k-space interpolation for all three acceleration factors.

Number of Cascaded Blocks: Table 3b indicates that a cascade of 3 blocks or 5 blocks yields the best performance. Increasing the number of blocks is in general expected to improve performance, but this comes at the cost of significantly increased number of network parameters, reducing network trainability with limited sample size.

5   Conclusion

In this paper, we presented a graph convolutional network to accelerate high-resolution 3D MRF along the slice-encoding direction. Our results suggest that improved tissue quantification accuracy can be achieved with reduced post-processing time. Therefore, our method improves the feasibility of 3D MRF clinical settings.

References

1. Akçakaya, M., Moeller, S., Weingärtner, S., Uğurbil, K.: Scan-specific robust artificial-neural-networks for k-space interpolation (RAKI) reconstruction: database-free deep learning for fast imaging. Magn. Reson. Med. **81**(1), 439–453 (2019)
2. Badve, C., et al.: MR fingerprinting of adult brain tumors: initial experience. Am. J. Neuroradiol. **38**(3), 492–499 (2017)
3. Mehta, B.B., et al.: Magnetic resonance fingerprinting: a technical review. Magn. Reson. Med. **81**(1), 25–46 (2019)
4. Chen, Y., Fang, Z., Hung, S.C., Chang, W.T., Shen, D., Lin, W.: High-resolution 3D MR fingerprinting using parallel imaging and deep learning. NeuroImage **206**, 116329 (2020)
5. Chen, Y.: Three-dimensional MR fingerprinting for quantitative breast imaging. Radiology **290**(1), 33–40 (2019)
6. Fang, Z., et al.: Deep learning for fast and spatially constrained tissue quantification from highly accelerated data in magnetic resonance fingerprinting. IEEE Trans. Med. Imaging **38**(10), 2364–2374 (2019)
7. Fang, Z., Chen, Y., Nie, D., Lin, W., Shen, D.: RCA-U-Net: residual channel attention U-Net for fast tissue quantification in magnetic resonance fingerprinting. In: Shen, D., et al. (eds.) MICCAI 2019. LNCS, vol. 11766, pp. 101–109. Springer, Cham (2019). https://doi.org/10.1007/978-3-030-32248-9_12
8. Hong, Y., Kim, J., Chen, G., Lin, W., Yap, P.T., Shen, D.: Longitudinal prediction of infant diffusion MRI data via graph convolutional adversarial networks. IEEE Trans. Med. Imaging **38**(12), 2717–2725 (2019)
9. Kipf, T.N., Welling, M.: Semi-supervised classification with graph convolutional networks. arXiv preprint arXiv:1609.02907 (2016)
10. Ma, D., et al.: Magnetic resonance fingerprinting. Nature **495**(7440), 187–192 (2013)
11. Ma, D., et al.: Fast 3D magnetic resonance fingerprinting for a whole-brain coverage. Magn. Reson. Med. **79**(4), 2190–2197 (2018)
12. Majumdar, S., Orphanoudakis, S., Gmitro, A., O'donnell, M., Gore, J.: Errors in the measurements of T2 using multiple-echo MRI techniques. I. effects of radiofrequency pulse imperfections. Magn. Reson. Med. **3**(3), 397–417 (1986)
13. Song, P., Eldar, Y.C., Mazor, G., Rodrigues, M.R.: Magnetic resonance fingerprinting using a residual convolutional neural network. In: ICASSP 2019–2019 IEEE International Conference on Acoustics, Speech and Signal Processing (ICASSP), pp. 1040–1044. IEEE (2019)
14. Stikov, N., Boudreau, M., Levesque, I.R., Tardif, C.L., Barral, J.K., Pike, G.B.: On the accuracy of T1 mapping: searching for common ground. Magn. Reson. Med. **73**(2), 514–522 (2015)
15. Yu, A.C., et al.: Development of a combined MR fingerprinting and diffusion examination for prostate cancer. Radiology **283**(3), 729–738 (2017)

Deep Attentive Wasserstein Generative Adversarial Networks for MRI Reconstruction with Recurrent Context-Awareness

Yifeng Guo[1], Chengjia Wang[2], Heye Zhang[1(✉)], and Guang Yang[3]

[1] Sun Yat-sen University, Guangzhou, China
zhangheye@mail.sysu.edu.cn
[2] University of Edinburgh, Edinburgh EH16 4TJ, UK
[3] National Heart and Lung Institute, Imperial College London,
London SW7 2AZ, UK

Abstract. The performance of traditional compressive sensing-based MRI (CS-MRI) reconstruction is affected by its slow iterative procedure and noise-induced artefacts. Although many deep learning-based CS-MRI methods have been proposed to mitigate the problems of traditional methods, they have not been able to achieve more robust results at higher acceleration factors. Most of the deep learning-based CS-MRI methods still can not fully mine the information from the k-space, which leads to unsatisfactory results in the MRI reconstruction. In this study, we propose a new deep learning-based CS-MRI reconstruction method to fully utilise the relationship among sequential MRI slices by coupling Wasserstein Generative Adversarial Networks (WGAN) with Recurrent Neural Networks. Further development of an attentive unit enables our model to reconstruct more accurate anatomical structures for the MRI data. By experimenting on different MRI datasets, we have demonstrated that our method can not only achieve better results compared to the state-of-the-arts but can also effectively reduce residual noise generated during the reconstruction process.

Keywords: Recurrent neural network · Wasserstein generative adversarial networks · MRI reconstruction

1 Introduction

Compressed sensing magnetic resonance imaging (CS-MRI) [1,2] has been proposed for accelerating MRI process. This technique uses a small fraction of data to reconstruct images from sub-Nyquist sampling. Assuming the raw data is compressible, CS-MRI performs nonlinear optimisations on the undersampling data without sacrificing the quality of the reconstructed images significantly.

However, it is still very challenging to consolidate the speed of reconstruction and robustness of image quality maintenance in one CS-MRI based framework.

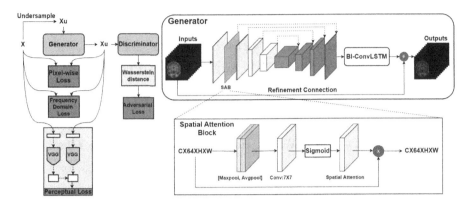

Fig. 1. The overall network architecture of our DAWGAN framework. Left: the workflow of our proposed model. Right: the details of our proposed generator with Bi-ConvLSTM and flow diagram of the proposed spatial attention block (SAB).

On the one hand, CS-MRI tries to solve underdetermined equations to perceive the original signals from the limited undersampled data. This requires nonlinear optimisation solvers for a common non-convex system that usually involve iterative computations, which can result in prolonged reconstruction time [3]. On the other hand, CS-MRI may produce images with degraded image quality and low signal-to-noise ratio (SNR) from randomly highly undersampled k-space data [4]. Moreover, in addition to a large amount of computation needed for the nonlinear optimisation, CS-MRI also requires that the acquisition matrix and the sparse transformation matrix are unrelated. Based on the above limitations, the acceleration factor of CS-MRI is generally between 2 and 6.

Recently, deep learning-based CS-MRI methods have emerged as an effective way to solve the problems of slow and unstable MRI reconstruction [5–11,18–26]. For example, a conditional Generative Adversarial Networks-based model (DAGAN) was proposed to achieve fast CS-MRI [6], but still, this end-to-end training neglected the correlation between adjacent 2D slices. Thus, although DAGAN can achieve fast MRI reconstruction, it may lose image quality without using a priori information. For another example, DC-CNN [7] applied cascades of convolutional neural networks with a residual connection for CS-MRI. Besides, DC-CNN also used a data consistency (DC) step to ensure that the output of each cascade was consistent with the original k-space information. However, DC-CNN approach was not able to effectively utilise the full temporal domain information. In contrast, a convolutional recurrent neural network (CRNN) method was proposed to incorporate a bidirectional convolutional recurrent unit for a faster and more stable reconstruction [8]. However, such an approach was not able to effectively exploit the k-space information from individual images.

In this study, we propose a GAN based architecture that works on continuous sequential data for CS-MRI. This intuitively mimics the way reporting clinicians scrutinise the 3D data by scrolling up and down to fully sense the information above and below the current 2D slice. Our method can not only overcome the shortcomings of slow reconstruction but can also maintain higher reconstructed image quality by combining the characteristics in time and frequency domains. To the best of our knowledge, this is the first study to combine Recurrent Neural Networks (RNN) with GAN in the field of MRI reconstruction. In particular, we design a novel generator with bidirectional convolutional long short-term memory (Bi-ConvLSTM) that can encode the a priori frequency and time-domain information. Besides, another significant contribution of our work is that we propose a spatial attention-based model that the attention unit in our model can distinguish between significant and non-significant features in terms of the MRI reconstruction task. In addition, we utilise WGAN with gradient penalty (WGAN-GP) as a critic function, which can significantly improve the stability of GAN. We also couple the adversarial loss with pixel-wise mean square error (MSE) and the perceptual loss [12] to achieve better reconstruction details with superior perceptual image quality.

2   Method

2.1   Problem Formulation of CS-MRI

Deep Learning-Based CS-MRI Reconstruction. Let $\mathbf{x} \in \mathbb{C}^D$ represents the slice of 2D images to be reconstructed, which consists of $\sqrt{N} \times \sqrt{N}$ pixels for one image, and let $\mathbf{y}$ denotes the undersampled measurements in k-space. For deep learning-based methods, previous studies such as [5] and [7] incorporated a CNN into CS-MRI reconstruction, transformed the unconstrained optimisation problem into:

$$\min_{\mathbf{x}} \lambda \|\mathbf{y} - \mathbf{F}_u \mathbf{x}\|_2^2 + \mathcal{R}(\mathbf{x}) + \zeta \left\|\mathbf{x} - f_{\mathrm{cnn}}\left(\mathbf{X}_u | \hat{\theta}\right)\right\|_2^2, \tag{1}$$

in which f_{cnn} denotes the forward propagation of data through the CNN paramet-rised by θ, and ζ is a regularisation parameter. $\mathcal{R}$ expresses regularisation terms on $\mathbf{x}$. $\mathbf{X}_u$ is the reconstruction from the zero-filled undersampled k-space measurements.

In the reconstruction network selection, many previous studies, e.g., [6,11,18, 19], relied on an encoder-decoder structure. Nevertheless, our preliminary experiments indicated that these single structures performed poorly in the PSNR. Moreover, there are also methods [7,8] that developed for the dynamic MR reconstruction, but they did not perform well at higher k-space undersampling.

2.2   DAWGAN for CS-MRI

In this study, we propose a Deep Attentive Wasserstein Generative Adversarial Networks (DAWGAN) method to reconstruct MRI images from highly under-sampled data with continuous sequential data. It contains three key components: a Bi-ConvLSTM block, a spatial attention block (SAB) and a WGAN-GP as the critic function. The workflow of our DAWGAN is summarised in Fig. 1.

Image Domain Feature Extraction via a Sequential Learning. To achieve more aggressive undersampling, one way is to encode the a priori frequency and time-domain information in sequential data, e.g., 2D MRI slices of a 3D volumetric data. We assumed $\mathbf{X}$ as the feature representation of our 2D sequential MRI data slices throughout the 3D volume. Here $\mathbf{X}_l^{(i)}$ denoted the representation at slice l and iteration i. We needed to take into account $\mathbf{X}_{l-1}^{(i)}$ and $\mathbf{X}_{l+1}^{(i)}$ in the reconstruction process to provide information for $\mathbf{X}_l^{(i)}$. To that end, we proposed a Bi-ConvLSTM subnetwork to exploit both temporal and iteration dependencies jointly. The Bi-ConvLSTM subnetwork can be formulated as:

$$
\begin{aligned}
\overrightarrow{\mathbf{X}}_{l,t}^{(i)} &= \sigma \left(\mathbf{W}_l * \mathbf{X}_{l-1,t}^{(i)} + \mathbf{W}_t * \overrightarrow{\mathbf{X}}_{l,t-1}^{i} + \mathbf{W}_i * \mathbf{X}_{l,t}^{(i-1)} + \overrightarrow{\mathbf{B}}_l \right) \\
\overleftarrow{\mathbf{X}}_{l,t}^{(i)} &= \sigma \left(\mathbf{W}_l * \mathbf{X}_{l-1,t}^{(i)} + \mathbf{W}_t * \overleftarrow{\mathbf{X}}_{l,t+1}^{(i)} + \mathbf{W}_i * \mathbf{X}_{l,t}^{(i-1)} + \overleftarrow{\mathbf{B}}_l \right) \\
\mathbf{X}_{l,t}^{(i)} &= \overrightarrow{\mathbf{X}}_{l,t}^{(i)} + \overleftarrow{\mathbf{X}}_{l,t}^{(i)}
\end{aligned}
\tag{2}
$$

where $\overrightarrow{\mathbf{X}}_{l,t}^{(i)}$ denoted the forward direction and $\overleftarrow{\mathbf{X}}_{l,t}^{(i)}$ denoted the backward direction. Through Bi-ConvLSTM layer, our model can learn the differences and correlations of successive MRI data slices. The output of the Bi-ConvLSTM layer then took a refinement connection to prevent data shifting.

Spatial Attention Block (SAB). The main aim of the designed SAB was to increase representation power by using attention mechanism: focusing on important features and suppressing unnecessary ones. Details about the SAB are shown in Fig. 1. Inspired by [13], we set the SAB after the first convolution block, which was also propagated to the up-sampling layers with the skip-connections. We conducted average-pooling and max-pooling operations on the feature map obtained from the upper layer to generate an efficient feature descriptor. Then we utilised a convolution layer to generate a feature map that could encode where to emphasize or suppress. The SAB took all the features extracted by the upper layer to calculate the attention map.

We assumed that the 2D maps generated by pooling operations were $\mathbf{F}_{\text{avg}} \in \mathbb{R}^{1 \times H \times W}$ and $\mathbf{F}_{\max} \in \mathbb{R}^{1 \times H \times W}$. Each denoted average-pooled features and max-pooled features across the feature map. The two maps were then stacked and convolved by a standard convolution layer to produce the 2D spatial attention map. Hence, our spatial attention map was computed as

$$
\mathbf{M}_{\mathbf{s}}(\mathbf{F}) = \sigma \left(f^{7 \times 7} \left([\mathbf{F}_{\text{avg}} ; \mathbf{F}_{\max}] \right) \right)
\tag{3}
$$

where $f^{7 \times 7}$ represented the convolution operation with the filter size of 7×7 and σ denoted the sigmoid function according to [13]. The spatial attention calculated the feature correlation across the channel domain to find the cardinal features across the entire spatial domain.

Loss Function and Training. Our loss function consisted of content loss and adversarial loss. The content loss function was basically made up of three parts, i.e., a pixel-wise image domain mean square error (MSE) loss, a frequency domain MSE loss and a perceptual VGG loss. The whole loss function could be formulated as

$$\mathcal{L}_{\text{TOTAL}} = \alpha \mathcal{L}_{\text{iMSE}} + \beta \mathcal{L}_{\text{fMSE}} + \gamma \mathcal{L}_{\text{VGG}} + \mathcal{L}_{\text{GEN}} \tag{4}$$

where α, β, γ represented the hyper-parameters.

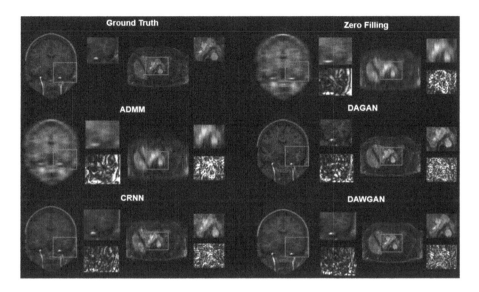

Fig. 2. Qualitative visualisation using 10% of the k-space data. For each subfigure, left: brain MRI data; right: cardiac MRI data.

Most of the GAN based CS-MRI studies used vanilla GAN objective [14], which applied the Kullback-Leibler (KL) divergence, as the adversarial loss function. However, during the training of the generator, when the generator deviated from the optimal solution, the parameters of the generator might not be updated continuously, which could then lead to complicated training process and *model collapse* [15]. In this study, we introduced WGAN-GP [16] as an alternative strategy of using Wasserstein distance to displace the KL divergence for solving the potential complications in training the GAN. WGAN-GP also introduced the *gradient penalty* to better solve the common gradient vanishing problem. We used a loss that was calculated as the following

$$\mathcal{L}_{\text{GEN}} = -E_{x \sim p_g} \left[f_{\text{model}}(x) \right]. \tag{5}$$

Table 1. Comparison study results using different CS-MRI methods.

Brain MRI Data						
Methods	10%		30%		50%	
	PSNR	MOS	PSNR	MOS	PSNR	MOS
Zero-filling	28.16(3.33)	1.02(0.13)	34.83(2.78)	1.12(0.21)	39.36(2.61)	1.09(0.34)
ADMM	28.20(3.36)	1.21(0.35)	35.21(4.03)	1.22(0.31)	39.99(4.08)	1.28(0.37)
DAGAN	33.25(4.10)	2.52(0.88)	38.12(3.56)	3.08(0.68)	45.41(4.13)	3.27(0.73)
CRNN	33.57(3.16)	2.78(0.62)	38.26(3.86)	2.98(0.43)	46.10(2.29)	3.58(0.61)
DAWGAN	**34.31(3.01)**	**3.01(0.69)**	**40.74(3.57)**	**3.23(0.69)**	**46.43(2.19)**	**3.98(0.72)**
Cardiac MRI Data						
Zero-filling	27.08(0.84)	1.02(0.13)	31.49(0.88)	1.12(0.21)	35.13(0.92)	1.09(0.34)
ADMM	27.20(1.64)	1.25(0.28)	31.88(1.72)	1.38(0.21)	35.54(1.73)	1.98(0.45)
DAGAN	29.35(1.33)	2.21(0.54)	33.85(1.62)	2.49(0.41)	37.86(1.22)	2.81(0.53)
CRNN	29.62(2.15)	2.68(0.61)	34.29(2.33)	2.83(0.71)	38.12(2.29)	2.96(0.67)
DAWGAN	**31.06(1.71)**	**2.92(0.72)**	**35.97(1.77)**	**3.06(0.59)**	**39.66(1.79)**	**3.42(0.56)**

In order to improve the perceptual quality, we also incorporated the content loss with three different combinations of the loss functions:

$$\min_{\theta_G} \mathcal{L}_{\text{iMSE}}(\theta_G) = \frac{1}{2} \|x_t - \hat{x}_u\|_2^2$$

$$\min_{\theta_G} \mathcal{L}_{\text{fMSE}}(\theta_G) = \frac{1}{2} \|y_t - \hat{y}_u\|_2^2 \quad (6)$$

$$\min_{\theta_G} \mathcal{L}_{\text{VGG}}(\theta_G) = \frac{1}{2} \|f_{\text{Vgg}}(x_t) - f_{\text{vgg}}(\hat{x}_u)\|_2^2.$$

We used normalised MSE (NMSE) as the optimisation cost function. However, the use of NMSE as content loss alone might lead to perceptually uneven reconstruction, resulting in a lack of coherent image details. Therefore, to consider the perceptual similarity of images, we also added NMSE of the frequency domain data and VGG loss ($\mathcal{L}_{\text{VGG}}$) as additional constraints.

3   Experiments and Results

3.1   Experiments

Datasets. Our experiments were performed on two datasets (1) Brain MRI dataset: We trained and tested our model using a MICCAI 2013 grand challenge dataset. In total, we included 726 3D data for our study. We randomly used 503 data for training, 173 for validation and 50 for testing. Each 2D slice had a shape of 256×256, and we normalised the intensities into a range of $[-1\ 1]$. [6] (2) Cardiac MRI dataset: A population of 100 3D LGE MRI patient data, which were made available through the 2018 Atrial Segmentation Challenge, were used in this work. The scanner used for this clinical study was a whole-body MRI scanner, within an image acquisition resolution of $0.625\,\text{mm}^3$. The studied data were randomly divided into 80 for training, 10 for validation and 10 for testing. Similarly, we normalised each slice into a range of $[-1\ 1]$.

Experiments Setup. For all the input data, we applied data augmentation on the input 2D image slices. Besides, we used raw k-space data with different undersampling ratios to simulate the corresponding acceleration factors. In particular, $10\%, 30\%$ and 50% retained raw k-space data were simulated representing $10\times, 3.3\times, 2\times$ accelerations assuming that the preparation time of MRI scanning is insignificant. All our comparison studies were carried out using different CS-MRI reconstruction algorithms using these three levels of undersampling ratios. Our studies were mainly divided into three experiments: First, We compared the performance of our method with that of other SOTA at three different acceleration factors. In addition to the traditional metrics of PSNR, we also introduced the mean opinion scores (MOS) to take human perception into account, which was the results of domain experts evaluating the reconstructions and averaging their perceptual quality. Then, at different acceleration factors, we tested the noise reduction effect of all models at different noise level to prove that our model could significantly suppress the residual noise. To test the noise tolerance of different CS-MRI methods, we added white Gaussian noises to the k-space data before applying the undersampling. Inspired by [17], we conducted a noise level estimation for all the reconstruction results. Finally, we tested the effectiveness of various network configurations of our proposed framework. In the final quantification, we used PSNR, SSIM and NMSE as the evaluation metrics.

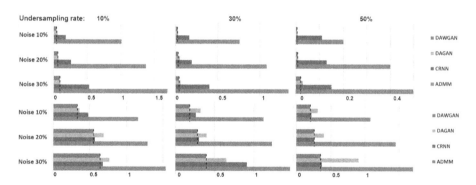

Fig. 3. Estimated residual noise from the reconstructed images with respect to different noise levels. Upper panel: brain MRI data; Lower panel: cardiac MRI data.

3.2   Results

Image Quality Comparison. Our method has demonstrated the best performance by comparing with four SATO methods (on both brain and cardiac MRI datasets). Table 1 shows that the results of our proposed DAWGAN performed best in PSNR and MOS. At the $10\times$ and $3.3\times$ acceleration factors, the PSNR and MOS achieved by our DAWGAN were significantly higher than the other methods. In addition, Fig. 2 shows that DAWGAN produced less noise-induced artefacts in all the simulation studies, while the other methods had

Table 2. Ablation study results of our framework with various network configurations.

Brain MRI Data

Comparison Study	10%				30%				50%			
	PSNR	NMSE	SSIM		PSNR	NMSE	SSIM		PSNR	NMSE	SSIM	
WGAN-GP+RNN	34.11(3.91)	0.16(0.03)	0.96(0.02)		40.95(4.13)	0.07(0.02)	0.98(0.01)		46.08(4.37)	0.03(0.01)	0.99(0.00)	
WGAN-GP+Attention	34.22(2.01)	0.17(0.03)	0.96(0.02)		39.90(3.57)	0.08(0.02)	0.98(0.01)		46.29(3.81)	0.03(0.01)	0.99(0.00)	
Attention+RNN	33.71(3.30)	0.16(0.03)	0.96(0.02)		40.74(3.57)	0.08(0.02)	0.98(0.01)		46.29(3.81)	0.03(0.01)	0.99(0.00)	
DAWGAN	**34.31(3.01)**	**0.16(0.03)**	**0.96(0.02)**		**40.74(3.57)**	**0.07(0.02)**	**0.98(0.01)**		**46.43(2.19)**	**0.03(0.01)**	**0.99(0.00)**	

Cardiac MRI Data

	PSNR	NMSE	SSIM		PSNR	NMSE	SSIM		PSNR	NMSE	SSIM	
WGAN-GP+RNN	30.62(2.33)	0.24(0.03)	0.89(0.02)		34.51(2.40)	0.15(0.02)	0.94(0.01)		38.44(2.51)	0.09(0.01)	0.97(0.00)	
WGAN-GP+Attention	30.68(2.61)	0.23(0.03)	0.89(0.02)		34.81(2.13)	0.14(0.02)	0.94(0.01)		37.87(2.54)	0.10(0.01)	0.96(0.00)	
Attention+RNN	30.51(2.85)	0.23(0.03)	0.87(0.02)		34.35(2.70)	0.15(0.02)	0.93(0.01)		37.67(2.10)	0.10(0.01)	0.96(0.00)	
DAWGAN	**31.06(1.71)**	**0.23(0.02)**	**0.90(0.02)**		**35.97(1.77)**	**0.13(0.01)**	**0.96(0.01)**		**39.66(1.79)**	**0.09(0.01)**	**0.97(0.00)**	

more noise-induced artefacts. Although CRNN and DAGAN could also suppress some artefacts, the reconstructions of the brain area were less detailed than those DAWGAN reconstructed. Moreover, ADMM and zero-filling could not effectively inhibit remaining aliasing artefacts.

Noise Suppression Comparison. In terms of reconstruction details, we demonstrated that DAWGAN could effectively reduce residual noise in the reconstructed images. As shown in Fig. 3, DAWGAN suppressed the noise effectively at different noise levels. Figure 4 shows the PSNR results with respect to different noise levels and various undersampling patterns. Our proposed DAWGAN also demonstrated considerable noise tolerance at different noise levels, and the mean value of the PSNR was higher than other methods.

Ablation Studies. The performance of our framework with various network components was shown by our ablation studies. The sub-models we compared were WGAN-GP+RNN, WGAN-GP+Attention and Attention+RNN. Table 2 shows that the DAWGAN full model was superior to other sub-model variations using all three metrics, which indicated that the current configurations in our proposed network architecture are effective.

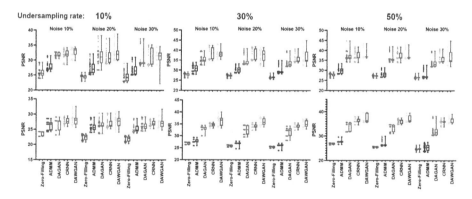

Fig. 4. PSNR with respect to different noise levels at various undersampling ratios, i.e., 10%, 30% and 50%, respectively. The upper panel shows the results of the brain MRI dataset, and the lower panel shows the results of the cardiac MRI dataset.

4   Conclusion

In this paper, we proposed the DAWGAN to reconstruct MRI images from highly undersampled k-space data. Our DAWGAN employed WGAN-GP to improve the stability of vanilla GAN. The incorporated Bi-ConvLSTM block can make full use of the relationships among successive MRI slices to improve the reconstruction results. In addition, the proposed SAB can distinguish between significant and non-significant features for our MRI reconstruction task. Our ablation studies have demonstrated the effectiveness of the key components of our framework. The comprehensive comparison studies on both brain and cardiac MRI datasets have corroborated that our method can not only achieve better reconstruction results but can also effectively reduce residual noise generated during the reconstruction process.

References

1. Donoho, D.L., et al.: Compressed sensing. IEEE Trans. Inf. Theory **52**(4), 1289–1306 (2006)
2. Lustig, M., Donoho, D., Pauly, J.M.: Sparse MRI: the application of compressed sensing for rapid MR imaging. Magn. Reson. Med. Off. J. Int. Soc. Magn. Reson. Med. **58**(6), 1182–1195 (2007)
3. Hollingsworth, K.G.: Reducing acquisition time in clinical MRI by data undersampling and compressed sensing reconstruction. Phys. Med. Biol. **60**(21), R297 (2015)
4. Ma, S., Yin, W., Zhang, Y., Chakraborty, A.: An efficient algorithm for compressed MR imaging using total variation and wavelets. In: 2008 IEEE Conference on Computer Vision and Pattern Recognition (2008)
5. Wang, S., et al.: Accelerating magnetic resonance imaging via deep learning. In: 2016 IEEE 13th International Symposium on Biomedical Imaging, pp. 514–517. IEEE (2016)
6. Yang, G., et al.: DAGAN: deep de-aliasing generative adversarial networks for fast compressed sensing MRI reconstruction. IEEE Trans. Med. Imaging **37**(6), 1310–1321 (2017)
7. Schlemper, J., Caballero, J., Hajnal, J.V., Price, A.N., Rueckert, D.: A deep cascade of convolutional neural networks for dynamic MR image reconstruction. IEEE Trans. Med. Imaging **37**(2), 491–503 (2017)
8. Qin, C., Schlemper, J., Caballero, J., Price, A.N., Hajnal, J.V., Rueckert, D.: Convolutional recurrent neural networks for dynamic MR image reconstruction. IEEE Trans. Med. Imaging **38**(1), 280–290 (2018)
9. Sun, J., Li, H., Xu, Z., et al.: Deep ADMM-Net for compressive sensing MRI. In: Advances in Neural Information Processing Systems, pp. 10–18 (2016)
10. Mardani, M., et al.: Deep generative adversarial neural networks for compressive sensing MRI. IEEE Trans. Med. Imaging **38**(1), 167–179 (2018)
11. Lee, D., Yoo, J., Ye, J.C.: Deep residual learning for compressed sensing MRI. In: 2017 IEEE 14th International Symposium on Biomedical Imaging, pp. 15–18. IEEE (2017)
12. Johnson, J., Alahi, A., Fei-Fei, L.: Perceptual losses for real-time style transfer and super-resolution. In: Leibe, B., Matas, J., Sebe, N., Welling, M. (eds.) ECCV 2016. LNCS, vol. 9906, pp. 694–711. Springer, Cham (2016). https://doi.org/10.1007/978-3-319-46475-6_43
13. Woo, S., Park, J., Lee, J.Y., Kweon, I.S.: CBAM: convolutional block attention module. In: Proceedings of the European Conference on Computer Vision (ECCV), pp. 3–19 (2018)

14. Goodfellow, I., et al.: Generative adversarial nets. In: Advances in Neural Information Processing Systems, pp. 2672–2680 (2014)
15. Arjovsky, M., Chintala, S., Bottou, L.: Wasserstein GAN. arXiv preprint arXiv:1701.07875 (2017)
16. Gulrajani, I., Ahmed, F., Arjovsky, M., Dumoulin, V., Courville, A.C.: Improved training of wasserstein GANs. In: Advances in Neural Information Processing Systems, pp. 5767–5777 (2017)
17. Liu, X., Tanaka, M., Okutomi, M.: Single-image noise level estimation for blind denoising. IEEE Trans. Image Process. **22**(12), 5226–5237 (2013)
18. Han, Y., Yoo, J., Kim, H.H., Shin, H.J., Sung, K., Ye, J.C.: Deep learning with domain adaptation for accelerated projection-reconstruction MR. Magn. Reson. Med. **80**(3), 1189–1205 (2018)
19. Wang, S., Huang, N., Zhao, T., Yang, Y., Ying, L., Liang, D.: 1D partial fourier parallel MR imaging with deep convolutional neural network. In: Proceedings of the 25st Annual Meeting of ISMRM, Honolulu, HI, USA, vol. 1, p. 2 (2017)
20. Quan, T.M., Jeong, W.-K.: Compressed sensing dynamic MRI reconstruction using GPU-accelerated 3D convolutional sparse coding. In: Ourselin, S., Joskowicz, L., Sabuncu, M.R., Unal, G., Wells, W. (eds.) MICCAI 2016. LNCS, vol. 9902, pp. 484–492. Springer, Cham (2016). https://doi.org/10.1007/978-3-319-46726-9_56
21. Seitzer, M., et al.: Adversarial and perceptual refinement for compressed sensing MRI reconstruction. In: Frangi, A.F., Schnabel, J.A., Davatzikos, C., Alberola-López, C., Fichtinger, G. (eds.) MICCAI 2018. LNCS, vol. 11070, pp. 232–240. Springer, Cham (2018). https://doi.org/10.1007/978-3-030-00928-1_27
22. Zhang, P., Wang, F., Xu, W., Li, Y.: Multi-channel generative adversarial network for parallel magnetic resonance image reconstruction in k-space. In: Frangi, A.F., Schnabel, J.A., Davatzikos, C., Alberola-López, C., Fichtinger, G. (eds.) MICCAI 2018. LNCS, vol. 11070, pp. 180–188. Springer, Cham (2018). https://doi.org/10.1007/978-3-030-00928-1_21
23. Duan, J., et al.: VS-Net: variable splitting network for accelerated parallel MRI reconstruction. In: Shen, D., et al. (eds.) MICCAI 2019. LNCS, vol. 11767, pp. 713–722. Springer, Cham (2019). https://doi.org/10.1007/978-3-030-32251-9_78
24. Schlemper, J., et al.: Stochastic deep compressive sensing for the reconstruction of diffusion tensor cardiac MRI. In: Frangi, A.F., Schnabel, J.A., Davatzikos, C., Alberola-López, C., Fichtinger, G. (eds.) MICCAI 2018. LNCS, vol. 11070, pp. 295–303. Springer, Cham (2018). https://doi.org/10.1007/978-3-030-00928-1_34
25. Quan, T.M., Nguyen-Duc, T., Jeong, W.K.: Compressed sensing MRI reconstruction using a generative adversarial network with a cyclic loss. IEEE Trans. Med. Imaging **37**(6), 1488–1497 (2018)
26. Hammernik, K., et al.: Learning a variational network for reconstruction of accelerated MRI data. Magn. Reson. Med. **79**(6), 3055–3071 (2018)

Learning MRI k-Space Subsampling Pattern Using Progressive Weight Pruning

Kai Xuan[1,2], Shanhui Sun[3], Zhong Xue[1], Qian Wang[2(✉)], and Shu Liao[1(✉)]

[1] Shanghai United Imaging Intelligence Co. Ltd., Shanghai, China
shu.liao@united-imaging.com
[2] Institute for Medical Imaging Technology, School of Biomedical Engineering,
Shanghai Jiao Tong University, Shanghai, China
wang.qian@sjtu.edu.cn
[3] United Imaging Intelligence, Cambridge, MA, USA

Abstract. Magnetic resonance (MR) imaging is widely used in clinical scenarios, while the long acquisition time is still one of its major limitations. An efficient way to accelerate the imaging process is to subsample the k-space, where MR signal is physically acquired, and then estimate the fully-sampled MR image from subsampled signal with a learned reconstruction model. In this work, we are inspired from the idea of neural network pruning and propose a novel strategy to learn the k-space subsampling pattern and the reconstruction model alternately in a data-driven fashion. More specifically, in each iteration of learning, we first greedily eliminate a few phases that are considered less important in the k-space according to their assigned weights, and then fine-tune the reconstruction model. In our pilot study, experiments demonstrated the robustness and superiority of our proposed method in both single- and multi-modal MRI settings.

Keywords: Image reconstruction · Learning-based subsampling · Magnetic resonance imaging

1 Introduction

The long acquisition time, as a consequence of many factors, is always a main bottleneck of magnetic resonance (MR) imaging, impeding its more widely usage in clinical scenarios. One of the most commonly used acceleration strategies is to subsample the k-space, where MR signals are physically acquired, and then reconstruct the MR image (MRI) from subsampled data. In the context of 2D MRI with Cartesian trajectories, which is widely used in clinical practice, the k-space is a 2D matrix with one axis named read-out direction and the other one called phase-encoding direction with different phase shifts. Also, k-space data is the Fourier transform of corresponding MR image thus smaller phase shift is equivalent to lower frequency, and vice versa. In such circumstance a

A. L. Martel et al. (Eds.): MICCAI 2020, LNCS 12262, pp. 178–187, 2020.
https://doi.org/10.1007/978-3-030-59713-9_18

valid subsampling pattern will be a 1D binary vector because only subsampling in phase-encoding direction can actually save acquisition time, determined by physical proprieties. There are two fundamental factors affecting the final reconstruction quality. One is the image reconstruction method, while the other is the design of the subsampling pattern.

A lot of image reconstruction methods have been successfully applied to MRI. Originally, simple techniques including linear filtering, reflecting, or zero-filling are used to recover MR images [3,6]. Sooner, more complex algorithms like Bayesian-based reconstruction are also adopted [13,14]. A milestone to this solution is the introduction of compressed sensing (CS) [2] when Lusting et al. [11] proposed CS-MRI. More recent years have witnessed a rapid improvement of reconstruction quality with more powerful machine learning algorithms like sparse encoding and deep learning [17,20].

With the fast development of reconstruction methods, the design of the k-space subsampling pattern is however less explored. Compared to manually designed subsampling trajectories like random sampling with uniform distribution or variable density [12,18], optimized patterns tend to perform better [24]. To optimize such a binary vector in a data-driven way, Zijlstra et al. [24] used Monte Carlo optimization and Liu et al. [9] adopted the greedy strategy. Computational cost is a large difficulty for these methods because to evaluate a pattern, images have to be reconstructed before quality assessment, which is time-consuming. As an alternative, the value of a pattern is sometimes approximated with indirect measurements like moment-based spectral analysis [8] or Cramér–Rao lower bound (CRLB) [4].

The computation becomes much more expensive when a reconstruction model has many learnable parameters. The subsampling operation is non-differentiable and is thus incompatible with gradient-descent-based algorithms like back propagation (BP). To optimize the pattern and a deep neural network jointly, Bahadir et al. [1] relaxed the sampling operation with a "soft" threshold, Zhang et al. [23] proposed to approximate the importance of each phase adversely with a discriminator, and Jin et al. [7] used reinforcement learning to solve this problem.

The task of neural network (NN) pruning is similar to the optimization of the subsampling pattern. First, they both aim at eliminating less important parts (i.e., neurons or channels for NN, phases for MRI) with minimal loss in performance. Then, the pruning process and the subsampling operation are both non-differentiable. Also, the measurement of exact importance is extremely expensive, and it is usually approximated, e.g. with the magnitude of weight [10], with sparsity of activation [5], and with Taylor expansion [15].

In this work, we propose to optimize the k-space subsampling pattern and the deep-learning-based reconstruction model jointly in a data-driven fashion, while with a computational complexity comparable to training only the reconstruction model with a fixed pattern. More specifically, the pruning of less important phases and optimization of the neural network are processed alternatively until the sampling sparsity requirement in the k-space is met. To measure the important of each phase, inspired by neural network pruning, we multiply each phase with a learnable weight before reconstruction. Experiments on large open-accessible dataset proved the robustness and effectiveness of our proposed method, and with this tool we also analyzed the benefit brought by multi-modal MR images.

2   Methods

In this section, we will first mathematically formulate the MRI reconstruction problem, and then introduce our strategy to optimize the subsampling pattern and the reconstruction model jointly. An overview of our proposed framework and the flowchart can be found in Fig. 1.

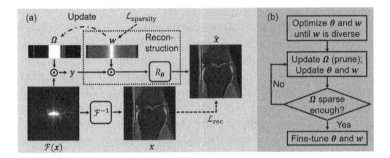

Fig. 1. The framework (a) and flowchart (b) of our proposed method. (a) The k-space signal $\mathcal{F}(\boldsymbol{x})$ is subsampled according to pattern $\boldsymbol{\Omega}$. Then, along phase-encoding direction, k-space signal is multiplied by weight vector $\boldsymbol{w}$, which will later guide the update of $\boldsymbol{\Omega}$. At last, a deep-learning-based reconstruction model R_θ is used to reconstruct $\boldsymbol{x}$ from subsampled signal. (b) To orchestrate the optimization process, first $\boldsymbol{\theta}$ and $\boldsymbol{w}$ are trained until $\boldsymbol{w}$ is diverse enough. Then $\boldsymbol{\Omega}$ is pruned according to $|\boldsymbol{w}|$ while $\boldsymbol{\theta}$ and $\boldsymbol{w}$ are optimized with BP. These two optimization processes continue alternatively until $\boldsymbol{\Omega}$ is sparse enough, after which $\boldsymbol{\theta}$ and $\boldsymbol{w}$ are fine-tuned without sparsity constraint.

2.1   MRI Reconstruction

With 2D MR image $\boldsymbol{x} \in \mathbb{C}^{N \times N}$ reconstructed from a fully sampled k-space, 2D discrete Fourier transformation (DFT) $\mathcal{F}(\cdot)$, the binary subsampling pattern $\boldsymbol{\Omega}$ with length N where $\|\boldsymbol{\Omega}\|_0 = M$, the subsampling process can be formulated as

$$\boldsymbol{y} = \text{diag}(\boldsymbol{\Omega})\mathcal{F}(\boldsymbol{x}), \tag{1}$$

where $\boldsymbol{y} \in \mathbb{C}^{N \times N}$ is the zero-padded k-space sample with at most M non-zeros columns (phase direction). Without loss of generality, the MRI reconstruction process can be notated as $R_\theta(\cdot)$, where R is the composition of reconstruction operations and $\boldsymbol{\theta}$ is the set of corresponding learnable parameters (if any). For example, in the context of reconstructing MRI in image space with a deep neural network, the reconstruction process can be the cascaded inverse Fourier transformation and deep neural network G with parameters $\boldsymbol{\theta}$, i.e. $R_\theta(\cdot) = G_\theta\left(\mathcal{F}^{-1}(\cdot)\right)$.

2.2   Subsampling by Pruning

In this paper, following Liu *et al.* [10], we propose to assign a weight w for each phase, and approximate the importance with magnitude of weight. Also, L1 loss is imposed on w during the training of the reconstruction model in favor of sparsity. Now, the loss function for the reconstruction model is

$$\theta^*, \boldsymbol{w}^* = \arg\min_{\theta,\boldsymbol{w}} \underbrace{\left\| \boldsymbol{x} - G_\theta\left(\mathcal{F}^{-1}\left(\text{diag}\left(\boldsymbol{w} \odot \boldsymbol{\Omega}\right) \mathcal{F}\left(\boldsymbol{x}\right)\right)\right)\right\|_1}_{\mathcal{L}_{\text{rec}}} + \lambda \underbrace{\|\boldsymbol{w}\|_1}_{\mathcal{L}_{\text{sparsity}}} \quad (2)$$

where λ is a hyper-parameter balancing two terms. $\|\cdot\|_1$ is L1 norm minimizing reconstruction error ($\mathcal{L}_{\text{rec}}$) and encouraging sparsity ($\mathcal{L}_{\text{sparsity}}$).

The subsampling pattern and reconstruction model are optimized alternately. In the ith iteration, the subsampling pattern is gradually updated following

$$\boldsymbol{\Omega}^i = \arg\max_{\Omega} \|(|\boldsymbol{w}| + \boldsymbol{n}) \odot \boldsymbol{\Omega}\|_1 \quad \text{s.t.} \quad \|\boldsymbol{\Omega} \odot \boldsymbol{\Omega}^{i-1}\|_0 = M^i, \quad (3)$$

where $\boldsymbol{n}$ is additive noise. The idea behind adding this noise term is that it helps the optimizer avoid a greedy pruning in a local phase region. The greedy pruning tends to yield a sub-optimal solution. M^i is the desired sparsity in i^{th} iteration with $M^0 = N$, $M^i = M^{i-1} - K$ and similarly, $\boldsymbol{\Omega}^0 = 1$. This equation can simply be interpreted as eliminating K phases with smallest noisy absolute weight $|w|$ in each iteration. When noise level is zero, (3) will act in a greedy way. More analysis regarding the motivation of using the additive noise term can be found in experiment and discussion section.

Figure 1(b) depicts a brief description about the proposed algorithm using a flowchart. To make major phases distinguishable enough from less important ones, first θ and $\boldsymbol{w}$ are trained according to (2) with $\boldsymbol{\Omega}^0$ until $\boldsymbol{w}$ is visually diverse enough. In each iteration, $\boldsymbol{\Omega}$ is pruned according to (3) while θ and $\boldsymbol{w}$ are optimized following (2). These two optimization processes continue alternatively until $\boldsymbol{\Omega}$ meets desired sparsity. Finally, θ and $\boldsymbol{w}$ are fine-tuned without $\mathcal{L}_{\text{sparsity}}$ since $\boldsymbol{\Omega}$ already meets our desired sparsity.

2.3   Multi-modal MRI

In addition to single modal MRI reconstruction, we extend our algorithm to multi-modal MRI reconstruction. MR images acquired in different modalities contain highly related information as well as complementary information. The utilization of multi-modal data for MRI reconstruction demonstrated a better image quality in previous work [21]. It is also clinically desirable, since multi-modal images are often acquired for a single visit of a patient to facilitate the diagnosis and treatment. With our proposed method, we can analyze the benefit brought by an additional modality $\boldsymbol{x}' \in \mathbb{C}^{N \times N}$ by comparing $\boldsymbol{w}$ trained with and without $\boldsymbol{x}'$. To be concrete, if some phases can be recovered from another modality, their importance, approximated by $\boldsymbol{w}$, will most probably decrease. And also, the reconstructed image may hold better quality since $\boldsymbol{\Omega}$ is freed from acquiring redundant information that x' already have. Figure 2 illustrates the multi-modal reconstruction procedure. The auxiliary modality is fused with subsampled MR signal by feeding channel-wise-concatenated x' and $\mathcal{F}^{-1}(\mathrm{diag}(w)y)$ to a neural network, which is a cascade of residual blocks and heading/tailing convolution layers adjusting channel number. The reconstruction process is the same for single-modal MRI except the non-existing x' will be set to all zeros.

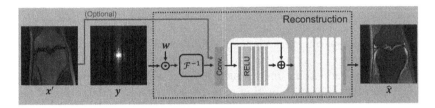

Fig. 2. A close view of MRI reconstruction. For multi-modal MRI, an auxiliary modality x' is concatenated with MR image subsampled in k-space as the input of the neural network reconstructing in the image space. In case multi-modal MR images are not available, x' is replaced with $O_{N \times N}$ and it reduces to the single-model setting.

3 Experiments and Discussion

In this section, first we validate our proposed algorithm on a single modal MRI reconstruction task. We compared our method to different subsampling methods. In addition, we perform the evaluation on multi-modal MRI reconstruction task. Furthermore, we perform an ablation study to understand the insights behind our algorithm design.

Experiments were conducted on raw k-space measurement from the large, open-accessible knee joint MRI dataset released by fastMRI [22]. Although there's no official guarantee of multi-modal MRI, we found in training and validation set, both proton-density (PD) and PD with fat-suppression (PDFS) modalities were acquired and they are almost aligned. Further, noisy or mismatched image pairs were eliminated by manually checking. At last, 227 and 45 pairs of PD and PDFS volumes were selected out of the whole fasMRI knee training/validation set, respectively. We used a total of 8332 pairs of 2D slices for training and 1665 for validation.

Figure 2 illustrates our neural network structure. The number of feature maps in each layer was 64 except in the input/output layer. Learning rate was 0.001 for θ and 0.0001 for w and they were optimized with Adam optimizer with default settings. θ and w were trained for 8 epochs before the gradual pruning in Fig. 1(b). And in each iteration, Ω was updated by pruning 2 phases ($K = 2$) and θ, w were optimized over 50 batches of training data. All MR images were center cropped to 320×320 ($N = 320$) for training and batch size was set to 10 in all settings. Our implementation was based on PyTorch [16] and SciPy [19].

In our experiments, we compared our method to other k-space sampling methods in terms of peak signal-to-noise ratio (PSNR). Note that PSNR is a sufficient quantitative index for comparing quality of sampling pattern taking the role in MRI reconstruction tasks.

3.1 Single-Modal MRI Reconstruction

We compared our algorithm to two widely used subsampling methods: random (Random) and equispaced (Equispaced) sampling, as well as LOUPE [1], a learning based method. Table 1 reports quantitative comparison of all methods on PD and PDFS MRI with an acceleration rate of 4 (4×) and 8 (8×). Overall, from Table 1 we can observe that methods using learned patterns produce higher PSNR results compared with those non-learned subsampling patterns given by Random and Equispaced methods. Among methods with learned subsampling patterns, our proposed method without noise (w/o Noise) is better than LOUPE. In addition, adding noise (w/ Noise) further improves its performance.

Table 1. Results of comparison of different sampling patterns in terms of PSNR. Note that w/ Noise and w/o Noise represent our proposed method pruning with and without noise n in (3), respectively.

PSNR	Random	Equispaced	LOUPE	w/o Noise	w/ Noise
PDFS (4×)	31.51	31.47	31.74	31.91	31.97
PDFS (8×)	29.54	29.76	30.46	30.66	30.83
PD (4×)	35.15	35.82	35.88	36.96	37.11
PD (8×)	31.95	32.32	33.91	34.62	34.64

Figure 3 shows examples of MRI reconstruction results using compared algorithms for both PDFS and PD based on 4× and 8× acceleration rates. Reconstructed images with learned subsampling patterns contain more details and less artifacts. It is worth noticing that our proposed method results a low-pass-like pattern when learned a sampling pattern without noise (See (3)). This result demonstrates a drawback of utilizing greedy weight pruning method. It turns out that greedy algorithm cannot fully utilize the coherence between phase shifts. In contrast to the greedy weight pruning, adding noise in (3) gives the pruning algorithm chance to explore more regions and mitigates the coherence problem.

3.2 Multi-modal MRI Reconstruction

In the context of multi-modal MRI reconstruction, we have 4× accelerate rate k-space sampling for PDFS input of the neural network and a fully-sampled PD as another input (Fig. 2). Our proposed method of learning subsampling pattern with noise achieves 32.07 dB in terms of PSRN. The multi-modality with

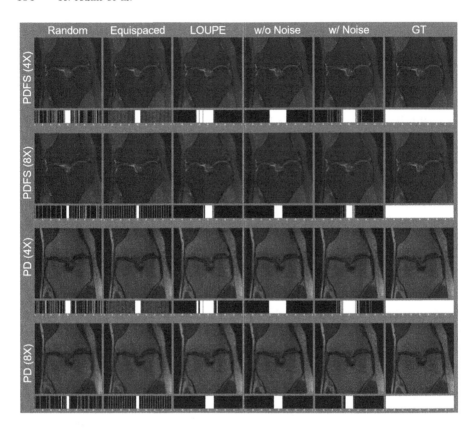

Fig. 3. Visual comparison of reconstructed image and subsampling pattern of different methods and data. From left to right are uniformly random (Random) sampling, equally spaced (Equispaced) sampling, LOUPE [1], our proposed method without (w/o Noise) and with (w/ Noise) noise, and fully sampled image (GT).

our method brings 0.1 dB increase in PSNR than single modal PDFS reconstruction (31.97 dB, Table 1). Figure 4(b) shows a comparison example using learned subsampling pattern for both single modal reconstruction and multimodal reconstruction. With a complementary modality, more textural information is reconstructed in the multi-modal setting than single one. From Fig. 4(b), we observe that the learned subsampling pattern pays less attention to middle frequencies than single one. Actually, w can explain this observation. As shown in Fig. 4(a), the weight for middle frequency is significantly lower that w trained in single-modal setting, even lower than the high frequency band. As a result, the learned subsampling pattern is mostly concentrated on the low frequency band, which is always the most important part, and then is the high frequency band, and nothing in middle frequency. We heuristically interpenetrate the change of weight as the information brought by another modality mostly comes from middle frequency. One may also note that weight of highest frequency decreases extremely fast in both settings, because sometimes phases with largest shift are just ignored during acquisition. In contrast, weight for lowest frequency barely declines during training, indicating they are the most important phase encoding.

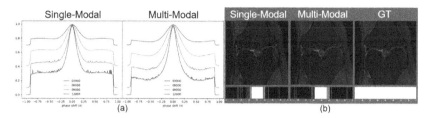

Fig. 4. Visual comparison of (a) w trained for 3000, 6000, 9000, and 12000 batches, (b) reconstructed image and subsampling pattern with single- and multi-modal MRI. With the existence of another modality, more details can be reconstructed, and weight for middle frequency decreases, leading to subsampling pattern paying less attention to this band.

3.3   Effect of Sparsity Constraint

Sparsity constraint $\mathcal{L}_{\text{sparsity}}$ in (2) is essential in our proposed method. The hyper-parameter λ is to encourage w to be diverse. We set it to 0.01 as a rule-of-thumb so that the important phase shifts are distinguishable enough from minor ones. Effect of different λ in (2) can be found in Fig. 5. Figure 5 indicates that when λ is too large (e.g. 0.1), there is not much difference between weights for high frequency and middle frequency. If it is a too small (e.g. 0.001), λ barely decreases any weight.

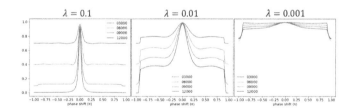

Fig. 5. Comparison of w after trained for 3000, 6000, 9000, and 12000 batches under different λ. w is not diverse enough without proper λ.

4   Conclusion

In this work, we proposed a new method to optimize the k-space subsampling pattern using a progressive weight pruning method. In our method, we simultaneously optimize the subsampling pattern and the deep learning based reconstruction model directly driven by the training data. Moreover, we introduced a random exploration strategy during the optimization which let the learner avoid performing greedy weight pruning. Experiments demonstrated the robustness and superiority of our proposed method. In addition to the single modal MRI reconstruction application, we found that the proposed sampling method is beneficial for a multi-modal MRI reconstruction application. Further research directions include broader application of our proposed method and better importance measurement, and the design of loss functions is also worth discussing.

Acknowledgment. This work was partially supported by the National Key Research and Development Program of China (2018YFC0116400), Science and Technology Commission of Shanghai Municipality (19QC1400600), and China Scholarship Council.

References

1. Bahadir, C.D., Dalca, A.V., Sabuncu, M.R.: Learning-based optimization of the under-sampling pattern in MRI. In: Chung, A.C.S., Gee, J.C., Yushkevich, P.A., Bao, S. (eds.) IPMI 2019. LNCS, vol. 11492, pp. 780–792. Springer, Cham (2019). https://doi.org/10.1007/978-3-030-20351-1_61
2. Donoho, D.L.: Compressed sensing. IEEE Trans. Inf. Theory **52**(4), 1289–1306 (2006)
3. Feinberg, D.A., Hale, J.D., Watts, J.C., Kaufman, L., Mark, A.: Halving MR imaging time by conjugation: demonstration at 3.5 kG. Radiology **161**(2), 527–531 (1986)
4. Haldar, J.P., Kim, D.: OEDIPUS: an experiment design framework for sparsity-constrained MRI. IEEE Trans. Med. Imaging **38**(7), 1545–1558 (2019)
5. Hu, H., Peng, R., Tai, Y.W., Tang, C.K.: Network trimming: a data-driven neuron pruning approach towards efficient deep architectures. arXiv:1607.03250 [cs], July 2016
6. Jackson, J., Meyer, C., Nishimura, D., Macovski, A.: Selection of a convolution function for Fourier inversion using gridding (computerised tomography application). IEEE Trans. Med. Imaging **10**(3), 473–478 (1991)
7. Jin, K.H., Unser, M., Yi, K.M.: Self-supervised deep active accelerated MRI. arXiv:1901.04547 [cs], January 2019
8. Levine, E., Hargreaves, B.: On-the-fly adaptive k-space sampling for linear MRI reconstruction using moment-based spectral analysis. IEEE Trans. Med. Imaging **37**(2), 557–567 (2018)
9. Liu, D.D., Liang, D., Liu, X., Zhang, Y.T.: Under-sampling trajectory design for compressed sensing MRI. In: 2012 Annual International Conference of the IEEE Engineering in Medicine and Biology Society, pp. 73–76, August 2012
10. Liu, Z., Li, J., Shen, Z., Huang, G., Yan, S., Zhang, C.: Learning efficient convolutional networks through network slimming. In: 2017 IEEE International Conference on Computer Vision (ICCV), pp. 2755–2763, October 2017
11. Lustig, M., Donoho, D.L., Santos, J.M., Pauly, J.M.: Compressed sensing MRI. IEEE Signal Process. Mag. **25**(2), 72–82 (2008)
12. Lustig, M., Donoho, D., Pauly, J.M.: Sparse MRI: the application of compressed sensing for rapid MR imaging. Magn. Reson. Med. **58**(6), 1182–1195 (2007)
13. Marseille, G.J., de Beer, R., Fuderer, M., Mehlkopf, A.F., van Ormondt, D.: Nonuniform phase-encode distributions for MRI scan time reduction. J. Magn. Reson., Ser. B **111**(1), 70–75 (1996)

14. McGibney, G., Smith, M.R., Nichols, S.T., Crawley, A.: Quantitative evaluation of several partial fourier reconstruction algorithms used in MRI. Magn. Reson. Med. **30**(1), 51–59 (1993)
15. Molchanov, P., Mallya, A., Tyree, S., Frosio, I., Kautz, J.: Importance estimation for neural network pruning. In: 2019 IEEE/CVF Conference on Computer Vision and Pattern Recognition (CVPR), pp. 11256–11264, June 2019
16. Paszke, A., et al.: PyTorch: An imperative style, high-performance deep learning library. In: Advances in Neural Information Processing Systems 32, pp. 8024–8035. Curran Associates, Inc. (2019)
17. Ravishankar, S., Bresler, Y.: MR image reconstruction from highly undersampled k-space data by dictionary learning. IEEE Trans. Med. Imaging **30**(5), 1028–1041 (2011)
18. Tsai, C.M., Nishimura, D.G.: Reduced aliasing artifacts using variable-density k-space sampling trajectories. Magn. Reson. Med. **43**(3), 452–458 (2000)
19. Virtanen, P., et al.: SciPy 1.0: fundamental algorithms for scientific computing in python. Nat. Methods **17**, 261–272 (2020)
20. Wang, S., et al.: Accelerating magnetic resonance imaging via deep learning. In: 2016 IEEE 13th International Symposium on Biomedical Imaging (ISBI), pp. 514–517, April 2016
21. Xiang, L., et al.: Ultra-fast T2-weighted MR reconstruction using complementary T1-weighted information. In: Frangi, A.F., Schnabel, J.A., Davatzikos, C., Alberola-López, C., Fichtinger, G. (eds.) MICCAI 2018. LNCS, vol. 11070, pp. 215–223. Springer, Cham (2018). https://doi.org/10.1007/978-3-030-00928-1_25
22. Zbontar, J., et al.: fastMRI: an open dataset and benchmarks for accelerated MRI. arXiv:1811.08839 [physics, stat], December 2019
23. Zhang, Z., Romero, A., Muckley, M.J., Vincent, P., Yang, L., Drozdzal, M.: Reducing uncertainty in undersampled MRI reconstruction with active acquisition. In: Proceedings of the IEEE Conference on Computer Vision and Pattern Recognition, pp. 2049–2058 (2019)
24. Zijlstra, F., Viergever, M.A., Seevinck, P.R.: Evaluation of variable density and data-driven K-space undersampling for compressed sensing magnetic resonance imaging. Invest. Radiol. **51**(6), 410–419 (2016)

Model-Driven Deep Attention Network for Ultra-fast Compressive Sensing MRI Guided by Cross-contrast MR Image

Yan Yang, Na Wang, Heran Yang, Jian Sun$^{(\boxtimes)}$, and Zongben Xu

Xi'an Jiaotong University, Xi'an 710049, China
{yangyan92,nawang2018,yhr.7017}@stu.xjtu.edu.cn,
{jiansun,zbxu}@xjtu.edu.cn

Abstract. Speeding up Magnetic Resonance Imaging (MRI) is an inevitable task in capturing multi-contrast MR images for medical diagnosis. In MRI, some sequences, e.g., in T2 weighted imaging, require long scanning time, while T1 weighted images are captured by short-time sequences. To accelerate MRI, in this paper, we propose a model-driven deep attention network, dubbed as MD-DAN, to reconstruct highly under-sampled long-time sequence MR image with the guidance of a certain short-time sequence MR image. MD-DAN is a novel deep architecture inspired by the iterative algorithm optimizing a novel MRI reconstruction model regularized by cross-contrast prior using a guided contrast image. The network is designed to automatically learn cross-contrast prior by learning corresponding proximal operator. The backbone network to model the proximal operator is designed as a dual-path convolutional network with channel and spatial attention modules. Experimental results on a brain MRI dataset substantiate the superiority of our method with significantly improved accuracy. For example, MD-DAN achieves PSNR up to 35.04 dB at the ultra-fast 1/32 sampling rate.

Keywords: Multi-contrast MRI · Prior learning · Model-driven net

1 Introduction

Magnetic Resonance Imaging (MRI) is a leading biomedical imaging technology, which depicts both anatomical and functional information for disease diagnosis. A typical MRI protocol comprises multi-contrast sequences of same anatomy which provide complementary information to enhance clinical diagnosis. For

---

Y. Yang, N. Wang—Contributed equally to this work.

Electronic supplementary material The online version of this chapter (https://doi.org/10.1007/978-3-030-59713-9_19) contains supplementary material, which is available to authorized users.

© Springer Nature Switzerland AG 2020
A. L. Martel et al. (Eds.): MICCAI 2020, LNCS 12262, pp. 188–198, 2020.
https://doi.org/10.1007/978-3-030-59713-9_19

instance, T1 Weighted Imaging (T1WI) is useful for delineation of morphological information including assessing the gray and white matter and identifying
fatty tissue. T2 Weighted Imaging (T2WI) is useful for delineation of edema
and inflammation. Fluid Attenuated Inversion Recovery (FLAIR) is useful for
suppressing cerebrospinal fluid (CSF) effects on the image to detect the periventricular hyperintense lesions in brain imaging. However, a major challenge in
MRI is its long data acquisition time, for example, typical scanning time for
T1WI, T2WI and FLAIR is about 20 min, limiting its clinical applications in
fetal imaging or dynamic imaging. In clinical routines, some contrast sequences
such as T1WI, requiring short repetition time (TR) and echo time (TE), allow
full sampling, while others, such as T2WI and FLAIR, requiring long TR and
TE, can be accelerated by under-sampling. In this paper, we aim to reconstruct
highly under-sampled long-time sequence MR image (e.g., T2WI) with the guidance of a certain short-time sequence MR image (e.g., T1WI) [26].

Compressive sensing MRI (CS-MRI) [16] is a predominant approach to accelerate MRI by under-sampling in k-space. Traditional model-based CS-MRI
methods rely on regularization related to image prior to improve reconstruction
quality, e.g., Total Variation (TV) [3,10], wavelet regularization [10,15], non-
local regularization [7,20] and dictionary learning [21,28]. However, it is challenging to handcraft an optimal regularizer. Recently, deep learning method has been
applied to CS-MRI. Wang et al. [24] first trained a deep CNN to learn the mapping to high-quality reconstructed images. [13] proposed a multi-scale residual
learning network (i.e., U-net) for image reconstruction by removing aliasing artifacts. [6,9,17,23,27] introduced imaging model or data consistency term to deep
networks to learn the priors of images from training data and greatly improve
reconstruction accuracy. All these CS-MRI methods consider reconstruction of
MRI images with a single contrast (e.g., T1WI, T2WI or FLAIR).

An alternative way to accelerate MRI is to synthesize missing contrast MR
image from other contrast with fully-sampled acquisitions. They either learn a
dictionary or sparse representation of source contrast patches for target contrast [11,22], or directly learn a mapping from source to target contrast by a
deep neural network [4,5,12,14]. However, such methods suffer from low-quantity
reconstruction without requiring samples in k-space. Recently, Xiang et al. [26]
proposed a Dense-Unet to accelerate MRI by reconstruction using both undersampled k-space data and guided MR image. They use a deep network to fuse
under-sampled T2WI image and guided fully-sampled T1WI image, and output
the reconstructed high quality T2WI MR image.

To reconstruct MR image from its under-sampled k-space acquisitions with
guided cross-contrast MR image, we propose a novel interpretable deep attention
network by integrating the k-space data constraint and cross-modality relations
into a single deep architecture. Specifically, we first propose a novel MR image
reconstruction model consisting of a data fidelity term based on k-space data
and a cross-contrast prior term modeling relations between contrasts. Then we
design an iterative algorithm based on half-quadratic splitting to minimize the
model by alternately performing guided image fusion and image reconstruction.

To learn the cross-contrast prior from training data, we substitute its proximal operator in the iterative algorithm by a novel backbone network, namely DPA-FusionNet, which is a dual-path convolutional network with channel and spatial attention. Finally, we unfold the iterative algorithm to be a deep architecture, dubbed as *model-driven deep attention network (MD-DAN)*, as shown in Fig. 1. Experiments on a brain MRI dataset show that our proposed MD-DAN can effectively reconstruct MR image with state-of-the-art reconstruction accuracy even in ultra-fast 1/32 sampling rate.

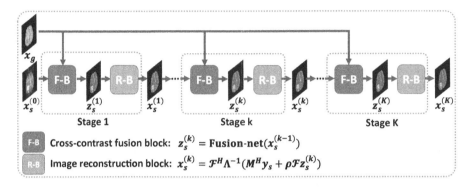

Fig. 1. Illustration of model-driven deep attention network, i.e., MD-DAN. Each stage consists of a cross-contrast fusion block (i.e., F-B) and an image reconstruction block (i.e., R-B). Given a reconstructed image by zero-filling (i.e., $x_s^{(0)}$) and a guided image (i.e., x_g), it outputs reconstructed MR image after K stages.

2   Method

Given under-sampled k-space data $\{y_s \in \mathbb{C}^U\}$ of long-time sequence MR image (e.g., T2WI), and fully sampled reconstructed MR image $\{x_g \in \mathbb{C}^N\}$ of short-time sequence (e.g., T1WI) as guidance, we aim to reconstruct the MR image $\{x_s \in \mathbb{C}^N\}$ from k-space data $\{y_s \in \mathbb{C}^U\}$, where N and U denote the cardinality of MR image and sampled k-space data. Based on the MR imaging mechanism, we design the following MRI reconstruction model:

$$x_s^* = \arg\min_{x_s} \frac{1}{2}||M\mathcal{F}x_s - y_s||_2^2 + \lambda f(x_s, x_g), \tag{1}$$

where $\mathcal{F} \in \mathbb{C}^{N \times N}$ is the Fourier transform and $M \in \mathbb{C}^{U \times N}$ ($U < N$) is the sampling matrix in k-space. The first term is a data term that enforces data consistency between reconstructed MR image x_s and its under-sampled data y_s in k-space. The second term is a *cross-contrast prior* that models the correlation between MR image x_s and guided image x_g. For example, $f(x_s, x_g)$ can be taken as joint TV, group wavelet-sparsity [10], weighted similarity of intensity [25] or similarity of image patches between multi-contrast MR images [20].

In this work, we aim to learn this prior instead of handcrafting it. To this end, we first design an iterative optimization algorithm for solving Eq. (1), where the prior will be transformed to be a proximal operator in the iterative algorithm. We then design a novel fusion network with two contrasts as inputs to replace this proximal operator to implicitly learn the cross-contrast prior. The iterative algorithm can be taken as a deep network and trained end-to-end.

2.1 Model Optimization

The reconstruction model of Eq. (1) can be solved efficiently by half-quadratic splitting (HQS) algorithm [8]. By introducing an auxiliary MR image z_s, s.t., $z_s = x_s$, Eq. (1) is equivalent to optimizing the following energy model:

$$\arg\min_{z_s, x_s} \frac{1}{2}||M\mathcal{F}x_s - y_s||_2^2 + \lambda f(z_s, x_g) + \frac{\rho}{2}||x_s - z_s||_2^2, \tag{2}$$

where penalty coefficient $\rho \to \infty$ during optimization. The energy model of Eq. (2) can be minimized by iteratively estimating the unknown variables z_s and x_s. At k-th iteration, we solve the following two subproblems.

Estimating z_s by Proximal Operator: Given MR image $x_s^{(k-1)}$ at iteration $k-1$, auxiliary MR image z_s can be updated with guided MR image x_g by:

$$z_s^{(k)} = \arg\min_{z_s} \frac{\rho}{2}||x_s^{(k-1)} - z_s||_2^2 + \lambda f(z_s, x_g). \tag{3}$$

By definition of proximal operator of a regularizer term $g(\cdot)$, i.e., $\text{Prox}_{\eta g}(v) = \arg\min_u \frac{1}{2}||v - u||_2^2 + \eta g(u)$, auxiliary image z_s is updated by optimizing Eq. (3):

$$z_s^{(k)} = \text{Prox}_{\frac{\lambda}{\rho} f(\cdot, x_g)}(x_s^{(k-1)}), \tag{4}$$

where proximal operator $\text{Prox}_{\frac{\lambda}{\rho} f(\cdot, x_g)}$ is a nonlinear mapping determined by cross-contrast prior $f(z_s, x_g)$, and maps input MR image $x_s^{(k-1)}$ to $z_s^{(k)}$ with guidance of guided MR image x_g. x_s is initialized by zero-filling: $x_s^{(0)} = \mathcal{F}^H M^H y_s$.

Estimating x_s by Image Reconstruction: Given updated auxiliary MR image $z_s^{(k)}$ at iteration k, reconstructed MR image x_s can be updated as:

$$x_s^{(k)} = \arg\min_{x_s} \frac{1}{2}||M\mathcal{F}x_s - y_s||_2^2 + \frac{\rho}{2}||x_s - z_s^{(k)}||_2^2. \tag{5}$$

This sub-problem has a closed-form solution:

$$x_s^{(k)} = \mathcal{F}^H \Lambda^{-1}(M^H y_s + \rho \mathcal{F} z_s^{(k)}), \tag{6}$$

where $\Lambda = M^H M + \text{diag}(\rho)$ is a diagonal matrix.

By iteratively updating auxiliary and reconstructed MR images using Eqs. (4) and (6), cross-contrast prior is imposed on x_s by proximal operator in Eq. (4) with guidance of MR image x_g.

2.2   Unfolded Network for Cross-contrast Prior Learning

We unfold the iterative algorithm (Eqs. (4), (6)) to be a deep architecture, dubbed *MD-DAN*. As illustrated in Fig. 1, the whole network contains K stages, each of which corresponds to one iteration with two blocks, i.e., **cross-contrast fusion block (F-B)** and **image reconstruction block (R-B)**, implementing updates of variable z_s and x_s in Eq. (4) and Eq. (6) respectively.

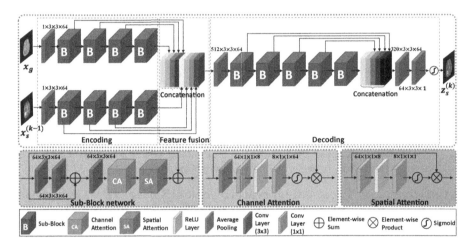

Fig. 2. The network structure of cross-contrast fusion network (DPA-FusionNet).

Cross-contrast Fusion Block. Instead of handcrafting cross-contrast prior $f(\cdot, \cdot)$, we substitute its proximal operator in Eq. (4) by a learnable network:

$$z_s^{(k)} = \mathrm{Prox}_{\frac{\lambda}{\rho} f(\cdot, x_g)}(x_s^{(k-1)}) \triangleq \textbf{Fusion-net}(x_s^{(k-1)}, x_g). \qquad (7)$$

The Fusion-net is to fuse two contrast images as inputs and output updated auxiliary MR image. As shown in Fig. 2, it is designed as a dual-path net including encoding, feature fusion and decoding blocks. Encoding block extracts features from guided image x_g and reconstructed image in $(k-1)$-th stage (i.e., $x_s^{(k-1)}$) by two sub-encoders, each of which consists of a convolution and four cascaded Sub-Block networks. Then features of different layers of encoding networks from different contrasts are concatenated and further fused by a decoding network consisting of a convolution, five cascaded Sub-Blocks, a multi-scale feature fusion (i.e., feature concatenation) followed by two convolution and a sigmoid operation, and the decoder outputs the updated auxiliary MR image $z_s^{(k)}$. We name this cross-contrast fusion network as *Dual Path Attention Fusion Network (DPA-FusionNet)*. Please see Fig. 2 for detailed network hyper-parameters.

Each Sub-Block network, as shown in Fig. 2, is designed with skip connection, channel attention (CA) [29] and spatial attention (SA) [19] modules. The two attention modules are introduced as follows.

Channel Attention (CA). Channel attention module is to make network pay more attention to important channel features from different contrasts. Firstly, we convert features denoted by F into a channel descriptor denoted by G through average pooling channel-by-channel: $G_c = \frac{1}{H \times W} \sum_i \sum_j F_c(i,j)$, where F_c is feature in c-th channel, and H and W are the height and width of the features. Then the channel descriptor G goes through the cascaded layers: Conv $\rightarrow$ ReLu $\rightarrow$ Conv $\rightarrow$ Sigmoid, and we get weights of different channels W^{CA}. Finally, weights are applied to the input features by element-wise product: $F_c^{out} = W_c^{CA} \otimes F_c$.

Spatial Attention (SA). Spatial attention module is to force network to pay more attention to important spatial regions of image, such as high-frequency and heavy artifact areas. For the features F, we get weights of different pixels W^{SA} by cascaded layers: Conv $\rightarrow$ ReLu $\rightarrow$ Conv $\rightarrow$ Sigmoid. Then weights are applied to the input features by element-wise product: $F_c^{out} = W^{SA} \otimes F_c$.

2.3   Network Training

We use mean squared error (MSE) loss to train the net: $\mathcal{L}(\Theta) = \sum_n ||x_n^{(K)}(\Theta) - x_n^{gt}||_2$, where $x_n^{(K)}(\Theta)$ is the network output for n-th data, Θ are network parameters of MD-DAN with backbone DPA-FusionNets in different stages initialized by Xavier random initialization. x_n^{gt} is targeting MR image reconstructed from fully-sampled k-space data. ρ is initialized as 1e−5. The backbone network at different stages does not share parameters. We implement and train MD-DAN by PyTorch using Adam with learning rate of 1e−4, on an Ubuntu 16.04 system with GTX 1080 Ti GPU. We conduct stage-wise training followed by an end-to-end fine tuning, and we train the network in each stage for 120 epochs.

3   Experiments

Experimental Settings. We evaluate MD-DAN on brain MR images from BraTS 2019 dataset[1] [1,2,18], providing clinically-acquired multi-contrast MRI scans of glioblastoma patients from 19 institutions. We respectively take 190 and 189 subjects as training and test set. We train our MD-DAN to reconstruct T2WI MR image from its under-sampled k-space data with guidance of T1WI image. Successive 2D slices are used to train MD-DAN and test accuracy is performed on whole 3D volume in size of $240 \times 240 \times 115$. Preprocessing steps including N4 corrected and peak normalization were applied. We used 1D Cartesian sampling with sampling rates of $1/8$, $1/16$ and $1/32$ to under-sample T2WI k-space data.

Compared Methods. We compare MD-DAN ($K = 4$) with five single contrast CS-MRI methods that reconstruct T2WI MR image from under-sampled k-space data including Zero-filling (ZF), DC-CNN [23], Dense-Unet-R, ResNet-R

---

[1] https://ipp.cbica.upenn.edu/.

Table 1. Comparison of average performance on the testing dataset.

Type	Method	1/8 rate PSNR	nRMSE	1/16 rate PSNR	nRMSE	1/32 rate PSNR	nRMSE	Time
T2→T2	ZF	26.0883	0.3312	25.9763	0.3357	25.0999	0.3719	0.010 s
	DC-CNN	35.8136	0.1095	32.2087	0.1665	28.0524	0.2667	0.031 s
	Dense-Unet-R	32.1378	0.1698	30.6749	0.1983	28.0846	0.2661	0.014 s
	ResNet-R	33.5167	0.1425	31.2695	0.1851	28.4251	0.2565	0.016 s
	DPA-FusionNet-R	33.5520	0.1420	31.3854	0.1825	28.5896	0.2516	0.026 s
T1→T2	Dense-Unet-S	29.6061	0.2235	29.6061	0.2235	29.6061	0.2235	0.014 s
	ResNet-S	30.2969	0.2079	30.2969	0.2079	30.2969	0.2079	0.015 s
	DPA-FusionNet-S	30.3193	0.2077	30.3193	0.2077	30.3193	0.2077	0.027 s
T1+T2	FCSA-MT	35.8089	0.1118	32.5788	0.1626	27.6125	0.2844	4.001 s
	Dense-Unet	35.4403	0.1147	34.2353	0.1314	32.9811	0.1520	0.015 s
→T2	ResNet	36.1856	0.1047	35.0296	0.1264	33.8310	0.1378	0.016 s
	MD-DAN	**40.6069**	**0.0639**	**37.9372**	**0.0868**	**35.0364**	**0.1206**	0.108 s

and DPA-FusionNet-R. We also compare with image synthesis methods including Dense-Unet-S, ResNet-S and DPA-FusionNet-S which synthesize T2WI MR image from T1WI MR image. Dense-Unet-R(S), ResNet-R(S) are respectively based on Dense-Unet (2 down and up sampling operations and 5 dense blocks [26]) and ResNet (9 convolution residual blocks [30]). DPA-FusionNet-R(S) is variant of our DPA-FusionNet in Fig. 2 without upper (for '-R') or lower (for '-S') path respectively for single modality reconstruction and cross-contrast synthesis. We further compare with three reconstruction methods guided by cross-contrast image including model-based method FCSA-MT [10] and two state-of-the-art deep learning networks of Dense-Unet [26] and ResNet [30]. The two deep networks take reconstructed under-sampled T2WI image and fully sampled T1WI image as input, and output the reconstructed T2WI image.

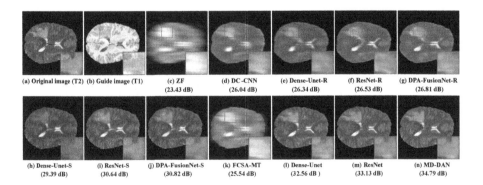

(a) Original image (T2) (b) Guide image (T1) (c) ZF (23.43 dB) (d) DC-CNN (26.04 dB) (e) Dense-Unet-R (26.34 dB) (f) ResNet-R (26.53 dB) (g) DPA-FusionNet-R (26.81 dB)

(h) Dense-Unet-S (29.39 dB) (i) ResNet-S (30.64 dB) (j) DPA-FusionNet-S (30.82 dB) (k) FCSA-MT (25.54 dB) (l) Dense-Unet (32.56 dB) (m) ResNet (33.13 dB) (n) MD-DAN (34.79 dB)

Fig. 3. Results for a brain MR image using 1/32 1D Cartesian sampling.

Results. Table 1 shows the quantitative results on test dataset with 1/8, 1/16 and 1/32 sampling rates. Our MD-DAN significantly outperforms all synthesis-based methods and single contrast reconstruction methods, especially at higher sampling rates 1/16 and 1/32, which verify the effectiveness of joint reconstruction using k-space data and guided image. For example, in 1/16 sampling rate, MD-DAN outperforms DPA-FusionNet-S by 7.62 dB and DC-CNN by 5.73 dB. Compared with Dense-Unet and ResNet in the same setting as ours, our MD-DAN achieves better reconstruction accuracy in PSNR and nRMSE using 50% less sampled data. Examples of reconstructed images by different methods in 1/32 sampling rate are shown in Fig. 3. More examples are in supplementary material. Our method yields higher-quality MR images without obvious artifacts.

Table 2. Comparison of average performance with sampling rate of 1/8.

	MD-Dense-Unet	MD-ResNet	MD-DAN
PSNR	39.6798	40.2295	**40.6069**
nRMSE	0.0708	0.0666	**0.0639**
Training time	5.5 days	6 days	7 days

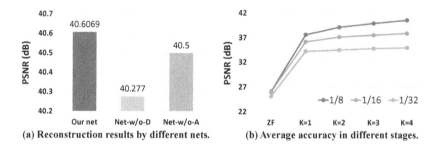

Fig. 4. Performance comparison of different network settings

Comparisons of Different Backbone Networks. We use a dual-path DPA-FusionNet in Fig. 2 as backbone to replace proximal operator in our model-driven network. To verify superiority of DPA-FusionNet, we also respectively use Dense-Unet and ResNet to replace the proximal operator in our framework, denoted as MD-Dense-Unet and MD-ResNet. Table 2 shows performance with sampling rate of 1/8 in 4 stages. Our MD-DAN with DPA-FusionNet as proximal operator works best and improves the PSNR at least 0.37 dB than others.

Effectiveness of Backbone Network Design. To explore effectiveness of dual-path structure and attention modules in backbone network DPA-FusionNet, we respectively discard either of them in MD-DAN, and the corresponding networks are denoted as Net-w/o-D and Net-w/o-A. The average PSNR values of trained Net-w/o-D and Net-w/o-A with $K = 4$ and $1/8$ sampling rate are shown in Fig. 4(a). Dual-path structure and attention modules are both beneficial for improving performance, e.g., the result of MD-DAN is higher in 0.33 dB and 0.11 dB than Net-w/o-D and Net-w/o-A respectively.

Effect of Number of Stages. To explore influence of the number of stages (i.e., K) on reconstruction accuracy, we train MD-DAN from stage 1 to stage 4 in a greedy way, and the average accuracies on test dataset at different sampling rates and stages are shown in Fig. 4(b). With the increase of stages, the accuracy of reconstruction is improved gradually at all three sampling rates.

4   Conclusion

We proposed a novel model-driven deep attention network to reconstruct highly under-sampled long-time sequence MR image with guidance of a short-time sequence MR image. We design a new reconstruction model with cross-contrast prior, inspiring us to design a novel deep architecture, in which cross-contrast prior is learned by replacing its proximal operator with a deep DPA-FusionNet. Our method can be extended to other contrast MR image (e.g., FLAIR). We also plan to extend it to simultaneously reconstruct multiple contrast MRI images.

Acknowledgement. This work was supported in part by NSFC under Grants 11971373, 61976173, 11690011, 61721002, U1811461, and in part by the National Key Research and Development Program of China under Grant 2018AAA0102201.

References

1. Bakas, S., et al.: Advancing the cancer genome atlas glioma MRI collections with expert segmentation labels and radiomic features. Nat. Sci. data **4**, 170117 (2017)
2. Bakas, S., et al.: Identifying the best machine learning algorithms for brain tumor segmentation, progression assessment, and overall survival prediction in the brats challenge. arXiv (2018)
3. Block, K.T., Uecker, M., Frahm, J.: Undersampled radial MRI with multiple coils. Iterative image reconstruction using a total variation constraint. Magn. Reson. Med. **57**(6), 1086–1098 (2007)
4. Chartsias, A., Joyce, T., Giuffrida, M.V., Tsaftaris, S.A.: Multimodal MR synthesis via modality-invariant latent representation. IEEE Trans. Med. Imaging **37**(3), 803–814 (2017)
5. Dar, S.U., Yurt, M., Karacan, L., Erdem, A., Erdem, E., Çukur, T.: Image synthesis in multi-contrast MRI with conditional generative adversarial networks. IEEE Trans. Med. Imaging **38**(10), 2375–2388 (2019)
6. Duan, J., et al.: VS-Net: variable splitting network for accelerated parallel MRI reconstruction. In: Shen, D., et al. (eds.) MICCAI 2019. LNCS, vol. 11767, pp. 713–722. Springer, Cham (2019). https://doi.org/10.1007/978-3-030-32251-9_78
7. Eksioglu, E.M.: Decoupled algorithm for MRI reconstruction using nonlocal block matching model: BM3D-MRI. J. Math. Imaging Vis. **56**(3), 430–440 (2016). https://doi.org/10.1007/s10851-016-0647-7

8. Geman, D., Yang, C.: Nonlinear image recovery with half-quadratic regularization. IEEE Trans. Image Process. **4**(7), 932–946 (1995)
9. Hammernik, K., et al.: Learning a variational network for reconstruction of accelerated MRI data. Magn. Reson. Med. **79**(6), 3055–3071 (2018)
10. Huang, J., Chen, C., Axel, L.: Fast multi-contrast MRI reconstruction. Magn. Reson. Imaging **32**(10), 1344–1352 (2014)
11. Huang, Y., Shao, L., Frangi, A.F.: Cross-modality image synthesis via weakly coupled and geometry co-regularized joint dictionary learning. IEEE Trans. Med. Imaging **37**(3), 815–827 (2017)
12. Joyce, T., Chartsias, A., Tsaftaris, S.A.: Robust multi-modal MR image synthesis. In: Descoteaux, M., Maier-Hein, L., Franz, A., Jannin, P., Collins, D.L., Duchesne, S. (eds.) MICCAI 2017. LNCS, vol. 10435, pp. 347–355. Springer, Cham (2017). https://doi.org/10.1007/978-3-319-66179-7_40
13. Lee, D., Yoo, J., Ye, J.C.: Deep residual learning for compressed sensing MRI. In: IEEE ISBI, pp. 15–18 (2017)
14. Li, H., et al.: DiamondGAN: unified multi-modal generative adversarial networks for MRI sequences synthesis. In: Shen, D., et al. (eds.) MICCAI 2019. LNCS, vol. 11767, pp. 795–803. Springer, Cham (2019). https://doi.org/10.1007/978-3-030-32251-9_87
15. Lustig, M., Donoho, D., Pauly, J.M.: Sparse MRI: the application of compressed sensing for rapid MR imaging. Magn. Reson. Med. **58**(6), 1182–1195 (2007)
16. Lustig, M., Donoho, D.L., Santos, J.M., Pauly, J.M.: Compressed sensing MRI. IEEE Signal Process. Mag. **25**(2), 72–82 (2008)
17. Meng, N., Yang, Y., Xu, Z., Sun, J.: A prior learning network for joint image and sensitivity estimation in parallel MR imaging. In: Shen, D., et al. (eds.) MICCAI 2019. LNCS, vol. 11767, pp. 732–740. Springer, Cham (2019). https://doi.org/10.1007/978-3-030-32251-9_80
18. Menze, B.H., et al.: The multimodal brain tumor image segmentation benchmark (BRATS). IEEE Trans. Med. Imaging **34**(10), 1993–2024 (2014)
19. Qin, X., Wang, Z., Bai, Y., Xie, X., Jia, H.: FFA-Net: feature fusion attention network for single image dehazing. In: AAAI (2020)
20. Qu, X., Hou, Y., Lam, F., Guo, D., Zhong, J., Chen, Z.: Magnetic resonance image reconstruction from undersampled measurements using a patch-based nonlocal operator. Med. Image Anal. **18**(6), 843–856 (2014)
21. Ravishankar, S., Bresler, Y.: MR image reconstruction from highly undersampled k-space data by dictionary learning. IEEE Trans. Med. Imaging **30**(5), 1028–1041 (2010)
22. Roy, S., Carass, A., Prince, J.L.: Magnetic resonance image example-based contrast synthesis. IEEE Trans. Med. Imaging **32**(12), 2348–2363 (2013)
23. Schlemper, J., et al.: A deep cascade of convolutional neural networks for dynamic MR image reconstruction. IEEE Trans. Med. Imaging **37**(2), 491–503 (2018)
24. Wang, S., Su, Z., Ying, L., Xi, P., Dong, L.: Accelerating magnetic resonance imaging via deep learning. In: IEEE ISBI, pp. 514–517 (2016)
25. Weizman, L., Eldar, Y.C., Ben, B.D.: Reference-based MRI. Med. Phys. **43**(10), 5357 (2016)
26. Xiang, L., et al.: Ultra-fast T2-weighted MR reconstruction using complementary T1-weighted information. In: Frangi, A.F., Schnabel, J.A., Davatzikos, C., Alberola-López, C., Fichtinger, G. (eds.) MICCAI 2018. LNCS, vol. 11070, pp. 215–223. Springer, Cham (2018). https://doi.org/10.1007/978-3-030-00928-1_25
27. Yang, Y., Sun, J., Li, H., Xu, Z.: Deep ADMM-Net for compressive sensing MRI. In: NIPS, pp. 10–18 (2016)

28. Zhan, Z., Cai, J.F., Guo, D., Liu, Y., Chen, Z., Qu, X.: Fast multiclass dictionaries learning with geometrical directions in MRI reconstruction. IEEE Trans. Biomed. Eng. **63**(9), 1850–1861 (2015)
29. Zhang, Y., Li, K., Li, K., Wang, L., Zhong, B., Fu, Y.: Image super-resolution using very deep residual channel attention networks. In: Ferrari, V., Hebert, M., Sminchisescu, C., Weiss, Y. (eds.) ECCV 2018. LNCS, vol. 11211, pp. 294–310. Springer, Cham (2018). https://doi.org/10.1007/978-3-030-01234-2_18
30. Zhu, J.Y., Park, T., Isola, P., Efros, A.A.: Unpaired image-to-image translation using cycle-consistent adversarial networks. In: ICCV, pp. 2223–2232 (2017)

Simultaneous Estimation of X-Ray Back-Scatter and Forward-Scatter Using Multi-task Learning

Philipp Roser[1,4]([✉]), Xia Zhong[2], Annette Birkhold[3], Alexander Preuhs[1], Christopher Syben[1], Elisabeth Hoppe[1], Norbert Strobel[5], Markus Kowarschik[3], Rebecca Fahrig[3], and Andreas Maier[1,4]

[1] Pattern Recognition Lab, Friedrich-Alexander-Universität Erlangen-Nürnberg (FAU), Erlangen, Germany
`philipp.roser@fau.de`
[2] Diagnostic Imaging, Siemens Healthcare GmbH, Erlangen, Germany
[3] Advanced Therapies, Siemens Healthcare GmbH, Forchheim, Germany
[4] Erlangen Graduate School in Advanced Optical Technologies (SAOT), Erlangen, Germany
[5] Institute of Medical Engineering, University of Applied Sciences Würzburg-Schweinfurt, Schweinfurt, Germany

Abstract. Scattered radiation is a major concern impacting X-ray image-guided procedures in two ways. First, back-scatter significantly contributes to patient (skin) dose during complicated interventions. Second, forward-scattered radiation reduces contrast in projection images and introduces artifacts in 3-D reconstructions. While conventionally employed anti-scatter grids improve image quality by blocking X-rays, the additional attenuation due to the anti-scatter grid at the detector needs to be compensated for by a higher patient entrance dose. This also increases the room dose affecting the staff caring for the patient. For skin dose quantification, back-scatter is usually accounted for by applying predetermined scalar back-scatter factors or linear point spread functions to a primary kerma forward projection onto a patient surface point. However, as patients come in different shapes, the generalization of conventional methods is limited. Here, we propose a novel approach combining conventional techniques with learning-based methods to simultaneously estimate the forward-scatter reaching the detector as well as the back-scatter affecting the patient skin dose. Knowing the forward-scatter, we can correct X-ray projections, while a good estimate of the back-scatter component facilitates an improved skin dose assessment. To simultaneously estimate forward-scatter as well as back-scatter, we propose a multi-task approach for joint back- and forward-scatter estimation by combining X-ray physics with neural networks. We show that, in theory, highly accurate scatter estimation in both cases is possible. In addition, we identify research directions for our multi-task framework and learning-based scatter estimation in general.

Keywords: X-ray scatter · Skin dose · Multi-task learning

A. L. Martel et al. (Eds.): MICCAI 2020, LNCS 12262, pp. 199–208, 2020.
https://doi.org/10.1007/978-3-030-59713-9_20

1 Introduction

X-ray fluoroscopic guidance enables minimally-invasive interventions. Unfortunately, photons scattered by the patient impair X-ray image quality and increase the X-ray dose affecting both patient as well as staff. There are two major types of scatter in X-ray imaging: back-scatter and forward-scatter.

Back-scatter contributes up to 30% to 60% of the total skin dose [18]. Unless properly accounted for, it impairs accurate online monitoring of skin dose, which is a critical means in dose management for interventional fluoroscopic imaging. By providing constant feedback on accumulated skin dose values to the physicians, they can spread the dose using table movements and C-arm rotations to avoid excessive peak skin dose values. Therefore, most X-ray imaging systems are equipped with a dose chamber measuring the kinetic energy released per unit mass in air (typically referred to as air kerma). These measured values can be used to calibrate either on-site Monte Carlo (MC) simulations [15], U-Net-accelerated dose simulations [21,22], or kerma forward projection (KFP) onto a digital patient model [4]. Simulation approaches can yield more accurate dose estimates if a precise model of the actual imaging setting is available. Unfortunately, prior knowledge, such as the exact patient anatomy, is not available in general. Therefore, current state-of-the-art systems rely on patient shape models, KFP, and several correction terms accounting for the inverse square law, skin absorption, and back-scatter [4,12,20]. Previous studies have shown that these back-scatter factors (BSF), determined experimentally or using MC simulations, have the potential to increase the accuracy of skin dose estimation [18,20]. Yet, BSFs are highly dependent on the imaging setting and specific patient. Despite this fact, they are usually pre-computed using MC simulation or measured empirically yielding large tables of BSFs to cover the whole patient population, anatomic regions, and X-ray characteristics. These tables are cumbersome to obtain and maintain. Furthermore, the rich source of information reflecting patient as well as X-ray beam characteristics contained in the X-ray images themselves is not used in any form by these dose estimation methods.

Forward-scatter, on the other hand, causes uneven exposure and loss of contrast in X-ray images. This is why, hardware- and/or software-based scatter correction methods are used to enhance image quality [25,26]. Typically, anti-scatter grids are used to physically block scattered X-rays. However, since they also absorb some primary radiation, a higher patient entrance dose is needed to maintain the desired X-ray exposure level at the detector [6]. With an increased focus on X-ray dose reduction, protection, and risk management, grid-less X-ray imaging is desirable. In particular, in pediatrics, where patient X-ray dose plays a crucial role and where patients are usually smaller, anti-scatter grids are commonly removed [9,28]. Software-only approaches, such as the recently introduced single-task U-Net-based deep scatter estimation (DSE) [16], might render conventional approaches [14,17,27] obsolete.

In this work, we apply convolutional neural networks (CNN) to estimate back-scatter as well as forward-scatter. To this end, we propose a multi-task framework to calculate back- and forward-scatter in a one-step procedure by combining X-ray physics' models with modern learning-based methods. Since back- and forward-scatter share the same mathematical description, we can leverage learning-based forward-scatter estimation to also infer back-scatter directly from an X-ray projection image and a patient shape model.

Contribution. To the best of our knowledge, this paper presents several novel ideas not yet published elsewhere: (1) learning-based back-scatter estimation using a patient model, (2) deriving back-scatter from an X-ray image projection, and (3) simultaneous back-scatter and forward-scatter estimation. Finally, we propose a lightweight network architecture to efficiently implement it reaching an accuracy comparable to outcomes obtained using a multi-task U-Net.

2   Materials and Methods

Figure 1 depicts the outline of the proposed method. The basic principle is to exploit our rich understanding of the photon interactions causing X-ray forward-scatter and back-scatter, respectively. Since both are caused by same underlying scattering interactions, it is reasonable to estimate their effects on image formation and skin dose together. Unfortunately, analytic and stochastic scatter estimation using established physics models is time-consuming and relies on accurate prior knowledge on the patient anatomy [3,19,29]. This can, however, be done faster and with a comparable accuracy, if we combine the underlying physics with a data-driven multi-task CNN. In particular, our approach involves KFP and a multi-task scattering model. Details are described below.

Kerma Forward Projection. As in conventional dose-monitoring systems, a 3-D patient shape model $\mathcal{M}$ is at the heart of our method. Potential sources of such a digital twin are a pre-operative computed tomography (CT) scan, a point cloud reconstruction based on 3-D-capable camera systems, or an active shape model based on meta-parameters such as weight, height, and age [12,15,31]. In the following, we assume that we are given an already registered patient shape model. Based on the patient shape model and X-ray system geometry, we calculate the distance per detector pixel, where the X-ray enters the patient $Z_{\mathrm{in}} \in \mathbb{R}^{w \times h}$, with the width w and height h of the detector. To make our method robust and flexible in a clinical situation, we use an efficient grid traversal algorithm [1], which allows for patient shapes defined by either a mesh, implicit or explicit analytical functions, or tomographic data. For each detector pixel, a ray iteratively traverses a cubic grid from the X-ray source position to the respective pixel. Once the ray intersects the patient model, the traveled (source-patient-surface) distance is stored in the patient entrance map Z_{in}. Once Z_{in} is known, the primary component of the skin dose per detector pixel $D_{\mathrm{p}} \in \mathbb{R}^{w \times h}$ (patient entrance dose) is computed by applying the inverse square law [4]. The air-kerma measured at the interventional reference point (IRP) K_{IRP} in mGy or $\mathrm{J\,g^{-1}}$ is projected onto the skin surface yielding

$$D_{\mathrm{p}} = K_{\mathrm{IRP}} \cdot d_{\mathrm{IRP}}^2 / Z_{\mathrm{in}}^2 \cdot f, \tag{1}$$

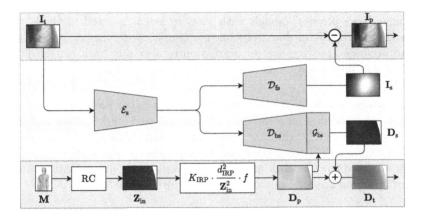

Fig. 1. Overview of the proposed method with a dedicated dual autoencoder-like architecture (DAE). Based on the patient shape model and the imaging geometry of the X-ray system, the ray-caster (RC) estimates an entrance map Z_{in}. Based on Z_{in}, the primary skin dose D_{p} is calculated using conventional KFP. An encoder $\mathcal{E}_{\text{s}}$ extracts the latent representation of the underlying scatter distribution. Using two independent decoders $\mathcal{D}_{\text{fs}}$ and $\mathcal{D}_{\text{bs}}$ forward-scatter I_{s} and back-scatter D_{s} are estimated, respectively. To account for the domain transfer to dose, $\mathcal{D}_{\text{bs}}$ is extended by an additional convolutional block $\mathcal{G}_{\text{bs}}$ with D_{p} as second input. Knowing the I_{s}, we can estimate the primary image I_{p}, while D_{s} can be used to calculate the total skin dose D_{t}.

where d_{IRP} is the distance between the X-ray source and the IRP and f is a unit-less tissue-conversion factor which can be pre-calculated for any X-ray spectrum. Unfortunately, for back-scatter D_{s}, no such analytical model exists without extensive prior knowledge on the patient anatomy. However, we can make use of the rich information encoded in the measured X-ray projection image.

Multi-task Scattering Model. Since, in the medical X-ray energy regime, the incoherent Compton scatter dominates over the coherent Rayleigh scatter, we can safely neglect the latter in the following considerations. The occurrence probability of scattering interactions can be expressed in terms of cross-sections (CS) σ. The Compton scattering model for an infinitesimal volume element $\partial\Omega$ is based on the differential Klein-Nishina (KN) CS $\frac{\partial\sigma_{\text{KN}}}{\partial\Omega}$, given by

$$\frac{\partial\sigma_{\text{KN}}}{\partial\Omega} = 0.5\, r_e^2\, P(E,\theta)^2 \left[P(E,\theta) + P(E,\theta)^{-1} - \sin^2\theta \right], \tag{2}$$

with the classical electron radius r_e, the scattering angle θ, and the ratio of photon energy E after and before the interaction $P(E,\theta)$. The ratio $P(E,\theta)$ is defined as

$$P(E,\theta) = \frac{1}{1 + (\frac{E}{m_e c^2})(1 - \cos\theta)}, \tag{3}$$

where m_e denotes the electron rest mass and c the speed of light, respectively. As Fig. 2 shows, the KN model has several useful properties.

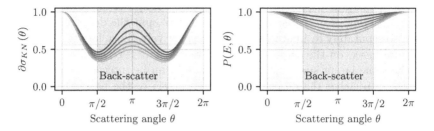

Fig. 2. Normalized KN CS $\partial\sigma_{KN}(\theta)$, left, and energy ratio $P(E, \theta)$, right, plotted against the scattering angle θ for $E \in \{20, 40, \ldots, 120\}$ in keV (dark blue to light blue). (Color figure online)

First, since Eq. 2 only depends on energy ratios and symmetric trigonometric functions, we can establish a functional relationship between back-scatter and forward-scatter. In other words, given the forward-scatter distribution, the back-scatter distribution is for infinitesimal volumes analytically described by the KN formula. From recent studies [16], we know that forward-scatter $\boldsymbol{I}_s = \mathcal{U}_{ST}(\boldsymbol{I}_t)$ in J cm^{-2} can be extracted from the measured X-ray projection $\boldsymbol{I}_t$ using a single-task U-Net $\mathcal{U}_{ST}$. Hence, we can relate forward- and back-scatter via

$$\left(\frac{\mu_{en}}{\rho}\right)^{-1} \boldsymbol{D}_s \sim \boldsymbol{I}_s = \mathcal{U}_{ST}(\boldsymbol{I}_t), \tag{4}$$

with the mass energy absorption coefficient $\left(\frac{\mu_{en}}{\rho}\right)$ in cm^2 g^{-1}. Since $\left(\frac{\mu_{en}}{\rho}\right)$ relates to a simple linear scaling, we can omit it below. Unfortunately, we do not know a simple yet accurate model of this relationship for arbitrary patient anatomies. However, being based on the same physical effects, it can be concluded that both forward- and back-scatter can be encoded by similar features or latent variables. Consequently, it appears attractive to learn both in a multi-task fashion. Since the U-Net has yielded promising results for forward-scatter estimation, we investigate its applicability to the task at hand. Therefore, by supplying the primary skin dose estimate $\boldsymbol{D}_s$ we extend the U-Net to an multi-task function $(\boldsymbol{D}_s, \boldsymbol{I}_s) = \mathcal{U}_{MT}(\boldsymbol{D}_p, \boldsymbol{I}_t)$. The U-Net comprises six levels, two convolutional layers per block, average pooling, 16 initial feature maps doubled after each pooling operation, rectified linear units (ReLU) [10] as activations, and 31 M parameters to train in total. Although the U-Net has proven to be a powerful function approximator for numerous tasks, its high parameter complexity and degree of connectivity is not easy to fully comprehend making it a potentially risky tool for clinical image processing.

Especially the U-Net's skip-connections lead to outputs covering the whole frequency spectrum. However, a Fourier analysis of Eqs. 2 and 3 for the diagnostic X-ray energy regime reveals mostly low-frequency characteristics. This observation is substantiated by the low amplitude and smoothness of the corresponding plots in Fig. 2. Therefore, we propose to degenerate the U-Net to a dual autoencoder-like CNN (DAE) without skip-connections to constrain its output frequency, as depicted in Fig. 1. While we keep the encoding path $\mathcal{E}_s$ to extract the latent scatter distribution from $\boldsymbol{I}_t$, we split the decoding paths $\mathcal{D}_{bs}$ and $\mathcal{D}_{fs}$ to separately estimate $\boldsymbol{D}_s$ and $\boldsymbol{I}_s$, respectively. Since forward- and back-scatter are based on the same particle interactions, it is reasonable for both to share the same latent space, while the decoders can be interpreted as opposing projections on either the detector (forward-scatter) or the patient skin surface (back-scatter). To further enforce low-frequency characteristics via a compact latent space, the number of feature maps is not doubled per down-sampling operation as it is typically done for the U-Net. Similar to $\mathcal{U}_{MT}$, the encoder $\mathcal{E}_s$ consists of six convolutional blocks (two layers with 16 3×3 kernels and ReLU activations) with average pooling operations in between. The decoders resemble $\mathcal{E}_s$ with bilinear up-sampling instead of average pooling. In addition, we extend $\mathcal{D}_{bs}$ by an additional convolutional block $\mathcal{G}_{bs}$ with $\boldsymbol{D}_p$ as second input to account for the domain transfer to skin dose. The number of parameters to train is with 105 k two magnitudes lower than the U-Net's. In general, $\boldsymbol{D}_s$ and $\boldsymbol{I}_s$ have both different co-domains and SI units. To circumvent error-prone loss weighting, we used the mean absolute percentage error (MAPE) cost function.

3   Results and Discussion

Data. As it is technically not feasible to extract matching pairs of X-ray projections in a clinical setup, we based our experiments on synthetic data. To this end, we simulated data using a MC photon transport code yielding X-ray images, Compton and Rayleigh scatter, and 3-D kerma distributions [2,23]. As patient models, we selected 22 head scans from the HNSCC-3DCT-RT data set [5] and 17 thorax scans from the CT-Lymph-Nodes data set [24], both provided by The Cancer Imaging Archive (TCIA) [7]. We used a 100 kVp spectrum and simulated 5×10^{10} primary photons for each X-ray image. All images comprise 256×384 pixels with an area of $1.16\,\mathrm{mm}^2 \times 1.03\,\mathrm{mm}^2$. For each patient, we computed 60 projections with three source positions, 20 projection angles ($0°$ to $95°$, $5°$ sampling), and fixed source-to-detector distance (130 cm). In total, 2340 data points were available. We separated one patient for validation and two patients for testing for both data sets.

Experimental Setup. To thoroughly assess our results and provide for an ablation study, we compared our method to several baseline networks: (a) a lean and straightforward single-task autoencoder-like network similar to one path of our DAE network (AE, six levels, two convolutional layers per block, average pooling, 16 feature maps per convolution, ReLU activation, 49 k parameters to train) and (b) a single-task U-Net (six levels, two convolutional layers per block, average pooling, 16 initial feature maps doubled after each pooling operation, ReLU activation, 31 M parameters to train). As inputs, we either considered the

Table 1. Expected error rates $\mu_\varepsilon(\sigma_\varepsilon)$ for all network settings. The AE and the U-Net are trained to either infer I_s or D_s. For dose estimation, $\mu_\varepsilon(\sigma_\varepsilon)$ we also compare to results obtained using back-scatter factors (BSFs) which were optimized for the specific settings.

	Method	Map	Head $\mu_\varepsilon(\sigma_\varepsilon)$ [%]		Thorax $\mu_\varepsilon(\sigma_\varepsilon)$ [%]	
			D_s	I_s	D_s	I_s
Baseline (single)	BSF	$D_p \mapsto D_s$	15.57(217)	–	8.04(208)	–
	AE	$I_t \mapsto I_s$	–	5.61(164)	–	10.12(401)
		$D_p \mapsto D_s$	24.56(380)	–	12.66(620)	–
	U-Net	$I_t \mapsto I_s$	–	5.61(255)	–	8.20(371)
		$D_p \mapsto D_s$	11.26(244)	–	9.46(553)	–
Ours (multi)	AE	$D_p, I_t \mapsto D_s, I_s$	21.98(401)	6.84(159)	10.12(427)	9.46(277)
	U-Net	$D_p, I_t \mapsto D_s, I_s$	9.11(220)	**4.22(166)**	**5.22(156)**	**6.70(280)**
	DAE	$D_p, I_t \mapsto D_s, I_s$	**8.09(142)**	7.87(226)	6.95(333)	9.06(324)

measured X-ray projection I_t or the primary dose distribution D_p. As outputs, we either considered the X-ray forward-scatter I_s or the back-scatter skin dose D_s, respectively. In the training phase, we fixed the hyper-parameters for all networks, including our multi-task learning approach. Therefore, we also minimized the MAPE for both single-task networks. To optimize the network weights, we used adaptive moments [13] with a learning rate of 10^{-4} and a batch size of four. Since this is the first time where back-scatter is estimated in a data-driven manner, we also provide error metrics for BSFs. To highlight the minimum error bounds achievable using conventional back-scatter correction, we calculated the BSFs specifically for each X-ray projection, which compares to an over-fitting scenario.

Results. Table 1 summarizes expected error rates for all network settings for the testing data. For the head data set, both, the single-task AE and the single-task U-Net extracted I_s from X-ray projections with similar and high accuracy. Nevertheless, both single-task networks leave room for improvement concerning skin dose estimation. Overall the multi-task U-Net and DAE approaches clearly outperformed the single-task ones, especially in terms of back-scatter dose estimation. However, the multi-task AE approach appears to lack the capacity to estimate D_s and I_s simultaneously. Instead it focused on one quantity with high accuracy (see head data) or both with lower accuracy (see thorax data). Overall, the multi-task U-Net found the best mapping for all test cases. Due to its high number of parameters, the multi-task U-Net can map the relationship between both scatter types easily. With our modifications of the AE, the resulting DAE network, however, performed almost on par with the multi-task U-Net, but it achieved this with a at much lower parameter complexity. Figure 3 combines exemplary qualitative results of both U-Nets and the DAE. The depicted error

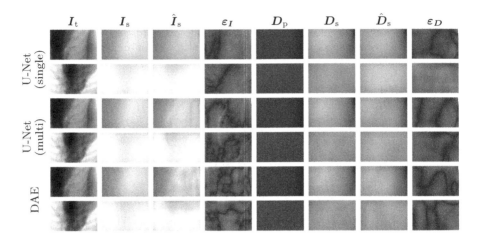

Fig. 3. Two qualitative examples of both U-Nets and the DAE for the thorax data set. From left to right: The measured X-ray projection I_t, corresponding forward-scatter ground truth I_s with associated network estimates $\hat{I}_s$ and relative error maps ε_I, and the primary skin dose D_p, corresponding back-scatter ground truth D_s with associated network estimates $\hat{D}_s$ and relative error maps ε_D. Error maps are in the range of 0% to 30% (dark blue over green to yellow). (Color figure online)

maps substantiate the quantitative results we observed. They also reveal current limitations of our approach. Especially the U-Net preserved spurious structural information in the scatter distributions due to its high number of parameters to train, and skip connections. Although the DAE performed worse in terms of quantitative results, it did not hallucinate high-frequency details (edges) to the same degree. Besides, with an average runtime of 100 ms, it was more than twice as fast as the U-Net with a runtime of 240 ms (CPU, Intel(R) i7-8850H), suggesting that it can be used in a clinical setup once current limitations are solved. Promising counter-measures include the incorporation of more prior knowledge, such as deriving scatter probabilities based on the patient shape model, or the combination with a first-order scatter estimation algorithm [8,11,30].

4   Conclusion

We presented an multi-task learning-based framework to (a) estimate the back-scatter contribution to the total skin dose and (b) estimate the forward-scatter in the X-ray projection image. To the best of our knowledge, this is the first paper investigating back-scatter estimation in an learning-based fashion and to then combine it with forward-scatter calculation. For the AE and the U-Net, we showed that, by estimating back- and forward-scatter in an multi-task fashion, the accuracy in both cases benefits compared to their respective single-task versions. We identified limitations in our approach and proposed appropriate counter-measures for future work. In addition, with only minor adjustments of the AE architecture, we achieved almost the same accuracy as the multi-task U-Net, while decreasing the parameter complexity and thus increasing the computational efficiency more than twofold. Especially in an interventional environment, where scatter is a critical aspect, our approach has the potential to facilitate dose reduction while maintaining or even improving image quality.

Disclaimer. The concepts and information presented in this paper are based on research and are not commercially available.

References

1. Amanatides, J., Woo, A.: A fast voxel traversal algorithm for ray tracing. In: Proceedings of the Eurographics (1987)
2. Badal, A., Badano, A.: Accelerating Monte Carlo simulations of photon transport in a voxelized geometry using a massively parallel graphics processing unit. Med. Phys. **36**(11), 4878–4880 (2009)
3. Baer, M., Kachelrieß, M.: Hybrid scatter correction for CT imaging. Phys. Med. Biol. **57**(21), 6849–6867 (2012)
4. Balter, S.: Methods for measuring fluoroscopic skin dose. Pediatr. Radiol. **36**(2), 136–140 (2006). https://doi.org/10.1007/s00247-006-0193-3
5. Bejarano, T., De Ornelas Couto, M., Mihaylov, I.: Head-and-neck squamous cell carcinoma patients with CT taken during pre-treatment, mid-treatment, and post-treatment dataset. The Cancer Imaging Archive (2018)
6. Chan, H.P., Doi, K.: Investigation of the performance of antiscatter grids: Monte Carlo simulation studies. Phys. Med. Biol. **27**(6), 785–803 (1982)
7. Clark, K., et al.: The cancer imaging archive (TCIA): maintaining and operating a public information repository. J. Digit. Imaging **26**(6), 1045–1057 (2013). https://doi.org/10.1007/s10278-013-9622-7
8. Freud, N., Duvauchelle, P., Pistrui-Maximean, S., Létang, J.M., Babot, D.: Deterministic simulation of first-order scattering in virtual x-ray imaging. Nucl. Instrum. Methods Phys. Res. B **222**(1), 285–300 (2004)
9. Fritz, S., Jones, A.K.: Guidelines for anti-scatter grid use in pediatric digital radiography. Pediatr. Radiol. **44**(3), 313–321 (2013). https://doi.org/10.1007/s00247-013-2824-9
10. Glorot, X., Bordes, A., Bengio, Y.: Deep sparse rectifier neural networks. In: Proceedings of the Fourteenth International Conference on Artificial Intelligence and Statistics, pp. 315–323 (2011)
11. Ingleby, H.R., Lippuner, J., Rickey, D.W., Li, Y.L., Elbakri, I.A.: Fast analytical scatter estimation using graphics processing units. J. X-Ray Sci. Technol. **23**(2), 119–133 (2015)
12. Johnson, P.B., Borrego, D., Balter, S., Johnson, K., Siragusa, D., Bolch, W.E.: Skin dose mapping for fluoroscopically guided interventions. Med. Phys. **38**(10), 5490–5499 (2011)
13. Kingma, D., Ba, J.: Adam: a method for stochastic optimization. In: Proceedings of the International Conference on Learning Representations (ICLR), December 2014
14. Li, H., Mohan, R., Zhu, X.R.: Scatter kernel estimation with an edge-spread function method for cone-beam computed tomography imaging. Phys. Med. Biol. **53**(23), 6729–6748 (2008)

15. Loy Rodas, N., Padoy, N.: Seeing is believing: increasing intraoperative awareness to scattered radiation in interventional procedures by combining augmented reality, Monte Carlo simulations and wireless dosimeters. Int. J. Comput. Assist. Radiol. Surg. **10**(8), 1181–1191 (2015). https://doi.org/10.1007/s11548-015-1161-x
16. Maier, J., et al.: Real-time scatter estimation for medical CT using the deep scatter estimation: Method and robustness analysis with respect to different anatomies, dose levels, tube voltages, and data truncation. Med. Phys. **46**(1), 238–249 (2019)
17. Ohnesorge, B., Flohr, T., Klingenbeck-Regn, K.: Efficient object scatter correction algorithm for third and fourth generation CT scanners. Eur. Radiol. **9**(3), 563–569 (1999). https://doi.org/10.1007/s003300050710
18. Petoussi-Henss, N., Zankl, M., Drexler, G., Panzer, W., Regulla, D.: Calculation of backscatter factors for diagnostic radiology using Monte Carlo methods. Phys. Med. Biol. **43**(8), 2237–2250 (1998)
19. Poludniowski, G., Evans, P.M., Hansen, V.N., Webb, S.: An efficient Monte Carlo-based algorithm for scatter correction in keV cone-beam CT. Phys. Med. Biol. **54**(12), 3847–3864 (2009)
20. Rana, V.K., Rudin, S., Bednarek, D.R.: A tracking system to calculate patient skin dose in real-time during neurointerventional procedures using a biplane x-ray imaging system. Med. Phys. **43**(9), 5131–5144 (2016)
21. Ronneberger, O., Fischer, P., Brox, T.: U-Net: convolutional networks for biomedical image segmentation. In: Navab, N., Hornegger, J., Wells, W.M., Frangi, A.F. (eds.) MICCAI 2015. LNCS, vol. 9351, pp. 234–241. Springer, Cham (2015). https://doi.org/10.1007/978-3-319-24574-4_28
22. Roser, P., et al.: Physics-driven learning of x-ray skin dose distribution in interventional procedures. Med. Phys. **46**(10), 4654–4665 (2019)
23. Roser, P., et al.: Fully-automatic CT data preparation for interventional x-ray skin dose simulation. In: Bildverarbeitung für die Medizin 2020. I, pp. 125–130. Springer, Wiesbaden (2020). https://doi.org/10.1007/978-3-658-29267-6_26
24. Roth, H., et al.: A new 2.5 d representation for lymph node detection in CT. The Cancer Imaging Archive (2018)
25. Rührnschopf, E.P., Klingenbeck, K.: A general framework and review of scatter correction methods in cone-beam CT. Part 2: scatter estimation approaches. Med. Phys. **38**(9), 5186–5199 (2011)
26. Rührnschopf, E.P., Klingenbeck, K.: A general framework and review of scatter correction methods in x-ray cone-beam computerized tomography. Part 1: scatter compensation approaches. Med. Phys. **38**(7), 4296–4311 (2011)
27. Sun, M., Star-Lack, J.M.: Improved scatter correction using adaptive scatter kernel superposition. Phys. Med. Biol. **55**(22), 6695–6720 (2010)
28. Ubeda, C., Vano, E., Gonzalez, L., Miranda, P.: Influence of the antiscatter grid on dose and image quality in pediatric interventional cardiology x-ray systems. Catheter. Cardio. Inte. **82**(1), 51–57 (2013)
29. Wang, A., et al.: Acuros CTS: a fast, linear Boltzmann transport equation solver for computed tomography scatter - Part II: system modeling, scatter correction, and optimization. Med. Phys. **45**(5), 1914–1925 (2018)
30. Yao, W., Leszczynski, K.W.: An analytical approach to estimating the first order scatter in heterogeneous medium. II. A practical application. Med. Phys. **36**(7), 3157–3167 (2009)
31. Zhong, X., Strobel, N., Kowarschik, M., Fahrig, R., Maier, A.: Comparison of default patient surface model estimation methods. In: Bildverarbeitung für die Medizin 2017. I, pp. 281–286. Springer, Heidelberg (2017). https://doi.org/10.1007/978-3-662-54345-0_64

Prediction and Diagnosis

MIA-Prognosis: A Deep Learning Framework to Predict Therapy Response

Jiancheng Yang[1,2,3], Jiajun Chen[3], Kaiming Kuang[3], Tiancheng Lin[1],
Junjun He[1], and Bingbing Ni[1,2,4(✉)]

[1] Shanghai Jiao Tong University, Shanghai, China
nibingbing@sjtu.edu.cn
[2] MoE Key Lab of Artificial Intelligence, AI Institute,
Shanghai Jiao Tong University, Shanghai, China
[3] Dianei Technology, Shanghai, China
[4] Huawei Hisilicon, Shanghai, China

Abstract. Predicting clinical outcome is remarkably important but challenging. Research efforts have been paid on seeking significant biomarkers associated with the therapy response or/and patient survival. However, these biomarkers are generally costly and invasive, and possibly dissatifactory for novel therapy. On the other hand, multi-modal, heterogeneous, unaligned temporal data is continuously generated in clinical practice. This paper aims at a unified deep learning approach to predict patient prognosis and therapy response, with easily accessible data, *e.g.*, radiographics, laboratory and clinical information. Prior arts focus on modeling single data modality, or ignore the temporal changes. Importantly, the clinical time series is asynchronous in practice, *i.e.*, recorded with irregular intervals. In this study, we formalize the prognosis modeling as a **multi-modal asynchronous** time series classification task, and propose a *MIA*-Prognosis framework with **Measurement, Intervention** and **Assessment** (MIA) information to predict therapy response, where a Simple Temporal Attention (SimTA) module is developed to process the asynchronous time series. Experiments on synthetic dataset validate the superiory of SimTA over standard RNN-based approaches. Furthermore, we experiment the proposed method on an in-house, retrospective dataset of real-world non-small cell lung cancer patients under anti-PD-1 immunotherapy. The proposed method achieves promising performance on predicting the immunotherapy response. Notably, our predictive model could further stratify low-risk and high-risk patients in terms of long-term survival.

Keywords: Asynchronous time series · Prognosis · Immunotherapy

1 Introduction

Modeling patient prognosis is a challenging but important topic in clinical research, where researchers analyze and predict clinical outcomes including

J. Yang and J. Chen–Contributed equally

© Springer Nature Switzerland AG 2020
A. L. Martel et al. (Eds.): MICCAI 2020, LNCS 12262, pp. 211–220, 2020.
https://doi.org/10.1007/978-3-030-59713-9_21

response to certain therapy (*e.g.*, radiotherapy, chemotherapy, surgery, immunotherapy for oncology), patient progression-free survival (PFS) and overall survival (OS). Research efforts have been paid on seeking significant biomarkers, *e.g.*, EGFR mutation for EGFR-TKI therapy [23], PD-L1 expression and tumor mutational burden (TMB) for immunotherapy [5]. However, these biomarkers are generally costly and invasive, and possibly dissatisfactory for novel therapy, *e.g.*, anti-PD-1 and anti-PD-L1 immunotherapy [14]. With more novel revolutionary therapy (including combination therapy [11]) available, a unified analytic framework for modeling patient prognosis is urged.

We address this issue via emerging deep learning technology by mining clinical data, *e.g.*, electronic health records (EHR) [13]. Specifically, we focus on a unified approach to model patient prognosis under certain therapy. Prior arts are generally developed on a single data modality [10,16]. Besides, only a few studies [18] take into account the temporal/serial information. In clinical practice, multi-modal temporal data is continuously generated with numerous kinds of sensors and records. It is remarkably valuable to mine the easily accessible information to develop the prognosis prediction system, *e.g.*, radiographics, laboratory and clinical information. We formalize the prognosis modeling as a **multi-modal asynchronous** time series classification task, and propose a *MIA*-Prognosis framework with **Measurement, Intervention** and **Assessment** (MIA) information, where Measurement and Intervention information are treated as inputs of multi-modal asynchronous time series to predict the Assessment as ground truth (details in Sect. 2.1).

An algorithmic challenge is how to effectively and efficiently process multimodal asynchronous time series like clinical information. Binkowski *et al.* [1] propose a gated CNN for asynchronous time series analysis, where asynchronous time intervals are regarded as input features. This approach might not be suitable for the clinical scenario since it is not essentially asynchronous; data is needed to learn representation for time intervals. What we need for real-world clinical data processing is a natively asynchronous model, which is flexible and light-weight to learn from expensive clinical data. Inspired by recent advances in natural language processing (NLP), *e.g.*, attention transformers [3,17,21] and relative position encoding [15], we propose a Simple Temporal Attention (SimTA) module to process asynchronous time series, where attention matrix is learned simply from the time intervals of asynchronous time series (details in Sect. 2.2).

The SimTA module is proven to be superior to standard RNN-based approaches in a synthetic asynchronous time series prediction dataset. Moreover, we experiment the proposed *MIA*-Prognosis framework on an in-house retrospective dataset of real-world non-small cell lung cancer (NSCLC) under anti-PD-1 immunotherapy. Our predictive model achieves promising performance on predicting the immunotherapy response after 90 days. Notably, this model could further stratify low-risk and high-risk patients in terms of long-term survival.

2   Methods

2.1   MIA Prognosis: The Framework

Categorizing Clinical Information. In clinical practice, data of a numerous variety of modalities is collected. Most medical data is unaligned in time steps, which means that it has varying intervals between adjacent steps in time series. Such limitations call for a unified framework that integrates asynchronous data of different modalities. To address this issue, we first divide clinical information into three categories according to data sources: **measurement**, **intervention** and **assessment**, which defined our *MIA*-Prognosis framework. Measurement data comes from medical examinations such as imaging data (computed tomography, ultrasound, X-ray), laboratory and genetic tests. Measurement is the main information in our *MIA*-Prognosis framework. Interventions include actions such as injections and operations. Assessment evaluates the effectiveness of interventions, *e.g.*, Response Evaluation Criteria in Solid Tumors (RECIST) [4], or 1-year overall survival rates. In this study, we use RECIST to obtain the "ground truth"[1] of therapy response, where complete response (CR), partial response (PR), stable disease (SD) are regarded as response (R), and PD (progressive disease) is regarded as non-response (non-R). Note that measurement, intervention and assessment can also be categoried into either serial or static data, which depends on its status over time. These categorizations are the basis of our framework's capability of integrating heterogeneous multi-modal data.

Model Overview. We propose a framework that integrates multi-modal data in asynchronous time series, named *MIA*-Prognosis. Figure 1 gives an overall description of our framework. Due to the fact that clinical data of different modalities is usually unaligned in time, it is impractical to simply concatenate these vectors together and pad zeroes at the time step where a certain modality is missing. Therefore, we process each modality independently in our framework. We pass serial data of each modality through its own SimTA module, which outputs a summary vector. The summary vector is added with a temporal encoding (adapted from position encoding [17]) of time intervals between the assessment time and the last time stamp. Static data goes through a multi-layer perceptron (MLP) that encodes it in high-dimensional embedding. We then concatenate summary vectors of serial data with static embedding, and input the concatenated vector into another MLP to give the final prediction of therapy response (R: response/non-R: non-response) after an unobserved period.

Existing deep sequential models, such as recurrent neural network (RNN), assume that time series data is synchronous in nature. However, this assumption does not hold in the context of clinical practice. Here we introduce a new module to help us process asynchronous data, named SimTA. Inspired by recent advances in natural language processing (NLP), *e.g.*, attention transformers [3,17] and relative position encoding [15], SimTA utilizes time interval information of asynchronous series to generate attention matrix, capturing temporal relationships

---

[1] RECIST is not theoretecally perfect. We refer to "ground truth" in a clinical sense.

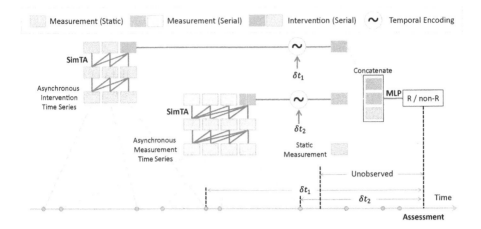

Fig. 1. The *MIA*-Prognosis framework, with Measurement, Intervention and Assessment information. The asynchronous time series is encoded by the proposed Simple Temporal Attention (SimTA) module into a summary vector. The summary vector is further added with a temporal encoding of time intervals between the assessment time and the last time stamp (δt_1 and δt_2). Together with static information, these features predict the therapy response (R/non-R) after an unobserved period.

between asynchronous time steps. It is worth noting that the latest steps in time series of different modalities are not likely to coincide with each other. In such cases, we use temporal encoding to make use of this information.

2.2   Simple Temporal Attention for Asynchronous Time Series

Simple Temporal Attention (SimTA) is the key factor that enables our framework to process asynchronous times series. Let $X \in \mathbb{R}^{T \times C}$ be an asynchronous time series of length T, and $\boldsymbol{\tau} = [\tau_1, \tau_2, \ldots, \tau_{T-1}] \in \mathbb{R}^{T-1}$ be the time interval vector between any adjacent time steps. A general formula of a single SimTA module can be described as:

$$SimTA(X, \boldsymbol{\tau}) = softmax(A)\sigma(f(X)), \tag{1}$$

where σ denotes an activation function of our choice, and f is a fully-connected layer. $A = S(X, \tau) \in \mathbb{R}^{T \times T}$ is the attention matrix that encodes relations between any two time steps. Matrix A can be calculated using any edge-aware attention mechanism, *e.g.*, multi-head self attention [17] with relative position encoding [15]. In this study, we use an extremely simplified version that only encodes linearly time intervals, which is validated effective in our experiments:

$$A = S(\boldsymbol{\tau}) = \begin{bmatrix} 0, & -\infty, & -\infty, & \dots -\infty \\ -\lambda\tau_1 + \beta, & 0, & -\infty, & \dots -\infty \\ -\lambda(\tau_1 + \tau_2) + \beta, & -\lambda\tau_2 + \beta, & 0, & \dots -\infty \\ \vdots & \vdots & \vdots & \ddots & \vdots \\ -\lambda\sum_1^{T-1}\tau_i + \beta, & -\lambda\sum_2^{T-1}\tau_i + \beta, & -\lambda\sum_3^{T-1}\tau_i + \beta, & \dots & 0 \end{bmatrix},$$

$$(2)$$

where $\lambda \in \mathbb{R}^+$ and $\beta \in \mathbb{R}$ are trainable parameters that apply a linear transformation on $\boldsymbol{\tau}$. The simplicity of this attention mechanism can help us cope with overfitting as well, considering we only have limited amount of data. The formula is based on the assumption that the more recent time steps should have stronger correlations with the current time step than further ones. Complex temporal information can be captured by stacking multiple SimTA modules. The complete SimTA model pipeline is a SimTA block of one or multiple SimTA modules, which outputs a summary vector, followed by an MLP that outputs the final prediction with softmax activation. To the best of our knowledge, the proposed SimTA is the first study to introduce attention mechanism to asynchronous time series analysis with proven effectiveness.

2.3 Counterpart Approaches

To demonstrate SimTA's capability of learning asynchronous temporal relations, we bring in LSTM (Long Short-term Memory) [7,9] as comparison. LSTM is a special type of RNN designed to capture relations over extended time intervals in sequences. In our experiments, we use LSTM models with a comparable size and the same training configuration as SimTA. Let X_i be the value of the time series at time step t_i. Three LSTM approaches are tested, which differ in their input: (1) Only X_i; (2) X_i and time intervals τ_i; (3) X_i and time stamps t_i. We mark them as LSTM, LSTM(i) and LSTM(s), respectively.

3 Experiments

3.1 Proof of Concept on Synthetic Dataset

We first validate the superiority of the proposed SimTA over RNN on asynchronous time series using a synthetic dataset.

Dataset and Experiment Settings. The synthetic dataset consists of the summation of N trigonometric functions with random periods. Each time series is computed as:

$$X_t = \sum_{j=1}^{N}[\alpha_j \sin(\omega_j \pi t + b_j) + \beta_j] + \eta\epsilon \quad (3)$$

where N is the number of trigonometric functions involved. ϵ is a white noise following the standard normal distribution, whose magnitude is controlled by a

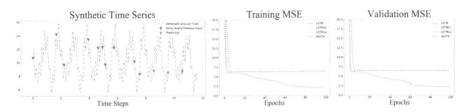

Fig. 2. The synthetic time series and MSE loss curves of LSTM, LSTM(i), LSTM(s) and the proposed SimTA. **Left**: The illustration of a data sample. **Middle**: Training MSE loss curves. **Right**: Validation MSE loss curves. We clip the y axis in both loss curves for the sake of better visualization.

constant η. In our experiments, we choose $N = 10$ and $\eta = 0.5$. We generate 10,000 such asynchronous series, which are split in 80/20 for training/validation, respectively. The training/validation data is sampled once from the predefined distribution and fixed throughout the experiment.

During training, 10 different X_t are randomly sampled from each series. The time intervals between any two adjacent points follow a uniform distribution between 0 and a maximum interval level I. The model is tasked to predict the next 3 points $(+1, +2, +3)$ following the last one in the input. Figure 2 shows a sample series of the synthetic dataset. In our experiments, we compare performances of SimTA and three LSTM models, which are LSTM, LSTM(i) and LSTM(s). For LSTM(i) and LSTM(s), X_t and the time information are concatenated into one vector. SimTA follows the model described in Sect. 2.2.

Results. We train all four models for 100 epochs. Figure 2 shows the training and validation mean squared error over the training phase. SimTA outperforms all three LSTM models on both training and validation data. It achieves significantly lower MSE (2.197 on SimTA and 6.427 on all LSTM approaches) compared with LSTM. The time information does help LSTM(i) and LSTM(s) to converge faster than vanilla LSTM, but all three end up with errors at the same level. It is worth noting that the LSTM model variants underfit the training set. From the observations above, we conjecture that the proposed SimTA outperforms existing standard sequential models such as LSTM for asynchronous time series.

3.2 Predicting Response to Anti-PD-1 Immunotherapy for Non-Small Cell Lung Cancer (NSCLC)

Background. Lung cancer is the most commonly diagnosed cancer worldwide. According to [2], lung cancer accounts for 18.4% of the global cancer deaths in 2018. NSCLC makes up 80%–85% of these cases. Deep learning has shown its potential in precision medicine for lung cancer [19,22,24,25]. Recently, immunotherapy has been proven to remarkably increase the overall survival and the life quality of patients with a variaty of cancers, including NSCLC. However, only a small percentage of patients benefit from immunotherapy and show lasting responses. There has been research on the identification of response predictors, whereas most of the effort are focused on biopsy analyses and serum biomarkers, *e.g.*, PD-L1 expression and tumor mutation burden (TMB) for first-line immunotherapy. These methods are expensive, invasive, and not always consistently associated with tumor responses. Furthermore, no biomarker is available for predicting second-line NSCLC immunotherapy outcome so far . Such limitations emphasize the necessity for convenient, economical and non-invasive indicators, especially for second-line immunotherapy treatment.

Dataset and Experiment Settings. In this retrospective study, 99 patients with advanced or metastatic stage IIIB and IV NSCLC under second-line immunotherapy are included. The dataset includes 793 CT scans, 1335 laboratory blood tests, 99 clinical data, and 320 response evaluations as per RECIST1.1 [4]. All data is further categorized into serial data and static data.

The CT scans are labelled by one radiologist with 8 years of experience, by manually segmenting the volume of interest (VOI) of the target lesion in each scan. An oncologist with 30 years of experience reviewed and confirmed the segmentation. CT volumes and segmentation masks are resampled to uniform spacing ($1\,mm \times 1\,mm \times 1\,mm$), with B-spline interpolation for CT volumes and nearest-neighbor interpolation for VOI masks. We use radiomics features [6] to represent the radiological features due to limited number of samples. With large data available, a fully end-to-end CNN could also be used as the feature extractor. 107 radiomics features are extracted from each VOI using PyRadiomics [8]. Radiomics features are treated as serial data unless there is only one CT examination. The serial blood test features are in 22 dimensions and static clinical information features are in 18 dimensions. All categorical features are encoded in one-hot vectors. Numeric features are normalized by removing the mean and scaling to unit variance to ensure stable training and faster convergence. Intervention information is one-hot encoded (*i.e.*, a binary flag at a time step). Serial radiomics, laboratory blood test and intervention is asynchronous in time.

In our experiments, models are tasked to output binary predictions of R (response) or Non-R (non-response) of each response evaluation, with static data and all serial data before 90 days prior to time of response assessment. We use binary cross entropy (BCE) as the loss function. SimTA model is optimized with Adam optimizer [12]. We split the 99 patients into 3-fold (33 patients in each fold), and perform 3-fold cross validation for evaluating our method. Hyperparameters of model structure and training configuration are chosen using a grid search with bootstrap on the training dataset in each cross validation fold. To verify the effectiveness of SimTA on asynchronous time series, we include LSTM, LSTM(i) and LSTM(s) in our ablation study as comparison. LSTM models of comparable parameters are trained under similar setting.

We further associate model prediction with clinical survival benefits, specifically, overall survival (OS) and progression-free survival (PFS). A cutoff value of 0.5 is used for stratifying patients into high-risk and low-risk groups. The serial data before 90 days prior to time of response assessment is used as input to the trained model. Kaplan-Meier analysis and log-rank test for the survival analysis validate the effectiveness of our method in terms of patient survival.

Table 1. Model performance of on predicting immunotherapy response, including standard RNN approaches (LSTM, LSTM(i), LSTM(s)) instead of the proposed SimTA, and our methods with multi-modal and single-modal inputs.

Methods	LSTM	LSTM(i)	LSTM(s)	Ours	Ours w/o radiomics	Ours w/o lab
AUC	0.71	0.71	0.70	0.80	0.47	0.58

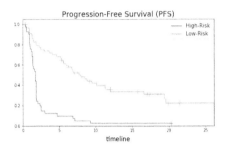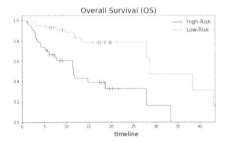

Fig. 3. Model performance on predicting NSCLC patient survival under anti-PD-1 immunotherapy. **Left**: Patient survival curve visualized by Kaplan-Meier (K-M) plot of progression-free survival (PFS), p-value of log-rank test to high/low-risk groups is < 0.01. **Right**: K-M plot of overall survival (OS), with $p < 0.01$.

Results. As depicted in Table 1, promising results are observed in predicting immunotherapy outcome using the proposed framework. The area under curve (AUC) of receiver operating characteristic (ROC) curve is 0.80 with our *MIA*-Prognosis framework, whereas vanilla LSTM, LSTM(i) and LSTM(s) are achieving 0.71, 0.71 and 0.70 AUC respectively. The LSTM counterparts does not totally fail in this case because the patients are taking CT scans and blood tests on a fairly regular schedule, the interval variance is mostly smaller than seven days. Still, SimTA outperforms LSTM by a large margin in this "mildly" asynchronous serial data modelling task. We also validate the necessity of multi-modal input. Without radiomics feature or laboratory blood test results, our framework reaches very low AUC of 0.47 and 0.58 respectively, suggesting the significance of multi-modal model in this task. Moreover, as shown in Fig. 3, the p-values for Kaplan-Meier analysis are significant in both PFS and OS tests. Therefore, our predictive model could further stratify the low- and high-risk patients in terms of patient survival.

4   Conclusion and Further Work

In this paper, we focus on a unified deep learning framework to predict therapy response, with easily accessible clinical data. The proposed framework named *MIA*-Prognosis utilizes clinical information including Measurement, Intervention and Assessment to model patient prognosis. We also propose a Simple

Temporal Attention (SimTA) module to process the asynchronous time series. The proof-of-concept experiments validate the superiority of SimTA over standard RNN approaches in asynchronous time series analysis. Moreover, our method is proven effective on an in-house dataset on predicting response to anti-PD-1 immunotherapy for real-world non-small cell lung cancer (NSCLC) patients. Importantly, our predictive model is associated with long-term patient survival in terms of progression-free survival (PFS) and overall survival (OS).

In future studies, it is valuable to apply the proposed *MIA*-Prognosis framework on other therapy and diseases. On the other hand, it is also important to design efficient and effective non-linear temporal attention module to enhance temporal relation learning of SimTA. Besides, a fully end-to-end model with CNN-based Radiomics [20] to encode the signature of radiographic features is worth exploring. Furthermore, it is interesting to explain what the *MIA*-Prognosis models from data-driven approaches.

Acknowledgment. This work was supported by National Science Foundation of China (61976137, U1611461). Authors would like to appreciate the Student Innovation Center of SJTU for providing GPUs.

References

1. Binkowski, M., Marti, G., Donnat, P.: Autoregressive convolutional neural networks for asynchronous time series. In: ICML (2018)
2. Bray, F., Ferlay, J., Soerjomataram, I., Siegel, R.L., Torre, L.A., Jemal, A.: Global cancer statistics 2018: globocan estimates of incidence and mortality worldwide for 36 cancers in 185 countries. CA Cancer J. Clini. **68**(6), 394–424 (2018)
3. Devlin, J., Chang, M.W., Lee, K., Toutanova, K.: Bert: pre-training of deep bidirectional transformers for language understanding. In: NAACL-HLT (2019)
4. Eisenhauer, E.A., et al.: New response evaluation criteria in solid tumours: revised recist guideline. Eur. J. Cancer **45**(2), 228–47 (2009)
5. Gibney, G.T., Weiner, L.M., Atkins, M.B.: Predictive biomarkers for checkpoint inhibitor-based immunotherapy. Lancet Oncol. **17**(12), e542–e551 (2016)
6. Gillies, R.J., Kinahan, P.E., Hricak, H.: Radiomics: images are more thanpictures, they are data. Radiology **278**(2), 563–577 (2016)
7. Greff, K., Srivastava, R.K., Koutník, J., Steunebrink, B., Schmidhuber, J.: Lstm: a search space odyssey. IEEE Trans. Neural Netw. Learn. Syst. **28**, 2222–2232 (2017)
8. van Griethuysen, J.J., et al.: Computational radiomics system to decode the radiographic phenotype. Cancer Res. **77**(21), e104–e107 (2017)
9. Hochreiter, S., Schmidhuber, J.: Long short-term memory. Neural Comput. **9**, 1735–1780 (1997)
10. Hosny, A., et al.: Deep learning for lung cancer prognostication: a retrospective multi-cohort radiomics study. PLoS Med. **15**(11), e1002711 (2018)
11. Jain, R.K.: Normalizing tumor vasculature with anti-angiogenic therapy: a new paradigm for combination therapy. Nature Med. **7**, 987–989 (2001)
12. Kingma, D.P., Ba, J.: Adam: a method for stochastic optimization. arXiv preprint arXiv:1412.6980 (2014)
13. Rajkomar, A., et al.: Scalable and accurate deep learning with electronic health records. NPJ Digit. Med. **1**(1), 18 (2018)

14. Sacher, A.G., Gandhi, L.: Biomarkers for the clinical use of pd-1/pd-l1 inhibitors in non-small-cell lung cancer: a review. JAMA Oncol. **2**(9), 1217–22 (2016)
15. Shaw, P., Uszkoreit, J., Vaswani, A.: Self-attention with relative position representations. In: NAACL-HLT (2018)
16. Sun, R., et al.: A radiomics approach to assess tumour-infiltrating cd8 cells and response to anti-pd-1 or anti-pd-l1 immunotherapy: an imaging biomarker, retrospective multicohort study. Lancet Oncol. **19**(9), 1180–1191 (2018)
17. Vaswani, A., et al.: Attention is all you need. In: NIPS (2017)
18. Xu, Y., et al.: Deep learning predicts lung cancer treatment response from serial medical imaging. Clin. Cancer Res. **25**(11), 3266–3275 (2019)
19. Yang, J., Deng, H., Huang, X., Ni, B., Xu, Y.: Relational learning between multiple pulmonary nodules via deep set attention transformers. In: ISBI (2020)
20. Yang, J., Fang, R., Ni, B., Li, Y., Xu, Y., Li, L.: Probabilistic radiomics: ambiguous diagnosis with controllable shape analysis. In: Shen, D., et al. (eds.) MICCAI 2019. LNCS, vol. 11769, pp. 658–666. Springer, Cham (2019). https://doi.org/10.1007/978-3-030-32226-7_73
21. Yang, J., et al.: Modeling point clouds with self-attention and gumbel subset sampling. In: CVPR, pp. 3323–3332 (2019)
22. Yang, Y., Yang, J., Ye, Y., Xia, T., Lu, S.: Development and validation of a deep learning model to assess tumor progression to immunotherapy. ASCO **37**, e20601–e20601 (2019)
23. Yu, H.A., et al.: Analysis of tumor specimens at the time of acquired resistance to egfr-tki therapy in 155 patients with egfr-mutant lung cancers. Clin. Cancer Res. **19**(8), 2240–2247 (2013)
24. Zhao, W., et al.: Toward automatic prediction of egfr mutation status in pulmonary adenocarcinoma with 3d deep learning. Cancer Med. **8**(7), 3532–3543 (2019)
25. Zhao, W., et al.: 3d deep learning from ct scans predicts tumor invasiveness of subcentimeter pulmonary adenocarcinomas. Cancer Res. **78**(24), 6881–6889 (2018)

M²Net: Multi-modal Multi-channel Network for Overall Survival Time Prediction of Brain Tumor Patients

Tao Zhou[1], Huazhu Fu[1(✉)], Yu Zhang[2], Changqing Zhang[3], Xiankai Lu[1], Jianbing Shen[1(✉)], and Ling Shao[1,4]

[1] Inception Institute of Artificial Intelligence, Abu Dhabi, UAE
huazhu.fu@inceptioniai.org , shenjianbingcg@gmail.com
[2] Department of Bioengineering, Lehigh University, Bethlehem, PA 18015, USA
[3] College of Intelligence and Computing, Tianjin University, Tianjin, China
[4] Mohamed bin Zayed University of Artificial Intelligence, Abu Dhabi, UAE

Abstract. Early and accurate prediction of overall survival (OS) time can help to obtain better treatment planning for brain tumor patients. Although many OS time prediction methods have been developed and obtain promising results, there are still several issues. *First*, conventional prediction methods rely on radiomic features at the local lesion area of a magnetic resonance (MR) volume, which may not represent the full image or model complex tumor patterns. *Second*, different types of scanners (*i.e.*, multi-modal data) are sensitive to different brain regions, which makes it challenging to effectively exploit the complementary information across multiple modalities and also preserve the modality-specific properties. *Third*, existing methods focus on prediction models, ignoring complex data-to-label relationships. To address the above issues, we propose an end-to-end OS time prediction model; namely, Multi-modal Multi-channel Network (M²Net). Specifically, we first project the 3D MR volume onto 2D images in different directions, which reduces computational costs, while preserving important information and enabling pre-trained models to be transferred from other tasks. Then, we use a modality-specific network to extract implicit and high-level features from different MR scans. A multi-modal shared network is built to fuse these features using a bilinear pooling model, exploiting their correlations to provide complementary information. Finally, we integrate the outputs from each modality-specific network and the multi-modal shared network to generate the final prediction result. Experimental results demonstrate the superiority of our M²Net model over other methods.

1   Introduction

Brain tumors are the most common and difficult-to-treat malignant neurologic tumors, with the highest mortality rate. Gliomas account for about 70% of primary brain tumors in adults [23]. Gliomas can be less aggressive (*i.e.*, low grade), yielding a life expectancy of several years, or more aggressive (*i.e.*, high grade),

© Springer Nature Switzerland AG 2020
A. L. Martel et al. (Eds.): MICCAI 2020, LNCS 12262, pp. 221–231, 2020.
https://doi.org/10.1007/978-3-030-59713-9_22

with a life expectancy of at most two years [7]. The prognosis of gliomas is often measured by the overall survival (OS) time, which varies largely across individuals. Thus, timely and accurate OS time prediction for brain tumor patients is of great clinical importance and could benefit individualized treatment care.

Recently, magnetic resonance imaging (MRI) has played an important role in the study of glioma prognosis [3,16,20,22,26,34,35]. By using informative imaging phenotypes from multi-modal MR scans (*e.g.*, native (T1), T1 contrast enhanced (T1ce), T2-weighted (T2), and Fluid Attenuated Inversion Recovery (FLAIR)), several methods have been proposed for survival prediction. For example, Pope *et al.* [20] found that non-enhancing tumor and infiltration areas are helpful for OS prediction by analyzing 15 MRI features. The study in [17] also used all kinds of informative radiomic features extracted from raw images for OS time prediction. Complementary information from multi-modal data has also played an important role in helping improve prediction performance. For example, Zhou *et al.* [29] extracted informative radiomic features based on segmented tumors from multi-modal data and then used the XGBoost as classifier. Feng *et al.* [4] proposed a linear model for survival prediction using imaging and non-imaging features extracted from multi-modal data. Isensee *et al.* [9] conducted OS prediction by training an ensemble of a random forest regressor. Nie *et al.* [19] used a 3D deep learning model to extract multi-modal multi-channel features and then feeding them into a support vector machine (SVM) classifier for OS prediction.

Although several methods have been developed for this task, there still exist several issues. *First*, conventional prediction approaches rely on radiomic features at the local lesion area of an MR volume. However, these shallow and low-level features might not fully characterize the image or model the tumor's complex patterns. *Second*, most methods simply concatenate features from different modalities into a single feature vector without considering the underlying correlations across multi-modal data. Moreover, rare methods exploit the correlations while also preserving the modality-specific attributes. However, both of aspects are important for improving model performance in multi-modal/view learning [27,32]. *Third*, many methods first extract features from multi-modal data, and then feed these features into a subsequent classification model (*e.g.*, SVM). Due to the possible heterogeneity between the features and the model, ignoring the correlation between the two could lead to sub-optimal results. Intuitively, integrating both components into a unified framework could improve the prediction performance.

To tackle the above issues, we propose an end-to-end **Multi-modal Multi-channel Network** (**M²Net**) for OS time prediction. M²Net first adopts a modality-specific network to automatically extract implicit and high-level features from different modalities (*i.e.*, different MR scans), and then applies a multi-modal shared network to fuse them using a bilinear pooling model. Here, the modality-specific network is used to preserve the modality-specific attributes, while the multi-modal shared network is used to exploit the correlations. Further, we add another layer that automatically weights the outputs from each

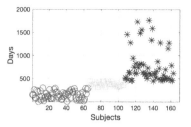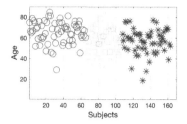

Fig. 1. Illustration of subjects' survival time in days (left) and age information (right), where red, green and blue denote the information of short-term, mid-term, and long-term, respectively. (Color figure online)

modality-specific network and the multi-modal shared network. Compared with conventional classification-based OS prediction methods, M²Net can not only extract high-level features to model complex tumor patterns, but also integrates feature learning and prediction model training into a unified framework. The unified framework enables label information to guide the feature learning more effectively. Besides, both the correlations across multi-modal data and the modality-specific properties can be seamlessly exploited. Finally, the experimental results demonstrate that our M²Net outperforms several classification-based OS prediction methods.

2   Data Description and Processing

The BraTS 2018 was organized using multi-institutional pre-operative MRI scans for the segmentation of intrinsically heterogeneous brain tumor sub-regions [7, 18]. In this dataset, each patient includes 1) T1, 2) T1ce, 3) T2, and 4) FLAIR volumes. In this study, we focus on OS time prediction; thus, we have 163 subjects with survival information (in days) for this task.

Definition of OS Time. Similar to several previous studies [29], we formulate the OS time prediction as classification-based task. Inspired by [29], we define OS time as the duration from the date the patient was first scanned, to the date of the patient's tumor-related death. To construct a classification task, the continuous survival time can be divided into three classes: (1) short-term survivors (*i.e.*, ≤10 months) (2) mid-term survivors (*i.e.*, between 10 and 15 months), and (3) long-term survivors (*i.e.*, ≥15 months). Figure 1 shows subjects' survival time in days (left) and ages (right). As shown in Fig. 1, it can clearly be seen that there are large intra-class differences, especially in the long-term class. Therefore, it is very challenging to achieve accurate prediction for OS time.

Image Patch Extraction. For our OS prediction task, we first locate the tumor region according to the tumor mask and extract an image patch that is centered at the tumor region. Then we resize the extracted image patch of each subject to a predefined size (*i.e.*, 124 × 124 × 124).

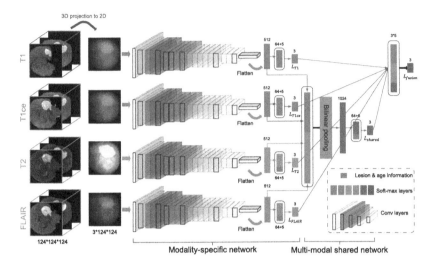

Fig. 2. Overview of the proposed M^2Net for OS time prediction. *First*, we project the 3D MR volumes onto three-channel 2D images in different directions. *Second*, we propose an end-to-end OS time prediction model with multi-modal data.

Lesion and Age Information. Because tumor size and age information may also be related to brain status, they can affect the prediction performance for OS time. Thus, we utilize these as supplemental features. Specifically, for every type of tumor, we calculate the size of each tumor using: $s_i = n_i / \sum n_i$ ($i = 1, 2, 3$), where n_i is the number of the i-th type of tumor. There are three types of tumors, *i.e.*, the necrotic and non-enhancing tumor core (label = 1), the peritumoral edema (label = 2), and the gadolinium-enhancing tumor (label = 4). We also calculate the total size using $s_{total} = \sum n_i / n_{non-zero}$, where $n_{non-zero}$ is the total number of the non-zero elements in the original MR volume. Further, we add the age information (denoted as "s_{age}") into the additional feature vector. Finally, we have the supplemental feature vector as $s = [s_{total}, s_1, s_2, s_3, s_{age}]^\top$.

3   The Proposed Method

In this study, we effectively exploit the correlations among multi-modalities as well as preserve the model-specific attribute, for improving the OS prediction performance. We detail the three main parts: modality-specific network, multi-modal shared network, and multi-modal multi-channel network below.

3.1 Modality-Specific Network

Convolutional neural networks (CNNs) are a class of deep neural networks that have been widely applied for analyzing visual tasks. Deep learning based CNN models can learn a hierarchy of features, in which high-level features are built upon low-level image features using a layer-wise strategy. Many pre-trained models have been developed and obtain very promising performances due to their being trained on large-scale datasets [8]. In this study, in order to reduce computational costs while preserving important information and enabling a pre-trained model to be transferred from another related task, we project 3D MR volumes onto multi-channel 2D images, in different directions (as shown in Fig. 2). Specifically, 3D images can be projected onto 2D images along each axis by averaging the sum of all slices from each volume. Using this strategy, we can obtain a three-channel 2D image. Then, we use the ResNet34 [8] as the base network to extract modality-specific attributes from each modality (*i.e.*, T1, T1ce, T2, and FLAIR). Note that, in ResNet34, the second last layer is a 7×7 max-pooling, which we change to a 4×4 max-pooling layer for our specific task. Thus, we have a fully-connected (FC) layer with 512 nodes (while ResNet34 has 1024 nodes in its FC layer). Finally, for each modality-specific network, we have the corresponding losses $\mathcal{L}_{T1}$, $\mathcal{L}_{T1ce}$, $\mathcal{L}_{T2}$, and $\mathcal{L}_{FLAIR}$, respectively.

3.2 Multi-modal Shared Network

Bilinear models [13] have been widely applied in fine-grained recognition, semantic segmentation, face recognition, and other fields, and have obtained impressive performance. Bilinear pooling often forms a descriptor by calculating $B(\mathcal{X}) = \sum_{s \in \mathcal{S}} x_s x_s^\top$, where $\mathcal{X} = (x_1, \ldots, x_{|\mathcal{S}|}; x_s \in \mathbb{R}^c)$ denotes a set of local descriptors, and $\mathcal{S}$ denotes the set of spatial locations. Using this equation, we can obtain a $c \times c$ matrix. However, the classic bilinear features are often high-dimensional, making them impractical for many applications. To address this issue, a compact bilinear pooling method [5] has been developed to obtain the same discriminative ability as the full bilinear representation but with only few thousand dimensions. Further, given two sets of local descriptors, $\mathcal{X}$ and $\mathcal{Y}$, a linear kernel machine can be used as follows:

$$\langle B(\mathcal{X}), B(\mathcal{Y}) \rangle = \langle \sum_{s \in \mathcal{S}} x_s x_s^\top, \sum_{u \in \mathcal{U}} y_u y_u^\top \rangle = \sum_{s \in \mathcal{S}} \sum_{u \in \mathcal{U}} \langle x_s, y_u \rangle^2, \tag{1}$$

where $\mathcal{U}$ denotes another set of spatial locations. If we could find some low dimensional projection function $\varphi(x) \in \mathbb{R}^d$, where $d \ll c^2$, that satisfies $\langle \varphi(x), \varphi(y) \rangle \approx k(x, y)$, we could approximate the inner product of (1) by:

$$\langle B(\mathcal{X}), B(\mathcal{Y}) \rangle = \sum_{s \in \mathcal{S}} \sum_{u \in \mathcal{U}} \langle x_s, y_u \rangle^2 \approx \sum_{s \in \mathcal{S}} \sum_{u \in \mathcal{U}} \langle \varphi(x), \varphi(y) \rangle^2 = \langle C(\mathcal{X}), C(\mathcal{Y}) \rangle, \tag{2}$$

where $C(\mathcal{X}) = \sum_{s \in \mathcal{S}} \phi(x_s)$ is the compact bilinear feature. It is clear from this analysis that any low-dimensional approximation of the polynomial kernel can be used to obtain compact bilinear features.

In this study, to learn the compact discriminative representations for multi-modal data, we utilize the compact bilinear pooling method to effectively fuse them and exploit the correlations across multiple modalities (as shown in Fig. 2). Thus, for the multi-modal shared network, we have the loss $\mathcal{L}_{\text{shared}}$.

3.3 M²Net: Multi-modal Multi-channel Network

To effectively exploit the correlations among multi-modal data, as well as preserve modality-specific attributes for each modality, we propose a Multi-modal Multi-channel Network (*i.e.*, M²Net). As shown in Fig. 2, we can obtain multiple outputs (*i.e.*, $\text{out}_v^{sp}, v = 1, 2, 3, 4$) for all modality-specific networks and one output (*i.e.*, out^{sh}) from the multi-modal shared network. A direct way to fuse the various outputs would be to weight them, *i.e.*, $\text{out} = \sum_v w_v^{sp} * \text{out}_v^{sp} + w^{sh} * \text{out}^{sh}$, where w_v^{sp} and w^{sh} denote the corresponding weights ($v = 1, 2, 3, 4$). However, in order to construct an end-to-end prediction model, we can cascade all the outputs (*i.e.*, out_v^{sp} and out^{sh}), and then add a soft-max layer after the cascaded layer. It is worth noting that this strategy automatically weights the various outputs. More importantly, it enables our model to automatically learn the contributions of each modality and the multi-modal fusion. Subsequently, the total loss of our M²Net can be obtained as

$$\mathcal{L} = \lambda_1(\mathcal{L}_{\text{T1}} + \mathcal{L}_{\text{T1ce}} + \mathcal{L}_{\text{T2}} + \mathcal{L}_{\text{FLAIR}} + \mathcal{L}_{\text{shared}}) + \lambda_2 \mathcal{L}_{\text{fusion}}, \qquad (3)$$

where $\mathcal{L}_{\text{fusion}}$ denotes the loss of the final fusion for multiple outputs, and λ_1 and λ_2 are trade-off parameters.

4 Experiments

4.1 Experimental Settings

In our experiments, we adopt a 10-fold cross-validation strategy for performance evaluation. We further split the training set into two groups as: training set (80%) and validation set (20%). The best model for the validation set is used to evaluate the testing set. We adopt four evaluation metrics, including accuracy, precision, recall, and F-score [21]. We calculate the precision and recall using a one-class-versus-all-other-classes strategy, and then calculate F-score using F-score $= \frac{2*\text{Precision}*\text{Recall}}{\text{Precision}+\text{Recall}}$. Finally, we report the mean and standard deviation.

We compare M²Net with the following methods: • **Baseline** method. Similar to the work [19], we first extract features from the outputs of the first FC layer (*i.e.*, with 512 nodes) for each modality, and then cascade the features from the four modalities and the supplemental features into a single feature vector (*i.e.*, with $4 \times 512 + 5$ dimension). Sparse learning [15] is applied to the cascaded features, and then the selected features are fed to an SVM classifier. • **3D CNN fusion** method. We use a 3D CNN [11] for each modality (*i.e.*, the extracted 3D patches) and then fuse them using a bilinear model (denoted as "3D CNN + fusion"). • **TPCNN** [29]. This method uses a CNN model to extract features

Table 1. Prediction results (mean ± standard deviation) of different methods. The best results are highlighted in **bold**.

Method	Accuracy	Precision	Recall	F-score
Baseline [19]	0.544 ± 0.140	0.478 ± 0.177	0.506 ± 0.135	0.489 ± 0.152
3D CNN fusion [11]	0.587 ± 0.093	0.479 ± 0.114	0.533 ± 0.071	0.501 ± 0.141
TPCNN [29]	0.636 ± 0.000	–	–	–
Multi-channel ResNet	0.624 ± 0.084	0.556 ± 0.167	0.572 ± 0.091	0.559 ± 0.136
Modality-specific nets	0.621 ± 0.069	0.580 ± 0.177	0.576 ± 0.067	0.569 ± 0.110
Multi-modal shared net	0.611 ± 0.071	0.449 ± 0.071	0.566 ± 0.045	0.499 ± 0.056
M^{2}Net	$\mathbf{0.664 \pm 0.061}$	$\mathbf{0.574 \pm 0.141}$	$\mathbf{0.613 \pm 0.075}$	$\mathbf{0.589 \pm 0.102}$

from multi-modal data and then employs XGBoost to build the regression model for OS time prediction. • **Multi-channel ResNet** method. We cascade the three-channel 2D images from all the four modalities into a $12 \times 124 \times 124$ image, and then we apply the pre-trained ResNet34 model for prediction. Here, a 1×1 convolutional layer must be added before the standard ResNet (since three-channel images are used as inputs). • **Modality-specific nets** method. This is a degraded version of our proposed model, which only uses the modality-specific network and then fuses all the outputs (*i.e.*, soft-max layers) to obtain the final results. • **Multi-modal shared net** method. This is another degraded version of our proposed model, which employs the multi-modal shared network using a bilinear model and then obtains the final prediction result. Note that, for fair comparison, we use the supplemental features for all comparison methods.

Implementation Details: The proposed M^{2}Net is implemented using PyTorch, and the model is trained using Adam optimizer. The decay rate is set to 10^{-4}, and the maximum epoch is 100. The batch size is set to 32. Besides, the training data are augmented by vertical and horizontal flips, and rotation. The trade-off parameters are set to $\lambda_1 = 0.1$ and $\lambda_2 = 0.5$.

4.2   Results Comparison

Table 1 shows the prediction performance of all the comparison methods. As can clearly be seen, our proposed method performs consistently better than all the comparison methods in four metrics. Compared with the baseline method, our method effectively improves the prediction performance. Note that the baseline method conducts feature learning and classification model learning using a two-steps strategy, while our method integrates them into an end-to-end framework. In contrast to the "3D CNN fusion" method, our model first projects the 3D MR volume onto 2D images in different directions and then transfers a pre-trained model to the prediction task. The results clearly validate the effectiveness of this strategy. Our method is also shown to have a better prediction performance than the TPCNN method. Besides, it is worth noting that the "Multi-channel

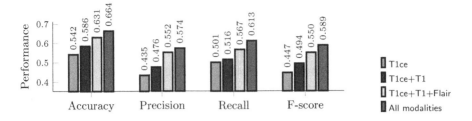

Fig. 3. Results comparison using multi-modal data fusion and single modality.

ResNet" method obtains a even better performance than Baseline and "3D CNN fusion" methods. Furthermore, the improvement of our model over "Modality-specific nets" and "Multi-modal shared net" clearly validates the importance of fusing multiple outputs from the modality-specific networks and the multi-modal shared network. Note that, while several other studies [4] also investigate OS time prediction, they tend to primarily focus on tumor segmentation and then directly conduct a regression task, reporting a mean square error. Thus, it is not possible to directly compare our method with these methods. Overall, our method obtains a relatively better performance than other comparison methods.

Multi-modal Data Fusion. To analyze the effectiveness of the multi-modal data fusion, we first apply a modality-specific network for each modality, and we report the classification results for the best modality (*i.e.*, T1ce in this experiment, as shown in Fig. 3). Then, we conduct the experiment using a two-modality fusion, *i.e.*, fusing T1ce and the three remaining modalities. We obtain the best performance when fusing the T1ce and T1 modalities. Next, we fuse the three modalities (*i.e.*, T1ce+T1+FLAIR), and obtain an overall better performance that the two-modality fusion. From the results, it can be seen that the prediction performance of our method improves when using more modalities, which also verifies the effectiveness of multi-modal data fusion.

4.3 Discussion

As reported in Table 1, our proposed model obtains much better prediction performance than all the comparison methods, the overall performance is still not very high. There could be several reasons for this. *First*, for the OS time prediction task, over-fitting is very likely to occur since a very small dataset is used. We also note that the prediction task is different from the tumor segmentation task, because the segmentation task can be conducted on each slice from a 3D volume, providing significantly more samples. *Second*, as seen from Fig. 1, there are large intra-class deviations in OS time, especially in the long-term class, and it is difficult to learn a reliable model to bridge these deviations. We also note that there is a small amount of subjects between short-term and middle-term classes that are very close together, which leads to ambiguous classification results.

In future work, we will focus on the exploration of features (including genetic and lesion information [10,24]) through clinical collaboration, to improve the prediction performance. Additionally, modeling brain tumor growth is in itself meaningful, enabling us to better understand the mechanisms behind the disease progression and formulate better treatment planning. Thus, the mathematical modeling [6,14] is a future direction. Besides, to deal with small sample issues, transfer learning [12,25,33] can be introduced to our OS prediction task by borrowing prior knowledge from related tasks to improving prediction performance. We also use multi-stage fusion strategy [31] to integrate multi-modal MR images for improving prediction performance. Further, we also use the generative adversarial networks [1,2,28,30] to synthesize more samples, which can be regarded as a form of data augmentation to enhance the prediction performance.

5 Conclusion

In this paper, we propose a novel M^{2}Net model to predict the OS time for brain tumor patients. Our M^{2}Net model preserves the attributes of different modalities using a modality-specific network while also exploiting the correlations across multi-modal data using a multi-modal shared network. Further, the multiple outputs can be automatically fused to obtain the final prediction result. Experimental results demonstrate that our M^{2}Net model improves OS time prediction accuracy, which also indicate the effectiveness of our proposed end-to-end prediction framework.

Acknowledgement. This research was supported in part by NSF of China (No: 61973090) and NSF of Tianjin (No: 19JCYBJC15200).

References

1. Chartsias, A., Joyce, T., Giuffrida, M.V., Tsaftaris, S.A.: Multimodal MR synthesis via modality-invariant latent representation. IEEE TMI **37**(3), 803–814 (2017)
2. Fan, J., Cao, X., Wang, Q., Yap, P.T., Shen, D.: Adversarial learning for mono-or multi-modal registration. Med. Image Anal. **58**, 101545 (2019)
3. Fan, J., Cao, X., et al.: BIRNet: brain image registration using dual-supervised fully convolutional networks. Med. Image Anal. **54**, 193–206 (2019)
4. Feng, X., Tustison, N., Meyer, C.: Brain tumor segmentation using an ensemble of 3 D U-Nets and overall survival prediction using radiomic features. In: International MICCAI Brainlesion Workshop, pp. 279–288. Springer (2018)
5. Gao, Y., Beijbom, O., et al.: Compact bilinear pooling. In: CVPR, pp. 317–326. IEEE (2016)
6. Gevertz, J.L., Torquato, S.: Modeling the effects of vasculature evolution on early brain tumor growth. J. Theor. Biol. **243**(4), 517–531 (2006)
7. Havaei, M., et al.: Brain tumor segmentation with deep neural networks. Med. Image Anal. **35**, 18–31 (2017)
8. He, K., Zhang, X., Ren, S., Sun, J.: Deep residual learning for image recognition. In: CVPR, pp. 770–778. IEEE (2016)

9. Isensee, F., Kickingereder, P., Wick, W., Bendszus, M., Maier-Hein, K.H.: Brain tumor segmentation and radiomics survival prediction: contribution to the BRATS 2017 challenge. In: Crimi, A., Bakas, S., Kuijf, H., Menze, B., Reyes, M. (eds.) BrainLes 2017. LNCS, vol. 10670, pp. 287–297. Springer, Cham (2018). https://doi.org/10.1007/978-3-319-75238-9_25

10. Kao, P.Y., Ngo, T., Zhang, A., Chen, J.W., Manjunath, B.S.: Brain tumor segmentation and tractographic feature extraction from structural MR images for overall survival prediction. In: Crimi, A., Bakas, S., Kuijf, H., Keyvan, F., Reyes, M., van Walsum, T. (eds.) BrainLes 2018. LNCS, vol. 11384, pp. 128–141. Springer, Cham (2019). https://doi.org/10.1007/978-3-030-11726-9_12

11. Khvostikov, A., Aderghal, K., Benois-Pineau, J., Krylov, A., Catheline, G.: 3D CNN-based classification using sMRI and MD-DTI images for alzheimer disease studies. arXiv preprint arXiv:1801.05968 (2018)

12. Li, W., Zhao, Y., Chen, X., Xiao, Y., Qin, Y.: Detecting alzheimer's disease on small dataset: a knowledge transfer perspective. IEEE JBHI **23**(3), 1234–1242 (2018)

13. Lin, T.Y., RoyChowdhury, A., Maji, S.: Bilinear cnn models for fine-grained visual recognition. In: ICCV, pp. 1449–1457. IEEE (2015)

14. Lipkova, J., et al.: Personalized radiotherapy design for glioblastoma: integrating mathematical tumor models, multimodal scans, and bayesian inference. IEEE TMI **38**(8), 1875–1884 (2019)

15. Liu, J., Ji, S., Ye, J.: SLEP: sparse learning with efficient projections. Ariz. State Univ. **6**(491), 7 (2009)

16. Liu, L., et al.: Overall survival time prediction for high-grade glioma patients based on large-scale brain functional networks. Brain Imaging Behav. **13**(5), 1333–1351 (2019)

17. Liu, Y., Xu, X., Yin, L., Zhang, X., Li, L., Lu, H.: Relationship between glioblastoma heterogeneity and survival time: an MR imaging texture analysis. Am. J. Neuroradiol. **38**(9), 1695–1701 (2017)

18. Menze, B.H., et al.: The multimodal brain tumor image segmentation benchmark (BRATS). IEEE TMI **34**(10), 1993–2024 (2015)

19. Nie, D., Lu, J., Zhang, H., Adeli, E., Wang, J., et al.: Multi-channel 3 D deep feature learning for survival time prediction of brain tumor patients using multimodal neuroimages. Sci. Rep. **9**(1), 1103 (2019)

20. Pope, W.B., Sayre, J., Perlina, A., Villablanca, J.P., Mischel, P.S., Cloughesy, T.F.: MR imaging correlates of survival in patients with high-grade gliomas. Am. J. Neuroradiol. **26**(10), 2466–2474 (2005)

21. Provost, F., Fawcett, T.: Robust classification for imprecise environments. Mach. Learn. **42**(3), 203–231 (2001)

22. Razek, A., El-Serougy, L., Abdelsalam, M., Gaballa, G., Talaat, M.: Differentiation of residual/recurrent gliomas from postradiation necrosis with arterial spin labeling and diffusion tensor magnetic resonance imaging-derived metrics. Neuroradiology **60**(2), 169–177 (2018)

23. Ricard, D., Idbaih, A., Ducray, F., Lahutte, M., Hoang-Xuan, K., Delattre, J.Y.: Primary brain tumours in adults. Lancet **379**(9830), 1984–1996 (2012)

24. Tang, Z., et al.: Pre-operative overall survival time prediction for glioblastoma patients using deep learning on both imaging phenotype and genotype. In: Shen, D., et al. (eds.) MICCAI 2019. LNCS, vol. 11764, pp. 415–422. Springer, Cham (2019). https://doi.org/10.1007/978-3-030-32239-7_46

25. Wang, Y.X., Hebert, M.: Learning to learn: model regression networks for easy small sample learning. In: Leibe, B., Matas, J., Sebe, N., Welling, M. (eds.) ECCV 2016. LNCS, vol. 9910, pp. 616–634. Springer, Cham (2016). https://doi.org/10.1007/978-3-319-46466-4_37

26. Zacharaki, E.I., Morita, N., Bhatt, P., Orourke, D., Melhem, E.R., Davatzikos, C.: Survival analysis of patients with high-grade gliomas based on data mining of imaging variables. Am. J. Neuroradiol **33**(6), 1065–1071 (2012)

27. Zhang, C., Liu, Y., Fu, H.: Ae2-nets: autoencoder in autoencoder networks. In: VPR, pp. 2577–2585. IEEE (2019)

28. Zhang, T., et al.: SkrGAN: sketching-rendering unconditional generative adversarial networks for medical image synthesis. In: Shen, D., et al. (eds.) MICCAI 2019. LNCS, vol. 11767, pp. 777–785. Springer, Cham (2019). https://doi.org/10.1007/978-3-030-32251-9_85

29. Zhou, F., Li, T., Li, H., Zhu, H.: TPCNN: two-phase patch-based convolutional neural network for automatic brain tumor segmentation and survival prediction. In: Crimi, A., Bakas, S., Kuijf, H., Menze, B., Reyes, M. (eds.) BrainLes 2017. LNCS, vol. 10670, pp. 274–286. Springer, Cham (2018). https://doi.org/10.1007/978-3-319-75238-9_24

30. Zhou, T., Fu, H., Chen, G., Shen, J., Shao, L.: Hi-net: hybrid-fusion network for multi-modal MR image synthesis. In: TMI. IEEE (2020)

31. Zhou, T., Thung, K.H., et al.: Effective feature learning and fusion of multimodality data using stage-wise deep neural network for dementia diagnosis. Hum. Brain Mapp. **40**(3), 1001–1016 (2019)

32. Zhou, T., Zhang, C., Peng, X., Bhaskar, H., Yang, J.: Dual shared-specific multi-view subspace clustering. IEEE Transactions on Cybernetics (2019)

33. Zhou, T., Fu, H., Gong, C., Shen, J., Shao, L., Porikli, F.: Multi-mutual consistency induced transfer subspace learning for human motion segmentation. In: CVPR, pp. 10277–10286. IEEE (2020)

34. Zhou, T., et al.: Deep multi-modal latent representation learning for automated dementia diagnosis. In: Shen, D., et al. (eds.) MICCAI 2019. LNCS, vol. 11767, pp. 629–638. Springer, Cham (2019). https://doi.org/10.1007/978-3-030-32251-9_69

35. Zhu, H., Yuan, M., Qiu, C., et al.: Multivariate classification of earthquake survivors with post-traumatic stress disorder based on large-scale brain networks. Acta Psychiatr. Scand. **141**(3), 285–298 (2020)

Automatic Detection of Free Intra-abdominal Air in Computed Tomography

Oliver Taubmann[1(✉)], Jingpeng Li[2], Felix Denzinger[1,2], Eva Eibenberger[1], Felix C. Müller[3,4], Mathias W. Brejnebøl[5], and Andreas Maier[2]

[1] Computed Tomography, Siemens Healthcare GmbH, Forchheim, Germany
`oliver.taubmann@siemens-healthineers.com`
[2] Pattern Recognition Lab, Friedrich-Alexander-Universität Erlangen-Nürnberg (FAU), Erlangen, Germany
[3] Department of Radiology, Herlev and Gentofte Hospital, Herlev, Denmark
[4] Siemens Healthcare A/S, Ballerup, Denmark
[5] Department of Radiology, Bispebjerg and Frederiksberg Hospital, Copenhagen, Denmark

Abstract. Pneumoperitoneum, the presence of air within the peritoneal cavity, is a comparatively rare but potentially urgent critical finding in patients presenting with acute abdominal pain. When prior laparoscopic treatment can be ruled out as a cause, it can indicate perforation of the wall of a hollow organ, which typically necessitates immediate surgery. Computed tomography (CT) is the gold standard for detecting free intra-abdominal air, yet subtle cases are easy to miss. More crucially though, if there is no initial suspicion of pneumoperitoneum, the scans may not be read immediately as other emergency patients take precedence. Therefore, fully automatic detection would provide a direct clinical benefit. In this work, an algorithm for this purpose is proposed which follows a sliding-window approach and has a deep-learning based classifier at its core. In addition to the baseline method, variants that rely on multi-scale inputs and recurrent layers to increase robustness are presented. In a five-fold cross validation on the training data, consisting in abdominal CT scans of 110 affected patients and 29 controls, our method achieved an area under the receiver-operating characteristic curve of 89% for case-level classification. Due to its high specificity at reasonable detection rates, it shows potential for use in triage, where false alerts are considered particularly harmful as they may disrupt the clinical workflow.

Keywords: Pneumoperitoneum · Free intra-abdominal air · CT

1 Introduction

The peritoneum is a membrane that tightly surrounds the majority of the abdominal organs. The folds of the peritoneum create an intra-abdominal space,

---

O. Taubmann and J. Li—Contributed equally to this work.

© Springer Nature Switzerland AG 2020
A. L. Martel et al. (Eds.): MICCAI 2020, LNCS 12262, pp. 232–241, 2020.
https://doi.org/10.1007/978-3-030-59713-9_23

the peritoneal cavity. During the diagnosis of patients with severe abdominal pain, the identification of air inside the peritoneal cavity (pneumoperitoneum) is an important and time-critical task for radiologists. The most common non-surgical cause of pneumoperitoneum is a perforation of the viscus. More precisely, free intra-abdominal air indicates the rupturing of a hollow organ, mostly of the gastrointenstinal tract. Such a rupture is a surgical emergency and requires immediate intervention in order to prevent contamination within the peritoneal cavity [9,10]. As the perforation itself often cannot be recognized directly with the imaging modalities commonly used in emergency care, the air acts as a surrogate radiological sign. Computed tomography (CT) is considered to be the standard for discrimination of extraluminal and intraluminal gas [1,5,9]. However, depending on the location and amount of free intra-abdominal air—ranging from sub-centimeter air bubbles in close proximity to the bowel wall to large quantities that freely move within the peritoneal cavity—its diagnosis can be challenging and inconspicuous cases might be missed, especially when there is no initial suspicion of pneumoperitoneum. In addition, other cases may occupy the higher-priority spots of the emergency reading worklist, causing easily identifiable cases to go unnoticed until the scan is read at a later time, at which the patient may already have faced serious consequences. An automatic detection algorithm would be useful [13] for both scenarios, i.e. identification and prioritized reading (triage) of potentially incidentally found cases.

While prior work has addressed detection of pneumoperitoneum on chest X-rays [7], to the best of our knowledge this paper marks the first time an algorithm for the automatic detection of free intra-abdominal air on CT scans is described in the literature. In our method, we tackle the detection task using a conventional sliding-window approach relying on deep-learning based classification of 3-D image patches. We demonstrate two variants to enhance this approach employing multi-scale inputs and recurrent learning in order to increase the spatial context of the input patch without compromising the association between the label and the location of the finding that caused it. In addition to the default setup that requires only trivial preprocessing, we also show results for a variant that relies on a prior segmentation of the abdominopelvic cavity being available. To assess the performance of our algorithm while also taking into account the heterogeneity and varying difficulty of the cases, we perform sub-group analysis based chiefly on the amount of free intra-abdominal air observable in the scan.

2   Methods

2.1   Preprocessing

The input to our model is any abdominal CT scan. The image volume is first resampled in z to obtain a standardized slice thickness of 3 mm. The in-plane resolution may vary. Next, a simple body segmentation is performed to obtain a binary mask of the image region occupied by the patient, which will serve as the search space unless explicitly stated otherwise. Alternatively, the search space can be restricted further by using a segmentation of the abdominopelvic cavity instead, which is obtained from a deep-learning model originally described by Yang et al. for liver segmentation [15].

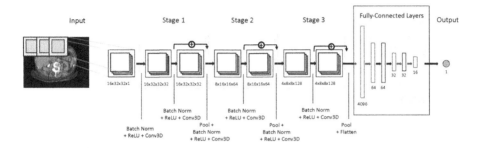

Fig. 1. Baseline convolutional neural network architecture for sliding-window detection, which maps each 3-D patch to a binary (sigmoid) output variable stating a probability for whether the patch contains free intra-abdominal air.

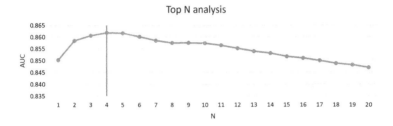

Fig. 2. Case-level AUCs for a baseline model (cf. Sects. 2.4, 3.1) for different values of N, i.e. how many of the highest individual patch predictions are used to determine the case-level prediction (cf. Sect. 2.2).

2.2 Sliding-Window Detection Approach

Inference Workflow. Subsequently, the volume is subdivided into regular 3-D patches of size $(s_x, s_y, s_z) = (64, 64, 16)$ pixels with 25% overlap in all dimensions. The patches are downsampled in the axial plane to a size of $(s'_x, s'_y, s_z) = (32, 32, 16)$ pixels and pixel intensities are normalized. Each patch is then separately fed to the convolutional neural network classifier depicted in Fig. 1. The model is trained to classify whether the patch contains free intra-abdominal air or not. The case-level prediction for a patient is then determined as the average of the output probability scores of the top N patches, i.e. the N boxes that achieved the highest probabilities, where $N = 4$ is determined empirically once and then kept fixed (cf. Fig. 2). Only patches that have overlap with the search space are considered. Below, this model will be referred to as the baseline model.

Model Training. The ground truth annotation consists in markers placed manually in the original scans by radiologists, indicating positions inside the air regions. (For a brief discussion of the chosen mode of annotation, please see Sect. 3.2 below.) If an image patch contains any markers, it is labeled as containing free intra-abdominal air, and vice versa. In order to avoid patches with off-center detections while at the same time ensuring complete coverage of all markers, a margin of half the size of the overlap region is excluded for this process of determining the labels. For training, mini batches of size 32 are augmented on the fly with random translations, small-angle rotations and flips, then fed to an Adam optimizer [6] to minimize binary cross-entropy on the training data. Frequency weighting is utilized to cope with the label imbalance. We use early stopping with a patience of 30 epochs monitoring Matthews correlation coefficient [8] calculated on validation data.

2.3   Enhancing Spatial Context

The obvious downside of sliding window classifiers is the limited spatial context available to the model. However, there is a trade-off involved that prevents one from the trivial solution of choosing ever larger patches to overcome this limitation: With more context, the association between the label and the finding that caused it grows weaker. As an extreme example, consider the case of whole-volume classification: While technically all information is present, even deep learning usually has a hard time identifying the relevant structures due to the enormous input space, unless training data on the order of thousands of cases is available. Below, we present two ways to cope with this trade-off.

Multi-scale Variant. The first is a simple extension that borrows its core idea from multi-scale/resolution analysis [14]. For each patch, a second patch centered at the same position, but with twice the (physical) size within the axial plane is extracted. It is downsampled to the same size in pixels as the original patch such that it has half its resolution. This second patch is fed to the model as a separate input and goes through a duplicated (but not parameter-tied) version of the convolutional stages of Fig. 1, with the resulting feature maps of both legs being concatenated before the first fully-connected layer. All other aspects of the training procedure are kept the same.

Recurrent Variant. Inspired loosely by approaches common in text recognition [12], the other variant is a way for the model to incorporate information from previous and subsequent patches by using recurrent units. More precisely, we treat all sliding-window patches with the same position along the axial direction, i.e. a z-slice of 3-D patches which corresponds to an axial slab in the input volume, as a sequence. Within the sequence, the patches are arranged to follow a zig-zag order (left-to-right, then right-to-left, etc.) in order to avoid "jumps." We then design a model to bi-directionally traverse such patch sequences and map them to the binary sequences of their labels. In order to have fixed-length sequences, we include the patches outside the search space as well and zero their predictions after inference. As training this model from scratch proves to be challenging, we adopt a transfer learning approach by taking the baseline model described in Sect. 2.2, freezing the trained weights and adding three layers of gated recurrent units (GRU) [2] after the first, third and fifth dense layer, each with the same output tensor dimensions as the preceding layer. These are then trained for ten more epochs, effectively fine-tuning the model to the sequence representation.

2.4 Experiments

Data. Abdominal CT scans from a total of $N_{pat} = 139$ patients were selected for a retrospective study, of which $N_{air} = 110$ were consecutive pneumoperitoneum cases, while the remaining $N_{ctrl} = 29$ were controls. The control cases are equivalently acquired scans of other emergency patients with an abdominal pathology that subsequently required surgery, rendering them directly comparable to those of the pneumoperitoneum patients as well as more difficult to distinguish from them than healthy individuals. For training and evaluation purposes, all N_{pat} patients are pooled and five random repetitions of five-fold cross validation (CV) are performed to achieve reliable performance estimates. From four out of five CV subsets used for training, one is set aside as the validation set.

Figures of Merit. Receiver-operating characteristic (ROC) curve analysis [4] is carried out in a micro-averaging fashion, i.e. with all out-of-sample predictions for one CV repetition combined in one curve, for two classification tasks: a) Patch-wise binary classification of all 3-D patches of all patients (window-level evaluation) and b) patient-wise binary classification according to the top-4 weighting procedure of patch predictions as described in Sect. 2.2 (case-level evaluation). While the latter represents a clinically more meaningful assessment, the former allows for a more fine-grained analysis of the model performance due to the much larger number of samples. In both evaluations, the same training/test splits were performed on the patient level.

Sub-group Analysis. The difficulty of the detection task varies substantially; the largest air volumes are immediately obvious even to a layperson, whereas a single, tiny bubble of non-physiologic air in the abdomen might be the proverbial needle in a haystack. In order to get a better intuition of the level of heterogeneity and how it affects the model performance, a simple sub-group analysis is employed: We treat the number of markers placed per patient as a surrogate for the amount of air, seeing as large air volumes are somewhat densely covered with markers to ensure that all sliding windows are properly labeled. Hence, the N_{air} pneumoperitoneum data sets are sorted w.r.t. the number of markers and then split into two partitions: $\mathcal{P}_{<t}$ ($\mathcal{P}_{\geq t}$) with less than (at least) t markers for $t \in \{50, 100\}$. The sizes of the sub-groups are $|\mathcal{P}_{<50}| = 49$, $|\mathcal{P}_{\geq 50}| = 61$ and $|\mathcal{P}_{<100}| = 76$, $|\mathcal{P}_{\geq 100}| = 34$. For evaluation of the individual sub-groups, each is separately pooled with all control cases for a robust assessment. See Fig. 3d for a histogram of the marker distribution.

3   Results and Discussion

3.1   Evaluation

Table 1 shows the areas under the curve (AUC) for the ROC curves of the baseline, multi-scale and recurrent variants for case-level and window-level evaluations. On the window level, there is a continuous improvement from the baseline to the multi-scale to the recurrent variant. While the changes may seem minimal, the difference between the baseline and the recurrent variant is statistically significant with $p < 10^{-4}$ in a two-tailed paired-sample t-test over the CV repetitions. This is due to the AUC being highly reproducible, as reflected by the small standard deviations, owing to the large total number of patches. It can also be seen that this improvement does not translate directly to the case-level. Here, the AUC even decreases a little for the enhanced variants. However, all of these differences are statistically insignificant: All models can be assumed to perform equally well on case level. In essence, the patches that were classified more accurately by the enhanced models were not influential for the case-level decisions, at least in the limited number of cases available. For the sake of conciseness, remaining results are shown only for the best-performing, recurrent model.

Table 1. AUC values for window-level and case-level evaluation, given in percent as mean ± standard deviation over five random repetitions of five-fold CV.

Evaluation	Baseline	Multi-scale	Recurrent
Window level	$94.8 \pm 0.2\%$	$95.0 \pm 0.5\%$	$95.6 \pm 0.2\%$
Case level	$86.2 \pm 1.5\%$	$84.8 \pm 2.8\%$	$85.3 \pm 2.2\%$

In Figs. 3a–c, ROC curves for one exemplary CV run are shown. Figures 3a, b summarize the results of the sub-group analysis. For the groups $\mathcal{P}_{\geq 50}$ and $\mathcal{P}_{\geq 100}$, AUCs of 96% and 99% are achieved even though these still contain 55% and 31% of all data sets, respectively. In other words, while there are very difficult cases that limit our system's overall sensitivity, close to a third of the pneumoperitoneum cases can be classified almost perfectly. Note that although any amount of free intra-abdominal air can indicate an emergency, there is evidence that local occurrences are clinically less significant [3]. Figure 3c shows the change in performance when using the abdominopelvic cavity segmentation instead of the body mask, which suggests that elimination of candidates outside of the abdomen is beneficial. Figure 4 demonstrates examples of model mistakes, i.e. false positive/negative predictions. Despite being a false alert, the saliency map visualized in Fig. 4b illustrates that the model behaves reasonably by paying attention to spots of air that a human reader would also inspect. Conversely,

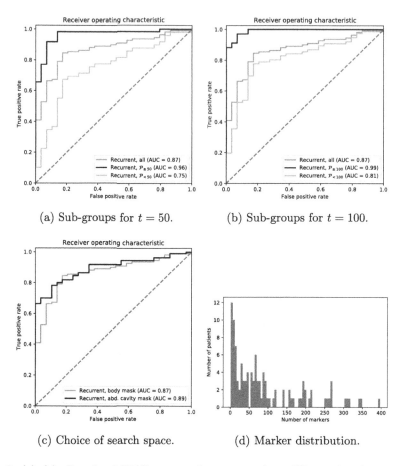

(a) Sub-groups for $t = 50$.   (b) Sub-groups for $t = 100$.

(c) Choice of search space.   (d) Marker distribution.

Fig. 3. (a)–(c): Case-level ROC curves of one exemplary CV run for the recurrent model. (d): Histogram of the distribution of ground truth markers.

Figs. 4c, d show a missed detection that is similarly reasonable as it has a subtle appearance and, unlike most, does not occur right below the body surface, but is trapped below the bowel.

3.2   Alternative Approaches

Besides the presented, marker-based mode of annotation, others which might lend themselves better to training a detection model were originally considered as well. At first glance, it may seem trivial to obtain a pixel-wise labeling of the regions as air in CT can usually be segmented by mere thresholding. This appears particularly promising as such ground truth masks would enable the use of image-to-image segmentation architectures [11], which make up the majority of the current state of the art. In our experience, this proved problematic for two reasons. First, soft reconstruction kernels and limited image resolution can

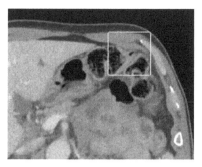

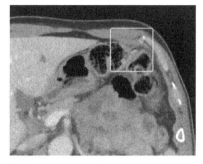

(a) False positive. The outlined patch contains no free intra-abdominal air.

(b) Image from (a) with a heatmap indicating salient spots for the model.

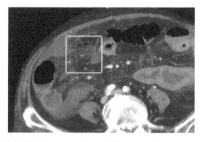

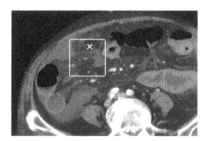

(c) False negative. The patch shows undetected free intra-abdominal air.

(d) Image from (c) with the marker indicating free intra-abdominal air.

Fig. 4. Examples of misclassifications by our model.

cause the bowel walls to become partially invisible, obscuring the transition from intra- to extraluminal space. Second, and more crucially, even with sophisticated segmentation tools it would be prohibitively time-consuming to segment all individual air bubbles in a patient, which can number in the dozens and may be interspersed with physiological air bubbles that prevent one from simply marking all air in a given region. In contrast, in our approach, only some representative bubbles need to be marked to obtain correct labels for each patch. An annotation of bounding boxes, which could be used to train region proposal networks or similar modern detection models [16], did not seem feasible either. This is in large part due to the many tiny, distributed air bubbles for which it would be difficult to cover them with boxes in a consistent manner, whereas huge air volumes can sometimes occupy about half of the abdomen.

4   Conclusion

In summary, we present the first algorithm described in literature for detecting pneumoperitoneum in CT scans. It follows a sliding-window approach with a deep recurrent neural network classifier at its core. When relying on an abdominopelvic cavity segmentation, it achieves a ROC AUC of 89% in a five-fold CV. Currently, its performance is primarily limited by subtle, under-represented cases preventing very high degrees of sensitivity. However, high specificity can be achieved while maintaining reasonable detection rates. Such an operating point is suitable for adjusting the reading order of emergency scans, where an unwarranted prioritization pre-empting more urgent cases is typically considered more harmful than a missed opportunity for earlier detection. This is particularly relevant as less than 1% of emergency patients present with pneumoperitoneum [9]. As the next steps, we plan to perform a more in-depth clinical validation as well as to collect and evaluate data from additional sites.

References

1. Bulas, D.I., Taylor, G.A., Eichelberger, M.R.: The value of CT in detecting bowel perforation in children after blunt abdominal trauma. Am. J. Roentgenol. **153**(3), 561–564 (1989)

2. Cho, K., et al.: Learning phrase representations using RNN encoder-decoder for statistical machine translation. arXiv preprint arXiv:1406.1078 (2014)

3. Cho, S.-J., et al.: Clinical significance of intraperitoneal air on computed tomography scan after endoscopic submucosal dissection in patients with gastric neoplasms. Surg. Endosc. **28**(1), 307–313 (2013). https://doi.org/10.1007/s00464-013-3188-9

4. Fawcett, T.: An introduction to ROC analysis. Pattern Recogn. Lett. **27**(8), 861–874 (2006)

5. Hainaux, B., et al.: Accuracy of MDCT in predicting site of gastrointestinal tract perforation. Am. J. Roentgenol. **187**(5), 1179–1183 (2006)

6. Kingma, D.P., Ba, J.: Adam: a method for stochastic optimization. arXiv preprint arXiv:1412.6980 (2014)

7. Luo, J.W., Lie, J.L., Chong, J.: Pneumoperitoneum on chest X-ray: a DCNN approach to automated detection and localization utilizing salience and class activation maps. In: SIIM Conference on Machine Intelligence in Medical Imaging (2018)

8. Matthews, B.W.: Comparison of the predicted and observed secondary structure of T4 phage lysozyme. Biochimica et Biophysica Acta (BBA)-Protein Struct. **405**(2), 442–451 (1975)

9. Nazerian, P., et al.: Accuracy of abdominal ultrasound for the diagnosis of pneumoperitoneum in patients with acute abdominal pain: a pilot study. Critical Ultrasound J. **7**(1), 1–7 (2015). https://doi.org/10.1186/s13089-015-0032-6

10. Paster, S.B., Brogdon, B.G.: Roentgenographic diagnosis of pneumoperitoneum. JAMA **235**(12), 1264–1267 (1976). https://doi.org/10.1001/jama.1976.03260380058035

11. Ronneberger, O., Fischer, P., Brox, T.: U-Net: convolutional networks for biomedical image segmentation. In: Navab, N., Hornegger, J., Wells, W.M., Frangi, A.F. (eds.) MICCAI 2015. LNCS, vol. 9351, pp. 234–241. Springer, Cham (2015). https://doi.org/10.1007/978-3-319-24574-4_28

12. Shi, B., Bai, X., Yao, C.: An end-to-end trainable neural network for image-based sequence recognition and its application to scene text recognition. IEEE Trans. Pattern Anal. Mach. Intell. **39**, 2298–2304 (2016)

13. Summers, R.M.: Progress in fully automated abdominal CT interpretation. Am. J. Roentgenol. **207**(1), 67–79 (2016)

14. Tompson, J.J., Jain, A., LeCun, Y., Bregler, C.: Joint training of a convolutional network and a graphical model for human pose estimation. In: Advances in Neural Information Processing Systems, pp. 1799–1807 (2014)
15. Yang, D., et al.: Automatic liver segmentation using an adversarial image-to-image network. In: Descoteaux, M., Maier-Hein, L., Franz, A., Jannin, P., Collins, D.L., Duchesne, S. (eds.) MICCAI 2017. LNCS, vol. 10435, pp. 507–515. Springer, Cham (2017). https://doi.org/10.1007/978-3-319-66179-7_58
16. Zhao, Z.Q., Zheng, P., Xu, S.T., Wu, X.: Object detection with deep learning: a review. IEEE Trans. Neural Netw. Learn. Syst. 1–21 (2019). https://doi.org/10.1109/TNNLS.2018.2876865

Prediction of Pathological Complete Response to Neoadjuvant Chemotherapy in Breast Cancer Using Deep Learning with Integrative Imaging, Molecular and Demographic Data

Hongyi Duanmu[1,2], Pauline Boning Huang[1], Srinidhi Brahmavar[2],
Stephanie Lin[1], Thomas Ren[1], Jun Kong[4], Fusheng Wang[2,3(✉)],
and Tim Q. Duong[1(✉)]

[1] Department of Radiology, Stony Brook School of Medicine, New York, USA
hongyi.duanmu@stonybrook.edu, Tim.Duong@stonybrookmedicine.edu
[2] Department of Computer Science, Stony Brook University, New York, USA
[3] Department of Biomedical Informatics, Stony Brook University, New York, USA
fusheng.wang@stonybrook.edu
[4] Department of Mathematics and Statistics, Georgia State University,
Atlanta, USA

Abstract. Neoadjuvant chemotherapy is widely used to reduce tumor size to make surgical excision manageable and to minimize distant metastasis. Assessing and accurately predicting pathological complete response is important in treatment planing for breast cancer patients. In this study, we propose a novel approach integrating 3D MRI imaging data, molecular data and demographic data using convolutional neural network to predict the likelihood of pathological complete response to neoadjuvant chemotherapy in breast cancer. We take post-contrast T1-weighted 3D MRI images without the need of tumor segmentation, and incorporate molecular subtypes and demographic data. In our predictive model, MRI data and non-imaging data are convolved to inform each other through interactions, instead of a concatenation of multiple data type channels. This is achieved by channel-wise multiplication of the intermediate results of imaging and non-imaging data. We use a subset of curated data from the I-SPY-1 TRIAL of 112 patients with stage 2 or 3 breast cancer with breast tumors underwent standard neoadjuvant chemotherapy. Our method yielded an accuracy of 0.83, AUC of 0.80, sensitivity of 0.68 and specificity of 0.88. Our model significantly outperforms models using imaging data only or traditional concatenation models. Our approach has the potential to aid physicians to identify patients who are likely to respond to neoadjuvant chemotherapy at diagnosis or early treatment, thus facilitate treatment planning, treatment execution, or mid-treatment adjustment.

Keywords: Artificial intelligence · Convolutional neural network · Magnetic resonance imaging

© Springer Nature Switzerland AG 2020
A. L. Martel et al. (Eds.): MICCAI 2020, LNCS 12262, pp. 242–252, 2020.
https://doi.org/10.1007/978-3-030-59713-9_24

1   Introduction

Neoadjuvant chemotherapy (NAC) [1] is widely used to reduce tumor size to make surgical excision manageable and to minimize distant metastasis. Pathological complete response (PCR) [2], defined as the absence of any residual diseases by pathology obtained from tumor tissue at the end of NAC, is used to assess treatment response. Patients with PCR are more likely to be candidates for breast-conserving surgery, sparing a full mastectomy, and are associated with longer recurrence free survival and overall survival [2]. The ability that determines individual patient's likelihood to respond to NAC using non-invasive imaging is likely to be clinically useful, as it avoids or minimizes unnecessary toxic chemotherapy or enables modification of regimens mid-treatment to achieve better efficiency. Unfortunately, early imaging and non-imaging clinical metrics generally do not have sufficient accuracy to predict eventual PCR.

Breast MRI is a standard of care for diagnosis, cancer staging, and treat monitoring because it can reliably identify in-breast disease (positive predictive value of 93%) [3]. Breast MRI is, however, less successful at predicting PCR [4]. Many studies have reported the use of radiological staging, tumor volume, and radiomic features from pre-treatment MRI to predict PCR [5,6]. Although promising, identifying reliable imaging metrics to predict eventual PCR remains an active area of research.

More recently, deep learning has become increasingly popular for image segmentation and classification [7], in particular, the convolutional neural network (CNN) [8]. CNN takes an input image, learns important features and saves these parameters as weights and bias to differentiate types of images [9,12]. Many studies relied on extracted radiomic features (such as volume, sphericity, DCE signal washout, etc.), not images per se, as inputs to predict PCR [10,11,14]. Several studies took MRI images themselves as inputs to predict PCR [13,16,17,21,22], however, none of them combined imaging with non-imaging data such as molecular subtypes and demographic information (race/age).

Molecular subtypes of breast cancer have been defined based on gene expression patterns, including estrogen receptor (ER), progesterone receptor (PR), and human epidermal growth factor 2–neu (HER2), which are tested routinely for clinical decision support. The expression of Ki67, a proliferation marker to measure the growth fraction of tumor cells,is commonly used in routine pathology [20,23]. The status of these subtypes can present demographic and clinical differences. In this paper, we take an integrative approach by consolidating all these molecular data and demographic data (race and age) with imaging data to provide more knowledge for PCR prediction.

The goal of this study is to build a CNN based model to predict the likelihood of PCR to neoadjuvant chemotherapy in breast cancer. The inputs include post-contrast T1-weighted MRI 3D volume as whole images without manual tumor segmentation, and clinical data (the molecular subtypes and demographical data). This approach uses a model in which MRI data and non-imaging data "convolved" to inform each other, instead of a typical concatenation of multiple data sources. This is achieved by channel-wise multiplication of the intermediate

Table 1. Non-imaging clinical data used in the prediction models.

Parameter	Description	Data type
Age	Patient age (Years)	Demographic
Race	1 = Caucasian 2 = African American 3 = Asian 4 = Native Hawaiian/Pacific Islander 5 = American Indian/Alaskan Native 6 = Multiple race	Demographic
Estrogen Receptor (ER)	0 = Negative 1 = Positive 2 = Indeterminate	Molecular
Progesterone Receptor (PR)	0=Negative 1 = Positive 2 = Indeterminate	Molecular
HER2 status	0 = Negative 1 = Positive −1 = indeterminate or not done	Molecular
3-level HR/HER2 category	1 = HR Positive, HER2 Negative 2 = HER2 Positive 3 = Triple Negative	Molecular
Ki-67	1 = <10% 2 = 10–20% 3 = >20%	Molecular

results of imaging and non-imaging clinical data branches. We take full advantage of the 3D MRI data, instead of typical 2D multi-slice MRI approaches. For comparison, prediction performance metrics are evaluated for CNN models using MRI data alone and CNN model using MRI data and non-image data by concatenation. We also generate heatmaps to understand the regions on the breast MRI that these models considered most important in the prediction.

2 Methods

2.1 Patient Cohort

Level 3 curated data from the I-SPY-1 TRIAL (2002–2006) are used [4,15]. All patients are diagnosed with stage 2 or 3 breast cancer with breast tumors at least 3 cm in size and underwent anthracycline-cyclophosphamide (AC) with or without taxane treatment. A total of 112 patients from the ISPY1 Clinical Trial dataset are used in this study. MRI data are the T1 post-contrast breast MR images obtained at pre-neoadjuvant chemotherapy. Non-imaging clinical data used in prediction include demographic data (age, race), Estrogen Receptor Status (ER), Progesterone Receptor Status (PR), human epidermal growth factor receptor 2 (HER2) Status, 3-level hormonal receptor (HR)/HER2 category, and Ki67 (Table 1). The outcome for prediction is tumor PCR (0 or 1) to NAC.

2.2   Methodology

We take a 3D centric approach and use the 3D MRI image volume as input instead of the commonly used 2D multi-slice based approaches. All MRI volumes are resized to $256 * 256 * 60$. We build three CNN models (Fig. 1) based on i) MRI volumes only (Imaging-Only-Model), ii) MRI data and non-imaging clinical data trained with parallel branches and concatenated toward the end (Parallel-Model), and iii) MRI data and non-imaging clinical data trained with one informing the other at the intermediate stages of the CNN (Interactive-Model).

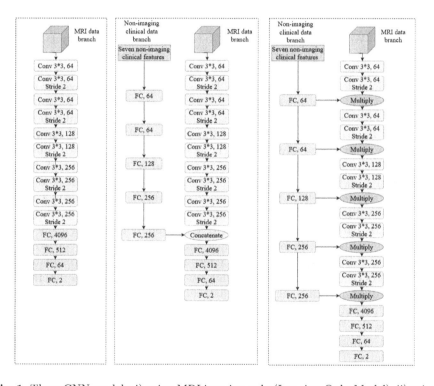

Fig. 1. Three CNN models: i) using MRI imaging only (Imaging-Only-Model), ii) using MRI data with concatenated non-imaging clinical data (Parallel-Model), and iii) using MRI data with non-imaging clinical data with interaction at the intermediate stages of the CNN (Interactive-Model).

The *Imaging-Only-Model* is a standard CNN model which processes images only. Ten convolutional layers after the input layer are used for potential image

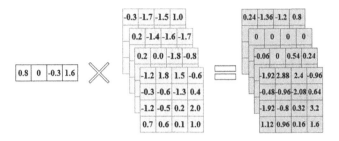

Fig. 2. One example of channel-wise multiplication. (Color figure online)

patterns extraction, which is inherited from the classical VGG model [9]. Since a 3D model is computationally demanding, we select VGG-13 as the skeleton model instead of other complex architectures such as ResNet [18] and Inception [25]. Considering the balance between efficiency and computing burden, we reduce the number of channels from originally 64-128-256-512-512 to 64-64-128-256-256. We replace max-pooling layers with convolutional layers of stride equal to 2, which has better capability in image pattern extraction [24]. All convolutional layers utilize $3 * 3$ kernels. One batch norm layer and one ReLU activation layer follow all convolutional layers, which has been proven to improve the robustness of the system [19].

The *Parallel-Model* includes MRI data and non-imaging clinical data trained as parallel branches and concatenated toward the end. The MRI data branch is identical to Imaging-Only-Model. The non-imaging data branch consists of, in total, 5 fully connected layers, while the trend of the number of channels remains the same with the MRI data branch. The final output of non-imaging clinical data branch, a 256 vector, is concatenated with the output of MRI data branch and fed into the last 4 fully connected layers for the final prediction.

The *Interactive-Model* includes MRI data and non-imaging clinical data interactively trained with one informing the other. Instead of concatenation, this new approach uses channel-wise multiplication at the intermediate levels of the network. This is achieved by channel-wise multiplying the intermediate results from the non-imaging clinical data branch (light yellow array) with the corresponding intermediate results from the MRI data branch (orange array) to generate the "interactive feature maps" (purple array) illustrated in Fig. 2. In essence, non-imaging clinical data "inform" the image data at intermediate CNN networks. This approach will incur minimal computational cost.

Channel-wise multiplication essentially involves kernel selection and weighted averaging. For example, if one feature from clinical side is zero, when it multiplies with corresponding feature maps convolved by one specific kernel, the output of that kernel is zero, which means based on the knowledge discovered from the clinical information, the system decides to ignore such pattern. Similarly, positive features from the non-imaging clinical data side emphasize that feature map (image patterns found by that specific kernel) and negative features restrain corresponding kernel's contribution. This approach effectively integrates MRI image and non-imaging clinical information using CNN networks.

112 patients in total, are randomly splitted into training and testing set with 80% (90 patients) and 20% (22 patients) respectively. All three models are trained for 100 epochs with stochastic gradient descent (SGD) and fully fine-tuned training parameters to obtain the best performance separately. In terms of testing speed, our system can make predictions for 20 patients per second in one NVIDIA Tesla V100 GPU. Time efficiency shows high potential in clinical use. Five-fold cross validation is used to account for outlying accuracies. The performance of the prediction model is evaluated with area under the curve (AUC), accuracy, sensitivity, specificity, and F1 score.

Table 2. Prediction performance comparison.

	Imaging-Only-Model	Parallel-Model	Interactive-Model
Accuracy	0.7407	0.7846	0.8300
AUC	0.5758	0.5871	0.8035
Sensitivity	0.2229	0.4000	0.6822
Specificity	0.9222	0.8929	0.8822
F1 score	0.3590	0.5525	0.7694

To further visualize the spatial location on the images that the CNN networks are paying attention to, heatmaps are generated with the class activation maps algorithm [26]. By adding global average pooling into CNN and calculating gradient back propagation, the class activation maps to one specific class can be obtained. The class activation maps indicates the discriminative image regions CNN paid attention to in the prediction. While this algorithm was originally proposed for more robust CNN training, it is widely used in visualization of deep learning systems.

3   Results

The comparison of the performance of the three models is shown in Table 2. The Interactive-Model significantly outperforms other models on AUC, F1 score, and accuracy, and is on a par with other models on specificity. The addition of non-imaging clinical data to image data in the CNN by concatenation improves PCR prediction performance indices. The Interactive-Model with novel method of incorporating clinical data and image data further improves these prediction performance indices. Of note, Interactive-Model yields higher sensitivity with minimal cost of specificity, compared favorably to Image-Only-Model and Parallel-Model in predicting PCR.

To evaluate the spatial location on the images that the model considered essential for the prediction, we also generate heatmaps showing the location that the systems paid attention to in the prediction (Fig. 3). With the Image-Only-Model, there are many regions outside the tumor that appear to be mistakenly considered to be important. With the Parallel-Model, there are fewer regions appearing to be mistakenly considered. With our Interactive-Model, attention regions inside and around the breast tumor regions are most relevant in predicting PCR.

4 Discussion

In our predictive model for PCR, we take post-contrast T1-weighted MRI as 3D volumes without the need of manual tumor contouring, and incorporate non-imaging clinical data (i.e., molecular subtypes and demographic information). Our model uses channel-wise multiplication of MRI data and non-imaging clinical data with interactions in the CNN architecture. The Interactive-Model significantly outperforms Imaging-Only-Model using CNN, and the Parallel-Model which concatenates MRI data and non-image clinical data in CNN.

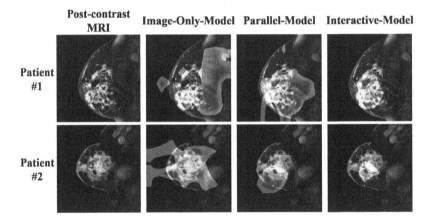

Fig. 3. Heatmaps of two patient cases showing the location that the three models paid attention to in the prediction.

There have only been a few studies that have applied deep learning on MRI images only to predict PCR [13,16,17,21,22] and there is only one study used non-imaging clinical data in the CNN model to predict PCR.

Braman et al. [13] applied deep learning to predict NAC in HER2 patients from pre-treatment 2D DCE MRI in a retrospective study. They explored both pre-contrast and late post-contrast phases of DCE MRI and found AUC of 0.77 to 0.93 respectively. They also reported that deep learning-based response prediction model exceeded both a multivariable model incorporating predictive clinical variables (AUC < 0.65) and a model utilizing semi-quantitative DCE-MRI pharmacokinetic measurements (AUC < 0.60). Their system is in 2D and needs radiologists' manual contouring, making it less practical.

Ha et al. [17] used convolutional neural networks (CNN) to predict neoadjuvant chemotherapy (NAC) response based on a breast MRI tumor dataset prior to initiation of chemotherapy. Breast tumor was identified on first T1 postcontrast dynamic images and segmented. A three-class neoadjuvant prediction model was evaluated (complete pathological response, partial response and no pathological response). The CNN achieved an overall accuracy of 88% in three-class prediction of neoadjuvant treatment response. They concluded that CNN architectures could predict NAC treatment response from pre-treatment breast MRI. Their study differed from our work in that they did not include non-imaging clinical data in their model and the input data used breast tumor area that was segmented instead of information from the whole images.

Ravichandran et al. [22] reported the use of CNN to predict PCR from pre-treatment dynamic contrast-enhanced MRI on a 2D basis. They found that a CNN utilizing a combination of both pre- and post-contrast images best distinguished responders, yielding, respectively, an AUC of 0.77 and an accuracy of 0.82. The study differed from our work in that they required breast tumors to be manually segmented and 2D multi-slices were used. In addition, a multinomial logistic regression model to incorporate HER2 status boosted performance for AUC to 0.85. Ravichandran et al. evaluated both pre- and post-contrast MRI, whereas we only used post contrast MRI.

El Adoui et al. [16] reported that their CNN model was able to classify patients who were responsive or non-responsive to chemotherapy. They used a two-branch CNN architecture, taking two breast tumor MRI slices acquired before and after the first round of chemotherapy as inputs, and reported an accuracy of 92.72% and an AUC of 0.96. The study differed from ours in that they did not include non-imaging clinical data in their model and the input data relied on segmented breast tumors.

Qu et al. [21] built a CNN model using pre- and post-NAC. Tumor regions were manually delineated on each slice by two expert radiologists on enhanced T1-weighted images. They found an AUC of 0.553 for pre-NAC, 0.968 for post-NAC and 0.970 for the combined data. It is not surprising that post-NAC has high AUC because in patients with PCR there are no residual breast tumor left, i.e, the volume from the manual segmentation is likely zero or very small which itself is predictive of PCR without sophisticated CNN analysis. It is more clinically relevant to predict PCR at pre-treatment or early treatment. It is also more clinically relevant to use information from whole images as input instead of segmented tumor, especially for post NAC data. Qu et al. also included molecular subtypes as a separate channel, but found that inclusion of molecular subtypes did not significantly increase in the AUC value for PCR prediction, in contrast to our findings. This is surprising in that many studies have shown that molecular receptor subtypes (Estrogen Receptor Status, Progesterone Receptor Status) and the cancer antigen Ki67, amongst others, are important for clinical diagnosis and treatment planning, in particular, for informing PCR [20,23].

One future work will be using a larger dataset to improve generalizability of the model. We also plan to incorporate multiple time points during NAC instead of only pre-treatment time point. In addition, we will explore multiple longitudinal MRI data and other data types during early treatment. We are also considering using Multiparametric MRI (i.e., T2-weighted MRI and diffusion-weighted MRI) for our future studies.

5   Conclusion

Incorporation of convolutional neural network of standard post-contrast breast MRI and non-imaging clinical data from pre-treatment or early treatment can provide high accurate prediction of complete pathological response associated with breast cancer. Our approach operates on the entire three-dimensional MRI images with most complete information without the need to manually segment tumor volume. With further development and testing on larger multi-institutional sample size, our approach has the potential to non-invasively identify patients who are likely to respond to neoadjuvant chemotherapy at diagnosis or early treatment, thus provides an opportunity for computer aided support for treatment planning, treatment execution, or mid-treatment adjustment.

Acknowledgements. This work was supported in part by a pilot grant from the Stony Brook Cancer Center, a Carol Baldwin pilot grant through the Stony Brook University School of Medicine, and the Biomedical Imaging Research Center (Radiology), and grants from NIH National Cancer Institute 1U01CA242936, and NSF ACI 1443054 and IIS 1350885.

References

1. Curigliano, G., et al.: De-escalating and escalating treatments for early-stage breast cancer: the St. Gallen international expert consensus conference on the primary therapy of early breast cancer 2017. Ann. Oncol. **28**(8), 1700–1712 (2017)
2. Cortazar, P., et al.: Pathological complete response and long-term clinical benefit in breast cancer: the CTNeoBC pooled analysis. Lancet **384**(9938), 164–172 (2014)
3. Fowler, A.M., Mankoff, D.A., Joe, B.N.: Imaging neoadjuvant therapy response in breast cancer. Radiology **285**(2), 358–375 (2017)
4. Hylton, N.M., et al.: Locally advanced breast cancer: MR imaging for prediction of response to neoadjuvant chemotherapy-results from ACRIN 6657/I-SPY trial. Radiology **263**(3), 663–672 (2012)
5. Marinovich, M.L., et al.: Early prediction of pathologic response to neoadjuvant therapy in breast cancer: systematic review of the accuracy of MRI. Breast **21**(5), 669–677 (2012)
6. Lindenberg, M.A., et al.: Imaging performance in guiding response to neoadjuvant therapy according to breast cancer subtypes: a systematic literature review. Crit, Rev. Oncol./Hematol. **112**, 198–207 (2017)
7. LeCun, Y., Bengio, Y., Hinton, G.: Deep learning. Nature **521**(7553), 436–444 (2015)

8. LeCun, Y., Haffner, P., Bottou, L., Bengio, Y.: Object recognition with gradient-based learning. Shape, Contour and Grouping in Computer Vision. LNCS, vol. 1681, pp. 319–345. Springer, Heidelberg (1999). https://doi.org/10.1007/3-540-46805-6_19

9. Simonyan, K., Zisserman, A.: Very deep convolutional networks for large-scale image recognition. arXiv preprint arXiv:1409.1556 (2014)

10. Mani, S., et al.: Machine learning for predicting the response of breast cancer to neoadjuvant chemotherapy. J. Am. Med. Inform. Assoc. **20**(4), 688–695 (2013)

11. Tahmassebi, A., Gandomi, A.H., Fong, S., Meyer-Baese, A., Foo, S.Y.: Multi-stage optimization of a deep model: a case study on ground motion modeling. PloS One **13**(9), e0203829–e0203829 (2018)

12. Arel, I., Rose, D.C., Karnowski, T.P.: Deep machine learning-a new frontier in artificial intelligence research [research frontier]. IEEE Comput. Intell. Mag. **5**(4), 13–18 (2010)

13. Braman, N., et al.: Deep learning-based prediction of response to HER2-targeted neoadjuvant chemotherapy from pre-treatment dynamic breast MRI: a multi-institutional validation study. arXiv preprint arXiv:2001.08570 (2020)

14. Cain, E.H., Saha, A., Harowicz, M.R., Marks, J.R., Marcom, P.K., Mazurowski, M.A.: Multivariate machine learning models for prediction of pathologic response to neoadjuvant therapy in breast cancer using MRI features: a study using an independent validation set. Breast Cancer Res. Treat. **173**(2), 455–463 (2019)

15. Clark, K., et al.: The cancer imaging archive (TCIA): maintaining and operating a public information repository. J. Digit. Imaging **26**(6), 1045–1057 (2013)

16. El Adoui, M., Drisis, S., Benjelloun, M.: A PRM approach for early prediction of breast cancer response to chemotherapy based on registered MR images. Int. J. Comput. Assist. Radiol. Surg. **13**(8), 1233–1243 (2018)

17. Ha, R., et al.: Prior to initiation of chemotherapy, can we predict breast tumor response? Deep learning convolutional neural networks approach using a breast MRI tumor dataset. J. Digit. Imaging **32**(5), 693–701 (2019)

18. He, K., Zhang, X., Ren, S., Sun, J.: Deep residual learning for image recognition. In: Proceedings of the IEEE Conference on Computer Vision and Pattern Recognition, pp. 770–778 (2016)

19. Ioffe, S., Szegedy, C.: Batch normalization: accelerating deep network training by reducing internal covariate shift. arXiv preprint arXiv:1502.03167 (2015)

20. Kalinowski, L., Saunus, J.M., McCart Reed, A.E., Lakhani, S.R.: Breast cancer heterogeneity in primary and metastatic disease. In: Ahmad, A. (ed.) Breast Cancer Metastasis and Drug Resistance. AEMB, vol. 1152, pp. 75–104. Springer, Cham (2019). https://doi.org/10.1007/978-3-030-20301-6_6

21. Qu, Y.-H., Zhu, H.-T., Cao, K., Li, X.-T., Ye, M., Sun, Y.-S.: Prediction of pathological complete response to neoadjuvant chemotherapy in breast cancer using a deep learning (DL) method. Thoracic Cancer **11**, 651–658 (2020)

22. Ravichandran, K., Braman, N., Janowczyk, A., Madabhushi, A.: A deep learning classifier for prediction of pathological complete response to neoadjuvant chemotherapy from baseline breast DCE-MRI. In: Medical Imaging 2018: Computer-Aided Diagnosis, vol. 10575, p. 105750C. International Society for Optics and Photonics (2018)

23. Schettini, F., et al.: HER2-enriched subtype and pathological complete response in HER2-positive breast cancer: a systematic review and meta-analysis. Cancer Treat. Rev. **84**, 101965 (2020)

24. Springenberg, J.T., Dosovitskiy, A., Brox, T., Riedmiller, M.: Striving for simplicity: the all convolutional net. arXiv preprint arXiv:1412.6806 (2014)
25. Szegedy, C., Vanhoucke, V., Ioffe, S., Shlens, J., Wojna, Z.: Rethinking the inception architecture for computer vision. In: Proceedings of the IEEE Conference on Computer Vision and Pattern Recognition, pp. 2818–2826 (2016)
26. Zhou, B., Khosla, A., Lapedriza, A., Oliva, A., Torralba, A.: Learning deep features for discriminative localization. In: Proceedings of the IEEE Conference on Computer Vision and Pattern Recognition, pp. 2921–2929 (2016)

Geodesically Smoothed Tensor Features for Pulmonary Hypertension Prognosis Using the Heart and Surrounding Tissues

Johanna Uthoff[1]([✉]), Samer Alabed[2], Andrew J. Swift[2,3], and Haiping Lu[1,3]

[1] Department of Computer Science, The University of Sheffield, Sheffield, UK
j.uthoff@sheffield.ac.uk
[2] Sheffield Teaching Hospitals NHS Foundation Trust, Sheffield, UK
[3] INSIGNEO, Institute for In Silico Medicine, The University of Sheffield, Sheffield, UK

Abstract. Cardiac magnetic resonance imaging (CMRI) provides non-invasive characterization of the heart and surrounding tissues. It is an important tool for the prognosis of pulmonary arterial hypertension (PAH), a disease with heterogeneous presentation that makes survival likelihood prediction a challenging task. In this paper, we propose a **G**eodesically **S**mooothed **T**ensor feature learning method (GST) that utilizes not only the heart but also its surrounding tissues to characterize disease severity for improving prognosis. Specifically, GST includes structures surrounding the heart by geodesic rings which were incrementally smoothed with Gaussian filters. This provides additive insight while modulating for patient positional differences for a subsequent tensor-based feature learning pipeline. We performed evaluation on Four Chamber and Short Axis CMRI from 150 individuals with confirmed PAH and 1-year mortality census (27 deceased, 123 alive). The proposed GST method improved AUC and Cox difference at 4-years post-imaging (Cox4YD) over the standardized measurement of right ventricular end systolic volume index (RVESVi: AUC: 0.58; Cox4YD: 0.18) on the Four Chamber protocol (AUC: 0.77; Cox4YD: 0.35). Only AUC was improved over RVESVi in the Short Axis scans (AUC: 0.77; Cox4YD: 0.16).

1 Introduction

Pulmonary arterial hypertension (PAH) is a severe disease affecting the cardiopulmonary system with a recorded 1-year mortality rate of 15% [3]. The heterogeneous nature of disease presentation and the diversity in disease progression at time-of-diagnosis makes predicting survival likelihood and proper treatment planning a challenging task for PAH [9]. Guidelines from the European Society of Cardiology and the European Respiratory Society assigns categorical risk of 1-year mortality to subjects based on assessment of symptom severity, 6-minute walk test, and right ventricle function [6]. There is evidence that prognostic indicators including clinical, echo-cardiogram imaging, and right heart catheterization (RHC) can be used to predict a subject's likelihood of mortality [2].

© Springer Nature Switzerland AG 2020
A. L. Martel et al. (Eds.): MICCAI 2020, LNCS 12262, pp. 253–262, 2020.
https://doi.org/10.1007/978-3-030-59713-9_25

Recently, standardized cardiac magnetic resonance imaging (CMRI) measures calculated from user-delineated contours were shown to have prognostic value in PAH subjects when combined with clinical measures (AUC $= 0.70 - 0.78$) [8].

Prior studies of artificial intelligence technologies for PAH prognosis have focused on segmentation-based ventricular volume and motion analysis and achieved a moderate performance accuracy (AUC $= 0.75$) [1]. From related literature, while feature extraction methods typically focus on specific regions of interest within a dataset, there are potentially features in non-target organ regions which could provide additional disease risk classification [15]. Inspired by this, we note that features from the surrounding tissues of the heart has the potential to benefit PAH prognosis. For example, PAH subjects with worse outcomes tend to have increased pulmonary vasculopathy [7] which could potentially be detected as features in the lungs. However, the automatic extraction of meaningful features from areas external to the heart is challenged by the diversity in patient positioning during scanning. As the priority of the imaging technician is the location of heart structures, the positioning of secondary structures (lungs, liver, appendages) may be inconsistent or difficult to align to a standard.

Moreover, temporal-spatial medical scans, such as CMRIs have high dimensionality characterizing the in-vivo tissue conditions of target organs and surrounding structures. However, the number of available samples is much smaller relevant to such high dimensionality, making machine learning challenging on such data. Recently, a tensor-based dimensionality reduction method named as multilinear principal component analysis (MPCA) [10] has shown promising results in automated diagnosis of PAH by detecting interpretable tensorial CMRI features [13]. It will be interesting to explore its application in prognosis.

Contributions: This paper proposes a **G**eodesically **S**moooothed **T**ensor feature learning method (GST) for PAH prognosis. More specifically, we use a simple, regional Gaussian smoothing to include tissues surrounding the heart incrementally. This has the potential to alleviate the challenges associated with positional discrepancies and promote a more comprehensive assessment of mortality risk. The geodescially smoothed CMRI is then passed to a tensor feature learning pipeline to predict the mortality. To the best of our knowledge, this is the first work to employ incremental geodesic-based image smoothing and investigate the prognostic value of tensor-based features from such smoothed images.

2   Methodology

Figure 1 shows an overview of the proposed GST pipeline for PAH prognosis. Source code is made available at https://github.com/pykale.

CMRI Data Preprocessing. Image preprocessing is consisted of (1) CMRI unit standardization, (2) magnetic field in-homogeneity bias correction, (3) inter-subject scan alignment, (4) incremental geodesic distance smoothing, and

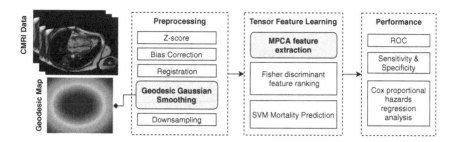

Fig. 1. Overview of the proposed pipeline for PAH prognosis including geodesic smoothing preprocessing, tensor feature learning, and performance analysis. Gray highlighted regions are described in more detail in the Methods.

(5) in-plan resolution down-sampling. CMRI units were standardized using z-score normalization to allow for meaningful comparison between subjects [11]. CMRI field in-homogeneity bias correction was performed using the N4ITK method to correct for acquisition artifacts [14]. Affine registration of three landmarks points placed by an expert in each of the datasets was performed to align subject hearts to the same image space; specific landmarks are detailed in the CMRI Scan Data section of the Experiments. Incremental geodesic distance smoothing is described in the next section. The in-plane scaling was done by max-pool at 2, 4, 8, 16 times resulting in down-sampled resolution images of 256×256, 128×128, 64×64, and 32×32 respectively.

Geodesic Smoothing. Our GST method performs incremental smoothing based on geodesic distance from the region of the heart using iterative mask dilation and Gaussian filters. Let H_i be the previous iteration i's mask with diameter d_{H_i}, ρ be the width of the concentric shell, s be the structuring element $4\pi\rho^2$, and X_m be indicating a phase in subject's cardiac cycle.

First we calculate the geodesic distance using dilation procedure from the edge of the initial mask H_i to the concentric edge at an expanding distance ρ towards the image bounds, hereby H_{i+1} follows the formulation:

$$H_{i+1} = H_i \oplus s. \tag{1}$$

The initial mask, H_i is subtracted from the dilated mask and element-wise multiplication is performed with image X_m. This results in a shell of image about the H_i, V_i, with inner diameter of d_{H_i} and outer diameter of $d_{H_i} + 2 \times \rho$, calculated as:

$$V_i = (H_{i+1} - H_i) \odot X_m. \tag{2}$$

The 2D Gaussian of V_i is calculated as W_i. Let σ be the smoothing factor. The formulation of W_i is as follows:

$$W_i = G[V_i(r,c)|\sigma] = \frac{1}{2\pi\sigma^2} e^{\frac{-r^2+c^2}{2\sigma^2}}, \tag{3}$$

which is updated for each i with $\Delta\sigma$ as the defined incremental increase in smoothing factor as:

$$\sigma = \sigma + \Delta\sigma. \tag{4}$$

Equations (1) to (4) are repeated until H_{i+1} has reached the image boundaries $(i = I)$. Finally, the incrementally smoothed image X'_m is calculated as the sum of all W_i as follows:

$$X'_m = \sum_{i=0}^{I} W_i. \tag{5}$$

The resulting X'_m is an image of the same size as X_i. The original masked heart region is unchanged from the original image and voxels along tangential arrays from the mask become increasingly more blurred. The lower left of Fig. 1 shows a typical mask and Fig. 2B shows an example on a Four Chamber scan.

Tensor Feature Learning. The learning of prognostic tensor features involves (1) MPCA feature extraction, (2) Fisher discriminant feature ranking, (3) support vector machine (SVM) training, and (4) cross validation. After CMRI data preprocessing, MPCA is applied to the aligned dataset for spatial-temporal feature extraction - see next section for details on the parameters for *Baseline*, *Masked*, *Surrounding Smoothing*, and *Geodesic Smoothing*. After MPCA, the extracted features are ranked using Fisher's discriminant score. Those top-ranked features were utilized to train a linear SVM through 10-fold *stratified* cross validation with class imbalance preserving fold generation. This was repeated 10 times for a 10×10-fold cross validation to help alleviate bias from random fold generation and increase the estimation reliability for small sample size.

Spatial-Temporal Feature Extraction. We identify prognostic characteristics by using MPCA [10] to learn multilinear bases from image stacks to obtain low-dimensional tensor features. Here, we represent our M CMRI samples as third-order tensors in the form $\{X_1, ..., X_M \in \mathbb{R}^{I_1 \times I_2 \times I_3}\}$. MPCA utilizes these inputs to extract low-dimensional tensor features $\{Y_1, ..., Y_M \in \mathbb{R}^{P_1 \times P_2 \times P_3}\}$ by learning three (order $N = 3$) projection matrices $\{\mathbf{U}^{(n)} \in \mathbb{R}^{I_n \times P_n}, n = 1, 2, 3\}$ as follows:

$$Y_m = X_m \times_1 \mathbf{U}^{(1)^\top} \times_2 \mathbf{U}^{(2)^\top} \times_3 \mathbf{U}^{(3)^\top}, m = 1, ..., M, \tag{6}$$

where $P_n < I_n$, thereby reducing the dimension of the input tensor to $P_1 \times P_2 \times P_3$ from $I_1 \times I_2 \times I_3$. The projection matrices $\{\mathbf{U}^{(n)}\}$ are optimized through maximizing the total scatter $\Psi_Y = \sum_{m=1}^{M} ||Y_m - \overline{Y}||_F^2$, where $\overline{Y} = \frac{1}{M} \sum_{m=1}^{M} Y_m$ is the mean tensor and $||\cdot||_F$ is the Frobenius norm [10]. MPCA has one hyperparameter Q determining the tensor subspace dimensions $\{P_1, P_2, P_3\}$ and its default setting takes only one iteration.

Table 1. Demographic and clinical information of study population. *N: number of subjects; IPAH: idiopathic PAH; WHO: World Health Organization.*

Attribute	Deceased	Alive	p
N	27	123	–
IPAH	14 (52%)	55 (44%)	0.62
Female	17 (63%)	88 (72%)	0.56
Age (Mean±STD)	70.3 ± 10.4	62.2 ± 12.7	**< 0.01**
WHO-2	0 (0%)	7 (6%)	
WHO-3	23 (85%)	103 (84%)	0.49
WHO-4	4 (15%)	13 (10%)	

Performance Assessment. Model performance was evaluated using 10×10-fold cross validation. The primary metric is the area under the receiver-operator characteristic curve (AUC). We computed both sensitivity and specificity. We also performed Cox proportional-hazards regression [4,12], a type of non-parametric survival analysis which relates variables to survival time; the measured effect is the *Hazard rate* which is the expected number of events/deaths per unit time. Categorical variables were assessed with Fisher's Exact Test or Chi-squared test and continuous variables were assessed with Wilcoxon Rank Sum test, (R, https://www.r-project.org/).

3 Experiments

For the task of PAH prognosis assessment, the proposed pipeline takes as input CMRI data represented by volumetric slices (spatial) of intensity units over 20 phases of the cardiac cycle (temporal).

Study Population. Subjects diagnosed with PAH and imaged prior to treatment with CMRI between December 2014 and February 2017 were included in this study following institutional review board approval and ethics committee review. In total, 150 subjects were included in this study; diagnosis of PAH was made following RHC within 48 h of imaging. Table 1 lists key demographics and clinical information for the study subjects. Subjects were censused in November 2019 at which time 80 subjects (52%) were alive. One-year mortality was calculated from date of imaging, with a total 27 subjects (18%) deceased within one year of CMRI assessment. The median survival time of those who died within 1-year was 173 days. Subjects who were deceased after 1-year tended to be older (70 years average) compared to those who were alive at 1-year follow-up (62 years average).

CMRI Scan Data. Two CMRI protocols - Short Axis and Four Chamber - were utilized in this study. All scans were performed on a 1.5 5 T GE HDx (GE Healthcare, Milwaukee, USA) using an 8-channel cardiac coil and retrospective electrocardiogram gating. Acquisition parameters followed clinical standards with a cardiac gated multi-slice balanced SSFP sequence (20 frames per cardiac cycle, slice thickness 8 mm, FOV 48, matrix 512×512, BW 125 125 kHz/pixel, TR/TE 3.7/1.6 ms). Expert reader defined landmarks were selected as the inferior hinge point, superior hinge point, and interolateral inflection point of right ventricular free wall for the Short Axis scan. For the Four Chamber the left ventricular apex, mitral annuli, and tricuspid annuli were used.

CMRI Tensor Experimental Setup. Four experimental levels were explored on both the Four Chamber and Short Axis datasets: (1) *Baseline*, (2) *Masked*, (3) *Gaussian Surroundings*, and (4) *Gaussian Geodesic*.

– The *Baseline Tensor* experiment took as input the complete unmasked CMRI and tensor feature extraction was performed on all voxels.
– The *Masked Tensor* experiment took as input CMRI with a user-defined ellipse circumscribing the heart and tensor feature extraction was performed on voxels within the masked heart region.
– The *Surrounding Smoothing Tensor (SST)* experiment took as input CMRI Gaussian smoothing performed on all voxels exterior to the ellipsoid mask with two levels of $\sigma = [0.5, 1.0]$ and tensor feature extraction was performed on all voxels.
– The *Geodesic Smoothing Tensor (GST)* experiment took as input CMRI with Gaussian smoothing performed with incremental increases on all voxels exterior to the ellipsoid mask with two levels of $\sigma = [0.5, 1.0]$, one level of incremental $\sigma_i = 0.1$, and two levels of concentric size $bw = [5, 10]$ and tensor feature extraction was performed on all voxels.

For all the experiments, in-plane scaling was done by max-pool at 2, 4, 8, 16 times resulting in down-sampled resolution images of 256×256, 128×128, 128×128, and 32×32 respectively.

Comparison Studies. Contour-based measurements of the right ventricle end-diastolic volume index (RVEDVi) and right ventricle end-systolic volume index (RVESVi) were made by an expert reader and thresholded based on published categories [8].

Classification Accuracy. Table 2 shows the performance results from experiments. The RVEDVi and RVESVi measures both achieved relatively low performance (AUC $= 0.577 - 0.581$). Performance improvement was seen with application of all tensor-based experiments over *Baseline Tensor*. In Four Chamber *Masked Tensor*, binary masking improved prediction quality by a difference of 0.12 in AUC. Minimal difference ($\Delta AUC < 0.01$) was seen by applying a constant Gaussian smoothing over the surrounding structures (*SST*). However, use

Table 2. Experimental classification settings and results. The best results were in bold for the last three columns. Precision is shown as two decimal places therefore some tied values in the table differ in the third digit (e.g. 0.58 and 0.77). *AUC: area-under-curve for 1-year mortality; SENS: sensitivity for 1-year mortality; Cox4YD: Cox Proportional-Hazards Regression difference at 4-years after imaging; RVEDVi: right ventricular end-diastolic volume index; RVESVi: right ventricular end-systolic volume index; SST: Surrounding Smoothed Tensor; GST: Geodesic Smoothed Tensor.*

	Experiment	Mask	Gaussian	Geodesic	AUC	SENS	Cox4YD
Standard	RVEDVi	–	–	–	0.58	0.52	0.02
	RVESVi	–	–	–	**0.58**	**0.74**	**0.18**
Four Chamber	Baseline	–	–	–	0.60 ± 0.03	0.52	0.16
	Masked	YES	–	–	0.70 ± 0.03	0.59	0.18
	SST	YES	0.5	–	0.71 ± 0.02	0.67	0.19
	SST	YES	1.0	–	0.70 ± 0.03	0.67	0.22
	GST	YES	0.5	5	**0.77 ± 0.03**	**0.70**	**0.35**
	GST	YES	1.0	10	0.77 ± 0.03	0.70	0.34
Short Axis	Baseline	–	–	–	0.56 ± 0.04	0.44	0.02
	Masked	YES	–	–	0.75 ± 0.03	0.70	0.15
	SST	YES	0.5	–	0.71 ± 0.02	0.70	0.11
	SST	YES	1.0	–	0.63 ± 0.03	0.59	0.08
	GST	YES	0.5	5	**0.76 ± 0.03**	**0.78**	**0.16**
	GST	YES	1.0	10	0.73 ± 0.03	0.70	0.15

of *GST* to incrementally increase the level of σ farther away from the heart improved performance by up to 0.07 difference in AUC. The best performing ($AUC = 0.774$) prognostic tool was the Four Chamber *GST* which extracted tensor features from the heart and concentric rings of 5-voxel steps increasing the level of blur sigma from 0.5 by 0.1 each step. On the Short Axis *Masked Tensor*, an improvement of 0.20 points in ΔAUC was achieved. Only modest improvement ($\Delta AUC = 0.01$) was seen in the Short Axis *GST*.

Cox Proportional-Hazards Regression Analysis. Cox regression analysis was performed on the top prognostic predictions - with Kaplan-Meier survival curves shown in Fig. 2A [5]. The Four Chamber *GST* had the hazards ratio (HR) of 2.89 (CI: [1.81, 4.63], $p < 0.01$), with good discrimination between survival curves beyond 365 days. Similar results were seen for the Short Axis *GST* (HR 1.81, [1.13, 2.90], $p = 0.01$). The CMRI standardized score of RVESVi categorized into low and intermediate-high score achieved a non-significant hazards (HR 1.70, [1.01, 2.88], $p = 0.05$). The WHO functional class clinical score also achieved high odds (HR 2.00, [1.17, 3.43], $p = 0.01$), however this is likely unrealistically biased by the absence of deceased subjects at time of census who were WHO category-2. A comparison of the difference at 4-years survival rates for predicted-deceased and predicted-survived, shown graphically in Fig. 2A, is included in Table 2 as column Cox4YD.

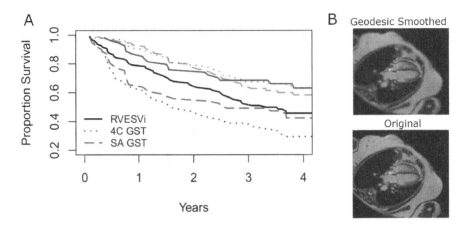

Fig. 2. A) Kaplan–Meier survival plots with standard right ventricular end-systolic volume index (RVESVi), and geodesic smoothing tensor Four Chamber (4C GST) and Short Axis (SA GST) predictions of survival through four years census. Higher curves with 50% transparency are subjects predicted-survival and lower curves at 0% transparency are subjects predicted-deceased. B) Example of geodesic smoothing (top) on a Four Chamber scan (bottom).

Table 3. Multivariate Cox proportional-hazards regression analysis. *WHO: World Health Organization; HR: Hazards Ratio; GST: Geodesic Smoothed Tensor.*

Attribute	Beta Coefficient	HR	p
Age	0.046 ± 0.011	1.05	**<0.01**
WHO Functional Class	0.635 ± 0.301	1.89	**0.04**
Four Chamber GST	0.937 ± 0.246	2.55	**<0.01**
Short Axis GST	0.374 ± 0.245	1.45	0.13

Uni-variate Cox indicated statistical significance ($p < 0.05$) in (1) Age, (2) WHO functional class, (3) Four Chamber *GST* 1-year mortality prediction, and (4) Short Axis *GST* 1-year mortality prediction. Multi-variate Cox proportional-hazards regression analysis was performed (Table 3) on these four variables demonstrated a high concordance of 0.732; this indicates a high probability for any two randomly selected subjects, the subject with the shorter survival time has a larger risk score.

Limitations. The current study was limited by the relatively small cohort (150 subjects, 27 cases of 1-year mortality). Mortality is affected not only by a subject's disease but also by lifestyle habits and treatment decisions, therefore it is possible there are levels of mortality risk (i.e. a novel virus, accident) that cannot ever be fully accounted in a prognostic assessment. We selected to extract features from two common scanning protocols (Four Chamber and Short Axis); in clinic there are additional CMRI acquisitions typically acquired which could be explored for additional or supplemental prognostic features.

4   Conclusions

This paper proposed a **G**eodesically **S**mooothed **T**ensor feature learning method (GST) which uses both the heart and its surrounding tissues for PAH prognosis. We have demonstrated that 1) tensor-based CMRI features can achieve higher PAH mortality prediction compared to standardized measures of ventricular volume, and 2) the full GST pipeline including surrounding tissues can further improves the performance, particularly on Four Chamber scans of the heart. Further study is therefore warranted to investigate if this improvement persists in a larger cohort and for other scanning protocols.

Acknowledgements. This work was supported by the Wellcome Trust: 215799/Z/19/Z and 205188/Z/16/Z; and EPSRC: EP/R014507/1.

References

1. Bello, G.A., et al.: Deep-learning cardiac motion analysis for human survival prediction. Nat. Mach. Intel. **1**(2), 95–104 (2019)
2. Benza, R.L., et al.: The reveal registry risk score calculator in patients newly diagnosed with pulmonary arterial hypertension. Chest **141**(2), 354–362 (2012)
3. Benza, R.L., Miller, D.P., Barst, R.J., Badesch, D.B., Frost, A.E., McGoon, M.D.: An evaluation of long-term survival from time of diagnosis in pulmonary arterial hypertension from the reveal registry. Chest **142**(2), 448–456 (2012)
4. Cox, D.R.: Regression models and life-tables. J. Roy. Stat. Soc. Ser. B (Methodol.) **34**(2), 187–202 (1972)
5. Efron, B.: Logistic regression, survival analysis, and the Kaplan-Meier curve. J. Am. Stat. Assoc. **83**(402), 414–425 (1988)
6. Galiè, N., et al.: 2015 ESC/ERS guidelines for the diagnosis and treatment of pulmonary hypertension: the joint task force for the diagnosis and treatment of pulmonary hypertension of the European Society of Cardiology (ESC) and the European Respiratory Society (ERS): endorsed by: Association for European Paediatric and Congenital Cardiology (AEPC), International Society for Heart and Lung Transplantation (ISHLT). Eur. Heart J. **37**(1), 67–119 (2016)
7. Guillevin, L.: Vasculopathy and pulmonary arterial hypertension. Rheumatology **48**(suppl-3), iii54–iii57 (2006)
8. Lewis, R.A., et al.: Identification of cardiac magnetic resonance imaging thresholds for risk stratification in pulmonary arterial hypertension. Am. J. Respir. Crit. Care Med. **201**(4), 458–468 (2020)
9. Ling, Y., et al.: Changing demographics, epidemiology, and survival of incident pulmonary arterial hypertension: results from the pulmonary hypertension registry of the United Kingdom and Ireland. Am. J. Respir. Crit. Care Med. **186**(8), 790–796 (2012)

10. Lu, H., Plataniotis, K.N., Venetsanopoulos, A.N.: MPCA: multilinear principal component analysis of tensor objects. IEEE Trans. Neural Netw. **19**(1), 18–39 (2008)

11. Reinhold, J.C., Dewey, B.E., Carass, A., Prince, J.L.: Evaluating the impact of intensity normalization on MR image synthesis. In: Medical Imaging 2019: Image Processing, vol. 10949, p. 109493H. International Society for Optics and Photonics (2019)

12. Rich, J.T., Neely, J.G., Paniello, R.C., Voelker, C.C., Nussenbaum, B., Wang, E.W.: A practical guide to understanding Kaplan-Meier curves. Otolaryngol.-Head Neck Surg. **143**(3), 331–336 (2010)

13. Swift, A.J., et al.: A machine learning cardiac magnetic resonance approach to extract disease features and automate pulmonary arterial hypertension diagnosis. Eur. Heart J.-Cardiovasc. Imaging (2020)

14. Tustison, N.J., et al.: N4ITK: improved N3 bias correction. IEEE Trans. Med. Imaging **29**(6), 1310–1320 (2010)

15. Uthoff, J., Sieren, J.C.: Information theory optimization based feature selection in breast mammography lesion classification. In: 2018 IEEE 15th International Symposium on Biomedical Imaging (ISBI 2018), pp. 817–821. IEEE (2018)

Ovarian Cancer Prediction in Proteomic Data Using Stacked Asymmetric Convolution

Cheng Yuan[1], Yujin Tang[1], and Dahong Qian[2(✉)]

[1] School of Biomedical Engineering, Shanghai Jiao Tong University, Shanghai, China
c.yuan@sjtu.edu.cn
[2] Institute of Medical Robotics, Shanghai Jiao Tong University, Shanghai, China
dahong.qian@sjtu.edu.cn

Abstract. Prediction of high grade ovarian cancer on proteomic data is a clinical challenge. Besides, it offers the potential for earlier intervention to increase overall survival, as well as guides the prophylactic ovarian removal to avoid unnecessary early menopause. In this work, we propose a model that learns how to detect ovarian cancer on images from uterine liquid proteomic data. The contributions of this work are two-fold. First, we propose an original method to use proteomic data without direct matching with the existing protein libraries as in the traditional method. The gray-scale peptide image generated by our method contains almost all information from mass spectrometry. Second, we pioneer in analyzing the uterine liquid proteomic data with deep convolutional neural networks. Specifically, we design a feature extractor consisting of stacked asymmetric convolutional layers, which could pay more attention to multiple compounds in different retention times and isotopes in similar mass/charge than symmetric convolutions. Another novelty is trying to find the patches contributing more in improving both sensitivity and specificity. In addition, we add an auxiliary classifier module near the end of training to push useful gradients into the lower layers and to improve the convergence during training. Compared with traditional proteome analysis, experimental results demonstrate the effectiveness and superiority of our model in high grade ovarian cancer prediction.

Keywords: High grade ovarian cancer · Prediction · Peptide image · Mass spectrometry · Auxiliary classifier

1 Introduction

The 5-year overall survival of high grade ovarian cancer (HGOC) is extremely low [1]. Thus, prediction of HGOC in high-risk population is meaningful and vitally needed. On the one hand, it offers the potential of earlier intervention for HGOC patients. On the other hand, benign patients avoid to undergo unnecessary prophylactic removal of ovaries and fallopian tubes, reducing morbidity of early menopause [2].

In present, diagnosis of HGOC relies on proteome screening methods of biomarkers. Multiple blood-based biomarkers have been proposed and tested over the years, however even the most established makers, namely CA125 and HE4, have not proven to

© Springer Nature Switzerland AG 2020
A. L. Martel et al. (Eds.): MICCAI 2020, LNCS 12262, pp. 263–271, 2020.
https://doi.org/10.1007/978-3-030-59713-9_26

be effective in improving survival [3, 4]. Several recent large-scale screening trials on blood CA125-based monitoring showed sensitivity and specificity lower than 60% [5, 6]. Due to the fact that plasma proteins can easily mask the proteins secreted from tumor cells, blood-based testing hampers biomarker identification. In contrast, Geiger *et al.* [7] performed a plasma microparticle screening method of uterine liquid biopsy. Intralumi-nal body fluids are expected to contain the putative biomarkers. Without interference of plasma proteins, they further found a 9-protein classifier with high-resolution mass spectrometry [8]. They pioneered in micro-particle analysis of uterine liquid biopsy and prevented biomarkers from interferences. However, their statistical analysis is insuffi-cient and based on regular matching between fragments they got and protein libraries, which loses lots of potentially valuable information. The complicated proteome exper-iments for traditional analysis not only bring risks to the result credibility, but also consume time.

Deep convolutional neural networks (CNNs) are widely used in clinical computer-aided diagnosis. Classic CNN architectures for prediction make full use of symmetric convolution kernels (such as the size of 3 * 3, 5 * 5, etc.) to learn correlations of neighbor pixels and extract spatial features commonly in images of radiology and pathology. In this work, we utilize a public proteomic dataset submitted by Geiger *et al*. We process proteomic information without matching with existing peptide libraries, and convert peptide signals from mass spectrometers into gray-scale images. Next, we design a feature extractor based on stacked asymmetric convolutional layers, which likely attend to multiple compounds in different retention time and isotopes in similar mass/charge. Furthermore, we use the module of auxiliary classifiers to improve the convergence of the network and achieve a better prediction.

2   Method

The proposed network intends to extract features from peptide images that are most related to HGOC. In our framework, we first integrate peaks and map them to the image according to the corresponding retention time and mass/charge. Next, the generated gray-scale peptide images are fed into the feature extractor based on stacked asymmetric convolution and then the auxiliary classifier, which trains a model to classify the testing data. Figure 1 shows the flowchart of our method.

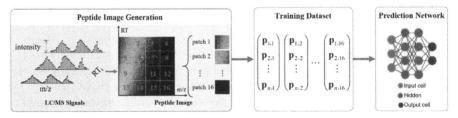

Fig. 1. The pipeline of our proposed method. Every patch cropped from the peptide images is a matrix in training dataset, which is named $\mathbf{p}_{i,j}$. The architecture of the prediction network is demonstrated in Sect. 2.2.

2.1 Peptide Image Generation

The flowchart of peptide image generation is exhibited in Fig. 2.

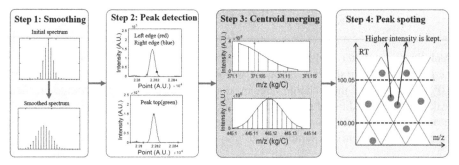

Fig. 2. The four-step pipeline of peptide image generation. In step 3, the charts display two main kinds of centroid merging. The peak with higher intensity is kept when the number of peaks exceed one at intervals of less than 0.05 s of the retention time.

Smoothing. At the start of the processing, we apply the linear smoothing average with respect to retention time and mass/charge. For every peak, namely $f(x)$, we set $2n$ as the window width of the filter. The smoothed peak is obtained by solving the following objective function.

$$f(x)_{new} = \frac{\sum_{i=-n}^{n}(n - |i|)f(x + i)}{n^2} \tag{1}$$

Peak Detection. The basic concept of peak detection algorithm contains differential calculus and noise estimation [9]. Peak detection is performed after the smoothing for retention time and mass/charge. To evaluate the noise, we determine three threshold values: the maximum amplitude differences between two adjacent points, the maxima of the first derivatives and the maxima of the second derivatives, which is respectively shown in Eq. (2), (3) and (4).

$$D(x) = f(x) - f(x - 1) \tag{2}$$

$$f'(x) = \frac{-2f(x - 2) - f(x - 1) + f(x + 1) + 2f(x + 2)}{10} \tag{3}$$

$$f''(x) = \frac{2f(x - 2) - f(x - 1) - 2f(x) - f(x + 1) + 2f(x + 2)}{7} \tag{4}$$

The value below 5% of each maximum and median of amplitude differences, first derivatives and second derivatives are calculated as the threshold values for peak detection. Zero median is set to 0.0001 to avoid errors.

The left edge of the peak is recognized when the amplitude and the first-order derivative both exceed the threshold values in two adjacent points. The peak top is recognized when the sign of the first-order derivative changes and the second-order derivative is less than it threshold. The right edge is recognized by the same way as the left.

Centroid Merging. After peak detection, the centroid merging is performed following the Eq. (5): the ions (point:n) in the defined region between the peak's left and right edges are averaged.

$$Centroid = \frac{\sum_n (ion_{count}(n) * masscharge(n))}{\sum_n ion_{count}(n)} \tag{5}$$

Peak Spotting. After applying algorithms above to the base peak chromatogram, the merged peak tops are shown as 'spots'. Two spots with the close retention time (less than 0.05 s) and the same mass/charge value in adjacent bins are selected by comparing their peak heights. In traditional proteome analysis, unwanted peaks are simply excluded by means of self-defined exclusion mass list. In contrast, we have no exclusion to avoid the experience interference. A sample of our gray-scale peptide image is shown in Fig. 3.

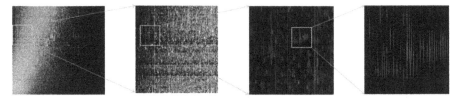

Fig. 3. A sample of the generated gray-scale peptide images. Details are zoomed out.

2.2 Proposed Model

In this section, we introduce our network in detail. We first describe the feature extractor based on stacked asymmetric convolution. The module takes the patches of the peptide image as the inputs and produces higher-level presentation of image features. Then we introduce the auxiliary classifier which can push useful gradients to the lower layers and improve the convergence. An overall illustration of the proposed network architecture is shown in Table 1.

Stacked Asymmetric Convolution. The input images contain information of multiple compounds in different retention time and isotopes in similar mass/charge. Asymmetric convolution kernels with a variety of sizes perceive the anisotropic characteristics of multiple fields, which has superiority in proteomics feature extraction. Let k_i and d_i be the size of convolution filter. Instead of employing the $k_i \times k_i$ convolution operation, we apply the asymmetric convolution operation in size of $k_i \times d_i$ to the input matrix. We design a triply stacked asymmetric convolution as shown in Fig. 4(a). Specific value selection of k_i and d_i are inspired by suggested design principles [10]. After every asymmetric convolution, two outputs are concatenated.

Table 1. The outline of the proposed network architecture.

Type	Kernel size/stride	Padding	Input size
asymmetric conv	As shown in Fig. 4(a)	same	$299 \times 299 \times 1$
pool	$3 \times 3/2$	valid	$147 \times 147 \times 64$
asymmetric conv	As shown in Fig. 4(a)	same	$73 \times 73 \times 64$
$3 \times$ aux_4	As shown in Fig. 4(b)	valid	$35 \times 35 \times 192$
$5 \times$ aux_5	As shown in Fig. 4(c)	valid	$35 \times 35 \times 288$
$3 \times$ aux_6	As shown in Fig. 4(d)	valid	$17 \times 17 \times 768$
pool	8×8	valid	$8 \times 8 \times 2048$
linear	logits	valid	$1 \times 1 \times 2048$
softmax	classifier	valid	$1 \times 1 \times 2$

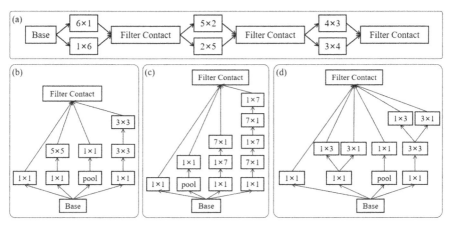

Fig. 4. The structure of stacked asymmetric convolution module and the auxiliary classifiers named aux_4, aux_5 and aux_6.

Auxiliary Classifier. After stacked asymmetric convolution layers, we add the auxiliary classifier module before softmax to achieve more accurate prediction as well as improve the convergence of the deep network. Lee *et al.* [11] suggested utilizing the auxiliary classifier to promote a more robust learning and a better convergence. The specific structure of the auxiliary classifiers named aux_4, aux_5 and aux_6 are shown in Fig. 4(b), (c) and (d), which are in similar structures with Inception module [10]. They can reduce the grid-size while expand the filter banks.

3 Experimental Setup

3.1 Materials

We used HGOC (n = 46) and control (n = 121, such as begin ovarian mass, endometrial polyp, normal endometrium, etc.) samples as raw data, which are from the publicly available proteomic dataset (http://proteomecentral.proteomexchange.org/cgi/Get Dataset?ID=PXD009655). After the steps containing collection, microvesicle isolation and in-solution digestion of utero-tubal lavage samples, peptides are analyzed by liquid-chromatography using the EASY-nLC1000 HPLC coupled to the Q-Exactive Plus or Q-Exactive HF mass spectrometry.

3.2 Implementation Details

The size of the generated peptide image is 1024×1024. To ensure that the input information is not overly compressed, we equally cropped each image into $4 \times 4 = 16$ patches marked through 1 to 16. In every training, the network learns from the patches with the same mark. In every testing, the network outputs prediction confidence scores of every patch. Then the image score is the mean of 16 patch scores.

There is no specifically designed network for classification task based on proteomic data in previous work. We designed two networks as baselines to make the comparison correlative and reliable. As the first baseline, we kept the encoder of stacked asymmetric convolution, added the full connection as a simple classifier and named it AC-CNN. As the second baseline, we trained the Inception network. The model has an encoder of 6-layer 3×3 convolution and uses the same auxiliary classifiers (aux_4, aux_5 and aux_6) as our proposed model.

All the models are trained with cross entropy loss and stochastic gradient descent optimizer with a learning rate of 0.01. Padding operation is applied before the concatenation in stacked asymmetric convolution. Limited by the number of samples in the public dataset, we applied a 7-fold cross-validation experiment to verify the reliability of the results. All the models was implemented using Python 3.6.8 on a DELL workstation with an NVIDIA GeForce GTX 1060 6-GB GPU.

4 Results and Discussion

4.1 Prediction Performance

In clinical diagnosis, physicians mainly focus on the sensitivity, specificity and f1-score. Prediction performance on the test dataset is reported in Table 2. Compared with two baseline networks, our proposed method has superiority in sensitivity, specificity and f1-score, since the asymmetry of image features and the convergence of auxiliary classifiers are jointly considered. In addition, we exhibit the prediction performance of the 9-protein classifier [8] and further prove the effectiveness of our proposed network. Our sensitivity and specificity is respectively 1.7% and 0.8% higher than the 9-protein classifier. It indicates that our method is more sensitive to positive samples, while the 9-protein classifier tends to find out negative samples because of the prior knowledge. Receiver operating characteristic (ROC) curves for method comparison are shown in Fig. 5(a). As for the 7-fold cross-validation, we compared the mean ROC curves of three models. Our network achieves the best ROC performance in all competing networks, which shows the advantages for ovarian cancer diagnosis based on proteomic data. Smaller standard deviations in Table 2 and smoother ROC curves require training and testing on more HGOC samples. Furthermore, Fig. 5(b) displays the distribution of samples with confidence scores in fold 2 testing. It provides another clear representation for the prediction performance.

Table 2. Prediction performances of competing methods.

Method	Sensitivity	Specificity	F1-score
Geiger *et al.* [8]	70.0%	76.2%	/
AC-CNN	55.4% ± 33.5%	51.7% ± 30.0%	38.0% ± 21.8%
Inception [10]	64.3% ± 22.4%	52.6% ± 25.6%	49.8% ± 15.8%
Our proposed	**71.7% ± 22.5%**	**77.0% ± 28.1%**	**66.8% ± 18.2%**

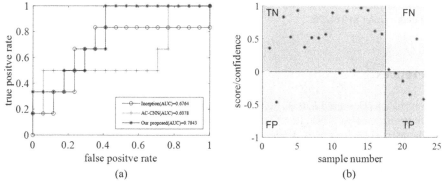

Fig. 5. (a) Mean ROC curves of our proposed network and two baselines. (b) A clear representation for the prediction performance, which is shown on fold 2 testing of our network.

4.2 Possible Domain Containing Biomarker

Here, an explorative research is to find the patches contributing more in improving both sensitivity and specificity. That means such patches contain important information for ovarian cancer prediction and could be recognized as possible domains containing biomarkers. Let m_1 and m_2 respectively represents the number of HGOC samples and control samples in test dataset. The final confidence score of the patch is defined as bellow.

$$final\ score = -m_1 \sum\nolimits_{HGOC} score(i) + m_2 \sum\nolimits_{control} score(i) \qquad (6)$$

Thus, the higher final score represents the more positive impact. Through averaging, we will further explore possible domains containing biomarkers for ovarian cancer diagnosis.

5 Conclusion

In this paper, we utilize a series of operations as smoothing, peak detection, centroid merging and peak spotting to map proteomic data into gray-scale images at first. In addition, we propose a deep neural network consisting of stacked asymmetric convolution and auxiliary classifier module for ovarian cancer prediction. Asymmetric convolution kernels with a variety of sizes perceive the anisotropic characteristics of multiple fields, which shows superiority in proteomics feature extraction. The auxiliary classifier named Inception replaces the full connection layer to push useful gradients to the lower layers and to improve the convergence. Using proteomic data from a public dataset, extensive experiments indicate that our method has the best prediction performance among the traditional analysis and two demonstrated baselines. Future directions include finding the specific locations of biomarkers and optimizing our network with consideration of individual protein expression in age and menopause.

Acknowledgement. The study was supported by the National Nature Science Foundation of China under Grant 81974276.

References

1. Vaughan, S., Coward, J.I., Bast, R.C., et al.: Rethinking ovarian cancer: recommendations for improving outcomes. Nat. Rev. Cancer **11**, 719–725 (2011)
2. Harmsen, M.G., Jong, M.A., Hoogerbrugge, N., et al.: Early salpingectomy (TUbectomy) with delayed oophorectomy to improve quality of life as alternative for risk-reducing salpingo-oophorectomy in BRCA1/2 mutation carriers (TUBA study): a prospective nonrandomized multicentre study. BMC Cancer **15**, 593–601 (2015)
3. Karlan, B.Y., Thorpe, J., Watabayashi, K., et al.: Use of CA125 and HE4 serum markers to predict ovarian cancer in elevated-risk women. Cancer Epidemiol. Biomarkers Prev. **23**, 1383–1393 (2014)
4. Sölétormos, G., Duffy, M.J., Suher, O.A.H., et al.: Clinical use of cancer biomarkers in epithelial ovarian cancer: updated guidelines from the european group on tumor markers. Int. J. Gynecol. Cancer **26**, 43–51 (2015)
5. Jacobs, I.J., Menon, U., Ryan, A., et al.: Ovarian cancer screening and mortality in the UK collaborative trial of ovarian cancer screening (UKCTOCS): a randomised controlled trial. Lancet **387**, 945–956 (2016)
6. Rosenthal, A.N., Fraser, L., Philpott, S., et al.: Evidence of stage shift in women diagnosed with ovarian cancer during phase II of the United Kingdom familial ovarian cancer screening study. J. Clin. Oncol. **35**, 1411–1420 (2017)
7. Harel, M., Oren-Giladi, P., Kaidar-Person, O., et al.: Proteomics of microparticles with SILAC quantification (PROMIS-Quan): a novel proteomic method for plasma biomarker quantification. Mol. Cell. Proteomics **14**, 1127–1136 (2015)

8. Georgina, D.B., Keren, B., Stav, S., et al.: Microvesicle proteomic profiling of uterine liquid biopsy for ovarian cancer early detection. Mol. Cell. Proteomics **18**(5), 865 (2019)

9. Hiroshi, T., Tomas, C., Tobias, K., et al.: MS-DIAL: data-independent MS/MS deconvolution for comprehensive metabolome analysis. Nature Meth. **12**(6), 523–526 (2015)

10. Christian, S., Vincent, V., Sergey, L., et al.: Rethinking the inception architecture for computer vision. In: IEEE Conference on Computer Vision and Pattern Recognition, pp 2818–2826 (2016)

11. Chen-Yu, L., Saining, X., Patrick, G., et al.: Deeply-supervised nets. In: Artificial Intelligence and Statistics, pp. 562–570 (2015)

DeepPrognosis: Preoperative Prediction of Pancreatic Cancer Survival and Surgical Margin via Contrast-Enhanced CT Imaging

Jiawen Yao[1]([✉]), Yu Shi[2], Le Lu[1], Jing Xiao[3], and Ling Zhang[1]

[1] PAII Inc., Bethesda, MD 20817, USA
yaojiawen076@paii-labs.com
[2] Shengjing Hospital of China Medical University, Shenyang, China
[3] Ping An Technology, Shenzhen, China

Abstract. Pancreatic ductal adenocarcinoma (PDAC) is one of the most lethal cancers and carries a dismal prognosis. Surgery remains the best chance of a potential cure for patients who are eligible for initial resection of PDAC. However, outcomes vary significantly even among the resected patients of the same stage and received similar treatments. Accurate preoperative prognosis of resectable PDACs for personalized treatment is thus highly desired. Nevertheless, there are no automated methods yet to fully exploit the contrast-enhanced computed tomography (CE-CT) imaging for PDAC. Tumor attenuation changes across different CT phases can reflect the tumor internal stromal fractions and vascularization of individual tumors that may impact the clinical outcomes. In this work, we propose a novel deep neural network for the survival prediction of resectable PDAC patients, named as 3D Contrast-Enhanced Convolutional Long Short-Term Memory network (CE-ConvLSTM), which can derive the tumor attenuation signatures or patterns from CE-CT imaging studies. We present a multi-task CNN to accomplish both tasks of outcome and margin prediction where the network benefits from learning the tumor resection margin related features to improve survival prediction. The proposed framework can improve the prediction performances compared with existing state-of-the-art survival analysis approaches. The tumor signature built from our model has evidently added values to be combined with the existing clinical staging system.

Keywords: Pancreatic Ductal Adenocarcinoma (PDAC) ·
3D Contrast-Enhanced Convolutional LSTM (CE-ConvLSTM) ·
Preoperative survival prediction

1 Introduction

Pancreatic ductal adenocarcinoma (PDAC) is one of the most lethal of all human cancers, which has an extremely very poor 5-year survival rate of 9% [15].

© Springer Nature Switzerland AG 2020
A. L. Martel et al. (Eds.): MICCAI 2020, LNCS 12262, pp. 272–282, 2020.
https://doi.org/10.1007/978-3-030-59713-9_27

Surgical resection, in combination with neoadjuvant chemotherapy, is the only potentially curative treatment for PDAC patients. Offering surgery to those who would most likely benefit (e.g., a high chance of long-term survival) is thus very important for improving life expectancy. Computed tomography (CT) remains the primary initial imaging modality of choice for the pancreatic cancer diagnosis. Previous work adopts image texture analysis for the survival prediction of PDACs [1]. However, the representation power of hand-crafted features on only venous phase CT might be limited. More recently, deep learning based approaches have shown good performances not only in traditional medical imaging diagnosis [20,21], but also in prognosis models like outcome prediction of lung cancer [12,23,25] and gliomas [11,13]. The success of 3DCNNs contributes to capture deep features not only in the 3D gross tumor volume but also in peritumoral regions. However, such models may not generalize well for PDAC because some important predictive information is not necessarily existing in isolated imaging modality/phase. Contrast-enhanced computed tomography (CE-CT) imaging plays a major role in the depiction, staging, and resectability evaluations of PDAC. Typical characteristics of pancreatic cancer include a hypo-attenuating mass during the pancreatic and venous phases observed in CE-CT. Enhancement variations among multiple CT phases are known to reflect the differences in internal stromal fractions and vascularization of individual PDACs that impact clinical outcomes [14]. However, the use of multi-phase CE-CT for quantitative analysis of PDAC prognosis has not been well investigated in the literature. There is also no established deep learning survival models to incorporate enhancement variations in multi-phase CE-CT. The margin resection status, which is known to be associated with the overall survival (OS) of PDAC patients [10], has been incorporated into the previous radiomics-based PDAC prognosis model [1]. However, the margin status is only available after the surgery is conducted. A preoperative prediction of the margin resection is desired but has not been investigated.

In this paper, we propose a novel 3D Contrast-Enhanced Convolutional Long Short-Term Memory (CE-ConvLSTM) network to learn the enhancement dynamics of tumor attenuation from multi-phase CE-CT images. This model can capture the tumor's temporal changes across several phases more effectively than the early fusion of input images. Furthermore, to allow the tumor resection margin information to contribute to the survival prediction preoperatively, we present a multi-task learning framework to conduct the joint prediction of margin status and outcome. The jointly learning of tumor resectability and tumor attenuation in a multi-task setting can benefit both tasks and derive more effective/comprehensive prognosis related deep image features. Extensive experimental results verify the effectiveness of our presented framework. The signature built from the proposed model remains strong in multivariable analysis adjusting for established clinical predictors, and can be combined with the established criteria for risk stratification and management of PDAC patients.

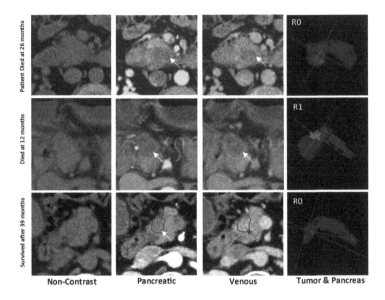

Fig. 1. Example of the multi-phase CE-CT images and PDAC tumor enhancement patterns. Tumor and pancreas masks show the spatial relationship between the pancreas and tumor. The white arrow depicts a hypo-attenuating tumor; blue arrow indicates an iso-attenuating mass. (Color figure online)

2   Methods

The preoperative multi-phase CE-CT pancreatic imaging used in this study have been scanned at three time points for PDACs located at the pancreas head and uncinate. After the non-contrast phase, average imaging time delays are 40–50 s for the pancreatic phase and 65–70 s for the portal venous phase. Figure 1 shows three examples to illustrate different tumor attenuation and resection margins of PDAC patients. Tumor attenuation in specific CT phases are very important characteristics to identify and detect the tumor. Each row in Fig. 1 represents one PDAC patient, and red boundaries are the tumor annotations. The white arrow indicates a typical hypo-attenuating tumor, while blue arrow shows an iso-attenuating tumor. In previous studies, Kim et al. reported that visually iso-attenuating PDACs are associated with better survival rates after surgery, as opposed to typical hypo-attenuating PDACs [9]. Hypo-attenuating mass can be clearly observed in both pancreatic and venous phases of the first and second patients, indicating low stromal fractions (worse clinical outcomes). For the third patient, even though tumor hypo-attenuating is observed in pancreatic phase, it then reflects iso- or even hyper-attenuating in the venous phase compared with its adjacent pancreas regions, indicating high stromal fractions (better survival). This reminds us that tumor enhancement changes across phases is a very useful marker to reflect tumor internal variations and could benefit prognosis. Besides tumor attenuation, another very important factor is the resection margin indicating the margin of apparently non-tumorous tissue around a tumor that has been surgically removed. More specifically, the resection margin is character- ized as R0 when no evidence of malignant glands was identified microscopically at the primary tumour site. R1 resections have malignant glands infiltrating at least one of the resection margins on the permanent section and are usually asso- ciated with poor overall survival [10]. From the Fig. 1, both tumors from the first and second patient display hypo-attenuating appearances, but it is clear to see that the second tumor has infiltrated out of the pancreas shown in tumor and pancreas masks. The pathological evidence indicates the second patient has the PDAC with R1 resections, and a follow-up study shows this patient has worse outcome than the first patient. Radiological observations about tumor attenua- tion and surgical margins status from CE-CT imaging motivate us to develop a preoperative PDAC survival model.

We use time points 1, 2, 3 to represent non-contrast, pancreatic, and venous phases, respectively. A radiologist with 18 years of experience in pancreatic imag- ing manually delineates the tumors slice by slice on the pancreatic phase. The segmentation of the pancreas is performed automatically by a nnUNet model [7] trained on a public pancreatic cancer dataset with annotations [16]. We use DEEDS [6] to register the non-contrast and venous phases to the pancreatic phase. Three image feature channels are derived: 1) CT-image of pancreas and PDAC (background removed) in the soft-tissue window $[-100, 200HU]$ and are normalized as zero mean and unit variance; 2) binary tumor segmentation mask and 3) binary pancreas segmentation mask. A multi-phase sequence of image subvolumes of $64 \times 64 \times 64$ pixels^3 centered at the tumor 3D centroid are cropped to cover the entire tumor and its surrounding pancreas regions. The dataset is prepared for every tumor volume from each phase scan, to build the 4D CE-CT data for both training and testing (as $X_t = \{X_t^{CT}, X_t^{M_T}, X_t^{M_P}\}, t \in \{1, 2, 3\}$, M_T: tumor mask, M_P: pancreas mask). Our joint learning network architecture is shown in Fig. 2. This network has two branches for predicting both the resec- tion margins and survival outcomes. The branch of resection margins uses one 3D-CNN model with six convolutional layers equipped with Batch Normaliza- tion and ReLu. Similar 3D architecture has shown good prediction performance for lung cancer [12]. Input of this branch is the concatenation of CT volumes at different time points and the corresponding tumor and pancreas masks: e.g., $X \in \mathbf{R}^{5 \times 64^3}$. This branch will try to learn the CT intensity attenuation vari- ations and the relationships between tumor and surrounding pancreas regions, which help classify the tumor into different resection status. Note that R0/R1 can only be obtained after the surgery and pathology. Our model can be applied preoperatively in real scenarios to offer PDAC patients with the appropriate advice regarding surgical decisions.

The branch of predicting outcomes use CT volumes at each phase (each phase is CT-M_T-M_P three-channel input, $X_t \in \mathbf{R}^{3 \times 64^3}$). The aim of this branch is to capture the tumor attenuation patterns across phases. Tumor attenuation usually means the contrast differences between the tumor and its surrounding pancreas regions so that we introduce both the tumor and pancreas masks into

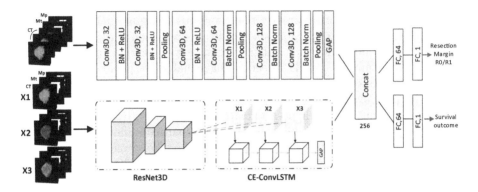

Fig. 2. An overview of the proposed multi-task model with CE-ConvLSTM.

input volumes. The core part of this branch is a recurrence module that allows the network to retain what it has seen and to update the memory when it sees a new phase image. A naive approach is to use a vanilla LSTM or ConvLSTM network. Conventional ConvLSTM is capable of modeling 2D spatio-temporal image sequences by explicitly encoding the 2D spatial structures into the temporal domain [2]. A more recent ST-ConvLSTM simultaneously learns both the spatial consistency among successive image slices and the temporal dynamics across different time points for the tumor growth prediction [24]. Instead of using adjacent 2D CT slices and motivated by 3D object reconstruction [4], we propose to use Contrast-Enhanced 3D Convolutional LSTM (CE-ConvLSTM) network to capture the temporally enhanced patterns from CE-CT sequences. CE-ConvLSTM can model 4D spatio-temporal CE-CT sequences by explicitly encoding their 3D spatial structures into the temporal domain. The main equations of ConvLSTM are as follows:

$$f_t = \sigma(W_f^X * X_t + W_f^H * H_{t-1} + b_f) \tag{1}$$
$$i_t = \sigma(W_i^X * X_t + W_i^H * H_{t-1} + b_i)$$
$$o_t = \sigma(W_o^X * X_t + W_o^H * H_{t-1} + b_o)$$
$$C_t = f_t \odot C_{t-1} + i_t \odot tanh(W_C^X * X_t + W_C^H * H_{t-1} + b_C)$$
$$H_t = o_t \odot tanh(C_t)$$

where X_t is the CE-CT sequences at time t, $*$ denotes the convolution operation, and $\odot$ denotes the Hadamard product. All the gates f, i, o, memory cell C, hidden state H are 4D tensors. We use a $3 \times 3 \times 3$ convolutional kernel and 128 as the channel dimension of hidden states for the LSTM unit. We employ 3D-ResNet18 [3,5] as the encoder to encode each three-channel input to the lower-dimensional feature maps for CE-ConvLSTM.

After the concatenation of feature maps from both tasks, the channel number of this common representation is 256. Then two separate fully-connected networks will use the common representation for each prediction task. In the

training phase, labels of the resection status and patient overall survival information (OS time and censoring status) are known for each input CE-CT sequence.
The weighted binary cross-entropy (BCE) loss is applied to the resection margin
prediction task, while the negative log partial likelihood [8,22] is used to predict
the survival outcome $\mathbf{y}_i$ of this patient which is a continuous score. This loss is
summarized as $L(\mathbf{y}_i) = \sum_i \delta_i(-\mathbf{y}_i + \log \sum_{j:t_j>=t_i} \exp(\mathbf{y}_j))$ where j is from the
set whose survival time is equal or larger than t_i ($t_j \geq t_i$).

3 Experiments

Dataset. Pancreatic CT scans of 205 patients (with resectable PDACs, mean
tumor size $= 2.5\,\mathrm{cm}$) were undertaken preoperatively during non-contrast, pancreatic, and portal venous phases (i.e., 615 CT volumes). Only 24 out of 205
patients have R1 resection margins, and the imbalanced class weighting in the
loss is considered. All images were resampled to an isotropic $1\,\mathrm{mm}^3$ resolution.
We adopt nested 5-fold cross-validation (with training, validation, and testing
sets in each fold) to evaluate our model and other competing methods. To augment the training data, we rotate the volumetric tumors in the axial direction
around the tumor center with the step size of $90°$ to get the corresponding 3D CT
image patches and their mirrored patches. We also randomly select the cropped
regions with random shifts for each iteration during the training process. This
data augmentation can improve the network's ability to locate the desired translational invariants. The batch sizes of our method and other models are the same
as 8. The maximum iteration is set to be 500 epochs, and the model with the
best performance on the validation set during training is selected for testing.

We first validate the performance from each branch shown in Fig. 2 for
single survival prediction task. Then we incorporate each model with CE-
ConvLSTM and report the results as shown in Table 1. C-index value is
adopted as our main evaluation metric for survival prediction. The C-index
quantifies the ranking quality of rankings and is calculated as follows $c =
\frac{1}{n} \sum_{i\in\{1...N|\delta_i=1\}} \sum_{t_j>t_i} I[\mathbf{y}_i > \mathbf{y}_j]$ where n is the number of comparable pairs
and $I[.]$ is the indicator function. t_i is the actual time observation of patient i and
$\mathbf{y}_i$ denotes the corresponding risk. The value of C-index ranges from 0 to 1. The
averaged C-index using the single pancreatic phase is 0.659 ± 0.075 from 3DCNN
model. We can see that the prediction is improved when using early fused CE-
CT images (CE-3DCNN, 0.662 vs. 0.659). When CE-ConvLSTM is adopted by
incorporating all phases, C-index is increased by 3% (3DCNN-CE-ConvLSTM)
compared with the model only using the pancreatic phase. 3D-ResNet18 with the
pre-trained weights [3] is used as an advanced model compared to the conventional 3D ConvNets. From Table 1, ResNet3D with CE-ConvLSTM has improved
performance versus ResNet3D with early fusion. Results in this table illustrate that the dynamic enhancement CT imaging patterns learned/captured by
CE-ConvLSTM can help achieve significant prediction improvements compared
against early fusion CNNs.

Table 1. Validation of CE-ConvLSTM with different CNN backbones.

Model	Res-CE-ConvLSTM	CE-ResNet3D	3DCNN-CE-ConvLSTM	CE-3DCNN
C-index	0.683 ± 0.047	0.675 ± 0.050	0.679 ± 0.052	0.662 ± 0.038

Table 2. Results of different methods. Mul: multi-task; cls: single classification.

	Task	Survival	Resection Margin: R0/R1		
		C-index	Balanced-ACC	Sensitivity	Specificity
Proposed	Mul	**0.705 ± 0.015**	**0.736 ± 0.141**	**0.813 ± 0.222**	0.659 ± 0.118
Tang et al. [17]	Mul	0.683 ± 0.056	0.574 ± 0.090	0.573 ± 0.336	0.575 ± 0.232
Lou et al. [12]	Mul	0.649 ± 0.070	0.682 ± 0.091	0.673 ± 0.236	0.690 ± 0.080
CE-ResNet3D	cls	–	0.604 ± 0.169	0.446 ± 0.347	**0.762 ± 0.157**
CE-3DCNN	cls	–	0.583 ± 0.145	0.460 ± 0.312	0.706 ± 0.161

To further evaluate the performance of multi-task models, we report the results in comparison to recent multi-task deep prediction methods [12,17], as illustrated in Table 2. Tang et al. propose using separate branches of 3D CNNs to predict both OS time and tumor genotype for glioblastoma (GBM) patients [17]. Lou et al. present a multi-task training model on shared hidden representations from single model [12]. We replace the RMSE loss from the original implementation of [17] with the negative log partial likelihood loss because it can handle alive patients (whereas authors discarded some patients who are still alive in [17]). Those models cannot capture tumor temporal enhancement changes. Classification performances are evaluated by the metrics of Balanced-Accuracy, Sensitivity, and Specificity. Single classification task uses CE-ResNet3D and CE-3DCNN can be found in the last two rows in Table 2. We can see the proposed method that uses CE-ConvLSTM to capture tumor enhancement patterns indeed achieves better performance in both tasks than the baseline multi-task models with early fusion.

In Table 3, univariate and multivariate cox proportional-hazards models are used to evaluate Hazard Ratio (HR) and log-rank test p-value for each factor including deep signature, radiomics signature and other clinicopathologic factors. To obtain the deep signature on all 205 patients, we use the mean and standard deviation of risk predictions on the training folds to normalize the risk scores of the corresponding testing fold. Then we combine results from each testing fold together. We also build radiomics signature using conventional radiologic features of input CE-CT images using Pyradiomics [19]. Each phase will provide 338 features, and thus each patient contains 1014 features in total. Features are then refined by Lasso-based Cox feature selection [18]. Finally, the selected features are fed to the Cox regression model to get the signature. Each feature selection is performed on each training fold, and the number of selected features varies across different folds. From the statistic analysis (Table 3), the proposed signature is the strongest prognostic factor in univariate analysis with the

Table 3. Univariate and multivariate cox regression analysis. HR: hazard ratio.

Factors	Univariate analysis			Multivariate cox	
	HR (95% CI)	C-index	p-value	HR (95% CI)	p-value
Stromal fraction	0.057(0.019−0.17)	0.659	6.38e−7	0.105(0.031−0.0.36)	3.39e−4
Clear fat plane	3.771(2.235−6.364)	0.559	1.39e−5	2.010(1.157−3.490)	0.013
Tumor size	1.693(1.218−2.352)	0.588	0.002	1.568(1.108−2.22)	0.011
Resection margin	4.557(2.772−7.492)	0.577	2.99e−7	3.704(2.147−6.390)	2.51e−6
TNM stage	1.475(1.107−1.966)	0.546	0.008	1.319(0.970−1.790)	0.078
CA19.9	1.001(1.000−1.001)	0.557	0.0207	1.000(0.999−1.001)	0.281
Radiomics signature	1.181(1.109−1.257)	0.645	1.17e−4	1.085(1.004−1.171)	0.038
Signature of [17]	1.704(1.426−2.035)	0.677	4.96e−9	1.035(0.824−1.30)	0.767
Our signature	1.764(1.478−2.105)	0.691	2.81e−10	1.435(1.142−1.803)	0.002

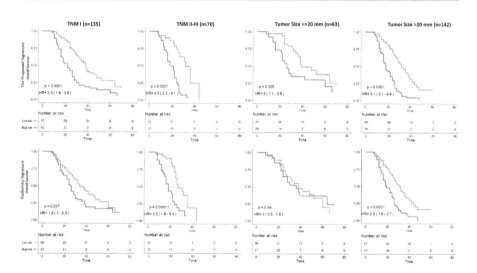

Fig. 3. Kaplan-Meier analyses of overall survival according to the proposed (1st row) and radiomics signature (2nd row) in patients with two different stratifications (TNM staging and Tumor size). The proposed signature significantly stratifies all subgroups (TNM I vs II-III; or Tumor size ≤20 mm vs >20 mm).

highest concordance index. The proposed signature remains strong in multivariable analysis (HR = 1.435, p = 0.002) adjusting for established clinicopathologic prognostic markers, e.g.: stromal fractions (HR = 0.105, $p < 0.001$), resection margins (HR = 3.704, $p < 0.0001$), and TNM stage (HR = 1.319, $p = 0.078$). Our proposed one is even stronger than any other CT-derived signature using radiomics analysis or deep learning model without encoding tumor enhancement patterns.

To demonstrate the added value of the proposed signature to the current staging system, we plot Kaplan-Meier survival curves in Fig. 3 for patients with further stratification by our signature after grouping by TNM and Tumor size, respectively, which are two well-established stratification criteria. The cut-off for our and radiomics signature is the median score of each signature. We study two subgroups of patients: 1) patients divided by TNM staging I vs. II-III, 2) patients divided by the primary PDAC tumor size $\leq$20 mm vs. >20 mm. It is shown that the proposed signature remains the most significant log-rank test outcome in the subgroup of patients, while the radiomics signature does not reach the statistical significance within the patient sub-population of PDAC tumor $\leq$ 20 mm. Results shown in Fig. 3 demonstrate that after using the current clinicopathologic TNM staging system or tumor size, our proposed multi-phase CT imaging based signature can indeed further provide the risk stratification with significant evidence. This described novel deep signature could be combined with the established clinicopathological criteria to refine the risk stratification and guide the individualized treatment of resectable PDAC patients.

4   Conclusion

In this paper, we propose a new multi-task CNN framework for cancer survival prediction by simultaneously predicting the tumor resection margins for resectable PDAC patients. The use of CE-ConvLSTM to consider and encode the dynamic tumor attenuation patterns of PDAC boost the whole framework, to significantly outperform the early fusion deep learning models and conventional radiomics-based survival models. Our results also validate that the proposed signature can serve as both a prognostic and predictive biomarker for subgroup patients after staging by the well-established pathological TNM or tumor size. This makes our model very promising in future clinical usage to refine the risk stratification and guide the surgery treatment of resectable PDAC patients. Future work will consider to integrate with other non-radiological factors (e.g. age, gender) to achieve better results.

References

1. Attiyeh, M.A., et al.: Survival prediction in pancreatic ductal adenocarcinoma by quantitative computed tomography image analysis. Ann. Surg. Oncol. **25**(4), 1034–1042 (2018)
2. Chen, J., Yang, L., Zhang, Y., Alber, M., Chen, D.Z.: Combining fully convolutional and recurrent neural networks for 3D biomedical image segmentation. In: Advances in Neural Information Processing Systems, pp. 3036–3044 (2016)
3. Chen, S., Ma, K., Zheng, Y.: Med3D: transfer learning for 3D medical image analysis. arXiv preprint arXiv:1904.00625 (2019)
4. Choy, C.B., Xu, D., Gwak, J.Y., Chen, K., Savarese, S.: 3D-R2N2: a unified approach for single and multi-view 3D object reconstruction. In: Leibe, B., Matas, J., Sebe, N., Welling, M. (eds.) ECCV 2016. LNCS, vol. 9912, pp. 628–644. Springer, Cham (2016). https://doi.org/10.1007/978-3-319-46484-8_38

5. Hara, K., Kataoka, H., Satoh, Y.: Can spatiotemporal 3D CNNs retrace the history of 2D CNNs and ImageNet? In: Proceedings of the IEEE Conference on Computer Vision and Pattern Recognition, pp. 6546–6555 (2018)
6. Heinrich, M.P., Jenkinson, M., Papież, B.W., Brady, S.M., Schnabel, J.A.: Towards realtime multimodal fusion for image-guided interventions using self-similarities. In: Mori, K., Sakuma, I., Sato, Y., Barillot, C., Navab, N. (eds.) MICCAI 2013. LNCS, vol. 8149, pp. 187–194. Springer, Heidelberg (2013). https://doi.org/10.1007/978-3-642-40811-3_24
7. Isensee, F., et al.: nnU-Net: self-adapting framework for U-Net-based medical image segmentation. arXiv preprint arXiv:1809.10486 (2018)
8. Katzman, J.L., Shaham, U., Cloninger, A., Bates, J., Jiang, T., Kluger, Y.: Deep-Surv: personalized treatment recommender system using a Cox proportional hazards deep neural network. BMC Med. Res. Methodol. 18(1), 24 (2018)
9. Kim, J.H., et al.: Visually isoattenuating pancreatic adenocarcinoma at dynamic-enhanced CT: frequency, clinical and pathologic characteristics, and diagnosis at imaging examinations. Radiology 257(1), 87–96 (2010)
10. Konstantinidis, I.T., et al.: Pancreatic ductal adenocarcinoma: is there a survival difference for R1 resections versus locally advanced unresectable tumors? what is a "true" R0 resection? Ann. Surg. 257(4), 731–736 (2013)
11. Liu, Z., Sun, Q., Bai, H., Liang, C., Chen, Y., Li, Z.C.: 3D deep attention network for survival prediction from magnetic resonance images in Glioblastoma. In: 2019 IEEE International Conference on Image Processing (ICIP), pp. 1381–1384. IEEE (2019)
12. Lou, B., et al.: An image-based deep learning framework for individualising radiotherapy dose: a retrospective analysis of outcome prediction. Lancet Digital Health 1(3), e136–e147 (2019)
13. Nie, D., Zhang, H., Adeli, E., Liu, L., Shen, D.: 3D deep learning for multi-modal imaging-guided survival time prediction of brain tumor patients. In: Ourselin, S., Joskowicz, L., Sabuncu, M.R., Unal, G., Wells, W. (eds.) MICCAI 2016. LNCS, vol. 9901, pp. 212–220. Springer, Cham (2016). https://doi.org/10.1007/978-3-319-46723-8_25
14. Prokesch, R.W., Chow, L.C., Beaulieu, C.F., Bammer, R., Jeffrey Jr., R.B.: Isoattenuating pancreatic adenocarcinoma at multi-detector row CT: secondary signs. Radiology 224(3), 764–768 (2002)
15. Siegel, R.L., Miller, K.D., Jemal, A.: Cancer statistics. CA Cancer J. Clin. 69(1), 7–34 (2019)
16. Simpson, A.L., et al.: A large annotated medical image dataset for the development and evaluation of segmentation algorithms. arXiv preprint arXiv:1902.09063 (2019)
17. Tang, Z., et al.: Pre-operative overall survival time prediction for glioblastoma patients using deep learning on both imaging phenotype and genotype. In: Shen, D., et al. (eds.) MICCAI 2019. LNCS, vol. 11764, pp. 415–422. Springer, Cham (2019). https://doi.org/10.1007/978-3-030-32239-7_46
18. Tibshirani, R.: The lasso method for variable selection in the Cox model. Stat. Med. 16(4), 385–395 (1997)
19. Van Griethuysen, J.J., et al.: Computational radiomics system to decode the radiographic phenotype. Cancer Res. 77(21), e104–e107 (2017)
20. Wang, X., Peng, Y., Lu, L., Lu, Z., Bagheri, M., Summers, R.M.: Chestx-ray8: hospital-scale chest x-ray database and benchmarks on weakly-supervised classification and localization of common thorax diseases. In: Proceedings of the IEEE Conference on Computer Vision and Pattern Recognition, pp. 2097–2106 (2017)

21. Wang, Y., et al.: Weakly supervised universal fracture detection in pelvic X-rays. In: Shen, D., et al. (eds.) MICCAI 2019. LNCS, vol. 11769, pp. 459–467. Springer, Cham (2019). https://doi.org/10.1007/978-3-030-32226-7_51

22. Yao, J., Zhu, X., Huang, J.: Deep multi-instance learning for survival prediction from whole slide images. In: Shen, D., et al. (eds.) MICCAI 2019. LNCS, vol. 11764, pp. 496–504. Springer, Cham (2019). https://doi.org/10.1007/978-3-030-32239-7_55

23. Yao, J., Zhu, X., Zhu, F., Huang, J.: Deep correlational learning for survival prediction from multi-modality data. In: Descoteaux, M., Maier-Hein, L., Franz, A., Jannin, P., Collins, D.L., Duchesne, S. (eds.) MICCAI 2017. LNCS, vol. 10434, pp. 406–414. Springer, Cham (2017). https://doi.org/10.1007/978-3-319-66185-8_46

24. Zhang, L., et al.: Spatio-temporal convolutional LSTMs for tumor growth prediction by learning 4D longitudinal patient data. IEEE Trans. Med. Imaging **39**(4), 1114–1126 (2019)

25. Zhu, X., Yao, J., Zhu, F., Huang, J.: WSISA: making survival prediction from whole slide histopathological images. In: CVPR, pp. 7234–7242 (2017)

Holistic Analysis of Abdominal CT for Predicting the Grade of Dysplasia of Pancreatic Lesions

Konstantin Dmitriev$^{(\boxtimes)}$ and Arie E. Kaufman

Department of Computer Science, Stony Brook University, Stony Brook, NY, USA
{kdmitriev,ari}@cs.stonybrook.edu

Abstract. Diagnosis of various pancreatic lesions in CT images is a challenging task owing to a significant overlap in their imaging appearance. An accurate diagnosis of pancreatic lesions and the assessment of their malignant progression, or the grade of dysplasia, is crucial for optimal patient management. Typically, the grade of dysplasia is confirmed histologically via biopsy, yet certain radiological findings, including extrapancreatic, can serve as diagnostic clues of the disease progression. This work introduces a novel method of transforming intermediate activations for processing intact imaging data of varying sizes with convnets with linear layers. Our method allows to efficiently leverage the 3D information of the entire abdominal CT scan to acquire a holistic picture of all radiological findings for an improved and more precise classification of pancreatic lesions. Our model outperforms current state-of-the-art methods in classifying four most common lesion types (by 2.92%), while additionally diagnosing the grade of dysplasia. We conduct a set of experiments to illustrate the effects of a holistic CT analysis and the auxiliary diagnostic data on the accuracy of the final diagnosis.

1 Introduction

Pancreatic cancer (PC) is one of the most aggressive types of cancer and is currently the fourth most common cause of cancer-related deaths in the United States [1]. PC is often asymptomatic but is associated with distinct precursor cystic lesions, such as intraductal papillary mucinous neoplasms (IPMNs) and mucinous cystic neoplasms (MCNs). Early detection and diagnosis of such lesions offer an opportunity to prevent the progression of the disease. However, cystic pancreatic lesions are a heterogeneous group of lesions, which also include serous cystadenomas (SCAs) and solid-pseudopapillary neoplasms (SPNs), which are considered to have a low malignant potential [23]. Despite the widespread use of high-resolution imaging, such as computed tomography (CT), the non-invasive diagnosis and characterization of pancreatic lesions is still a challenge even for an experienced radiologist due to the overlapping demographic and appearance characteristics (Fig. 1) of lesions [10]. The potential complications and risks associated with surgical resection make an accurate non-invasive diagnostic assessment critically important.

© Springer Nature Switzerland AG 2020
A. L. Martel et al. (Eds.): MICCAI 2020, LNCS 12262, pp. 283–293, 2020.
https://doi.org/10.1007/978-3-030-59713-9_28

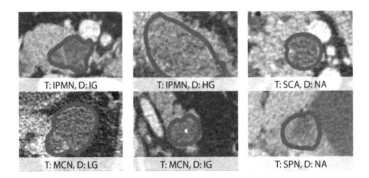

Fig. 1. Types (T) of different pancreatic lesions in CT and associated grades of dysplasia (D), namely, low- (LG), intermediate- (IG), or high-grade (HG) dysplasia.

Optimal management of pancreatic lesions also relies on the malignant progression of a lesion to estimate which patients would benefit the most from a surgical resection as only a small portion of lesions progress to PC during the lifetime of a patient. The malignant progression of a lesion is associated with the grade of dysplasia, namely, low- (LG), intermediate- (IG), or high-grade (HG) dysplasia. Current guidelines recommend to observe LG and IG precursor lesions and to resect HG lesions to decrease the risk of invasive PC [2]. Traditionally, the grade of dysplasia is determined pathologically via a biopsy procedure. However, some studies show that certain radiological features indicate an association with the HG dysplasia in pancreatic lesions. These features include the dilation of the main pancreatic duct [6] or the common bile duct [19], extrapancreatic malignant neoplasms [4], and lesion size and location [20].

Previous Work. Significant progress has been made in the development of computer-aided diagnosis (CAD) systems to aid the clinicians in the process of differentiation of various abnormalities in radiological images [5,14,17,18,27,29, 30,33], including for classification of pancreatic lesions [9,13,15,24,25,32]. Latter methods mainly focus on the classification of the pre-segmented lesions into the four most common lesion types, or their sub-groups (benign vs. malignant, SCN vs. non-SCN, etc.), ignoring the grades of dysplasia for lesions with malignant potential. In addition, due to the significant variations in the size of the original 3D regions with lesions, previous methods often divide them into smaller 3D or 2D subvolumes or patches of fixed size. Each patch is then analyzed individually, and the final probabilities are generated as the average between all patches. In such an approach, the spatial relationship between the patches is neglected, leading to the potential inconsistencies between individual classifications and an overall limited diagnostic performance. A similar in spirit, yet different, issue is common in the analysis of gigapixel pathology images [22,26]. Finally, another common limitation of these works is the diagnosis of pancreatic lesions in isolation from the surrounding vascular structures and organs, including the pancreas, which leads to the loss of additional potential diagnostic clues.

In this paper, we attempt to approach the diagnosis of pancreatic lesions along with the associated grade of dysplasia in a holistic, rather than an isolated way. We hypothesize that while the radiological appearance of lesions provides some diagnostic clues, a more accurate diagnosis is possible by analyzing the entire peritoneal region in a CT scan, and obtaining a complete picture of all clinical findings, including extrapancreatic. The **contributions** of our paper are the following: a method for efficient transformation of intermediate activations in convolutional networks (convnets) with linear layers to process CT volumes of varying sizes; improved pancreatic lesion classification results (by 2.92%), along with the prediction of the associates grade of dysplasia; and, to the best of our knowledge, this work is the first to describe a lesion classification method based on the holistic analysis of the entire intact abdominal CT scan. The importance of the holistic approach is illustrated by a comprehensive set of experiments.

2   Method

Let $\mathbf{X}^i$ be the available diagnostic records of a patient i with a pathologically confirmed pancreatic lesion of type t^i with the associated grade of dysplasia d^i. These records $\mathbf{X}^i$ include an abdominal CT scan I^i, and the optional binary segmentation masks of the pancreas M_p^i and the lesion M_l^i, and the patient's age a_i at the acquisition time and their gender g_i. The goal is to model a function $\mathcal{F}_\theta(\mathbf{X}^i) = (\hat{\mathbf{t}}^i, \hat{\mathbf{d}}^i)$ using a convnet, where $\hat{\mathbf{t}}^i \in T = \{\text{IPMN, MCN, SCA, SPN}\}$ and $\hat{\mathbf{d}}^i \in D = \{\text{NA, LG, IG, HG}\}$ are the estimated probabilities of the lesion type and dysplasia. Each abdominal scan I^i of size $Z_i \times H_i \times W_i$ is a stack of Z_i axial 2D slices of size $H_i \times W_i$. While the dimensions of each 2D slice are typically unvarying and are equal to 512×512 regardless of the scanning equipment, the number of slices Z_i in each scan varies between patients depending on patient's size and positioning inside the scanner. Unlike the traditional solutions, our proposed model allows us to avoid dividing the input scan I_i into subvolumes of predefined size or individual 2D slices, but keep it intact to preserve the spatial information and to obtain a holistic picture of all pathological findings which may be present in the CT scans along with the pancreatic lesions.

Base Model: The overview of our model is illustrated in Fig. 2. It consists of the convolutional backbone $\mathcal{F}_\theta^1(\mathbf{X}^i) = \mathbf{A}^i$ for feature extraction (Fig. 2(a)), and the classification head $\mathcal{F}_\theta^2(\mathbf{A}^i) = (\hat{\mathbf{t}}^i, \hat{\mathbf{d}}^i)$ (Fig. 2(c)). In other words, our target function can be expressed as $\mathcal{F}_\theta(\mathbf{X}^i) = \mathcal{F}_\theta^2(\mathcal{F}_\theta^1(\mathbf{X}^i)) = (\hat{\mathbf{t}}^i, \hat{\mathbf{d}}^i)$. The backbone $\mathcal{F}_\theta^1$ consists of three convolutional layers of various kernel sizes and strides, each followed by a Leaky ReLU activation function, proceeded by nine ResNet bottleneck blocks [11]. The classification head $\mathcal{F}_\theta^2$ includes an additional $1 \times 1 \times 1$ convolutional layer to reduce the size of the activation maps generated by the backbone and two linear layers which perform the final classification.

The convolutional backbone $\mathcal{F}_\theta^1$ is agnostic to the size $Z_i \times H_i \times W_i$ of the input volumes in $\mathbf{X}^i$, and generates the activation maps $\mathbf{A}^i$ of size $\frac{Z_i}{16} \times \frac{H_i}{16} \times \frac{W_i}{16}$, or more specifically, $\frac{Z_i}{16} \times H' \times W'$, as the dimensions $H_i \times W_i$ of 2D slices are

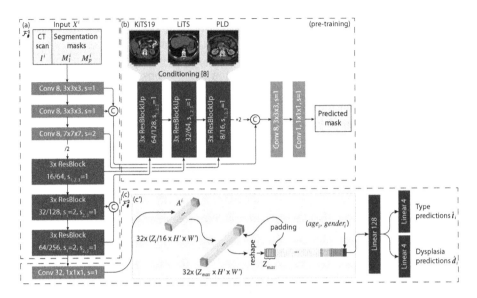

Fig. 2. Schematic view of our classification framework. The main classification model consists of (a) a feature extractor $\mathcal{F}_\theta^1$, or backbone, and (c) a classification head $\mathcal{F}_\theta^2$. The backbone processes a full intact 3D CT scan of arbitrary size $Z_i \times H \times W$ with optional segmentation masks of the pancreas M_p^i and lesion(s) M_l^i provided as a second channel, and generates a tensor of activations maps of size $Z_i/16 \times H' \times W'$. These activation maps are (c') adaptively padded to a predefined size $Z_{max} \times H' \times W'$, flattened, and passed to the classification head which generated the final output. During training, padded activation maps can be organized into batches for faster processing (see Sec. 2). To improve the classification performance, the backbone was augmented with a (b) conditioned decoder and pre-trained for segmentation.

constant, but the number of slices Z_i varies between each volume. As a result, the size of the input $\mathbf{A}^i$ to the classification head $\mathcal{F}_\theta^2$ constantly varies. However, traditionally organized networks with linear layers require the input to be of fixed size. Common workarounds include zero-padding of the original 3D volumes in $\mathbf{X}^i$ and adaptive pooling of the final activations to a fixed size. These approaches come with limitations. The former significantly increases memory requirements and computational complexity due to a larger size of the input and the intermediate activation maps. On the other hand, adaptive pooling can potentially lead to a severe loss of spatial information encoded in the activation maps. Instead, we propose a simple, yet effective way of transforming the activation maps to a fixed size, which can be processed by the linear layers. Particularly, we propose to adaptively pad the final activations $\mathbf{A}^i$ generated by the convolutional backbone $\mathcal{F}_\theta^1$ to a pre-defined size Z_{max} before reshaping, or flattening, them to a 1D tensor and passing them to the classification head $\mathcal{F}_\theta^2$. Our proposed approach allows one to process input data $\mathbf{X}^i$ of varying size, while avoiding the increase of computational complexity and the loss of the important spatial information.

However, as the original data samples $\mathbf{X}^i$ are left unpadded and can significantly vary in size, they cannot be easily collated into batch tensors for more efficient processing that takes advantage of the parallelism. To accelerate the training process, we propose the following: during the forward step, each sample $\mathbf{X}^i$ from a desired batch of size N is first processed individually by the backbone $\mathcal{F}_\theta^1$. Then, the padded activations maps $\mathbf{A}^i$ are combined into a batch tensor of size N and are processed jointly by the classification head $\mathcal{F}_\theta^2$. Consequently, during the backward step, gradients can be accumulated across all N samples in the batch, and the gradient updates can be applied to both $\mathcal{F}_\theta^2$ and $\mathcal{F}_\theta^1$. The efficiency of the proposed framework is evaluated and compared against the traditional input padding in Sec. 3.

Due to many medical datasets being often rather small, training a classification model from scratch might lead to overfitting. Moreover, it is unlikely that the model will learn to reliably detect and use extrapancreatic abnormalities, such as renal lesions, for the diagnosis of pancreatic lesions. Instead, the model will likely use irrelevant features. To alleviate this issue, we propose to pretrain our feature extractor $\mathcal{F}_\theta^1$ for a segmentation task using an additional decoder branch. The architecture of the decoder (Fig. 2(b)) is similar to the backbone, yet significantly narrower. Such an architecture puts more load on the backbone and forces it to learn robust feature representations, potentially at the cost of a limited segmentation performance. Additionally, we condition the decoder [8] to relax the dataset requirements, and to allow us to train a single multi-class segmentation model on multiple binary or ternary, rather than multi-class, segmentation datasets.

3   Experiments

Datasets: To train and evaluate the proposed model, we utilize three datasets of CT scans with various abdominal pathologies. Specifically, during the initial segmentation pre-training step, we use the publicly available Liver Tumor Segmentation (LiTS) dataset [3], the Kidney Tumor Segmentation (KiTS19) dataset [12], and a dataset with pancreatic lesions (PLD). The latter is also used during the final classification training and evaluation steps. The CT images in each dataset were minimally preprocessed by being downsampled from $Z_i \times 512 \times 512$ to $Z_i \times 128 \times 128$ ($Z_i \in [76, 190]$). The information loss due to this step is minimal. Furthermore, this is optional and is not a principal part of our framework. Additionally, we normalize the CT intensities to $[-1, 1]$ range after clipping the original values around $[-300, 300]$ range. This procedure imitates the standard radiologists' reading protocol for pancreatic lesion evaluation (similar to "window/level" adjustments), while also preserving intensity consistency between scans.

LiTS dataset consists of 131 manually annotated contrast-enhanced abdominal CT volumes with various hepatic tumors collected from several clinical sites. The gender and age distribution of the subjects in the dataset, as well as the used imaging equipment, are unknown.

KiTS19 dataset consists of 210 manually annotated contrast-enhanced CT scans (60% males and 40% females; mean age, 60) with renal lesions collected from several clinical sites. The imaging equipment is unknown.

PLD dataset contains 141 contrast-enhanced abdominal CT scans (44 males and 97 females; mean age, 59 ± 17) collected with Siemens SOMATOM scanner (Siemens Medical Solutions, Malvern, PA), using venous phase protocol. The dataset exhibits examples of the four most common pancreatic lesions: 74 cases of IPMNs, 17 cases of MCNs, 33 cases of SCAs, and 17 cases of SPNs. Each scan was accompanied with the segmentation outlines of the pancreas and the lesion(s) generated by a semi-automatic framework [7]. The histopathological diagnosis for each subject was confirmed by a pancreatic pathologist based on the subsequent resection. The grade of dysplasia on IMPNs and MCNs was noted on a three-grade scale: LG (IPMNs: 18, MCNs: 11), IG (IPMNs: 29, MCNs: 5), HG (IPMNs: 27, MCNs: 1). Examples of various lesions from this dataset are illustrated in Fig. 1.

Data Augmentation: The data augmentation routine plays an important role in our method. Each image was augmented with random rotations within $\pm 25°$ range, random vertical and horizontal translations within ± 15 pixels range, and random scaling within $\pm 5\%$ range, during both segmentation pre-training and the final classification training stages. More importantly, as padding the activations extracted by the backbone might introduce an additional source for data overfitting (the model might get biased to the original size of the tensor), we perform random volume clippings (± 15 voxels) along Z direction to alleviate such detrimental effects.

Table 1. Comparison of the segmentation accuracy (mean DSC, %) of our *single* pretrained conditioned model against previous works evaluated on the same or similar datasets. (Note that outperforming SOTA results was not the goal of this paper.)

Method	LiTS [3]		KiTS19 [12]		Misc.	
	Liver	Tumor	Kidneys	Tumor	Pancreas	Lesion
Ours	83.53	65.45	81.48	75.69	66.05	85.93
Myronenko et al. [31]	-	-	97.42	81.03	-	-
Isensee et al. [16]	95.8	72.5	-	-	-	-
Zhou et al. [34]	-	-	-	-	86.65	63.94

Pre-training: During the initial segmentation pre-training step, our backbone network $\mathcal{F}_\theta^1$ and the additional decoder were trained to minimize a loss function, based on the common Dice Similarity Coefficient (DSC) metric, $\mathcal{L}(M, \hat{M}) = 1 - \frac{2 \sum M \odot \hat{M}}{\sum M + \sum \hat{M}}$. We use AdaBound optimizer [28] with the initial rate of $1e-4$, and the final rate of 0.01. Prior to training, we defined a lookup table of conditional values for the decoder for each of the six segmentation classes (i.e., liver, liver tumors, kidneys, etc.) with random values sampled from $[-20, 20]$, as suggested by Dmitriev et al. [8]. A mix of all three datasets was randomly split into training, validation, and testing sets using a 70/10/20 ratio. The results were evaluated using the DSC metric and are presented in Table 1. We additionally compare our results against previous works evaluated on the same or similar datasets. It is important to emphasize that updating the state-of-the-art (SOTA) segmentation results was not our goal during this pre-training step, especially given the decoder architecture deliberately bounded in performance.

Classification Results and Discussion: Our experiments are conducted on the PLD dataset using stratified 5-fold cross-validation with similar type and dysplasia distributions in training and testing folds. Each training fold was further split into training and validation sets using a 90/10 ratio for early stopping to prevent overfitting. Each experiment considered different key parameters. Particularly, we studied the effects of various combinations of input data X_i on the final performance, namely, the raw CT scan I^i, binary masks of the pancreas M_p^i and the lesion(s) M_l^i, and the age and gender of the patient. Additionally, we compared the impact of segmentation pre-training on classification performance. In each experiment, the model was trained to minimize the joint class-balanced cross-entropy loss (for lesion type and dysplasia) using Adam optimizer [21] with the initial learning rate of $1e - 4$. We recorded the overall accuracy of predicting lesion type, the associated grade of dysplasia on 4-point (LG, IG, HG, NA) and 3-point (LG or IG, HG, NA) scales, and both (lesion type and dysplasia together). The final classification results are reported in Table 2.

Table 2. Classification performance comparison across different experiments and competitive models. Different inputs X^i to our model are in parenthesis: raw CT scan; binary lesion mask, M_p; patient's age and gender, AG.

	Type	Dysplasia (4-point scale)	Both	Dysplasia (3-point scale)	Both
Li et al. [25]	72.80%	-	-	-	-
Dmitriev et al. [9]	83.60%	-	-	-	-
Ours (CT)	48.23%	30.49%	19.15%	47.52%	27.66%
Ours (CT + M_p)	85.81%	71.63%	61.70%	72.34%	62.41%
Ours (CT + M_p + AG)	**86.52%**	**75.17%**	**67.37%**	**75.88%**	**68.08%**
Ours (CT + M_p + AG) without pre-training	73.75%	59.57%	35.46%	62.41%	38.29%

The model trained only on the raw CT scans performed poorly and did not surpass the 50% accuracy mark for predicting lesion's type, and performed even worse on predicting the grade of dysplasia. A likely explanation of such performance is the model's inability to deduct the target structure for classification, namely, pancreatic lesions. To bring attention of the model to the pancreas and its lesions, we experimented with augmenting the raw CT scan with binary masks of the pancreas M_p^i, lesion(s) M_l^i, and both, as a second channel. This simply helps to bring the attention of the model to the pancreas and pancreatic lesions within the CT scan rather than to "mask out" areas outside of the mask. We observed significant, yet compatible, performance improvements in each experiment, and the best results were achieved using only binary masks of the lesions. Finally, we examined if these results could be improved by feeding the age and gender of the patient to the classification head, as these demographical features are considered important diagnostic clues [10]. In this experiment, we slightly increased the number of units in the first linear layer of the classification head from 128 to 130, to accommodate the extra input, and combined the activations generated by the backbone with patient's age and gender, which were normalized and binarized, respectively. We observed an additional improvement in the performance, outperforming previous solutions [9] by 2.92% in predicting the type, while also predicting the grade of dysplasia with >67% accuracy.

Additionally, we examined the importance of our holistic diagnostic approach, namely, analysis of the entire CT volume without splitting it into smaller pieces and identification of all clinical findings, such as extrapancreatic abnormalities. In our approach, the latter was encouraged through pre-training the backbone $\mathcal{F}_\theta^1$ to detect and segment various lesions, such as renal and hepatic tumors. To evaluate the importance of this step, we compared the performance of the pre-trained model and the model trained from scratch, using the raw CT scan, a binary mask of the lesion, and the age and gender of the patient as the input data. The results are reported in Table 2. Notably, model ability to diagnose the lesion type was affected significantly less than its ability to diagnose the grade of dysplasia. We believe this can be attributed to the model being able to learn to utilize demographical and radiological features of the lesions to correctly predict their type, but not being able to reliably identify additional diagnostic clues outside of the pancreas to diagnose the grade of dysplasia. The results of the experiment support the importance of our holistic diagnostic approach.

Finally, we conducted another experiment to assess and compare the benefits of our proposed adaptive padding of the activation maps against the traditional padding of input data. Particularly, we measured the differences in the memory requirements and computation time required to complete one forward and backward pass between these two approaches. For this experiment, we set $Z_{max} = 12$, which equals to $Z_i = 16 \times 12 = 192$ for traditional padding, and we fixed the number of samples per batch $N = 5$ (batching of the padded activation maps passed to the classification head vs. batching of the padded original input volumes). All experiments were performed on NVidia RTX6000. The proposed method required 2.3 GB of GPU memory as opposed to 6.1 GB when padding the original volumes. While the proposed method was 56.8% slower in this experiment, one can speed up the processing with larger N. The same approach might be impossible when using the traditional padding technique as the memory requirements will increase significantly, especially for very large volumes (e.g., $192 \times 512 \times 512$).

4   Conclusion

We presented a simple, yet effective, method of adaptive padding of intermediate activations for processing intact imaging data of varying size with convnets with linear layers, and an efficient training procedure for such a setup. Our method applied for a holistic diagnosis of the pancreatic lesion showed improved performance, while also predicting the associated grades of dysplasia. Despite extensive evaluation and experiments with various sources of input data, our study has several limitations. Specifically, a loss function which explicitly addresses the fact that benign lesions do not have associated grade of dysplasia could be an interesting direction for future work. Furthermore, potential performance gains could be achieved by using or combining images of different modalities. Additionally, a large, potentially multi-center, clinical study is needed to verify the robustness of our system.

Acknowledgments. This research was supported in part by NSF grants NRT1633299, CNS1650499, OAC1919752, and ICER1940302.

References

1. Cancer Facts & Figures. American Cancer Society (2020)
2. Basturk, O., Hong, S.M., et al.: A revised classification system and recommendations from the Baltimore consensus meeting for neoplastic precursor lesions in the pancreas. Am. J. Surg. Pathol. **39**(12), 1730–1741 (2015)
3. Bilic, P., et al.: The liver tumor segmentation benchmark (LiTS). arXiv preprint arXiv:1901.04056 (2019)
4. Buerke, B., Domagk, D., Heindel, W., Wessling, J.: Diagnostic and radiological management of cystic pancreatic lesions: important features for radiologists. Clin. Radiol. **67**(8), 727–737 (2012)
5. Chen, T., et al.: Multi-view learning with feature level fusion for cervical dysplasia diagnosis. In: Shen, D., et al. (eds.) MICCAI 2019. LNCS, vol. 11764, pp. 329–338. Springer, Cham (2019). https://doi.org/10.1007/978-3-030-32239-7_37
6. Del Chiaro, M., et al.: Main duct dilatation is the best predictor of high-grade dysplasia or invasion in intraductal papillary mucinous neoplasms of the pancreas. Ann. Surg. (2019)
7. Dmitriev, K., Gutenko, I., Nadeem, S., Kaufman, A.: Pancreas and cyst segmentation. In: Medical Imaging 2016: Image Processing, vol. 9784, p. 97842C (2016)
8. Dmitriev, K., Kaufman, A.E.: Learning multi-class segmentations from single-class datasets. In: Proceedings of IEEE Conference Computer Vision Pattern Recognition, June 2019
9. Dmitriev, K., et al.: Classification of pancreatic cysts in computed tomography images using a random forest and convolutional neural network ensemble. In: Descoteaux, M., Maier-Hein, L., Franz, A., Jannin, P., Collins, D.L., Duchesne, S. (eds.) MICCAI 2017. LNCS, vol. 10435, pp. 150–158. Springer, Cham (2017). https://doi.org/10.1007/978-3-319-66179-7_18
10. Farrell, J.J., Fernández-del Castillo, C.: Pancreatic cystic neoplasms: management and unanswered questions. Gastroenterology **144**(6), 1303–1315 (2013)

11. He, K., Zhang, X., Ren, S., Sun, J.: Deep residual learning for image recognition. In: Proceedings of IEEE Conference Computer Vision Pattern Recognition, pp. 770–778 (2016)
12. Heller, N., et al.: The KiTS19 challenge data: 300 kidney tumor cases with clinical context, CT semantic segmentations, and surgical outcomes. arXiv preprint arXiv:1904.00445 (2019)
13. Hu, H., Li, K., Guan, Q., Chen, F., Chen, S., Ni, Y.: A multi-channel multi-classifier method for classifying pancreatic cystic neoplasms based on ResNet. In: International Conference on Artificial Neural Networks, pp. 101–108 (2018)
14. Hussain, M.A., Hamarneh, G., Garbi, R.: ImHistNet: learnable image histogram based DNN with application to noninvasive determination of carcinoma grades in CT scans. In: Shen, D., et al. (eds.) MICCAI 2019. LNCS, vol. 11769, pp. 130–138. Springer, Cham (2019). https://doi.org/10.1007/978-3-030-32226-7_15
15. Hussein, S., Kandel, P., Bolan, C.W., Wallace, M.B., Bagci, U.: Lung and pancreatic tumor characterization in the deep learning era: novel supervised and unsupervised learning approaches. IEEE Trans. Med. Imaging 38(8), 1777–1787 (2019)
16. Isensee, F., Petersen, J., Kohl, S.A., Jäger, P.F., Maier-Hein, K.H.: nnU-Net: breaking the spell on successful medical image segmentation. arXiv preprint arXiv:1904.08128 (2019)
17. Jiménez-Sánchez, A., et al.: Medical-based deep curriculum learning for improved fracture classification. In: Shen, D., et al. (eds.) MICCAI 2019. LNCS, vol. 11769, pp. 694–702. Springer, Cham (2019). https://doi.org/10.1007/978-3-030-32226-7_77
18. Kanayama, T., et al.: Gastric cancer detection from endoscopic images using synthesis by GAN. In: Shen, D., et al. (eds.) MICCAI 2019. LNCS, vol. 11768, pp. 530–538. Springer, Cham (2019). https://doi.org/10.1007/978-3-030-32254-0_59
19. Kawamoto, S., Horton, K.M., Lawler, L.P., Hruban, R.H., Fishman, E.K.: Intraductal papillary mucinous neoplasm of the pancreas: can benign lesions be differentiated from malignant lesions with multidetector CT? RadioGraphics 25(6), 1451–1468 (2005)
20. Khashab, M.A., et al.: Tumor size and location correlate with behavior of pancreatic serous cystic neoplasms. Am. J. Gastroenterol. 106(8), 1521–1526 (2011)
21. Kingma, D.P., Ba, J.: Adam: a method for stochastic optimization. arXiv preprint arXiv:1412.6980 (2014)
22. Kong, B., Wang, X., Li, Z., Song, Q., Zhang, S.: Cancer metastasis detection via spatially structured deep network. In: Niethammer, M., et al. (eds.) IPMI 2017. LNCS, vol. 10265, pp. 236–248. Springer, Cham (2017). https://doi.org/10.1007/978-3-319-59050-9_19
23. Kowalski, T., et al.: Management of patients with pancreatic cysts: analysis of possible false-negative cases of malignancy. J. Clin. Gastroenterol. 50(8), 649 (2016)
24. LaLonde, R., et al.: INN: inflated neural networks for IPMN diagnosis. In: Shen, D., et al. (eds.) MICCAI 2019. LNCS, vol. 11768, pp. 101–109. Springer, Cham (2019). https://doi.org/10.1007/978-3-030-32254-0_12
25. Li, H., et al.: Differential diagnosis for pancreatic cysts in CT scans using densely-connected convolutional networks. arXiv preprint arXiv:1806.01023 (2018)
26. Li, Y., Ping, W.: Cancer metastasis detection with neural conditional random field. arXiv preprint arXiv:1806.07064 (2018)

27. Liang, D., et al.: Combining convolutional and recurrent neural networks for classification of focal liver lesions in multi-phase CT images. In: Frangi, A.F., Schnabel, J.A., Davatzikos, C., Alberola-López, C., Fichtinger, G. (eds.) MICCAI 2018. LNCS, vol. 11071, pp. 666–675. Springer, Cham (2018). https://doi.org/10.1007/978-3-030-00934-2_74

28. Luo, L., Xiong, Y., Liu, Y., Sun, X.: Adaptive gradient methods with dynamic bound of learning rate. In: Proceedings of ICLR, May 2019

29. Luo, L., et al.: Deep angular embedding and feature correlation attention for breast MRI cancer analysis. In: Shen, D., et al. (eds.) MICCAI 2019. LNCS, vol. 11767, pp. 504–512. Springer, Cham (2019). https://doi.org/10.1007/978-3-030-32251-9_55

30. Manvel, A., Vladimir, K., Alexander, T., Dmitry, U.: Radiologist-level stroke classification on non-contrast CT scans with deep U-Net. In: Shen, D., et al. (eds.) MICCAI 2019. LNCS, vol. 11766, pp. 820–828. Springer, Cham (2019). https://doi.org/10.1007/978-3-030-32248-9_91

31. Myronenko, A., Hatamizadeh, A.: 3D kidneys and kidney tumor semantic segmentation using boundary-aware networks. arXiv preprint arXiv:1909.06684 (2019)

32. Wei, R., et al.: Computer-aided diagnosis of pancreas serous cystic neoplasms: a radiomics method on preoperative MDCT images. Technol. Cancer Res. Treat. 18 (2019)

33. Zhao, Z., Lin, H., Chen, H., Heng, P.-A.: PFA-ScanNet: pyramidal feature aggregation with synergistic learning for breast cancer metastasis analysis. In: Shen, D., et al. (eds.) MICCAI 2019. LNCS, vol. 11764, pp. 586–594. Springer, Cham (2019). https://doi.org/10.1007/978-3-030-32239-7_65

34. Zhou, Y., et al.: Hyper-pairing network for multi-phase pancreatic ductal adenocarcinoma segmentation. In: Shen, D., et al. (eds.) MICCAI 2019. LNCS, vol. 11765, pp. 155–163. Springer, Cham (2019). https://doi.org/10.1007/978-3-030-32245-8_18

Feature-Enhanced Graph Networks for Genetic Mutational Prediction Using Histopathological Images in Colon Cancer

Kexin Ding[1], Qiao Liu[2], Edward Lee[3], Mu Zhou[4], Aidong Lu[1],
and Shaoting Zhang[5(✉)]

[1] Department of Computer Science, UNC Charlotte, Charlotte, NC, USA
[2] Department of Automation, Tsinghua University, Beijing, China
[3] Department of Radiology, Stanford University, Stanford, CA, USA
[4] SenseBrain Research, Santa Clara, CA, USA
[5] SenseTime Research, Shanghai, China
szhang16@uncc.edu

Abstract. Mining histopathological and genetic data provides a unique avenue to deepen our understanding of cancer biology. However, extensive cancer heterogeneity across image- and molecular-scales poses technical challenges for feature extraction and outcome prediction. In this study, we propose a feature-enhanced graph network (FENet) for genetic mutation prediction using histopathological images in colon cancer. Unlike conventional approaches analyzing patch-based feature alone without considering their spatial connectivity, we seek to link and explore non-isomorphic topological structures in histopathological images. Our FENet incorporates feature enhancement in convolutional graph neural networks to aggregate discriminative features for capturing gene mutation status. Specifically, our approach could identify both local patch feature information and global topological structure in histopathological images simultaneously. Furthermore, we introduced an ensemble strategy by constructing multiple subgraphs to boost the prediction performance. Extensive experiments on the TCGA-COAD and TCGA-READ cohort including both histopathological images and three key genes' mutation profiles (APC, KRAS, and TP53) demonstrated the superiority of FENet for key mutational outcome prediction in colon cancer.

Keywords: Histopathological image analysis · Graph convolutional networks · Gene mutation prediction

1 Introduction

Colon cancer [1,2] is the third common cancer worldwide that accounts for 13% of all new cancer incidence and approximately 8% of all cancer deaths. Especially

This study has been partially supported by fund of STCSM (19511121400).

A. L. Martel et al. (Eds.): MICCAI 2020, LNCS 12262, pp. 294–304, 2020.
https://doi.org/10.1007/978-3-030-59713-9_29

colon adenoma and carcinoma are known to occur through a series of histopathological changes due to key genetic alterations [3–5]. Therefore, the prediction of genetic mutations over the course of cancer evolution is highly desired towards accurate detection and diagnosis of colon cancer.

Mining histopathological and genetic profiles provides a unique avenue to deepen our understanding of cancer biology across scales. Particularly, genetic mutations play key roles in colon cancer evolution at all clinical stages (Fig. 1). For example, KRAS gene mutation has been proven to be an independent prognostic factor in patient with advanced colon cancer [5]. Also, APC gene mutation triggers chain of molecular and histological changes, leading to increased growth of colon cancer cells [6]. In parallel, whole slide images (WSIs) permit extensive cell-level characterization of individual patients. Thus increasing emphasis has been placed on extracting quantitative features from WSIs for outcome assessment [7,8]. Preliminary evidences suggested the usefulness of quantitative features extracted from large-scale image patches [9–13]. However, conventional approaches were merely focused on image patches and thus unable to consider rich topological structures as shown in WSIs. Especially local and spatial connectivity of image-based findings, critical to understand characteristics of cancer heterogeneity [14], has not been explicitly analyzed.

To explore the topological structure of WSIs, we introduce a graph-based analysis with a goal to capture both spatial and local histopatholgical variations. Specifically, we focus on analyzing spatial-based graph convolutional networks (Spatial-GCNs) due to their advances in network flexibility [15–17]. For example, spatial-GCNs allows convolutional operations locally on each node with weight sharing across locations and structures [18]. Thus it is more convenient to integrate node features with their neighborhood information in spatial-GCNs, compared with Spectral-based graph convolutional networks (Spectral-GCNs) that commonly requires entire graph Laplacian embedding. In addition, spatial-GCNs offers mechanisms of aggregators for feature integration. For example, GraphSage [15], GCN [19] and GIN [17] models demonstrate their learning ability using max-, mean-, or sum-pooling aggregators. However, it remains uncertain about how can we implement an efficient graph structure to characterize histopathological images, especially for the classification of non-isomorphic topological structures in our study.

In this study, we proposed the feature-enhanced graph network (FENet), as a novel graph-driven approach for gene mutation prediction in patients with colon cancer (Fig. 1). To tackle the challenge of cancer heterogeneity, our FENet model consists of multiple subgraphs construction from histopathological image patches. Conceptually, FENet can be considered as an ensemble learning solution for predicting genetic signal variations with enhanced feature integration. Our major contributions are summarized as follows.

- We introduced an efficient transformation method between WSIs and graph structured data. Focusing on generating spatially-connected graphs, our approach can link and explore local feature and global topological structure of WSIs simultaneously.

– We developed a feature-enhanced model to underscore discriminative feature learning. In this architecture, we improved the ability of distinguishing non-isomorphic topological structure, and adaptively selected the node representations from the different graph convolution layers.
– Our ensemble strategy of network models alleviates cancer heterogeneity so that integration of multiple subgraph outcomes leads to a significant improvement of prediction performance.

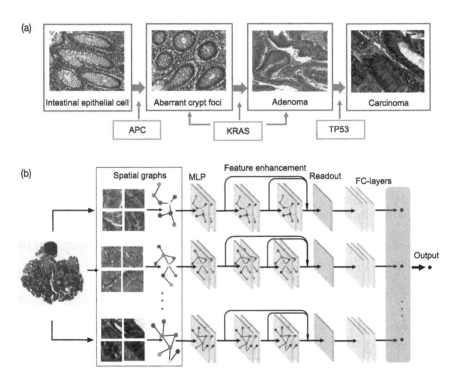

Fig. 1. (a) Illustration of colon cancer evolution in histopathology over time and its key genetic mutations. (b) the proposed FENet networks architecture.

2   Methodology

Overview of FENet Model. Our FENet is constructed by multiple subgraphs using image patches from each patient's WSI. We particularly underscored the spatially-connected subgraph construction strategy that was seldom addressed in prior studies [9–13]. To increase feature learning performance, we then described our feature-enhanced mechanism to aggregate the features of neighboring patches and combine the aggregated feature as the updated central node representation. Moreover, our feature-enhanced mechanism could adaptively select the node representations from different graph convolution layers. Finally, an ensemble strategy was introduced to combine the prediction results of subgraphs for predicting important mutation statuses (i.e., mutated and non-mutated classes).

Spatially-Connected Subgraph Construction. For each whole slide image (WSI), we randomly selected a set of sampled patches $P = \{P_1, P_2, \ldots, P_N\}$ from all patches generated from WSIs (N is the number of patches). Random selection always maintains the distribution of the original feature underlying WSI, which provides a comprehensive description of WSIs. In our study, we defined patches as graph nodes in each subgraph, and the spatial distance between two patches determines whether there exists a graph edge. We emphasized the analysis of non-isomorphic graphs because informative image patches caused by cancer mutation can be always in various spatial locations in WSIs [14]. This property is opposed to isomorphic graphs which strictly share the same adjacency neighborhood matrix. Therefore, we first used a pre-trained ResNet18 model to extract high-level features (node attribute) of all individual patches within a subgraph. We constructed graph representation for each subgraph by calculating its adjacent matrix with criterion that the edge exists if the spatial distance is below a fixed threshold. Precisely, we measured the spatial distance by directly calculating the Euclidean distance of two patches mapped on the original WSI. Algorithm 1 describes the details of the subgraph construction.

Algorithm 1. Spatially-connected Subgraph Construction Algorithm

---

Input: A set of selected patches in one slide $P = P_1, P_2, ..., P_N$; a set of top-left corresponding coordinates of the selected patches $C_{p_i} = (x_1, y_1), (x_1, y_1), ..., (x_N, y_N)$;
$\quad$ **N** is the number of patches.
$\quad$ Threshold **t** is the distance
Output: Adjacency Matrix **A** and Feature Matrix **X**

1: **function** CONSTRUCTING(A, P, C_{p_i}, N, t)
2: $\quad$ $X \leftarrow$ Feature_extractor(P)
3: $\quad$ **for** $i \leftarrow 1$ to N **do**
4: $\quad\quad$ $(x_j, y_j) \leftarrow c_{p_j}$
5: $\quad\quad$ $s_{ij} \leftarrow \sqrt{(x_i - x_j)^2 + (y_i - y_j)^2}$
6: $\quad\quad$ **if** $s_{ij} < t$ **then**
7: $\quad\quad\quad$ $A_{ij} \leftarrow 1$
8: $\quad\quad$ **end if**
9: $\quad$ **end for**
10: $\quad$ **return** A, X
11: **end function**

Feature-Enhanced Mechanism. Spatial-based convolutional networks utilize a neighborhood aggregation strategy that iteratively updates the representation of a central node by AGGREGATE and COMBINE [15,16]. The AGGREGATE can aggregate neighboring node representations of the center node, while the COMBINE combines the neighborhood node representation with the center node representation to obtain the updated center node representation. After k iterations of aggregation and combination, a node's representation captures the structural information within its k-hop network neighborhood. Formally, the kth layer spatial-based graph convolutional network can be represented as

$$a_v^{(k)} = \text{AGGREGATE}(\{h_u^{(k-1)} : u \in \mathcal{N}(v)\}) \tag{1}$$

$$h_v^{(k)} = \text{COMBINE}(h_{v-1}^{(k)}, a_v^{(k)}) \tag{2}$$

Where $h_v^{(k)}$ is the feature vector of node v at the kth layer. $\mathcal{N}(v)$ is a set of nodes adjacent to v. As illustrated in [17], the selection of AGGREGATE and COMBINE is critical to capture the graph topological structures. The traditional aggregation strategy using max-pooling (GraphSage) considers multiple node representations as one node representation while ignoring multiplicities. Alternatively, the mean-pooling (GCN) captures the statistical and distributional information within the graph rather than the exaction of the entire graph structure. These measurements can be useful if the graphs are isomorphic (i.e., with a strictly same adjacency matrix). However, the topological structure of subgraphs are non-isomorphisms in our study, improving the ability to distinguish topological non-isomorphism is a critical issue. As proved in [17], sum-pooling captures the full structural information of the entire graph representation. Additionally, we use multi-layer perceptrons (MLPs) with Rectified Linear Unit (ReLU) and batch normalization [20] to model in each graph neural network layers. Therefore, FENet updates node feature representation $h_v^{(k)}$ as:

$$h_v^{(k)} = \text{MLP}^{(k)}(h_{v-1}^{(k)} + \sum_{u \in \mathcal{N}(v)} h_u^{(k-1)}) \tag{3}$$

It is known that nodes in the central and boundary regions of a graph requiring different frequencies of aggregation to achieve optimal performance [21]. Our WSI-based graph contains both central and marginal nodes since each WSI has a unique spatial distribution of cancerous regions. To consider all topological structural information, we emphasized the useful information from all depths of network layers. Therefore, we aggregated node representation from each previous layer to the last layer. By this design, the network could adaptively select the most meaningful information during the training process and find the desired representation for each node. To achieve the subgraph classification, we introduced the READOUT function to converts node representations into a graph representation. To guarantee that each part in the network is injective, we also selected GlobalAddPooling as a READOUT function to aggregate information from all nodes and to convert it to the entire graph representation. Formally,

FENet aggregates layers information and converts node representation to graph representation h_G by:

$$h_G = \text{READOUT}(h_v^1 \| h_v^2 \| ... \| h_v^k, v \in G) \tag{4}$$

where $\|$ is the feature concatenation. The feature concatenation could aggregate node representation in different graph convolutional layers. After READOUT representation, the final prediction outcome for each subgraph is classified by fully-connected layers.

Ensemble Strategy. Our ensemble strategy utilized the majority vote to aggregate all subgraphs' prediction outcomes derived from the same WSI. The ensemble strategy is motivated by two major findings. First, since a WSI can contain a large number of patches (e.g., 10k) that allows us to explore the diversity of WSI characteristics via individual subgraphs. Second, analyzing individual patches creates a sizeable computational burden as shown in conventional CNN-based approaches. The ensemble strategy, therefore, achieves a good trade-off between informative representation of WSI and computational burden. We highlighted that ensemble learning allowed FENet to increase its generalization power by exploiting the advance of multiple spatial subgraphs of which the predictive error can be reduced by majority vote.

3 Experiments and Results

Dataset. We collected whole slide images (WSIs) from The Cancer Genome Atlas Colon Adenocarcinoma (TCGA-COAD) dataset [22], which contains 421 WSIs with colon tumors. We identified the associated colon cancer genetic mutational profiles from Cbioportal [23]. Data exclusion criteria included that we removed 40 WSIs with noisy stained annotation. We also removed patient data without key mutational profiles. We finally analyzed the total number of 274 patients' WSIs with a resolution of 40X (0.25 microns/pixel). For each type of mutational profile in each patient, we assigned the outcome label as a positive class if mutated and as a negative class if non-mutated. We focus on three key genes (APC, KRAS, and TP53) that are significantly associated with colon cancer evolution over various clinical stages with treatment impact [22]. We found that 70% of samples contain APC mutation, 60% of samples contain TP53 mutation, and 40% of samples contain KRAS mutation. We also collected 30 WSIs from The Cancer Genome Atlas Rectum Adenocarcinoma (TCGA-READ) as an external dataset. We found that 75% of samples contain APC mutation, 70% of samples contain TP53 mutation, and 36% of samples contain KRAS mutation.

Experimental Settings and Implementations. In data preprocessing, all slides are color-normalized by Macenko's method [24], and the foreground is segmented using OTSU [25]. To obtain a tumor region in WSI, we trained a tumor detection model by performing a pre-trained ResNet18 on the NCT-CRC-HE-100K dataset [26]. A fault-tolerant tumor region delineation is acceptable due to our patch-focused analysis. We then generated non-overlapping patches with a size of 224 * 224 that was resized from raw 512 * 512 patches within the tumor region. We then randomly generated five different subsets of patches to build subgraphs. The number of selected patches in each subset was set to 1,000 due to the consideration of computational efficiency [27].

For graph construction, we used a pre-trained ResNet18 to extract features from patches as the feature matrix. For the adjacency matrix, we calculated the Euclidean distance between patches' coordinate values recorded in their raw WSI and determine the connection by a fixed spatial distance threshold. Precisely, we calculated the mode value from the statistical distribution of the spatial Euclidean distances among all pairs of patches to determine the threshold value. Finally, the feature matrix and the adjacency matrix were used for the non-isomorphic graph representation of WSI.

In the experiments, we evaluated our FENet with multiple competing methods on TCGA-COAD cohort. The performances are achieved from 10-fold cross-validation and reported the mean accuracy and AUC among 10 times experiments for each prediction task. We guarantee patches in the same WSI will not appear in different sets simultaneously. Moreover, to show the generalization power, we trained FENet on TCGA-COAD and tested it on TCGA-READ. To facilitate model training, we augmented patch samples from the minority class in the training set to balance the number of positive and negative samples. Notably, we always kept the real positive and negative ratios at the testing stage. In the training set, we dropped a fraction of edges (drop rate = 0.3) to reduce potential overfitting as commonly done in [28]. The optimal hyperparameters were obtained by a grid search. For a fair comparison, we always used the same hyperparameters setting in all experiments to ensure differences only come from the variants of the methods. All models were trained with initial learning rate $1e-3$, batch size 64 by Adam optimizer [29] with a weight decay $5e-3$ and the cross-entropy loss.

Method Comparison and Ablation Study. We compared our approach with multiple state-of-the-art models. GCN [19] is a spectral-GCNs with mean-pooling aggregation. GraphSAGE [15] is a spatial-GCNs model applying node embedding with max-pooling aggregation. GAT [18] is a spatial-GCNs model that leverages masked self-attentional layers to graph convolutions. We used VGG16 [31] and ResNet18 [32] as a baseline CNN-based model. Additionally, we designed several ablation studies. First, we introduced the Perceptron-3 model by replacing model's MLPs by 1-layer perceptrons with sum-pooling as the aggregator. Meanwhile, Perceptron-3 comes without ensemble strategy and the operation feature concatenation (i.e., aggregating node representation from each layer to the last layer). Second, MLP-n model removes both the ensemble strategy and feature concatenation in the FENet-n, where n is the number of MLP in the model. Third, the FENet-n w/o ensemble model keeps all components in the

FENet (e.g., feature concatenation, MLPs and sum-pooling) except the ensemble strategy. Finally, the FENet-3 is our proposed approach with full components.

Table 1. The prediction performance of mean accuracy and variance of FENet with competing approaches and ablation studies.

Network architecture	Network type	TP53	KRAS	APC
VGG16 [30]	Deep CNN	59.89 ± 0.84	58.28 ± 1.12	56.36 ± 0.83
ResNet18 [31]	Deep CNN	62.49 ± 2.22	61.33 ± 2.51	70.86 ± 0.18
GCN [19]	Spectral-based	60.98 ± 0.71	62.21 ± 0.89	70.85 ± 0.24
ChebNet [32]	Spectral-based	60.47 ± 0.86	61.62 ± 0.93	68.70 ± 0.76
GraphSAGE [15]	Spatial-based	60.31 ± 0.95	63.98 ± 1.27	71.39 ± 1.14
GAT [16]	Spatial-based	62.47 ± 0.86	60.47 ± 1.42	68.79 ± 0.92
Perceptron-3	Ablation analysis	64.50 ± 0.25	65.75 ± 0.62	68.89 ± 0.68
MLP-2	Ablation analysis	66.03 ± 0.36	64.21 ± 0.29	70.37 ± 0.25
FENet-2 w/o essemble	Ablation analysis	65.45 ± 0.56	67.57 ± 0.52	71.72 ± 0.22
MLP-3	Ablation analysis	65.63 ± 0.14	66.29 ± 0.08	71.77 ± 0.21
FENet-3 w/o essemble	Ablation analysis	67.44 ± 0.36	69.87 ± 0.25	73.18 ± 0.31
FENet-3	Our method	**77.00 ± 0.01**	**80.38 ± 0.01**	**79.93 ± 0.01**

Results and Analysis. In Table 1, our FENet demonstrated leading performance across all three key mutational prediction tasks. Even viewing our perceptron-3 model without ensemble learning or feature concatenation, it is remarkable that Perceptron-3 was quite competitive to the graph-based baseline models (e.g., GCN, GraphSAGE and GAT). This key finding can be ascribed to the contribution of our design of sum-pooling aggregator followed by [17], leading to improved ability to distinguish different non-isomorphic topological structures. Notably, our task is challenging because traditional CNN-based approaches working on patches alone may not be able to perform well (e.g., VGG16). In Fig. 2, we observed that the feature concatenation performed strongly for improving the performance by comparing MLP-3 model and FENet-3 without ensemble. A similar finding was confirmed by viewing MLP-2 model and FENet-2 without ensemble. Furthermore, we recognized the boosted performance with our ensemble strategy on all three prediction tasks (Fig. 2). Overall, the ensembled strategy greatly increased the generalization power of our FENet probably due to its ability to capture diverse topological information of WSI derived from our multiple spatially-connected subgraphs. Besides the superior performance on TCGA-COAD cohort, the proposed model also showed a relatively good performance on TCGA-READ. The test accuracy of TP53 on TCGA-READ is 0.7325, and AUC is 0.7522. Both test accuracy and AUC of KRAS and APC are higher than 0.6. The comparative performance on TCGA-READ shows a potential generalization power of the proposed model.

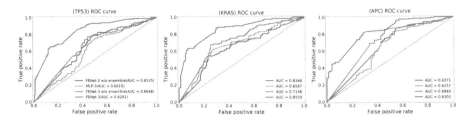

Fig. 2. The ROC-curve comparison of our approach with or without ensemble strategy. FENet with ensemble strategy brings the boosted performance over other methods.

4   Conclusion

In this study, we have proposed the FENet to tackle the challenging problem of predicting gene mutational status in colon cancer using histopathological images. Our findings supported that the convergence of multi-scale cancer data leads to novel insights into modeling cancer biology. To address the cancer heterogeneity, we highlighted the importance of exploring spatial and local connections of image patches via graph construction, which offers an efficient means for important molecular outcome prediction. Extensive experiments suggested that our model with a subgraph ensemble strategy outperformed current state-of-the-art approaches. In the future work, we plan to perform and validate a large-scale of our analysis across different types of cancers and thus gain more insights into multi-scale cancer data integration.

References

1. Mármol, I., Sánchez-de-Diego, C., Pradilla, D.A.: Colorectal carcinoma: a general overview and future perspectives in colorectal cancer. Int. J. Mol. Sci. **18**, 197 (2017)
2. Bray, F.: GLOBOCAN estimates of incidence and mortality worldwide for 36 cancers in 185 countries. CA Cancer J. Clin. **68**, 394–424 (2018)
3. Iacopetta, B.: TP53 mutation in colorectal cancer. In: Human Mutation (2003)
4. Armaghany, T., Wilson, J.D., Chu, Q., Mills, G.: Genetic alterations in colorectal cancer. Gastrointest. Cancer Res. **5**, 19–27 (2012)
5. Jancik, S., Drabek, J., Radzioch, D., Hajduch, M.: Clinical relevance of KRAS in human cancers. J. Biomed. Biotechnol. **2010**, 150960 (2010)
6. Fodde, R.: The APC gene in colorectal cancer. Eur. J. Cancer **38**(7), 867–71 (2002)
7. Kather, J.N.: Deep learning can predict microsatellite instability directly from histology in gastrointestinal cancer. Nat. Med. **25**, 1054–1056 (2019)
8. Coudray, N., et al.: Classification and mutation prediction from non-small cell lung cancer histopathology images using deep learning. Nat. Med. **24**(10), 1559–1567 (2018)
9. Li, Z., Zhang, X., Muller, H., Zhang, S.: Large-scale retrieval for medical image analytics: a comprehensive review. Med. Image Anal. **43**, 66–84 (2018)
10. Ghaznavi, F., Evans, A., Madabhushi, A., Feldman, M.: Digital imaging in pathology: whole-slide imaging and beyond. Annu. Rev. Pathol. **8**, 331–359 (2013)

11. Mobadersany, P., et al.: Predicting cancer outcomes from histology and genomics using convolutional networks. Proc. Nat. Acad. Sci. U.S.A. **115**, E2970–E2979 (2018)
12. Zhang, X., Su, H., Yang, L., Zhang, S.: Fine-grained histopathological image analysis via robust segmentation and large-scale retrieval. In: IEEE Conference on Computer Vision and Pattern Recognition (2015)
13. Duan, Q., et al.: SenseCare: a research platform for medical image informatics and interactive 3D visualization. https://arxiv.org/abs/2004.07031
14. Heindl, A., Nawaz, S., Yuan, Y.: Mapping spatial heterogeneity in the tumor microenvironment: a new era for digital pathology. Lab. Investig. **95**, 377–384 (2015)
15. Hamilton, W.L., Ying, R., Leskovec, J.: Inductive representation learning on large graphs. In: Neural Information Processing Systems (2017)
16. Velickovic, P., Cucurull, G., Casanova, A., Romero, A., Lio, P., Bengio, Y.: Graph attention networks. In International Conference on Learning Representations (2018)
17. Xu, K, Hu, W., Leskovec, J., Jegelka, S.: How powerful are graph neural networks. In: International Conference on Learning Representations (2019)
18. Wu, Z., Pan, S., Chen, F., Long, G., Zhang, C., Yu, P.S.: A comprehensive survey on graph
19. Kipf, T.N., Welling, M.: Semi-supervised classification with graph convolutional networks. In: International Conference on Learning Representations (2016)
20. Ioffe, S., et al.: Batch normalization: accelerating deep network training by reducing internal covariate shift. In: International Conference on Machine Learning (2015)
21. Xu, K., Li, C., Tian, Y., Sonobe, T., Kawarabayashi, K., Jegelka, S.: Representation learning on graphs with jumping knowledge networks. In: International Conference on Machine Learning (2018)
22. Kirk, S., et al.: Radiology data from the cancer genome atlas colon adenocarcinoma [TCGA-COAD] collection. In: The Cancer Imaging Archive. https://doi.org/10.7937/K9/TCIA.2016.HJJHBOXZ
23. Cbioportal Homepage. https://www.cbioportal.org/
24. Macenko, M., et al.: A method for normalizing histology slides for quantitative analysis. In: Proceedings of IEEE International Symposium on Biomedical Imaging (2011)
25. Otsu, N.: A threshold selection method from gray-level histogram. IEEE Trans. Syst. Man Cybern. B Cybern. **9**, 62–66 (1979)
26. Kather, J.N., Halama, N, Marx, A.: 100,000 histological images of human colorectal cancer and healthy tissue (Version v0.1) [Data set]. Zenodo (2018). https://doi.org/10.5281/zenodo.1214456
27. Li, R., Yao, J., Zhu, X., Li, Y., Huang, J.: Graph CNN for survival analysis on whole slide pathological images. In: Medical Image Computing and Computer Assisted Intervention (2018)
28. Rong, Y., Huang, W., Xu, T., Huang, J.: DropEdge: towards deep graph convolutional networks on node classification. In: International Conference on Learning Representations (2020)
29. Kingma, D., Jimmy B.: Adam: a method for stochastic optimization. In: International Conference on Learning Representations (2015)
30. Simonyan, K., Zisserman, A.: Very deep convolutional networks for large-scale image recognition. In: International Conference on Learning Representations (2015)

31. He, K., Zhang, X., Ren, S., and Sun, J. : Deep residual learning for image recognition. In: IEEE Conference on Computer Vision and Pattern Recognition (2016)
32. Defferrard, M., Bresson, X., Vandergheynst, P.: Convolutional neural networks on graphs with fast localized spectral filtering. In: Neural Information Processing Systems (2016)

Spatial-And-Context Aware (SpACe) "Virtual Biopsy" Radiogenomic Maps to Target Tumor Mutational Status on Structural MRI

Marwa Ismail[1][(✉)], Ramon Correa[1], Kaustav Bera[1], Ruchika Verma[1],
Anas Saeed Bamashmos[2], Niha Beig[1], Jacob Antunes[1], Prateek Prasanna[3],
Volodymyr Statsevych[4], Manmeet Ahluwalia[2], and Pallavi Tiwari[1]

[1] Case Western Reserve University, Cleveland, OH, USA
mxi125@case.edu
[2] Brain Tumor and Neuro-Oncology Center, Cleveland Clinic, Cleveland, OH, USA
[3] Stony Brook University, New York, USA
[4] Imaging Institute, Cleveland Clinic, Cleveland, OH, USA

Abstract. With growing emphasis on personalized cancer-therapies, radiogenomics has shown promise in identifying target tumor mutational status on routine imaging (i.e. MRI) scans. These approaches largely fall into two categories: (1) deep-learning/radiomics (context-based) that employ image features from the entire tumor to identify the gene mutation status, or (2) atlas (spatial)-based to obtain likelihood of gene mutation status based on population statistics. While many genes (i.e. EGFR, MGMT) are spatially variant, a significant challenge in reliable assessment of gene mutation status on imaging is the lack of available co-localized ground truth for training the models. We present Spatial-And-Context aware (SpACe) "virtual biopsy" maps that incorporate context-features from co-localized biopsy sites along with spatial-priors from population atlases, within a Least Absolute Shrinkage and Selection Operator (LASSO) regression model, to obtain a per-voxel likelihood of the presence of a mutation status (M^+ vs M^-). We then use probabilistic pair-wise Markov model to improve the voxel-wise likelihood. We evaluate the efficacy of SpACe maps on MRI scans with co-localized ground truth obtained from biopsy, to predict the mutation status of 2 driver genes in Glioblastoma (GBM): (1) $EGFR^+$ versus $EGFR^-$, (n = 91), and (2) $MGMT^+$ versus $MGMT^-$, (n = 81). When compared against state-of-the-art deep-learning (DL) and radiomic models, SpACe maps obtained training and testing accuracies of 90% (n = 70) and 90.48% (n = 21) in identifying EGFR amplification status, compared to 80% and 71.4% via radiomics, and 74.28% and 65.5% via DL. For MGMT methylation status, training and testing accuracies using SpACe were 88.3% (n = 60) and 71.5% (n = 21), compared to 52.4% and 66.7% using radiomics, and 79.3% and 68.4% using DL. Following validation, SpACe maps could provide surgical navigation to improve localization of sampling sites for targeting of specific driver genes in cancer.

© Springer Nature Switzerland AG 2020
A. L. Martel et al. (Eds.): MICCAI 2020, LNCS 12262, pp. 305–314, 2020.
https://doi.org/10.1007/978-3-030-59713-9_30

Keywords: Tumor mutation status · Radiogenomics · Spatial prior

1 Introduction

With treatments for solid cancers transitioning towards personalized therapy, mutational profiling for identifying target gene mutation status or gene amplification status using tissue biopsies is becoming *status-quo* for most cancers. However, a significant challenge in reliable assessment of gene mutation or amplification status, is the underlying genomic heterogeneity which makes it challenging to identify the "true" mutational status based on random tissue sampling [1]. Multiple studies have shown that certain gene mutations (e.g. MGMT promoter methylation, EGFR) have varying expressions across different parts of the tumor or between primary or secondary metastatic sites [2,3]. There is hence a need for developing "virtual biopsy" techniques on imaging that can comprehensively capture the gene mutation heterogeneity of solid tumors, and potentially assist in surgical navigation to identify sampling sites for biopsy targeting.

The field of radiogenomics has provided a surrogate mechanism to predict gene mutational status on routine imaging (i.e. MRI) by training machine-learning models. Most of the existing radiogenomic models fall in two categories: (1) deep learning [4–6]/radiomics [7–9], and (2) atlas-based probabilistic approaches [10,11]. In the absence of co-localized biopsy sites on MRI, deep learning/radiomic approaches employ features from the entire tumor to predict the gene mutational status. In contrast, atlas-based approaches obtain the likelihood of the mutational status of driver genes such as MGMT and EGFR at different spatial locations by creating probabilistic radiographic atlases obtained from a large population. These population-based approaches, however, do not leverage any tumor-specific information in the model.

In this work, we present an attempt at creating "virtual biopsy" radiogenomic maps for predicting gene mutational status on MRI, by combining two complementary attributes that capture mutational heterogeneity at: (1) population-level via *spatial-priors* for presence or absence of mutation status (M^+, M^-) using probabilistic atlases from a retrospective cohort, and (2) local tumor-level by incorporating *context-priors* that capture mutational heterogeneity via radiomic attributes obtained from a stereotactically co-localized biopsy site within the tumor. The spatial and context priors are combined within a Least Absolute Shrinkage and Selection Operator (LASSO) regression model to obtain a per-voxel likelihood of increased expression of the gene mutation (M^+, M^-) at that location. The prediction of per-voxel likelihood of the presence of a specific gene mutation is further improved using probabilistic pairwise Markov models. In this work, we evaluate these Spatial and Context Aware (SpACe) maps in the context of two problems in Glioblastoma (GBM): (1) predicting EGFR status (amplified $(EGFR^+)$, non-amplified $(EGFR^-)$), and (2) predicting MGMT status (methylated $(MGMT^+)$, non-methylated $(MGMT^-)$), from routine MRI scans. The pipeline of the entire workflow is illustrated in Fig. 1.

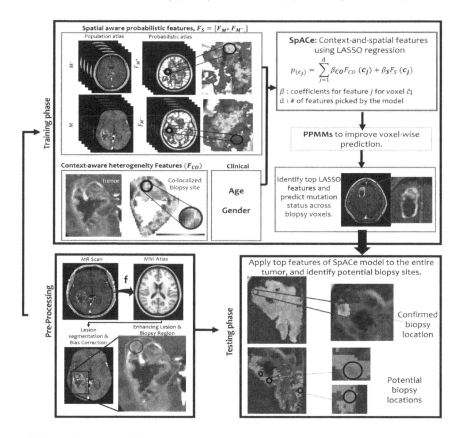

Fig. 1. Overview of the workflow of **SpACe** to create "virtual biopsy" maps.

2   Methods

2.1   Notation

We define an image scene I as $I = (C, f)$, where I is a spatial grid C of voxels $c \in C$, in a 3D space, $\mathbb{R}^3$. Each voxel, $c \in C$ is associated with an intensity value $f(c)$. I_B represents the co-localized biopsy location on MRI scans, such that $I_B \subset I$. $\mathbb{F}(c)$ denotes the feature set obtained for every $c \in C_B$. For gene M, M^+ defines amplified/methylated, while M^- defines non-amplified/unmethylated.

2.2   Computing Context-Aware Mutational Heterogeneity from Stereotactic Biopsy Locations ($\mathbb{F}_{\mathbb{CO}}$)

We define "context" as local heterogeneity attributes computed from the co-localized biopsy site on imaging, using radiomic features including Haralick features (capture image heterogeneity [12]), Gabor features (capture structural details at different orientations and scales), Laws (capture spots and ripples-like patterns), and CoLlAGe features (capture localized gradient orientation changes [13]). Specifically, for every $c \in C_B$, we extract a set of 3D radiomic features (i.e. Haralick, Gabor, Laws, CoLlAGe). We define $\mathbb{F}_\theta^k(c)$, where θ is the type of feature family (e.g. Haralick, Gabor), and $k \in \{1, ..., n\}$, where n is the number of feature attributes for every feature family. Feature pruning is then conducted on the extracted features using Spearman's correlation metric, to eliminate redundant features. The pruned "context-aware" features (152 for EGFR cohort, 149 for MGMT cohort) are finally aggregated into one feature descriptor $\mathbb{F}_{CO}$.

2.3 Computing Spatially-Aware Priors ($\mathbb{F}_S$) for Likelihood of Gene Mutation Status (M^+, M^-) Using Probabilistic Atlases

Using the lesion segmentation obtained for every patient in the training set, two different population atlases for gene M are constructed using subjects that belong to either M^+ or M^-. This is done to quantify the frequency of occurrence of every voxel across M^+ and M^-, and compute voxel-wise probability values, $P_w(c)$, $w \in (M^+, M^-)$. All scans need to first be registered to an isotropic reference atlas (i.e. MNI152; Montreal Neurological Institute). The intensity values are then averaged across $c \in C$ across all the annotated binary images of all patients involved in the study. This means that for $c \in C$, two probability values from these two atlases could be obtained, that characterize the probability of a voxel c being M^+ or M^-. The 2 probability values (P_{M^+}, P_{M^-}) for every voxel $c \in C$ are finally aggregated in the spatial feature descriptor $\mathbb{F}_S = [P_{M^+}, P_{M^-}]$.

2.4 Creating SpACe Maps for Predicting Voxel-Wise Mutational Heterogeneity in the Tumor

In order to obtain a voxel-wise prediction $p(c)$ of the gene mutation status, the context descriptor ($\mathbb{F}_{CO}$), spatial descriptor ($\mathbb{F}_S$), age ($\mathbb{F}_A$), and gender ($\mathbb{F}_G$) of every patient in the training set, are incorporated within a LASSO model [14]. LASSO model is selected to obtain a likelihood score using a parsimonious feature set by utilizing its capability in reducing variance when shrinking features, while simultaneously not increasing the bias. The model optimizes weights for all feature variables simultaneously across cross validation, with enforcing balance of the two classes. We designed the LASSO model to perform regularization of feature parameters as follows: $[\hat{\beta}] = argmin\{|y - F_{CO}\beta_{CO}|^2 + \lambda_{CO}|\beta_{CO}| + |y - F_S\beta_S|^2 + \lambda_S|\beta_S| + |y - F_A\beta_A|^2 + \lambda_A|\beta_A| + |y - F_G\beta_G|^2 + \lambda_G|\beta_G|\}$, where $[\hat{\beta}] = \{\hat{\beta}_1, ..., \hat{\beta}_d\}$ is the shrinked set of d coefficients obtained after regularization, and $y \in [M^+, M^-]$. Feature shrinkage was handled by optimizing the penalty term λ, by minimizing mean cross-validated error across 10 folds. The voxel-wise likelihood is then computed as the weighted sum of the selected features for the set of coefficients $[\hat{\beta}]$, as follows: $p_{SpACe}(c) = \sum_{j=1}^d \hat{\beta}_j \mathbb{F}_j(c)$, where $j \in \{1, ..., d\}$, and d is the number of features selected by LASSO. After obtaining the likelihood map for every $c \in C_B$, we incorporate probabilistic pairwise Markov models (PPMMs) to improve voxel-wise gene mutation prediction [15].

PPMMs are adopted from Markov Random Fields, through formulating priors in terms of probability density functions, hence allowing the creation of more robust prediction models. The input to this model is the voxel-wise likelihood values obtained from LASSO model. Interaction between neighboring sites is then modelled, to improve voxel-wise likelihood scores, and to finally obtain $\mathbb{F}_{SpACe}$ maps.

2.5 Applying SpACe Maps on Testing Sets for Predicting Voxel-Wise Mutational Heterogeneity Within the Tumor

The top features selected on the training set are applied to the entire tumor on the test set, for obtaining voxel-wise likelihood scores for predicting the mutation status. For the purpose of computing accuracy of our model, we predict the mutation status $[M^+, M^-]$ based on pooled likelihood values for an already known biopsy site, and compare the prediction with the known mutation status. As an additional qualitative analysis, we threshold the likelihood values obtained from the entire tumor (threshold obtained empirically), followed by connected component analysis and PPMM, to obtain 2–3 hot-spots of high likelihood mutation sites. These hot-spots prospectively could be used to drive surgical navigation as potential sites for biopsy localization.

3 Experimental Design

3.1 Data Description and Preprocessing

We employed a unique retrospective dataset of a total of 100 GBM patients who underwent CT-guided biopsy for disease confirmation, since surgical resection was not feasible (due to location or other clinical reasons) for these patients. Segmentation of the enhancing lesion was conducted by an experienced radiologist on the MR scans. The biopsy site was co-localized by co-registering CT images with the MRI scans, followed by expert evaluation for confirmation. All scans were then registered to an MNI152 atlas and then bias-corrected using N4 bias correction [16]. These studies were then divided into two cohorts: (a) S_1: EGFR amplified ($EGFR^+$) versus non-amplified ($EGFR^-$) studies, and (2) S_2: MGMT methylated ($MGMT^+$) versus unmethylated ($MGMT^+$). For S_1, we had a total of 91 subjects of which 70 were used for training (35 amp, 35 non-amp), and the remaining 21 (6 $EGFR^+$, 15 $EGFR^-$) were used for validation. For S_2, of a total of 81 subjects, 60 (28 $MGMT^+$, 32 $MGMT^-$) were used for training, while 21 subjects were used for validation (5 $MGMT^+$, 16 $MGMT^-$).

3.2 Implementation Details

Two experiments were set-up using cohorts S_1 (**Experiment 1**: $EGFR^+$ versus $EGFR^-$) and S_2 (**Experiment 2**: $MGMT^+$ versus $MGMT^-$), respectively. For both experiments, we extracted a total of 316 3D context-features for every

$c \in C_B$ (where C_B was a 1-cm diameter sphere in our case), including 1 raw feature, 8 gray features, 13 gradient features, 26 Haralick features, 64 Gabor features, 152 Laws features, and 52 CoLlAGe features, extracted using 2 window sizes $w = 3 \times 3$ and $w = 5 \times 5$. These features are pruned as detailed in Sect. 2.2 to obtain $\mathbb{F}_{CO}$. In addition, population atlases were constructed to quantify the frequency of occurrence of $EGFR^+$ versus $EGFR^-$ and $MGMT^+$ versus $MGMT^-$ as detailed in Sect. 2.3. Similarly, for both experiments, $\mathbb{F}_{SpACe}$ was created following 10 runs of 10-fold cross validation, as detailed in Sect. 2.4. The median value of the likelihood values across all voxels from the biopsy sites of all subjects was used as a threshold to determine M^+, M^- for every voxel. Finally, majority voting was used to obtain the mutation status for every biopsy site.

3.3   Comparative Strategies

In order to evaluate the efficacy of SpACe model, we compared our results with two state-of-the-art methods, radiomic-based, and deep-learning-based that employ features from the entire tumor to predict amplification/methylation status. For the radiomic experiment ($\mathbb{F}_{Rad}$), a total of 316 radiomic features were extracted from the entire enhancing lesion of every subject (same attributes that were extracted from biopsy sites), and the feature vector was constructed from the 4 statistics: median, variance, skewness, and kurtosis values that were computed for every feature across all voxels for this patient, for a total of 1264 features. After feature pruning, 283 features were fed to the LASSO model to compute patient-wise scores that determined their gene status.

For the DL approach to predict the mutation status, we used a deep residual neural network (ResNet) as described in [5]. ResNet has previously been used to predict EGFR and MGMT mutation in GBM and other cancers [5,17]. Specifically, patches of size 128×128 were sampled from the center of the selected MRI slices and augmented using horizontal flips and random rotations to enlarge the limited training data. Following patch sampling, we trained separate deep ResNet networks with 18 layers (ResNet-18) for each of the two experiments ($EGFR^+$ vs. $EGFR^-$ and $MGMT^+$ vs. $MGMT^-$) using stochastic gradient optimization. In order to train the networks on MRI scans, we used a pre-trained model on ImageNet and performed transfer learning using the sampled patches from the MRI scans. We trained each model for 25 epochs with dropout values ranging from 0.1–0.5 based on minimum validation loss, to avoid overfitting. Models with minimum loss were locked down to test patches obtained from the test set.

4   Results and Discussion

4.1   Experiment 1: Determining EGFR Amplification Status

Using $\mathbb{F}_{CO}$ alone, training and testing accuracies were reported as 61.43% and 66.67% respectively. Combination of $\mathbb{F}_{CO}$ features with $\mathbb{F}_S$ features into LASSO

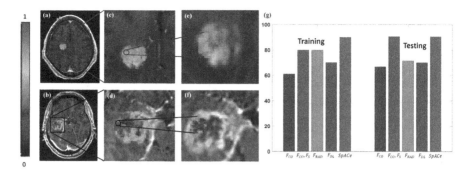

Fig. 2. $EGFR^+$ (a) and $EGFR^-$ (b) cases with voxel-wise probabilities calculated using SpACe maps. Heatmaps with voxel-wise probabilities for the entire tumor area for $EGFR^+$ (c) and $EGFR^-$ (d) are shown, where "red" represents amplified and "blue" represents non-amplified status. Confirmed biopsy sites are enclosed by a circle. (e), (f) show biopsy region heatmaps, which confirm the mutation status of the tumor. Tumor heatmaps in (c), (d) show other clusters that could be potential candidates for biopsy sites. The prediction accuracies for predicting mutation status in the two patients in (a) and (b) using SpACe were 92.5% and 96% respectively. (g) shows a bar graph with accuracies for both training and testing sets for $EGFR^+$ versus $EGFR^-$, using $\mathbb{F}_{CO}$, $[\mathbb{F}_{CO}. \mathbb{F}_S]$, $\mathbb{F}_{Rad}$, $\mathbb{F}_{DL}$, and $\mathbb{F}_{SpACe}$. (Color figure online)

model yielded training and testing accuracies of 80% and 90.48% respectively, using 8 $\mathbb{F}_{CO}$ (1 raw, 1 gray, 2 gradient, 1 Haralick, 3 Gabor) and 2 $\mathbb{F}_S$ features. Next, we evaluated the efficacy of including $\mathbb{F}_{CO}$, $\mathbb{F}_S$, and clinical features ($\mathbb{F}_A$, $\mathbb{F}_G$) into our model to predict the mutation status. Clinical features did not improve accuracy of the model. PPMMs were then employed, and successfully corrected the amplification status for 7 subjects from the training set, yielding final training and testing accuracies of 90% (1 amp, 6 non-amp subjects were misclassified out of 70) and 90.48% (2 non-amp subjects are misclassified out of 21) respectively. Apart from computing the area under curve (AUC), we also computed the area under the precision-recall curve (AUPRC), for both training and testing sets, to handle imbalance of classes in our test set [18]. AUCs for EGFR amplification prediction on training and test sets were 0.78 and 0.8 respectively, while AUPRCs were 0.76 and 0.7 respectively. Using radiomic features from the entire tumor to predict mutation status yielded training and testing accuracies of 80% and 71.43% respectively. Further, the Res-Net model to predict EGFR status yielded training and testing accuracies of 74.28% and 65.52%, significantly underperforming in comparison to the SpACe model. Results of SpACe model on 2 different patients are illustrated in Fig. 2.

4.2   Experiment 2: Determining MGMT Methylation Status

Using $\mathbb{F}_{CO}$ alone, the model achieved training and testing accuracies of 76.67% and 57.14% respectively. When combining $\mathbb{F}_{CO}$ features with $\mathbb{F}_S$ features into LASSO model, we obtained training and testing accuracies of 81.67% and 61.9%

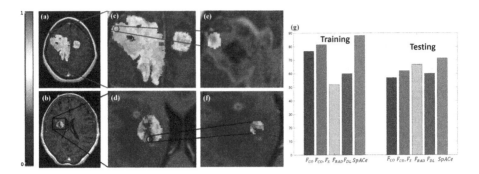

Fig. 3. $MGMT^+$ (a) and $MGMT^-$ (b) cases with voxel-wise probabilities calculated using SpACe maps. Heatmaps with voxel-wise probabilities for the entire tumor area for $MGMT^+$ (c) and $MGMT^-$ cases (d) are shown, where "red" represents methylated and "blue" represents unmethylated status. Confirmed biopsy sites are enclosed by a circle. (e), (f) show biopsy region heatmaps, which confirm the mutation status of the tumor. Tumor heatmaps in (c), (d) show other clusters that could be potential candidates for biopsy sites. The prediction accuracies for cases (a) and (b) using SpACe are 98% and 99% respectively. (g) shows a bar graph with accuracies for both training and testing sets using $\mathbb{F}_{CO}$, [$\mathbb{F}_{CO}$. $\mathbb{F}_S$], $\mathbb{F}_{Rad}$, $\mathbb{F}_{DL}$, and $\mathbb{F}_{SpACe}$. (Color figure online)

respectively. Next, using $\mathbb{F}_{CO}$, $\mathbb{F}_S$, and clinical features, the model picked a set of 12 features that included 8 $\mathbb{F}_{CO}$ features; 1 gray, 3 Haralick, and 4 Gabor features, in addition to P_{MGMT^+}, P_{MGMT^-}, $\mathbb{F}_A$, and $\mathbb{F}_G$. This model yielded training and testing accuracies of 83.3% and 66.67% respectively. Applying PPMMs for predicting methylation status on these results corrected the mutation status for 3 training subjects as well as 1 testing subject, with final accuracies of 88.3% and 71.5% respectively, compared to training and testing accuracies of 82% and 67% without applying PPMMs. AUCs for MGMT methylation prediction on training and test sets were 0.75 and 0.71 respectively, and AUPRCs were 0.72 and 0.69 respectively. When using radiomic features from the entire tumor to predict methylation status, training and testing accuracies were 76.67% and 52.38%. In addition, the DL model that was trained to predict methylation status gave training and testing accuracies of 79.37% and 68.40%, suggesting that results obtained using SpACe maps outperformed both comparative approaches. Results on 2 different patients are illustrated in Fig. 3.

5   Concluding Remarks

In this work, we presented the first-attempt at creating "virtual biopsy" radiogenomic maps for predicting gene mutational status on MRI, by combining two complementary attributes: (1) *spatial-priors* for presence or absence of mutation status via probabilistic atlases from a retrospective cohort, and (2) *context-priors* to capture mutational heterogeneity using radiomic attributes obtained from a stereotactically co-localized biopsy site within the tumor. These spatial-and-context aware (SpACe) maps were evaluated in the context of two experiments: predicting (1) EGFR amplification status, and (2) MGMT methylation status, on Glioblastoma. Our results demonstrated that SpACe outperformed state-of-the-art radiomic and deep learning approaches that were performed on the entire tumor, instead of learning features from the co-localized biopsy site. The virtual biopsy maps created using SpACe could not only improve prediction of gene mutation status of the tumor, but could also serve as surgical navigation to guide potential biopsy sites for specific gene mutations.

References

1. Koljenović, S., et al.: Discriminating vital tumor from necrotic tissue in human glioblastoma tissue samples by Raman spectroscopy. Lab. Invest. **82**(10), 1265–1277 (2002)
2. Qazi, M., et al.: Intratumoral heterogeneity: pathways to treatment resistance and relapse in human glioblastoma. Ann. Oncol. **28**(7), 1448–1456 (2017)
3. Della Puppa, A., et al.: MGMT expression and promoter methylation status may depend on the site of surgical sample collection within glioblastoma: a possible pitfall in stratification of patients? J. Neurooncol. **106**(1), 33–41 (2012). https://doi.org/10.1007/s11060-011-0639-9
4. Chang, P., et al.: Deep-learning convolutional neural networks accurately classify genetic mutations in gliomas. Am. J. Neuroradiol. **39**(7), 1201–1207 (2018)
5. Korfiatis, P., Kline, T.L., Lachance, D.H., Parney, I.F., Buckner, J.C., Erickson, B.J.: Residual deep convolutional neural network predicts MGMT methylation status. J. Digit. Imaging **30**(5), 622–628 (2017). https://doi.org/10.1007/s10278-017-0009-z
6. Li, Z., Wang, Y., Yu, J., Guo, Y., Cao, W.: Deep learning based radiomics (DLR) and its usage in noninvasive IDH1 prediction for low grade glioma. Sci. Rep. **7**(1), 1–11 (2017)
7. Li, Z.-C., et al.: Multiregional radiomics features from multiparametric MRI for prediction of MGMT methylation status in glioblastoma multiforme: a multicentre study. Eur. Radiol. **28**(9), 3640–3650 (2018). https://doi.org/10.1007/s00330-017-5302-1
8. Parker, N.R., et al.: Intratumoral heterogeneity identified at the epigenetic, genetic and transcriptional level in glioblastoma. Sci. Rep. **6**, 22477 (2016)
9. French, P.J., et al.: Defining EGFR amplification status for clinical trial inclusion. Neuro-oncology **21**(10), 1263–1272 (2019)
10. Ellingson, B., et al.: Probabilistic radiographic atlas of glioblastoma phenotypes. Am. J. Neuroradiol. **34**(3), 533–540 (2013)
11. Bilello, M., et al.: Population-based MRI atlases of spatial distribution are specific to patient and tumor characteristics in glioblastoma. NeuroImage Clin. **12**, 34–40 (2016)
12. Haralick, R.M., et al.: Textural features for image classification. IEEE Trans. Syst. Man Cybern. **3**(6), 610–621 (1973)
13. Prasanna, P., et al.: Co-occurrence of local anisotropic gradient orientations (collage): a new radiomics descriptor. Sci. Rep. **6**, 37241 (2016)
14. Tibshirani, R.: Regression shrinkage and selection via the lasso. J. Roy. Stat. Soc. Ser. B (Methodol.) **58**(1), 267–288 (1996)

15. Monaco, J.P., et al.: High-throughput detection of prostate cancer in histological sections using probabilistic pairwise markov models. Med. Image Anal. **14**(4), 617–629 (2010)
16. Tustison, N.J., et al.: N4ITK: improved N3 bias correction. IEEE Trans. Med. Imaging **29**(6), 1310–1320 (2010)
17. Xiong, J., et al.: Implementation strategy of a CNN model affects the performance of CT assessment of EGFR mutation status in lung cancer patients. IEEE Access **7**, 64583–64591 (2019)
18. Keilwagen, J., Grosse, I., Grau, J.: Area under precision-recall curves for weighted and unweighted data. PLoS ONE **9**(3), e92209 (2014)

CorrSigNet: Learning CORRelated Prostate Cancer SIGnatures from Radiology and Pathology Images for Improved Computer Aided Diagnosis

Indrani Bhattacharya[1]([⊠]), Arun Seetharaman[2], Wei Shao[1], Rewa Sood[2], Christian A. Kunder[4], Richard E. Fan[3], Simon John Christoph Soerensen[3,5], Jeffrey B. Wang[1], Pejman Ghanouni[1,3], Nikola C. Teslovich[3], James D. Brooks[3], Geoffrey A. Sonn[1,3], and Mirabela Rusu[1]

[1] Department of Radiology, School of Medicine, Stanford University, Stanford, CA 94305, USA
ibhatt@stanford.edu
[2] Department of Electrical Engineering, Stanford University, Stanford, CA 94305, USA
[3] Department of Urology, School of Medicine, Stanford University, Stanford, CA 94305, USA
[4] Department of Pathology, School of Medicine, Stanford University, Stanford, CA 94305, USA
[5] Department of Urology, Aarhus University, Aarhus, Denmark

Abstract. Magnetic Resonance Imaging (MRI) is widely used for screening and staging prostate cancer. However, many prostate cancers have subtle features which are not easily identifiable on MRI, resulting in missed diagnoses and alarming variability in radiologist interpretation. Machine learning models have been developed in an effort to improve cancer identification, but current models localize cancer using MRI-derived features, while failing to consider the disease pathology characteristics observed on resected tissue. In this paper, we propose CorrSigNet, an automated two-step model that localizes prostate cancer on MRI by capturing the pathology features of cancer. First, the model learns MRI signatures of cancer that are correlated with corresponding histopathology features using Common Representation Learning. Second, the model uses the learned correlated MRI features to train a Convolutional Neural Network to localize prostate cancer. The histopathology images are used only in the first step to learn the correlated features. Once learned, these correlated features can be extracted from MRI of new patients (without histopathology or surgery) to localize cancer. We trained and validated our framework on a unique dataset of 75 patients with 806 slices who underwent MRI followed by prostatectomy surgery. We tested our method on an independent test set of 20 prostatectomy patients (139

M. Rusu—We thank the Departments of Radiology and Urology at Stanford University, for their support for this work.

© Springer Nature Switzerland AG 2020
A. L. Martel et al. (Eds.): MICCAI 2020, LNCS 12262, pp. 315–325, 2020.
https://doi.org/10.1007/978-3-030-59713-9_31

slices, 24 cancerous lesions, 1.12M pixels) and achieved a per-pixel sensitivity of 0.81, specificity of 0.71, AUC of 0.86 and a per-lesion AUC of 0.96 ± 0.07, outperforming the current state-of-the-art accuracy in predicting prostate cancer using MRI.

Keywords: Computer aided diagnosis · Common representation Learning · MRI · Histopathology images · Prostate cancer

1 Introduction

Early localization of prostate cancer from MRI is crucial for successful diagnosis and local therapy. However, subtle differences between benign conditions and cancer on MRI often make human interpretation challenging, leading to missed diagnoses and an alarming variability in radiologist interpretation. Human interpretation of prostate MRI suffers from low inter-reader agreement (0.46–0.78) [1] and high variability in reported sensitivity (58–98%) and specificity (23–87%) [2].

Predictive models can help standardize radiologist interpretation, but current models [3–8] often learn from MRI only, without considering the disease pathology characteristics. These approaches derive MRI features that are agnostic to the biology of the tumor. Moreover, current predictive models mostly use inaccurate labels (either from biopsies [6] that suffer from sampling errors, or cognitive registration of pre-operative MRI with digital histopathology images of surgical specimens, where a radiologist retrospectively outlines the lesions on MRI [4]). MRI under-estimates the tumor size [9], making outlines on MRI alone insufficient to capture the entire extent of disease. Furthermore, it is challenging to outline the ~20% of tumors that are not clearly seen on MRI, even when using histopathology images as reference [1]. These MRI-based models use a variety of techniques including traditional classifiers with hand-crafted and radiomic features [3,5,7], as well as deep learning based models [4,8]. The current state-of-the-art approach [4] to predict a cancer probability map for the entire prostate uses a Holistically Nested Edge Detection (HED) [10] algorithm.

In this paper, we propose CorrSigNet, a two-step approach for predicting prostate cancer using MRI. First, CorrSigNet leverages spatially aligned radiology and histopathology images of prostate surgery patients to learn MRI cancer signatures that correlate with features extracted from the histopathology images. Second, CorrSigNet uses these correlated MRI signatures to train a predictive model for localizing cancer when histopathology images are not available, e.g. before surgery. This approach enables learning MRI signatures that capture tumor biology information from surgery patients with histopathology images, and then translating those learned signatures for prediction in patients without surgery/biopsy. Prior studies lack such correlation analysis of the two modalities. Our approach shows improved prostate cancer prediction compared to the current state-of-the-art method [4].

2 Proposed Method

2.1 Dataset

We used 95 prostate surgery patients with pre-operative multi-parametric MRI (T2-weighted and Apparent Diffusion Coefficient) and post-operative digitized histopathology images. Custom 3D printed molds were used to ensure that excised prostate tissue was sectioned in the same plane as the T2-weighted (T2W) MRI. An expert pathologist annotated cancer on the histopathology images. We spatially aligned the pre-operative MRI and digitized histopathology images of the excised tissue via the state-of-the-art RAPSODI registration platform [11]. RAPSODI achieved a Dice similarity coefficient of 0.98 ± 0.01 for the prostate, prostate boundary Hausdorff distance of $1.71 \pm 0.48\,$mm, and urethra deviation of $2.91 \pm 1.25\,$mm between registered histopathology images and MRI. Such careful registration of radiology and pathology images of the prostate enabled (1) correlation analysis of the two modalities at a pixel-level, and (2) accurate mapping of cancer labels from pathology to radiology images. We considered multiple slices per patient (average 7 slices/patient) irrespective of cancer size. Slices with missing cancer annotations were discarded during training. The dataset included some patients with cancer that had extra prostatic extensions, but our analysis was focused only on cancers inside the prostate.

2.2 Data Pre-processing

We smoothed the histopathology images with a Gaussian filter with $\sigma = 0.25$ to prevent downsampling artifacts, padded and then downsampled them to 224×224, resulting in an X-Y resolution of $0.29 \times 0.29\,$mm^2. We projected and resampled the T2W and ADC images, prostate masks, and cancer labels on the corresponding downsampled histopathology images, such that they also had the same X-Y resolution of $0.29 \times 0.29\,$mm^2. This ensured that each pixel in each modality represented the same physical area.

Since MRI intensities vary significantly between scanners and scanning protocols, we standardized the T2W and ADC intensities using the histogram alignment approach proposed by Nyúl et al. [12]. We used prostate masks to standardize intensities within the prostate, and then applied the learned transformation to the image region beyond the prostate. After intensity standardization, we normalized the intensities to have zero mean and standard deviation of 1.

We randomly split the 95 patients to create our train, validation, and test sets with 66, 9, and 20 patients respectively. After horizontal flipping based data augmentation, the train and validation sets had 700 and 106 slices respectively. The test set included 139 slices, 24 cancerous lesions, 1.12M pixels in the prostate with 9% cancer pixels. We performed MRI scale standardization on the train set, and used the learned histograms to standardize the validation and test sets. We followed a similar strategy for MRI intensity normalization.

2.3 Learning Correlated Features

Feature Extraction: We extracted features from the T2W, ADC, and histopathology images by passing them through the first two convolutional layers of a pre-trained VGG-16 architecture [13]. Thus, each 224×224 image yielded a $224 \times 224 \times 64$ representation, generating 64 features per pixel. We sampled pixels from within the prostate, and concatenated the T2W and ADC features to form the MRI representation per pixel. Thus, for each pixel, we had the MRI representation $\mathcal{R}_i \in \mathbb{R}^{128}$ and the histopathology representation $\mathcal{P}_i \in \mathbb{R}^{64}$.

Common Representation Learning: We trained a Correlational Neural Network architecture (CorrNet) [14] to learn common representations from MRI and histopathology features per pixel. Given N pixels, each pixel input Z_i to the CorrNet model had two views: the MRI feature representation for pixel i, $\mathcal{R}_i$, and the histopathology feature representation for pixel i, $\mathcal{P}_i$. We used a fully-connected CorrNet model with a single hidden layer, where the hidden layer $H(Z_i) \in \mathbb{R}^k$ was computed as:

$$H(Z_i) = W\mathcal{R}_i + V\mathcal{P}_i + b \tag{1}$$

where $W \in \mathbb{R}^{k \times 128}$, $V \in \mathbb{R}^{k \times 64}$ and $b \in \mathbb{R}^{k \times 1}$. The reconstructed output Z_i' was computed from the hidden layer as:

$$Z_i' = [W'H(Z_i), V'H(Z_i)] + b' \tag{2}$$

where $W' \in \mathbb{R}^{128 \times k}$, $V' \in \mathbb{R}^{64 \times k}$ and $b' \in \mathbb{R}^{(128+64) \times 1}$. In contrast to the original CorrNet model, we did not use any non-linear activation function. We learned the parameters $\theta = \{W, V, W', V', b, b'\}$ of the system by minimizing the following objective function, as detailed in [14]:

$$J(\theta) = \sum_{i=1}^{N} [L(Z_i, H(Z_i)) + L(Z_i, H(R_i)) + L(Z_i, H(P_i)) - \lambda corr(H(R_i), H(P_i))]$$

$$\tag{3}$$

$$corr(H(R_i), H(P_i))$$

$$= \frac{\sum_{i=1}^{N}[(H(R_i) - \overline{H(R)})(H(P_i) - \overline{H(P)}]}{\sqrt{\sum_{i=1}^{N}(H(R_i) - \overline{H(R)})^2 \sum_{i=1}^{N}(H(P_i) - \overline{H(P)})^2}} \tag{4}$$

where L is the reconstruction error, λ is the scaling parameter to determine the relative weight of the correlation error with respect to the reconstruction errors, $\overline{H(R)}$ is the mean hidden representation of the 1st view and $\overline{H(P)}$ is the mean hidden representation of the 2nd view. Thus, the CorrNet model (i) minimizes the self and cross reconstruction errors, and (ii) maximizes the correlation between the hidden representations of the two views. Training CorrNet using pixel representations from within the prostate gave ample training samples to optimize the model, and to learn differences between cancer and non-cancer pixels.

After the CorrNet model was trained, we used the learned weights W, b to project the MRI feature representations $\mathcal{R}_i$ onto the k dimensional hidden space to form *CorrNet representations* of the input MRI. The *CorrNet representations* are correlated with the corresponding histopathology features, and once trained, can be constructed even in the absence of histopathology images. Figure 1 shows the pipeline for learning common representations.

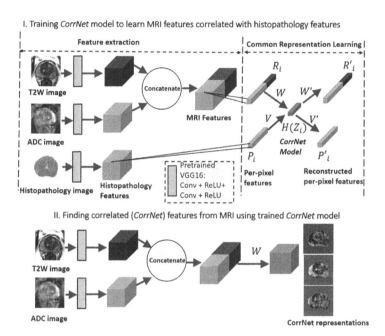

Fig. 1. Learning correlated representations from spatially aligned MRI and histopathology images, and then constructing the correlated (*CorrNet*) representations from MRI alone using learned weights.

Fig. 2. Five-dimensional *CorrNet representations* for one example MRI slice

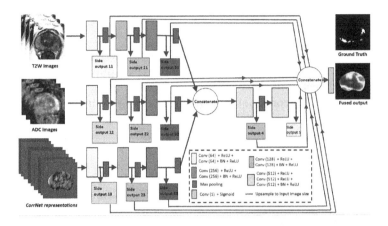

Fig. 3. HED-branch-3 model for predicting cancer probability maps.

Training: From the 66 patients in the training cohort, we sampled all the cancer pixels from within the prostate, and randomly sampled an equal number of non-cancer pixels, also from within the prostate, thereby generating a training set of $\approx 1.2M$ pixels, with equal number of cancer and non-cancer pixels. This ensured that we train the CorrNet with a balanced dataset of two classes. We used $\lambda = 2$ to weigh the cross-correlation error higher than the reconstruction errors. We chose a squared error loss L for the reconstruction errors. We trained the CorrNet model with varying hidden layer dimensions, namely: $k \in \{1, 3, 5, 15, 30\}$. For each k, we used a learning rate $\eta = 10^{-5}$, and 300 training epochs. Figure 2 shows *CorrNet* representations of an example MRI slice, with $k = 5$.

2.4 Prediction of Prostate Cancer Extent

We modified the Holistically Nested Edge Detection (HED) architecture [10] to predict cancer probability maps for the entire prostate. We considered two modified versions of HED: (1) HED-3, and (2) HED-branch-3. The HED-3 model evaluates how well *CorrNet representations* alone perform in predicting cancer, while the HED-branch-3 model evaluates how well *CorrNet representations* combined with T2W and ADC images perform in predicting cancer. We represent our model using correlated feature learning and HED-3 as CorrSigNet(k), and our model with correlated feature learning and HED-branch-3 as CorrSigNet(T2W, ADC, k), where k is the *CorrNet* feature dimension. For example, CorrSigNet(5) uses only 5 correlated features for prediction, whereas CorrSigNet (T2W, ADC, 5) uses the normalized T2W and ADC intensities in addition to 5 correlated features for prediction. We chose a prediction model similar to the HED architecture because it is known to learn and combine multi-scale and multi-level features, and has been successfully applied to anatomy segmentations from CT scans [15–17], and to prostate cancer prediction [4].

In HED-3, we input three adjacent *CorrNet slice representations* of the prostate and output predictions for only the central slice. This ensured that the 2D-HED model learned the 3D volumetric continuity from MRI/ histopathology/ correlated features. This also helped in reducing false positive rates.

In HED-branch-3 (shown in Fig. 3), we combined the *CorrNet slice representations* together with the normalized T2W and ADC images as inputs to the model. Similar to HED-3, we considered three adjacent slices for each input sequence (T2W, ADC, *CorrNet representations*), and predicted cancer probability maps for the central slice only. However, in HED-branch-3 model, we processed each input sequence independently using the first three blocks, concatenated the three outputs from the three independent blocks, and processed the concatenated output using the next 2 blocks. Since the input sequences are processed independently in the first three blocks, we had a total of 11 side outputs, which were fused together using a Conv-1D layer to form the weighted fused output. We computed balanced cross-entropy losses for each of the 12 outputs (11 side outputs and 1 fused output) while training the architecture, but computed evaluation metrics only on the fused output. We used 3×3 kernels for all convolution layers except the last Conv-1D layer. The number of filters in each layer is stated in the legend in Fig. 3. For both the HED-3 and HED-branch-3 models, we added Batch Normalization in each block, before ReLU activation, as opposed to the HED model used by [4] which used Batch Normalization in each layer. No post-processing steps were performed on the prediction maps.

Training: We trained both models using an Adam optimizer with an initial learning rate $\eta = 10^{-3}$, weight decay $\alpha = 0.1$, epochs = 200 and early stopping.

3   Experimental Results

Quantitative Evaluation: We quantitatively evaluated our models on a per-pixel and a per-lesion basis, with ground truth labels derived from pathologist cancer annotations on registered histopathology images. For a direct comparison, we reproduced the current state-of-the-art model [4] to the best of our understanding, and computed both pixel-level and lesion-level evaluation metrics of this model on our test data (20 patients, 139 slices, 24 cancerous lesions, 1.12M pixels in the prostate). It may be noted that the AUC numbers reported in [4] are computed on a lesion level, and not on a pixel-level. Our pixel-level metrics including all pixels within the prostate provide a more rigorous evaluation.

Pixel Level Analysis: We tested the performance of the CorrSigNet models with different inputs and varying *CorrNet* feature dimension k using the following pixel-level evaluation metrics (computed using 1.12M pixels in the prostate): sensitivity, specificity, and AUC of the ROC curve, with a probability threshold of 0.5. We note from Table 1 that CorrSigNet performs better than [4], with consistently higher AUC numbers in pixel-level analysis. The sensitivity and specificity numbers vary within the models. Our tests showed that at least 3 CorrNet features were necessary for improved performance over MRI alone.

Table 1. Pixel-level quantitative evaluation of CorrSigNet models

Model	Sensitivity	Specificity	AUC
HED [4] (current state-of-the-art)	0.75	0.74	0.80
CorrSigNet(1)	0.72	0.78	0.81
CorrSigNet(3)	0.82	0.71	**0.86**
CorrSigNet(5)	0.77	0.77	**0.86**
CorrSigNet(15)	0.76	0.78	0.84
CorrSigNet(30)	0.75	0.81	0.85
CorrSigNet(T2W, ADC, 1)	0.73	0.79	0.83
CorrSigNet(T2W, ADC, 3)	0.70	**0.85**	**0.86**
CorrSigNet(T2W, ADC, 5)	*0.81*	*0.72*	***0.86***
CorrSigNet(T2W, ADC, 15)	0.78	0.78	**0.86**
CorrSigNet(T2W, ADC, 30)	**0.83**	0.71	**0.86**

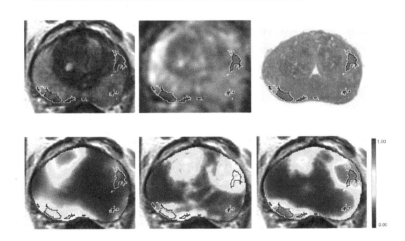

Fig. 4. Spatially aligned (a) T2W, (b) ADC, and (c) histopathology images. Data obtained and processed as detailed in Sect. 2.1. Prediction results using (d) the current state-of-the-art method [4], (e) our model CorrSigNet(5), and (f) our model CorrSigNet(T2W, ADC, 5).

We chose CorrSigNet (T2W, ADC, 5) as the optimum model, because it had high sensitivity, specificity and AUC, with an optimum number of parameters. Between false positives and false negatives, we note that a false negative is more detrimental than a false positive in the task of cancer prediction.

Lesion level analysis: We performed lesion-level analysis using the evaluation method detailed in [4] and found that CorrSigNet(T2W, ADC, 5) achieved a per-lesion AUC of 0.96 ± 0.07 compared to a per-lesion AUC of 0.92 ± 0.09 by [4] on the same test set.

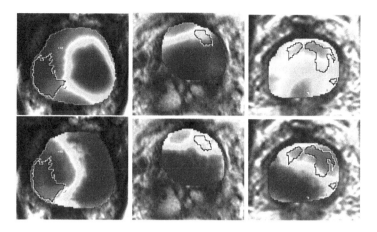

Fig. 5. (Top) Prediction results using the current state-of-the-art method [4]. (Bottom) Prediction results from our model CorrSigNet(T2W, ADC, 5).

Qualitative Evaluation: Figure 4 shows the same slice as in Fig. 2 with aligned T2W, ADC, and histopathology images, and prediction results using current state-of-the-art method [4], our CorrSigNet(5) and CorrSigNet(T2W, ADC, 5) models. It may be noted that [4] fails to detect the cancerous regions on the left and right of the images, while the *CorrNet* representations alone can identify the cancer regions, and when combined with T2W and ADC images, they predict the cancer regions with high probability. It may also be noted that CorrSigNet(T2W, ADC, 5) shows fewer false positives than [4]. This example shows the strength of learning correlated MRI signatures in identifying subtle, and sometimes MRI-invisible cancers. Figure 5 shows more example slices from different patients, comparing the state-of-the-art approach [4] and our prediction results with CorrSigNet(T2W, ADC, 5). We note that our model with correlated features (1) can identify subtle and smaller cancer regions, (2) have better overlap with ground truth cancer labels, and (3) have fewer false positives.

4   Conclusion

In this paper, we presented a novel method to learn correlated signatures of cancer from spatially aligned MRI and histopathology images of prostatectomy surgical specimens, and then use these learned correlated signatures in predicting prostate cancer extent from MRI. Quantitatively, our method improved performance of automated prostate cancer localization (per-pixel AUC of 0.86, per-lesion AUC of 0.96 ± 0.07), as compared to the current state-of-the-art method [4] (per-pixel AUC 0.80, per-lesion 0.92 ± 0.09). Qualitatively, we found that correlated features could capture subtle cancerous regions and sometimes MRI-invisible cancers, had better overlap with ground truth labels, and fewer false positives. Correlated features have the capability of capturing tumor biology information from histopathology images in an unprecedented way, and these features, once learned, can be extracted in patients without histopathology images. In future work, we intend to conduct experiments with augmented datasets and in a cross-validation framework to boost the performance of our models.

References

1. Barentsz, J.O., et al.: Synopsis of the PI-RADS v2 guidelines for multiparametric prostate magnetic resonance imaging and recommendations for use. Eur. Urol. **69**(1), 41 (2016)
2. Ahmed, H.U., et al.: Diagnostic accuracy of multi-parametric MRI and TRUS biopsy in prostate cancer (PROMIS): a paired validating confirmatory study. Lancet **389**(10071), 815–822 (2017)
3. Viswanath, S.E., et al.: Central gland and peripheral zone prostate tumors have significantly different quantitative imaging signatures on 3 Tesla endorectal, in vivo T2-weighted MR imagery. J. Magn. Reson. Imaging **36**(1), 213–224 (2012)
4. Sumathipala, Y., Lay, N., Turkbey, B., Smith, C., Choyke, P.L., Summers, R.M.: Prostate cancer detection from multi-institution multiparametric MRIs using deep convolutional neural networks. J. Med. Imaging **5**(4), 044507 (2018)
5. Litjens, G., Debats, O., Barentsz, J., Karssemeijer, N., Huisman, H.: Computer-aided detection of prostate cancer in MRI. IEEE Trans. Med. Imaging **33**(5), 1083–1092 (2014)
6. Armato, S.G., et al.: PROSTATEx challenges for computerized classification of prostate lesions from multiparametric magnetic resonance images. J. Med. Imaging **5**(4), 044501 (2018)
7. Viswanath, S.E., et al.: Comparing radiomic classifiers and classifier ensembles for detection of peripheral zone prostate tumors on T2-weighted MRI: a multi-site study. BMC Med. Imaging **19**(1), 22 (2019)
8. Cao, R., et al.: Joint prostate cancer detection and gleason score prediction in mp-MRI via FocalNet. IEEE Trans. Med. Imaging **38**(11), 2496–2506 (2019)
9. Priester, A., et al.: Magnetic resonance imaging underestimation of prostate cancer geometry: use of patient specific molds to correlate images with whole mount pathology. J. Urol. **197**(2), 320–326 (2017)
10. Xie, S., Tu, Z.: Holistically-nested edge detection. In: Proceedings of the IEEE International Conference on Computer Vision, pp. 1395–1403 (2015)
11. Rusu, M., et al.: Registration of pre-surgical MRI and histopathology images from radical prostatectomy via RAPSODI. Med. Phys. (2020, in press)
12. Nyúl, L.G., Udupa, J.K., Zhang, X.: New variants of a method of MRI scale standardization. IEEE Trans. Med. Imaging **19**(2), 143–150 (2000)
13. Simonyan, K., Zisserman, A.: Very deep convolutional networks for large-scale image recognition. arXiv preprint arXiv:1409.1556 (2014)
14. Chandar, S., Khapra, M.M., Larochelle, H., Ravindran, B.: Correlational neural networks. Neural Comput. **28**(2), 257–285 (2016)
15. Harrison, A.P., Xu, Z., George, K., Lu, L., Summers, R.M., Mollura, D.J.: Progressive and multi-path holistically nested neural networks for pathological lung segmentation from CT images. In: Descoteaux, M., Maier-Hein, L., Franz, A., Jannin, P., Collins, D.L., Duchesne, S. (eds.) MICCAI 2017. LNCS, vol. 10435, pp. 621–629. Springer, Cham (2017). https://doi.org/10.1007/978-3-319-66179-7_71

16. Roth, H.R., Lu, L., Farag, A., Sohn, A., Summers, R.M.: Spatial aggregation of holistically-nested networks for automated pancreas segmentation. In: Ourselin, S., Joskowicz, L., Sabuncu, M.R., Unal, G., Wells, W. (eds.) MICCAI 2016. LNCS, vol. 9901, pp. 451–459. Springer, Cham (2016). https://doi.org/10.1007/978-3-319-46723-8_52

17. Nogues, I., et al.: Automatic lymph node cluster segmentation using holistically-nested neural networks and structured optimization in CT images. In: Ourselin, S., Joskowicz, L., Sabuncu, M.R., Unal, G., Wells, W. (eds.) MICCAI 2016. LNCS, vol. 9901, pp. 388–397. Springer, Cham (2016). https://doi.org/10.1007/978-3-319-46723-8_45

Preoperative Prediction of Lymph Node Metastasis from Clinical DCE MRI of the Primary Breast Tumor Using a 4D CNN

Son Nguyen[1,2], Dogan Polat[2], Paniz Karbasi[1], Daniel Moser[1], Liqiang Wang[1], Keith Hulsey[2], Murat Can Çobanoğlu[1], Basak Dogan[2], and Albert Montillo[1,2(✉)]

[1] Lyda Hill Department of Bioinformatics, UT Southwestern Medical Center, Dallas, TX 75390, USA
[2] Department of Radiology, UT Southwestern Medical Center, Dallas, TX 75390, USA
{Son.Nguyen,Albert.Montillo}@UTSouthwestern.edu

Abstract. In breast cancer, undetected lymph node metastases can spread to distal parts of the body for which the 5-year survival rate is only 27%, making accurate nodal metastases diagnosis fundamental to reducing the burden of breast cancer, when it is still early enough to intervene with surgery and adjuvant therapies. Currently, breast cancer management entails a time consuming and costly sequence of steps to clinically diagnose axillary nodal metastases status. The purpose of this study is to determine whether preoperative, clinical DCE MRI of the primary tumor alone may be used to predict clinical node status with a deep learning model. If possible then many costly steps could be eliminated or reserved for only those with uncertain or probable nodal metastases. This research develops a data-driven approach that predicts lymph node metastasis through the judicious integration of clinical and imaging features from preoperative 4D dynamic contrast enhanced (DCE) MRI of 357 patients from 2 hospitals. Innovative deep learning classifiers are trained from scratch, including 2D, 3D, 4D and 4D deep convolutional neural networks (CNNs) that integrate multiple data types and predict the nodal metastasis differentiating nodal stage N0 (non metastatic) against stages N1, N2 and N3. Appropriate methodologies for data preprocessing and network interpretation are presented, the later of which bolster radiologist confidence that the model has learned relevant features from the primary tumor. Rigorous *nested* 10-fold cross-validation provides an unbiased estimate of model performance. The best model achieves a high sensitivity of 72% and an AUROC of 71% on held out test data. Results are strongly supportive of the potential of the combination of DCE MRI and machine learning to inform diagnostics that could substantially reduce breast cancer burden.

Electronic supplementary material The online version of this chapter (https://doi.org/10.1007/978-3-030-59713-9_32) contains supplementary material, which is available to authorized users.

A. L. Martel et al. (Eds.): MICCAI 2020, LNCS 12262, pp. 326–334, 2020.
https://doi.org/10.1007/978-3-030-59713-9_32

Keywords: Breast cancer · Nodal metastases · DCE MRI · Deep learning

1 Introduction

Breast cancer is the most common cancer among women in many countries including the USA and causes more premature deaths than any cancer other than lung cancer. Among women with undetected breast lymph node metastasis, the 5 year survival rate is only 27% [3]. The presence of lymph node metastasis is the single most important prognostic factor in breast cancer [1]. Beyond prognosis, the detection of nodal metastases is used for cancer staging and to determine the course of surgical treatment, and is an important index for postoperative chemotherapy and radiotherapy.

The management of breast cancer entails a time consuming sequence of costly steps to diagnose whether the patient has axillary nodal metastases. In many hospitals this process entails: (1) ultrasound (US) imaging of the axilla along with tumor diagnosis costing $750, (2) breast DCE MRI (if US is negative MRI may still be positive) at $3,500, (3) axillary US again to reidentify the sentinel node $750, (4) US guided biopsy at $1,500, (5) pathology evaluation of the biopsy specimen costing $1,200. The earlier steps, typically (1) and (2), are used towards the clinical diagnosis of the lymph node status, cNode, which has 4 levels:N0 (no nodal metastasis), and increasing levels of nodal metastasis: N1, N2, and N3.

The purpose and clinical value of this study is to determine whether preoperative, clinical DCE MRI of the primary tumor may be used to predict node status with a deep learning model. If possible then steps 1,3 and 4 could be eliminated or reserved for only those patients with uncertain or probable metastases. Relying upon clinical MRI (1.5T) enables research results to have potential for direct clinical impact without costly upgrades to 3.0T. Furthermore, methods developed using the primary tumor will be applicable to most patients given the current standard protocol that images the tumor, while axillary nodes may not be in the field of view (FOV).

Dynamic contrast enhanced (DCE) MRI contains abundant information about the vascularity and structure of the tumor and is valuable to quantify cancer aggressiveness. Recently, preliminary results demonstrated that hand crafted tumoral features from 2D DCE MRI are can help predict nodal metastasis using classical machine learning SVM [6]. While promising these results relied upon research grade 3.0T MRI and were restricted to data from just 100 subjects acquired at one hospital. Furthermore validation performance is reported rather than more rigorous test set performance.

Deep learning has been shown to find useful patterns for image analysis [4]. It has been used to detect breast cancer with near expert radiologist accuracy [7]. However to the best of our knowledge, deep learning has not been used to predict axillary nodal metastasis from DCE MRI.

The strengths and contributions of this work are four-fold: (1) A 4D CNN model is proposed that automatically learns to fuse information from 4D DCE

MRI (3D over time) and non-imaging clinical information. (2) The model relies exclusively on the primary tumor and does not require nodes to be within the FOV nor high field strength imaging which may not be available. (3) The model achieves a promising 72% accuracy while using rigorous nested cross-fold validation and while training and testing upon an extensive dataset of 357 subjects from two hospital sites using two distinct image acquisition protocols and hardware. (4) Saliency mapping demonstrates that the proposed model correctly learns to utilize primary tumor voxels when identifying both metastatic and non-metastatic subjects.

2   Materials and Methods

2.1   Materials

Clinical 1.5T DCE MRI was obtained from 357 breast cancer patients, whose characteristics are summarized in Table 1. Data for each subject includes a single precontrast and four serial dynamic image volumes acquired at a temporal resolution of 90s/phase obtained before and immediately after intravenous bolus infusion of a contrast agent. 221 subjects were obtained from *Parkland Hospital, Dallas, Texas* where dynamic VIBRANT sagittal images were acquired with a GE Optima MR450w 1.5T scanner using 0.1 mmol/kg gadopentetate dimeglumine contrast medium. The remaining 136 subjects came from *UT Southwestern Medical Center, Dallas, Texas* where dynamic FSPGR (THRIVE) axial images are acquired with a Philips Intera 1.5T scanner using 0.1 mmol/kg Gadavist contrast medium. Additionally, *four clinical features* were obtained including age (yrs), estrogen receptor status (ER), human epidermal growth factor receptor-2 (HER2), and a marker for proliferation (Ki-67). *Clinical node status* (cNode) ground truth was determined by one of 13 board-certified radiologists, fellowship-trained in breast imaging and breast MRI, who assess the 4D DCE MRI and ultrasound imaging information, clinical measures, clinical history available at the time of the image reading.

2.2   Methods

Each subject has five 3D MRI volumes which are denoted as time1, time2, time3, time4, time5. Board certified radiologists traced the boundary of the primary tumor of each subject on the time3 volume. Then a 3D cuboidal bounding box encompassing the tumor region of interest (ROI) and peri-tumoral area was defined and used to crop each subject data to consistently-sized 3D volumes. Three difference images were then defined for each subject by subtracting voxelwise the cropped time3-time1, time4-time1 and time5-time1. These difference volumes are used to train the deep learning model and the processing steps are illustrated in Fig. 1. Next the intensities for the difference images from each hospital are harmonized by: (1) clipping the values of the lowest and highest 0.5%

Table 1. Demographics and disease characteristic of subjects included in this analysis.

Variable	Age							cNode status				Tumor stage			
Category	21–30	31–40	41–50	51–60	61–70	71–80	81–90	N0	N1	N2	N3	T1	T2	T3	T4
Percentage	2	16	34	23	19	5	1	62	28	4	6	30	44	19	7
Number of patients	7	57	121	82	68	18	4	221	101	13	22	107	157	68	25

intensities, and (2) computing the mean and standard deviation of the intensities per hospital and transforming the intensities to have zero mean and unit variance.

The data is partitioned using nested, stratified group 10-fold cross-validation with the test data held out and not used during training nor validation. This ensured a subject's data appears in only 1 fold and each fold has the same ratio of patients from each hospital, and the same ratio of negative (cNode = N0) and positive (N1, N2, N3) labels. Data is augmented 27× by small random translation and rotation of the cropping volume.

In this work, we develop two-category classifiers to predict whether or not there is lymph node metastasis. Our classifiers are predominantly CNNs [2] because they automatically learn a hierarchy of intensity features from difference volumes. For brevity we explain the most complex 4D hybrid CNN model that we build which takes as input a set of difference volumes and the clinical data. The remaining models (e.g. 2D CNN, 3D CNN) are simpler and their architectural details can be found in the supplemental file. The 4D hybrid CNN model consist of four 4D convolutional layers and three fully connected layers. Each convolutional layer is followed by max-pooling. Then the output of the convolutional layers are concatenated with the clinical features to form the final classifier and all layers use batch normalization [5], except the last which outputs the metastasis diagnosis. Figure 2 visualizes the model architecture and how the data dimensions change in each layer.

The output of the model can be summarized in Eq. (1)

$$\mathbf{p}^k = f(\mathbf{x}^k, \mathbf{c}^k) \tag{1}$$

where $\mathbf{x}^k$ is the set of 3D difference images for the kth patient, $\mathbf{c}^k$ is their clinical data, and 2-dimensional $\mathbf{p}$ is the predicted probabilities that the model $f()$ assigns to the two output categories. The threshold is $>= 0.5$ for positive prediction and < 0.5 for negative prediction.

Since the dataset has moderate imbalance (62% of subjects are non-metastatic) and we would like the model to focus more on positive than negative cases, we apply a weighted cost function where the cost for positive cases is twice that of the negative cases:

$$E = \frac{1}{N} \sum_{k=1}^{N} \sum_{c=1}^{2} (p_c^k - l_c^k)^2 \cdot w_c \tag{2}$$

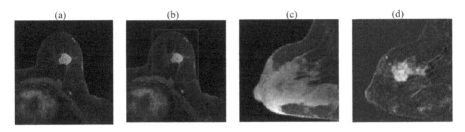

Fig. 1. Preprocessing the volumetric DCE MRI. (a) primary tumor is radiologist delineated at time3 in each slice (green contour), (b) MRI is cropped to a cuboidal volume around tumor, (c) sagittal view showing breast at time1, (d) tumor is enhanced by computing difference images, shown here: time3-time1. (Color figure online)

where E is the cost function, N is the number of subjects, $\mathbf{l}^k$ and $\mathbf{p}^k$ are the label and prediction of the kth patient respectively, and $\mathbf{w}$ is the weight assigned according to class label where: $w_1 = 1$ and $w_2 = 2$. This MSE cost function works well on classification tasks when using a softmax output [10]. Model fitting is trained from scratch and model weights are learned with the Adam [9] optimizer using: an adaptive learning rate initialized to 0.001, beta1 = 0.9, beta2 = 0.999, and a batch size of 36. Models are implemented in Tensorflow and trained on a Linux workstation with 2 NVIDIA V100s GPUs.

3   Results

Five types of input feature sets were tested: clinical features alone, 2D images alone, 3D images alone, 4D images (3D + time), and 4D with clinical features. Results are summarized in Table 2. *With clinical data only,* an XGBoost classifier [11] attained a performance just above chance accuracy with an AUC = 0.55. Dense feedforward neural networks did not perform better. The relative importance of the 4 clinical features computed as f scores, are shown in Fig. 3. **Ki67** and **HER2** are top ranked and have approximately equal importance while **age** and **ER** status are less important. *Using 2D difference images* and a deep 2D CNN, performance improved to an AUC = 0.61. *Using the additional context of 3D difference images* performance further improved to AUC = 0.66. Each 3D difference image is input independently, however group stratification is used so that a subject's data appears in either train, validation or test. Subsequently, all three 3D difference images of a patient (time3-time1, 4-1, and 5-1) are concatenated forming an input 4D tensor to the 4D CNN. This produced the best results including an AUC = 0.67 and, when combined with clinical data an AUC of 0.71 and true positive rate of 0.72. This compares favorably with the results of a related study [13] in which texture features were manually specified.

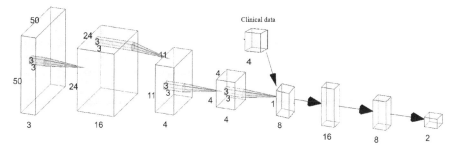

Fig. 2. The 4D CNN model architecture. The model consists of 4 convolutional layers (red pyramids) followed by 3 fully connected layers (horz. arrows). Input and feature maps are 4D tensors; to visualize them they are rendered as 3D volumes by omitting *one dimension*, e.g. the input layer (left) is $50 \times 50 \times \underline{50} \times 3$ but rendered as $50 \times 50 \times 3$. Four clinical features are concatenated with the 4 outputs of last conv. layer, creating an 8-vector as input to the dense layers. One-hot encoded output layer predicts probability of nodal metastasis and no metastasis. Illustration generated using [12]. (Color figure online)

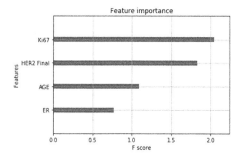

Fig. 3. Relative importance of each clinical feature. Features with higher scores are more valuable for predicting metastasis.

Table 2. Comparative performance across input feature sets on the held out test set. Increasing prediction performance was observed across the five sets of inputs evaluated including: only clinical features, 2D images, 3D images, 4D images (3D + time), and 4D with clinical features.

Clinical only		2D image		3D image		4D image		4D img & clinical	
AUC	TPR	AUC	TPR	AUC	TPR	AUC	TPR	AUC	TPR
0.55	0.24	0.61	0.35	0.66	0.84	0.67	0.72	0.71	0.72

4   Discussion

The results (Table 2) suggest the more data the model has the better its performance. This makes sense: 4D provides spatiotemporal context that 3D and 2D lack, while clinical features provide biomolecular information not accessible

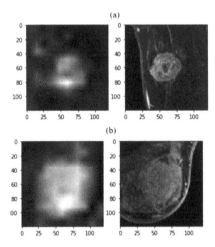

Fig. 4. Important voxels revealed through saliency mapping. Saliency mapping with Grad-CAM (left column) demonstrates that the primary active tumor voxels in MRI (right column) are those most valuable for predicting boht (a) non-metastatic cancer, and (b) axillary metastatic cancer.

Table 3. Comparison prediction accuracy and stability across feature sets. Both accuracy and stability increases when clinical features are added to the 4D CNN.

Fold	4D image AUC	TPR	TNR	4D & clinical AUC	TPR	TNR
0	0.61	0.71	0.45	0.77	0.71	0.73
1	0.78	0.79	0.68	0.85	0.71	0.76
2	0.71	0.93	0.23	0.83	0.86	0.64
3	0.82	0.71	0.77	0.67	1.00	0.00
4	0.37	1.00	0.00	0.58	0.29	0.87
5	0.65	0.64	0.59	0.62	0.71	0.55
6	0.82	1.00	0.00	0.8	0.83	0.61
7	0.62	0.38	0.64	0.54	0.62	0.41
8	0.75	0.62	0.82	0.8	0.92	0.48
9	0.55	0.46	0.55	0.56	0.54	0.52
mean	0.67	0.72	0.47	**0.71**	**0.72**	**0.56**
std	0.14	0.21	0.3	0.11	0.21	0.24

through MRI. To reveal which regions the top model learned as important Grad-CAM [8] was used to generate saliency maps. The most important voxels to predict metastasis free (Fig. 4a) or high probability of metastasis (Fig. 4b) are the primary tumor and its surround. More distal voxels are less important. That healthy breast tissue and non-breast tissue are of less value, makes sense biologically, and since the algorithm was not explicitly provided the non-linear tumor boundary this saliency map results suggests the model learned the segmentation on its own. Additional benefits are observed when adding clinical data to the 4D model including improving both the AUC and the true negative rate (TNR), while simultaneously reducing the variance across folds (Table 3), which further increases confidence in these performance estimates.

Our approach has some room for further improvement. Our reported AUC is 0.72 is very promising, but still needs improvement. We expect to improve the AUC to at least 0.8 to facilitate clinical adoption, while 0.9 is a long term target. There are several steps that can lead us there. Differences between hospital data is currently mitigated through preprocessing, however adversarial domain adaption could directly learn a model agnostic to these differences and may improve performance. Second, unsupervised pretraining on external data could improve performance. Third, other data combination methods beyond concatenation could be beneficial. Fourth, our preprocessing assumes no motion between MRI time frames, however patient movement can occur and could impede model performance. In the future we will apply motion correction to suppress any difference image artifact.

5   Conclusions

This work demonstrates that a deep 4D CNN has the potential to learn to predict axillary nodal metastases with high accuracy through the judicious fusion of spatiotemporal features of the primary tumor visible in DCE MRI, and that this further improves with the addition of clinical measures. Such high precision noninvasive methods that utilize standard clinical MRI (1.5T) and do not require complete imaging of the axillary nodes would fit well with clinical practices and could with further refinement, help patients avoid the costs associated with unnecessary lymph node surgery. The proposed method was tested on an extensive dataset of 357 subjects from 2 hospitals with distinct imaging protocols. Saliency mapping demonstrated that the model used tumor voxels to predict nodal metastasis which agrees with the expectation that aggressive tumors that spread to the nodes have appearance distinct from less invasive tumors. Such diagnostic methods hold the potential to streamline time consuming and costly steps currently used to clinically diagnose nodal metastases, improve doctor efficiency, and help select safe and effective treatments that reduce postoperative complications.

References

1. Fisher, B., et al.: Relation of number of positive axillary nodes to the prognosis of patients with primary breast cancer. An NSABP update. Cancer **52**(9), 1551–1557 (1983)
2. Soffer, S., Ben-Cohen, A., Shimon, O., Amitai, M.M., Greenspan, H., Klang, E.: Convolutional neural networks for radiologic images: a radiologist's guide. Radiology **290**(3), 590–606 (2019)
3. American Cancer Society Homepage, March 2020. https://www.cancer.org/cancer/breast-cancer/understanding-a-breast-cancer-diagnosis/breast-cancer-survival-rates.html
4. LeCun, Y., Bengio, Y., Hinton, G.: Deep learning. Nature **521**(7553), 436–444 (2015)
5. Ioffe, S., Szegedy, C.: Batch normalization: accelerating deep network training by reducing internal covariate shift (2015). arXiv preprint arXiv:1502.03167
6. Cui, X., et al.: Preoperative prediction of axillary lymph node metastasis in breast cancer using radiomics features of DCE-MRI. Sci. Rep. **9**(1), 1–8 (2019)
7. Zhou, J., et al.: Weakly supervised 3D deep learning for breast cancer classification and localization of the lesions in MR images. J. Magn. Reson. Imaging **50**(4), 1144–1151 (2019)
8. Selvaraju, R.R., Cogswell, M., Das, A., Vedantam, R., Parikh, D., Batra, D.: Grad-CAM: visual explanations from deep networks via gradient-based localization. In: Proceedings of the IEEE International Conference on Computer Vision, pp. 618–626 (2017)

9. Kingma, D.P., Ba, J.: Adam: a method for stochastic optimization (2014). arXiv preprint arXiv:1412.6980
10. Tyagi, K., Nguyen, S., Rawat, R., Manry, M.: Second order training and sizing for the multilayer perceptron. Neural Process. Lett. **51**(1), 963–991 (2019). https://doi.org/10.1007/s11063-019-10116-7
11. Chen, T., Guestrin, C.: XGBoost: a scalable tree boosting system. In: Proceedings of the 22nd ACM SIGKDD International Conference on Knowledge Discovery and Data Mining, pp. 785–794, August 2016
12. LeNail, A.: NN-SVG: publication-ready neural network architecture schematics. J. Open Source Softw. **4**(33), 747 (2019). https://doi.org/10.21105/joss.00747
13. Kulkarni, S., Xi, Y., Ganti, R., Lewis, M., Lenkinski, R., Dogan, B.: Contrast texture-derived MRI radiomics correlate with breast cancer clinico-pathological prognostic factors. In: Radiological Society of North America 2017 Scientific Assembly and Annual Meeting, Chicago, IL, 26 November–1 December 2017. archive.rsna.org/2017/17002173.html

Learning Differential Diagnosis of Skin Conditions with Co-occurrence Supervision Using Graph Convolutional Networks

Junyan Wu[1], Hao Jiang[2], Xiaowei Ding[1(✉)], Anudeep Konda[1], Jin Han[1], Yang Zhang[1], and Qian Li[1]

[1] Voxelcloud Inc., Los Angeles, CA, USA
xding@voxelcloud.io
[2] Northeastern University, Shenyang, Liaoning Province, China

Abstract. Skin conditions are reported the 4th leading cause of nonfatal disease burden worldwide. However, given the colossal spectrum of skin disorders defined clinically and shortage in dermatology expertise, diagnosing skin conditions in a timely and accurate manner remains a challenging task. Using computer vision technologies, a deep learning system has proven effective assisting clinicians in image diagnostics of radiology, ophthalmology and more. In this paper, we propose a deep learning system (DLS) that may predict differential diagnosis of skin conditions using clinical images. Our DLS formulates the differential diagnostics as a multi-label classification task over 80 conditions when only incomplete image labels are available. We tackle the label incompleteness problem by combining a classification network with a Graph Convolutional Network (GCN) that characterizes label co-occurrence and effectively regularizes it towards a sparse representation. Our approach is demonstrated on 136,462 clinical images and concludes that the classification accuracy greatly benefit from the Co-occurrence supervision. Our DLS achieves 93.6% top-5 accuracy on 12,378 test images and consistently outperform the baseline classification network.

Keywords: Graph Convolutional Networks (GCN) · Multi-label classification · Incomplete label · Skin differential diagnosis

1 Introduction

Skin problems and conditions are common health concerns and their diagnostics are largely based on visual clues. According to [8], 27% of the U.S. population were seen by a physician for skin disease in 2013 and the affected individuals

---

Electronic supplementary material The online version of this chapter (https://doi.org/10.1007/978-3-030-59713-9_33) contains supplementary material, which is available to authorized users.

A. L. Martel et al. (Eds.): MICCAI 2020, LNCS 12262, pp. 335–344, 2020.
https://doi.org/10.1007/978-3-030-59713-9_33

averaged 1.5 skin diseases. Diagnosing and treating skin conditions remains a challenge as a diverse set of skin diseases, with over 3000 entities identified in the literature [10], need to be differentiated by dermatologists who are in significant shortage relative to the rising demand. As a first step, deriving a group of possible causes from visual impression is deeply rooted in clinical practice, therefore often refereed to as "differential diagnosis" (see [1]).

Recent advances in computer vision promise an accessible and reliable solution to differential diagnosis on clinical skin images. Previous work has demonstrated the efficacy of Covolutional Neural Networks (CNN) powering decision support systems to radiologists, ophthalmologists and pathologists. In dermatology, dermoscopy images has attracted much attention in which the image modality is more standardized and target labels are multiple magnitudes less than the number of skin conditions. Recently, more efforts are cast on clinical images in a direct effort to target multiple skin conditions, for example in [9], a deep learning system(DLS) was trained to distinguish 26 disease classes. In their work, authors misrepresented differential diagnosis as a multi-class classification problem therefore inherently undermined the interpretability of predictions and correlation between labels. [4] took on a dataset of 129,450 clinical images consisting of 2032 skin conditions, however, it overlooked the differential diagnosis problem and evaluated their DLS mostly on binary classification tasks, *i.e.* cancer versus non-cancer.

In this paper, we propose a deep learning system (DLS) that may predict differential diagnosis of skin conditions using clinical images. Our DLS formulates the differential diagnostics as a multi-label classification task over 80 conditions when only incomplete image labels are available. We tackle the label incompleteness problem by combining a classification network with a Graph Convolutional Network (GCN) that characterizes label co-occurrence and effectively regularizes it towards a sparse representation. Our approach is demonstrated on 136,462 clinical images and concludes that the classification accuracy greatly benefit from the co-occurrence supervision.

Our GCN-CNN approach highlights three major advantages:

- By introducing co-occurrence supervision by means of GCN layers, we effectively handle correlated image labels even when annotations are incomplete.
- Our approach is end-to-end trainable and readily applicable to any CNN architecture.
- GCN can be flexibly initialized by either empirical or expert-provided inputs that may adapt well per applications.

Finally, we evaluate our DLS on 12,378 user taken images acquired through a telehealth platform. To the best of our knowledge, this is the first study to investigate the performance of a DLS to differentiate skin conditions outside clinic. We report a top-5 accuracy of 93.6% on test images and argue for the value of DLS in extending the reach of dermatological expertise with tremendous accessibility and accuracy.

2   Related Work

Graph Convolutional Networks (GCNs) were first introduced in [6]. In its original application, *i.e.* the problem of nodes classification, only a small subset of nodes had their labels available. By introducing a fast approximation to spectral graph convolutions, labels to unknown nodes can be effectively learned in a semi-supervised manner as information from labeled nodes propagates through GCN layers.

In [3], ML-GCN was proposed for multi-label classification task. Different from [6], a graph structure was constructed from data and ML-GCN may directly incorporate a representation learned from a convolutional network. The graph structure is a directed graph over object labels and an edge of "Label$_i$ → Label$_j$" means when Label$_i$ is present, Label$_j$ is likely to be present too.

Formally, the output of ML-GCN, *i.e.*, $\boldsymbol{W} = \{\boldsymbol{w}_i\}_{i=1}^{C}$, parameterizes a mapping from feature vectors, learned from a conventional convolution network, to C labels. As for the final prediction, a CNN-based model learns an image representation $\boldsymbol{x}$ and the predicted scores $\boldsymbol{y}$ can be derived as,

$$\text{CNN} \quad \widehat{\boldsymbol{y}} = f(\boldsymbol{W} \cdot \boldsymbol{x}) \tag{1}$$

$$\text{ML-CNN} \quad \widehat{\boldsymbol{y}} = f(\tilde{\boldsymbol{W}}(\boldsymbol{D}, \boldsymbol{Z}) \cdot \boldsymbol{x}), \tag{2}$$

where D is the directed graph derived from correlation, $Z = \{z_i\}_{i=1}^{C}$ is a set of semantic embeddings to each label, both of which are predetermined therefore fed into GCN based classifier $\tilde{\boldsymbol{W}}$. As authors claimed, $\tilde{\boldsymbol{W}}$ benefits from both conditional dependence characterized in $\boldsymbol{D}$ and semantic proximity embedded in $\boldsymbol{Z}$.

3   Method

Our GCN-CNN approach naturally extends ML-GCN and tailor it specifically to the differential diagnostics problem. The overall framework of GCN-CNN is presented in Fig. 1. An undirected graph replaces the directed graph in [3] given the symmetry of conditional probability when two conditions are discerned. We leverage spectral graph convolution in [6] between GCN layers. Finally, we demonstrate and evaluate the graph construction empirically from label co-occurrence and by medical expert.

Our GCN branch consists of two graph convolutional (GC) layers. We follow the work of [6] and use the k−th order filter of spectral graph convolution that propagates the neighboring nodes up to k steps. The first GC layer (GCN-1) takes order 1 and convolves directly on neighboring nodes that is equivalent to a $1-$st order spectral filter, similarly, the second layer (GCN-2) takes the same order and extends to indirect neighbors as a $2-$nd order filter on the original graph. Except for computational advantage which reduces the graph convolution complexity to linear to the number of edges, this characterization avoids over smoothing label nodes by a deep GCN.

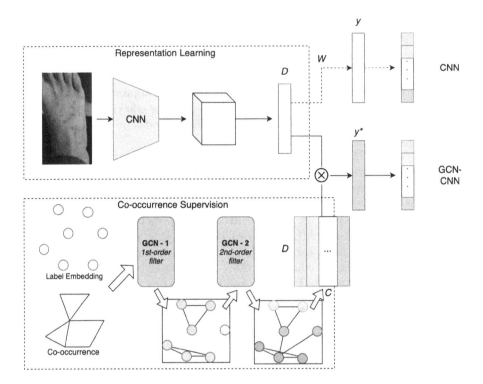

Fig. 1. Overview of GCN-CNN: the GCN branch propagates label co-occurrence and semantic embedding. A trainable representation network has its feature vectors dot product with GCN output and generate final predictions.

During the training time, we empirically estimate a co-occurrence graph using only training data. An undirected graph $\mathcal{G} = \{\mathcal{V}, \mathcal{E}\}$ encodes the conditional dependency between image labels, *i.e.* skin condition, and implicitly supervises the classification task. Node representations in $\mathcal{V}$ embed semantic meaning to labels which complements $\mathcal{G}$ particularly when one label includes another as a substring. As for edges, $e_{i,j} \in \mathcal{E}$ is sparsely constructed such that

$$e_{i,j} = \mathbb{1}(\frac{C(i,j)}{C(i) + C(j)} \geq t), \tag{3}$$

where $C(i,j)$ is the number of images that have both label i and j, $C(i)$ and $C(i)$ are the total number of images in class i and j respectively. In our experiment, a differential graph is also constructed with domain knowledge by board certified dermatologists. We ask two dermatologists to provide overlapped differential diagnoses groups, as many as possible, and connect an edge when two labels appear in at least one differential group by both dermatologists.

Since our approach is end-to-end trainable, we simply use multi-label cross entropy as loss function to the classification task.

4   Experiments

4.1   Dataset

A total of 136,462 user taken images were used for training and testing. In our dataset, the images are directly acquired by end users on a telehealth platform operating in China. All images are extracted randomly from consultations to physicians that involves skin problems. We split this dataset into a training set of 124,084 images and a test set of 12,378 images and conduct annotation differently.

For the training set, we obtain single reader annotation by randomly selecting a dermatologist from a group of 25. Due to the cost of annotating medical images and limited expertise available to the task, single reading on the image is commonly used in practice. To each dermatologist, we present the image and ask for their impression of skin conditions relative to the symptoms that manifests, as many as possible. Upon finish, 81.7% of training images carry a single label which outnumber doubly labeled images (15.5%) and triply labeled images (2.8%).

Multi-reader annotations were collected and aggregated on the test set. At least two dermatologists were blindly and independently presented with the same images and asked to conduct annotation the same as training. As the first two dermatologists fail to converge, a third dermatologist is involved and final labels are determined by majority voting. In contrast to the training set, testing images have 46.0% singly labeled, 38.1% doubly labeled and 12.7% triply labeled. The label distribution is significantly different between training and testing and it highlights the label incompleteness issues commonly encountered on single reader annotations.

Finally, we select the top 80 frequent conditions and a complete list is available in Appendix A. Differential graph contributed by domain experts takes into account lesion morphology, configuration and distribution, more details are provided in Appendix B.

4.2   Experiment Setting

The node representation of our GCN takes dimensions of 700 (GCN-0), 1024 (GCN-1) and 2048 (GCN-2). The initial label embedding input at GCN-0 utilizes BioSentVec [2] specifically trained on biomedical corpus. We investigate two undirected graph construction: empirical label co-occurrence and differential graph by domain experts.

All input images are downsized to 448×448 and we use a Resnet-101 [5] as the classification backbone. A linear layer was added after FC-2048 to conduct dot product between image features and GCN node features.

During training time, CNN backbone is first trained with 300 epochs at an initial learning rate of 0.1 with step decay. Thereafter, we randomly initialize GCN branch and train the whole GCN-CNN architecture end-to-end for another 300 epochs at a learning rate of 0.0003.

To formally evaluate our approach, we consider Resnet-101 alone as the baseline and benchmark our method against Li, Y et al. [7], which improved multi-label classification by means of a novel loss function for pairwise ranking. More efforts focus on the GCN branch as multiple constructions are formally compared.

4.3   Evaluation Metrics

Top 1/3/5 accuracy and mAP across all labels are major metrics of interest in evaluation. Additionally, we report Hamming Loss, Ranking Loss and Ranking One Error that are particularly suited for multi-label classification problems. Formal definitions are available in Appendix D.

It is worth mentioning that top-n accuracy considers a prediction true positive when the top-n predictions overlap with GT label set. Therefore, it characterizes the relevance of model prediction rather than the completeness of differential label sets. Authors recommend multi-label metrics since they complement top-n metrics and hint on the comprehensiveness of model predictions while adjusting for the prediction ranking.

5   Results

5.1   Performance Gain on Co-occurrence Graph

As shown in Table 1, GCN-CNN with empirical co-occurrence graph outperforms its competitors by all measures. It also predicts differential diagnoses more accurately and comprehensively based on multi-label classification metrics in Table 2. Random graph initialization leads to inferior performance to baseline, not surprisingly, given the limited depth of GCN constrains its representativeness of label dependency.

Table 1. Performance comparison: classification metrics

Method	top1 acc	top3 acc	top5 acc	mAP
Resnet-101	0.682	0.866	0.918	0.5067
Li, Y et al. [7]	0.689	0.866	0.916	0.498
Ours (random graph)	0.653	0.855	0.912	0.469
Ours (co-occurrence graph)	**0.703**	**0.885**	**0.936**	**0.546**

We exemplify model predictions in Fig. 2. The baseline Resnet makes fewer predictions given the imbalanced labels between training and testing. The label incompleteness also deteriorates the performance of Li, Y et al. which is incapable of leverage co-occurrence information effectively. The number of false positive

Table 2. Performance comparison: multi-label metrics

Method	Hamming Loss	Ranking Loss	Ranking One Error
Resnet-101	0.164	0.475	0.326
Li, Y et al. [7]	0.171	0.492	0.341
Ours (random graph)	0.274	0.496	0.340
Ours (co-occurrence graph)	**0.093**	**0.456**	**0.287**

predictions by Li, Y et al. also increases as label ranking is adjusted based upon incomplete supervision.

Given the sparsity of co-occurring labels in our training data, we did not investigate the sensitivity of model performance to the choice of threshold t in Eq. (3).

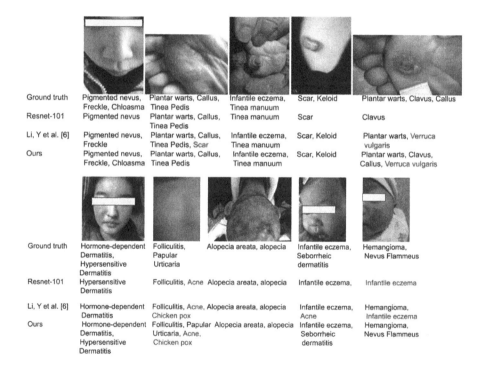

Fig. 2. Sample predictions: false positives are colored red. (Color figure online)

5.2 Graph Output from GCN

As a byproduct of our GCN branch, the proximity of nodes after GCN-2 may hint on a refined label dependency. We consider the difference between GCN-0 nodes proximity and GCN-2 nodes proximity as the learned label dependency. In Fig. 3, BioSentVec encoded label proximity is mainly driven by their semantic meaning. However in Fig. 4, label proximity has reduced to isolated clusters that highly correlate with differential groups, e.g. Appendage lesions, Perithyroid disease, Ulcerative changes, etc. This observation explains how GCN-CNN has improved classification accuracy and introduces extra interpretability of the system.

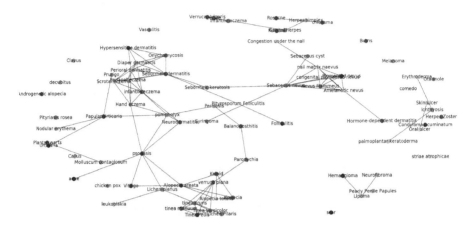

Fig. 3. Nodes proximity at GCN-0: BioSentVec embeddings.

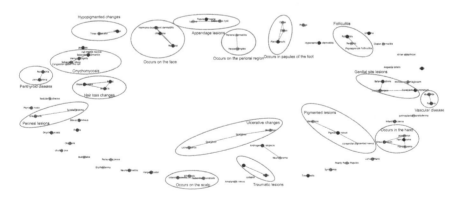

Fig. 4. Nodes proximity at GCN-2: isolated clusters are highlighted and interpreted by dermatologists.

5.3 Differential Graph by Domain Experts

We also evaluate an undirected graph construction by domain experts. As shown in Table 3, differential graph achieves comparable performance against empirical co-occurrence graph. Marginal gains are observed on top 1, 3 and 5 accuracy but none is statistically significant. Considering the laborious task to construct such graph when the label set grows, empirical co-occurrence graph is an strongly recommended.

Table 3. Performance comparison for different graph construction

Method	top1 acc	top3 acc	top5 acc	mAP
Ours (co-occurrence graph)	0.703	0.885	0.936	**0.546**
Ours (knowledge graph)	**0.708**	**0.891**	**0.940**	0.545

6 Conclusion

We introduce label co-occurrence supervision via a GCN branch for the problem of differential diagnosis of skin conditions. Our approach has significantly improved classification accuracy and completeness even when trained on incomplete labels that are commonly seen in medical imaging applications. By testing on user taken images of skin issues, we report a top-5 accuracy of 93.6%. This deep learning system is promising to be used as clinical decision support to medical professional with limited training in dermatology, as well as an accessible self-diagnostic tool directly to consumers.

Besides, the GCN branch leads to explainable visualization of label proximity that can be readily utilized for interpretation and debugging. Moreover, our GCN-CNN approach is end-to-end trainable, may adapt to any classification backbone and add zero effort during inference time. Furthermore, the GCN branch may extend to a diversity of features, e.g. patient demographics and medical history, therefore leverage multi-modal information in a classification task.

References

1. Ashton, R., Leppard, B.: Differential Diagnosis in Dermatology. CRC Press, Boca Raton (2014)
2. Chen, Q., Peng, Y., Lu, Z.: BioSentVec: creating sentence embeddings for biomedical texts. In: 2019 IEEE International Conference on Healthcare Informatics (ICHI), pp. 1–5. IEEE (2019)
3. Chen, Z.M., Wei, X.S., Wang, P., Guo, Y.: Multi-label image recognition with graph convolutional networks. In: Proceedings of the IEEE Conference on Computer Vision and Pattern Recognition, pp. 5177–5186 (2019)
4. Esteva, A., et al.: Dermatologist-level classification of skin cancer with deep neural networks. Nature **542**(7639), 115–118 (2017)
5. He, K., Zhang, X., Ren, S., Sun, J.: Deep residual learning for image recognition. In: Proceedings of the IEEE Conference on Computer Vision and Pattern Recognition, pp. 770–778 (2016)

6. Kipf, T.N., Welling, M.: Semi-supervised classification with graph convolutional networks. arXiv preprint arXiv:1609.02907 (2016)
7. Li, Y., Song, Y., Luo, J.: Improving pairwise ranking for multi-label image classification. In: Proceedings of the IEEE Conference on Computer Vision and Pattern Recognition, pp. 3617–3625 (2017)
8. Lim, H.W., et al.: The burden of skin disease in the United States. J. Am. Acad. Dermatol. **76**(5), 958–972 (2017)
9. Liu, Y., et al.: A deep learning system for differential diagnosis of skin diseases. arXiv preprint arXiv:1909.05382 (2019)
10. Segre, J.A., et al.: Epidermal barrier formation and recovery in skin disorders. J. Clin. Investig. **116**(5), 1150–1158 (2006)

Cross-Domain Methods
and Reconstruction

Unified Cross-Modality Feature Disentangler for Unsupervised Multi-domain MRI Abdomen Organs Segmentation

Jue Jiang and Harini Veeraraghavan[✉]

Department of Medical Physics, Memorial Sloan Kettering Cancer Center,
New York, USA
veerarah@mskcc.org

Abstract. Our contribution is a unified cross-modality feature disentagling approach for multi-domain image translation and multiple organ segmentation. Using CT as the labeled source domain, our approach learns to segment multi-modal (T1-weighted and T2-weighted) MRI having no labeled data. Our approach uses a variational auto-encoder (VAE) to disentangle the image content from style. The VAE constrains the style feature encoding to match a universal prior (Gaussian) that is assumed to span the styles of all the source and target modalities. The extracted image style is converted into a latent style scaling code, which modulates the generator to produce multi-modality images according to the target domain code from the image content features. Finally, we introduce a joint distribution matching discriminator that combines the translated images with task-relevant segmentation probability maps to further constrain and regularize image-to-image (I2I) translations. We performed extensive comparisons to multiple state-of-the-art I2I translation and segmentation methods. Our approach resulted in the lowest average multi-domain image reconstruction error of 1.34 ± 0.04. Our approach produced an average Dice similarity coefficient (DSC) of 0.85 for T1w and 0.90 for T2w MRI for multi-organ segmentation, which was highly comparable to a fully supervised MRI multi-organ segmentation network (DSC of 0.86 for T1w and 0.90 for T2w MRI).

Keywords: Multi-domain translation · Disentagled networks · Unsupervised multi-modal MRI segmentation · abdominal organs

1 Introduction

Magnetic resonance imaging guided radiation therapy treatments require accurate segmentation, which is currently done by clinicians [1]. Despite the availability

Electronic supplementary material The online version of this chapter (https://doi.org/10.1007/978-3-030-59713-9_34) contains supplementary material, which is available to authorized users.

of in-treatment-room-imaging, due to lack of fast segmentation methods, these treatments cannot be used to adapt treatments and precisely target tumors. Deep learning methods cannot be directly applied to MRI as large expert-labeled datasets are lacking. Therefore, we developed an unsupervised multi-domain MRI (T1-weighted, T2-weighted) segmentation approach by using CT as a source domain. Our approach performs parameter efficient multi-domain adaptation without requiring multiple one-to-one domain adaptation networks.

Cross-domain adaptation has been successfully used for one-to-one domain adaptation and medical image segmentation [2–4]. These methods produce image-to-image (I2I) translations using the generative adversarial networks (GAN) [5], including the cycle GAN [6]. GANs map random noise into output images by modeling the intensity distribution of a target domain. But GANs make two unrealistic assumptions for model convergence; the discriminators have infinite capacity to drive the generators, and very large number of training samples are available [7]. The work in [8] showed that the use of appropriate priors like Lipschitz densities are needed to constrain GAN training and avoid mode collapse.

Bijection constraints, which regularize GAN training by matching a joint distribution of an image and a latent distribution [9,10], have shown feasibility to model diverse intensity and color variations within the same modality. However, these constraints do not specify the dependency (or correlation) structure between the image and the latent distribution and are not guaranteed to cover the different modes in the required target task [11,12]. Prior methods that combined image and feature-level constraints as in [3,4,6,13] are intrinsically one-to-one mappers, and extension to multi-domain mapping would require $\mathbb{O}(n \times (n-1))$ networks, which is not computationally feasible for practical applications.

Disentangling methods [14,15], which extract a domain-invariant content and domain specific style have shown generalizable classification performance on target domains while reducing mode collapse. Example approaches include, task-relevant losses [11], domain classifier losses [16–18], and domain generalization methods that match style features with a known prior assumed to span across a number of seen and unseen domains [11,12]. However, in addition to requiring domain-specific style encoders [16,17], these methods use image-level matching losses, which is insufficient to model translation of multiple organs that transform differently with respect to one another across the imaging domains.

Key Improvements and Differences: We used one universal content encoder and one variational auto-encoder to extract image content and style code from multiple domains. The style code is converted into a vector of latent style scales that modulate the generator filters processing content features; target domain code is injected into the generator for synthesizing target modality. Our approach requires no additional networks for encoding domain-specific style features, and requires similar number of networks and parameters as a one-to-one mapper. To our best knowledge, ours is the first approach to perform such scalable multi cross-modality adaptation for disparate medical imaging modalities. We improve upon mode-seeking constraints [19] to reduce mode collapse by introducing a joint distribution discriminator that combines images with their generated segmentation probability maps to compute domain mismatches. This preserves organs geometry and appearance.

Contributions:

1. A compact cross-modality feature disentangling approach for diverse medical imaging modalities adaptation,
2. An end-to-end multi-domain translation and unsupervised cross-modality segmentation network, and
3. A new joint distribution (image, segmentation map) discriminator to force preservation of multiple organ appearance on generated images.

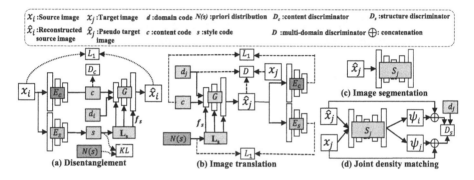

Fig. 1. Many-to-many domain translation and multi-domain MRI segmentation with joint distribution discriminator.

2 Method

2.1 Notation:

Bold letters denote a matrix, $\mathbf{x}, \mathbf{X}$; mapping functions are denoted by non-bold letters, e.g., $E_c : \mathbf{x_i} \rightarrow \mathbf{c}$, vectors are indicated by italicized letters, e.g., d. $\mathbf{X} = \cup_{k=1}^{N} \mathbf{X}_k \in \mathbb{R}^{H \times W}$ are the set of N domains, H is the height and W is the width of the images. The domain code is represented using one-hot-coding, and is denoted by $\mathbf{d} = \cup_{k=1}^{N} \mathbf{d}_k$. Source domain is indicated by i, the target domain by j, and the transformed image by $\hat{x}_j$. Domain-invariant content encoder is denoted by $E_c : x_i \rightarrow \mathbf{c}$, where $c \in \mathbb{R}^{H' \times W' \times C}$, and $\{H', W', C\}$ are the height, width of the convolved feature image and number of content features, respectively. The style encoder is denoted by $E_s : x_i \rightarrow s$, where $s \in \mathbb{R}^{1 \times C_s}$, C_s is the number of style features. G is the generator. L_s is the latent scale layer that maps style code s into a vector of latent style scales $f_s \in \mathbb{R}^{1 \times F}$, where F is the number of filters in the generator. D_c and D are content and multi-modality GAN discriminators. D_s is the joint distribution (image, segmentation probability map) discriminator. S_i is the segmentator for modality i.

2.2 Feature Disentanglement and Image Translation

Style and Content Feature Disentanglement. Disentangled image content and style features (Fig. 1a) are computed using a sequence of convolutional layers and a variational auto-encoder (VAE) [20], respectively. Assuming that a latent Gaussian prior ($z \in \mathcal{N}(0, I)$) spans the styles of all the domains, the encoder E_s extracts a style code that matches with this prior using KL-divergence. The style code is then transformed into latent style scale f_s by a latent scale (LS) layer [21]. The latent style scale for each domain is learned. It modulates (as a multiplier) the strength of the various generator filters processing the image content features to produce the desired I2I translation. The vector of latent style scale outputs are shown as multiple outputs for LS in Fig. 1(a). These are combined with generator filters using residual blocks. The domain code is injected through channel-wise concatenation with the content code. A self-reconstruction loss is combined with the KL-divergence loss in the VAE as:

$$L_{VAE} = \sum_i \mathop{E}_{x_i \sim X_i} [D_{KL}(E_s(x_i)||z)] + \|\hat{x}_i - x_i\|_1. \tag{1}$$

where $\hat{x}_i = G(E_c(x_i), L_s(E_s(x_i)), d_i)$. The domain invariant content features are produced using an adversarial training by:

$$L_{adv}^c = \sum_i \sum_{j;j \neq i} \mathop{E}_{\substack{x_i \sim X_i \\ x_j \sim X_j}} [log(D_c(E_c(x_i)))] + \mathbb{E}[1 - log(D_c(E_c(x_j)))], \tag{2}$$

Image Translation Losses: We compute content reconstruction, latent code regression, domain adversarial loss and mode-seeking losses to constrain multi-domain I2I translation (Fig. 1 b).

Content reconstruction loss forces content preservation in the transformation $\hat{x}_j = G(E_c(x_i), L_s(E_s(x_j)), d_j)$ upon transforming an image x_i from domain i to j. This is computed as:

$$L_c = \sum_i \sum_{j;j \neq i} \mathop{E}_{\substack{x_i \sim X_i \\ x_j \sim X_j}} \|E_c(x_i) - E_c(\hat{x}_j)\|_1. \tag{3}$$

Latent code regression loss constrains the generator to produce unique mappings of image for a given latent code z. This is accomplished through a reverse mapping that produces point-wise estimates of latent code as done in [9,22]. The latent regression error is then computed as:

$$L_{lr} = \sum_i \sum_{j;j \neq i} \mathop{E}_{x \sim X_i} \|z - E_s(G(E_c(x), L_s(z), d_j))\|_1. \tag{4}$$

Domain adversarial loss is computed to distinguish the generated images from the distribution of the individual domains using styles coded from x_j and sampled from $\mathcal{N}(0, I)$. This is computed as:

$$\min_G \max_D L_{GAN} = \sum_i \sum_{j;j\neq i} \{ \mathbb{E}_{x_j \sim X_j} [log(D(x_j, d_j))] +$$
$$\mathbb{E}_{x_i \sim X_i} [\frac{1}{2} log(1 - D(G(E_c(x_i), L_s(E_s(x_j)), d_j), d_j))] + \quad (5)$$
$$\mathbb{E}_{z \sim N(0,1)} [\frac{1}{2} log(1 - D(G(E_c(x_i), L_s(z), d_j), d_j))\}$$

Multi-domain translation is additionally stabilized by employing bias center instance normalization (BCIN) in both the generator and discriminators, as it has been shown to improve the consistency and diversity in the generated I2I transformations [23]. BCIN also allows the domain labels $d_j \in \mathbf{d}$ to be directly injected into G and D, which eliminates additional computations otherwise needed for calculating domain classification loss via the discriminator as done in [17,22].

Mode Seeking Loss: As proposed in [19], we compute a mode-seeking loss to prevent mode collapse such that the chances of producing the same image from two different latent vectors is reduced. Briefly, given two random latent style codes $z_1, z_2 \in \mathcal{N}(0, I)$, an image $x_i \sim X_i$, target domain code d_j and a generator G, mode seeking regularization maximizes the ratio of the distance between the generated images and the corresponding latent vectors as:

$$L_{ms} = \max_G \left(\sum_i \sum_{j;j\neq i} \frac{d_I(G(E_c(x_i), L_s(z_1), d_j), G(E_c(x_i), L_s(z_2), d_j))}{d_z(z_1, z_2)} \right) \quad (6)$$

Finally, a new joint distribution discriminator is employed as described in the next Subsection to constrain the multi-domain translation for segmentation.

2.3 Segmentation

CT is the source domain containing expert segmentations, $\{X_i, Y_i\}$, and the target MRI domain only contains the image sequences for training $\mathbf{X} = \cup_{j=1, j\neq i}^N \mathbf{X}_j$. Separate multi-organ segmentation networks S_j are trained for each target modality j (Fig. 1(c)) by using transformed images $\hat{x}_j = G(E_c(x_i), L_s(z), d_j)$, obtained via randomly sampled style z during segmentation training. Cross entropy loss was used to optimize these networks as:

$$L_{seg} = \sum_{j,j\neq i} \mathbb{E}_{\hat{x}_j \sim \hat{X}_j, l_i \sim Y_i} [log P(l_i | S_j(\hat{x}_j))]. \quad (7)$$

Joint Distribution Structure Discriminator. Prior work [4] has shown that modality hallucination occurs while transforming highly disparate modalities when GAN training is not regularized to model the structures of interest. Therefore, we introduce a new adversarially trained joint distribution structure discriminator D_s (Fig. 1(d)) that explicitly conditions image generation by focusing domain mismatch detection within the structures of interest. This is done by treating images and their segmentation probabilities as a joint distribution, and implemented as channel-wise concatenation. Voxel-wise segmentation probability maps (channel-wise accumulation except the 0-th channel that corresponds to background label) $\hat{\psi}_j$ for the CT to MRI translated and the unrelated real MRIs ψ_j are obtained from the *SoftMax* operation of the segmentation network S_j. D_s uses the domain code d_j to compute domain mismatch as:

$$L_{st} = \sum_{j:j\neq i} \{\mathbb{E}_{x_j \sim X_j}[log(D_s(x_j, \psi_j, d_j))] + \mathbb{E}_{\hat{x}_j \approx X_j}[log(1 - D_s(\hat{x}_j, \hat{\psi}_j, d_j))]\} \quad (8)$$

The total loss is computed as:

$$L_{total} = L_{GAN} + \lambda_{vae}L_{VAE} + \lambda_c L_{adv}^c + \lambda_{lr}L_{lr} + \lambda_{ms}L_{ms} + \lambda_{st}L_{st} + \lambda_{seg}L_{seg} \quad (9)$$

2.4 Implementation Details and Network Structure

All networks were implemented using the Pytorch library and were trained on Nvidia GTX V100 with 16 GB memory. The ADAM algorithm [24] with an initial learning rate of 2e-4 and batch size of 1 was used during training. We set $\lambda_{vae}=1$, $\lambda_c=1$, $\lambda_{lr}=10$, $\lambda_{ms}=1$, $\lambda_{st}=1$ and $\lambda_{seg}=5$ in the training. The learning rate was kept constant for the first 50 epochs and decayed to zero in the next 50 epochs. In order to ensure stable training, the encoders, G and VAE networks are trained cooperatively and optimized at a different iteration than the discriminators and segmentors.

The content encoder E_c is a fully convolutional network for projecting the images into a spatial feature map. The feature map retains the spatial structure by using a small output stride of 2. The style encoder E_s is composed of several convolution and pooling layers followed by global pooing and fully connected layers, with the output layer implemented using a reparameterization trick as done in [20]. The latent scale module L_s consists of 5 fully connected layers with *tanh* operation to produce style scale features within a range of $[-1, 1]$. The generator G is composed of 5 residual blocks with BCIN used for domain code injection. The segmentation networks S are implemented using a standard Unet [25]. Content discriminator D_c is implemented using PatchGAN. The joint density structure discriminator D_s and multi-domain discriminator D are implemented using PatchGAN with BCIN used for domain injection. Details of all networks are included in the supplementary documents.

3   Experiments and Results

Dataset: We used 20 MRIs (T1w and T2w MRI) from the Combined Healthy Abdominal Organ Segmentation (CHAOS) challenge data [26] and a completely different set of 30 patients with expert-segmented CT scans from [27] for the analysis. CHAOS CT scans only have liver segmentations. Ten MRIs were used in training (without expert segmentations) and validation (with expert segmentations) while the remaining 10 MRIs were held out for independent testing.

Networks were trained using 256×256 pixels image patches obtained from 14038 individual CT slices, and from 8000 T1w and 7872 T2w MRI slices, respectively. All MRIs were acquired from 1.5T Philips scanners, with a resolution of 256×256, pixel spacing [1.36 mm to 1.89 mm] and slice thickness [5.5 mm to 9 mm], and processed with bias field correction. CTs had a resolution of 512×512 pixels, a pixel resolution [0.7 to 0.8 mm], and a slice thickness [3 mm to 3.2 mm]. CT and MR image sets were normalized in range −1 to +1 prior to training and testing.

Experiments: Performance comparisons were done against state-of-the-art multi-domain disentanglement based translation methods, DRIFT++ [22] and starGAN [17]. Translation accuracy was evaluated using cyclic reconstruction [8] of CT going through T1w and T2w MRI using mean absolute error (MAE) and through T-SNE cluster distances between translated and real MRIs.

Segmentation performance was compared against multiple translation based segmentation methods including Cycada [13], CycleGAN [6], variational auto-encoder based method UNIT [28], synergistic feature encoder (SIFA) [29], and the SynSeg [3] methods. Dice similarity coefficient (DSC) was used to measure accuracy. All methods were trained from scratch with identical image sets and subject to reasonable hyperparameter optimization for equitable comparisons.

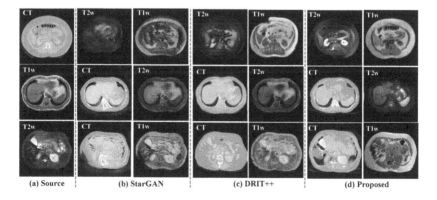

Fig. 2. Multi-domain many-to-many translation.

3.1 Results and Discussion

Image Translation. Fig. 2 shows example translation from all three modalities using our and other methods. As shown, both StarGAN [17] and DRIT++[22] produce less accurate translation for all three modalities compared with our method. Figure 3 shows T-SNE clusterings computed from the signal intensities of the various organs of interest using multiple methods and overlayed with the distribution of real T1w and T2w MRIs. The average cluster distance to the corresponding real MRI using our method was 5.05 and 14.00 for T2w and T1w MRIs, respectively. The average cluster distances of comparison methods were much higher with CycleGAN [6] of 73.90 for T2w, 101.37 for T1w, Star-GAN [17] of 73.39 for T2w and 77.49 for T1w, and DRIT++[22] of 87.32 for T2w and 70.73 for T1w MRIs, respectively. Furthermore, as shown in Table 1, our method produced the lowest reconstruction error for one step and two-step reconstructions.

Segmentation. Table 2 shows the segmentation accuracies achieved by our and multiple methods. We also evaluated the performance when a network trained on CT was directly applied to T1w MRI and T2w MRIs. As seen, the performance worsens when using the network trained on CT dataset alone. We evaluated the performance of the approach without the joint density structure discriminator in order to evaluate its use for segmentation performance. As shown, adding the joint density discriminator improved the segmentation performance. Overall, our approach produced an average accuracy of 0.85 for T1w and 0.90 for T2w MRI, which was higher than all other compared methods and only slightly lower than fully supervised training with T1w MRI, and slightly improved over supervised training for T2w MRI. Figure 4 shows example segmentations produced for T1w and T2w images using all the compared methods against expert segmentation.

Ablation Tests. We evaluated the impact of mode-seeking loss [19] and joint distribution matching losses introduced in this work for T2w MRI segmentation. The two step reconstruction error increased 27% and 19%, while segmentation accuracy dropped by 8% and 7% when removing mode seeking loss and structure discriminator loss, respectively.

Table 1. Reconstruction Error.

Method	One-step reconstruction		Two-step reconstruction	
	CT →T1w→CT	CT→ T2w→CT	CT→T1w→T2w→CT	CT→T2w→T1w→CT
StarGAN	2.30±1.37	2.56±1.30	5.56±2.44	6.55±2.07
DRIT++	2.32±1.00	2.53±1.08	6.13±1.40	6.04±2.23
Proposed	**1.34±0.40**	**1.20±0.39**	**2.36±0.44**	**2.27±0.67**

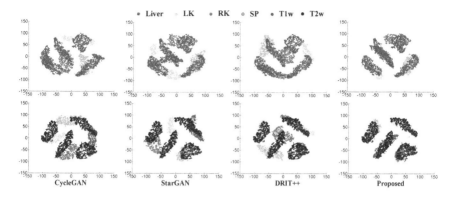

Fig. 3. TSNE clusters computed from the generated and real T1w, T2w MRI of organs.

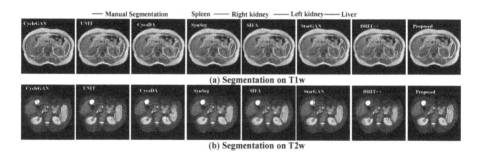

Fig. 4. Segmentation results comparing the various methods to the proposed method.

Discussion. We developed a multi cross-modality adaptation-based unsupervised segmentation approach that uses only single encoder/decoder for producing one to many mapping. Prior feature disentanglement medical image analysis methods used separate one-to-one mapping based segmentors [29–31], wherein each additional modality would require one additional encoder and decoder. Also, multi-domain adaptation was often done to handle scanner-related imaging variation in that same modality [30–32]. Our work also extends unsupervised image-level classification [14,33] to unsupervised multi-domain, multiple organ segmentation. Our results showed improved segmentation and reconstruction performance against the compared methods.

Table 2. Overall segmentation accuracy on CHAOS dataset (In-Phase). Liver-LV, Spleen-SP, Left kidney-LK, Right kidney-RK.

Method	T1w					T2W				
	LV	SP	LK	RK	Avg	LV	Sp	LK	RK	Avg
MRI supervised	0.91	0.86	0.82	0.83	0.86	0.92	0.87	0.91	0.90	0.90
CT only	0.00	0.00	0.00	0.00	0.00	0.01	0.08	0.29	0.22	0.15
CycleGAN [6]	0.82	0.83	0.63	0.61	0.72	0.86	0.75	0.88	0.87	0.84
UNIT [28]	0.89	0.81	0.64	0.62	0.74	0.87	0.76	0.91	0.88	0.86
CycaDA [13]	0.85	0.73	0.71	0.70	0.75	0.88	0.68	0.86	0.86	0.82
SynSeg [3]	0.89	**0.85**	0.73	0.70	0.79	0.88	0.77	0.89	0.85	0.85
SIFA [29]	**0.90**	**0.85**	0.77	0.78	0.83	0.89	0.77	0.90	0.89	0.86
StarGAN [17]	0.76	0.60	0.56	0.67	0.65	0.83	0.69	0.69	0.73	0.74
DRIT++[22]	0.83	0.63	0.61	0.66	0.68	0.73	0.81	0.81	0.82	0.79
Proposed - D_s	0.89	0.79	0.74	0.73	0.79	0.87	0.86	0.91	0.89	0.88
Proposed + D_s	**0.90**	0.84	**0.81**	**0.83**	**0.85**	**0.90**	**0.89**	**0.92**	**0.90**	**0.90**

4   Conclusion

We developed a multi-domain adversarial translation and segmentation method applied to unsupervised multiple MRI sequences segmentation. We showed that a universal multi-domain disentanglement using content and style extractors can produce reasonably accurate multi-domain translation and reasonably accurate segmentation of multiple organs.

Acknowledgement. This work was supported by the MSK Cancer Center support grant/core grant P30 CA008748.

References

1. Kupelian, P., Sonke, J.: Magnetic-resonance guided adaptive radiotherapy: a solution to the future. Semin Radiat Oncol **24**(3), 227–32 (2014)
2. Chartsias, A., Joyce, T., Dharmakumar, R., Tsaftaris, S.A.: Adversarial image synthesis for unpaired multi-modal cardiac data. In: Tsaftaris, S.A., Gooya, A., Frangi, A.F., Prince, J.L. (eds.) SASHIMI 2017. LNCS, vol. 10557, pp. 3–13. Springer, Cham (2017). https://doi.org/10.1007/978-3-319-68127-6_1
3. Huo, Y., et al.: SynSeg-Net: synthetic segmentation without target modality ground truth. IEEE Trans. Med. Imaging **38**(4), 1016–1025 (2018)
4. Jiang, J., et al.: Tumor-Aware, adversarial domain adaptation from CT to MRI for lung cancer segmentation. In: Frangi, A.F., Schnabel, J.A., Davatzikos, C., Alberola-López, C., Fichtinger, G. (eds.) MICCAI 2018. LNCS, vol. 11071, pp. 777–785. Springer, Cham (2018). https://doi.org/10.1007/978-3-030-00934-2_86
5. Goodfellow, I., et al.: Generative adversarial nets. In: Advances in Neural Information Processing Systems (NeurIPS), pp. 2672–2680 (2014)

6. Zhu, J.Y., Park, T., Isola, P., Efros, A.: Unpaired image-to-image translation using cycle-consistent adversarial networks. In: International Conference on Computer Vision (ICCV), pp. 2223–2232 (2017)
7. Arora, S., Ge, R., Liang, Y., Ma, T., Zhang, Y.: Generalization and equilibrium in generative adversarial nets GANs. In: International Conference on Machine Learning (ICML) (2017)
8. Qi, G.J.: Loss-sensitive generative adversarial networks on Lipschitz densities. Int. J. Comput. Vis. **128**(5), 1118–1140 (2020)
9. Zhu, J.Y., et al.: Toward multimodal image-to-image translation. In: International Conference on Neural Information Processing Systems (NeurIPS), pp. 465–476 (2017)
10. Donahue, J., Krähenbühl, P., Darrell, T.: Adversarial feature learning. CoRR abs/1605.09782 (2016)
11. Peng, X., Huang, Z., Sun, X., Saenko, K.: Domain agnostic learning with disentangled representations. In: International Conference on Machine Learning, pp. 5102–5112 (2019)
12. Liu, A., Liu, Y.C., Yeh, Y.Y., Wang, Y.C.F.: A unified feature disentangler for multi-domain image translation and manipulation. In: International Conference on Neural Information Processing Systems (NeurIPS), pp. 2595–2604 (2018)
13. Hoffman, J., et al.: CyCADA: cycle-consistent adversarial domain adaptation. In: International Conference on Machine Learning, pp. 1989–1998 (2018)
14. Huang, X., Liu, M.Y., Belongie, S., Kautz, J.: Multimodal unsupervised image-to-image translation. In: European Conference on Computer Vision (ECCV), pp. 172–189 (2018)
15. Liu, Y., Yeh, Y., Fu, T., Wang, S., Chiu, W., Wang, Y.F.: Detach and adapt: learning cross-domain disentangled deep representation. In: IEEE Confernce on Computer Vision and Pattern Recognition(CVPR), pp. 8867–8876 (2018)
16. Lee, H.Y., Tseng, H.Y., Huang, J.B., Singh, M., Yang, M.H.: Diverse image-to-image translation via disentangled representations. In: European Conference on Computer Vision (ECCV), pp. 35–51 (2018)
17. Choi, Y., Choi, M., Kim, M., Ha, J.W., Kim, S., Choo, J.: StarGAN: Unified generative adversarial networks for multi-domain image-to-image translation. In: IEEE Confernce on Computer Vision and Pattern Recognition (CVPR), pp. 8789–8797 (2018)
18. Liu, Y., Wang, Z., Jin, H., Wassell, I.: Multi-task adversarial network for disentangled feature learning. In: IEEE Confernce on Computer Vision and Pattern Recognition (CVPR), pp. 3743–3751 (2018)
19. Mao, Q., Lee, H., Tseng, H., Ma, S., Yang, M.: Mode seeking generative adversarial networks for diverse image synthesis. In: IEEE Confernce on Computer Vision and Pattern Recognition (CVPR), pp. 1429–1437 (2019)
20. Kingma, D.P., Welling, M.: Auto-encoding variational Bayes. arXiv preprint arXiv:1312.6114 (2013)
21. Alharbi, Y., Smith, N., Wonka, P.: Latent filter scaling for multimodal unsupervised image-to-image translation. In: IEEE Confernce on Computer Vision and Pattern Recognition (CVPR), pp. 1458–1466 (2019)
22. Lee, H.Y., et al.: DRIT++: diverse image-to-image translation via disentangled representations. Int. J. Comput. Vis. 1–16 (2020)
23. Yu, X., Ying, Z., Li, T., Liu, S., Li, G.: Multi-mapping image-to-image translation with central biasing normalization. arXiv preprint arXiv:1806.10050 (2018)
24. Kingma, D.P., Ba, J.: Adam: a method for stochastic optimization. In: International Conference on Learning Representations (ICLR) (2014)

25. Ronneberger, O., Fischer, P., Brox, T.: U-Net: convolutional networks for biomedical image segmentation. In: Navab, N., Hornegger, J., Wells, W.M., Frangi, A.F. (eds.) MICCAI 2015. LNCS, vol. 9351, pp. 234–241. Springer, Cham (2015). https://doi.org/10.1007/978-3-319-24574-4_28

26. Ali, E., Alper, S.M., Oğuz, D., Mustafa, B., Sinem, N.G.: CHAOS - combined (CT-MR) healthy abdominal organ segmentation challenge data

27. Landman, B., Xu, Z., Igelsias, J., Styner, M., Langerak, T., Klein, A.: MICCAI multi-atlas labeling beyond the cranial vault-workshop and challenge (2015)

28. Hou, L., Agarwal, A., Samaras, D., Kurc, T.M., Gupta, R.R., Saltz, J.H.: Unsupervised histopathology image synthesis. arXiv preprint arXiv:1712.05021 (2017)

29. Chen, C., Dou, Q., Chen, H., Qin, J., Heng, P.A.: Synergistic image and feature adaptation: Towards cross-modality domain adaptation for medical image segmentation. In: Proceedings of the (AAAI) Conference on Artificial Intelligence, vol. 33, pp. 865–872 (2019)

30. Yang, J., et al.: Unsupervised domain adaptation via disentangled representations: application to cross-modality liver segmentation. In: Shen, D., et al. (eds.) MICCAI 2019. LNCS, vol. 11765, pp. 255–263. Springer, Cham (2019). https://doi.org/10.1007/978-3-030-32245-8_29

31. Chartsias, A., et al.: Factorised spatial representation learning: application in semi-supervised myocardial segmentation. In: Frangi, A.F., Schnabel, J.A., Davatzikos, C., Alberola-López, C., Fichtinger, G. (eds.) MICCAI 2018. LNCS, vol. 11071, pp. 490–498. Springer, Cham (2018). https://doi.org/10.1007/978-3-030-00934-2_55

32. Ouyang, C., Kamnitsas, K., Biffi, C., Duan, J., Rueckert, D.: Data Efficient Unsupervised Domain Adaptation For Cross-modality Image Segmentation. In: Shen, D., Liu, T., Peters, T.M., Staib, L.H., Essert, C., Zhou, S., Yap, P.-T., Khan, A. (eds.) MICCAI 2019. LNCS, vol. 11765, pp. 669–677. Springer, Cham (2019). https://doi.org/10.1007/978-3-030-32245-8_74

33. Li, H., Pan, S.J., Wang, S., Kot, A.C.: Domain generalization with adversarial feature learning. In: IEEE Conference on Computer Vision and Pattern Recognition, pp. 5400–5409 (2018)

Dynamic Memory to Alleviate Catastrophic Forgetting in Continuous Learning Settings

Johannes Hofmanninger[1], Matthias Perkonigg[1], James A. Brink[2],
Oleg Pianykh[2], Christian Herold[1], and Georg Langs[1(✉)]

[1] Department of Biomedical imaging and Image-guided Therapy, Computational
Imaging Research Lab, Medical University of Vienna, Vienna, Austria
{johannes.hofmanninger,matthias.perkonigg,georg.langs}@meduniwien.ac.at
[2] Department of Radiology, Massachusetts General Hospital, Harvard Medical
School, Boston, USA

Abstract. In medical imaging, technical progress or changes in diagnostic procedures lead to a continuous change in image appearance. Scanner manufacturer, reconstruction kernel, dose, other protocol specific settings or administering of contrast agents are examples that influence image content independent of the scanned biology. Such domain and task shifts limit the applicability of machine learning algorithms in the clinical routine by rendering models obsolete over time. Here, we address the problem of data shifts in a continuous learning scenario by adapting a model to unseen variations in the source domain while counteracting catastrophic forgetting effects. Our method uses a dynamic memory to facilitate rehearsal of a diverse training data subset to mitigate forgetting. We evaluated our approach on routine clinical CT data obtained with two different scanner protocols and synthetic classification tasks. Experiments show that dynamic memory counters catastrophic forgetting in a setting with multiple data shifts without the necessity for explicit knowledge about when these shifts occur.

Keywords: Continuous learning · Domain adaptation · Dynamic memory

1 Introduction

In clinical practice, medical images are produced with continuously changing policies, protocols, scanner hardware or settings resulting in different visual appearance of scans despite the same underlying biology. In most cases, such a shift in visual appearance is intuitive for clinicians and does not lessen their ability to assess a scan. However, the performance of machine-learning based

---

J. Hofmanninger and M. Perkonigg—Authors contributed equally.

© Springer Nature Switzerland AG 2020
A. L. Martel et al. (Eds.): MICCAI 2020, LNCS 12262, pp. 359–368, 2020.
https://doi.org/10.1007/978-3-030-59713-9_35

methods can deteriorate significantly when the data distribution changes. *Continuous learning* adapts a model to changing data and new tasks by sequentially updating the model on novel cases. The ground truth labels of these cases can be, for instance, acquired by corrective labelling. However, *catastrophic forgetting* is a major undesired phenomenon affecting continuous learning approaches [9]. That is, when a model is continuously updated on a new task or a different data distribution, the model performance will deteriorate on the preceding tasks. Alleviating catastrophic forgetting is one of the major challenges in continuous learning.

Here, we propose an approach for a scenario where new data are sequentially available for model training. We aim to utilize such a data stream to frequently update an existing model without the requirement of keeping the entire data available. We assume a real-world clinical setting where changing visual domains and appearance of classification targets can occur gradually over time and where the information about such an eventual shift in data is not available to the continuous learning system. Figure 1 illustrates the scenario and the experimental setup. A trained model (base-model) has been trained to perform well on a certain classification task. Subsequently, a continuous data stream is used to update the model with the aim to learn variations of the initial task given new data. This model should become accurate on new data, while at the same time staying accurate on data generated by previous technology. Accordingly, the final model is then evaluated on all tasks to assess the effect of catastrophic forgetting and the classification performance. Note that here we use the term *task* to denote the detection of the same target but on shifted visual domains and not for additional target classes.

Related Work. There are various groups of methods dealing with the problem of continuously updating a machine learning model over time such as continuous learning, continuous domain adaptation or active learning. They operate on similar but different assumptions about the problem settings, the data available and the level of supervision required. For example, continuous domain adaption assumes novel data to be shifted but closely related to the previous data. Continuous learning makes no assumptions about domain shifts and is not limited to a specific task to perform in new data (e.g. incremental learning). Active learning is characterized by the task of automatically selecting examples for which supervision is beneficial. In this work, we propose a continuous learning technique which can also be categorized as a supervised continuous domain adaptation method.

Various methods for continuous learning have been proposed to alleviate catastrophic forgetting in scenarios where multiple tasks are learned sequentially.

A popular method to retain previous knowledge in a network is elastic weight consolidation (EWC) [7]. EWC is a regularization technique aiming to constrain parameters of the model which are critical for performing previous tasks during the training of new tasks. Alternative methods attempt to overcome catastrophic forgetting by rehearsing past examples [8] or proxy information (pseudorehearsal) [10] when new information is added [11]. In the field of medical imaging, continuous learning has been demonstrated to reduce catastrophic forgetting on segmentation and classification tasks. Karani et al. proposed domain-specific batch norm layers to adapt to new domains (different MR protocols) while learning segmentations of various brain regions [5]. Baweja et al. applied EWC to sequentially learn normal brain structure and white matter lesion segmentation [1]. Ravishankar et al. propose a pseudorehearsal technique and training of task-specific dense layers for pneumothorax classification [10]. These current approaches expect that the information about the domain to which a training example belongs, is available to the learning system. In real world medical imaging data, such information may not be available at the image level (e.g. a change in treatment policies, or different hardware updates across departments). At the same time, changes in protocol or scanner manufacturer may not automatically lead to a loss of performance of the model and considering each protocol as a novel domain may lead do adverse effects such as overfitting.

Contribution. We propose an approach for continuous learning of continuously or repeatedly shifting domains. This is in contrast to most previous methods treating updates as sequentially adding well specified tasks. In contrast to existing rehearsal methods, we propose a technique that automatically infers data shifts without explicit knowledge about them. To this end, the method maintains a diverse memory of previously seen examples that is dynamically updated based on high level representations in the network.

2   Method

We continuously update the parameters of an already trained model with new training data. Our approach composes this training data to capture novel data characteristics while sustaining the diversity of the overall training corpus. It chooses examples from previously seen data (*dynamic memory* $\mathcal{M}$) and new examples (*input-mini-batch* $\mathcal{B}$) to form the training data (*training-mini-batch* $\mathcal{T}$) for the model update.

Our approach is a *rehearsal method* to counter catastrophic forgetting in continuous learning. We adopt a *dynamic memory (DM)*

$$\mathcal{M} = \{\langle \mathbf{m}_1, n_1 \rangle, \ldots, \langle \mathbf{m}_M, n_M \rangle\} \tag{1}$$

of a fixed-size M holding image-label pairs $\langle \mathbf{m}, n \rangle$ that are stored and eventually replaced during continuous training. To alleviate catastrophic forgetting, a subset of cases of $\mathcal{M}$ is used for rehearsal during every update step. It is critical that the diversity of $\mathcal{M}$ is representative of the visual variation across all tasks, even without explicit knowledge about the task membership of training examples. As the size of $\mathcal{M}$ is fixed, the most critical step of such an approach is to decide which samples to keep in $\mathcal{M}$ and which to replace with a new sample. To this end, we define a memory update strategy based on following rules: (1) every novel case will be stored in the memory, (2) a novel case can only replace a case in memory of the same class and (3) the case in memory that will be replaced is

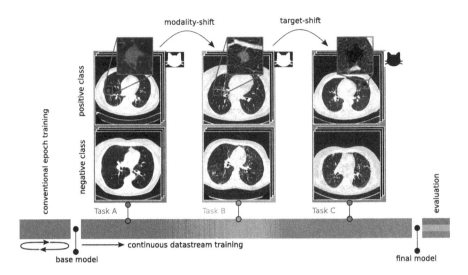

Fig. 1. Experimental setup for a continuous learning scenario: we assume a learning scenario for which a conventionally trained model (e.g. multi-epoch training) performing well on Task A is available. This model is continuously updated on a data stream with a shift in overall image appearance caused by scanner parameters (modality-shift) and a shift in the appearance of the classification target (target-shift). The timing of shifts is not known a priori. For evaluation, the final model is evaluated on a test set of all three tasks.

close according to a high level metric. Rule 1 allows the memory to dynamically adapt to changing variation. Rule 2 prevents class imbalance in the memory and rule 3 prevents the replacement of previous cases if they are visually distant. The metric used in rule 3 is critical as it ensures that cases of previous tasks are kept in memory and not fully replaced over time. We define a high-level metric based on the gram matrix $G^l \in \mathbb{R}^{N_l \times N_l}$ where N_l is the number of feature maps in layer l. $G^l_{ij}(\mathbf{x})$ is defined as the inner product between the vectorized activations $\mathbf{f}_{il}(\mathbf{x})$ and $\mathbf{f}_{jl}(\mathbf{x})$ of two feature maps i and j in a layer l given a sample image $\mathbf{x}$:

$$G^l_{ij}(\mathbf{x}) = \frac{1}{N_l M_l} \mathbf{f}_{il}(\mathbf{x})^\top \mathbf{f}_{jl}(\mathbf{x}) \tag{2}$$

where M_l denotes the number of elements in the vectorized feature map (width $\times$ height). For a set of convolution layers $\mathcal{L}$ we define a gram distance $\delta(\mathbf{x}, \mathbf{y})$ between two images $\mathbf{x}$ and $\mathbf{y}$ as:

$$\delta(\mathbf{x}, \mathbf{y}) = \sum_{l \in \mathcal{L}} \frac{1}{N_l^2} \sum_{i=1}^{N_l} \sum_{j=1}^{N_l} (G^l_{ij}(\mathbf{x}) - G^l_{ij}(\mathbf{y}))^2 \tag{3}$$

The rationale behind using the gram matrix is the fact, that the gram matrix encodes high level style information. Here, we are interested in this style information to maintain a diverse memory not only with respect to the content but also with respect to different visual appearances. Similar gram distances have been used in computer vision methods in the area of neural style transfer as a way to compare the style of two natural images [2].

During continuous training, a memory update is performed after an input-mini-batch $\mathcal{B} = \{\langle \mathbf{b}_1, c_1 \rangle, \dots, \langle \mathbf{b}_B, c_B \rangle\}$ of B sequential cases (image $\mathbf{b}$ and label c) is taken from the data stream. Sequentially, each element of $\mathcal{B}$ replaces an element of $\mathcal{M}$. More formally, given an input sample $\langle \mathbf{b}_i, c_i \rangle$, the sample will replace the element in $\mathcal{M}$ with index

$$\xi(i) = \arg \min_{j} \delta(\mathbf{b}_i, \mathbf{m}_j) \mid c_i = n_j, \ j \in \{1, \dots, M\}. \tag{4}$$

During the initial phase of continuous training, the memory is filled with elements of the data stream. Only after the desired proportion of a class in the memory is reached, the replacement strategy is applied. After the memory is updated, a model update is done by assembling a training-mini-batch $\mathcal{T} = \{\langle \mathbf{t}_1, u_1 \rangle, \dots, \langle \mathbf{t}_T, u_T \rangle\}$ of size T. Each misclassified element of $\mathcal{B}$ for which the model predicted the wrong label is added to $\mathcal{T}$ and additional cases are randomly drawn from $\mathcal{M}$ until $|\mathcal{T}| = T$. Finally, using the training-mini-batch $\mathcal{T}$, a forward and backward pass is performed to update the parameters of the model.

3 Experiments and Results

We evaluated and studied the DM method in a realistic setting using medical images from clinical routine. To this end, we collected a representative dataset described in Sect. 3.1. Based on these data, we designed a learning scenario with three tasks (A, B and C) in which a classifier pre-trained on task A is continuously updated with changing input data over time. Figure 1 illustrates the learning scenario, the data, and the experimental setup. Within the continuous dataset, we created two shifts, (1) a *modality-shift* between scanner protocols and (2) a *target-shift* by changing the target structure from high to low intensity.

3.1 Dataset

In total, we collected 8883 chest CT scans from individual studies and for each extracted an axial slice at the center of the segmented [4] lung. Each scan was performed on a Siemens scanner with either B3 reconstruction kernel and 3 mm slice-thickness (B3/3) or B6 reconstruction kernel and 1 mm slice-thickness (B6/1). We collected 3784 cases with B3/3 protocol and 5099 cases with the B6/1

Table 1. Data: splitting of the data into a base, continuous, validation, and test set. The number of cases in each split are shown.

	Task A	Task B	Task C	Total
Protocol	B3/3	B6/1	B6/1	
Target	High	High	Low	
Base	1513	0	0	1513
Continuous	1513	1000	2398	4911
Validation	377	424	424	1225
Test	381	427	426	1234

protocol. We imprinted a synthetic target structure in the form of a cat on random locations, rotations and varying scale at 50% of the cases (see also Fig. 1). The high-intensity target structures were engraved by randomly adding an offset between 200 and 400 hounsfield units (HU) and the low-intensity target structures by subtracting between 200 and 400 HU. A synthetic target was chosen to facilitate data set collection, as well as to create a dataset without label noise. Table 1 lists the data collected and shows the partitioning into base, continuous, validation and test split and the stratification into the three tasks.

3.2   Experiments

We created the base model by fine-tuning a pre-trained Res-Net50 [3] model, as provided by the pytorch-torchvision library[1] on the *base* training set (Task A). Given this base model, we continuously updated the model parameters on the *continuous* training set using different strategies:

- **Naive**: The naive approach serves as a baseline method by training sequentially on the data stream, without any specific strategy to counter catastrophic forgetting.
- **Elastic Weight Consolidation (EWC)**: As a reference method, we used EWC as presented in [7]. EWC regularizes weights that are critical for previous tasks based on the fisher information. We calculated the fisher information after training on the base set (Task A) so that in further updates, weights that are important for this task are regularized.
- **EWC-fBN**: In preliminary experiments, we found, that EWC is not suitable for networks using batch norm layers in scenarios where a modality-shift occurs. The reason for that is, that the regularization of EWC does not effect batch norm parameters (mean and variance) and that these parameters are constantly updated by novel data. Thus, to quantify this effect, we show results where we fixed the batch norm layers once the base training was completed.
- **Dynamic Memory (DM)**: The method as described in this paper. The input-mini-batch size B and training-mini-batch size T have been set to 8 for all experiments. If not stated differently, results have been computed with a memory size M of 32. The gram matrices have been calculated on feature maps throughout the network covering multiple scales. Specifically, we used the feature maps of the last convolution layers before the number of channels is increased. That is, for Res-Net50 we calculate the matrices on four maps with 256, 512, 1024 and 2048 features.

In addition to the continuous scenario we trained an upper bound network in a conventional, epoch-based way, using all training data at once (*full training*). All training processes were run five times to assess the variability and robustness and the results were averaged. All models were trained using an Adam optimizer [6] and binary cross entropy as a loss function for the classification task. With the described methods we,

---

[1] https://pytorch.org.

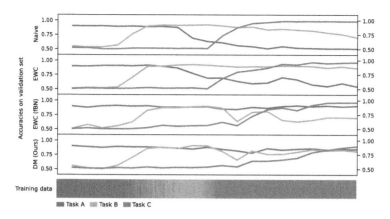

Fig. 2. Validation accuracy during training: the curves show the change of validation accuracy for each of the tested approaches. Accuracies are computed on the validation sets for the three tasks during training. The last row show how the composition of the training data stream is changing over time.

- studied the **dynamics during training** by calculating the classification accuracies on the validation set every 30 iterations.
- calculated **quantitative results** after training on the test set for each task separately. Classification accuracy, backward transfer (BWT) and forward transfer (FWT), as defined in [8] were used. BWT and FWT measure the influence that learning a new task has on the performance of previously learned, respectively future tasks. Thus, larger BWT and FWT values are preferable. Specifically, negative BWT quantifies catastrophic forgetting.
- studied the **influence of memory size** M on the performance of our method during the training process and after training. We included experiments with M set to 16, 32, 64, 80, 128 and 160.

3.3   Results

Dynamics during training are shown in Fig. 2. As expected, the naive approach shows catastrophic forgetting for task A (drop of accuracy from 0.91 to 0.51) and B (drop from 0.93 to 0.70) after new tasks are introduced. Without any method to counteract forgetting, knowledge of previous tasks is lost after sequential updates to the model. EWC exhibits catastrophic forgetting for task A (0.92 to 0.54) after the introduction of task B and C data into the data stream. The shift from task A to B is a modality shift, the batch norm layers of the network adapt to the second modality and knowledge about the first modality is lost. Although EWC protected the weights that are important for task A, those weights were not useful after the batch norm layers were adapted. EWC-fBN avoids this problem by fixing the batch norm layers together with the weights that are relevant for task A. In this setting a forgetting effect for task B (accuracy drop from 0.90 to 0.70) can be observed after the target-shift to task C.

This effect is due to the requirement of EWC to know when shifts occur. Thus, in our scenario, EWC only regularizes weights that are important for task A. As described previously, DM does not have this requirement by dynamically adapting to changing data. Using the DM approach only mild forgetting occurs and all three tasks reach a comparable performance of about 0.85 accuracy after training.

Quantitative results are shown in Table 2. The large negative BWT values for the naive approach (−0.32) and EWC (−0.20) indicate that these methods suffer from catastrophic forgetting. Using EWC-fBN mitigates the forgetting for task A, but the model is forgetting part of the knowledge for task B when task C is introduced (observable in Fig. 2). Both, DM and EWC-fBN show comparable backward and forward transfer capabilities. The accuracy values in Table 2 show that DM performs equally well on all tasks, while the other approaches show signs of forgetting of task A (naive and EWC) and task B (naive and EWC-fBN).

The influence of memory size during training is shown in Fig. 3a For a small M of 16, adapting to new tasks is fast. However, such a limited memory can only store a limited data variability leading to catastrophic forgetting effects. Increasing memory size decreases catastrophic forgetting effects but slows adaption of new tasks. For the two largest investigated sizes 128 and 160 adaption, especially after the target-shift (Task C), is slower. The reason is, that more elements of task A are stored in the memory and more iterations are needed to fill the memory with samples of task C. This reduces the probability that elements from task C are drawn and in turn, slows down training for task C. For the same reason, with higher memory sizes there is an increase in task A accuracy, since examples from the first task are more often seen by the network. Results indicate that setting the memory size is a trade-off between faster adaption to novel data and more catastrophic forgetting. In our setting, training worked comparably well on M of 32, 64, and 80. Considering memory sizes between 16 and 160, setting $M = 32$ resulted in the highest average accuracy of 0.87 on the validation set (Fig. 3b). Therefore, we used this model for comparison to other continuous learning methods on the test set.

Table 2. Accuracy and BWT/FWT values for our dynamic memory (DM) method compared to baseline methods. Results were calculated on the test set after continuous training. Lower values marked with $\otimes$ indicate forgetting.

	ACC Task A	ACC Task B	ACC Task C	BWT	FWT
Naive	$\otimes$ 0.51 ± 0.00	$\otimes$ 0.71 ± 0.02	0.98 ± 0.01	−0.32 ± 0.01	0.05 ± 0.00
EWC	$\otimes$ 0.57 ± 0.00	0.83 ± 0.00	0.91 ± 0.00	−0.20 ± 0.00	0.04 ± 0.00
EWC-fBN	0.89 ± 0.02	$\otimes$ 0.74 ± 0.05	0.94 ± 0.04	−0.08 ± 0.02	0.06 ± 0.00
DM (Ours)	0.81 ± 0.02	0.85 ± 0.02	0.92 ± 0.04	−0.07 ± 0.02	0.06 ± 0.01
Full training	0.92 ± 0.01	0.91 ± 0.02	0.97 ± 0.00	-	-

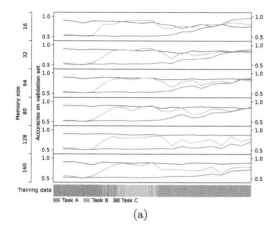

M	Task A	Task B	Task C	Avg
16	0.73	0.83	0.96	0.84
32	0.82	0.86	0.93	0.87
64	0.85	0.81	0.87	0.84
80	0.86	0.82	0.87	0.85
128	0.87	0.75	0.71	0.78
160	0.88	0.79	0.69	0.79

(a)   (b)

Fig. 3. Memory size: (a) the effect of memory size during training with DM. Small memory sizes tend to catastrophic forgetting, while large memory sizes lead to slow training of later tasks. **(b)** classification accuracy on the validation set after training for varying memory sizes.

4   Conclusion

Here, we presented a continuous learning approach to deal with modality- and tasks-shifts induced by changes in protocols, parameter setting or different scanners in a clinical setting. We showed that maintaining a memory of diverse training samples mitigates catastrophic forgetting of a deep learning classification model. We proposed a memory update strategy that is able to automatically handle shifts in the data distribution without explicit information about domain membership and the moment such a shift occurs.

Acknowledgments. This work was supported by Austrian Science Fund (FWF) I 2714B31 and Novartis Pharmaceuticals Corporation.

References

1. Baweja, C., Glocker, B., Kamnitsas, K.: Towards continual learning in medical imaging (2018). https://arxiv.org/abs/1811.02496
2. Gatys, L., Ecker, A., Bethge, M.: A neural algorithm of artistic style. J. Vis. **16**(12), 326 (2016). https://doi.org/10.1167/16.12.326
3. He, K., Zhang, X., Ren, S., Sun, J.: Deep residual learning for image recognition. In: Proceedings of the IEEE Conference on Computer Vision and Pattern Recognition (2015). http://image-net.org/challenges/LSVRC/2015/

4. Hofmanninger, J., Prayer, F., Pan, J., Röhrich, S., Prosch, H., Langs, G.: Automatic lung segmentation in routine imaging is primarily a data diversity problem, not a methodology problem. Eur. Radiol. Exp. **4**(50), 1–13 (2020). https://doi.org/10.1186/s41747-020-00173-2

5. Karani, N., Chaitanya, K., Baumgartner, C., Konukoglu, E.: A lifelong learning approach to brain MR segmentation across scanners and protocols. In: Frangi, A.F., Schnabel, J.A., Davatzikos, C., Alberola-López, C., Fichtinger, G. (eds.) MICCAI 2018. LNCS, vol. 11070, pp. 476–484. Springer, Cham (2018). https://doi.org/10.1007/978-3-030-00928-1_54

6. Kingma, D.P., Ba, J.L.: Adam: a method for stochastic optimization. In: 3rd International Conference on Learning Representations. (ICLR 2015) (2015)

7. Kirkpatrick, J., et al.: Overcoming catastrophic forgetting in neural networks. Proc. Natl. Acad. Sci. U.S.A. **114**(13), 3521–3526 (2017). https://doi.org/10.1073/pnas.1611835114

8. Lopez-Paz, D., Ranzato, M.: Gradient episodic memory for continual learning. In: Advances in Neural Information Processing Systems, pp. 6468–6477 (2017)

9. McCloskey, M., Cohen, N.J.: Catastrophic Interference in Connectionist Networks: The Sequential Learning Problem. Psychol. Learn. Motiv. Adv. Res. Theory **24**(C), 109–165 (1989). https://doi.org/10.1016/S0079-7421(08)60536-8

10. Ravishankar, H., Venkataramani, R., Anamandra, S., Sudhakar, P., Annangi, P.: Feature transformers: privacy preserving lifelong learners for medical imaging. In: Shen, D., et al. (eds.) MICCAI 2019. LNCS, vol. 11767, pp. 347–355. Springer, Cham (2019). https://doi.org/10.1007/978-3-030-32251-9_38

11. Robins, A.: Catastrophic forgetting, rehearsal and pseudorehearsal. Connection Sci. **7**(2), 123–146 (1995). https://doi.org/10.1080/09540099550039318

Unlearning Scanner Bias for MRI Harmonisation

Nicola K. Dinsdale[1(✉)], Mark Jenkinson[1,2,3], and Ana I.L. Namburete[4]

[1] Wellcome Centre for Integrative Neuroimaging, FMRIB, University of Oxford, Oxford, UK
nicola.dinsdale@dtc.ox.ac.uk
[2] Australian Institute for Machine Learning (AIML), Department of Computer Science, University of Adelaide, Adelaide, Australia
[3] South Australian Health and Medical Research Institute (SAHMRI), North Terrace, Adelaide, Australia
[4] Institute of Biomedical Engineering, University of Oxford, Oxford, UK

Abstract. Combining datasets is vital for increased statistical power, especially for neurological conditions where limited data is available. However, variance due to differences in acquisition protocol and hardware limits our ability to combine datasets. We propose an iterative training scheme based on domain adaptation techniques, aiming to create scanner-invariant features while simultaneously maintaining overall performance on the main task. We demonstrate this on age prediction, but expect that our proposed training scheme will be applicable to any feedforward network and classification or regression task. We show that not only can we harmonise three MRI datasets from different studies, but can also successfully adapt the training to work with very biased datasets. The training scheme should, therefore, be applicable to most real-world data scenarios, enabling harmonisation for the task of interest.

Keywords: Harmonisation · Joint domain adaptation · MRI

1 Introduction

Whilst a few very large projects, such as the UK Biobank [14] now exist, the majority of dataset sizes in neuroimaging studies remain relatively small. Therefore, combining datasets from multiple sites and scanners is vital to give improved statistical power. However, this leads to greater variance in the data, largely due to differences in acquisition protocol and hardware [6]. Thus, harmonisation is required to achieve joint unbiased analysis of data from different scanners.

One popular harmonisation method is ComBat [6], which performs post-hoc normalisation using a linear model, making image-derived values comparable. This was extended to incorporate a nonlinear model in [12] and explicitly to encode bias caused by nonbiological variance in the model in [15]. The majority

© Springer Nature Switzerland AG 2020
A. L. Martel et al. (Eds.): MICCAI 2020, LNCS 12262, pp. 369–378, 2020.
https://doi.org/10.1007/978-3-030-59713-9_36

of other methods for MRI harmonisation focus on making images produced on one scanner look as if they came from another, with recent studies using deep learning methods (eg. [4]). CycleGANS [18] have also been used to transform images between domains [17].

Instead of harmonising the images, we propose to harmonise the features extracted by deep learning networks using a joint domain adaptation approach. Domain adaptation assumes that we have a source domain D_s with learning task T_s and a target domain D_t with learning task T_t and either $D_s \neq D_t$ or $T_s \neq T_t$ [3]; the success of the domain adaptation depends on the existence of a similarity between the two domains [16]. For harmonisation, we consider the case where $D_s \neq D_t$; that is, when the data was collected on distinct scanners. One of the most successful methods for domain adaptation is the DANN network [8] which uses a gradient reversal layer [7] to train a discriminator adversarially, creating a feature representation which is discriminative for the main task but indiscriminate as to the domains. There is, however, little exploration of the effect of domain adaptation on the performance on the *source domain* data, whereas for harmonization it is vital that the network performs well across all the datasets.

In [9] a method is proposed to solve both domain transfer and task transfer simultaneously. Similarly to DANN, they complete domain adaptation using adversarial methods but, rather than using a gradient reversal layer to update the domain predictor in opposition to the task, they use an iterative training scheme: they alternate between learning the best domain classifier with a given feature representation and then minimise a confusion loss which aims to force the domain predictions to become closer to a uniform distribution so it 'maximally confuses' the domain classifier [9]. Compared to DANN-style networks it is also better at ensuring an equally uninformative classifier across the domains [2] because of the confusion loss, which is desirable for the harmonisation scenario. This network is applied in [2] where iterative unlearning creates classifiers that are blind to spurious variations in the data. These together form the inspiration for this work.

In this work, we apply a framework similar to that introduced in [9] for harmonisation, by posing the problem as a joint domain adaptation problem. We aim to create a feature representation that is invariant to the scanner from which the data were acquired and show that this network is still able to perform the task of interest. By taking a joint domain adaptation approach, we also require that the network is successful at all tasks and is not just driven by the larger dataset; thus, we explore the effect of training with different datasets. We show that scanner information can be successfully 'unlearned' in realistic scenarios, allowing us to harmonise data for the main task of interest. The code is available at: https://github.com/nkdinsdale/Unlearning_for_MRI_harmonisation.

2   Method

We explore two different data regimes and show that, by adapting the loss functions, the same learning framework can be used in all scenarios. We consider

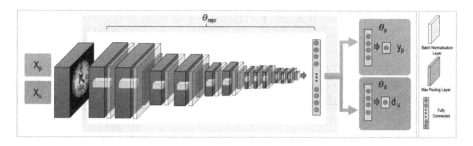

Fig. 1. Network architecture. $\boldsymbol{X}_p$ and $\boldsymbol{X}_u$ represent the input data for the network, $\boldsymbol{X}_p$ for the main task and $\boldsymbol{X}_u$ for unlearning. These can be the same data, subsets of each other, or different datasets. For $\boldsymbol{X}_p$ the labels $\boldsymbol{y}_p$ are the labels for the main task and for $\boldsymbol{X}_u$ the labels are the domain labels $\boldsymbol{d}_u$. $\boldsymbol{\Theta}_{repr}$ are the parameters of the convolutional layers and first fully connected layer that form the encoder; $\boldsymbol{\Theta}_p$ are the parameters of the fully connected layers which predict the main task; $\boldsymbol{\Theta}_d$ are the parameters of the fully connected layers in the domain predictor.

both training with fully supervised data with similar distributions, and training when the data distributions are biased for the main task label both for the task of age prediction.

2.1   Standard Supervised Training

This data regime corresponds to the scenario where we have training labels for the main task available for data from all scanners and the data distributions for the main task are similar. In this case, $\boldsymbol{X}_p$ and $\boldsymbol{X}_u$ are a single dataset $\boldsymbol{X}$ which is used to evaluate all of the training iterations.

The aim of the 3D network shown in Fig. 1 is to find a feature representation $\boldsymbol{\Theta}_{repr}$ that maximises the performance on a primary task while minimising the performance of a discriminator, which aims to predict the site of origin of the data. In this case, we use age prediction (based on T1-weighted MRI scans) as an example task, but the training procedure should generalise to any feedforward architecture and task. $\boldsymbol{\Theta}_{repr}$ represents the parameters of the encoder which are shared between the two output branches; $\boldsymbol{\Theta}_p$ represents the parameters for the primary age prediction task, and $\boldsymbol{\Theta}_d$ represents the parameters of the domain prediction branch. We consider the case of three datasets, each with input images $\boldsymbol{X} \in \mathbb{R}^{W \times H \times D \times 1}$ and task labels $\boldsymbol{y} \in \mathbb{R}$, with different domains $\boldsymbol{d}$, representing scans acquired from three distinct scanners.

Three loss functions are used in the training of the network. The first loss is for the main task and is conditioned on each domain:

$$L_p(\boldsymbol{X}, \boldsymbol{y}, \boldsymbol{d}; \boldsymbol{\Theta}_{repr}, \boldsymbol{\Theta}_p) = \sum_{n=1}^{N} \frac{1}{S_n} \sum_{j=1}^{S_n} L_n(\boldsymbol{y}_{j,n}, \hat{\boldsymbol{y}}_{j,n}) \tag{1}$$

where N is the number of domains and S_n is the number of subjects from domain n such that $\boldsymbol{y}_{j,n}$ is the true task label for the jth subject from the nth

domain and L_n is the loss function for the main task evaluated for the data from domain n. This loss takes the form of mean squared error (MSE) for the age prediction task. The loss is calculated for each domain separately to prevent the performance being driven by the larger dataset, especially when there is a large imbalance in sample numbers. The domain information is then unlearned using two loss functions in combination. The domain loss is simply the categorical cross entropy:

$$L_d(\boldsymbol{X}, \boldsymbol{d}, \boldsymbol{\Theta}_{repr}; \boldsymbol{\Theta}_d) = -\sum_{n=1}^{N} \mathbb{1}[d = n] log(p_n) \tag{2}$$

which assesses how much domain information remains in Θ_{repr}. p_n are the softmax outputs of the domain classifier and are used in the confusion loss to remove information, by penalising deviations from a uniform distribution:

$$L_{conf}(\boldsymbol{X}, \boldsymbol{d}, \boldsymbol{\Theta}_d; \boldsymbol{\Theta}_{repr}) = -\sum_{n=1}^{N} \frac{1}{N} log(p_n) \tag{3}$$

Therefore the overall method minimises the total loss function $L = L_p + \alpha L_d + \beta L_{conf}$ where α and β represent weights of the relative contributions of the different loss functions. Eqs. (2) and (3) directly oppose each other and therefore cannot be optimised in a single step. Therefore, we iterate through updating the different loss functions, resulting in three forward passes per batch.

2.2 Biased Domains

Finally, we consider the scenario where there exists a large difference between the two domains such that the main task label is highly indicative of the scanner and, thus, unlearning scanner information leads to unlearning important information for the main task. For example, consider the scenario where the age distributions for two studies are only slightly overlapping or the scenario where nearly all the subjects with a given condition were collected on one of the scanners: we show that this problem can be reduced by unlearning the domain, using different data to that used to train the main task. This is simple to train within the same learning framework, only requiring Eqs. (2) and (3) to be evaluated across a different dataset or subset of the data. For instance, in the case of slightly overlapping age distributions, the domain information would be unlearned, only using the overlapping section. For the case of a dataset being biased with more subjects with a given pathology being from one scanner, the unlearning would only be completed with the controls. In this case, the equations become:

$$L_d(\boldsymbol{X}_u, \boldsymbol{d}_u, \boldsymbol{\Theta}_{repr}; \boldsymbol{\Theta}_d) = -\sum_{n=1}^{N} \mathbb{1}[d = n] log(p_n) \tag{4}$$

$$L_{conf}(\boldsymbol{X}_u, \boldsymbol{d}_u, \boldsymbol{\Theta}_d; \boldsymbol{\Theta}_{repr}) = -\sum_{n=1}^{N} \frac{1}{N} log(p_n) \tag{5}$$

where X_u and d_u are the input images and domain labels for the intersection data, and can either be a subset of X_p or a further dataset of controls only used to unlearn scanner information. It should be noted that task labels are not required for this data, making it much easier to obtain. The main task is still evaluated across the whole dataset X_p.

3   Experimental Setup

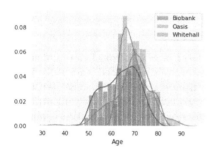

Fig. 2. Data distributions for the three datasets, normalised so that the distributions of the smaller datasets can be seen.

For the experiments in this work, T1 weighted MRI scans from three datasets were used: UK Biobank [14] (Siemens Skyra 3T) which had been processed using the UK Biobank Pipeline [1] (5508 training, 1377 testing); healthy subjects from the OASIS dataset [11] (Siemens Tesla Vision 1.5T) at multiple time points, split into training and test sets at the subject level (813 training, 217 testing), and subjects from the Whitehall II study [5] (Siemens Magnetom Verio 3T) (452 training, 51 testing). The input images for all datasets were resized to $128 \times 128 \times 128$ voxels and then every fourth slice was retained, leaving 32 slices in the z direction, chosen so as to maximise coverage across the whole brain whilst minimising redundancy, and allowing a larger batch size and number of filters to be used. The inputs were also normalised to have zero mean and unit standard deviation. We chose to investigate the age prediction task as this is a task for which accurate task labels are easy to obtain. The data distributions can be seen in Fig. 2.

The network was implemented in Python 3.6 using PyTorch (1.0.1) and is based on the VGG-16 architecture [13]; however, our proposed training procedure is applicable to any feedforward network. A batch size of 16 was used throughout, with each batch constrained to contain at least one example from each dataset, increasing the stability during training. To achieve this, the smaller datasets were oversampled. α and β were empirically set for the different experiments, taking values between 1 and 20.

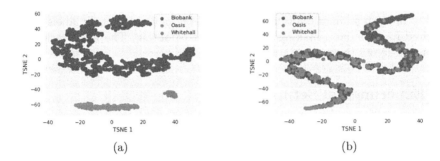

(a)   (b)

Fig. 3. a) T-SNE [10] plot of the fully connected layer in Θ_{repr} from before unlearning. It can be seen that the domains can be almost entirely separated, except for two datapoints grouped incorrectly, showing that data from each scanner has its own distinct distribution. b) T-SNE plot of the fully connected layer in Θ_{repr} after unlearning. It can be seen that, through the unlearning, the distributions become entirely jointly embedded.

Table 1. Results comparing unlearning to training the network in different combinations on the datasets using Mean Absolute Error (MAE) as the metric for the task of age regression. Scanner accuracy is the accuracy achieved by a domain predictor given the fixed feature representation at convergence, evaluating only for the datasets the network was trained on. The number in brackets indicates random chance. B = Biobank, O = OASIS, W = Whitehall

Training data			Biobank MAE	OASIS MAE	Whitehall MAE	Scanner classification accuracy (%)
B	O	W				
Normal training						
1. ✓	✗	✗	3.25 ± 2.36	16.50 ± 6.77	13.81 ± 5.42	–
2. ✗	✓	✗	5.61 ± 3.52	4.27 ± 3.79	6.73 ± 4.82	–
3. ✗	✗	✓	5.61 ± 3.65	5.22 ± 4.83	3.15 ± 2.81	–
4. ✓	✓	✗	3.30 ± 2.50	4.00 ± 2.78	4.71 ± 3.42	98 (50)
5. ✓	✗	✓	3.31 ± 2.49	4.45 ± 3.53	3.05 ± 2.84	100 (50)
6. ✗	✓	✓	5.71 ± 3.59	4.05 ± 3.71	3.21 ± 2.94	100 (50)
7. ✓	✓	✓	3.24 ± 2.47	4.19 ± 3.50	2.89 ± 2.70	96 (33)
Unlearning						
8. ✓	✓	✗	3.41 ± 2.04	3.79 ± 2.99	4.60 ± 3.47	48 (50)
9. ✓	✗	✓	3.41 ± 2.58	4.07 ± 4.12	2.81 ± 2.57	52 (50)
10. ✗	✓	✓	3.38 ± 2.64	3.91 ± 3.53	2.82 ± 2.65	50 (50)
11. ✓	✓	✓	3.38 ± 2.64	3.90 ± 3.53	2.56 ± 2.47	34 (33)

4   Results

4.1   Supervised Unlearning

We compared our method to standard training on all three datasets individually and on combinations of datasets and compare the mean absolute errors (MAEs) between methods. The results can be seen in Table 1. It can first be seen that training on all three datasets using normal training gives the best overall performance of the different regimes for standard training as would be expected (row 7), giving the lowest MAE overall across the datasets. This, however, produces a feature representation Θ_{repr} in which the three datasets can be separated as is shown in Fig. 3a, so information relating to the scanner is being used to inform the age prediction. This would be particularly problematic if there was a large correlation between task label and scanner as is shown in Sect. 4.2. On the other hand, it can be seen from the results of unlearning on all three datasets (row 11) that we are able to remove scanner information successfully, from the fact that the scanner accuracy is approximately random chance. Simultaneously, there is little decrease in performance across the datasets, showing that the unlearning is not detrimental to performance. In fact, a lower MAE is achieved for the two small datasets (OASIS and Whitehall) and only the performance on Biobank decreases, probably because the network is no longer driven by its much larger size. For reference, standard training using the loss function conditioned on each dataset led to MAEs of 3.55 ± 2.68, 3.90 ± 3.53 and 2.62 ± 2.65 years, respectively, and so we are probably seeing improvement due to the unlearning process. This is because, in essence, this method is a domain adaptation approach and so we are harnessing information from each dataset to boost the network's overall performance by removing scanner differences. Figure 3b) confirms the success of the harmonisation as the scanner domains can no longer be separated.

The comparison for training on two datasets and testing the trained network on the third unseen dataset (e.g. comparing row 6 and 10) also shows that the unlearning procedure helps the network to generalise better: the MAE values for the unseen dataset in both cases are improved by unlearning. This shows that by removing scanner information, preventing it from influencing the prediction, the network learns features that are more applicable to other datasets.

4.2   Biased Datasets

We also assessed network performance when training with biased subsets of the OASIS and Biobank datasets using: (i) a standard training regime and, (ii) naïve unlearning (unlearning on the whole of both datasets) and (ii) unlearning on just the overlap set. We considered three degrees of overlap: 5, 10 and 15 years. The resulting networks were then tested on the full testing dataset, spanning across the whole age range. Figure 4 shows the resulting errors. As expected, it can be seen that the normal training regime produces large errors, especially outside of the range of the Biobank training data, and is entirely driven by the larger Biobank training data.

Table 2. MAE results for Biobank and OASIS data from training with datasets with varying degrees of overlap as shown in the first column. Scanner accuracy is calculated by training a domain classifier on the fixed feature representation. Random chance is given in brackets.

Method	Biobank MAE	OASIS MAE	Scanner Classification Accuracy (%)
5 Years			
1. Standard	16.5 ± 5.94	15.5 ± 6.95	100 (50)
2. Naïve Unlearning	6.11 ± 3.99	4.44 ± 4.20	58 (50)
3. Unlearning on Overlap	5.49 ± 3.67	4.37 ± 4.05	53 (50)
10 Years			
4. Standard	9.66 ± 5.83	13.6 ± 6.58	100 (50)
5. Naïve Unlearning	4.20 ± 2.90	4.29 ± 4.01	56 (50)
6. Unlearning on Overlap	3.93 ± 2.81	4.04 ± 3.86	52 (50)
15 Years			
7. Standard	8.91 ± 5.31	10.4 ± 5.55	100 (50)
8. Naïve Unlearning	3.82 ± 2.84	4.39 ± 4.07	57(50)
9. Unlearning on Overlap	3.75 ± 2.78	3.99 ± 3.52	50 (50)

With naïve unlearning, the network is not able to correct for both scanners and the results for the OASIS dataset are poor, whereas by using unlearning on just the overlap subjects, the error is reduced across both datasets. The only region which performs worse is the lower end of the OASIS dataset, probably because when the network was being driven only by the Biobank data, the network generalised to OASIS testing points from the same range. Naïve unlearning also performs slightly less well at removing scanner information on the testing domain. This probably indicates that the features learned also encode some age information and so generalise less well across the whole age range.

These results show the power of the network to remove strong bias from the data, with a large reduction in error across the dataset compared to standard learning. There is also a clear improvement compared to unlearning on the whole dataset, showing that information which was key to the age prediction task was being removed when the whole dataset was used for unlearning, and this is lessened by training on only the overlap dataset.

5   Discussion

We have shown that, through using an iterative training scheme to 'unlearn' scanner information, we can create features from which most scanner information has been removed and, thus, harmonise the data for a given task. The training regime is flexible and could be implemented with any feedforward network and likely data scenarios such as biased datasets, meaning that unlearning should be applicable for most real-world MRI harmonisation problems.

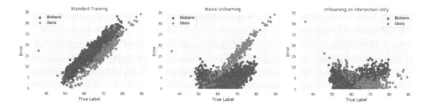

Fig. 4. Errors for the three different training regimes: standard training, naïve unlearning and unlearning only on the overlap data (10 year case). It can be seen that unlearning only on the overlap dataset leads to much lower losses across both datasets.

Acknowledgements. ND is supported by the Engineering and Physical Sciences Research Council (EPSRC) and Medical Research Council (MRC) [grant number EP/L016052/1]. MJ is supported by the National Institute for Health Research (NIHR), Oxford Biomedical Research Centre (BRC), and this research was funded by the Wellcome Trust [215573/Z/19/Z]. The Wellcome Centre for Integrative Neuroimaging is supported by core funding from the Wellcome Trust [203139/Z/16/Z]. AN is grateful for support from the UK Royal Academy of Engineering under the Engineering for Development Research Fellowships scheme.The computational aspects of this research were supported by the Wellcome Trust Core Award [Grant Number 203141/Z/16/Z] and the NIHR Oxford BRC. The views expressed are those of the author(s) and not necessarily those of the NHS, the NIHR or the Department of Health.

References

1. Alfaro-Almagro, F., et al.: Image processing and quality control for the first 10,000 brain imaging datasets from UK Biobank. bioRxiv **166**, 130385 (2017)
2. Alvi, M.S., Zisserman, A., Nellåker, C.: Turning a blind eye: explicit removal of biases and variation from deep neural network embeddings. In: ECCV Workshops (2018)
3. Ben-David, S., Blitzer, J., Crammer, K., Kulesza, A., Pereira, F., Vaughan, J.: A theory of learning from different domains. Mach. Learn. **79**, 151–175 (2010)
4. Dewey, B., et al.: DeepHarmony: a deep learning approach to contrast harmonization across scanner changes. Magn. Reson. Imaging **64**, 160–170 (2019)
5. Filippini, N., et al.: Study protocol: the Whitehall II imaging sub-study. BMC psychiatry **14**, 159 (2014). https://doi.org/10.1186/1471-244X-14-159
6. Fortin, J.P., et al.: Harmonization of cortical thickness measurements across scanners and sites. NeuroImage **167**, 104–120 (2017). https://doi.org/10.1016/j.neuroimage.2017.11.024
7. Ganin, Y., Lempitsky, V.S.: Unsupervised domain adaptation by back propagation. ArXiv (2014)
8. Ganin, Y., et al.: Domain-adversarial training of neural networks. J. Mach. Learn. Res. **17**, 59:1–59:35 (2015)
9. Hoffman, J., Tzeng, E., Darrell, T., Saenko, K.: Simultaneous deep transfer across domains and tasks. 2015 IEEE International Conference on Computer Vision (ICCV), pp. 4068–4076 (2015)

10. van der Maaten, L., Hinton, G.: Viualizing data using t-SNE. J. Mach. Learn. Res. **9**, 2579–2605 (2008)
11. Marcus, D., Wang, T., Parker, J., Csernansky, J., Morris, J., Buckner, R.: Open access series of imaging studies (OASIS): cross-sectional MRI data in young, middle aged, nondemented, and demented older adults. J. Cogn. Neurosci. **19**, 1498–507 (2007). https://doi.org/10.1162/jocn.2007.19.9.1498
12. Pomponio, R., et al.: Harmonization of large MRI datasets for the analysis of brain imaging patterns throughout the lifespan. NeuroImage **208**, 116450 (2019). https://doi.org/10.1016/j.neuroimage.2019.116450
13. Simonyan, K., Zisserman, A.: Very deep convolutional networks for large-scale image recognition, September 2014
14. Sudlow, C., et al.: UK Biobank: an open access resource for identifying the causes of a wide range of complex diseases of middle and old age. Plos Med. **12**, e1001779 (2015)
15. Wachinger, C., Rieckmann, A., Pölsterl, S.: Detect and correct bias in multi-site neuroimaging datasets. bioRxiv (2020)
16. Wilson, G., Cook, D.J.: A survey of unsupervised deep domain adaptation (2018)
17. Zhao, F., Wu, Z., Wang, L., Lin, W., Xia, S., Li, G.: Harmonization of infant cortical thickness using surface-to-surface cycle-consistent adversarial networks, pp. 475–483 (2019)
18. Zhu, J.Y., Park, T., Isola, P., Efros, A.A.: Unpaired image-to-image translation using cycle-consistent adversarial networks. In: 2017 IEEE International Conference on Computer Vision (ICCV), pp. 2242–2251 (2017)

Cross-domain Medical Image Translation by Shared Latent Gaussian Mixture Model

Yingying Zhu[1]([✉]), Youbao Tang[1], Yuxing Tang[1], Daniel C. Elton[1], Sungwon Lee[1], Perry J. Pickhardt[2], and Ronald M. Summers[1]

[1] Imaging Biomarkers and Computer-Aided Diagnosis Laboratory, Radiology and Imaging Sciences, National Institutes of Health, Clinical Center, Bethesda, MD 20892, USA
{yingying.zhu,rms}@nih.gov
[2] School of Medicine and Public Health, University of Wisconsin, Madison, WI 53706, USA

Abstract. Current deep learning based segmentation models generalize poorly to different domains due to the lack of sufficient labelled image data. An important example in radiology is generalizing from contrast enhanced CT to non-contrast CT. In real-world clinical applications, cross-domain image analysis tools are in high demand since medical images from different domains are generally used to achieve precise diagnoses. For example, contrast enhanced CT at different phases are used to enhance certain pathologies or internal organs. Many existing cross-domain image-to-image translation models show impressive results on large organ segmentation by successfully preserving large structures across domains. However, such models lack the ability to preserve fine structures during the translation process, which is significant for many clinical applications, such as segmenting small calcified plaques in the aorta and pelvic arteries. In order to preserve fine structures during medical image translation, we propose a patch-based model using shared latent variables from a Gaussian mixture. We compare our image translation framework to several state-of-the-art methods on cross-domain image translation and show our model does a better job preserving fine structures. The superior performance of our model is verified by performing two tasks with the translated images - detection and segmentation of aortic plaques and pancreas segmentation. We expect the utility of our framework will extend to other problems beyond segmentation due to the improved quality of the generated images and enhanced ability to preserve small structures.

1   Introduction

Developing deep learning based segmentation models which can generalize to different domains has been in high demand since different type of medical images are usually collected in real clinical practice to achieve a precisely diagnosis. For

A. L. Martel et al. (Eds.): MICCAI 2020, LNCS 12262, pp. 379–389, 2020.
https://doi.org/10.1007/978-3-030-59713-9_37

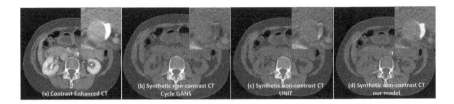

Fig. 1. (a) Real post-contrast CT scan. (b) Synthetic non-contrast CT using Cycle-GAN [21]. (c) using shared latent variables from a Gaussian distribution [4,8]. (d) Gaussian mixture model.

example, a patient might have a non-contrast and a contrast-enhanced CT scan generated by injecting an intravenous contrast agent to highlight different internal structures at different time points. For instance, the arteries are enhanced in early the early phase and the kidneys are enhanced in the late phase as the contrast agent is metabolized in the kidneys.

Although many existing works [5,13–17] have used image-to-image translation techniques to assist in medical image analysis tasks, less work been done to address cross-domain image segmentation due to the lack of sufficient labelled data from different domains. Only a few studies have used synthetic images generated by cross-modality image-to-image translation (e. g., CT and MRI) for cross-domain image (e. g., organ) segmentation [20,22]. However, image segmentation across CT and MRI scans is of less clinical importance due to its poor performance caused by the large difference between the two modalities. Moreover, CT scans are typically used to scan a large range (fast but low resolution) while MRI scans are often targeted at small regions (slow but high resolution). Paired CT and MRI scans for the same region of the body are rarely collected in clinical practice. Such works usually focus on large organ (e. g., heart) segmentation. However, hardly any work has been done for the small structure segmentation across domains, for example calcified plaque segmentation in the aorta and pelvic arteries under different contrast levels. This is a clinically important problem, since calcified plaque in the arteries is a strong predictor of heart attack [9].

Applying existing image translation models to improve calcified plaque segmentation on different domains is impractical since these model show inconsistent performance for preserving fine/tiny structures after image translation (as shown in Fig. 1, the calcified plaques are blurry and covered by neighboring structures by CycleGAN [21] and UNIT [4,8]). We hypothesize this is because UNIT assumes a shared latent Gaussian variable across domains and real medical images actually lie in a shared Gaussian mixture model since different internal structures (image patches) lies in different local clusters as shown on the right side of Fig. 1.

In order to address these problems, we proposed a patch-based domain invariant method using shared latent variables from a Gaussian mixture model for image translation [7,23]. In order to quantitatively evaluate the image translation performance of our model, we applied it to calcified plaque detection/segmentation and pancreas segmentation on both non-contrast and contrast enhanced CT images. We compared our model to several image translation networks. Experimental results showed that our model performs much better than the baseline (without image translation) and competing methods (e. g., Cycle-GAN [21] and UNIT [8]) by improving the performance of both tasks. It is worth noting that our model is trained using unpaired images across different domains, which means it can be easily adapted to real-world clinical practice where paired images are relatively rare.

2   Method

2.1   Unsupervised Image-to-Image Translation Networks (UNIT)

Let $x_1 \in \mathcal{X}_1$ and $x_2 \in \mathcal{X}_2$ be two images from non-contrast and contrast enhanced CT, respectively. The image size of x_1, x_2 is 512×512 in our dataset. Liu et al. [8] proposed two sets of variational auto-encoders for the two different domains and translate the images across the two domains via a shared latent space $\mathcal{Z}$. The shared latent space $\mathcal{Z}$ is conditionally independent from the two domains and is enforced to follow a Gaussian distribution with unit variance. Intuitively, this latent space $\mathcal{Z}$ encodes the underlying morphological structure of objects and is domain invariant. For example, the latent space $\mathcal{Z}$ may depend on the shape of internal organs which is invariant across image domains. They implemented this by sharing the latent space $\mathcal{Z}$ layers of two variational autoencoders. The UNIT model [8] trained using the whole image, which results in loss of detailed structures as shown in Fig. 1(b). Similarly, the CycleGAN method also translates imags between domains using a shared latent space and cycle consistent loss but can not preserve the detailed structures (see Fig. 1(c)).

2.2   Patch-Based Mixtures Gaussian Image-to-Image Translation

We proposed a patch based method and extracted many small randomly sampled patches from each image. The optimal patch size is determined using testing with the validation set. We used a patch size of 32×32 in all our experiments. We extracted random patches at the same location from x_1, x_2 respectively. We extracted the image features using a pre-trained CNN on these small patches and found that they lie in different local clusters as shown in Fig. 2. Intuitively, image patches from different organs or internal structures will be clustered into different local clusters. Based on these observations, we proposed to model the domain independent shared latent space $\mathcal{Z}$ using a mixture Gaussian model, $\mu_z = [\mu_{z,1}, \cdots, \mu_{z,k}], \Sigma_z = [\Sigma_{1,\mathbf{z}}, \cdots, \Sigma_{\mathbf{z},\mathbf{K}}]$ are the mean and variance for different Gaussian components.

$$z \sim \sum_{k=1}^{K} \pi_k \mathcal{N}(z|\mu_{k,z}, \Sigma_{k,z}), s.t. \sum_{k=1}^{K} \pi_k = 1, \tag{1}$$

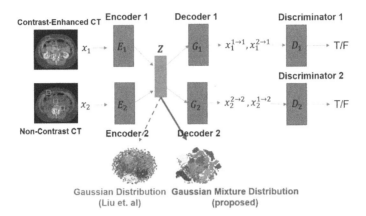

Fig. 2. The framework of our proposed model compared to UNIT [8] model. Our model assumes that the shared latent variable lies in a Gaussian mixed distribution (different patches lies in different local clusters and Liu et. al. [8] assume a single Gaussian distribution.

K is the number of Gaussian components following   [2]. It is worth noting that K is determined by the validation dataset on the downstream segmentation/detection task. We follow [8] and use 6 sub-networks: two domain image encoders: E_1, E_2, two domain image generators G_1, G_2, and two domain adversarial discriminators D_1, D_2 as shown in Fig. 2. We use VGG-16 as the encoder and a reversed VGG-16 structure as the decoder. We use the variational inference model from *Edward* [18] to solve the parameters in our model. The encoder outputs a set of mean vectors for each Gaussian component: $E_1(x_1, \theta_1), E_2(x_2, \theta_2)$, $\Sigma_1 = [\Sigma_{1,1}, \cdots, \Sigma_{1,K}], \Sigma_2 = [\Sigma_{2,1}, \cdots, \Sigma_{2,K}]$ are the variance matrix for each Gaussian component. $\Theta_1 = [\theta_{1,1}, \cdots, \theta_{1,K}]$, $\Theta_2 = [\theta_{2,1}, \ldots, \theta_{2,K}]$ are the weights for using a shared encoder to output the mean for different Gaussian components. The distribution of latent code z given x_1, x_2 are listed as,

$$q_1(z|x_1) \sim \sum_{i=1}^{K} \pi_k \mathcal{N}\left(z|E_1(x_1, \theta_{1,k}), \Sigma_{k,1}\right), \sum_{k=1}^{K} \pi_k = 1,$$

$$q_2(z|x_2) \sim \sum_{i=1}^{K} \pi_k \mathcal{N}\left(z|E_2(x_2, \theta_{2,k}), \Sigma_{k,2}\right), \sum_{k=1}^{K} \pi_k = 1.$$

The reconstructed image $\hat{x}_1^{1->1} = G_1(z \sim q_1(z|x_1))$ for two variational autoenconder (VAE) (E_1, G_1) and (E_2, G_2) are $\hat{x}_2^{2->2} = G_2(z \sim q_2(z|x_2))$, we have the VAE loss:

$$\mathcal{L}_{VAE_1}(E_1, G_1, \boldsymbol{\Theta}_1, \boldsymbol{\Sigma}_1, \boldsymbol{\Sigma}_z, \mu_z) = \lambda_1 \mathrm{KL}(q_1(z|x_1)||p(z)) \tag{2}$$
$$-\lambda_2 \mathbb{E}_{z \sim q_1(z|x_1)}[\log pG_1(x_1|z)]$$
$$\mathcal{L}_{VAE_2}(E_2, G_2, \boldsymbol{\Theta}_2, \boldsymbol{\Sigma}_2, \boldsymbol{\Sigma}_z, \mu_z) = \lambda_1 \mathrm{KL}(q_2(z|x_2)||p(z)) \tag{3}$$
$$-\lambda_2 \mathbb{E}_{z \sim q_2(z|x_2)}[\log pG_1(x_2|z)],$$

where λ_1, λ_2 are the parameters controls the weights for the objective terms and the KL divergence terms penalized derivation of the distribution if the latent variable from the prior distribution.

Our model also has two generative adversarial networks: $GAN_1 = \{G_1, D_1\}$, $GAN_2 = \{G_2, D_2\}$. D_1, D_2 are constrained to output true if images are sampled from the first or second domain respectively and output false if the images are generated from G_1, G_2 respectively. We have the following conditional GAN objective functions, which constrain the translated images to resemble images in their respective target domains:

$$\mathcal{L}_{GAN_1}(E_2, G_1, D_1, \boldsymbol{\Theta}_1, \boldsymbol{\Sigma}_1, \boldsymbol{\Sigma}_z, \mu_z) = \lambda_0 \mathbb{E}_{x_1 \sim P_{\mathcal{X}_1}}[\log D_1(x_1)] \tag{4}$$
$$+\lambda_0 \mathbb{E}_{z \sim q_2(z|x_2)}[\log D_1(G_1(z))]$$
$$\mathcal{L}_{GAN_2}(E_1, G_2, D_2, \boldsymbol{\Theta}_2, \boldsymbol{\Sigma}_2, \boldsymbol{\Sigma}_z, \mu_z) = \lambda_0 \mathbb{E}_{x_2 \sim P_{\mathcal{X}_2}}[\log D_2(x_2)] \tag{5}$$
$$+\lambda_0 \mathbb{E}_{z \sim q_1(z|x_1)}[\log D_2(G_2(z))],$$

Similarily to the previous equation, the hyperparameter λ_0 balances the impact of the GAN objective function. We also incorporate a cycle consistency constraint to ensure that twice translated images resemble the original image and a KL divergence term which penalizes the latent code from deviating too far from the prior distribution.

$$\mathcal{L}_{CC_2}(E_1, G_1, E_2, G_2, \boldsymbol{\Theta}_1, \boldsymbol{\Theta}_2, \boldsymbol{\Sigma}_1, \boldsymbol{\Sigma}_2, \boldsymbol{\Sigma}_z, \mu_z) \tag{6}$$
$$= \lambda_3 \mathrm{KL}(q_1(z|x_1)||p(z))$$
$$+\lambda_4 \mathrm{KL}(q_2(z|x_1^{1->2})||p(z)) - \lambda_4 \mathbb{E}_{z \sim q_2(z|x_1^{1->2})}[\log pG_1(x_1|z)]$$
$$\mathcal{L}_{CC_2}(E_2, G_2, E_1, G_1, \boldsymbol{\Theta}_1, \boldsymbol{\Theta}_2, \boldsymbol{\Sigma}_1, \boldsymbol{\Sigma}_2, \boldsymbol{\Sigma}_z, \mu_z)$$
$$= \lambda_3 \mathrm{KL}(q_2(z|x_2)||p(z)) \tag{7}$$
$$+\lambda_4 \mathrm{KL}(q_1(z|x_2^{2->1})||p(z)) - \lambda_4 \mathbb{E}_{z \sim q_1(z|x_2^{2->1})}[\log pG_2(x_2|z)]$$

Combining all the above objective functions, our final objective functions is:

$$\arg \min(E_1, E_2, G_1, G_2, \boldsymbol{\Theta}_1, \boldsymbol{\Theta}_2, \boldsymbol{\Sigma}_1, \boldsymbol{\Sigma}_2, \boldsymbol{\Sigma}_z, \mu_z) \max(D_1, D_2) \tag{8}$$
$$\mathcal{L}_{VAE_1}(E_1, G_1, \boldsymbol{\Theta}_1, \boldsymbol{\Sigma}_1, \boldsymbol{\Sigma}_z, \mu_z) + \mathcal{L}_{VAE_2}(E_2, G_2, \boldsymbol{\Theta}_1, \boldsymbol{\Sigma}_1, \boldsymbol{\Sigma}_z, \mu_z)$$
$$+\mathcal{L}_{CC1}(E_1, G_1, E_2, G_2, \boldsymbol{\Theta}_1, \boldsymbol{\Sigma}_1, \boldsymbol{\Theta}_2, \boldsymbol{\Sigma}_2, \boldsymbol{\Sigma}_z, \mu_z)$$
$$+\mathcal{L}_{CC2}(E_1, G_1, E_2, G_2, \boldsymbol{\Theta}_1, \boldsymbol{\Sigma}_1, \boldsymbol{\Theta}_2, \boldsymbol{\Sigma}_2, \boldsymbol{\Sigma}_z, \mu_z)$$
$$+\mathcal{L}_{GAN_1}(E_1, G_1, D_1, \boldsymbol{\Theta}_1, \boldsymbol{\Sigma}_1) + \mathcal{L}_{GAN_2}(E_2, G_2, D_2, \boldsymbol{\Theta}_2, \boldsymbol{\Sigma}_2)$$

3   Experiments

Evaluation Tasks. In order to evaluate the image translation model quantitatively, we evaluated the performance of our image translation models on two challenging image segmentation tasks: calcified plaque detection/segmentation and pancreas segmentation on both the non-contrast and contrast enhanced CT scans given labelled CT scans only from one domain. We also compared our model to two recent state-of-the art image translation models: CycleGAN [21] and UNIT [4,8], which have been broadly applied to image translation on natural images.

Image Translation Training. For training we used 140 unpaired CT scans (70 non-contrast, 70 contrast enhanced) taken from renal donor patients at the University of Wisconsin Medical Center. We applied our image translation model to generate synthetic contrast enhanced CT scans from the labelled non-contrast CT scans and used them as augmented data to improve our plaque segmentation/detection performance and pancreas segmentation on both contrast enhanced and non-contrast CT scans.

The image translation training dataset was separated into 10 folds and one fold was used as validation data for selecting hyperparameters including the number of Gaussian mixture components K and $\lambda_0, \lambda_1, \lambda_2, \lambda_3, \lambda_4$. The validation dataset was evaluated for downstream plaque detection tasks using the model trained from Sect. 3.1. The optimal value of K is very important for our final image translation results. A small K value can lead to blurring of local image structures and a large K value can add to computation cost and more uncertainty on the output images. We used $K = 25$ based on the validation dataset performance. It is worth noting that the setting of K can vary between datasets. In order to select the best generated image for cross-domain image segmentation task, we also train a quality control network using an independent CT dataset selected from DeepLesion data [19] to remove the synthetic CT images with artifacts.

3.1   Calcified Plaque Detection and Segmentation

Labelling calcified plaques in the CT scans is very time consuming since the plaques are very small, frequently on the order of just a few pixels in CT images. In our experiments, we only have labelled training CT images from low dose non-contrast CT scans. We trained a 2D detection and segmentation model [6] on 75 low dose CT scans which contained a total of 25,200 images (transverse cross sections/slices), including 2,119 with plaques. The training dataset was divided into 10 folds and we used 9 folds as the training dataset and 1 fold as the validation dataset for parameter selection. We shuffled the training and validation dataset and trained 10 2D Mask R-CNN [3] models and applied these models to our independent testing dataset. We report the mean and standard derivation across all 10 models in Table 1. For this work we labelled an additional testing dataset with 30 contrast enhanced CT scans and 30 non-contrast scans from a different dataset collected at University of Wisconsin Medical Center. It had plaque labeled manually (7/30 of these scans contained aortic plaques, with a total of 53 plaques overall).

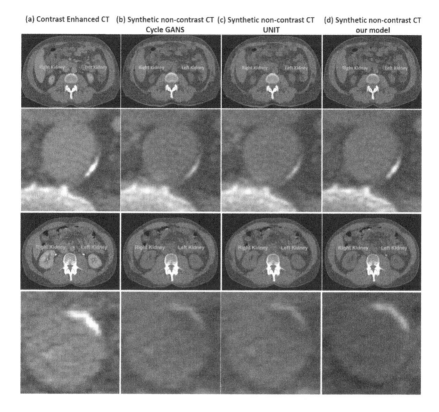

Fig. 3. Visualization of synthetic non-contrast CT scans from contrast enhanced CT (a) scans by (b) CycleGAN (c) UNIT (d) our model. We show two cases of median quality generated images by all competing methods.

We selected an synthetic image which shows the median performance on image translation (as shown in Fig. 3). Figure 3 (a) shows contrast enhanced CT scans in the late phase (the kidneys are enhanced with bright pixel values). There is a very small plaque in this image which was translated into non-contrast CT scans by cycleGAN [21], UNIT [8] and our model. The whole translated images looks very similar by all competing methods, however, the calcified plaque pixel brightness is better preserved by our approach.

Quantitative Results. The calcified plaque detection and segmentation results are shown in Table 1. Our model achieved similar plaque detection and segmentation performance to the real pre-contrast CT scans (precision decreased about $> 1.5\%$, recall decreased about $> 4\%$) and dice coefficients drops about > 0.1.

Table 1. Plaque detection and segmentation results. The first column gives detection results for the original data without image translation. The 2nd, 3rd, and 4th columns give results for non-contrast (NC) and contrast enhanced (CE) plaque detection after non-contrast to contrast enhanced image translation with different image translation models. SYN= synthetic images.

Method	Baseline	CycleGAN [1]	UNIT [4,8]	Ours
Training	NC	NC & SYN CE	NC & SYN CE	NC & SYN CE
Testing	NC	NC	NC	NC
Precision	78.4 ± 1.82%	79.5 ± 1.75%	80.2 ± 1.68%	**80.7 ± 1.65%**
Recall	82.4 ± 2.45%	83.1 ± 2.26%	84.7 ± 2.13%	**85.2 ± 2.03%**
Dice	0.724 ± 0.245	0.733 ± 0.223	0.751 ± 0.218	**0.756 ± 0.193**
Testing	CE	CE	CE	CE
Precision	48.6 ± 3.52%	61.5 ± 2.87%	64.7 ± 2.64%	**78.2 ± 2.58%**
Recall	54.3 ± 3.64%	64.3 ± 3.21%	69.8 ± 3.05%	**81.2 ± 2.87%**
Dice	0.452 ± 0.251	0.534 ± 0.236	0.566 ± 0.198	**0.676 ± 0.176**

The detection and segmentation model trained on synthetic images generated from UNIT [8] and CycleGAN [21], by contrast, shows a > 15% drop in precision, a > 13% drop in recall and drop > 0.18 in Dice coefficients caused by loss of fine structures.

3.2 Pancreas Segmentation

Pancreas segmentation is very important for the diagnosis of pancreas cancer and surgical planning. Pancreas segmentation is challenging since the pancreas is very small compared to other internal organs and has large variance in its shape and orientation. Most existing pancreas segmentation approaches focus on segmenting pancreas only in contrast enhanced CT where the pancreas structures are more enhanced and have clearer boundaries. Current public pancreas segmentation data are only labelled on contrast enhanced CT. We combined two public contrast enhanced CT datasets for pancreas segmentation. The first one includes 82 labelled contrast enhanced CT scans from the Cancer Imaging Archive database and second one has 281 contrast enhanced CT scans from the Medical Segmentation Decathlon database [10,11].

We use 10-fold cross validation and report the mean and standard derivation across the 10 folds. In order to improve pancreas segmentation on non-contrast CT images, we generated non-contrast CT from these contrast enhanced CT and used them to train a cross-domain 3D segmentation model. We use the multiple scale 3D Unet model proposed in [12] and compared with a 3D U-Net trained using synthetic non-contrast CT scans generated the different image translation models (CycleGAN [21], UNIT [4,8] and our model). We use 24 non-contrast CT scans annotated by an expert radiologist as the non-contrast pancreas segmentation testing dataset.

Table 2. Pancreas segmentation results with false positives pixel numbers and Dice scores on contrast enhanced CT and non-contrast CT. The first column show the baseline model which is only trained using contrast enhanced CT scans. Th second, third, and last column show the results trained by contrast enhanced CT and synthetic non-contrast CT generated by CycleGAN [21], UNIT [4,8] and our method respectively. NC = non-contrast CT. CE = contrast enhanced CT. SYN = synthetic.

Method	Baseline	CycleGAN [21]	UNIT [4,8]	Ours
Training	CE	CE & SYN NC	CE & SYN NC	CE & SYN NC
Testing	CE	CE	CE	CE
Precision	$85.7 \pm 1.82\%$	$87.2 \pm 1.62\%$	$88.1 \pm 1.53\%$	**$89.5 \pm 1.44\%$**
Recall	$86.8 \pm 2.1\%$	$88.7 \pm 1.97\%$	$89.5 \pm 1.83\%$	**$90.7 \pm 1.68\%$**
Dice	0.728 ± 0.173	0.728 ± 0.154	0.731 ± 0.142	**0.734 ± 0.136**
Testing	NC	NC	NC	NC
Precision	$78.1 \pm 2.83\%$	$82.8 \pm 2.67\%$	$83.2 \pm 2.45\%$	**$84.7 \pm 2.25\%$**
Recall	$81.5 \pm 3.0\%$	$84.3 \pm 2.81\%$	$86.2 \pm 2.75\%$	**$87.2 \pm 2.56\%$**
Dice	0.642 ± 0.183	0.684 ± 0.172	0.697 ± 0.163	**0.725 ± 0.153**

Quantitative Results. Table 2 shows the quantitative results of cross domain pancreas segmentation results on both contrast enhanced CT and non-contrast CT using different image translation methods (CycleGAN [21], UNIT [8] and our model). As shown in the top part of the table, the pancreas segmentation results on contrast enhanced CT scans are similar for all competing methods. Adding synthetic non-contrast CT scans in the training can also improve the pancreas segmentation on contrast enhanced CT, and our method shows slight improvement compared to all other methods. For pancreas segmentation on non-contrast CT, adding synthetic non-contrast CT images in the training can significantly improve the segmentation Dice score and reduce false positive pixels. For example, CycleGAN and UNIT show an average improvement of Dice score $> 0.04/ > 0.05$ and reduction in false positive pixels $> 4000/ > 4500$ compared to the baseline model. Our model shows the best performance and achieves a > 0.08 improvement in Dice score and a reduction of > 5000 false positive pixels on average compared to the baseline model.

4   Conclusion

In this work, we proposed an image translation model using shared latent variables from a Gaussian mixture distribution to preserve fine structures on medical images. We applied our method to two challenging medical imaging segmentation tasks: cross domain (non-contrast and contrast enhanced CT) calcified plaque detection/segmentation and pancreas segmentation. We demonstrated that our model can translate the medical images across different domains with better preservation of fine structures compared to two state-of-the-art image translation models for natural images. In the future work, we will explore the application of this model to translating medical images across multiple domains, for example contrast enhanced CT scans at different phases from non-contrast CT scans. Possible applications of this method are generating synthetic images to reduce radiation dose or creating 100% contrast enhanced CT scans from 10% dose contrast enhanced CT scans to reduce the dose of intravenous contrast agent to be used on patients.

Acknowledgments. This research was supported in part by the Intramural Research Program of the National Institutes of Health Clinical Center. We thank NVIDIA for GPU card donations.

References

1. Clark, K., et al.: The cancer imaging archive (TCIA): maintaining and operating a public information repository. J. Digit. Imaging **26**(6), 1045–1057 (2013)
2. Dilokthanakul, N., et al.: Deep unsupervised clustering with gaussian mixture variational autoencoders. arXiv e-prints arXiv:1611.02648 (2016)
3. He, K., Gkioxari, G., Dollár, P., Girshick, R.: Mask R-CNN. In: International Conference on Computer Vision (ICCV), pp. 2980–2988 (2017)
4. Huang, X., Liu, M.Y., Belongie, S., Kautz, J.: Multimodal unsupervised image-to-image translation. In: European Conference on Computer Vision (ECCV), pp. 172–189 (2018)
5. Jin, D., Xu, Z., Tang, Y., Harrison, A.P., Mollura, D.J.: CT-realistic lung nodule simulation from 3D conditional generative adversarial networks for robust lung segmentation. In: International Conference on Medical Image Computing and Computer-Assisted Intervention, pp. 732–740 (2018)
6. Liu, J., Yao, J., Bagheri, M., Sandfort, V., Summers, R.M.: A semi-supervised CNN learning method with pseudo-class labels for atherosclerotic vascular calcification detection. In: International Symposium on Biomedical Imaging (ISBI), pp. 780–783 (2019)
7. Liu, L., Nie, F., Wiliem, A., Li, Z., Zhang, T., Lovell, B.C.: Multi-modal joint clustering with application for unsupervised attribute discovery. IEEE Trans. Image Process. **27**(9), 4345–4356 (2018)
8. Liu, M.Y., Breuel, T., Kautz, J.: Unsupervised image-to-image translation networks. In: Advances in Neural Information Processing Systems (NeurIPS), pp. 700–708 (2017)
9. Parab, S.Y., Patil, V.P., Shetmahajan, M., Kanaparthi, A.: Coronary artery calcification on chest computed tomography scan-anaesthetic implications. Indian J. Anaesth. **63**(8), 663 (2019)
10. Roth, H., et al.: DeepOrgan: multi-level deep convolutional networks for automated pancreas segmentation. In: Navab, N., Hornegger, J., Wells, W.M., Frangi, A.F. (eds.) MICCAI 2015. LNCS, vol. 9349, pp. 556–564. Springer, Cham (2015). https://doi.org/10.1007/978-3-319-24553-9_68
11. Simpson, A.L., et al.: A large annotated medical image dataset for the development and evaluation of segmentation algorithms. arXiv e-prints arXiv:1902.09063 (2019)
12. Sriram, S.A., Paul, A., Zhu, Y., Sandfort, V., Pickhardt, P.J., Summers, R.: Multilevel U-Net for pancreas segmentation from non-contrast CT scans through domain adaptation. In: Hahn, H.K., Mazurowski, M.A. (eds.) Medical Imaging 2020: Computer-Aided Diagnosis. SPIE, March 2020

13. Tang, Y.B., Oh, S., Tang, Y.X., Xiao, J., Summers, R.M.: CT-realistic data augmentation using generative adversarial network for robust lymph node segmentation. In: Medical Imaging: Computer-Aided Diagnosis, vol. 10950, p. 109503V (2019)
14. Tang, Y., et al.: CT image enhancement using stacked generative adversarial networks and transfer learning for lesion segmentation improvement. In: International Workshop on Machine Learning in Medical Imaging, pp. 46–54 (2018)
15. Tang, Y., Tang, Y., Xiao, J., Summers, R.M.: XLSor: a robust and accurate lung segmentor on chest x-rays using criss-cross attention and customized radiorealistic abnormalities generation. In: International Conference on Medical Imaging with Deep Learning, pp. 457–467 (2019)
16. Tang, Y.X., Tang, Y.B., Han, M., Xiao, J., Summers, R.M.: Abnormal chest X-ray identification with generative adversarial one-class classifier. In: International Symposium on Biomedical Imaging, pp. 1358–1361 (2019)
17. Tang, Y., Tang, Y., Sandfort, V., Xiao, J., Summers, R.M.: TUNA-Net: task-oriented unsupervised adversarial network for disease recognition in cross-domain chest X-rays. In: International Conference on Medical Image Computing and Computer-Assisted Intervention, pp. 431–440 (2019)
18. Tran, D., Kucukelbir, A., Dieng, A.B., Rudolph, M., Liang, D., Blei, D.M.: Edward: a library for probabilistic modeling, inference, and criticism. arXiv e-prints arXiv:1610.09787 (2016)
19. Yan, K., Wang, X., Lu, L., Summers, R.M.: DeepLesion: automated mining of large-scale lesion annotations and universal lesion detection with deep learning. J. Med. Imaging **5**(3), 1–11 (2018)
20. Zhou, T., Fu, H., Chen, G., Shen, J., Shen, J., Shao, L.: Hi-net: Hybrid-fusion network for multi-modal MR image synthesis. IEEE Trans. Med. Imaging **PP(99)**, 1 (2020)
21. Zhu, J.Y., Park, T., Isola, P., Efros, A.A.: Unpaired image-to-image translation using cycle-consistent adversarial networks. In: International Conference on Computer Vision (ICCV), pp. 2223–2232 (2017)
22. Zhu, Y., Elton, D.C., Lee, S., Pickhardt, P., Summers, R.: Image translation by latent union of subspaces for cross-domain plaque detection. arXiv e-prints arXiv:2005.11384 (2020)
23. Zhu, Y., Huang, D., la Torre Frade, F.D., Lucey, S.: Complex non-rigid motion 3D reconstruction by union of subspaces. In: Proceedings of CVPR, June 2014

Self-supervised Skull Reconstruction in Brain CT Images with Decompressive Craniectomy

Franco Matzkin[1]([✉]), Virginia Newcombe[2], Susan Stevenson[2],
Aneesh Khetani[2], Tom Newman[2], Richard Digby[2], Andrew Stevens[2],
Ben Glocker[3], and Enzo Ferrante[1]

[1] Research Institute for Signals, Systems and Computational Intelligence, sinc(i),
CONICET, FICH-UNL, Santa Fe, Argentina
fmatzkin@sinc.unl.edu.ar
[2] Division of Anaesthesia, Department of Medicine, University of Cambridge,
Cambridge, UK
[3] BioMedIA, Imperial College London, London, UK

Abstract. Decompressive craniectomy (DC) is a common surgical procedure consisting of the removal of a portion of the skull that is performed after incidents such as stroke, traumatic brain injury (TBI) or other events that could result in acute subdural hemorrhage and/or increasing intracranial pressure. In these cases, CT scans are obtained to diagnose and assess injuries, or guide a certain therapy and intervention. We propose a deep learning based method to reconstruct the skull defect removed during DC performed after TBI from post-operative CT images. This reconstruction is useful in multiple scenarios, e.g. to support the creation of cranioplasty plates, accurate measurements of bone flap volume and total intracranial volume, important for studies that aim to relate later atrophy to patient outcome. We propose and compare alternative self-supervised methods where an encoder-decoder convolutional neural network (CNN) estimates the missing bone flap on post-operative CTs. The self-supervised learning strategy only requires images with complete skulls and avoids the need for annotated DC images. For evaluation, we employ real and simulated images with DC, comparing the results with other state-of-the-art approaches. The experiments show that the proposed model outperforms current manual methods, enabling reconstruction even in highly challenging cases where big skull defects have been removed during surgery.

Keywords: Skull reconstruction · Self-supervised learning ·
Decompressive craniectomy

Electronic supplementary material The online version of this chapter (https://
doi.org/10.1007/978-3-030-59713-9_38) contains supplementary material, which is
available to authorized users.

1 Introduction

Decompressive craniectomy (DC) is a surgical procedure performed for controlling the intracranial pressure (ICP) under some abnormal conditions which could be associated with brain lesions such as traumatic brain injury (TBI) [10]. In this procedure, a portion of the skull (bone flap) is removed, alleviating the risks associated with the presence of hematomas or contusions with a significant volume of blood [3]. In order to monitor the patient's condition and potential complications from the injury, computed tomography scans (CTs) of the affected area are acquired before and after this intervention [2].

Previous works which study the complications that can emerge after DC suggest that the volume of the skull defect is an important parameter to evaluate the decompressive effort [14,16]. A manual method to estimate such volume was proposed by Xiao and co-workers [17]. The authors developed a simple equation relying on three basic manual measurements which are multiplied and provide a good approximation of the real skull defect size. However, this method requires manual intervention and its accuracy is limited by the geometric approximation which does not take into account specific details of the skull shape.

Alternatively, the extracted bone flap volume could be estimated from a 3D model of the defect, which may be also useful for estimating materials and dimensions of eventual cranioplasty custom-made implants [5]. These can be used instead of the stored bone flap after DC, which has shown to carry potential complications if reused [4]. Different methods can estimate such shapes: one strategy is to take advantage of the symmetry present in the images [6]. However, it has the restriction of handling only unilateral DCs. Another simple and effective alternative could be the subtraction of the aligned pre- and post-operative CT scans, highlighting the missing part of the skull. Of course, this cannot be done if the provided data only contains post-operative images, which tends to be a common situation in real clinical scenarios.

We propose a bone flap reconstruction method which directly operates on post-operative CT scans, can handle any type of DC (not only unilateral) and is more accurate than current state-of-the-art manual methods. Our model employs encoder-decoder convolutional neural networks (CNN) and is trained following a self-supervised strategy, in the sense that it only requires images with complete skull for training, and avoids the need for annotated DC images.

Contributions: Our contributions are 3-fold: (i) to our knowledge, this is the first deep learning based model to perform skull reconstruction from brain CT images, (ii) the method outperforms the accuracy of manual and automatic state-of-the-art algorithms both in real and simulated DC and (iii) we introduce a self-supervised training procedure which enables learning skull reconstruction using only complete skull images which are more common than images with DC.

2 Self-supervised Skull Reconstruction

Our reconstruction method consists of a CNN which operates on binary skull images obtained after pre-processing the CT. We designed a virtual craniectomy

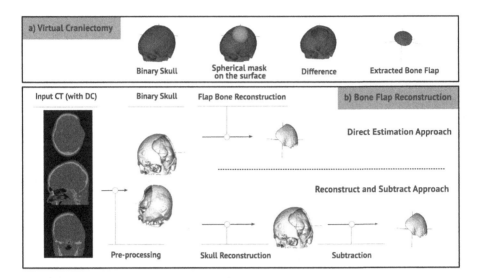

Fig. 1. a) Virtual Craniectomy process: given a skull, a spherical mask is applied in the surface for extracting a bone flap. b) In the direct estimation (DE) strategy, from the binary skull mask with DC, the bone flap is predicted by the network. In the reconstruct and subtract strategy (RS), the full skull is first reconstructed. Then, the binary mask with DC is subtracted from the complete skull, and the difference map is used as bone flap estimation.

(VC) procedure where full skulls are used to simulate DC patients by randomly removing bone flaps from specific areas. We used the VC to train various CNN architectures which follow alternative strategies: reconstructing only the missing flap or reconstructing the full skull and then subtracting. In the following, we describe in detail every stage of the reconstruction method.

2.1   Pre-processing

This stage extracts a binary skull mask from a CT and consists of three steps:

1. Registration: the images are registered to an atlas using rigid transformations, bringing all images into the same coordinate system. For registration, we use SimpleElastix [9], a state-of-the-art registration software publicly available. This pre-alignment encourages the model to focus on variations in the morphology of the skull, rather than learning features associated with its orientation and position.
2. Resampling: After registration, images are resampled to isotropic resolution (2 mm).
3. Thresholding: In CT scans, global thresholding [1] can be employed to extract the bones due to their high values in terms of Hounsfield units [15]. We used a threshold value of 90HU. As we can observe in Fig. 1b) a binary mask of the skull is obtained after pre-processing.

2.2 Virtual Craniectomy

We designed a virtual craniectomy procedure to simulate the effect of DC on full skulls. This enables the use of head CTs with the complete skull to self-supervise the learning process, avoiding the need of manually annotated DC images where the flap is segmented. This process implies extracting the intersection of the input skull with a spherical-shaped binary mask, which can be located in its upper part and have a variable size, and use such intersection as the ground truth during training.

We remove skull flaps from random locations, excluding the zone corresponding to the lower part (containing the bones between the jaw and the spine), where a craniectomy would not occur. The radius of the sphere was established so that the volume of the extracted bone flaps would match with standard surgeries. We defined a radius between 5 and 53 voxels to simulate craniectomies of 0.7 to 350 cm^3 of flap volume. This process is depicted in Fig. 1a).

2.3 Network Architectures

We implemented alternative encoder-decoder CNN architectures to address the flap reconstruction problem which are based on fully convolutional neural networks, but follow different reconstruction strategies, illustrated in Fig. 1b). Note that our contributions are not related to novel CNN architectures (we employ standard autoencoders and U-Net), but regarding the VC-based self-supervised strategy and its application to a new problem (i.e. skull reconstruction) where deep learning approaches have not been explored to date.

a) Reconstruct and subtract with autoencoder (RS-AE): The first model is a fully convolutional autoencoder (AE) trained to reconstruct the complete version of a DC skull (see the Supplementary Material for a detailed description of the AE architecture). Following an approach similar to that of Larrazabal et al. [7,8], we employ a denoising AE where the training process does not only include noise for data augmentation, but also virtual craniectomies. During training, we employ only full skulls: a random VC is applied before the skull enters the AE, which is trained to output its full version. Similar to previous strategies initially developed for unsupervised lesion detection [12], at test time, we reconstruct the bone flap by subtracting the original DC and its reconstructed full version to generate a difference map. The difference map constitutes the final bone flap 3D estimation, from which we can compute features like volume, etc.

b) Direct estimation with U-Net (DE-UNET): The second model directly estimates the bone flap, avoiding the full skull reconstruction and subtraction steps, which may introduce errors in the process. We employ the same encoder-decoder architecture used for the AE, but including skip connections, resulting in a 3D version of the standard U-Net architecture [13] (a detailed description of the architecture is given in the Supplementary Material). For training, instead of aiming to reconstruct the full skull, we learn to reconstruct the bone flap removed during the VC. Note that, similar to the previous model, we only require full skulls for training, enabling self-supervised learning without bone flap annotations.

c) Reconstruct and subtract with U-Net (RS-UNET): For completeness, we also explore the use of the U-Net following the reconstruct-and-subtract strategy, to evaluate the impact of the skip-connections in the resulting reconstruction.

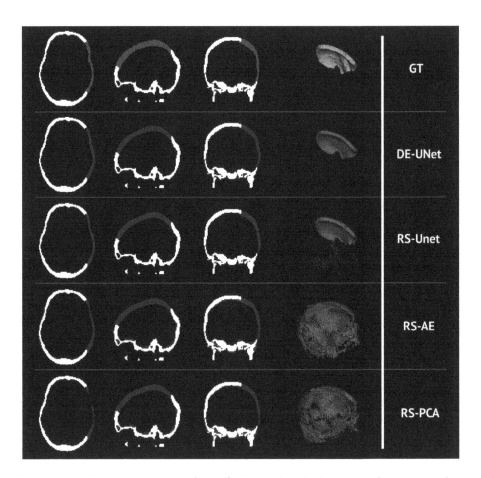

Fig. 2. Bone flap reconstruction (in red) obtained with the approaches compared in this work for a real decompressive craniectomy case from our test dataset. (Color figure online)

2.4 Training and Implementation

The CNN architectures were implemented in PyTorch 1.4 and trained on an NVIDIA TITAN Xp GPU[1]. During training, the images are fed to the network by adding salt and pepper noise and performing VC (with probability of 0.8), allowing the networks to see both intact and VC skulls. For all models the loss function L consists in a combination of the Dice loss L_{Dice} and the Binary Cross Entropy (BCE) Loss L_{BCE}. While cross-entropy loss optimizes for pixel level accuracy, the Dice loss function enhances the segmentation quality [11]. The compound loss function is defined as:

$$L = L_{Dice} + \lambda L_{BCE} \tag{1}$$

where the parameter $\lambda = 1$ was chosen by grid search. To improve generalization we incorporated dropout layers and use early stopping on validation data.

3 Experiments and Discussion

3.1 Database

The images used for this work were provided by the University of Cambridge (Division of Anaesthesia, Department of Medicine). They consist in 98 head CT images of 27 patients with Traumatic Brain Injury (TBI), including 31 images with DC and 67 cases with full skull. For training, we used full skull images only, excluding those patients who also have associated an image with DC. Patients which include pre and post-operative CT images were used for testing, since the difference between both images after registration was employed as ground-truth for the evaluation of bone flap estimation (an example is shown in green in Fig. 2). In this context, we employed 52 images for training (corresponding to 17 different patients) and 10 for testing (since there are only 10 patients with pre and post DC studies). The 36 images not included in the study were either pre-operative images of patients from the test split or post-operative without their corresponding pre-operative.

3.2 Baseline Models

We implemented a baseline model based on principal component analysis (PCA) for the task of flap bone estimation which follows the reconstruct and subtract strategy (RS-PCA). The principal components (see the Supplementary Material for visualizations of these components) were obtained by applying PCA to the vectorized version of the pre-processed complete skulls from the training fold. Similar to the RS-AE approach, the learnt latent representation provides a base for the space of complete skulls. Therefore, for reconstruction, we take the incomplete skull and project it to the learnt space to obtain its full version.

---

[1] The source code of our project is publicly available at: http://gitlab.com/matzkin/deep-brain-extractor.

For the task of flap bone volume estimation, we also compared our methods with the manual state-of-the-art ABC approach [17]. The ABC method requires to annotate manual measurements on the DC images (see the Supplementary Material for an example) and estimates the flap volume following simple geometric rules (a complete description of ABC can be found in the original publication [17]).

3.3 Experiments and Results

We performed experiments for bone flap reconstruction and volume estimation in real and simulated craniectomies. The simulations were done by performing 100 random virtual craniectomes to every complete skull from the test fold, resulting in a total of 1000 simulations for test. Figure 2 provides a qualitative comparison of the reconstructions (in red) obtained using the different approaches in a real DC. It can be observed that those based on the reconstruct and subtract strategy using AE and PCA produce spurious segmentations in areas far from the flap. The best reconstructions are achieved using the DE-UNet and RS-UNet, highlighting the importance of the skip connections.

The quantitative analysis is summarized in Figs. 3 and 4. Figure 3 shows Dice coefficient and Hausdorff distance between the ground-truth and reconstructed bone flaps for all the methods in real and simulated scenarios. Figure 4 includes scatter plots showing the accuracy of the bone flap volume estimation: we compare the predicted volume (x-axis) with the expected volume (y-axis). The closer the points to the identity, the more accurate are the predictions. For volume estimation we also include the manual ABC method. From these results, we observe that DE-UNet outperforms the other methods in both tasks, producing even

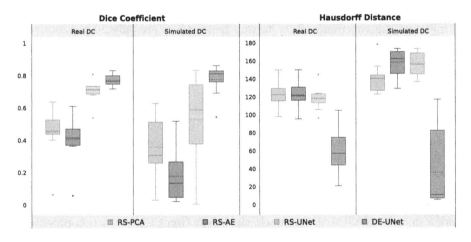

Fig. 3. Dice Coefficient and Hausdorff Distance (in mm) of the proposed methods output compared with the ground-truth (dashed line indicates the mean value). It can be seen that the DE-UNet outperforms the other discussed methods.

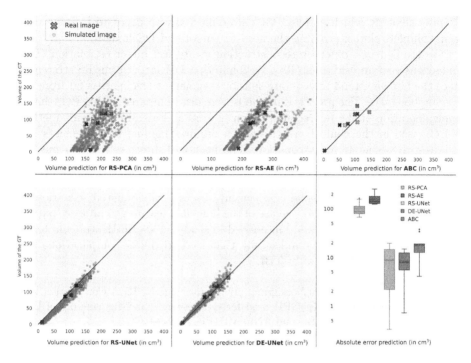

Fig. 4. Quantitative comparison for bone flap volume estimation with the different methods implemented in this study. The scatter plots show the estimated (x-axis) vs ground-truth bone flap volume. Note that RS-PCA, RS-AE, RS-UNet and DE-UNet show results for both real (cross markers in color) and simulated cases (circles in grey). For ABC, we only show results in real cases since the actual CT image is required for manual annotation (and virtual craniectomies for simulations are perform directly on the binary skull mask). (Color figure online)

better volume estimations than the manual ABC approach. We observed that Reconstruct and Subtract methods usually generate spurious pixels as prediction (as can be seen in Fig. 2) and a post-processing step may be needed after subtracting the pre and post-operative images (e.g. taking the biggest connected component, or applying morphological operations in the prediction). This does not tend to happen with Direct Estimation, what explains the gain in performance.

4   Conclusions

In this work, we propose and compare alternative self-supervised methods to estimate the missing bone flap on post-operative CTs with decompressive craniectomy. To our knowledge, this is the first study that tackles skull reconstruction and bone flap estimation using deep learning. We introduced a self-supervised training strategy which employs virtual craniectomy to generate training data from complete skulls, avoiding the need for annotated DC images.

We studied two different reconstruction strategies: direct estimation (DE) and reconstruct and subtract (RS). We found that DE outperforms RS strategies, since the last ones tend to generate spurious segmentations in areas far from the missing bone flap. The proposed methods were also compared with a PCA-based implementation of the RS reconstruction process and a state-of-the-art method (ABC) used in the clinical practise which requires manual measurements and relies on a geometric approximation. The proposed direct estimation method based on the U-Net architecture (DE-UNet) outperforms all the other strategies.

The performance of our method was measured in real cases (TBI patients who underwent decompressive craniectomy) as well as simulated scenarios. In the future, we plan to explore the use of the bone flap features to improve patient treatment. In this sense, we are interested in studying specific features in terms of volume and shape of a craniectomy that leads to fewer complications and improves patient outcome after TBI.

Acknowledgments. The authors gratefully acknowledge NVIDIA Corporation with the donation of the Titan Xp GPU used for this research, and the support of UNL (CAID-PIC-50220140100084LI) and ANPCyT (PICT 2018-03907).

References

1. van Eijnatten, M., van Dijk, R., Dobbe, J., Streekstra, G., Koivisto, J., Wolff, J.: CT image segmentation methods for bone used in medical additive manufacturing. Med. Eng. Phys. **51**, 6–16 (2018). https://doi.org/10.1016/j.medengphy.2017.10.008
2. Freyschlag, C.F., Gruber, R., Bauer, M., Grams, A.E., Thomé, C.: Routine postoperative computed tomography is not helpful after elective craniotomy. World Neurosurg. (2018). https://doi.org/10.1016/j.wneu.2018.11.079. http://www.sciencedirect.com/science/article/pii/S1878875018326299
3. Galgano, M., Toshkezi, G., Qiu, X., Russell, T., Chin, L., Zhao, L.R.: Traumatic brain injury: current treatment strategies and future endeavors. Cell Transplant. **26**(7), 1118–1130 (2017). https://doi.org/10.1177/0963689717714102. pMID: 28933211
4. Herteleer, M., Ectors, N., Duflou, J., Calenbergh, F.V.: Complications of skull reconstruction after decompressive craniectomy. Acta Chirurgica Belgica **117**(3), 149–156 (2016). https://doi.org/10.1080/00015458.2016.1264730
5. Hieu, L., et al.: Design for medical rapid prototyping of cranioplasty implants. Rapid Prototyping J. **9**(3), 175–186 (2003). https://doi.org/10.1108/13552540310477481
6. Huang, K.C., Liao, C.C., Xiao, F., Liu, C.C.H., Chiang, I.J., Wong, J.M.: Automated volumetry of postoperative skull defect on brain CT. Biomed. Eng. Appli. Basis Commun. **25**(03), 1350033 (2013). https://doi.org/10.4015/s1016237213500336
7. Larrazabal, A.J., Martínez, C., Glocker, B., Ferrante, E.: Post–DAE: anatomically plausible segmentation via post-processing with denoising autoencoders. IEEE Trans. Med. Imaging (2020). https://doi.org/10.1109/TMI.2020.3005297

8. Larrazabal, A.J., Martinez, C., Ferrante, E.: Anatomical priors for image segmentation via post-processing with denoising autoencoders. In: Shen, D., et al. (eds.) MICCAI 2019. LNCS, vol. 11769, pp. 585–593. Springer, Cham (2019). https://doi.org/10.1007/978-3-030-32226-7_65

9. Marstal, K., Berendsen, F., Staring, M., Klein, S.: Simpleelastix: a user-friendly, multi-lingual library for medical image registration. In: Proceedings of the IEEE Conference on Computer Vision and Pattern Recognition Workshops, pp. 134–142 (2016)

10. Moon, J.W., Hyun, D.K.: Decompressive craniectomy in traumatic brain injury: a review article. Korean J. Neurotrauma **13**(1), 1 (2017). https://doi.org/10.13004/kjnt.2017.13.1.1

11. Patravali, J., Jain, S., Chilamkurthy, S.: 2D-3D fully convolutional neural networks for cardiac MR segmentation. In: Pop, M., et al. (eds.) STACOM 2017. LNCS, vol. 10663, pp. 130–139. Springer, Cham (2018). https://doi.org/10.1007/978-3-319-75541-0_14

12. Pawlowski, N., et al.: Unsupervised lesion detection in brain CT using Bayesian convolutional autoencoders (2018)

13. Ronneberger, O., Fischer, P., Brox, T.: U-Net: convolutional networks for biomedical image segmentation. In: Navab, N., Hornegger, J., Wells, W.M., Frangi, A.F. (eds.) MICCAI 2015. LNCS, vol. 9351, pp. 234–241. Springer, Cham (2015). https://doi.org/10.1007/978-3-319-24574-4_28

14. Sedney, C., Julien, T., Manon, J., Wilson, A.: The effect of craniectomy size on mortality, outcome, and complications after decompressive craniectomy at a rural trauma center. J. Neurosci. Rural Pract. **5**(3), 212 (2014). https://doi.org/10.4103/0976-3147.133555

15. Seeram, E.: Computed Tomography - E-Book: Physical Principles, Clinical Applications, and Quality Control. Elsevier Health Sciences (2015). https://books.google.com.ar/books?id=DTCDCgAAQBAJ

16. Tanrikulu, L., et al.: The bigger, the better? about the size of decompressive hemicraniectomies. Clin. Neurol. Neurosurg. **135**, 15–21 (2015). https://doi.org/10.1016/j.clineuro.2015.04.019

17. Xiao, F., et al.: Estimating postoperative skull defect volume from CT images using the ABC method. Clin. Neurol. Neurosurg. **114**(3), 205–210 (2012). https://doi.org/10.1016/j.clineuro.2011.10.003.   http://www.sciencedirect.com/science/article/pii/S0303846711003076

X2Teeth: 3D Teeth Reconstruction from a Single Panoramic Radiograph

Yuan Liang$^{(\boxtimes)}$, Weinan Song, Jiawei Yang, Liang Qiu, Kun Wang, and Lei He

University of California, Los Angeles, CA 90095, USA
liangyuandg@ucla.edu, lhe@ee.ucla.edu

Abstract. 3D teeth reconstruction from X-ray is important for dental diagnosis and many clinical operations. However, no existing work has explored the reconstruction of teeth for a whole cavity from a single panoramic radiograph. Different from single object reconstruction from photos, this task has the unique challenge of constructing multiple objects at high resolutions. To conquer this task, we develop a novel ConvNet *X2Teeth* that decomposes the task into teeth localization and single-shape estimation. We also introduce a patch-based training strategy, such that *X2Teeth* can be end-to-end trained for optimal performance. Extensive experiments show that our method can successfully estimate the 3D structure of the cavity and reflect the details for each tooth. Moreover, *X2Teeth* achieves a reconstruction IoU of 0.681, which significantly outperforms the encoder-decoder method by 1.71× and the retrieval-based method by 1.52×. Our method can also be promising for other multi-anatomy 3D reconstruction tasks.

Keywords: Teeth reconstruction · Convolutional Neural Network · Panoramic radiograph

1 Introduction

X-ray is an important clinical imaging modality for dental diagnosis and surgical operations. Compared to Cone Beam Computed Tomography (CBCT), X-ray exceeds in lower cost and less absorbed dose of radiation. However, X-ray imagery cannot provide 3D information about teeth volumes or their spatial localization, which is useful for many dental applications [4,12], *e.g.*, micro-screws planning, root alignment assessment, and treatment simulations. Moreover, with volumetric radiation transport, the understanding and interpretation of X-ray imagery can be only apparent to experienced experts [8]. As such, the 3D visualization of X-rays can also be beneficial for applications related to patient education and physician training.

There have been several researches on the 3D reconstruction of a single tooth from its 2D scanning. For example, [10] models the volume of a tooth from X-rays by deforming the corresponding tooth atlas according to landmark aligning. [1,2] reconstruct a tooth from its crown photo by utilizing the surface

© Springer Nature Switzerland AG 2020
A. L. Martel et al. (Eds.): MICCAI 2020, LNCS 12262, pp. 400–409, 2020.
https://doi.org/10.1007/978-3-030-59713-9_39

reflectance model with shape priors. Despite those work, no one has explored the 3D teeth reconstruction of a whole cavity from a single panoramic radiograph. This task is more challenging than the single tooth reconstruction, since not only tooth shapes but also spatial localization of teeth should be estimated from their 2D representation. Moreover, all the existing methods of tooth reconstruction [1,2,10] utilize ad-hoc image processing steps and handcrafted shape features. Currently, Convolutional Neural Networks (ConvNet) provide an accurate solution for single-view 3D reconstruction by discriminative learning, and have become the state-of-the-art for many photo-based benchmarks [6,8,14]. However, the application of ConvNet on the teeth reconstruction has not yet been explored.

In this work, we pioneer the study of 3D teeth reconstruction of the whole cavity from a single panoramic radiograph with ConvNet. Different from most 3D reconstruction benchmarks [5,13], which target at estimating a single volume per low-resolution photo, our task has the unique challenge to estimate the shapes and localization of multiple objects at high resolutions. As such, we propose *X2Teeth*, an end-to-end trainable ConvNet that is compact for multi-object 3D reconstruction. Specifically, *X2Teeth* decomposes the reconstruction of teeth for a whole cavity into two sub-tasks of teeth localization and patch-wise tooth reconstruction. Moreover, we employ the random sampling of tooth patches during training guided by teeth localization to reduce the computational cost, which enables the end-to-end optimization of the whole network. According to experiments, our method can successfully reconstruct the 3D structure of the cavity, as well as restore the teeth with details at high resolutions. Moreover, we show *X2Teeth* achieves the reconstruction Intersection over Union (IoU) of 0.6817, outperforming the state-of-the-art encoder-decoder method by $1.71\times$ and retrieval-based method by $1.52\times$, which demonstrates the effectiveness of our method. To the best of our knowledge, this is the first work that explores 3D teeth reconstruction of the whole cavity from a single panoramic radiograph.

2 Methodologies

Figure 1 shows the overall architecture of our *X2Teeth*. We define the input of *X2Teeth* as a 2D panoramic radiograph (Fig. 1(1)), and the output as a 3D occupancy grid (Fig. 1(5)) of multiple categories for indicating different teeth. Different from the existing single-shape estimations [6,14] that mostly employ a single encoder-decoder structure for mapping the input image to one reconstructed object, *X2Teeth* decomposes the task into object localization (Fig. 1(b)) and patch-wise single tooth reconstruction (Fig. 1(c)). As such, the reconstruction can be carried out at high resolutions for giving more 3D details under the computational constraint, since tensor dimensions within the network can be largely reduced compared to directly reconstructing the whole cavity volume. Moreover, both sub-tasks share a feature extraction subnet (Fig. 1(a)), and the whole model can be end-to-end optimized by employing a sampling-based training strategy for the optimal performance. With the derived teeth localization

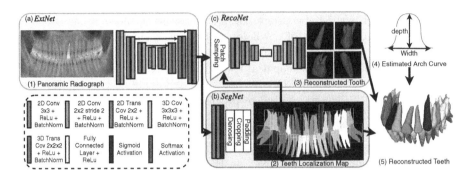

Fig. 1. Overall architecture of *X2Teeth*. *X2Teeth* consists of three subnets: (a) *ExtNet*, (b) *SegNet* and (c) *ReconNet*. *ExtNet* captures deep representations of teeth from the input panoramic radiograph. Based on the representations, *SegNet* performs pixel-wise classification followed by segmentation map denoising for localizing teeth. *ReconNet* samples tooth patches from the derived feature map and performs single-shape reconstruction. The final reconstruction of the whole cavity is the assembling of the reconstructed teeth according to the teeth localization and arch curve that estimated via β function model. The whole model can be end-to-end trained.

and tooth volumes, the final reconstruction of the cavity is derived by assembling different objects along the dental arch that is estimated via a β function model.

2.1 Model Architecture

Given the panoramic radiograph, our *X2Teeth* consists of three components: (1) a feature extracting subnet *ExtNet* for capturing teeth representations, (2) a segmentation subnet *SegNet* for estimating teeth localization, and (3) a patch-wise reconstruction subnet *ReconNet* for estimating the volume of a single tooth from the corresponding feature map patch. The detailed model configuration can be seen from the Fig. 1.

ExtNet. As shown in Fig. 1(a), *ExtNet* has an encoder-decoder structure consisting of 2D convolutions for capturing contexture features from the input panoramic radiograph (Fig. 1(1)). The extracted features are at high resolutions as the input image, and are trained to be discriminative for both *SegNet* and *ReconNet* to increase the compactness of the network. *ExtNet* utilizes strided convolutions for down-sampling and transpose convolutions for up-sampling, as well as channel concatenations between different layers for feature fusion.

SegNet. Given the feature map of *ExtNet*, *SegNet* maps it into a categorical mask $Y_{seg} \in \mathbb{Z}^{H \times W \times C}$, where H and W are image height and width, while C denotes the number of categories of teeth. Especially, a categorical vector $y \in Y_{mask}$ is multi-hot encoded, since nearby teeth can overlap in a panoramic radiograph because of the 2D projecting. With the categorical mask, *SegNet* further performs denoising by keeping the largest island of segmentation per tooth type, and localizes teeth by deriving their bounding boxes as shown in Fig. 1(2). As indicated in Fig. 1(b), *SegNet* consists of 2D convolutional layers followed by a *Sigmoid* transfer in order to perform the multi-label prediction. In our experiments, we set $C = 32$ for modeling the full set of teeth of an adult, including the four wisdom teeth that possibly exist for some individuals.

ReconNet. *ReconNet* samples the feature patch of a tooth, and maps the 2D patch into the 3D occupancy probability map $Y_{recon} \in \mathbb{R}^{H_p \times W_p \times D_p \times 2}$ of that tooth, where H_p, W_p, D_p are patch height, width and depth, respectively. The 2D feature patch is cropped from the feature map derived from *ExtNet*, while the cropping is guided by the tooth localization derived from *SegNet*. Similar to [6,8], *ReconNet* has an encoder-decoder structure consisting of both 2D and 3D convolutions. The encoder employs 2D convolutions, and its output is flattened into a 1D feature vector for the fully connected operation; while the decoder employs 3D convolutions to map this feature vector into the target dimension. Since our *ReconNet* operates on small image patches rather than the whole x-ray, the reconstruction can be done at high resolutions for restoring the details of teeth. In this work, we set $H_p = 120$, $W_p = 60$, $D_p = 60$ since all teeth fit into this dimension.

Teeth Assembling. By assembling the predicted tooth volumes according to their estimated localization from x-ray segmentation, we can achieve the 3D reconstruction of the cavity as a flat plane without the depth information about the cavity. This reconstruction is an estimation for the real cavity that is projected along the dental arch. Many previous work has investigated the modeling and prediction of the dental arch curve [11]. In this work, we employ the β function model introduced by [3], which estimates the curve by fitting the measurements of cavity depth and width (Fig. 1(4)). As the final step, our prediction of teeth for the whole cavity (Fig. 1(5)) can be simply achieved by bending the assembled flat reconstruction along the estimated arch curve.

2.2 Training Strategy

The loss function of *X2Teeth* is composed of two parts: segmentation loss L_{seg} and patch-wise reconstruction loss L_{recon}. For L_{seg}, considering that a pixel can be of multiple categories because of teeth overlaps on X-rays, we define the segmentation loss as the average of dice loss across all categories. Denote the segmentation output Y_{seg} at a pixel (i, j) to be a vector $Y_{seg}(i, j)$ of length C, where C is the number of possible categories, then

$$L_{seg}(Y_{seg}, Y_{gt}) = 1 - \frac{1}{C} \sum_C \frac{\sum_{i,j} 2Y_{seg}(i, j)Y_{gt}(i, j)}{\sum_{i,j} (Y_{seg}(i, j) + Y_{gt}(i, j))}, \quad (1)$$

where Y_{gt} is the multi-hot encoded segmentation ground-truth. For L_{recon}, we employ the 3D dice loss for defining the difference between the target and the predicted volumes. Let the reconstruction output Y_{recon} at a pixel (i, j, k) be a

Bernoulli distribution $Y_{recon}(i,j,k)$, then

$$L_{recon}(Y_{recon}, Y_{gt}) = 1 - 2\frac{\sum_{c=1}^{2}\sum_{i,j,k} Y_{recon}(i,j,k)Y_{gt}(i,j,k)}{\sum_{c=1}^{2}\sum_{i,j,k}(Y_{recon}(i,j,k) + Y_{gt}(i,j,k))}, \quad (2)$$

where Y_{gt} is the reconstruction ground-truth.

We employ a two-stage training paradigm. In the first stage, we train *ExtNet* and *SegNet* for the teeth localization by optimizing L_{seg}, such that the model can achieve an acceptable tooth patch sampling accuracy. In the second stage, we train the whole *X2Teeth* including *ReconNet* by optimizing the loss sum $L = L_{seg} + L_{recon}$ for both localization and reconstruction. Note that Adam optimizer is used for optimization. For each GPU, we set the batch size of panoramic radiograph as 1, and the batch size of tooth patches as 10. Besides, standard augmentations are employed for images, including random shifting, scaling, rotating and adding Gaussian noise. Finally, we implement our framework in Pytorch, and trained for the experiments on three NVidia Titan Xp GPUs.

3 Experiments

In this section, we validate and demonstrate the capability of our method for the teeth reconstruction from the panoramic radiograph. First, we introduce our in-house dataset of X-ray and panoramic radiograph pairs with teeth annotations from experts. Second, we validate *X2Teeth* by comparing with two state-of-the-art single view 3D reconstruction methods. Finally, we look into the performance of *X2Teeth* on the two sub-tasks of teeth localization and single tooth reconstruction.

3.1 Dataset

Ideally, we need paired data of the panoramic radiographs and CBCT scans captured from the same subject to train and validate *X2Teeth*. However, in order to control the radiation absorbed by subjects, such data pairs can rarely be collected in clinical settings. Therefore, we take an alternative approach by collecting high resolution CBCT scans and synthesize their corresponding panoramic radiographs. Such synthesis is valid since CBCT scans contain full 3D information of cavity, while panoramic radiographs are the 2D projections of them. Several previous work has demonstrated promising results for high quality synthesis, and in our work, we employ the method of Yun *et al.* [16] for building our dataset. Our in-house dataset contains 23 pairs of 3D CBCT scans and panoramic radiographs, each with a resolution ranging form 0.250 mm to 0.434 mm. All CBCT scans and panoramic radiographs are first labeled with pixel-wise tooth masks by 3 annotators, and then reviewed by 2 board-certificated dentists. Finally, we randomly split the dataset into 15 pairs for training, 1 pair for validation, and 7 pairs for testing.

3.2 Overall Evaluation of Teeth Reconstruction

Table 1. Comparison of reconstruction accuracy between *X2Teeth* and general purpose reconstruction methods in terms of IoU, detection accuracy (DA) and identification accuracy (FA). We report each metric in the format of *mean ± std*.

Method	IoU	DA	DF
3D-R2N2	0.398 ± 0.183	0.498 ± 0.101	0.592 ± 0.257
DeepRetrieval	0.448 ± 0.116	0.594 ± 0.088	0.503 ± 0.119
X2Teeth (ours)	**0.682 ± 0.030**	**0.702 ± 0.042**	**0.747 ± 0.038**

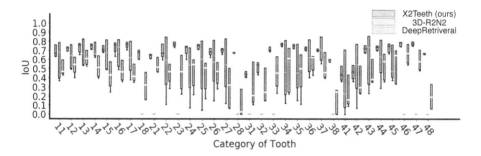

Fig. 2. IoU comparison of different tooth types between *X2Teeth*, 3D-R2N2, and Deep-Retrieval.

We compare our *X2Teeth* with two general purpose reconstruction methods that have achieved state-of-the-art performance as baselines: 3D-R2N2 [6] and Deep-Retrieval [15]. 3D-R2N2 employs an encoder-decoder network to map the input image to a latent representation, and reasons about the 3D structure upon it. To adapt 3D-R2N2 for high resolution X-rays in our tasks, we follow [15] by designing the output of the model to be 128^3 voxel grids, and up-sampling the prediction to the original resolution for evaluation. DeepRetrieval is a retrieval-based method that reconstructs images by deep feature recognition. Specifically, 2D images are embedded into a discriminative descriptor by using a ConvNet [9] as its representation. The corresponding 3D shape of a known image that shares the smallest Euclidean distance with the query image according to the representation is then retrieved as the prediction.

Quantitative Comparison. We evaluate the performance of models with intersection over union (IoU) between the predicted and the ground-truth voxels, as well as detection accuracy (DA) and identification accuracy (FA) [7]. The formulations of the metrics are:

$$IoU = \frac{|D \cap G|}{|D \cup G|}, \quad DA = \frac{|D|}{|D \cap G|} \quad and \quad FA = \frac{|D \cap G|}{|D|}, \qquad (3)$$

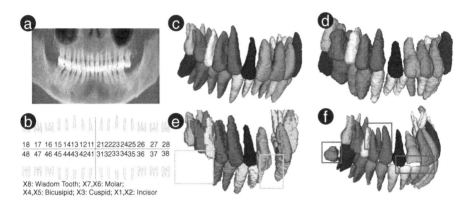

Fig. 3. Comparison of the reconstruction between (d) *X2Teeth* (ours), (e) 3D-R2N2, and (f) DeepRetrieval. (a) shows the input panoramic radiograph from the testing set, (c) shows the ground-truth of reconstruction, and (b) is the teeth numbering rule.

where G is the set of all teeth in ground-truth data, and D is the set of predicted teeth. As shown in Table 1, *X2Teeth* outperforms both baseline models significantly in terms of all three metrics. Specifically, *X2Teeth* achieves a mean IoU of 0.682, which outperforms 3D-R2N2 by 1.71×, and DeepRetrieval 1.52×. Similarly, Fig. 2 reveals IoUs for all the 32 types of tooth among the three methods, where our method has the highest median and the smallest likely range of variation (IQR) for all tooth types, which shows the consistent accuracy of *X2Teeth*. Yet, we also find that all algorithms have a lower accuracy for wisdom teeth (numbering 18, 28, 38, and 48) than the other teeth, indicating that the wisdom teeth are more subject-dependent, and thus difficult to predict.

Qualitative Comparison. Figure 3 visualizes the 3D reconstructions of a panoramic radiograph (Fig. 3(a)) from the testing set, which clearly shows our *X2Teeth* can achieve more appealing results than the other two methods. As for 3D-R2N2, its reconstruction (Fig. 3(e)) misses several teeth in the prediction as circled with green boxes, possibly because spatially small teeth can lose their representations within the deep feature map during the deep encoding process. The similar issue of missing tooth in predictions has also been previously reported in some teeth segmentation work [7]. Moreover, the reconstruction of 3D-R2N2 has coarse object surfaces that lack details about each tooth. This is because 3D-R2N2 is not compact enough and can only operate at the compressed resolution. As for DeepRetrieval, although the construction (Fig. 3(f)) has adequate details of teeth since its retrieved from high-resolution dataset, it fails to reflect the unique structure of individual cavity. The red boxes in Fig. 3(f) point out the significant differences in wisdom teeth, tooth root shapes, and teeth occlusion between the retrieved teeth and the ground-truth. Comparing to these two methods, *X2Teeth* has achieved a reconstruction (Fig. 3(d)) that can reflects both the unique structure of cavity and the details of each tooth, by formulating the task as the optimization of two sub-tasks for teeth localization and single tooth reconstruction.

3.3 Sub-task Evaluations

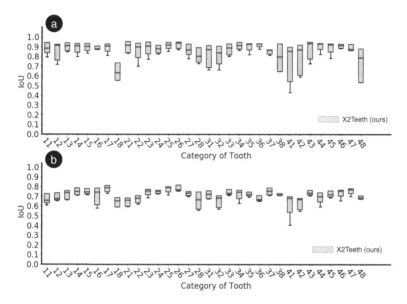

Fig. 4. (a) Segmentation IoUs of various teeth for the teeth localization sub-task. (b) Reconstruction IoUs of various teeth for the single tooth reconstruction sub-task.

For better understanding the performance of *X2Teeth*, we evaluate its accuracy on the two sub-tasks of teeth localization and single tooth reconstruction. Figure 4(a) shows the IoUs of different teeth for the 2D segmentation, where our method achieves an average IoU of 0.847 ± 0.071. The results validate that *X2Teeth* can accurately localize teeth, which enables the further sampling of tooth patches for the patch-based reconstruction. We also observe that the mean segmentation IoU for the 4 wisdom teeth (numbering X8) is 0.705 ± 0.056, which is lower than the other teeth. This is possibly because they have lower contrasts with surrounded bone structures, such that are more challenging to segment. Figure 4(b) demonstrates the IoUs of different types of teeth for the single tooth reconstruction, where our method achieves a mean IoU of 0.707 ± 0.044. Still, wisdom teeth have the significantly lower mean IoU of 0.668 ± 0.050, which can be contributed by the lower contrast with surroundings, less accurate localization, and the subject-dependent nature of their shapes. Moreover, incisor teeth (numbering X1 and X2) are observed to have less accurate reconstructions with the mean IoU of 0.661 ± 0.031. We argue the reason can be their feature vanishing in the deep feature maps considering their small spatial size.

4   Conclusion

In this paper, we initialize the study of 3D teeth reconstruction of the whole cavity from a single panoramic radiograph. In order to solve the challenges posed by the high resolution of images and multi-object reconstruction, we propose *X2Teeth* to decompose the task into teeth localization and single tooth reconstruction. Our *X2Teeth* is compact and employs sampling-based training strategy, which enables the end-to-end optimization of the whole model. Our experiments qualitatively and quantitatively demonstrate that *X2Teeth* achieves accurate reconstruction with tooth details. Moreover, our method can also be promising for other multi-anatomy 3D reconstruction tasks.

References

1. Abdelrahim, A.S., El-Melegy, M.T., Farag, A.A.: Realistic 3D reconstruction of the human teeth using shape from shading with shape priors. In: 2012 IEEE Computer Society Conference on Computer Vision and Pattern Recognition Workshops, pp. 64–69. IEEE (2012)
2. Abdelrehim, A.S., Farag, A.A., Shalaby, A.M., El-Melegy, M.T.: 2D-PCA shape models: application to 3D reconstruction of the human teeth from a single image. In: Menze, B., Langs, G., Montillo, A., Kelm, M., Müller, H., Tu, Z. (eds.) MCV 2013. LNCS, vol. 8331, pp. 44–52. Springer, Cham (2014). https://doi.org/10.1007/978-3-319-05530-5_5
3. Braun, S., Hnat, W.P., Fender, D.E., Legan, H.L.: The form of the human dental arch. Angle Orthod. **68**(1), 29–36 (1998)
4. Buchaillard, S., Ong, S.H., Payan, Y., Foong, K.W.: Reconstruction of 3D tooth images. In: 2004 International Conference on Image Processing, ICIP 2004, vol. 2, pp. 1077–1080. IEEE (2004)
5. Chang, A.X., et al.: ShapeNet: an information-rich 3D model repository. arXiv preprint arXiv:1512.03012 (2015)
6. Choy, C.B., Xu, D., Gwak, J.Y., Chen, K., Savarese, S.: 3D-R2N2: a unified approach for single and multi-view 3D object reconstruction. In: Leibe, B., Matas, J., Sebe, N., Welling, M. (eds.) ECCV 2016. LNCS, vol. 9912, pp. 628–644. Springer, Cham (2016). https://doi.org/10.1007/978-3-319-46484-8_38
7. Cui, Z., Li, C., Wang, W.: ToothNet: automatic tooth instance segmentation and identification from cone beam CT images. In: Proceedings of the IEEE Conference on Computer Vision and Pattern Recognition, pp. 6368–6377 (2019)
8. Henzler, P., Rasche, V., Ropinski, T., Ritschel, T.: Single-image tomography: 3D volumes from 2D cranial x-rays. In: Computer Graphics Forum, vol. 37, pp. 377–388. Wiley Online Library (2018)
9. Krizhevsky, A., Sutskever, I., Hinton, G.E.: ImageNet classification with deep convolutional neural networks. In: Advances in Neural Information Processing Systems, pp. 1097–1105 (2012)
10. Mazzotta, L., Cozzani, M., Razionale, A., Mutinelli, S., Castaldo, A., Silvestrini-Biavati, A.: From 2D to 3D: construction of a 3D parametric model for detection of dental roots shape and position from a panoramic radiograph–a preliminary report. Int. J. Dent. **2013** (2013)
11. Noroozi, H., Hosseinzadeh Nik, T., Saeeda, R.: The dental arch form revisited. Angle Orthod. **71**(5), 386–389 (2001)

12. Rahimi, A., et al.: 3D reconstruction of dental specimens from 2D histological images and μct-scans. Comput. Methods Biomech. Biomed. Eng. **8**(3), 167–176 (2005)
13. Sun, X., et al.: Pix3D: dataset and methods for single-image 3D shape modeling. In: Proceedings of the IEEE Conference on Computer Vision and Pattern Recognition, pp. 2974–2983 (2018)
14. Tatarchenko, M., Dosovitskiy, A., Brox, T.: Octree generating networks: efficient convolutional architectures for high-resolution 3D outputs. In: Proceedings of the IEEE International Conference on Computer Vision, pp. 2088–2096 (2017)
15. Tatarchenko, M., Richter, S.R., Ranftl, R., Li, Z., Koltun, V., Brox, T.: What do single-view 3D reconstruction networks learn? In: Proceedings of the IEEE Conference on Computer Vision and Pattern Recognition, pp. 3405–3414 (2019)
16. Yun, Z., Yang, S., Huang, E., Zhao, L., Yang, W., Feng, Q.: Automatic reconstruction method for high-contrast panoramic image from dental cone-beam CT data. Comput. Methods Programs Biomed. **175**, 205–214 (2019)

Domain Adaptation for Ultrasound Beamforming

Jaime Tierney$^{(\boxtimes)}$, Adam Luchies, Christopher Khan, Brett Byram,
and Matthew Berger

Vanderbilt University, Nashville, TN 37235, USA
`jaime.e.tierney@vanderbilt.edu`

Abstract. Ultrasound B-Mode images are created from data obtained
from each element in the transducer array in a process called beamforming. The beamforming goal is to enhance signals from specified spatial
locations, while reducing signal from all other locations. On clinical systems, beamforming is accomplished with the delay-and-sum (DAS) algorithm. DAS is efficient but fails in patients with high noise levels, so various adaptive beamformers have been proposed. Recently, deep learning
methods have been developed for this task. With deep learning methods, beamforming is typically framed as a regression problem, where
clean, ground-truth data is known, and usually simulated. For *in vivo*
data, however, it is extremely difficult to collect ground truth information, and deep networks trained on simulated data underperform when
applied to *in vivo* data, due to domain shift between simulated and *in
vivo* data. In this work, we show how to correct for domain shift by learning deep network beamformers that leverage both simulated data, and
unlabeled *in vivo* data, via a novel domain adaption scheme. A challenge
in our scenario is that domain shift exists both for noisy input, and clean
output. We address this challenge by extending cycle-consistent generative adversarial networks, where we leverage maps between synthetic
simulation and real *in vivo* domains to ensure that the learned beamformers capture the distribution of both noisy and clean *in vivo* data. We
obtain consistent *in vivo* image quality improvements compared to existing beamforming techniques, when applying our approach to simulated
anechoic cysts and *in vivo* liver data.

Keywords: Ultrasound beamforming · Domain adaptation · Deep
learning

1 Introduction

Ultrasound imaging is an indispensable tool for clinicians because it is real-
time, cost-effective, and portable. However, ultrasound image quality is often

Electronic supplementary material The online version of this chapter (https://
doi.org/10.1007/978-3-030-59713-9_40) contains supplementary material, which is
available to authorized users.

© Springer Nature Switzerland AG 2020
A. L. Martel et al. (Eds.): MICCAI 2020, LNCS 12262, pp. 410–420, 2020.
https://doi.org/10.1007/978-3-030-59713-9_40

suboptimal due to several sources of image degradation that limit clinical utility. Abdominal imaging is particularly challenging because structures of interest are beneath various tissue layers which can corrupt received channel signals due to phenomena like off-axis scattering or reverberation clutter [4].

Several advanced beamforming methods have been proposed to address this problem. Compared to conventional delay-and-sum (DAS) beamforming which applies constant delays and weights to the received channel data for a given spatial location, advanced methods aim to adaptively adjust the received channel data to enhance signals of interest and suppress sources of image degradation. This adaptive beamforming has been accomplished through coherence-based techniques [21,22], adaptive apodization schemes [15,33], as well as through model-based approaches [1,2,6,7]. Although effective, most of these advanced methods are computationally intensive and/or limited by user defined parameters. For example, despite achieving consistently superior image quality compared to DAS as well as other advanced techniques, a model-based approach called aperture domain model image reconstruction (ADMIRE) is exceedingly computationally complex, preventing real-time adjustable implementations [8].

More recently, there has been growing interest in using deep learning methods for ultrasound beamforming. These efforts generally fall under two categories, the first having the goal of using neural networks to reconstruct fully sampled data from some form of sub-sampled receive channel data [9,19,30,31,34]. The second has the goal of using neural networks to perform a form of adaptive beamforming using physical ground truth information during training [16,23,24,27,28,36]. The former is restricted to the desired fully sampled DAS or advanced beamforming output while the latter is theoretically capable of surpassing DAS or advanced beamformer performance. Other adaptive beamforming deep learning methods have been proposed that use an advanced beamformer as ground truth [32], which despite providing improvements to DAS, are constrained by the performance of the adaptive beamformer that they mimic.

One of the main limitations of deep learning for performing adaptive ultrasound beamforming is the lack of ground truth information *in vivo*. Previous efforts have primarily relied on simulations to generate labeled training data sets [16,23,27]. Unlike *in vivo* data, which is also costly to obtain, simulations can be controlled to generate unlimited amounts of realistic training data for which ground truth information is known. Network beamformers trained with simulated data have shown some success at generalizing to *in vivo* data [16,23,24]. However, despite sophisticated ultrasound simulation tools, a domain shift still exists between simulated and *in vivo* data, which ultimately limits network beamformer performance.

To address these limitations, we propose a novel domain adaptation scheme to incorporate *in vivo* data during training. Our approach uses cycle-consistent generative adversarial networks (CycleGANs) which learn maps between two data distributions absent of paired data [35]. GANs have been proposed previously in the context of ultrasound beamforming [28] but, to the best of our knowledge, have never been considered for performing domain adaption with real unlabeled *in vivo* data. Further, although CycleGANs have been leveraged to address domain shift in inputs for recognition tasks [14], in our scenario domain shift exists for *both* inputs and outputs. We mitigate both sources of domain shift by composing CycleGAN maps with domain-specific regressors to effectively learn deep *in vivo* beamformers. We develop and evaluate our approach using simulated anechoic cyst and *in vivo* liver data, and compare our approach to conventional DAS, deep neural networks (DNNs) trained using simulated data only, as well as established coherence and model-based advanced beamforming techniques.

2   Methods

2.1   Domain Adaptation

The basic intuition behind our approach is to, simultaneously, learn both *regressors* for beamforming, as well as *maps* that allow us to transform simulated channel data into corresponding *in vivo* data, and vice versa. More concretely, we denote S as our simulated domain, T as our *in vivo* domain, x_s and x_t refer to input channel data for simulated and *in vivo* data, respectively, and y_s and y_t refer to output channel data for their respective S and T domains. All channel data are d-length signals that we treat as d-dimensional vectors. We are provided a set of simulated input and output pairs, each denoted (x_s, y_s), as well as a set of *in vivo* inputs, but no corresponding outputs, denoted x_t. Our main goal is to learn a function $F_t : \mathbb{R}^d \to \mathbb{R}^d$ that serves as a beamformer for *in vivo* data. The challenge we address in this work is how to learn F_t absent of any *in vivo* outputs y_t, wherein our goal is to produce (x_t, y_t) pairs that approximate actual input/output *in vivo* pairs, and can be used to train F_t.

There are several ways to address this problem. Previous deep beamformer approaches [23] learn F_t from simulated pairs (x_s, y_s). However, applying F_t to *in vivo* inputs leads to *domain mismatch*, as the data distributions of x_s and x_t differ, due to simulation modeling assumptions that do not always capture *in vivo* physics. Thus, the starting point for our approach is to mitigate domain shift in the inputs. Specifically, our aim is to learn a mapping $G_{S \to T}$ that takes a given x_s and maps it into a corresponding *in vivo* input x_t. A common method for learning maps between domains is the use of generative adversarial networks [12], specifically for image translation tasks [17]. Such methods, however, assume paired data, where in our case simulated and *in vivo* data are not paired. For unpaired data, the CycleGAN approach of Zhu et al. [35] proposes to learn maps from S to T, $G_{S \to T}$, *and* from T to S, $G_{T \to S}$, enforcing cycle-consistency between maps. Specifically, for $G_{S \to T}$ we formulate the adverarial loss as follows:

$$L_{G_{S \to T}}(G_{S \to T}, D_T) = \mathbb{E}_{x_t \sim X_T}[\log D_T(x_t)] + \mathbb{E}_{x_s \sim X_S}[1 - \log D_T(G_{S \to T}(x_s))], \tag{1}$$

where D_T is the discriminator, tasked with distinguishing real *in vivo* data from fake *in vivo* data produced by $G_{S \to T}$, while X_S and X_T are the distributions for simulated and *in vivo* data respectively. We may define a similar adversarial

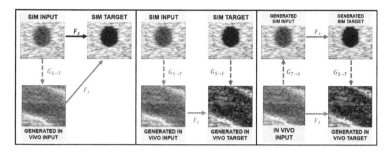

Fig. 1. Summary of input and output domain adaptation for training an *in vivo* beam-former, F_t. Green and orange arrows indicate *in vivo* input and output data generation, respectively. The left schematic summarizes previous efforts for which output domain adaptation was not considered [14]. In comparison, data used to compute $L_{F_{T1}}$ and $L_{F_{T2}}$ for the proposed method are indicated by the middle and right schematics, respectively. Data used to compute L_{F_S} is indicated by the black arrow on the left.

loss $L_{G_{T\to S}}$ for $G_{T\to S}$, with its own discriminator D_S. A cycle consistency regularization is also incorporated to ensure similarity between reconstructed signals and original data [35], defined as

$$L_{cyc}(G_{S\to T}, G_{T\to S}) = \mathbb{E}_{x_s\sim X_S}[||G_{T\to S}(G_{S\to T}(x_s)) - x_s||_1] \\ + \mathbb{E}_{x_t\sim X_T}[||G_{S\to T}(G_{T\to S}(x_t)) - x_t||_1]. \tag{2}$$

The above discriminators and maps can be jointly optimized with F_t, where we may provide generated, paired *in vivo* data via $(G_{S\to T}(x_s), y_s)$, highlighted in Fig. 1(left). This is at the core of the CyCADA method [14]. CyCADA is focused on recognition problems, e.g. classification and semantic segmentation, and thus the target output used (class label) can be easily leveraged from the source domain. However, for our scenario this is problematic, as domain shift still exists between y_s and y_t, and thus training on $(G_{S\to T}(x_s), y_s)$ necessitates F_t to *both* resolve domain gap, and beamform.

In contrast to CyCADA [14], we would rather have F_t focus *only* on beamforming. To address this, we leverage our domain maps for which we make the assumption that the domain shift in the inputs is identical to domain shift in the outputs, and also introduce a learned function F_s for beamforming simulated data, to arrive at the following *in vivo* beamforming losses, as illustrated in Fig. 1:

$$L_{F_S} = \mathbb{E}_{x_s\sim X_S}[||F_s(x_s) - y_s||_l] \tag{3}$$

$$L_{F_{T1}} = \mathbb{E}_{x_s\sim X_S}[||F_t(G_{S\to T}(x_s)) - G_{S\to T}(y_s)||_l], \tag{4}$$

$$L_{F_{T2}} = \mathbb{E}_{x_t\sim X_T}[||F_t(x_t) - G_{S\to T}(F_s(G_{T\to S}(x_t)))||_l]. \tag{5}$$

Intuitively, $L_{F_{T1}}$ ensures F_t can beamform generated *in vivo* data produced from paired simulated data. The term $L_{F_{T2}}$ ensures that F_t can beamform real *in vivo* data. In Fig. 1, example fully reconstructed simulated anechoic cyst and *in vivo*

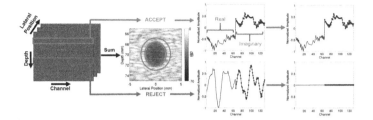

Fig. 2. Example labeled simulated anechoic cyst training data. An example channel data set and corresponding B-mode image is shown on the left to indicate accept and reject regions. A total of 10 depth locations were used as input and output to each network as illustrated by the red and green boxes on the channel data. These boxes represent 10 pixels within the corresponding red and green regions in the B-mode image. Example input and output aperture domain signals are shown on the right for a single depth. Real and imaginary components are stacked to form a 1D vector as the input and output to the DNN beamformer.

images are used, however our networks operate on aperture domain signals, as depicted in Fig. 2 and described in more detail in the following section.

Our full loss may be formulated as follows:

$$L = \underbrace{\lambda_s L_{G_{S \to T}} + \lambda_t L_{G_{T \to S}} + \lambda_c L_{cyc}}_{\text{GAN}} + \underbrace{\lambda_{F_S} L_{F_S} + \lambda_{F_T} (L_{F_{T1}} + L_{F_{T2}})}_{\text{Regressor}}, \quad (6)$$

where we simultaneously optimize for discriminators, generators, and regressors. We set GAN-related weights based on Hoffman et al. [14] (i.e., $\lambda_s = 2$, $\lambda_t = 1$, $\lambda_c = 10$), while the regressor loss weights were empirically chosen and both set to 1. We also regularize discriminators based on the approach of Mescheder et al. [26]. Furthermore, in order to ensure that the regressors utilize distinct, and shared, features from the two domains, we learn a single regressor $F : \mathbb{R}^d \times \mathbb{R}^d \times \mathbb{R}^d \to \mathbb{R}^d$ using the augmentation method of Daumé [5], such that $F_s(x_s) = F(x_s, x_s, 0)$ and $F_t(x_t) = F(x_t, 0, x_t)$, e.g. the first argument captures domain-invariant features, while the second and third arguments capture simulated and *in vivo* dependent features, respectively.

2.2 Data Summary

Our networks operate on time delayed aperture domain signals to perform a regression-based beamforming for each received spatial location. A Hilbert transform was applied to all received channel data prior to network processing to generate real and imaginary components. Training and test examples were generated from simulated anechoic cyst data as well as *in vivo* liver data. Although trivial, anechoic cysts provide clean and intuitive ground truth information and ensure an obvious domain shift between simulated and *in vivo* data.

Training Data. Field II [18] was used to simulate channel data of 12 5 mm diameter anechoic cyst realizations focused at 70mm using a 5.208 MHz center frequency, 20.832 MHz sampling frequency, 1540 m/s sound speed, and 65 active element channels with a pitch of 298 μm. These parameters were used to mimic the probe used for acquiring the *in vivo* data, as described below. Simulated training data were split into accept and reject regions depending on whether the aperture signals within a 0.5λ axial kernel (i.e., 10 depths) originated from a location outside or inside of the cyst, respectively. An output y_s is taken to be the input x_s if the signal is in the accept region, whereas the output y_s is a vector of zeros if x_s is in the reject region. Example simulated training data are shown in Fig. 2. Each aperture domain signal was concatenated through depth in addition to concatenating real and imaginary components. The number of accept and reject training examples was made equal (i.e., the full background was not used for training). For each simulated data set, 2,782 aperture domain examples (i.e., pre-reconstructed pixels) were used during training, resulting in a total of 33,384 total paired simulated examples.

A 36 year old healthy male gave informed written consent in accordance with the local institutional review board to acquire free-hand ultrasound channel data of his liver. A Verasonics Vantage Ultrasound System (Verasonics, Inc., Kirkland, WA) and ATL L7-4 (38 mm) linear array transducer were used to acquire channel data of 15 different fields of view of the liver, 6 of which were used for training. Acquisition parameters matched those used for simulations. For each of the 6 data sets used for training, similar to what was done for the simulations, aperture domain signals originating from spatial locations within a region around the focus were extracted. The same total number of examples were used from *in vivo* data as were used from simulations (i.e., 33,384 unpaired *in vivo* examples).

Test Data. For testing, 21 separate anechoic cyst realizations were simulated using the same parameters as above. White gaussian noise was added to the test realizations to achieve a signal-to-noise ratio of 50 dB. The remaining 9 *in vivo* examples not used during training were used for testing. For both the simulations and *in vivo* data, the full field of view was used for testing. A sliding window of 1 depth was used to select 10 depth inputs and overlapping depth outputs were averaged.

2.3   Evaluation

Network Details. Network hyperparameters corresponding to layer width, number of hidden layers, and regression losses (e.g. mean squared error, l_1, Huber loss) were varied. Models were tested on *in vivo* validation data, withheld from training and testing. The model that produced the highest CNR on the validation *in vivo* data was selected. Additional details are included in supplementary materials.

Both convolutional [16,27,28] and fully connected [23,36] neural networks have been investigated for the purposes of ultrasound beamforming, and it was shown previously that there was minimal difference between the two architectures [3]. Our networks are implicitly convolutional through depth, but fully connected across the transducer elements, which is consistent with the known signal coherence patterns of ultrasound channel data [25]. For this reason, and based on the network approach used for comparison in this work [23], all networks in this work – generators, discriminators, and regressors – are implemented as fully connected.

Comparison to Established Beamformers. Both frequency [23,36] and time [16,27] domain approaches have been considered for ultrasound deep learning methods. Given the added complexity of our proposed training architecture and the success of other time domain implementations [16,27], we use time domain data in this work. Therefore, as a direct baseline comparison to the proposed DA-DNN approach, a conventional DNN trained only on time domain simulated data, but with otherwise similar network parameters, was also evaluated. For this approach, the only loss used for optimization is summarized in Eq. 3. Additionally, for completeness, an established frequency-domain DNN approach [23] was also evaluated. This approach differs from the aforementioned conventional DNN approach in that it uses short time Fourier transformed (STFT) data and trains separate networks for individual frequencies. For this approach, model selection was performed as in [23] to highlight a best case scenario.

In addition to comparing the proposed DA-DNN approach to conventional and STFT DNN beamforming, performance was also evaluated in comparison to other established beamformers, including conventional DAS, the generalized coherence factor (GCF) [22], and aperture domain model image reconstruction (ADMIRE) [1]. For the GCF approach, as suggested by Li et al. [22], a cutoff spatial frequency of 3 frequency bins (i.e., $M_0 = 1$) was used to compute the weighting mask.

Performance Metrics. Contrast-to-noise ratio (CNR) and contrast ratio (CR) were used to evaluate beamformer performance as follows,

$$CNR = 20 \log_{10} \frac{|\mu_{background} - \mu_{lesion}|}{\sqrt{\sigma_{background}^2 + \sigma_{lesion}^2}} \; ; \; CR = -20 \log_{10} \frac{\mu_{lesion}}{\mu_{background}} \quad (7)$$

where μ and σ are the mean and standard deviation of the uncompressed envelope. Images were made for qualitative comparison by log compressing the envelope data and scaling to a 60dB dynamic range.

3   Results

DNN beamformers trained using only simulated data work well on simulations but often fail to generalize to *in vivo* data, as demonstrated in Fig. 3. Minimal difference was observed between the conventional DNN and DA-DNN approach

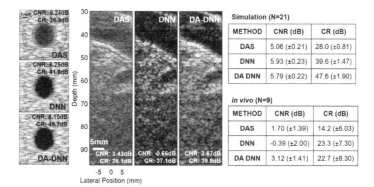

Fig. 3. Example DAS, DNN, and DA-DNN results for simulated anechoic cysts and *in vivo* data. All images are scaled to individual maximums and displayed with a 60dB dynamic range. Example regions of interest used to compute CNR and CR are indicated on the DAS images in red. Corresponding CNR and CR values are displayed on each image. Tables on the right indicate average CNR and CR (± standard deviation) for each method for simulations and *in vivo* data.

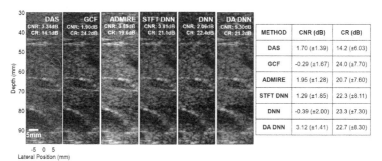

Fig. 4. Example *in vivo* B-mode images are shown for each beamformer. The regions of interest used to compute image quality metrics for the displayed example are shown in red on the DAS B-mode image. All images are scaled to individual maximums and a 60 dB dynamic range. Average CNR and CR (± standard deviation) across all 9 *in vivo* test examples are indicated in the table on the right.

when tested on simulated anechoic cysts in terms of CNR. In contrast, substantial improvements were observed both qualitatively and quantitatively when using the DA-DNN beamformer on *in vivo* data. Despite producing a high CR *in vivo*, the conventional DNN approach resulted in substantial drop out (i.e., extreme low amplitude pixels) in the background, resulting in a lower CNR than DAS. DA-DNN beamforming was able to maintain a higher CR compared to DAS while also improving CNR.

Qualitative and quantitative improvements in image quality with the DA-DNN beamformer were observed compared to the evaluated beamformers, as shown in Fig. 4. GCF and conventional DNN beamformers produce noticeably better contrast than DAS, but they also result in more drop out regions compared to ADMIRE and DA-DNN. These trends are described quantitatively by the corresponding tables, for which DA-DNN produced the highest average CNR overall while still maintaining higher CR than DAS.

4   Conclusion

Conventional DNN adaptive beamforming relies on ground truth training data which is difficult to acquire *in vivo*. To address this challenge, we propose a domain adaptation scheme to train an *in vivo* beamformer absent of any labeled *in vivo* data. We demonstrated substantial image quality improvements using our proposed approach compared to conventional DNN beamforming and to other established beamformers, including DAS, GCF, and ADMIRE. We show that DA-DNN beamforming achieved image quality consistent with or higher than state of the art ADMIRE without the same computational limitations. As stated throughout, an important fundamental assumption of our approach is that the domain shift between simulated and *in vivo* data is the same for the inputs and the outputs. Based on our results, this seems to be a reasonable baseline assumption, but it's worth investigating this further in future work.

References

1. Byram, B., Dei, K., Tierney, J., Dumont, D.: A model and regularization scheme for ultrasonic beamforming clutter reduction. IEEE Trans. Ultrason. Ferroelectr. Freq. Control **62**(11), 1913–1927 (2015)
2. Byram, B., Jakovljevic, M.: Ultrasonic multipath and beamforming clutter reduction: a chirp model approach. IEEE Trans. Ultrason. Ferroelectr. Freq. Control **61**(3), 428–440 (2014)
3. Chen, Z., Luchies, A., Byram, B.: Compact convolutional neural networks for ultrasound beamforming. In: 2019 IEEE International Ultrasonics Symposium (IUS), pp. 560–562. IEEE (2019)
4. Dahl, J.J., Sheth, N.M.: Reverberation clutter from subcutaneous tissue layers: simulation and in vivo demonstrations. Ultrasound Med. Biol. **40**(4), 714–726 (2014)
5. Daumé III, H.: Frustratingly easy domain adaptation. In: Proceedings of ACL (2007)
6. Dei, K., Byram, B.: The impact of model-based clutter suppression on cluttered, aberrated wavefronts. IEEE Trans. Ultrason. Ferroelectr. Freq. Control **64**(10), 1450–1464 (2017)
7. Dei, K., Byram, B.: A robust method for ultrasound beamforming in the presence of off-axis clutter and sound speed variation. Ultrasonics **89**, 34–45 (2018)
8. Dei, K., Schlunk, S., Byram, B.: Computationally efficient implementation of aperture domain model image reconstruction. IEEE Trans. Ultrason. Ferroelectr. Freq. Control **66**(10), 1546–1559 (2019)
9. Gasse, M., Millioz, F., Roux, E., Garcia, D., Liebgott, H., Friboulet, D.: High-quality plane wave compounding using convolutional neural networks. IEEE Trans. Ultrason. Ferroelectr. Freq. Control **64**(10), 1637–1639 (2017)

10. Glorot, X., Bengio, Y.: Understanding the difficulty of training deep feedforward neural networks. In: Proceedings of the Thirteenth International Conference on Artificial Intelligence and Statistics, pp. 249–256 (2010)

11. Glorot, X., Bordes, A., Bengio, Y.: Deep sparse rectifier neural networks. In: Proceedings of the Fourteenth International Conference on Artificial Intelligence and Statistics, pp. 315–323 (2011)

12. Goodfellow, I., et al.: Generative adversarial nets. In: Advances in Neural Information Processing Systems, pp. 2672–2680 (2014)

13. He, K., Zhang, X., Ren, S., Sun, J.: Delving deep into rectifiers: surpassing human-level performance on ImageNet classification. In: Proceedings of the IEEE International Conference on Computer Vision, pp. 1026–1034 (2015)

14. Hoffman, J., et al.: CyCADA: cycle-consistent adversarial domain adaptation. arXiv preprint arXiv:1711.03213 (2017)

15. Holfort, I.K., Gran, F., Jensen, J.A.: Broadband minimum variance beamforming for ultrasound imaging. IEEE Trans. Ultrason. Ferroelectr. Freq. Control **56**(2), 314–325 (2009)

16. Hyun, D., Brickson, L.L., Looby, K.T., Dahl, J.J.: Beamforming and speckle reduction using neural networks. IEEE Trans. Ultrason. Ferroelectr. Freq. Control **66**(5), 898–910 (2019)

17. Isola, P., Zhu, J.Y., Zhou, T., Efros, A.A.: Image-to-image translation with conditional adversarial networks. In: Proceedings of the IEEE Conference on Computer Vision and Pattern Recognition, pp. 1125–1134 (2017)

18. Jensen, J.A.: Field: a program for simulating ultrasound systems. Med. Biol. Eng. Comput. **34**, 351–353 (1996)

19. Khan, S., Huh, J., Ye, J.C.: Universal deep beamformer for variable rate ultrasound imaging. arXiv preprint arXiv:1901.01706 (2019)

20. Kingma, D.P., Ba, J.: Adam: a method for stochastic optimization. arXiv preprint arXiv:1412.6980 (2014)

21. Lediju, M.A., Trahey, G.E., Byram, B.C., Dahl, J.J.: Short-lag spatial coherence of backscattered echoes: imaging characteristics. IEEE Trans. Ultrason. Ferroelectr. Freq. Control **58**(7), 1377–1388 (2011)

22. Li, P.C., Li, M.L.: Adaptive imaging using the generalized coherence factor. IEEE Trans. Ultrason. Ferroelectr. Freq. Control **50**(2), 128–141 (2003)

23. Luchies, A.C., Byram, B.C.: Deep neural networks for ultrasound beamforming. IEEE Trans. Med. Imaging **37**(9), 2010–2021 (2018)

24. Luchies, A.C., Byram, B.C.: Training improvements for ultrasound beamforming with deep neural networks. Phys. Med. Biol. **64**, 045018 (2019)

25. Mallart, R., Fink, M.: The van Cittert-Zernike theorem in pulse echo measurements. J. Acoust. Soc. Am. **90**(5), 2718–2727 (1991)

26. Mescheder, L., Geiger, A., Nowozin, S.: Which training methods for GANs do actually converge? arXiv preprint arXiv:1801.04406 (2018)

27. Nair, A.A., Tran, T.D., Reiter, A., Bell, M.A.L.: A deep learning based alternative to beamforming ultrasound images. In: 2018 IEEE International Conference on Acoustics, Speech and Signal Processing (ICASSP), pp. 3359–3363. IEEE (2018)

28. Nair, A.A., Tran, T.D., Reiter, A., Bell, M.A.L.: A generative adversarial neural network for beamforming ultrasound images: invited presentation. In: 2019 53rd Annual Conference on Information Sciences and Systems (CISS), pp. 1–6. IEEE (2019)

29. Paszke, A., et al.: Automatic differentiation in PyTorch (2017)

30. Perdios, D., Besson, A., Arditi, M., Thiran, J.P.: A deep learning approach to ultrasound image recovery. In: 2017 IEEE International Ultrasonics Symposium (IUS), pp. 1–4. IEEE (2017)
31. Senouf, O., et al.: High frame-rate cardiac ultrasound imaging with deep learning. In: Frangi, A.F., Schnabel, J.A., Davatzikos, C., Alberola-López, C., Fichtinger, G. (eds.) MICCAI 2018. LNCS, vol. 11070, pp. 126–134. Springer, Cham (2018). https://doi.org/10.1007/978-3-030-00928-1_15
32. Simson, W., et al.: End-to-end learning-based ultrasound reconstruction. arXiv preprint arXiv:1904.04696 (2019)
33. Synnevag, J.F., Austeng, A., Holm, S.: Adaptive beamforming applied to medical ultrasound imaging. IEEE Trans. Ultrason. Ferroelectr. Freq. Control **54**(8), 1606–1613 (2007)
34. Yoon, Y.H., Khan, S., Huh, J., Ye, J.C.: Efficient B-mode ultrasound image reconstruction from sub-sampled RF data using deep learning. IEEE Trans. Med. Imaging **38**(2), 325–336 (2018)
35. Zhu, J.Y., Park, T., Isola, P., Efros, A.A.: Unpaired image-to-image translation using cycle-consistent adversarial networks. In: Proceedings of the IEEE International Conference on Computer Vision, pp. 2223–2232 (2017)
36. Zhuang, R., Chen, J.: Deep learning based minimum variance beamforming for ultrasound imaging. In: Wang, Q., et al. (eds.) PIPPI/SUSI -2019. LNCS, vol. 11798, pp. 83–91. Springer, Cham (2019). https://doi.org/10.1007/978-3-030-32875-7_10

CDF-Net: Cross-Domain Fusion Network for Accelerated MRI Reconstruction

Osvald Nitski[1,2], Sayan Nag[1,2], Chris McIntosh[2], and Bo Wang[1,2,3]

[1] University of Toronto, Toronto, Canada
{osvald.nitski,sayan.nag}@mail.utoronto.ca
[2] University Health Network, Toronto, Canada
chris.mcintosh@rmp.uhn.ca
[3] Vector Institute, Toronto, Canada
bowang@vectorinstitute.ai

Abstract. Accurate reconstruction of accelerated Magnetic Resonance Imaging (MRI) would produce myriad clinical benefits including higher patient throughput and lower examination cost. Traditional approaches utilize statistical methods in the frequency domain combined with Inverse Discrete Fourier Transform (IDFT) to interpolate the under-sampled frequency domain (referred as k-space) and often result in large artifacts in spatial domain. Recent advances in deep learning-based methods for MRI reconstruction, albeit outperforming traditional methods, fail to incorporate raw coil data and spatial domain data in an end-to-end manner. In this paper, we introduce a cross-domain fusion network (CDF-Net), a neural network architecture that recreates high resolution MRI reconstructions from an under-sampled single-coil k-space by taking advantage of relationships in both the frequency and spatial domains while also having an awareness of which frequencies have been omitted. CDF-Net consists of three main components, a U-Net variant operating on the spatial domain, another U-Net performing inpainting in k-space, and a 'frequency informed' U-Net variant merging the two reconstructions as well as a skip connected zero-filled reconstruction. The proposed CDF-Net represents one of the first end-to-end MRI reconstruction network that leverages relationships in both k-space and the spatial domain with a novel 'frequency information pathway' that allows information about missing frequencies to flow into the spatial domain. Trained on the largest public fastMRI dataset, CDF-Net outperforms both traditional statistical interpolation and deep learning-based methods by a large margin.

Keywords: MRI reconstruction · Deep learning · Accelerated acquisition

1 Introduction

Magnetic Resonance Imaging (MRI) is a popular non-invasive diagnostic imaging technique benefiting from its ability to produce detailed images of soft tissue

© Springer Nature Switzerland AG 2020
A. L. Martel et al. (Eds.): MICCAI 2020, LNCS 12262, pp. 421–430, 2020.
https://doi.org/10.1007/978-3-030-59713-9_41

structures and its lack of radiation [1]. However, a main drawback is the slow acquisition, increasing speed of acquisition has been a major field of ongoing research for decades, beginning with classical enhancements such as modification of pulse sequences and improving the magnetic field gradients [2,3].

MRIs acquire images by employing electromagnetic frequencies and recovering the signal strength of a given frequency and offset occurring in the spatial domain. The 2D complex-valued frequency domain information acquired by the MR is referred to as k-space, an Inverse Discrete Fourier Transform (IDFT) can convert the frequency information into a complex valued spatial domain image. MRI acquisition can be accelerated by omitting some portion of k-space samples; however, due to the violation of the Nyquist sampling theorem, ground truth full-resolution images cannot possibly be recovered without artifacts [4] which can often be too severe for clinical use.

Our approach, Cross-Domain Fusion Network (CDF-Net), is to combine frequency domain interpolation, which we refer to as k-space inpainting, and spatial domain up-sampling, which we refer to as image super-resolution, into a unified, non-iterative, end-to-end framework capable of being a component of a larger uncertainty reducing active acquisition framework. CDF-Net improves upon previous architectures with two primary contributions, first by leveraging both frequency and spatial domain relationships in an end-to-end manner, secondly by allowing cross-domain communication via frequency informed channel modulation in the spatial domain.

2   Related Work

Recent advancements in machine learning for accelerated acquisition MRI have shown promising results [5–7]; newer architectures see gains from adversarial training [8], attention mechanisms [9], and recurrent inference machines [10]. Complementary to these approaches is to focus on the unique dual domain nature of MR reconstruction; data-consistency layers can be used to preserve known information in k-space [11,12], KIKI-Net allows learning in both domains by iteratively transitioning between frequency and spatial domains and applying a domain-specific enhancement network in each [13].

In this paper, we have compared our proposed architecture (CDF-Net) with the following CNN based MRI reconstruction architectures:

2.1   U-Net

Based on fully convolutional neural network, U-Net is an end-to-end architecture which comprises of a contracting path for capturing attribute information along with a symmetric expanding path for facilitating accurate localization employing skip-connections from encoder to decoder to transmit information at various resolutions throughout the entire network [14]. A standard U-Net with an initial channel count of 256 is used as a baseline model for accelerated MRI reconstruction by super-resolution of the IDFT of a zero-filled under-sampled k-space. Reconstructions are performed on 2D slices and do not share information across slices.

2.2 KIKI-Net

Taejoon Eo et al. [13] proposed a CNN architecture called KIKI-Net for accelerated MR reconstruction specifically designed to leverage patterns in both k-space and the spatial domain [13]. The framework consists of 3 parts including a CNN for k-space completion called KCNN, a CNN for restoring images called ICNN and an interleaved data-consistency (IDC) module. IDC is a data consistency block which transfers the image to frequency domain and averages modified k-space lines with known ones. Two successive iterations of the KCNN, then ICNN & IDC, with appropriate domain transforms between, gives a KIKI-Net which has been reported as the best out of all the combinations of K-space and spatial networks [13].

CDF-Net offers an end-to-end alternative that has better communication between domains. To make comparisons fair, we train KIKI-Net[1] with a U-Net as the KCNN and our frequency informed U-Net variant as the ICNN.

3 Our Approach: CDF-Net

In this section, we discuss specific architecture design for an end-to-end accelerated MRI reconstruction model for the proposed CDF-Net as well as how to accelerate training using deep supervision. Next, we describe the novel 'frequency information pathway' which allows networks operating in the spatial domain to gain explicit knowledge of missing frequencies.

3.1 Cross Domain Fusion Network

The Basic Architecture. The proposed CDF-Net consists of three 2D U-Net variations that separately solve the problems of k-space inpainting, spatial domain super-resolution (SISR), and a multi-image super-resolution (MISR) combining the two previous outputs. The detailed structure of CDF-Net is shown in Fig. 1. First, a U-Net takes as input k_u and recovers k' while a Frequency-Informed U-Net, described in the next section, takes input $\mathcal{F}_{2D}^{-1}(I_u)$ and recovers I'_s. The k' output of the k-space network is converted to I'_k in the spatial domain via $\mathcal{F}_{2D}^{-1}(k')$. Another Frequency-Informed U-Net takes as input the channel-wise concatenation two aforementioned outputs, I'_k and I'_s, as well as a skip connection of the reconstructed zero-filled image, $\mathcal{F}_{2D}^{-1}(k_u)$, and resolves differences to recover a final I'_{CDF}.

---

[1] The original KIKI-Net is not open-source and the paper does not offer a full description of its CNN architectures. We do our best to recreate it and modify it such that comparisons are fair.

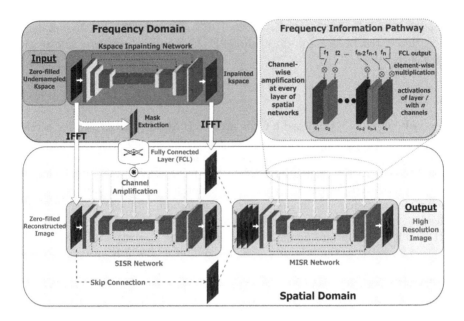

Fig. 1. Diagram of CDF-Net Architecture: Networks operate in both frequency and spatial domain with cross-communication. Under-sampled k-space input goes through a k-space inpainting network while the zero-filled reconstruction goes through a SISR network. The outputs of these two network are combined with a skip-connection from the input reconstruction and fed into the final MISR network. Networks operating in the spatial domain also receive information about which frequencies are under-sampled by passing masks through a fully connected layer (FCL) via our novel frequency-information pathways.

During training, gradients are back-propagated from the loss on the I'_{CDF} output through the earlier spatial and k-space inpainting networks as well. This allows the entire network to be trained in a single backwards pass. Though, three separate passes are required if deep supervision on I'_s and I'_k losses is desired.

The novelty of CDF-Net lies in the image fusion task of the third network; previous attempts to use both k-space and spatial domain information have been limited to iterative methods [13]. CDF-Net offers an end-to-end approach to using data in both k-space and the spatial domain enabled by differentiable FFT operations that propagate gradients between domains. This enabled faster training and inference than iterative methods.

Additionally, the sub-modules of CDF-Net can be replaced with newer networks under future advancements or networks more specified to the task at hand. This features make CDF-Net an attractive backbone for future specialized networks. Our implementation of CDF-Net solves these tasks using U-Net variants.

Frequency-Informed U-Net. Under the Cartesian k-space sub-sampling regime, in which all phase encodings of certain frequency are zero-filled, it is possible to retrieve a 1-dimensional bitmap of missing frequencies. The 'Frequency-Informed' U-Net accomplishes incorporates what information is under-sampled by feeding the extracted missing frequency bitmap from k-space through a feed forward neural network to output multipliers for each channel a CNN, similar to an attention mechanism [15,16] but with no normalization over channels.

Where b_m represents the 1-dimensional bitmap of sampled frequencies, W_{f_i}, b_{f_i} and W_i, b_i represent the weights and biases of the frequency-information fully connected layer and the weights and biases of the modulated CNN layer, respectively, σ the chosen activation function, and $\otimes$ element-wise product operator, the following equations define the frequency information channel update:

$$X_i = Relu(b_m W_{f_i} + b_{f_i}) \otimes \sigma(X_{i-1} W_i + b_i) \tag{1}$$

Deep Supervision. Because the k-space inpainting task and Single Image Super Resolution (SISR) task both output an image, in the frequency and spatial domain respectively, it is possible to optimize these outputs with auxiliary losses. In our implementation we set the weights of loss on both intermediary outputs equal to 0.1 times the weight of the final loss, both intermediary losses decay exponentially to zero midway through training. Enabling these losses quickens convergence but has a negligible effect on final performance, effects are shown in the ablation study. Smooth L1 loss is used for all objectives.

4 Results

In this section, CDF-Net is compared to a state of the art iterative MRI accelerated reconstruction method [13], a classic U-Net [14], and a popular non deep-learning approach, ESPIRiT [17]. Deep learning models are compared at two size with U-Nets of initial width of 64 and 128 channels. CDF-Net performs better than all other networks, even at its lower parameter count.

4.1 Dataset Overview

With the objective of advancing machine learning based reconstruction approaches, the fastMRI dataset [18] is the first large-scale dataset for accelerated MR Image reconstruction. It consists of raw MRI data of $1,594$ volumes, comprising of $56,987$ slices from a range of MRI systems. It contains both raw k-space data from multi-coil scanners, as well as ground truth images and emulated single-coil k-space data[19]. Since the test set and challenge set are only released in their under-sampled k-space form, we split the validation set randomly into validation and test sets, which both have ground truths available.

Under-sampling of the k-space data from a full acquisition is performed by masking k-space lines in the phase encoding direction. In $4\times$ under-sampling the central 8% of the bandwidth is sampled, in $8\times$ the central 4% is sampled, in both cases the remaining frequencies are randomly sampled to achieve the desired acceleration. We summarize the experimental set-up in Table 1.

Table 1. Number of volumes and slices assigned to each dataset. The training and validation datasets are used to fit model parameter values. The test dataset is used to compare the results across several reconstruction approaches.

Dataset overview				
Sets	Volumes		Slices	
	Multi-coil	Single-coil	Multi-coil	Single-coil
Training	973	973	34,742	34,742
Validation	99	99	3,550	3,550
Test	100	100	3,585	3,585

Table 2. NMSE, PSNR and SSIM values of popular models at small and large capacities (number representing initial U-Net channels) at $4\times$ and $8\times$ accelarations for Single-Coil Data.

Single-coil comparison study						
Methods	NMSE		PSNR		SSIM	
	$4\times$	$8\times$	$4\times$	$8\times$	$4\times$	$8\times$
CDF-128 (ours)	**0.0277**	**0.0465**	**32.94**	**29.59**	**0.7603**	**0.6663**
CDF-64 (ours)	0.0283	0.0488	32.72	29.26	0.7499	0.6495
KIKI-128	0.0281	0.0474	32.81	29.49	0.7567	0.6586
KIKI-64	0.0308	0.0492	32.54	29.22	0.7411	0.6409
U-Net256	0.0338	0.0489	32.08	29.31	0.7332	0.6334
ESPIRiT	0.0479	0.0795	31.69	27.12	0.6030	0.4690

4.2 Comparison to Existing Methods

In this section, we report detailed experimental results for two scenarios: single-coil and multi-coil MRI reconstructions. CDF-Net is also compared to existing state-of-the-art models at both $4\times$ and $8\times$ acceleration factors. We report the detailed comparisons at single-coil dataset in Table 2 and multi-coil dataset in Table 3. Note that, the proposed CDF-Net outperforms the existing methods in all metrics at both accelerations for both datasets. For example, CDF-Net outperforms the equivalent sized KIKI-Net, the iterative cross-domain state-of-the-art method, by 0.8807 to 0.8679 average SSIM, a 0.0128 improvement at multi-coil dataset. CDF-Net also outperforms the classic U-Net by 0.0355 average SSIM and the traditional ESPIRiT compressed sensing reconstruction by 0.2702 SSIM. Example outputs of each model at $4\times$ and $8\times$ accelerations are shown in Fig. 2 and Fig. 3, respectively, at multiple magnifications. Qualitative comparisons further demonstrate the superiority of the proposed method.

CDF-Net's advantage over the classic U-Net is its ability to make inferences in the frequency domain, while its advantage to KIKI-Net is better information flow and from end-to-end training and skip connections. Additionally, CDF-Net is a framework in which the U-Nets could be replaced with future improved CNNs.

Table 3. NMSE, PSNR and SSIM values of popular models at small and large capacities (number representing initial U-Net channels) at 4× and 8× accelerations for Multi-Coil Data.

Multi-coil comparison study						
Methods	NMSE		PSNR		SSIM	
	4×	8×	4×	8×	4×	8×
CDF-128 (ours)	**0.0088**	**0.0125**	**36.77**	**34.21**	**0.9003**	**0.8611**
CDF-64 (ours)	0.0106	0.0135	35.23	33.55	0.8837	0.8522
KIKI-128	0.0101	0.0132	35.66	33.92	0.8864	0.8571
KIKI-64	0.0114	0.0150	33.88	32.99	0.8686	0.8499
U-Net256	0.0130	0.0143	34.31	33.60	0.8752	0.8513
ESPIRiT	0.0503	0.0760	33.88	28.25	0.6280	0.5930

Table 4. Ablation Study Results (NMSE, PSNR and SSIM) for various ablations and both acceleration factors (4× and 8×). The results show that the images reconstructed by CDF-Net with 128 initial channels and a depth of 4 are better than all ablated variants.

Multi-coil ablation study						
Methods	NMSE		PSNR		SSIM	
	4×	8×	4×	8×	4×	8×
CDF-128	**0.0088**	**0.0125**	**36.77**	**34.21**	**0.9003**	**0.8611**
CDF-128 (no DS)	0.0090	0.0126	**36.77**	34.18	0.8997	0.8609
CDF-128 (no FI)	0.0095	0.0132	36.72	34.11	0.8998	0.8602
CDF-128 (no k-space)	0.0109	0.0149	33.37	32.96	0.8854	0.8540
CDF-128 (no spatial)	0.0301	0.0345	31.05	30.95	0.8168	0.7677
CDF-64	**0.0106**	**0.0135**	35.23	**33.55**	**0.8837**	**0.8522**
CDF-64 (no DS)	0.0107	**0.0135**	**35.24**	33.52	0.8833	0.8523
CDF-64 (no FI)	0.0110	0.0143	35.20	33.54	0.8831	0.8504
CDF-64 (no k-space)	0.0120	0.0159	33.45	32.74	0.8620	0.8481
CDF-64 (no spatial)	0.0351	0.0396	30.25	29.45	0.7751	0.7234

To make a fair comparison the KIKI-Net model included frequency information pathways as well.

4.3   Ablation Study

We report the results for ablation study in Table 4. As expected wider networks perform better across all ablations. The larger variant of CDF-Net, CDF-Net128, with each sub-module U-Net beginning with 128 channels and doubling to a depth of 4 performed better than the smaller model of equal depth but 64 base channels, CDF-Net64, without overfitting. Larger models were infeasible to train

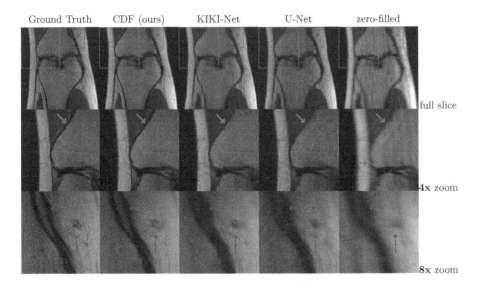

Fig. 2. Comparison of 4× Multi-Coil reconstructions for 4 different methods to ground truth as well as magnified sections to show detail. The qualitative superiority of our method over existing approaches is clearly demonstrated by the zoomed-in plots

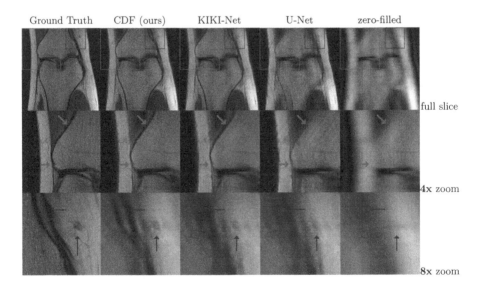

Fig. 3. Comparison of 8× Multi-Coil reconstructions for 4 different methods to ground truth as well as magnified sections to show detail. The qualitative superiority of our method over existing approaches is clearly demonstrated by the zoomed-in plots

due to GPU capacity, additionally U-Net baselines show a significant drop-off in performance improvements beyond 128 base channels [18].

Removing the frequency domain network but leaving the skip connection results in CDF-spatial64 and the larger CDF-spatial128. These architectures are similar to two U-Nets applied one after the other with a skip-connection from input to the first U-Nets output; both have frequency-information pathways. The converse is keeping only the k-space network, essentially a k-space CNN, with two size variants CDF-k-space64 and CDF-k-space128. Both ablated networks are outperformed by the full CDF-Net. The spatial networks outperform the k-space networks.

These ablations show that while CDF-Net's performance is primarily due to its spatial components, the k-space inpainting network indeed plays a role in improving performance

5   Conclusion

In this paper, we propose CDF-Net, a deep-learning model for accelerated MRI reconstruction. It is the first end-to-end network to use both raw k-space data and spatial data for inference, while prior networks either performed operations on both domains iteratively or only used one of them. We also introduce "frequency-information pathways" to selectively modulate spatial-domain CNN channels based on the instance's specific under-sampled frequencies, empirically we showed that this gives a moderate boost in performance.

Extensive experimental results on the large-scale fastMRI dataset, using both single-coil and multi-coil data at $4\times$ and $8\times$ acceleration factors, demonstrate that, CDF-Net outperforms both existing deep learning and traditional accelerated MRI reconstruction methods. Ablation studies show that each component of CDF-Net is contributing to its performance, with the spatial networks having the strongest effect, and the k-space inpainting network, and frequency information pathway giving a meaningful boost as well.

Future improvements will come from active acquisition methods that are guided to minimize CDF-Net's reconstruction error by selectively sampling k-space data. Other under-sampling masks besides zeroing out along the phase-encoding direction are also possible, though they must be constrained to physically possible acquisitions. We have defined the larger domain-fusion problem such that advancements can be made in any of the sub tasks. CDF-Net has the ability to be upgraded with advanced networks substituting any of the constituent U-Nets.

References

1. Zhu, B., Liu, J.Z., Cauley, S.F., Rosen, B.R.: Image reconstruction by domain transform manifold learning. Nature **555**(7697), 487 (2018)
2. Hennig, J., Nauerth, A., Friedburg, H.: Rare imaging: a fast imaging method for clinical MR. Magn. Reson. Med. **3**(6), 823–833 (1986)

3. Oppelt, A., Graumann, R., Barfuss, H., Fischer, H., Hartl, W., Schajor, W.: FISP - a new fast MRI sequence. Electromedica **54**(1), 15–18 (1986)

4. Moratal, D., Valles-Luch, A., Marti-Bonmati, L., Brummer, M.E.: k-space tutorial: an MRI educational tool for a better understanding of k-space. Biomed. Imaging Interv. J. **4**(1) (2008)

5. Knoll, F., et al.: Deep learning methods for parallel magnetic resonance image reconstruction. Preprint arXiv:1904 (2019)

6. Klatzer, T., et al.: Learning a variational network for reconstruction of accelerated MRI data. Magn. Reson. Med. **79**, 3055–3071 (2018)

7. Aggarwal, H.K., Mani, M.P., Jacob, M.: MoDL: model-based deep learning architecture for inverse problems. IEEE Trans. Med. Imaging **38**(2), 394–405 (2018)

8. Zhang, P., Wang, F., Xu, W., Li, Yu.: Multi-channel generative adversarial network for parallel magnetic resonance image reconstruction in K-space. In: Frangi, A.F., Schnabel, J.A., Davatzikos, C., Alberola-López, C., Fichtinger, G. (eds.) MICCAI 2018. LNCS, vol. 11070, pp. 180–188. Springer, Cham (2018). https://doi.org/10.1007/978-3-030-00928-1_21

9. Huang, Q., Yang, D., Wu, P., Qu, H., Yi, J., Metaxas, D.: MRI reconstruction via cascaded channel-wise attention network. In: 2019 IEEE 16th International Symposium on Biomedical Imaging, pp. 1622–1626 (2019)

10. Lønning, K., Putzky, P., Caan, M.W.A., Welling, M.: A deep cascade of convolutional neural networks for dynamic MR image reconstruction. Int. Soc. Magn. Reson. Med. (2018)

11. Schlemper, J., Caballero, J., Hajnal, J.V., Price, A.N., Rueckert, D.: A deep cascade of convolutional neural networks for dynamic MR image reconstruction. IEEE Trans. Med. Imaging **37**(2), 493–501 (2017)

12. Schlemper, J., Caballero, J., Price, A.N., Hajnal, J.V., Rueckert, D., Qin, C.: A deep cascade of convolutional neural networks for dynamic MR image reconstruction. IEEE Trans. Med. Imaging **38**(1), 280–290 (2019)

13. Eo, T., Jun, Y., Kim, T., Jang, J., Lee, H.-J., Hwang, D.: KIKI-net: cross-domain convolutional neural networks for reconstructing undersampled magnetic resonance images. Magn. Reson. Med. **80**(5), 2188–2201 (2018)

14. Ronneberger, O., Fischer, P., Brox, T.: U-Net: convolutional networks for biomedical image segmentation. In: Navab, N., Hornegger, J., Wells, W.M., Frangi, A.F. (eds.) MICCAI 2015. LNCS, vol. 9351, pp. 234–241. Springer, Cham (2015). https://doi.org/10.1007/978-3-319-24574-4_28

15. Chen, K., Wang, J., Chen, L.-C., Gao, H., Xu, W., Nevatia, R.: ABC-CNN: an attention based convolutional neural network for visual question answering. arXiv preprint arXiv:1511.05960 (2015)

16. Bello, I., Zoph, B., Vaswani, A., Shlens, J., Le, Q.V.: Attention augmented convolutional net-works. arXiv preprint arXiv:1904.09925 (2019)

17. Uecker, M., et al.: ESPIRiT-an eigenvalue approach to autocalibrating parallel MRI: where sense meets GRAPPA. Magn. Reson. Med. **71**(3), 990–1001 (2014)

18. Tygert, M., Zbontar, J.: fastMRI: an open dataset and benchmarks for accelerated MRI. Preprint arXiv:1811.08839 (2018)

19. Tygert, M., Zbontar, J.: Simulating single-coil MRI from the responses of multiple coils. Preprint arXiv:1811.08026 (2018)

Domain Adaptation

Improve Unseen Domain Generalization via Enhanced Local Color Transformation

Jianhao Xiong[1], Andre Wang He[1], Meng Fu[1], Xinyue Hu[1], Yifan Zhang[1],
Congxin Liu[1], Xin Zhao[1], and Zongyuan Ge[1,2(✉)]

[1] Airdoc LLC, Beijing, China
xiongjianhao@airdoc.com
[2] Monash eResearch Center, Monash University, Melbourne, Australia
zongyuan.ge@monash.edu

Abstract. Recent application of deep learning in medical image achieves expert-level accuracy. However, the accuracy often degrades greatly on unseen data, for example data from different device designs and population distributions. In this work, we consider a realistic problem of domain generalization in fundus image analysis: when a model is trained on a certain domain but tested on unseen domains. Here, the known domain data is taken from a single fundus camera manufacture, i.e. Canon. The unseen data are the image from different demographic population and with distinct photography styles. Specifically, the unseen images are taken from Topcom, Syseye and Crystalvue cameras. The model performance is evaluated by two objectives: age regression and diabetic retinopathy (DR) classification. We found that the model performance on unseen domain could decrease significantly. For example, the mean absolute error (MAE) of age prediction could increase by 57.7 %. To remedy this problem, we introduce an easy-to-use method, named enhanced domain transformation (EDT), to improve the performance on both seen and unseen data. The goal of EDT is to achieve domain adaptation without using labeling and training on unseen images. We evaluate our method comprehensively on seen and unseen data sets considering the factors of demographic distribution, image style and prediction task. All the results demonstrate that EDT improves the performance on seen and unseen data in the tasks of age prediction and DR classification. Equipped with EDT, the R^2 (coefficient of determination) of age prediction could be greatly improved from 0.599 to 0.765 (n = 29,577) on Crystalvue images, and AUC (area under curve) of DR classification increases from 0.875 to 0.922 (n = 1,015).

Keywords: Deep learning · Fundus image · Unseen domain

Electronic supplementary material The online version of this chapter (https://doi.org/10.1007/978-3-030-59713-9_42) contains supplementary material, which is available to authorized users.

© Springer Nature Switzerland AG 2020
A. L. Martel et al. (Eds.): MICCAI 2020, LNCS 12262, pp. 433–443, 2020.
https://doi.org/10.1007/978-3-030-59713-9_42

1 Introduction

Many applications of deep learning show an expert level accuracy. However, the performance and robustness of deep learning model are often challenged by unseen [11,23]. In realistic environments, the accuracy could greatly degrade on unseen data, e.g. data from different device designs and population distributions, which is usually considered as the problem of domain generalization [25]. In the scenario of medical image application, the problem is often caused by the fact that the training data are often obtained from few hospitals with specific machines but the trained model is required to perform well on a greater variety of machines [27].

Data augmentation are widely used as an effective solution to improve generalization. The most common targets of the augmentation are the appearance, quality and layout of image. Medical image quality could be manipulated by affecting the sharpness, blurriness and noise. Gaussian noise and Gaussian blur could be very useful to improve performance in CT scan [3,21]. Image appearance could be augmented by image intensities such as brightness, saturation and contrast. Gamma correction of saturation is found useful in segmenting retinal blood vessels segmentation [16]. Spatial transformations such as rotation, scaling and deformation is widely used in fundus image [17]. These augmentation methods could be helpful in improving model performance on the seen domain but its improvement on unseen domain is very limited. Mixup is a data augmentation method wherein new training data are created through by convex combination of pairs of images and existing labels [26]. An improvement is reproted in medical segmentation with the assist of mixup method [18]. This method has received a lot attention in medical image segmentation. Several application of mixup has been reported in the segmentation of tumors and prostate cancer [10,15], but mixup works under the condition that labels of target data is available. This method addresses well labeled data which is not the concern of this work.

Unseen data generalization could be increased by synthetic data augmentation which normally uses Generative Adversarial Network (GAN) to generate synthetic medical images. Fu et al. use a CycleGAN to generate synthetic 3D microscopy images, and their results show that there is a performance improvement of the segmentation model [6]. Guibas et al. propose a method of GAN to generate pairs of synthetic fundus images [8] to improve image fidelity. Huang et. al. use pix2pix network to transform an image of a healthy CXR into one with pathology as an augmentation method, which also improves segmentation performance [9]. The generative method is useful for unseen domain adaptation but one drawback here is that the training and prediction of the GAN are computationally intensive. In addition, another limitation is that the generative method normally requires a certain amount of unseen data for GAN model training.

For retinal fundus image, one of the major generalization problems is from the intrinsic design of fundus camera manufacturers. Designing a fundus camera requires the balanced combination of an imaging system and an illumination system [4]. The combination varies illumination uniformity, illumination ratio, backreflections, spectral sensitivity et. al. Another problem is the image quality and clarity which is highly related to the design of camera [14]. This has important implications for diagnosis like diabetic retinopathy (DR).

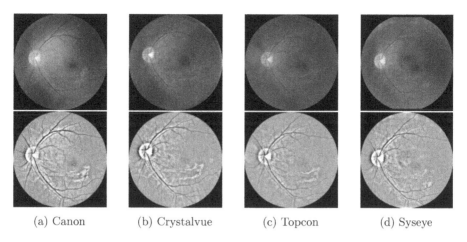

(a) Canon   (b) Crystalvue   (c) Topcon   (d) Syseye

Fig. 1. Raw images of a participant by using Canon CR2, Crystalvue FundusVue, Syseye retiCam 3100 and Topcon NW 400. Raw images are given in the fisrt row and the corresponding EDT transformed images are given in the second row.

To improve the problem, we propose a method named enhanced domain transformation (EDT). The motivation of the method is that if a transformed color space presents the unseen data with a distribution identical to that of training data, the model trained on the space should generalize better, because empirical risk minimization requires training and testing data to hvae independent and identical distributions [24]. The illustration of the transformation is given by Fig. 1. To achieve this transformation, we first considering use differential values for information representation and local adaptive enhancement to constrain the distribution of representation value. Furthermore, additional color space information from unseen domain is used in the augmentation in training by principal component analysis. These processes could reduce the difference between the seen and unseen data and adapt training model to unseen data. The EDT is an easy-to-use method without restriction on neural network architecture. Our results demonstrate that the method improves the unseen domain generalization in the tasks of both age regression and DR classification without the incorporation of unseen data during training. This method can be easily adapt to many machine learning architectures with only a few lines of code.

2 Data Sets and Model Development

All data sets are graded by ophthalmologists and trained professional graders. DR severity of mild, moderate, severe, and proliferative are all labeled as positive and the none DR images are labeled as negative. DR severity is graded according to the Chinese national retinopathy diagnosis guidance [28]. There are four DR conditions in our study, including mild, moderate, severe, and proliferative, which are all labeled as positive. In contrast, the non-symptom images

Table 1. Population and camera manufacture characteristics of training and validation data sets.

	Trainning	Validation
No. of images	316,863	16,677
No. of participants	158,432	8,338
Manufacture	Canon	Canon
Age, years (s.d.)	41.57 (13.63)	41.50 (13.56)
DR, % (no.)	0.522 % (1,653)	0.528 % (88)

are labeled as negatives. The age information is directly deduced from the birth date of participants. A variety of cameras are used to collect our data sets, including Canon CR1/CR2, Crystalvue FundusVue/TonoVue, Syseye retiCam 3100 and Topcon NW 400, where all fundus images are taken by using standard 45° fields of view. Details of development data sets and test data sets are given in Table 1 and Table 2. Dataset A is obtained from 33 participants. Each participant is taken 4 fundus images (two images for each eye) by each camera, so the images from each camera have same demographic characters. Dataset B and Dataset E are used for internal validation of the seen domain, and Dataset C, D and F are for the unseen domains. During the training, model only has access to color information of Crystalvue and Topcon images (domains of Dataset C and F). No information of Syseye image is used during training and validation of the model.

Table 2. Population and camera manufacture characteristics of test data sets for regression and classification performance evaluation.

Age regression			
Data set name	Manufacture	No. of images	Age, years (s.d.)
Dataset A	Canon, Crystalvue, Topcom, Syseye	528	31.29 (5.48)
Dataset B	Canon	17,512	41.51 (13.53)
Dataset C	Crystalvue	29,577	38.57 (12.98)
Dataset D	Syseye	9,737	38.01 (8.61)
DR classification			
Name	Manufacture	No. of images	DR, % (no.)
Dataset E	Canon	1,092	23.44 % (256)
Dataset F	Crystalvue	1,015	21.18 % (215)

3   Method

3.1   Model Development

In this study, we used the Inception-v3 deep learning network [22] as the deep learning backbone and follow the same process of [19] for the tasks of age prediction and DR classification. The predicted age and DR classification score are the two outputs of the network which are trained simultaneously. Loss functions of binary cross entropy and mean absolute error are use in training DR classification and age prediction respectively.

The input image size of the network is 299 × 299. A universal augmentation process of random crop, random horizontal flip, and ±15° rotation applied in the development of EDT and straching scale and adaptive contrast [1] enhancement (Adaptive Local Contrast Enhenacement, ALCE) models [20]. The models are trained on Keras platform v2.2.2 [2] and use the Adam method as the optimizer [12] with initial learning rate of 0.001. The training and testing is performed on NVIDIA GTX 1080Ti GPU ×1 on a batch size of 64.

3.2   Enhanced Domain Transformation

ENHANCED DOMAIN TRANSFORMATION (EDT) is an augmentation method designed to boost domain generalization on unseen images. The essential idea of EDT is that by applying image local average subtraction and contrast enhancement, we are able to project images with large pixel-value variations into an identical and independent distribution with vital pathology information preserved but background noise suppressed. In addition, we employ a special PCA which is designed to shift the training image color with the most present color component in a given image to imitate the unseen domain image color. In essence, the jittered color space would be useful to simulate the unseen domain conditions, thus improving the model generalization. The given image could be from the seen domain, unseen domain or non-fundus images (ImageNet, etc.).

The raw input image is a BGR image with a dimension of H × W × 3, where H and W are the height and width of the image respectively. The raw image is first subtracted the local average color from each pixel using methods described in Graham' Kaggle competation report [7]. The subtracted image f is then goes through an average blurring process. For a given channel of f_k ($k \in \{1,2,3\}$), the channel is applied as:

$$f_k'(x,y) = \frac{1}{(2c+1)^2} \sum_{i=-c}^{c} \sum_{j=-c}^{c} f_k(x+i, y+j), \qquad (1)$$

where $f_k(x,y)$ is the given channel of f, x and y are the x_{th} and y_{th} pixel of $g_k(x,y)$. $2c+1$ is the kernel size of the process, which we set as 5 for training and testing. The blurring shown by Eq. 1 is not included in the target of DR classification problem, because this process could remove the DR abnormal like microaneurysm and haemorrhage, which degrades the classification performance. ALCE is then applied on the returned image f' which is defined as

$$f' \rightarrow g = 255 \frac{\Psi(f') - \Psi(f'_{\min})}{\Psi(f'_{\max}) - \Psi(f'_{\min})}, \tag{2}$$

where $\Psi(f')$ is a sigmoid function. Then, we apply a special PCA on the set of BGR pixel values of the training data, which mixes the color space of seen and unseen images. In this process, the training data from the known domain is shifted by the main BGR components of the seen and unseen data. Only the eigenvectors and eigenvalues of unseen images are used here, but the unseen images themselves are excluded from training process. Specifically, we calculate the eigenvectors and eigenvalues of the 3×3 BGR covariance matrix of the Canon images and two images from manufacturers Crystalvue and Topcon. There are 148 images from each manufacturer. This mix of seen data and unseen data color component is implemented as:

$$g' = g + [\mathbf{p}_1, \mathbf{p}_2, \mathbf{p}_3][\alpha_1\lambda_1, \alpha_2\lambda_2, \alpha_3\lambda_3]^{\mathrm{T}}, \tag{3}$$

where $\mathbf{p}_l$ and α_l are the l_{th} eigenvector and eigenvalue of the 3×3 RGB covariance matrix and $l \in 1, 2, 3$. λ_1, λ_2 and λ_3 are random variables that follow the Gaussian distribution with mean zero and standard deviation 0.1. The returned g' is scaled to [0,1] through contrast stretching as:

$$g'_n = \frac{g' - g'_{\min}}{g'_{\max} - g'_{\min}}, \tag{4}$$

where $g'_{\max}$ and $g'_{\min}$ are the maximum and minimum of input image g' and the g'_n is the stretched output of g'. The EDT is applied in training, validation and testing, except that PCA augmentation described by Eq. 3 is not included in testing and validation. A step by step illustration of EDT's modification on image is given in supplementary materiel.

4   Results

First, we present the results of Dataset A in Table 3. We observe a degradation of model performance on all methods across all unseen data domains. The worst performance degradation occurs when testing images from Crystalvue camera. The largest performance difference is between Canon and Crystalvue (2.43 vs 4.84) on the ALCE benchmark. For the unseen Syseye images, EDT still outperform Contrast Stretching and ALCE. Our proposed EDT model achieves the best performance under both MAE and R2 metric on all camera domains. These results suggest that the EDT has the best age prediction performance on both seen and unseen data. It demonstrates that EDT's augmentation and transformation based on enhanced local contrast improving the generalization on regression model.

Table 3. The performance of EDT on Dataset A (performance measured by MAE and coefficient of determination (R^2). Test is performed on 132 images from same population for each manufacture's camera.

Manufacture	MAE			R^2		
	Contrast stretching	ALCE	EDT	Contrast stretching	ALCE	EDT
Canon	3.08	2.43	**2.13**	0.53	0.67	**0.71**
Crystalvue	4.73	4.83	**3.71**	−0.21	−0.25	**0.21**
Topcon	3.16	2.59	**2.26**	0.48	0.64	**0.73**
Syseye	3.25	2.77	**2.67**	0.36	0.54	**0.58**

Table 4. Performance of contrast stretching, ALCE and EDT method on seen Dataset B (Canon images), Dataset C (Crystalvue images) and unseen Dataset D (Syseye images).

	Dataset B		Dataset C		Dataset D	
	R^2 (95% CI)	MAE	R^2 (95% CI)	MAE	R^2 (95% CI)	MAE
Contrast Stretching	0.88 (0.87, 0.88)	3.49	0.60 (0.59, 0.61)	5.50	0.24 (0.21,0.27)	7.47
ALCE	0.88 (0.88, 0.89)	3.32	0.72 (0.71, 0.73)	4.76	0.24 (0.21 0.27)	7.47
EDT	**0.89** (0.89, 0.90)	**3.15**	**0.77** (0.76, 0.77)	**4.31**	**0.31** (0.27, 0.33)	**7.16**

The results of Dataset-A can be further validated by B and C as shown in Table 4. To show the statistic significance, we caculate 95% CIs by using 2,000 bootstrap samples [5]. Not surprisingly, we again found that EDT also performs significantly better than ALCE and contrast stretching on both data sets. For example, the MAE of EDT on Dataset B and C are the lowest values at 3.146 and 4.307 respectively. The EDT's degradation percentage between the two data sets is only 36.9% while contrast stretching and ALCE reach the percentages of 57.7% and 43.4% respectively, which all indicate that EDT has a statistically significant improvement than the others. EDT also outperforms the others on the unseen Syseye images. The MAE is reduced by 0.31 years on Dataset D. A further illustration of the age prediction is given in Fig. 2.

We also examined the effect of EDT on retinal eye DR classification. EDT without blurring performs better than others, although the gain is moderate. The highest AUC on Dataset F is 0.922 performed by omitted blurring EDT. There is a performance loss induced by EDT with the blurring process, which may be due to the removed details related to DR. The AUC improved from 0.912 to 0.922 when the blurring process is omitted.

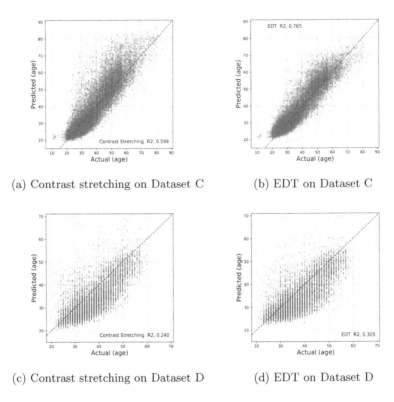

(a) Contrast stretching on Dataset C   (b) EDT on Dataset C

(c) Contrast stretching on Dataset D   (d) EDT on Dataset D

Fig. 2. Age prediction performance of EDT model and contrast stretching model. The red lines represent $y = x$. (Color figure online)

5   Discussion

Our results indicate that the usage of images from different camera manufactures causes the degradation in both age prediction and DR binary classification. These performance loss seems inevitable if the unseen data is not involved in training. In the case of age regression and binary DR classification, the age prediction tends to be more susceptible than the classification task, which may due to the fact that the numerical stability of regression task on a large range of the continuous value is more difficult than a binary classification of DR. Encouragingly, our results demonstrates that EDT could significantly improve the prediction accuracy on the unseen domain data without sacrificing the performance on the seen data. The processes such as PCA color jittering and feature space normalization involved in EDT could be considered as general methods on various tasks to enhance useful features and to remove the unnecessary noise. The noise may derive from several factors like camera imaging and illumination system, as well as the unique color style, enhanced pixel value and some image characters induced from these factors seems unnecessary for our prediction targets.

Table 5. DR classification AUCs of two validation data sets.

Method	AUC of dataset E (95% CI)	AUC of dataset F (95% CI)
Contrast Stretching	0.971 (0.962, 0.980)	0.913 (0.891, 0.935)
ALCE	0.959 (0.945, 0.972)	0.875 (0.851, 0.908)
EDT (No Bluring)	**0.972** (0.962, 0.980)	**0.922** (0.899, 0.944)
EDT	0.968 (0.954, 0.978)	0.912 (0.891 , 0.935)

As shown that if those unnecessary image features are removed by blurring, background extracting and contrast enhancing, the model can generalize better on many unseen settings on both age regression and DR classification tasks. The process like EDT is difficult to be automatically learned by deep algorithm because training data cannot cover all the possible diversity of real data.

Beside the promising results, the expense of EDT is very moderate. As the EDT only contains ordinary processes like blurring and contrast enhancement which are all computationally efficient. In addition, because we use the pre-calculated and stored PCA values of various image styles during in the training process, the PCA step in EDT is greatly sped up compared with the original design [13]. Furthermore, EDT is an easy-to-use method because there is no restriction on neural network architecture for the usage of the EDT. Ordinary machine learning image algorithm could be equipped with EDT by adding a few lines of code. Despite these advantages, there exists certain limitation in EDT. Certain modifications may be demanded for adapting this method. As shown in DR classification task, the blurring process imposes a penalty on the classification performance. Hence, users may only need to use parts of EDT for their specific learning targets, and the adaptation could be done by trial and error.

References

1. Jain, A.: Fundamentals of Digital Image Processing. Prentice-Hall, Upper Saddle River (1989)
2. Chollet, F., et al.: Keras. https://keras.io. Accessed 29 Feb 2020
3. Christ, P.F., et al.: Automatic liver and lesion segmentation in CT using cascaded fully convolutional neural networks and 3D conditional random fields. In: Ourselin, S., Joskowicz, L., Sabuncu, M.R., Unal, G., Wells, W. (eds.) MICCAI 2016. LNCS, vol. 9901, pp. 415–423. Springer, Cham (2016). https://doi.org/10.1007/978-3-319-46723-8_48
4. DeHoog, E., Schwiegerling, J.: Optimal parameters for retinal illumination and imaging in fundus cameras. Appl. Opt. **47**(36), 6769–6777 (2008)
5. Efron, B.: Better bootstrap confidence intervals. J. Am. Stat. Assoc. **82**(397), 171–185 (1987)
6. Fu, C., et al.: Three dimensional fluorescence microscopy image synthesis and segmentation. In: Proceedings of the IEEE Conference on Computer Vision and Pattern Recognition Workshops, pp. 2221–2229. IEEE (2018)

7. Graham, B.: Diabetic retinopathy detection competition report. University of Warwick (2015)
8. Guibas, J.T., Virdi, T.S., Li, P.S.: Synthetic medical images from dual generative adversarial networks (2017). arXiv preprint arXiv:1709.01872
9. Huang, X., Liu, M.Y., Belongie, S., Kautz, J.: Multimodal unsupervised image-to-image translation. In: Proceedings of the European Conference on Computer Vision, pp. 172–189 (2018)
10. Jung, W., Park, S., Jung, K.H., Hwang, S.I.: Prostate cancer segmentation using manifold mixup u-net. In Proceedings of the Medical Imaging with Deep Learning (MIDL), pp. 8–10 (2019)
11. Karimi, D., Dou, H., Warfield, S.K., Gholipour, A.: Deep learning with noisy labels: exploring techniques and remedies in medical image analysis. Med. Image Anal. **65**, 101759 (2020)
12. Kingma, D.P., Ba, J.: Adam: A method for stochastic optimization (2014). arXiv preprint arXiv:1412.6980
13. Krizhevsky, A., Sutskever, I., Hinton, G.E.: ImageNet classification with deep convolutional neural networks. In: Advances in Neural Information Processing Systems, pp. 1097–1105 (2012)
14. Li, H., Liu, W., Zhang, H.F.: Investigating the influence of chromatic aberration and optical illumination bandwidth on fundus imaging in rats. J. Biomed. Opt. **20**(10), 106010 (2015)
15. Li, Z., Kamnitsas, K., Glocker, B.: Overfitting of neural nets under class imbalance: analysis and improvements for segmentation. In: Shen, D., et al. (eds.) MICCAI 2019. LNCS, vol. 11766, pp. 402–410. Springer, Cham (2019). https://doi.org/10.1007/978-3-030-32248-9_45
16. Liskowski, P., Krawiec, K.: Segmenting retinal blood vessels with deep neural networks. IEEE Trans. Med. Imaging **35**(11), 2369–2380 (2016)
17. Maninis, K.-K., Pont-Tuset, J., Arbeláez, P., Van Gool, L.: Deep retinal image understanding. In: Ourselin, S., Joskowicz, L., Sabuncu, M.R., Unal, G., Wells, W. (eds.) MICCAI 2016. LNCS, vol. 9901, pp. 140–148. Springer, Cham (2016). https://doi.org/10.1007/978-3-319-46723-8_17
18. Panfilov, E., Tiulpin, A., Klein, S., Nieminen, M.T., Saarakkala., S.: Improving robustness of deep learning based knee MRI segmentation: mixup and adversarial domain adaptation. In: Proceedings of the IEEE International Conference on Computer Vision Workshops (2019)
19. Poplin, R., et al.: Prediction of cardiovascular risk factors from retinal fundus photographs via deep learning. Nat. Biomed. Eng. **2**(3), 158 (2018)
20. Sinthanayothin, C., Boyce, J., Cook, H., Williamson, T.: Automated localisation of the optic disc, fovea, and retinal blood vessels from digital colour fundus images. Brit. J. Ophthalmol. **83**(8), 902–910 (1999)
21. Sirinukunwattana, K., et al.: Gland segmentation in colon histology images: the glas challenge contest. Med. Image Anal. **35**, 489–502 (2017)
22. Szegedy, C., Vanhoucke, V., Ioffe, S., Shlens, J., Wojna, Z.: Rethinking the inception architecture for computer vision. In: Proceedings of the IEEE Conference on Computer Vision and Pattern Recognition, pp. 2818–2826 (2016)
23. Tajbakhsh, N., et al.: Embracing imperfect datasets: a review of deep learning solutions for medical image segmentation. Med. Image Anal. **63**, 101693 (2020)
24. Vapnik, V.: Statistical Learning Theory. Wiley, New York (1998)
25. Yasaka, K., Abe, O.: Deep learning and artificial intelligence in radiology: current applications and future directions. PLoS Med. **15**(11), e1002707 (2018)

26. Zhang, H., Cisse, M., Dauphin, Y.N., Lopez-Paz, D.: mixup: Beyond empirical risk minimization (2017). arXiv preprint arXiv:1710.09412
27. Zhang, L., et. al: When unseen domain generalization is unnecessary? Rethinking data augmentation (2019). arXiv preprint arXiv:1906.03347
28. 中华医学会眼科学会眼底病学组: 我国糖尿病视网膜病变临床诊疗指南. 中华眼科杂志**50**(11), 851–865 (2014)

Transport-Based Joint Distribution Alignment for Multi-site Autism Spectrum Disorder Diagnosis Using Resting-State fMRI

Junyi Zhang, Peng Wan, and Daoqiang Zhang[✉]

College of Computer Science and Technology,
Nanjing University of Aeronautics and Astronautics, Nanjing, China
dqzhang@nuaa.edu.cn

Abstract. Resting-state functional magnetic resonance imaging (rs-fMRI) has been a promising technique for computer-aided diagnosis of neurodevelopmental brain diseases, *e.g.*, autism spectrum disorder (ASD), due to its sensitivity to the progressive variations of brain functional connectivity. To overcome the challenge of overfitting resulting from small sample size, recent studies began to fuse multi-site datasets for improving model generalization. However, these existing methods generally simply combine multiple sites into single dataset, ignoring the heterogeneity (*i.e.*, data distribution discrepancy) among diverse sources. Actually, *distribution alignment*, alleviating the inter-site distribution shift, is the fundamental step for multi-site data analysis. In this paper, we propose a novel Transport-based Multi-site Joint Distribution Adaptation (TMJDA) framework to reduce multi-site heterogeneity for ASD diagnosis by aligning joint feature and label distribution using optimal transport (OT). Specifically, with the given target domain and multi-source domains, our TMJDA method concurrently performs joint distribution alignment in each pair of source and target domains, upon which, multiple domain-specific classifiers are further aligned by penalizing decision inconsistency among diverse classifiers for reducing inter-site distribution discrepancy. We evaluated our TMJDA model on the public Autism Brain Imaging Data Exchange (ABIDE) database. Experimental results demonstrate the effectiveness of our method in ASD diagnosis based on multi-site rs-fMRI datasets.

Keywords: Multi-site · Optimal transport · Autism spectrum disorder

1 Introduction

Autism spectrum disorder (ASD), characterized by deficits in social communication and the presence of restricted, repetitive behaviors [1], is the most

---

J. Zhang and P. Wan—These authors contributed equally to this work.

© Springer Nature Switzerland AG 2020
A. L. Martel et al. (Eds.): MICCAI 2020, LNCS 12262, pp. 444–453, 2020.
https://doi.org/10.1007/978-3-030-59713-9_43

prevalent neurodevelopmental disorder that ultimately seriously affects individual social and daily life. Considerable studies [2,3] have proven that the variations of brain functional connectivity pattern (e.g., local overconnectivity and long-distant underconnectivity) are associated with ASD and its progression, especially considering that this changing process occurs even earlier than the appearance of disconnection syndromes. Resting-state fMRI can non-invasively capture profound brain connectivity variations induced by the pathological process. Therefore, various learning-based methods [4–6] have been proposed for automatically mining the relevant local or global connectivity characteristics which are used as important biomarkers in ASD diagnosis.

Nonetheless, clinical practical applications of these studies are still faraway due to the limitation of reproducibility across diverse autism research centers [7]. This mainly arises from the potential variations across individuals, MR scanners, and data acquisition protocols [8]. In terms of statistical learning, the resulting data heterogeneity, i.e., distribution discrepancy, severely degrades the model generalization when applied to a new target imaging domain. In recent years, a number of techniques [9–11] have been developed to deal with the problem of distribution shift across multiple data sites, typically Unsupervised Domain Adaptation (UDA), which leverages knowledge learned from one or more related source domains to facilitate classifier construction in target domain with no annotated samples [12]. One of the common strategies is to seek a transformation $\mathcal{T}$ such that the new representations of input data are matching (i.e., $P_s(\mathcal{T}(X)) = P_t(\mathcal{T}(X))$) under some distribution measures (e.g., Bergman Divergence [13], Maximum Mean Discrepancy (MMD) [14] and et al. [15,16]). A particular case is the use of optimal transport (OT) [17,18], which seeks for an optimal transport plan γ by minimizing the global transportation cost between the empirical distributions of X_s and X_t, thereby interpolating X_s via barycentric mapping and ultimately bringing the source and target distribution closer. In this way, a shared embedding space could be learned while preserving the discriminative information [19], which is suitable to our task since the goal is to mine the common brain connectivity variation pattern across multiple ASD research centers.

In this paper, we propose a novel multi-source unsupervised domain adaptation framework using optimal transport, named as Transport-based Multi-site Joint Distribution Adaptation (TMJDA), to reduce the heterogeneity of multi-site resting-state fMRI data for improving ASD diagnostic performance. As shown in Fig. 1, our proposed TMJDA model admits a multiway parallel adaptation framework. To be specific, each source domain is adapted to the target domain using our proposed transport-based joint distribution alignment block (T-JDA block), meanwhile, a multi-classifier inconsistency cost is incorporated for reducing the mismatching of outputs of different classifiers, under the circumstance of knowledge propagation among diverse sources. The main innovations of our proposed TMJDA model can be summarized as follows, (1) each T-JDA block can propagate *feature/label* from source to target by minimizing the global transportation cost between empirical joint distribution of two domains; (2) the parallel adaptation architecture guided by an inconsistency loss can furtherly propagate knowledge transfer across all data sites.

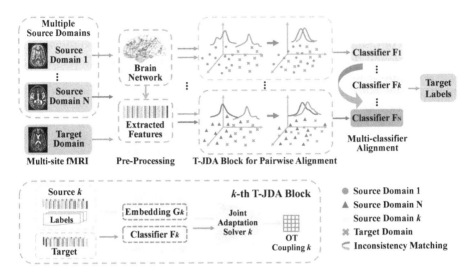

Fig. 1. An overview of the proposed Transport-based Multi-site Joint Distribution Adaptation (TMJDA) framework. Our framework receives labeled multi-source samples and a target domain with no annotated samples. Specific T-JDA block performs joint distribution alignment in each pair of source and target domains while a shared classifier F is learnt in both domains, upon which, multiple classifiers are aligned to classify target samples. (Best viewed in color)

2 Materials and Method

2.1 Datasets and Image Pre-processing

The Autism Brain Imaging Data Exchange [20] database is a qualified multi-site heterogeneous data collection. It contains resting-state fMRI and clinical data on altogether 1,112 subjects from 17 diverse sites. Considering that several sites suffer from a small sample size, five sites with more than 50 subjects contributes to our work: *Leuven, USM, UCLA, UM* and *NYU*. These sites contain 468 subjects, including 218 patients with ASD and 250 age-matched typical control individuals (TCs).

Considering the reproducibility of our work, we used the published version of this database in Preprocessed Connectome Project initiative. Firstly, Configurable Pipeline for the Analysis of Connectomes (C-PAC) [21] was used to preprocess resting-state fMRI, including slice-timing correction, image realignment and nuisance regression. After that, time series of average voxel signals were taken from each brain region extracted by the predefined Anatomical Automatic Labeling (AAL) atlas with 116 ROIs. Based on that, the correlation between each pair of ROIs is estimated using Pearson Correlation Coefficient, resulting a 116 × 116 symmetrical matrix representing functional connectivity for each subject. Finally, we utilized the lower-triangular matrix without 116 diagonal elements and flatten it to 6,670-dimensional features for each subject.

2.2 Transport-Based Multi-site Joint Adaptation

The standard learning paradigm of multi-source unsupervised domain adaptation (MUDA) assumes the existence of N source datasets $\{(X_s^k, Y_s^k)\}_{k=1}^N$, $Y_s^k = \{y_i^{sk}\}_{i=1}^{n_{sk}}$ is the associated class label for k-th source domain, and a target set of data $X_t = \{x_j^t\}_{j=1}^{n_t}$ with unknown labels. In the context of UDA, we assume that these joint probability distributions of source domain $P_s^k(X, Y)$ differ from that of target domain $P_t(X, Y)$. In order to determine the set of labels Y_t associated with X_t, the core task is to align empirical joint distribution $\widehat{P}_s^k(X, Y)$ and $\widehat{P}_t(X, Y)$, with expect to transfer knowledge from multiple sources to assist decision boundary learning in target domain.

Overall Adaptation Framework. Figure 1 shows a schematic diagram of our Transport-based Multi-site Joint Distribution Adaptation (TMJDA) framework, characterized by a multiway parallel adaptation architecture, which consists of 1) data preprocessing block, 2) transport-based joint distribution alignment block (T-JDA block) and 3) multi-classifier alignment. After data preprocessing, T-JDA block performs joint distribution alignment in each pair of source and target domains in parallel, accompanied by multi-classifier inconsistency loss to propagate knowledge transfer across all domains.

Transport-Based JDA Block. Given one pair of source and target domain, traditional assumption is that, the conditional distributions are kept unchanged under transformation $\mathcal{T}$, i.e., $P_s(Y|\mathcal{T}(X)) = P_t(Y|\mathcal{T}(X))$, but it remains controversial in practical problems. A more reasonable assumption should admit there exist shift in both marginal and conditional probabilities, that is, one should permit the direct alignment of joint distributions $P_s(X, Y)$ and $P_t(X, Y)$. Following the Kantorovich formulation [22], Courty et al. [23] proposed to seek a probabilistic coupling γ between empirical joint distributions, i.e., $\widehat{P}_s = \frac{1}{n_s} \sum_{i=1}^{n_s} \delta_{x_i^s, y_i^s}$ and $\widehat{P}_t = \frac{1}{n_t} \sum_{j=1}^{n_t} \delta_{x_j^t, y_j^t}$, which minimizes the transport cost:

$$\gamma_0 = \underset{\gamma \in \Pi(\widehat{P}_s, \widehat{P}_t)}{\arg\min} \sum_{ij} D(x_i^s, y_i^s; x_j^t, y_j^t)\gamma_{ij} \tag{1}$$

where $\Pi = \{\gamma \in (\mathbf{R}^+)^{n_s \times n_t} | \gamma \mathbf{1}_{n_s} = \widehat{P}_s, \gamma^{\mathrm{T}} \mathbf{1}_{n_t} = \widehat{P}_t\}$ and $\mathbf{1}_d$ is a d-dimensional one vector. The joint cost $D(x_i^s, y_i^s; x_j^t, y_j^t) = \alpha c(x_i^s, x_j^t) + \mathcal{L}(y_i^s, y_j^t)$ is a weighted combination of the distances in feature space and the classification loss $\mathcal{L}$ (e.g., hinge or cross-entropy) in label space. Since no label is available in the target domain, y_j^t is replaced by a proxy $f(x_j^t)$ using predictive function f.

Inspired by this, we propose a basic T-JDA block to learn a shared common space which shrinks the feature/label discrepancy between domains. The block consists of three components, $i.e.$, 1) shared feature embedding function $g_k(\cdot)$, 2) domain-specific classifier $f_k(\cdot)$ and 3) joint distribution adaptation solver $\mathcal{L}_{t-jda}^k$. Upon which, T-JDA block jointly aligns the feature space and label conditional distribution, such that the estimation of both mapping $g_k(\cdot)$ and classifier $f_k(\cdot)$ is embedded in the optimization of optimal transport coupling γ between $\widehat{P}_s^k(X,Y)$ and $\widehat{P}_t(X,Y)$.

We use Eq. (1) as the estimate of adaptation cost in each T-JDA block, where $c(\cdot,\cdot)$ is chosen as the squared Euclidean metric and $\mathcal{L}_t$ is a traditional cross-entropy loss. The T-JDA solver is reformulated as:

$$\mathcal{L}_{t-jda}^k = \sum_{i=1}^{n_{sk}} \sum_{j=1}^{n_t} \gamma_{ij}^k (\alpha \|g_k(x_i^{sk}) - g_k(x_j^t)\|^2 + \mathcal{L}_t(y_i^{sk}, f_k(g_k(x_j^t)))) \qquad (2)$$

Multi-classifier Inconsistency Matching. Since domain-specific classifiers in each T-JDA block are learned from diverse source-target domain pairs, the outputs of these classifiers may be inconsistent when predicting the same target sample, especially for samples near domain-specific decision boundary [24]. In order to reduce the prediction discrepancy among multiple classifiers while facilitate knowledge propagation among multiple diverse sources, we penalize decision inconsistency by minimizing the absolute value of the discrepancy in the outputs of all pairs of classifiers ($i.e.$, $f_k(\cdot)$ and $f_m(\cdot)$) when classifying target samples. The inconsistency loss is formulated as:

$$\mathcal{L}_{inc} = \frac{1}{N \times (N-1)} \sum_{k,m=1}^{N} \mathbb{E}_{x \sim X_t} [|f_k(g_k(x_j^t)) - f_m(g_m(x_j^t))|] \qquad (3)$$

Multi-site Joint Alignment and Classification. Our TMJDA model performs joint distribution alignment and ASD diagnostic based on multi-site rs-fMRI datasets. It optimizes the learnable parameters for each basic T-JDA block in parallel, while updating the classifier both in source and target domains. It is worth noting that, the mapping $g_k(\cdot)$ and classifier $f_k(\cdot)$ are shared by source-target domain pairs in k-th T-JDA block. As shown in Fig. 1, the total loss function $\mathcal{L}_{total}$ is a combination of adaptation cost in multiway T-JDA block ($i.e.$, $\mathcal{L}_{t-jda}$), source-domain classification loss ($i.e.$, $\mathcal{L}_{cls}$) and multi-classifier inconsistency cost in target domain ($i.e.$, $\mathcal{L}_{inc}$). To be specific, we use Eq. (2) as the adaptation cost in each T-JDA block, upon which, the multiway adaptation cost is reformulated as $\mathcal{L}_{t-jda} = \sum_{k=1}^{N} \mathcal{L}_{t-jda}^k$. Considering the possible performance degradation of classifier in source domain, $i.e.$, catastrophic forgetting [25], due to overemphasis on features and labels alignment, a classification loss using cross-entropy is incorporated here, which is formulated as:

$$\mathcal{L}_{cls} = \sum_{k=1}^{N} \mathbb{E}_{x \sim X_s^k} \mathcal{L}_s(y_i^{sk}, f_k(g_k(x_i^{sk}))) \qquad (4)$$

We define total loss function as:

$$\mathcal{L}_{total} = \mathcal{L}_{cls} + \mathcal{L}_{t-jda} + \lambda \mathcal{L}_{inc} \tag{5}$$

where λ is a parameter to control the contribution of multi-classifier inconsistency cost. Finally, we average the outputs of all domain-specific classifiers as the prediction of target sample by majority voting.

Optimization. The loss function is optimized by an alternative minimization approach [26]. We start with submodule (g_k, f_k) pre-trained on the corresponding labeled source domain. At the training stage, we randomly sample a batch of n_b samples from one of N source domains and a batch of n_b samples from target domain, and perform optimation as follows:

(1) With fixed parameters of submodule $(\widehat{g}_k, \widehat{f}_k)$, the optimization of T-JDA solver is a standard OT problem, and we can obtain coupling γ_k by minimizing the transport cost:

$$\min_{\gamma_k \in \Pi(\widehat{P}_s, \widehat{P}_t)} \sum_{i,j=1}^{n_b} \gamma_{ij}^k (\alpha \|\widehat{g}_k(x_i^{sk}) - \widehat{g}_k(x_j^t)\|^2 + \mathcal{L}_t(y_i^{sk}, \widehat{f}_k(\widehat{g}_k(x_j^t)))) \tag{6}$$

(2) With fixed couplings $\widehat{\gamma}_1 \dots \widehat{\gamma}_N$ obtained at previous step, we concurrently update multiple submodules $(g_1, f_1) \dots (g_N, f_N)$, which follows classical minibatch stochastic gradient descent (SGD) algorithm. We can obtain submodule (g_k, f_k) by minimizing the total loss:

$$\frac{1}{n_b} \sum_{i=1}^{n_b} \mathcal{L}_s(y_i^{sk}, f_k(g_k(x_i^{sk}))) + \sum_{i,j=1}^{n_b} \widehat{\gamma}_{ij}^k (\alpha \|g_k(x_i^{sk}) - g_k(x_j^t)\|^2$$
$$+ \mathcal{L}_t(y_i^{sk}, f_k(g_k(x_j^t)))) + \lambda \frac{1}{N-1} \sum_{m=1}^{N} \frac{1}{n_b} \sum_{j=1}^{n_b} (|f_k(g_k(x_j^t)) - f_m(g_m(x_j^t))|) \tag{7}$$

Implementation. Our TMJDA model was implemented using Python based on Keras. For simplicity, feature embedding function consists of only two fully connected layers of 1024 and 128 hidden units followed by a ReLU nonlinearity respectively, and domain-specific classifier consists of a fully connected layer of 64 hidden units follow by a softmax layer. Besides, we set the batch size to 32, learning rate to 0.01 and SGD momentum to 0.01. Instead of fixing the trade-off parameters, i.e., α in $\mathcal{L}_{t-jda}$ and λ of $\mathcal{L}_{inc}$, we set $\alpha_k = 1/\max \|g_k(x_i^{sk}) - g_k(x_j^t)\|^2$ to normalize the range of distances in domain-specific feature space [23], and set $\lambda_p = 1/(1 + \exp(-10p)) - 1$ to suppress noisy activations from multi-classifier at early training phases [27].

3   Experiments

3.1   Experimental Setup

We perform an application for multi-site datasets from the ABIDE database to evaluate the diagnostic performance of Transport-based Multi-site Joint Distribution Adaptation (TMJDA) model for ASDs and TCs. In our experiment, we alternately treat one site as target domain, and the rest are sources. The evaluation metrics including classification accuracy (ACC), sensitivity (SEN) and specificity (SPE). Let us note TP (*resp.*, TN) as true positives (*resp.*, negatives) and FP (*resp.*, FN) as false positives (*resp.*, negatives), we define ACC = (TP + TN)/(TP + FP + TN + FN), SEN = TP/(TP + FN) and SPE = TN/(TN + FP), where ACC corresponds to the ability of model to predict correctly, SEN corresponds to the abilty to detect healthy individuals and SPE corresponds to the ability to detect patients with ASD.

Considering there is a small amount of work involved in multi-site adaptation for disease diagnosis, in our experiment, we choose a recent method Multi-center Low-Rank Representation (MCLRR) [9] as multi-site adaptation baseline, and compare TMJDA model with multi-site combine methods, including Denoising Auto-Encoder (DAE) [6] and Optimal Transport Domain Adaptation (OTDA) [19]. Under this setting, we testify whether multi-site heterogeneity has an impact on ASD diagnosis, and demonstrates the effectiveness of our TMJDA model. Furthermore, we evaluate a variant TMJDA-1, *i.e.*, without multi-classifier inconsistency cost, to demonstrate the contribution of parallel domain-specific mapping and multi-classifier alignment to our model.

3.2   Results and Discussion

We compare TMJDA with or without inconsistency cost and baseline methods on five data sites and show the performance comparison of classification accuracy in Fig. 2. Furthermore, we report the detailed results on *UM* site in Table 1. From these results, we get the following insightful observations.

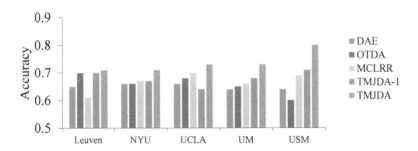

Fig. 2. Performance comparison of classification accuracy for ASDs and TCs on multiple sites from ABIDE database.

In Fig. 2, it shows that TMJDA outperforms all compared baseline methods. The average classification accuracy across multiple sites achieved by these methods are 65.00%, 65.82%, 66.76%, 68.11% and 73.29% respectively, which demonstrates the effectiveness of our method in ASD diagnosis based on multi-site rs-fMRI datasets.

Table 1. Results of classification accuracy, sensitivity and specificity on *UM* site.

	Method	ACC (%)	SEN (%)	SPE (%)
Multi-site combined	DAE [6]	64.00	66.00	62.00
	OTDA [19]	65.49	70.83	61.53
Multi-site adaptation	MCLRR [9]	66.30	75.16	59.53
	TMJDA-1	68.14	66.67	69.23
	TMJDA (ours)	**72.57**	75.00	**70.77**

In Table 1, comparing multi-site combine and multi-site adaptation methods, it is obvious that the multi-site heterogeneity can have a negative effect on ASD diagnosis, even the former one benefits from data expansion. TMJDA-1 outperforms MCLRR in most cases, it indicates that compared with the latter one, which tries to map all domains into a common latent space, TMJDA-1 is simpler and more efficient to concurrently performs joint distribution alignment in multiple source-target domain pairs to seek domain-specific mapping. Comparing TMJDA with or without inconsistency cost, TMJDA achieves much better performance than TMJDA-1, which indicates that aligning domain-specific classifiers by penalizing the decision inconsistency among diverse classifiers can furtherly propagate knowledge transfer across diverse domains.

4 Conclusions

In this paper, we propose a novel Transport-based Multi-site Joint Distribution Adaptation (TMJDA) framework to reduce the heterogeneity of multi-site resting-state fMRI data for improving ASD diagnostic performance. Our framework concurrently performs joint distribution alignment using optimal transport in diverse source-target domain pairs, and the parallel adaptation architecture guided by multi-classifier alignment to classify target samples. Experimental results on the public ABIDE database demonstrate the effectiveness of our method. Furthermore, the TMJDA framework holds the promise to be applied to other multi-site databases.

Acknowledgement. This work was supported by the National Natural Science Foundation of China (Nos. 61876082, 61732006, 61861130366), the National Key R&D Program of China (Grant Nos. 2018YFC2001600, 2018YFC2001602, 2018ZX10201002) and the Royal Society-Academy of Medical Sciences Newton Advanced Fellowship (No. NAF\R1\180371).

References

1. Amaral, D.G., et al.: Autism BrainNet: a network of postmortem brain banks established to facilitate autism research. Handb. Clin. Neurol. **150**, 31–39 (2018)
2. Kana, R.K., et al.: Disrupted cortical connectivity theory as an explanatory model for autism spectrum disorders. Phys. Life Rev. **8**(4), 410–437 (2011)
3. Maximo, J.O., et al.: The implications of brain connectivity in the neuropsychology of autism. Neuropsychol. Rev. **24**(1), 16–31 (2014)
4. Anderson, J.S., et al.: Functional connectivity magnetic resonance imaging classification of autism. Brain **134**(12), 3742–3754 (2011)
5. Wang, M., et al.: Multi-task exclusive relationship learning for Alzheimer's disease progression prediction with longitudinal data. Med. Image Anal. **53**, 111–122 (2019)
6. Heinsfeld, A.S., et al.: Identification of autism spectrum disorder using deep learning and the ABIDE dataset. NeuroImage Clin. **17**, 16–23 (2018)
7. Button, K.S., et al.: Power failure: why small sample size undermines the reliability of neuroscience. Nat. Rev. Neurosci. **14**(5), 365–376 (2013)
8. Abraham, A., et al.: Deriving reproducible biomarkers from multi-site resting-state data: an autism-based example. NeuroImage **147**, 736–745 (2017)
9. Wang, M., Zhang, D., Huang, J., Shen, D., Liu, M.: Low-rank representation for multi-center autism spectrum disorder identification. In: Frangi, A.F., Schnabel, J.A., Davatzikos, C., Alberola-López, C., Fichtinger, G. (eds.) MICCAI 2018. LNCS, vol. 11070, pp. 647–654. Springer, Cham (2018). https://doi.org/10.1007/978-3-030-00928-1_73
10. Wachinger, C., et al.: Domain adaptation for Alzheimer's disease diagnostics. NeuroImage **139**, 470–479 (2016)
11. Itani, S., et al.: A multi-level classification framework for multi-site medical data: application to the ADHD-200 collection. Expert Syst. Appl. **91**, 36–45 (2018)
12. Pan, S., et al.: A survey on transfer learning. IEEE Trans. Knowl. Data Eng. **22**(10), 1345–1359 (2009)
13. Si, S., et al.: Bregman divergence-based regularization for transfer subspace learning. IEEE Trans. Knowl. Data Eng. **22**(7), 929–942 (2009)
14. Long, M., et al.: Transfer feature learning with joint distribution adaptation. In: ICCV, pp. 2200–2207. IEEE, Piscataway (2013)
15. Sun, B., Saenko, K.: Deep CORAL: correlation alignment for deep domain adaptation. In: Hua, G., Jégou, H. (eds.) ECCV 2016. LNCS, vol. 9915, pp. 443–450. Springer, Cham (2016). https://doi.org/10.1007/978-3-319-49409-8_35
16. Xu, R., et al.: Deep cocktail network: Multi-source unsupervised domain adaptation with category shift. In: CVPR, pp. 3964–3973. IEEE, Piscataway (2018)
17. Courty, N., Flamary, R., Tuia, D.: Domain adaptation with regularized optimal transport. In: Calders, T., Esposito, F., Hüllermeier, E., Meo, R. (eds.) ECML PKDD 2014. LNCS (LNAI), vol. 8724, pp. 274–289. Springer, Heidelberg (2014). https://doi.org/10.1007/978-3-662-44848-9_18
18. Courty, N., et al.: Optimal transport for domain adaptation. IEEE Trans. Pattern Anal. Mach. Intell. **39**(9), 1853–1865 (2016)
19. Chambon, S., et al.: Domain adaptation with optimal transport improves EEG sleep stage classifiers. In: PRNI, pp. 1–4. IEEE, Piscataway (2018)
20. Di Martino, A., et al.: The autism brain imaging data exchange: towards a large-scale evaluation of the intrinsic brain architecture in autism. Mol. Psychiatry **19**(6), 659–667 (2014)

21. Craddock, C., et al.: Towards automated analysis of connectomes: the configurable pipeline for the analysis of connectomes (C-PAC). Front Neuroinform. **42** (2013)
22. Kantorovich, L.V.: On the translocation of masses. J. Math. Sci. **133**(4), 1381–1382 (2006)
23. Courty, N., et al.: Joint distribution optimal transportation for domain adaptation. In: NeurIPS, pp. 3730–3739. MIT Press, Cambridge (2017)
24. Zhu, Y., et al.: Aligning domain-specific distribution and classifier for cross-domain classification from multiple sources. In: AAAI, pp. 5989–5996. AAAI, Palo Alto (2019)
25. Li, Z., et al.: Learning without forgetting. IEEE Trans. Pattern Anal. Mach. Intell. **40**(12), 2935–2947 (2017)
26. Damodaran, B.B., Kellenberger, B., Flamary, R., Tuia, D., Courty, N.: DeepJDOT: deep joint distribution optimal transport for unsupervised domain adaptation. In: Ferrari, V., Hebert, M., Sminchisescu, C., Weiss, Y. (eds.) ECCV 2018. LNCS, vol. 11208, pp. 467–483. Springer, Cham (2018). https://doi.org/10.1007/978-3-030-01225-0_28
27. Ganin, Y., et al.: Unsupervised domain adaptation by backpropagation. In: ICML, pp. 1180–1189. ACM, New York (2015)

Automatic and Interpretable Model for Periodontitis Diagnosis in Panoramic Radiographs

Haoyang Li[1,2,3], Juexiao Zhou[1,4], Yi Zhou[5], Jieyu Chen[6(✉)], Feng Gao[6(✉)], Ying Xu[2,5(✉)], and Xin Gao[1(✉)]

[1] Computational Bioscience Research Center (CBRC), King Abdullah University of Science and Technology (KAUST), Thuwal, Saudi Arabia
xin.gao@kaust.edu.sa
[2] Cancer Systems Biology Center, China-Japan Union Hospital, Jilin University, Changchun, China
[3] College of Computer Science and Technology, Jilin University, Changchun, China
[4] Southern University of Science and Technology, Shenzhen, China
[5] Department of Biochemistry and Molecular Biology and Institute of Bioinformatics, University of Georgia, Athens, USA
xyn@uga.edu
[6] The Sixth Affiliated Hospital, Sun Yat-sen University, Guangzhou, China
{chenjy335,gaof57}@mail.sysu.edu.cn

Abstract. Periodontitis is a prevalent and irreversible chronic inflammatory disease both in developed and developing countries, and affects about 20%-50% of the global population. The tool for automatically diagnosing periodontitis is highly demanded to screen at-risk people for periodontitis and its early detection could prevent the onset of tooth loss, especially in local community and health care settings with limited dental professionals. In the medical field, doctors need to understand and trust the decisions made by computational models and developing interpretable machine learning models is crucial for disease diagnosis. Based on these considerations, we propose an interpretable machine learning method called Deetal-Perio to predict the severity degree of periodontitis in dental panoramic radiographs. In our method, alveolar bone loss (ABL), the clinical hallmark for periodontitis diagnosis, could be interpreted as the key feature. To calculate ABL, we also propose a method for teeth numbering and segmentation. First, Deetal-Perio segments and indexes the individual tooth via Mask R-CNN combined with a novel calibration method. Next, Deetal-Perio segments the contour of the alveolar bone and calculates a ratio for individual tooth to represent ABL. Finally, Deetal-Perio predicts the severity degree of periodontitis given the ratios of all the teeth. The entire architecture could not only outperform state-of-the-art methods and show robustness on two data sets in both periodontitis prediction, and teeth numbering and segmentation tasks, but also be interpretable for doctors to understand the reason why Deetal-Perio works so well.

ⓒ Springer Nature Switzerland AG 2020
A. L. Martel et al. (Eds.): MICCAI 2020, LNCS 12262, pp. 454–463, 2020.
https://doi.org/10.1007/978-3-030-59713-9_44

Keywords: Teeth segmentation and numbering · Interpretable machine learning · Periodontitis diagnosis · Panoramic radiograph

1   Introduction

Periodontitis is a chronic inflammatory disease of periodontium resulting in inflammation within the supporting tissues of the teeth, progressive attachment, and bone loss [11]. Periodontitis is prevalent in both developed and developing countries, and affects about 20%–50% of the global population which makes it a public health concern [12]. Thus, the tool for automatically diagnosing periodontitis is highly demanded to provide the invaluable opportunity to screen at-risk people for periodontitis and its early detection could prevent the onset of tooth loss, especially in local community and health care settings where dentists are not easily accessible [1]. The form of periodontitis is characterized by periodontal ligament loss and resorption of the surrounding alveolar bone caused by severe inflammatory events [14]. Cumulative alveolar bone loss (ABL) results in weakening of the supporting structures of the teeth, and predisposes the patient to tooth mobility and loss [2] (Fig. 1B). Thus ABL is a hallmark of the periodontal disease [17]. To calculate ABL, it is necessary to gather the contours of the individual tooth and the alveolar bone. In this situation, teeth numbering and segmentation are essential and fundamental tasks for periodontitis diagnosis. In addition, dentists usually need to serve numerous patients and read a large number of panoramic radiographs daily. Thus an automatic tool for teeth numbering and segmentation to enhance efficiency and improve the quality of dental care is highly needed [4, 18].

Several methods have been proposed to tackle the periodontitis prediction or teeth numbering and segmentation tasks. Joo et al. [8] proposed a classification method for the periodontal disease based on convolutional neural network (CNN) by using periodontal tissue images. This method classified four states of periodontitis and the accuracy on validation data was 0.81. Ozden et al. [13] tested three classification algorithms, artificial neural networks (ANN), support vector machine (SVM), and decision tree (DT) to diagnose periodontal diseases by using 11 measured variables of each patient as raw data. The results showed that DT and SVM were best to classify the periodontal diseases with high accuracy (0.98 of precision). It revealed that SVM and DT appeared to be practical as a decision-making aid for the prediction of periodontal disease.

As for the teeth numbering and segmentation, Wirtz et al. [16] proposed a coupled shape model in conjunction with a neural network by using panoramic radiographs. The network provided a preliminary segmentation of the teeth region which is used to initialize the coupled shape model. Then the 28 individual teeth (excluding wisdom teeth) were segmented and labeled using gradient image features in combination with the model's statistical knowledge. The experimental result showed an average dice of 0.744. Chen et al. [4] used faster regions with CNN features to detect and number teeth in dental periapical films. They proposed three post-processing techniques to improve the numbering performance. Results revealed that mean average precision (mAP) was 0.80 and the performance of this method was close to the level of junior dentists. Cui et al. [6] used deep CNN to achieve automatic and accurate tooth instance segmentation and identification from cone beam CT (CBCT) images. They extracted the edge map from the CBCT image to enhance image contrast along shape boundaries. Next, the edge map and input images were passed through 3D Mask R-CNN with encoded teeth spatial relationships. Their method produced accurate instance segmentation and identification results automatically.

The main limitations of methods mentioned above are as follows: (1) the bias of detecting and numbering teeth in some cases with severe periodontitis due to the disturbance of a large number of missing teeth, (2) the lacking capability of their methods on predicting the number of missing teeth in the shortage data volume of some individual classes, and (3) the lack of interpretability of predicting the severity degree of periodontitis.

In this paper, we try to overcome these limitations through the following contributions. (1) We propose an automatic and interpretable machine learning method called Deetal-Perio to predict the severity degree of periodontitis from dental panoramic radiographs. (2) As a subroutine of Deetal-Perio, we further propose a method for teeth numbering and segmentation which consists of a novel calibration algorithm. (3) Deetal-Perio outperforms state-of-the-art methods and shows the robustness on the two data sets from two hospitals. (4) Deetal-Perio uses ABL as the feature for periodontitis diagnosis and is thus fully interpretable.

2   Materials and Methods

2.1   Data Sets

Suzhou Stomatological Hospital supplied a total of 302 digitized panoramic radiographs (hereinafter referred to as the *Suzhou* dataset). Each radiograph has a high resolution of 1480 pixels by 2776 pixels and was annotated with the consensus taken by three professional dentists following the Fédération Dentaire Internationale (FDI) numbering system to get the contours of teeth and their labels as the ground truth. FDI numbering system divides all teeth into four quadrants where teeth are labeled as 11–18, 21–28, 31–38, 41–48, respectively (Fig. 1A). Among all radiographs, 298 were labeled with the severity degree of periodontitis by dentists, including four categories: 52 of no periodontitis, 189 of mild periodontitis, 43 of moderate periodontitis and 14 of severe periodontitis. These radiographs were randomly divided into two data sets, a training set with 270 images and a test set with 28 images. This random split was repeated for three times and the results were averaged.

To test the robustness and generality of Deetal-Perio, we also collected another dataset from the Sixth Affiliated Hospital, Sun Yat-sen University (hereinafter referred to as the *Zhongshan* dataset). This dataset includes 62 high-resolution panoramic radiographs which are categorized by four classes mentioned above (36 of no periodontitis, 12 of mild periodontitis, 6 of moderate periodontitis and 8 of severe periodontitis).

A

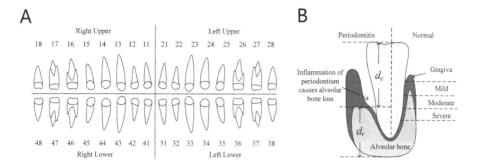

Fig. 1. A. FDI numbering system. It divides all teeth into four quadrants where teeth are labeled as 11–18, 21–28, 31–38, 41–48, respectively. B. The left of tooth shows the appearance of periodontitis and the representation of ABL. The right of tooth shows the appearance of a normal tooth.

2.2 Methods

The architecture of Deetal-Perio is as follows. First, Deetal-Perio segments and numbers individual tooth via Mask R-CNN combined with a novel calibration method. Next, Deetal-Perio segments the contour of the alveolar bone and calculates the ratio for individual tooth which could represent ABL as the key feature to predict periodontitis. Finally, Deetal-Perio uses XGBoost to predict the severity degree of periodontitis by given a vector of ratios from all the numbered teeth. The entire architecture is shown in Fig. 2.

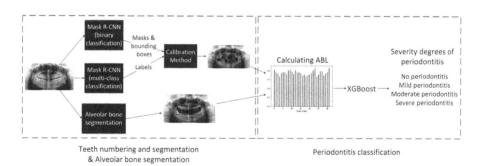

Fig. 2. The workflow of Deetal-Perio.

Teeth Segmentation and Numbering. Inspired by the state-of-the-art architecture in object classification and segmentation called Mask R-CNN [7], we tried to segment the teeth with binary classification via Mask R-CNN. That was, we wanted to differentiate teeth from the background image. The result revealed that Mask R-CNN could detect almost all of the teeth given a radiograph. Then, we tried to segment and number the teeth with multi-class classification via Mask R-CNN and this time, we wanted to identify the labels of these

A   Binary classification Mask R-CNN   **B**   Multi-class classification Mask R-CNN

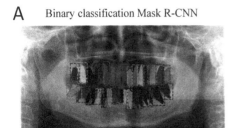 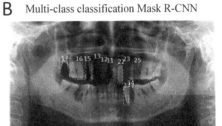

Fig. 3. A. The result of binary classification Mask R-CNN. B. The result of multi-class classification Mask R-CNN.

teeth. The result showed that only a minority of teeth could be detected and numbered due to the limitation of data from individual classes, but compared with the ground truth, most detected teeth were numbered correctly. Figure 3 shows these two results, which provide complementary information to each other. Thus, we proposed to combine the results from the binary and multi-class Mask R-CNN models together. We extracted the bounding boxes and masks from the binary classification results, and the labels of numbered teeth from the multi-class classification results. We further proposed a calibration method to integrate the results from the two classifiers, refine the labels of numbered teeth, and infer the labels of unnumbered teeth.

The calibration method is designed as follows: first, $B = \{B_1, B_2, ..., B_m\}$ and $M = \{M_1, M_2, ..., M_n\}$ represent m and n of center points of teeth's bounding boxes from the results of binary and multi-class classification Mask R-CNN, respectively. Then, we found the closest tooth to each tooth of M in B by calculating the Euclidean distance and assigned the labels of teeth from M to B. Next, we iterated each tooth in B to judge whether neighboring teeth are labeled and calibrate its own label until all the teeth in B satisfied two conditions: each tooth had been labeled and all labeled results followed the FDI numbering system. The details of algorithm are given in Algorithm 1. Finally, all teeth are labeled in B which are considered to be the results of teeth segmentation and numbering step.

The Representation of ABL. After applying the teeth segmentation and numbering method, we obtained the contours of individual teeth and their labels. We also acquired the contour of the alveolar bone by applying Mask R-CNN. Next, we calculated the ratio $d = d_c/(d_c + d_r)$ for each individual tooth, where d_c denotes the vertical distance from the alveolar bone to the top of the dental crown and d_r denotes the vertical distance from the bottom of the dental root to the alveolar bone (Fig. 1B). Due to the smoothness of the contour of the alveolar bone, we randomly selected two points on this contour to draw a line. Then, we randomly chose 50% points on the contour of the dental crown to calculate the vertical distance from these points to this line respectively and defined the largest

Algorithm 1. The calibration method for teeth numbering

Input: $B = \{B_1, ..., B_m\}$, $M = \{M_1, ..., M_n\}$, L_i : the label of tooth i, X_i : the x coordinate of tooth i
Output: calibrated $\{L_{B_1}, ..., L_{B_m}\}$
1: $L_{B_i}, ..., L_{B_m}$ are set to 0
2: **for** $i \leftarrow 1$ to n **do**
3:     $B_k \leftarrow$ the nearest point to M_i by calculating the Euclidean distance
4:     $L_{B_k} \leftarrow L_{M_i}$
5: **end for**
6: **while** exists tooth unlabeled in B or not following FDI numbering system **do**
7:     **for** each unlabeled tooth B_i **do**
8:         **if** $L_{B_{i-1}} = 0$ and $L_{B_{i+1}} = 0$ **then**
9:             continue
10:         **else if** $L_{B_{i-1}} \neq 0$ and $L_{B_{i+1}} \neq 0$ **then**
11:             find nearer tooth B_j to B_i between B_{i-1} and B_{i+1}, the other tooth is B_h
12:             $B_k \leftarrow$ the neighbor tooth of B_j (not B_i)
13:         **else if** only one of B_{i-1} and B_{i+1} is labeled **then**
14:             $B_j \leftarrow$ labeled tooth between B_{i-1} and B_{i+1}
15:             $B_k \leftarrow$ the neighbor tooth of B_j (not B_i)
16:         **end if**
17:         $z \leftarrow round(|X_{B_i} - X_{B_j}|/|X_{B_j} - X_{B_k}|)$
18:         infer L_{Bi} considering $z - 1$ of missing teeth bewteen B_j and B_i
19:     **end for**
20: **end while**

---

vertical distance as d_c. We estimated d_r in a similar way. In our case, d_c could be a good estimation to represent ABL showing the level of destruction of the alveolar bone and in order to normalize d_c, we divided it by $d_c + d_r$ which is the estimation of the perpendicular height of the tooth. Thus, d could represent the ABL to reflect the severity degree of periodontitis.

Periodontitis Prediction. After acquiring the ratio d of individual teeth, each radiograph would output a vector of ratios $D = \{d_1, d_2, ..., d_{32}\}$ where each radio corresponds to a label of tooth. Apparently, some radiographs do not have all the 32 teeth. In such cases, the ratios of teeth which do not exist in the radiograph are set to be the mean value of its neighboring teeth' ratios. We then solved the periodontitis classification task by XGBoost [5] which has gained attention as the algorithm of choice for many winning teams of machine learning competitions [15]. To tackle the class imbalance problem, we used Synthetic Minority Oversampling Technique (SMOTE) [3] for over-sampling the minority classes. After over-sampling, D was inputted as the feature to classify the four-class severity degree of periodontitis by XGBoost.

3   Result

Figure 4 shows two examples of teeth segmentation and numbering results on the *Suzhou* dataset and *Zhongshan* dataset respectively. We first set out to evaluate

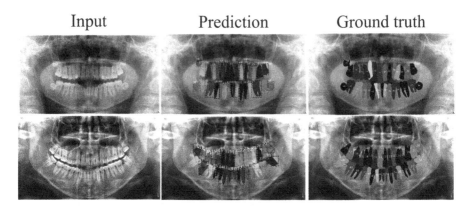

Fig. 4. Two examples from teeth segmentation and numbering results from the *Suzhou* (top) and *Zhongshan* (bottom) datasets respectively.

the performance of teeth numbering and segmentation, by using three metrics: (1) Dice (all) denotes the overall dice coefficient of teeth segmentation. (2) Dice (single) denotes the mean value of all dice coefficients from all labeled teeth respectively. (3) The mAP is calculated as the average of the average precisions for all object classes. We compared with the baseline multi-class classification Mask R-CNN, and the state-of-the-art methods, including the methods proposed by Chen et al. [4] and Wirtz et al. [16] which are trained on the *Suzhou* dataset for fair comparison (Table 1). Thanks to the calibration method in Deetal-Perio, we could number teeth much more correctly than other compared methods. Thus, the performance of Deetal-Perio in the segmentation of individual tooth is also better than other methods. To prove the robustness of Deetal-Perio, we tested our method directly on the *Zhongshan* dataset, without re-training the model, and the mAP, dice (all) and dice (single) reached 0.841, 0.852 and 0.748 respectively, which are close to the results from the *Suzhou* dataset. This cross-set validation demonstrates the robustness and generality of our method.

We then evaluated the performance of periodontitis prediction. Table 2 compares our method with several baseline machine learning methods and the method proposed by Joo et al. [8] which is trained on the *Suzhou* dataset for fair comparison. The metrics include F1-score and mean accuracy over the four classes. Our method has a stably good accuracy over all the four classes: 0.932 for no periodontitis, 0.891 for mild periodontitis, 0.952 for moderate periodontitis and 0.992 for severe periodontitis. We also tested it on the *Zhongshan* dataset directly, without re-training, and the F1-score and mean accuracy reached 0.454 and 0.817 respectively. The boosting algorithm of XGBoost made it a strong leaner to enhance the performance compared with the simple decision trees, and the regularization of XGBoost made it robust against the noise and thus outperforming Adaboost. Compared to Joo et al. [8], our method greatly reduced the feature dimension from 1480×2776 to a 1×32 vector instead of simply implementing a CNN for image classification with a large number of disturbing redundant features. In addition, the intermedia results of teeth segmentation and numbering as well as the geometrically calculated ABL provide dentists completely transparent and interpretable information, so that they not only know that our method works, but also understand how it works.

Table 1. Performance comparison tested on the *Suzhou* dataset between Deetal-Perio and other methods on teeth numbering/segmentation task, by mAP, Dice (all) and Dice (single). The method of Chen, et al. [4] only has tooth number, but not segmentation function.

Methods	mAP	Dice (all)	Dice (single)
Deetal-Perio	**0.826**	**0.868**	**0.778**
Mask R-CNN (multi-class)	0.372	0.496	0.377
Wirtz et al. [16]	0.410	0.763	0.517
Chen et al. [4]	0.195	—	—

Table 2. Performance comparison tested on the *Suzhou* dataset between Deetal-Perio and five methods by F1-score and accuracy on the periodontitis prediction task.

Methods	F1-score	Accuracy
Deetal-Perio	**0.878**	**0.884**
CNN	0.659	0.689
Decision Tree	0.792	0.806
SVM	0.655	0.675
Adaboost	0.652	0.657
Joo et al. [8]	0.310	0.641

4   Discussion and Conclusion

In this paper, we proposed a fully automatic and completely interpretable method, Deetal-Perio, for diagnosing the severity degrees of periodontitis from panoramic radiographs using ABL as the key feature. As the intermediate results, our method also accomplished teeth numbering and segmentation. Comprehensive experiments on two datasets show that Deetal-Perio not only dramatically outperforms other compared methods in both teeth segmentation and numbering, and periodontitis prediction, but is also robust and generalizable on independent datasets, which makes Deetal-Perio a suitable method for periodontitis screening and diagnostics.

Despite the success of Deetal-Perio, the performance of teeth numbering relies on the numbering results from the multi-class classification Mask R-CNN model in Deetal-Perio. This can cause issues when there are radiographs with severe periodontitis which have only few or abnormal shapes of teeth. To overcome this limitation, more data, especially the one with these special cases, need to be obtained to further improve the performance, for which few-shot learning might be a helpful way to deal with such special situations [9,10].

Acknowledgement. We thank He Zhang, Yi Zhang, and Yongwei Tan at Suzhou Stomatological Hospital for providing the data. The research reported in this publication was supported by the King Abdullah University of Science and Technology (KAUST) Office of Sponsored Research (OSR) under Award No. FCC/1/1976-04, FCC/1/1976-06, FCC/1/1976-17, FCC/1/1976-18, FCC/1/1976-23, FCC/1/1976-25, FCC/1/1976-26, and REI/1/0018-01-01.

References

1. Balaei, A.T., de Chazal, P., Eberhard, J., Domnisch, H., Spahr, A., Ruiz, K.: Automatic detection of periodontitis using intra-oral images. In: 2017 39th Annual International Conference of the IEEE Engineering in Medicine and Biology Society (EMBC), pp. 3906–3909 (2017). https://doi.org/10.1109/EMBC.2017.8037710

2. Bhatt, A.A., et al.: Contributors. In: Cappelli, D.P., Mobley, C.C. (eds.) Prevention in Clinical Oral Health Care, pp. v–vi. Mosby, Saint Louis (2008). https://doi.org/10.1016/B978-0-323-03695-5.50001-X, http://www.sciencedirect.com/science/article/pii/B978032303695550001X

3. Chawla, N.V., Bowyer, K.W., Hall, L.O., Kegelmeyer, W.P.: Smote: synthetic minority over-sampling technique. J. Artif. Intell. Res. **16**, 321–357 (2002)

4. Chen, H., et al.: A deep learning approach to automatic teeth detection and numbering based on object detection in dental periapical films. Sci. Rep. **9**(1), 3840 (2019). https://doi.org/10.1038/s41598-019-40414-y

5. Chen, T., Guestrin, C.: Xgboost: a scalable tree boosting system. In: Proceedings of the 22nd ACM Sigkdd International Conference on Knowledge Discovery and Data Mining, pp. 785–794. ACM (2016)

6. Cui, Z., Li, C., Wang, W.: ToothNet: automatic tooth instance segmentation and identification from cone beam CT images. In: Proceedings of the IEEE Conference on Computer Vision and Pattern Recognition, pp. 6368–6377 (2019)

7. He, K., Gkioxari, G., Dollár, P., Girshick, R.: Mask R-CNN. In: Proceedings of the IEEE International Conference on Computer Vision, pp. 2961–2969 (2017)

8. Joo, J., Jeong, S., Jin, H., Lee, U., Yoon, J.Y., Kim, S.C.: Periodontal disease detection using convolutional neural networks. In: 2019 International Conference on Artificial Intelligence in Information and Communication (ICAIIC), pp. 360–362 (2019). https://doi.org/10.1109/ICAIIC.2019.8669021

9. Li, H., et al.: Modern deep learning in bioinformatics. J. Mol. Cell Biol., June 2020. https://doi.org/10.1093/jmcb/mjaa030, mjaa030

10. Li, Y., Huang, C., Ding, L., Li, Z., Pan, Y., Gao, X.: Deep learning in bioinformatics: Introduction, application, and perspective in the big data era. Methods **166**, 4–21 (2019)

11. Lindhe, J., et al.: Consensus report: chronic periodontitis. Ann. Periodontol. **4**(1), 38 (1999)

12. Nazir, M.A.: Prevalence of periodontal disease, its association with systemic diseases and prevention. Int. J. Health Sci. **11**(2), 72–80 (2017). https://www.ncbi.nlm.nih.gov/pubmed/28539867, www.ncbi.nlm.nih.gov/pmc/articles/PMC5426403/

13. Ozden, F.O., Ozgonenel, O., Ozden, B., Aydogdu, A.: Diagnosis of periodontal diseases using different classification algorithms: a preliminary study. Niger. J. Clin. Practice **18**(3), 416–421 (2015)

14. de Pablo, P., Chapple, I.L.C., Buckley, C.D., Dietrich, T.: Periodontitis in systemic rheumatic diseases. Nature Rev. Rheumatol. **5**(4), 218–224 (2009). https://doi.org/10.1038/nrrheum.2009.28

15. Volkovs, M., Yu, G.W., Poutanen, T.: Content-based neighbor models for cold start in recommender systems. Proc. Recommender Syst. Challenge **2017**, 1–6 (2017)

16. Wirtz, A., Mirashi, S.G., Wesarg, S.: Automatic teeth segmentation in panoramic X-Ray images using a coupled shape model in combination with a neural network. In: Frangi, A.F., Schnabel, J.A., Davatzikos, C., Alberola-López, C., Fichtinger, G. (eds.) MICCAI 2018. LNCS, vol. 11073, pp. 712–719. Springer, Cham (2018). https://doi.org/10.1007/978-3-030-00937-3_81

17. Yang, M., Nam, G.E., Salamati, A., Baldwin, M., Deng, M., Liu, Z.J.: Alveolar bone loss and mineralization in the pig with experimental periodontal disease. Heliyon **4**(3), e00589 (2018)

18. Zhou, L., et al.: A rapid, accurate and machine-agnostic segmentation and quantification method for CT-based covid-19 diagnosis. IEEE Trans. Med. Imaging **99**, 1 (2020)

Residual-CycleGAN Based Camera Adaptation for Robust Diabetic Retinopathy Screening

Dalu Yang, Yehui Yang, Tiantian Huang, Binghong Wu, Lei Wang,
and Yanwu Xu$^{(\boxtimes)}$

Intelligent Health Unit, Baidu Inc., Beijing, China
{yangdalu, yangyehui01, huangtiantian01, wubinghong,
wanglei15, xuyanwu}@baidu.com

Abstract. There are extensive researches focusing on automated diabetic retinopathy (DR) detection from fundus images. However, the accuracy drop is observed when applying these models in real-world DR screening, where the fundus camera brands are different from the ones used to capture the training images. How can we train a classification model on labeled fundus images acquired from only one camera brand, yet still achieves good performance on images taken by other brands of cameras? In this paper, we quantitatively verify the impact of fundus camera brands related domain shift on the performance of DR classification models, from an experimental perspective. Further, we propose camera-oriented residual-CycleGAN to mitigate the camera brand difference by domain adaptation and achieve increased classification performance on target camera images. Extensive ablation experiments on both the EyePACS dataset and a private dataset show that the camera brand difference can significantly impact the classification performance and prove that our proposed method can effectively improve the model performance on the target domain. We have inferred and labeled the camera brand for each image in the EyePACS dataset and will publicize the camera brand labels for further research on domain adaptation.

Keywords: Domain adaptation · Diabetic Retinopathy screening · Camera brand

1 Introduction

Diabetic Retinopathy (DR) is a leading cause of blindness worldwide that affects approximately 30% of diabetes mellitus patients [1, 2]. There are extensive researches focusing on the automatic early detection of DR via color fundus photography, and they achieve remarkable classification performance [3–6].

D. Yang and Y. Yang—Equal contributions.

Electronic supplementary material The online version of this chapter (https://doi.org/10.1007/978-3-030-59713-9_45) contains supplementary material, which is available to authorized users.

A. L. Martel et al. (Eds.): MICCAI 2020, LNCS 12262, pp. 464–474, 2020.
https://doi.org/10.1007/978-3-030-59713-9_45

Despite that most of these high-performance results are reproducible with open-sourced code and specific public datasets, the generalizability of these researches is usually poor on fundus images with completely different distributions. As a result, the application of automatic DR detection in real-world DR screening programs is still limited. One of the key differences between experimental and real-world settings is the use of different brands of fundus cameras. It is common that the training dataset does not contain images captured by the same camera model used in the DR screening site.

By collecting extra data or applying extra limitations, the impact of domain shift in real-world settings can be avoided. Gulshan et al. collected fundus images from a large number of hospitals across the US and India, covering most mainstream camera brands and produce a model with good generalizability [3], but the data collection process is extremely costly. Abràmoff et al. took a different approach by restricting the usage of the software on only one single camera model (TRC-NW400, Topcon) [7]. The data distribution is effectively under control, but this approach limits the application of automated DR screening to places that can afford the specific camera model.

On the other hand, without extra data collection or limitations, domain adaptation is the technique that can counter the domain shift and increase model performance on different data distributions. In recent years, deep learning based domain adaptation techniques have been well developed [8, 9], and applied widely on different medical imaging tasks [10–12]. However, domain adaptation is rarely addressed in fundus photography images. Currently [13] is the only work on fundus image domain adaptation, where the authors adapt the optic disc and optic cup segmentation model trained on images from Zeiss Visucam 500 to images from Canon CR-2. However, their experiments are performed on only one source domain and one target domain, thus fail to address the domain shift problem of different camera brands in general.

In this paper, we propose a camera adaptation method to address a general camera adaptation problem: the performance of DR screening models fluctuates across images from various fundus camera brands. The key contributions of this paper are: 1) To our knowledge, we are the first to quantitatively verify the impact of fundus camera brands related domain shift on the performance of deep learning DR classification models, from an experimental perspective. 2) We propose a novel camera-oriented residual-CycleGAN which can mitigate the domain shift and achieve increased classification performance on target camera images, without retraining the base model. We perform extensive ablation experiments on both public datasets and private datasets for comparison of the results. 3) We will publicize our camera brand grouping label of the whole EyePACS dataset to enable further adaptation researches. EyePACS is so far the largest fundus dataset of multiple camera brands. Our camera brand domain label on EyePACS will both help the development of real-world DR screening systems and benefit the researchers in the field of domain adaptation in general.

2   Methods

2.1   Overview

In this paper, we characterize the camera adaptation as an unsupervised domain adaptation problem, under the setting with only one labeled source domain (camera brand) and multiple unlabeled target domains. Formally, we have source domain A with image set X_A and label set Y_A, and a set of target domain image set X_B, X_C, X_D, etc. Our domain adaptation process consists of three steps, as shown in Fig. 1: 1) We first train a classifier M with X_A and Y_A (Fig. 1a); 2) We use X_A and unlabeled X_B, X_C, X_D, etc. to obtain domain transformations G_B, G_C, G_D, etc., respectively (Fig. 1b); 3) We apply the transformations G_B, G_C, G_D, etc. on $x_B \in X_B$, $x_C \in X_C$, $x_D \in X_D$, etc. at test time, and use M to get the final classification results (Fig. 1c).

In general, M has a poor performance when applied directly on the target domain images X_B, X_C, X_D. With the domain transformation, although the classifier M is never trained on the target domain images, it has a better classification performance.

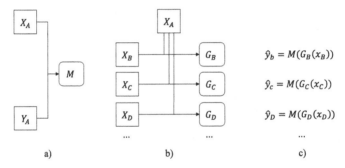

Fig. 1. Overview of the problem and domain adaptation process. X_A: source domain image set; Y_A: source domain label set; M: classifier trained with source domain images and labels; X_B, X_C, X_D: target domain image sets; G_B, G_C, G_D: domain transformation models; $\hat{y}_B$, $\hat{y}_C$, $\hat{y}_D$: predicted labels of target domain images x_B, x_C, x_D.

2.2   Camera Oriented Residual-CycleGAN

The overall network resembles a CycleGAN-like structure, with residual generator and customized discriminator. Figure 2 illustrates our proposed domain transformation network. The major components of the network include: the residual generator F and G, the task specific feature extractor f, and the domain discriminator D_A and D_B. For convenience, we denote the original images from domain A and B as a and b, and the transformed image as a_B and b_A. The residual domain transformation is:

$$\begin{aligned} a_B &= a + F(a), \\ b_A &= b + G(b). \end{aligned} \tag{1}$$

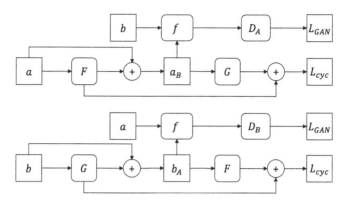

Fig. 2. The framework of residual-CycleGAN. F and G: residual generators applying on domain A and B; a and b: images from domain A and B; a_B and b_A: transformed image a to domain B and image b to domain A; f: camera-oriented feature extractor; D_B and D_A: discriminator for real domain B and A; L_{cyc}: cycle-consistency loss; L_{GAN}: generative-adversarial loss.

Intuitively, the generator F takes image from domain A to a residual transformation, which serves as an additive filter that can be applied to a and result in a fake a_B that has the distribution of domain B. Similarly, the generator G does the same thing for b to transform from domain B to domain A. In the remaining part of this section, we first formulate the domain transformation learning problem, then we present the details of each component in our network.

Formulation of Learning. Following the vanilla CycleGAN formulation, our total objective function includes three components, i.e., the adversarial loss, the cycle-consistency loss, and the identity loss. The description of each loss is as follows:

1. *Adversarial Loss.* The adversarial loss is defined as:

$$
\begin{aligned}
\mathcal{L}_{GAN}(F, D_B) &= \mathbb{E}_{b \in B}[\log D_B(f(b))] + \mathbb{E}_{a \in A}[\log(1 - D_B(f(a_B)))], \\
\mathcal{L}_{GAN}(G, D_A) &= \mathbb{E}_{a \in A}[\log D_A(f(a))] + \mathbb{E}_{b \in B}[\log(1 - D_A(f(b_A)))].
\end{aligned}
\tag{2}
$$

Notice a_B and b_A are the transformed images defined in (1). The adversarial loss regularizes the training through the minimax optimization, i.e., $\min_F \max_{D_B} \mathcal{L}_{GAN}(F, D_B)$ and $\min_G \max_{D_A} \mathcal{L}_{GAN}(G, D_A)$, to ensure that the transformed images are indistinguishable from the other domain.

2. *Cycle-consistency Loss.* Due to the residual nature of our network, the cycle-consistency loss can be expressed as:

$$
\mathcal{L}_{cyc}(F, G) = \mathbb{E}\big[\|F(a) + G(a_B)\|_2\big] + \mathbb{E}\big[\|G(b) + F(b_A)\|_2\big].
\tag{3}
$$

Intuitively, minimizing this cycle-consistency loss encourages the sum of residues added to the original image to be zero, when the original image is transformed from domain A to B and back to A.

3. *Identity Loss.* The identity loss in our network takes the form of:

$$\mathcal{L}_{idt}(F,G) = \mathbb{E}\big[\|F(b)\|_2\big] + \mathbb{E}\big[\|G(a)\|_2\big]. \tag{4}$$

The identity loss further encourages an identity mapping on real target domain image. [14] claims the application of identity loss prevents dramatic change of image hue, which is crucial when dealing with fundus images. We notice that in our experiment, using L2-norm in $\mathcal{L}_{cyc}$ and $\mathcal{L}_{idt}$ yields a faster convergence than using the L1-norm as stated in the original cycleGAN model.

Combining Eq. (2, 3, 4), the overall loss function is:

$$\mathcal{L}(F,G,D_A,D_B) \quad = \mathcal{L}_{GAN}(F,D_B) + \mathcal{L}_{GAN}(G,D_A) \\ + \lambda_1 \mathcal{L}_{idt}(F,G) + \lambda_2 \mathcal{L}_{cyc}(F,G) . \tag{5}$$

And the objective is to solve for F and G:

$$F^*, G^* = \arg\min_{F,G} \max_{D_A,D_B} \mathcal{L}(F,G,D_A,D_B). \tag{6}$$

Residual Generator. The architecture of generator F and G consists of 4 convolution layers with stride-2, 8 sequentially connected residual blocks, and 4 transposed convolutions to restore the original image size. All the (transposed) convolution layers are followed by LeakyReLU and instance normalization.

Discriminator. The discriminator between real and fake images of a specific domain consists of a two layer fully connected neural network. The non-linearity activation function in between is still LeakyReLU. The discriminator takes the task specific features as input, which is detailed below. We notice that during the adversarial training, the weights in the discriminators are updated, while the feature extractors are fixed to ensure that the transfer happens in the desired dimension.

Task Specific Feature Extractor. In our specific problem setting, we find it helpful to use predefined feature extractors, which guided the generator to capture camera brand and DR related information. The three feature extractors are:

1. *Mutual information score between color channels.* According to [14, 15], mutual information between channels can quantitatively capture lateral chromatic aberration information.
2. *Normalized color histogram.* Different camera brands can have different color temperatures of the flashlight or different post-processing or enhancing preferences. We characterize these differences using normalized color histogram.
3. *Deep Features from DR Classifier.* The final set of feature descriptor comes from the classification model M trained by the source domain fundus image. Specifically, it is the feature vector of size $1 \times n$ after the average pooling layer in ResNet models in our experiment.

See supplementary materials for a detailed description of these feature extractors.

3   Experiments and Results

The two purposes of our experiments are: 1) To quantitatively verify the impact of camera-brand-related domain shift on the performance of typical DR classification deep learning models. 2) To prove that our proposed adaptation model can effectively mitigate the domain shift and increase the classification performance of the model on target domain images.

We perform two sets of experiments: 1) We train a ResNet-50 model M with images of source domain camera brand, then directly use M to test images of both source and target domains. 2) We train the camera-oriented residual-CycleGAN for images of a target domain camera. We use the residual generator G to transform the target domain images, and then test M for performance. To further prove the effectiveness of camera-oriented residual-CycleGAN, we perform ablation studies on two SOTA adaptation methods, namely DANN [16] and vanilla CycleGAN [17].

3.1   Datasets

EyePACS Dataset. This publicly available dataset is released by Kaggle's Diabetic Retinopathy Detection Challenge [18]. We identify five camera brands in the dataset. See supplementary materials for a detailed process of camera brand identification and the brand labels for each image. The original split of EyePACS dataset was 35,126 images for training and 53,576 for testing, with five DR severity grades. Our new distribution of DR severity grades and camera brands is listed in Table 1. We notice that the images of camera A used for training comes both the original training set and the public part of the testing set, while images of camera A (for testing), B, C, D, and E come from the private part of the testing set.

Private Dataset. Our private dataset contains 55,326 historical fundus images collected from over 20 hospitals in China. The camera brand information of each image is available. Our collaborative licensed ophthalmologists graded the images for binary classification of positive (has referable DR) vs. negative (has no referable DR). Referable DR is defined as DR with grade II or above according to [19]. The distribution of DR label and domain is summarized in Table 2.

Table 1. Label and domain distribution of EyePacs dataset

Camera brand	Grade 0	Grade 1	Grade 2	Grade 3	Grade 4
A (training)	14601	1301	3237	532	499
A (testing)	13334	1305	2927	455	532
B	6628	540	1445	181	173
C	3755	457	570	71	65
D	3633	411	609	85	72
E	3831	317	709	173	121

Table 2. Label and domain distribution of the private dataset

Camera brand	No. of negative samples	No. of positive samples
I (for training)	26,656	12,534
I (for testing)	3,395	1,512
II	1,282	2,403
III	354	440
IV	380	683
V	912	558

3.2   Implementation Details

We first adjust the black margins (crop the horizontal direction and pad the vertical direction if necessary) to make the image a square. Then we resize the adjusted images to 512×512 resolution and normalize to 0 to 1 as the input of the network.

We use the Adam optimizer with a start learning rate of 1e−4 decaying linearly to 1e−5 over 200 epochs. The parameters λ_1 and λ_2 in the overall loss function are experimentally set to 0.2 and 5, respectively. In general, λ_1 and λ_2 should be set as large as possible, but not to trigger oscillation of loss during adversarial learning. The convergence is determined by visually check a fixed set of transformed images.

We implement the model with paddlepaddle architecture [20, 21]. The training process is relatively computationally intensive. With four NVIDIA Tesla-P40 GPUs, 200 epochs translate to about 80 h for our resplitted eyePACS dataset and 120 h for our private dataset, for each of the five camera brand adaptation tasks. We will publicize the camera brand list of the EyePACS dataset and code for training residual-CycleGAN for easier reproduction of the research.

For experiments performed on the EyePACS dataset (5-class grading task) and on the private dataset (binary classification task), we use quadratic weighted kappa [22] and the area under the receiver operating characteristics curve (AUC) as the evaluation criteria, respectively.

3.3   Quantitative Verification on the Impact of Domain Shift

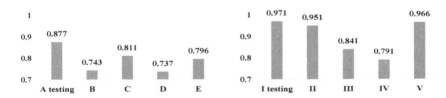

Fig. 3. Left: kappa value of the model trained on images taken by camera brand A and tested on various camera brands. Right: AUC value of the model trained on images taken by camera brand I and tested on various camera brands.

Impact of Domain Shift in EyePACS Dataset. The performance of the model on testing set of different camera brands is shown in the left chart of Fig. 3. The significant performance drop between source domain (camera brand A) and target domains (B, C, D, and E, especially B and D) indicates there is a non-trivial impact from difference in camera brands on the DR classification model.

Impact of Domain Shift in Private Dataset. Similarly, the right chart of Fig. 3 shows the drop between source domain (camera brand I) and target domains (especially III and IV), showing that the model is vulnerable to domain shift impact.

We observe several differences between camera brands that correlate with the performance decrease. 1) Overall sharpness: the sharpness of fundus image usually depends on post-processing settings and pixel geometry of the camera. In DR classification, sharpness of edge is an important discriminative feature between hard exudates and drusen. A shift in overall sharpness between source and target domain may confuse the model trained only on one domain. 2) Color distribution: the difference in color distribution or white balance can easily cause the model to confuse dust with a small hemorrhage, leading to the misclassification of the whole image.

3.4 Camera-Oriented Residual-CycleGAN Mitigates the Domain Shift

Domain Adaptation Performance on EyePACS Dataset: The quantitative results are listed in Table 3. Our method shows best performance in three out of four target domains and exceeds the model without adaptation by a large margin. This result shows that in this task our method can effectively transform the image to mitigate the domain shift and increase the classification performance on target domains.

Domain Adaptation Performance on Private Dataset: The results are summarized in Table 4. Our method shows best performance in two out of four target domains, which is as effective as DANN.

Besides the quantitative comparison of results, some examples of images in the source domain and target domain, with or without transformation are listed in the supplementary materials for a better qualitative impression.

Compared with the vanilla CycleGAN, our camera-oriented residual-CycleGAN performs better in most of the target domains in the two datasets. We notice that the vanilla CycleGAN performs worse than no adaptation on several of the tasks. Unlike vanilla CycleGAN, our residual-CycleGAN has explicit residual implementation that, combined with the cycle-consistency loss, in effect limits the range of the residue and prevents the generator from creating non-existing lesion or structures. Thus, the residual-CycleGAN almost always outperforms the no adaptation.

The performance of our adaptation approach is better than DANN only by a small margin. However, we notice that DANN, as well as ADDA [23] and other domain invariant feature extraction methods, requires retraining the baseline model when perform the adaptation. In real-world practice the "base model" is an ensemble of many models with complex results merging logic. In these situations, our approach has the advantage of easiness in deployment.

Gulshan et al. report remarkable AUCs of 0.991 and 0.990 on EyePACS-1 and Messidor-2 testing sets on referable DR detection task [3]. Their training set includes mixed camera brands, presumably covering the testing set brands. As a comparison, our baseline AUC (where training and testing sets are of one identical camera brand) is 0.971. Part of the gap may be due to model architecture difference (Inception-v3 vs. ResNet-50), or suggest that having mixed brands in the training set could assist the model to find robust features detecting referable DR. Since having multiple brands in the training set is quite common in real-world practice, a natural next step is to evaluate and improve the multi-to-one brand adaptation performance.

Table 3. Quadratic weighted kappa for DR grading, EyePACS dataset

Camera brand	No adaptation	Ours	DANN [14]	CycleGAN [14]
A (testing set)	0.877	–	–	–
B	0.743	0.740	0.764	**0.782**
C	0.811	**0.852**	0.806	0.790
D	0.737	**0.781**	0.695	0.754
E	0.796	**0.804**	0.780	0.729

Table 4. AUC for binary DR classification, private dataset

Camera brand	No adaptation	Ours	DANN [14]	CycleGAN [14]
I (testing set)	0.971	–	–	–
II	0.951	**0.956**	0.939	0.929
III	0.841	**0.882**	0.865	0.864
IV	0.791	0.875	**0.891**	0.804
V	0.966	0.965	**0.970**	0.956

4   Conclusions

In this paper, our quantitative evaluation show that without domain adaptation, the domain shift caused by different fundus camera brands can significantly impact the performance typical deep learning models, when performing DR classification tasks. Further, our proposed camera-oriented residual-CycleGAN can effectively mitigate camera brands domain shift, improving the classification performance of the original model on target domain images. With our proposed method, the adaptation can be performed directly before a trained classification network, enabling easy deployment.

One future step is to extend the method to a many-to-one or many-to-many setting, i.e. the training or testing set contain different camera brands. A major challenge is defining a proper cycle-consistency loss among multiple domain transformations. The computational cost is also a burden against the scaling of our current method, especially where high resolution is required for detection of certain diseases.

References

1. Ruta, L.M., Magliano, D.J., Lemesurier, R., Taylor, H.R., Zimmet, P.Z., Shaw, J.E.: Prevalence of diabetic retinopathy in Type 2 diabetes in developing and developed countries. Diabet. Med. **30**(4), 387–398 (2013)
2. Beagley, J., Guariguata, L., Weil, C., Motala, A.A.: Global estimates of undiagnosed diabetes in adults. Diabetes Res. Clin. Pract. **103**(2), 150–160 (2014)
3. Gulshan, V., et al.: Development and validation of a deep learning algorithm for detection of diabetic retinopathy in retinal fundus photographs. Jama **316**(22), 2402–2410 (2016)
4. Pratt, H., Coenen, F., Broadbent, D.M., Harding, S.P., Zheng, Y.: Convolutional neural networks for diabetic retinopathy. Procedia Comput. Sci. **90**, 200–205 (2016)
5. Gargeya, R., Leng, T.: Automated identification of diabetic retinopathy using deep learning. Ophthalmology **124**(7), 962–969 (2017)
6. Yang, Y., Li, T., Li, W., Wu, H., Fan, W., Zhang, W.: Lesion detection and grading of diabetic retinopathy via two-stages deep convolutional neural networks. In: Descoteaux, M., Maier-Hein, L., Franz, A., Jannin, P., Collins, D.L., Duchesne, S. (eds.) MICCAI 2017. LNCS, vol. 10435, pp. 533–540. Springer, Cham (2017). https://doi.org/10.1007/978-3-319-66179-7_61
7. Abràmoff, M.D., Lavin, P.T., Birch, M., Shah, N., Folk, J.C.: Pivotal trial of an autonomous AI-based diagnostic system for detection of diabetic retinopathy in primary care offices. NPJ Digit. Med. **1**(1), 1–8 (2018)
8. Csurka, G.: Domain adaptation for visual applications: a comprehensive survey. arXiv preprint arXiv:1702.05374 (2017)
9. Wang, M., Deng, W.: Deep visual domain adaptation: a survey. Neurocomputing **312**, 135–153 (2018)
10. Ghafoorian, M., et al.: Transfer learning for domain adaptation in MRI: application in brain lesion segmentation. In: Descoteaux, M., Maier-Hein, L., Franz, A., Jannin, P., Collins, D. L., Duchesne, S. (eds.) MICCAI 2017. LNCS, vol. 10435, pp. 516–524. Springer, Cham (2017). https://doi.org/10.1007/978-3-319-66179-7_59
11. Dong, N., Kampffmeyer, M., Liang, X., Wang, Z., Dai, W., Xing, E.: Unsupervised domain adaptation for automatic estimation of cardiothoracic ratio. In: Frangi, A.F., Schnabel, J.A., Davatzikos, C., Alberola-López, C., Fichtinger, G. (eds.) MICCAI 2018. LNCS, vol. 11071, pp. 544–552. Springer, Cham (2018). https://doi.org/10.1007/978-3-030-00934-2_61
12. Ren, J., Hacihaliloglu, I., Singer, Eric A., Foran, David J., Qi, X.: Adversarial domain adaptation for classification of prostate histopathology whole-slide images. In: Frangi, A.F., Schnabel, J.A., Davatzikos, C., Alberola-López, C., Fichtinger, G. (eds.) MICCAI 2018. LNCS, vol. 11071, pp. 201–209. Springer, Cham (2018). https://doi.org/10.1007/978-3-030-00934-2_23
13. Liu, P., Kong, B., Li, Z., Zhang, S., Fang, R.: CFEA: collaborative feature ensembling adaptation for domain adaptation in unsupervised optic disc and cup segmentation. In: Shen, D., et al. (eds.) MICCAI 2019. LNCS, vol. 11768, pp. 521–529. Springer, Cham (2019). https://doi.org/10.1007/978-3-030-32254-0_58
14. Piva, A.: An overview on image forensics. In: ISRN Signal Processing (2013)
15. Cozzolino, D., Verdoliva, L.: Noiseprint: a CNN-based camera model fingerprint. IEEE Trans. Inf. Forensics Secur. **15**, 144–159 (2019)
16. Ganin, Y., et al.: Domain-adversarial training of neural networks. J. Mach. Learn. Res. **17**(1), 2030–2096 (2016)

17. Zhu, J.Y., Park, T., Isola, P., Efros, A.A.: Unpaired image-to-image translation using cycle-consistent adversarial networks. In: Proceedings of the IEEE International Conference on Computer Vision, pp. 2223–2232 (2017)

18. Kaggle diabetic retinopathy competition (2015). https://www.kaggle.com/c/diabeticreti nopathy-detection

19. Chakrabarti, R., Harper, C.A., Keeffe, J.E.: Diabetic retinopathy management guidelines. Expert Rev. Ophthalmol. **7**(5), 417–439 (2012)

20. https://github.com/PaddlePaddle/Paddle/tree/1.6.2

21. https://github.com/PaddlePaddle/models/blob/develop/PaddleCV/PaddleGAN/network/CycleGAN_network.py

22. Cohen, J.: Weighted kappa: nominal scale agreement provision for scaled disagreement or partial credit. Psychol. Bull. **70**(4), 213 (1968)

23. Tzeng, E., Hoffman, J., Saenko, K., Darrell, T.: Adversarial discriminative domain adaptation. In: Proceedings of the IEEE Conference on Computer Vision and Pattern Recognition, pp. 7167–7176 (2017)

Shape-Aware Meta-learning for Generalizing Prostate MRI Segmentation to Unseen Domains

Quande Liu[1], Qi Dou[1,2(✉)], and Pheng-Ann Heng[1,3]

[1] Department of Computer Science and Engineering,
The Chinese University of Hong Kong, Shatin, Hong Kong SAR, China
{qdliu,qdou,pheng}@cse.cuhk.edu.hk
[2] T Stone Robotics Institute, The Chinese University of Hong Kong,
Shatin, Hong Kong SAR, China
[3] Guangdong Provincial Key Laboratory of Computer Vision and Virtual Reality
Technology, Shenzhen Institutes of Advanced Technology,
Chinese Academy of Sciences, Shenzhen, China

Abstract. Model generalization capacity at domain shift (*e.g.*, various imaging protocols and scanners) is crucial for deep learning methods in real-world clinical deployment. This paper tackles the challenging problem of domain generalization, *i.e.*, learning a model from multi-domain source data such that it can directly generalize to an unseen target domain. We present a novel shape-aware meta-learning scheme to improve the model generalization in prostate MRI segmentation. Our learning scheme roots in the gradient-based meta-learning, by explicitly simulating domain shift with virtual meta-train and meta-test during training. Importantly, considering the deficiencies encountered when applying a segmentation model to unseen domains (i.e., incomplete shape and ambiguous boundary of the prediction masks), we further introduce two complementary loss objectives to enhance the meta-optimization, by particularly encouraging the *shape compactness* and *shape smoothness* of the segmentations under simulated domain shift. We evaluate our method on prostate MRI data from six different institutions with distribution shifts acquired from public datasets. Experimental results show that our approach outperforms many state-of-the-art generalization methods consistently across all six settings of unseen domains (Code and dataset are available at https://github.com/liuquande/SAML).

Keywords: Domain generalization · Meta-learning · Prostate MRI segmentation

1 Introduction

Deep learning methods have shown remarkable achievement in automated medical image segmentation [12, 22, 30]. However, the clinical deployment of existing models still suffer from the performance degradation under the distribution shifts

© Springer Nature Switzerland AG 2020
A. L. Martel et al. (Eds.): MICCAI 2020, LNCS 12262, pp. 475–485, 2020.
https://doi.org/10.1007/978-3-030-59713-9_46

across different clinical sites using various imaging protocols or scanner vendors. Recently, many domain adaptation [5,13] and transfer learning methods [11,14] have been proposed to address this issue, while all of them require images from the target domain (labelled or unlabelled) for model re-training to some extent. In real-world situations, it would be time-consuming even impractical to collect data from each coming new target domain to adapt the model before deployment. Instead, learning a model from multiple source domains in a way such that it can directly generalize to an unseen target domain is of significant practical value. This challenging problem setting is *domain generalization (DG)*, in which no prior knowledge from the unseen target domain is available during training.

Among previous efforts towards the generalization problem [11,21,27], a naive practice of aggregating data from all source domains for training a deep model (called 'DeepAll' method) can already produce decent results serving as a strong baseline. It has also been widely used and validated in existing literature [4,8,26]. On top of DeepAll training, several studies added data augmentation techniques to improve the model generalization capability [24,29], assuming that the domain shift could be simulated by conducting extensive transformations to data of source domains. Performance improvements have been obtained on tasks of cardic [4], prostate [29] and brain [24] MRI image segmentations, yet the choices of augmentation schemes tend to be tedious with task-dependence. Some other approaches have developed new network architectures to handle domain discrepancy [15,25]. Kour *et al.* [15] developed an unsupervised bayesian model to interpret the tissue information prior for the generalization in brain tissue segmentation. A set of approaches [1,23] also tried to learn domain invariant representations with feature space regularization by developing adversarial neural networks. Although achieving promising progress, these methods rely on network designs, which introduces extra parameters thus complicating the pure task model.

Model-agnostic meta-learning [10] is a recently proposed method for fast deep model adaptation, which has been successfully applied to address the domain generalization problem [2,7,17]. The meta-learning strategy is flexible with independence from the base network, as it fully makes use of the gradient descent process. However, existing DG methods mainly tackle image-level classification tasks with natural images, which are not suitable for the image segmentation task that requires pixel-wise dense predictions. An outstanding issue remaining to be explored is how to incorporate the shape-based regularization for the segmentation mask during learning, which is a distinctive point for medical image segmentation. In this regard, we aim to build on the advantages of gradient-based meta-learning, while further integrate shape-relevant characteristics to advance model generalization performance on unseen domains.

We present a novel **shape-aware meta-learning (SAML)** scheme for domain generalization on medical image segmentation. Our method roots in the meta-learning episodic training strategy, to promote robust optimization by simulating the domain shift with meta-train and meta-test sets during model training. Importantly, to address the specific deficiencies encountered when applying a learned segmentation model to unseen domains (i.e., incomplete shape

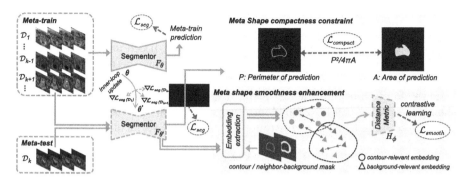

Fig. 1. Overview of our shape-aware meta-learning scheme. The source domains are randomly split into meta-train and meta-test to simulate the domain shift (Sect. 2.1). In meta-optimization: (1) we constrain the shape compactness in meta-test to encourage segmentations with complete shape (Sect. 2.2); (2) we promote the intra-class cohesion and inter-class separation between the contour and background embeddings regardless of domains, to enhance domain-invariance for robust boundary delineation (Sect. 2.3).

and ambiguous boundary of the predictions), we further propose two complementary shape-aware loss functions to regularize the meta optimization process. First, we regularize the *shape compactness* of predictions for meta-test data, enforcing the model to well preserve the complete shape of segmentation masks in unseen domains. Second, we enhance the *shape smoothness* at boundary under domain shift, for which we design a novel objective to encourage domain-invariant contour embeddings in the latent space. We have extensively evaluated our method with the application of prostate MRI segmentation, using public data acquired from six different institutions with various imaging scanners and protocols. Experimental results validate that our approach outperforms many state-of-the-art methods on the challenging problem of domain generalization, as well as achieving consistent improvements for the prostate segmentation performance across all the six settings of unseen domains.

2   Method

Let $(\mathcal{X}, \mathcal{Y})$ denote the joint input and label space in an segmentation task, $\mathcal{D} = \{\mathcal{D}_1, \mathcal{D}_2, ..., \mathcal{D}_K\}$ be the set of K source domains. Each domain $\mathcal{D}_k$ contains image-label pairs $\{(x_n^{(k)}, y_n^{(k)})\}_{n=1}^{N_k}$ sampled from domain distributions $(\mathcal{X}_k, \mathcal{Y})$, where N_k is the number of samples in the k-th domain. Our goal is to learn a segmentation model $F_\theta : \mathcal{X} \rightarrow \mathcal{Y}$ using all source domains $\mathcal{D}$ in a way such that it generalizes well to an unseen target domain $\mathcal{D}_{tg}$. Figure 1 gives an overview of our proposed shape-aware meta-learning scheme, which we will detail in this section.

2.1 Gradient-Based Meta-Learning Scheme

The foundation of our learning scheme is the gradient-based meta-learning algorithm [17], to promote robust optimization by simulating the real-world domain shifts in the training process. Specifically, at each iteration, the source domains $\mathcal{D}$ are randomly split into the meta-train $\mathcal{D}_{tr}$ and meta-test $\mathcal{D}_{te}$ sets of domains. The meta-learning can be divided into two steps. First, the model parameters θ are updated on data from meta-train $\mathcal{D}_{tr}$, using Dice segmentation loss $\mathcal{L}_{seg}$:

$$\theta' = \theta - \alpha \nabla_\theta \mathcal{L}_{seg}(\mathcal{D}_{tr}; \theta), \tag{1}$$

where α is the learning-rate for this inner-loop update. Second, we apply a meta-learning step, aiming to enforce the learning on meta-train $\mathcal{D}_{tr}$ to further exhibit certain properties that we desire on unseen meta-test $\mathcal{D}_{te}$. Crucially, the meta-objective $\mathcal{L}_{meta}$ to quantify these properties is computed with the updated parameters θ', but optimized towards the original parameters θ. Intuitively, besides learning the segmentation task on meta-train $\mathcal{D}_{tr}$, such a training scheme further learns how to generalize at the simulated domain shift across meta-train $\mathcal{D}_{tr}$ and meta-test $\mathcal{D}_{te}$. In other words, the model is optimized such that the parameter updates learned on virtual source domains $\mathcal{D}_{tr}$ also improve the performance on the virtual target domains $\mathcal{D}_{te}$, regarding certain aspects in $\mathcal{L}_{meta}$.

In segmentation problems, we expect the model to well preserve the complete shape (compactness) and smooth boundary (smoothness) of the segmentations in unseen target domains. To achieve this, apart from the traditional segmentation loss $\mathcal{L}_{seg}$, we further introduce two complementary loss terms into our meta-objective, $\mathcal{L}_{meta} = \mathcal{L}_{seg} + \lambda_1 \mathcal{L}_{compact} + \lambda_2 \mathcal{L}_{smooth}$ (λ_1 and λ_2 are the weighting trade-offs), to explicitly impose the shape compactness and shape smoothness of the segmentation maps under domain shift for improving generalization performance.

2.2 Meta Shape Compactness Constraint

Traditional segmentation loss functions, *e.g.*, Dice loss and cross entropy loss, typically evaluate the pixel-wise accuracy, without a global constraint to the segmentation shape. Trained in that way, the model often fails to produce complete segmentations under distribution shift. Previous study have demonstrated that for the compact objects, constraining the shape compactness [9] is helpful to promote segmentations for complete shape, as an incomplete segmentation with irregular shape often corresponds to a worse compactness property.

Based on the observation that the prostate region generally presents a compact shape, and such shape prior is independent of observed domains, we propose to explicitly incorporate the compact shape constraint in the meta-objective $\mathcal{L}_{meta}$, for encouraging the segmentations to well preserve the shape completeness under domain shift. Specifically, we adopt the well-established Iso-Perimetric Quotient [19] measurement to quantify the shape compactness, whose definition is $C_{IPQ} = 4\pi A/P^2$, where P and A are the perimeter and area of the shape, respectively. In our case, we define the shape compactness loss as the reciprocal form of this C_{IPQ} metric, and expend it in a pixel-wise manner as follows:

$$\mathcal{L}_{compact} = \frac{P^2}{4\pi A} = \frac{\sum_{i \in \Omega} \sqrt{(\nabla p_{u_i})^2 + (\nabla p_{v_i})^2 + \epsilon}}{4\pi (\sum_{i \in \Omega} |p_i| + \epsilon)}, \tag{2}$$

where p is the prediction probability map, Ω is the set of all pixels in the map; ∇p_{u_i} and ∇p_{v_i} are the probability gradients for each pixel i in direction of horizontal and vertical; ϵ ($1e^{-6}$ in our model) is a hyperparameter for computation stability. Overall, the perimeter length P is the sum of gradient magnitude over all pixels $i \in \Omega$; the area A is calculated as the sum of absolute value of map p.

Intuitively, minimizing this objective function encourages segmentation maps with complete shape, because an incomplete segmentation with irregular shape often presents a relatively smaller area A and relatively larger length P, leading to a higher loss value of $\mathcal{L}_{compact}$. Also note that we only impose $\mathcal{L}_{compact}$ in meta-test $\mathcal{D}_{te}$, as we expect the model to preserve the complete shape on unseen target images, rather than overfit the source data.

2.3 Meta Shape Smoothness Enhancement

In addition to promoting the complete segmentation shape, we further encourage smooth boundary delineation in unseen domains, by regularizing the model to capture domain-invariant contour-relevant and background-relevant embeddings that cluster regardless of domains. This is crucial, given the observation that performance drop at the cross-domain deployment mainly comes from the ambiguous boundary regions. In this regard, we propose a novel objective $\mathcal{L}_{smooth}$ to enhance the boundary delineation, by explicitly promoting the intra-class cohesion and inter-class separation between the contour-relevant and background-relevant embeddings drawn from each sample across all domains $\mathcal{D}$.

Specifically, given an image $x_m \in \mathbb{R}^{H \times W \times 3}$ and its one-hot label y_m, we denote its activation map from layer l as $M_m^l \in \mathbb{R}^{H_l \times W_l \times C_l}$, and we interpolate M_m^l into $T_m^l \in \mathbb{R}^{H \times W \times C_l}$ using bilinear interpolation to keep consistency with the dimensions of y_m. To extract the contour-relevant embeddings $E_m^{con} \in \mathbb{R}^{C_l}$ and background-relevant embeddings $E_m^{bg} \in \mathbb{R}^{C_l}$, we first obtain the binary contour mask $c_m \in \mathbb{R}^{H \times W \times 1}$ and binary background mask $b_m \in \mathbb{R}^{H \times W \times 1}$ from y_m using morphological operation. Note that the mask b_m only samples background pixels around the boundary, since we expect to enhance the discriminativeness for pixels around boundary region. Then, the embeddings E_m^{con} and E_m^{bg} can be extracted from T_m^l by conducting weighted average operation over c_m and b_m:

$$E_m^{con} = \frac{\sum_{i \in \Omega}(T_m^l)_i \cdot (c_m)_i}{\sum_{i \in \Omega}(c_m)_i}, \quad E_m^{bg} = \frac{\sum_{i \in \Omega}(T_m^l)_i \cdot (b_m)_i}{\sum_{i \in \Omega}(b_m)_i}, \tag{3}$$

where Ω denotes the set of all pixels in T_m^l, the E_m^{con} and E_m^{bg} are single vectors, representing the contour and background-relevant representations extracted from the whole image x_m. In our implementation, activations from the last two deconvolutional layers are interpolated and concatenated to obtain the embeddings.

Next, we enhance the domain-invariance of E^{con} and E^{bg} in latent space, by encouraging embeddings' intra-class cohesion and inter-class separation among samples from all source domains $\mathcal{D}$. Considering that imposing such regularization directly onto the network embeddings might be too strict to impede the convergence of $\mathcal{L}_{seg}$ and $\mathcal{L}_{compact}$, we adopt the contrastive learning [6] to achieve this constraint. Specifically, an embedding network H_ϕ is introduced to project the features $E \in [E^{con}, E^{bg}]$ to a lower-dimensional space, then the distance is computed on the obtained feature vectors from network H_ϕ as $d_\phi(E_m, E_n) = \|H_\phi(E_m) - H_\phi(E_n)\|_2$, where the sample pair (m, n) are randomly drawn from all domains $\mathcal{D}$, as we expect to harmonize the embeddings space of $\mathcal{D}_{te}$ and $\mathcal{D}_{tr}$ to capture domain-invariant representations around the boundary region. Therefore in our model, the contrastive loss is defined as follows:

$$\ell_{contrastive}(m, n) = \begin{cases} d_\phi(E_m, E_n), & \text{if } \tau(E_m) = \tau(E_n) \\ (max\{0, \zeta - d_\phi(E_m, E_n)\})^2, & \text{if } \tau(E_m) \neq \tau(E_n) \end{cases}. \quad (4)$$

where the function $\tau(E)$ indicates the class (1 for E being E^{con}, and 0 for E^{bg}) ζ is a pre-defined distance margin following the practice of metric learning (set as 10 in our model). The final objective $\mathcal{L}_{smooth}$ is computed within mini-batch of q samples. We randomly employ either E^{con} or E^{bg} for each sample, and the $\mathcal{L}_{smooth}$ is the average of $\ell_{contrastive}$ over all pairs of (m, n) embeddings:

$$\mathcal{L}_{smooth} = \sum_{m=1}^{q} \sum_{n=m+1}^{q} \ell_{contrastive}(m, n)/C(q, 2). \quad (5)$$

where $C(q, 2)$ is the number of combinations. Overall, all training objectives including $\mathcal{L}_{seg}(\mathcal{D}_{tr}; \theta)$ and $\mathcal{L}_{meta}(\mathcal{D}_{tr}, \mathcal{D}_{te}; \theta')$, are optimized together with respect to the original parameters θ. The $\mathcal{L}_{smooth}$ is also optimized with respect to H_ϕ.

3   Experiments

Dataset and Evaluation Metric. We employ prostate T2-weighted MRI from 6 different data sources with distribution shift (cf. Table 1 for summary of their sample numbers and scanning protocols). Among these data, samples of Site A, B are from NCI-ISBI13 dataset [3]; samples of Site C are from I2CVB dataset [16]; samples of Site D ,E, F are from PROMISE12 dataset [20]. Note that the NCI-ISBI13 and PROMISE12 actually include multiple data sources, hence we decompose them in our work. For pre-processing, we resized each sample to 384×384 in axial plane, and normalized it to zero mean and unit variance. We then clip each sample to only preserve slices of prostate region for consistent objective segmentation regions across sites. We adopt Dice score (Dice) and Average Surface Distance (ASD) as the evaluation metric.

Table 1. Details of our employed six different sites obtained from public datasets.

Dataset	Institution	Case num	Field strength (T)	Resolution (in/through plane) (mm)	Endorectal coil	Manufactor
Site A	RUNMC	30	3	0.6-0.625/3.6–4	Surface	Siemens
Site B	BMC	30	1.5	0.4/3	Endorectal	Philips
Site C	HCRUDB	19	3	0.67–0.79/1.25	No	Siemens
Site D	UCL	13	1.5 and 3	0.325–0.625/3–3.6	No	Siemens
Site E	BIDMC	12	3	0.25/2.2–3	Endorectal	GE
Site F	HK	12	1.5	0.625/3.6	Endorectal	Siemens

Implementation Details. We implement an adapted Mix-residual-UNet [28] as segmentation backbone. Due to the large variance on slice thickness among different sites, we employ the 2D architecture. The domains number of meta-train and meta-test were set as 2 and 1. The weights λ_1 and λ_2 were set as 1.0 and $5e^{-3}$. The embedding network H_ϕ composes of two fully connected layers with output sizes of 48 and 32. The segmentation network F_θ was trained using Adam optimizer and the learning rates for inner-loop update and meta optimization were both set as $1e^{-4}$. The network H_ϕ was also trained using Adam optimizer with learning rate of $1e^{-4}$. We trained 20 K iterations with batch size of 5 for each source domain. For batch normalization layer, we use the statistics of testing data for feature normalization during inference for better generalization performance.

Comparison with State-of-the-Art Generalization Methods. We implemented several state-of-the-art generalization methods for comparison, including a data-augmentation based method (BigAug) [29], a classifier regularization based method (Epi-FCR) [18], a latent space regularization method (LatReg) [1] and a meta-learning based method (MASF) [7]. In addition, we conducted experiments with 'DeepAll' baseline (i.e., aggregating data from all source domains for training a deep model) and 'Intra-site' setting (i.e., training and testing on the same domain, with some outlier cases excluded to provide general internal performance on each site). Following previous practice [7] for domain generalization, we adopt the leave-one-domain-out strategy, i.e., training on K-1 domains and testing on the one left-out unseen target domain.

As listed in Table 2, DeepAll presents a strong performance, while the Epi-FCR with classifier regularization shows limited advantage over this baseline. The other approaches of LatReg, BigAug and MASF are more significantly better than DeepAll, with the meta-learning based method yielding the best results among them in our experiments. Notably, our approach (cf. the last row) achieves higher performance over all these state-of-the-art methods across all the six sites, and outperforms the DeepAll model by 2.15% on Dice and $0.60mm$ on ASD, demonstrating the capability of our shape-aware meta-learning scheme to deal

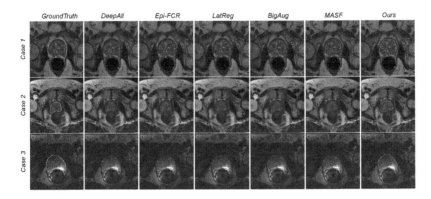

Fig. 2. Qualitative comparison on the generalization results of different methods, with three cases respectively drawn from different unseen domains.

Table 2. Generalization performance of various methods on Dice (%) and ASD (*mm*).

Method	Site A		Site B		Site C		Site D		Site E		Site F		Average	
Intra-site	89.27	1.41	<u>88.17</u>	1.35	<u>88.29</u>	1.56	83.23	3.21	83.67	2.93	85.43	1.91	86.34	2.06
DeepAll (baseline)	87.87	2.05	85.37	1.82	82.94	2.97	86.87	2.25	84.48	2.18	85.58	1.82	85.52	2.18
Epi-FCR [18]	88.35	1.97	85.83	1.73	82.56	2.99	86.97	2.05	85.03	1.89	85.66	1.76	85.74	2.07
LatReg [1]	88.17	1.95	86.65	1.53	83.37	2.91	87.27	2.12	84.68	1.93	86.28	1.65	86.07	2.01
BigAug [29]	88.62	1.70	86.22	1.56	83.76	2.72	87.35	1.98	85.53	1.90	85.83	1.75	86.21	1.93
MASF [7]	88.70	1.69	86.20	1.54	84.16	2.39	87.43	1.91	86.18	1.85	86.57	1.47	86.55	1.81
Plain meta-learning	88.55	1.87	85.92	1.61	83.60	2.52	87.52	1.86	85.39	1.89	86.49	1.63	86.24	1.90
+ $\mathcal{L}_{compact}$	89.08	1.61	87.11	1.49	84.02	2.47	87.96	1.64	86.23	1.80	87.19	1.32	86.93	1.72
+ $\mathcal{L}_{smooth}$	89.25	1.64	87.14	1.53	**84.69**	2.17	87.79	1.88	86.00	1.82	87.74	1.24	87.10	1.71
SAML (**Ours**)	**89.66**	**1.38**	**87.53**	**1.46**	84.43	2.07	**88.67**	**1.56**	**87.37**	**1.77**	**88.34**	**1.22**	**87.67**	**1.58**

with domain generalization problem. Moreover, Fig. 2 shows the generalization segmentation results of different methods on three typical cases from different unseen sites. We observe that our model with shape-relevant meta regularizers can well preserve the complete shape and smooth boundary for the segmentation in unseen domains, whereas other methods sometimes failed to do so.

We also report in Table 2 the cross-validation results conducted within each site, i.e., Intra-site. Interestingly, we find that this result for site D/E/F is relatively lower than the other sites, and even worse than the baseline model. The reason would be that the sample numbers of these three sites are fewer than the others, consequently intra-site training is ineffective with limited generalization capability. This observation reveals the important fact that, when a certain site suffers from severe data scarcity for model training, aggregating data from other sites (even with distribution shift) can be very helpful to obtain a qualified model. In addition, we also find that our method outperforms the Intra-site model in 4 out of 6 data sites, with superior overall performances on both Dice and ASD, which endorses the potential value of our approach in clinical practice.

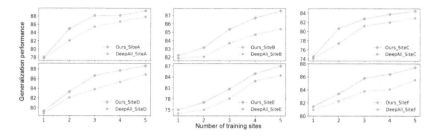

Fig. 3. Curves of generalization performance on unseen domain as the number of training source domain increases, using DeepAll method and our proposed approach.

Ablation Analysis. We first study the contribution of each key component in our model. As shown in Table 2, the plain meta-learning method only with $\mathcal{L}_{seg}$ can already outperform the DeepAll baseline, leveraging the explicit simulation of domain shift for training. Adding shape compactness constraint into $\mathcal{L}_{meta}$ yields improved Dice and ASD which are higher than MASF. Further incorporating L_{smooth} (SAML) to encourage domain-invariant embeddings for pixels around the boundary, consistent performance improvements on all six sites are attained. Besides, simply constraining L_{smooth} on pure meta-learning method ($+ L_{smooth}$) also leads to improvements across sites.

We further investigate the influence of training domain numbers on the generalization performance of our approach and the DeepAll model. Figure 3 illustrates how the segmentation performance on each unseen domain would change, as we gradually increase the number of source domains in range $[1, K-1]$. Obviously, when a model is trained just with a single source domain, directly applying it to target domain receives unsatisfactory results. The generalization performance progresses as the training site number increases, indicating that aggregating wider data sources helps to cover a more comprehensive distribution. Notably, our approach consistently outperforms DeepAll across all numbers of training sites, confirming the stable efficacy of our proposed learning scheme.

4   Conclusion

We present a novel shape-aware meta-learning scheme to improve the model generalization in prostate MRI segmentation. On top of the meta-learning strategy, we introduce two complementary objectives to enhance the segmentation outputs on unseen domain by imposing the shape compactness and smoothness in meta-optimization. Extensive experiments demonstrate the effectiveness. To our best knowledge, this is the first work incorporating shape constraints with meta-learning for domain generalization in medical image segmentation. Our method can be extended to various segmentation scenarios that suffer from domain shift.

Acknowledgement. This work was supported in parts by the following grants: Key-Area Research and Development Program of Guangdong Province, China (2020B010165004), Hong Kong Innovation and Technology Fund (Project No. ITS/426/17FP), Hong Kong RGC TRS Project T42-409/18-R, and National Natural Science Foundation of China with Project No. U1813204.

References

1. Aslani, S., Murino, V., Dayan, M., Tam, R., Sona, D., Hamarneh, G.: Scanner invariant multiple sclerosis lesion segmentation from MRI. In: ISBI, pp. 781–785. IEEE (2020)
2. Balaji, Y., Sankaranarayanan, S., Chellappa, R.: Metareg: towards domain generalization using meta-regularization. In: NeurIPS, pp. 998–1008 (2018)
3. Bloch, N., et al.: NCI-ISBI 2013 challenge: automated segmentation of prostate structures. The Cancer Imaging Archive 370 (2015)
4. Chen, C., et al.: Improving the generalizability of convolutional neural network-based segmentation on CMR images. arXiv preprint arXiv:1907.01268 (2019)
5. Chen, C., Dou, Q., Chen, H., Qin, J., Heng, P.A.: Unsupervised bidirectional cross-modality adaptation via deeply synergistic image and feature alignment for medical image segmentation. IEEE TMI (2020)
6. Chen, T., Kornblith, S., Norouzi, M., Hinton, G.: A simple framework for contrastive learning of visual representations. arXiv preprint arXiv:2002.05709 (2020)
7. Dou, Q., de Castro, D.C., Kamnitsas, K., Glocker, B.: Domain generalization via model-agnostic learning of semantic features. In: NeurIPS. pp. 6450–6461 (2019)
8. Dou, Q., Liu, Q., Heng, P.A., Glocker, B.: Unpaired multi-modal segmentation via knowledge distillation. IEEE TMI (2020)
9. Fan, R., Jin, X., Wang, C.C.: Multiregion segmentation based on compact shape prior. TASE **12**(3), 1047–1058 (2014)
10. Finn, C., Abbeel, P., Levine, S.: Model-agnostic meta-learning for fast adaptation of deep networks. ICML (2017)
11. Gibson, E., et al.: Inter-site variability in prostate segmentation accuracy using deep learning. In: Frangi, A.F., Schnabel, J.A., Davatzikos, C., Alberola-López, C., Fichtinger, G. (eds.) MICCAI 2018. LNCS, vol. 11073, pp. 506–514. Springer, Cham (2018). https://doi.org/10.1007/978-3-030-00937-3_58
12. Jia, H., Song, Y., Huang, H., Cai, W., Xia, Y.: HD-Net: hybrid discriminative network for prostate segmentation in MR images. In: Shen, D., et al. (eds.) MICCAI 2019. LNCS, vol. 11765, pp. 110–118. Springer, Cham (2019). https://doi.org/10.1007/978-3-030-32245-8_13
13. Kamnitsas, K., et al.: Unsupervised domain adaptation in brain lesion segmentation with adversarial networks. In: Niethammer, M., et al. (eds.) IPMI 2017. LNCS, vol. 10265, pp. 597–609. Springer, Cham (2017). https://doi.org/10.1007/978-3-319-59050-9_47
14. Karani, N., Chaitanya, K., Baumgartner, C., Konukoglu, E.: A lifelong learning approach to brain MR segmentation across scanners and protocols. In: Frangi, A.F., Schnabel, J.A., Davatzikos, C., Alberola-López, C., Fichtinger, G. (eds.) MICCAI 2018. LNCS, vol. 11070, pp. 476–484. Springer, Cham (2018). https://doi.org/10.1007/978-3-030-00928-1_54

15. Kouw, W.M., Ørting, S.N., Petersen, J., Pedersen, K.S., de Bruijne, M.: A cross-center smoothness prior for variational Bayesian brain tissue segmentation. In: Chung, A.C.S., Gee, J.C., Yushkevich, P.A., Bao, S. (eds.) IPMI 2019. LNCS, vol. 11492, pp. 360–371. Springer, Cham (2019). https://doi.org/10.1007/978-3-030-20351-1_27

16. Lemaître, G., Martí, R., Freixenet, J., Vilanova, J.C., Walker, P.M., Meriaudeau, F.: Computer-aided detection and diagnosis for prostate cancer based on mono and multi-parametric MRI: a review. CBM **60**, 8–31 (2015)

17. Li, D., Yang, Y., Song, Y.Z., Hospedales, T.M.: Learning to generalize: Meta-learning for domain generalization. In: AAAI (2018)

18. Li, D., Zhang, J., Yang, Y., Liu, C., Song, Y.Z., Hospedales, T.M.: Episodic training for domain generalization. In: ICCV, pp. 1446–1455 (2019)

19. Li, W., Goodchild, M.F., Church, R.: An efficient measure of compactness for two-dimensional shapes and its application in regionalization problems. IJGIS **27**(6), 1227–1250 (2013)

20. Litjens, G., et al.: Evaluation of prostate segmentation algorithms for MRI: the promise12 challenge. MIA **18**(2), 359–373 (2014)

21. Liu, Q., Dou, Q., Yu, L., Heng, P.A.: MS-Net: multi-site network for improving prostate segmentation with heterogeneous MRI data. IEEE TMI (2020)

22. Milletari, F., Navab, N., Ahmadi, S.A.: V-Net: fully convolutional neural networks for volumetric medical image segmentation. In: 3DV, pp. 565–571. IEEE (2016)

23. Otálora, S., Atzori, M., Andrearczyk, V., Khan, A., Müller, H.: Staining invariant features for improving generalization of deep convolutional neural networks in computational pathology. Front. Bioeng. Biotechnol. **7**, 198 (2019)

24. Paschali, M., Conjeti, S., Navarro, F., Navab, N.: Generalizability *vs.* robustness: investigating medical imaging networks using adversarial examples. In: Frangi, A.F., Schnabel, J.A., Davatzikos, C., Alberola-López, C., Fichtinger, G. (eds.) MICCAI 2018. LNCS, vol. 11070, pp. 493–501. Springer, Cham (2018). https://doi.org/10.1007/978-3-030-00928-1_56

25. Yang, X., et al.: Generalizing deep models for ultrasound image segmentation. In: Frangi, A.F., Schnabel, J.A., Davatzikos, C., Alberola-López, C., Fichtinger, G. (eds.) MICCAI 2018. LNCS, vol. 11073, pp. 497–505. Springer, Cham (2018). https://doi.org/10.1007/978-3-030-00937-3_57

26. Yao, L., Prosky, J., Covington, B., Lyman, K.: A strong baseline for domain adaptation and generalization in medical imaging. MIDL (2019)

27. Yoon, C., Hamarneh, G., Garbi, R.: Generalizable Feature learning in the presence of data bias and domain class imbalance with application to skin lesion classification. In: Shen, D., et al. (eds.) MICCAI 2019. LNCS, vol. 11767, pp. 365–373. Springer, Cham (2019). https://doi.org/10.1007/978-3-030-32251-9_40

28. Yu, L., Yang, X., Chen, H., Qin, J., Heng, P.A.: Volumetric convnets with mixed residual connections for automated prostate segmentation from 3d MR images. In: AAAI (2017)

29. Zhang, L., et al.: Generalizing deep learning for medical image segmentation to unseen domains via deep stacked transformation. In: IEEE TMI (2020)

30. Zhu, Q., Du, B., Yan, P.: Boundary-weighted domain adaptive neural network for prostate mr image segmentation. IEEE TMI **39**(3), 753–763 (2019)

Automatic Plane Adjustment of Orthopedic Intraoperative Flat Panel Detector CT-Volumes

Celia Martín Vicario[1,2], Florian Kordon[1,2,3] ⓘ, Felix Denzinger[1,2], Markus Weiten[2], Sarina Thomas[4], Lisa Kausch[4], Jochen Franke[5], Holger Keil[5], Andreas Maier[1,3] ⓘ, and Holger Kunze[2(✉)] ⓘ

[1] Pattern Recognition Lab, Friedrich-Alexander-Universität Erlangen-Nürnberg (FAU), Erlangen, Germany
[2] Siemens Healthcare GmbH, Forchheim, Germany
Holger.HK.Kunze@siemens-healthineers.com
[3] Erlangen Graduate School in Advanced Optical Technologies (SAOT), Friedrich-Alexander-Universität Erlangen-Nürnberg (FAU), Erlangen, Germany
[4] Division of Medical Image Computing, German Cancer Research Center, Heidelberg, Germany
[5] Department for Trauma and Orthopaedic Surgery, BG Trauma Center Ludwigshafen, Ludwigshafen, Germany

Abstract. Flat panel computed tomography is used intraoperatively to assess the result of surgery. Due to workflow issues, the acquisition typically cannot be carried out in such a way that the axis aligned multiplanar reconstructions (MPR) of the volume match the anatomically aligned MPRs. This needs to be performed manually, adding additional effort during viewing the datasets. A PoseNet convolutional neural network (CNN) is trained such that parameters of anatomically aligned MPR planes are regressed. Different mathematical approaches to describe plane rotation are compared, as well as a cost function is optimized to incorporate orientation constraints. The CNN is evaluated on two anatomical regions. For one of these regions, one plane is not orthogonal to the other two planes. The plane's normal can be estimated with a median accuracy of 5°, the in-plane rotation with an accuracy of 6°, and the position with an accuracy of 6 mm. Compared to state-of-the-art algorithms the labeling effort for this method is much lower as no segmentation is required. The computation time during inference is less than 0.05 s.

Keywords: Orthopedics · Flat planel CT · Multiplanar reconstruction

The authors gratefully acknowledge funding of the Erlangen Graduate School in Advanced Optical Technologies (SAOT) by the Bavarian State Ministry for Science and Art.

A. L. Martel et al. (Eds.): MICCAI 2020, LNCS 12262, pp. 486–495, 2020.
https://doi.org/10.1007/978-3-030-59713-9_47

1   Introduction

Intraoperative 3D acquisition is an important tool in trauma surgery for assessing the fracture reduction and implant position during a surgery and the result before releasing the patient out of the operating room [1]. While X-ray images help to assess the result of surgeries in standard cases, in complex anatomical regions like the calcaneus or ankle, 3D imaging provides a mean to resolve ambiguities. Recent studies have shown that, depending on the type of surgery, intraoperative correction rates are up to 40% when such systems are available [2]. If such a tool is not available, postoperative computed tomography is recommended. If a result is observed that should be revised (e.g. an intraarticular screw misplacement) a second surgery is necessary. Therefore, intraoperative 3D scans help to improve the outcome of the surgery and to avoid revision surgeries.

   To be able to understand the patient's anatomy, the volume should be oriented in a standardized way, as is customary in the volumes provided by the radiology department. For those acquisitions, the patient is typically carefully positioned, so that for example the axial slices of the computed tomography (CT) are aligned to the bone of interest. In the axis aligned multiplanar reconstructions (MPR), steps and gaps especially in intra-articular spaces can be analyzed without any further reformation. Also, with these carefully aligned slices, malpositioned screws can be diagnosed easily.

   However, intraoperative 3D acquisitions are performed with the patient lying on the table in the position that the surgery requires. Moreover, in some cases the imaging system cannot be aligned to the patient due to mechanical restrictions and the setup of the operating theater. Consequently, the scan often results in volumes which are not aligned to the patient anatomy. To obtain the correct presentation of the volume, it is essential to correct the rotation of the volume [2]. For the calcaneus region, the surgeon needs an average of 46 to 55 s for manual adjustment of the standard planes depending on his experience [3]. Additional time is spent as he gets unsterile and needs at least to change gloves.

   For 3D acquisitions often mobile and fixed mounted C-arm systems are used, as they combine the functionality of 2D X-ray imaging and 3D acquisition without adding another device to the operating theater, contrary to intraoperative CT systems. However, their 3D field of view is limited so that they typically cover the region of interest but not more. Often in such volumes only a restricted number of landmarks is visible, therefore landmark based approaches generally fail.

   Since the volumes are acquired after a trauma surgery, typically a larger number of metal implants and screws are inserted into the patient's body generating severe metal artifacts. An automation of anatomically correct alignment of axial, coronal, and sagittal MPRs given the above-mentioned restriction is desired.

   Literature review shows that automatic alignment is not a new topic. [4] already implemented an automatic rotation of axial head CT slices. [5, 6] covered the derivation of the brain midline using tools of pattern recognition. In [2] SURF features are extracted from the dataset and registered to an atlas with annotated MPRs. The quality is dependent on the choice of atlas and the feature extraction. Also, the registration needs to be carefully designed to support strongly truncated volumes [7] and typically the capture range of rotation is limited. An alternative approach to the present problem was followed by [8], in which shape models with attached labels for MPRs were registered to the volume. Although this solution solves the problem, in order to train the shape models for each body region the anatomical structures need to be segmented in the training datasets. Thus, the use of that algorithm requires a huge amount of manual work. For this approach, a time of 23 s was reported for the shape model segmentation and subsequent plane regression.

A first artificial intelligence approach for ultrasound images was presented in [9], in which the plane regression task was solved by using random forest trees. Recently, a convolutional neural network (CNN) was proposed, which extracts the standard slices from a stack of ultrasound images [10]. In that paper the rotation was described using different representations as quaternion or Euler angles where quaternions were showed to be superior. In [11] a rotation matrix was used of which only the first five or six values are calculated (5D or 6D method). This approach applies the definition of a proper rotation matrix, that states that the columns of a rotation matrix are orthonormal vectors and form a right-handed coordinate system. So, knowing the first vector and two coordinates of the second vector suffices to calculate the remaining four entries.

An inter-rater study was performed to evaluate the accuracy of two raters in the manual adjustment of the standard planes for the proximal femur regions in [12]. The found error estimations in that region are $6.3°$ for the normal and in-plane rotation and 9.3 mm for the translation. For this region, the planes are similarly well defined as for the upper ankle region and better defined compared to the calcaneus region.

In this paper we want to present a 3D CNN, with which the standard plane parameters are regressed directly from the volume. Since a plane regression network is used, no segmentation is needed, and the training of the proposed algorithm requires only the description of the planes. We compare three different approaches to describe the rotation and evaluate the optimal strategy for a regression of multiple MPRs.

To the best of our knowledge we present the first 3D CNN to directly estimate standard plane's parameters. Doing so we introduce the 6D method in medical imaging and compare it to well-established methods like Euler angles and quaternions.

In Sect. 2 we explain the methods of our approach, we introduce the employed mathematical description of planes, describe the normalization of the coordinate system, neural network, and the cost function. The implementation and the data we used for training and testing is described in Sect. 3 as well as the study design we followed. Thereafter we present the results of our experiments. Finally, we discuss the results in Sect. 5.

2 Methods

2.1 Plane Description

Mathematically a plane can be described by the point A and the vectors e_u and e_v showing right and upwards with increasing screen coordinate values. This description has the advantage that the image rotation is incorporated. The plane normal e_w is the cross product of e_u and e_v. The point A is the center of the plane.

The matrix $R = [e_u, e_v, e_w]$ can be interpreted as the rotation matrix from the volume to the plane. This rotation can be represented by rotation matrices (6D method), defined by Euler angles as well as by quaternions. As the coordinates of quaternions and matrices are in the range of $[-1, 1]$ these values are used directly. For the Euler angles however, the sine and cosine values of the angles are used. By this, we solve the problem of periodicity of the values and compress the values the range to the interval $[-1, 1]$. The angle value can be retrieved by the atan2 method.

2.2 Neural Network and Augmentation

To regress the parameters, we use an adapted version of the CNN proposed in [13] (Fig. 1). In contrast to the original work, the dropout layer was removed.

For each regressed value, the last fully connected layer has one output node. The number per MPR plane depends on the selected model for the rotation and varies between seven for quaternions and nine for sine and cosine values of the Euler angles and 6D representation. Thereby, 3 output nodes represent the translation whereas the remaining nodes describe the rotation.

During training, online augmentation of the volumes is employed. Since a neural network which combines rotation and translation parameters shall be trained using a combined loss function, we normalize the translation such that the origin of the volume is in the center of the volume, and that the volume edge length has a normalized length of 1. We apply a random rotation within the interval $[-45, 45]°$, random spatial scale of the volume by a factor in range $[0.95, 1.05]$, translation by $[-12, 12]$mm, a center crop, and sub-sampling. Additionally, mirroring in x-direction is added with a probability of 0.5 that allows to simulate left-right handedness of the volume.

For speedup and for reducing the number of interpolations, the augmentation operations are applied in a single step using their homogeneous matrices to create a composite matrix and therefore, the spatially augmented volume is interpolated just once during the sub-sampling.

To be robust to imperfect intensity calibration, the HU values added by 1000 HU are multiplied by a factor uniformly sampled of the range $[0.95, 1.05]$. Finally, a windowing function $w(x) = 1/\left(1 + e^{g(0.5-x)}\right)$ is applied after clipping the volume intensity values to the range of $[-490, 1040]$HU and rescaling it to $[0, 1]$. The gain parameter g is set dependent on the min/max values. This function helps to compress the values to the range $[0, 1]$. In contrast to min-max normalization, it reduces the signal variance of metal and air which typically contains little to no information about the plane's parameters. To speed up read-in, we down-sampled the original volumes to volumes of 128^3 voxels with length 160.25mm. The cost function is chosen to be

$$L = \alpha L_{rotation} + \beta L_{translation} + \gamma L_{orthogonality} \tag{1}$$

with $L_{translation}$ being the Euclidian distance of the normalized value of A, $L_{rotation}$ the Euclidian distance of the parameters describing the rotation, and $L_{orthogonality}$ the average of the cross products of the MPR normals.

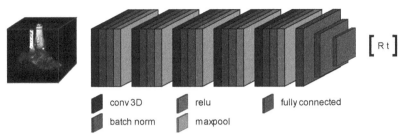

Fig. 1. Architecture of the regression CNN. 5 convolutional blocks are followed by 3 fully connected layers.

3   Experiments

3.1   Datasets

For the evaluation of the approach we use a dataset consisting of 160 volumes of the calcaneus region and 220 volumes of the ankle region. They were partly acquired intraoperatively to assess the result of a surgery and partly from cadavers, which were prepared for surgical training. The cadaver volumes were typically acquired twice: once without modifications and once with surgical instruments laid above, in order to produce metal artifacts without modifying the cadavers. The exact distribution of the datasets is listed in Table 1.

For each body region 5 data splits were created, taking care that volumes of the same patient belong to the same subset and that the distribution of dataset origin is approximately the same as in the total dataset. For all volumes, standard planes were defined according to [13].

For the ankle volumes axial, coronal, and sagittal MPRs, for the calcaneus datasets axial, sagittal, and semi-coronal planes were annotated by a medical engineer after training (Fig. 2). The labelling was performed using a syngo XWorkplace VD20 which was modified to store the plane description. Axial, sagittal, and coronal MPRs were adjusted with coupled MPRs. The semi-coronal plane was adjusted thereafter with decoupled planes.

Table 1. Origin and distribution of the used datasets.

	Cadaver			Clinical	Total
	Metal implants	Metal outside	No metal	Metal implants	
Calcaneus	9	63	62	26	160
Ankle	36	61	56	67	220

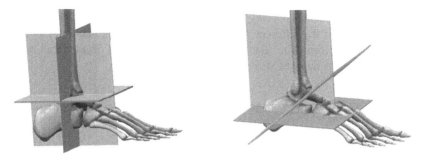

Fig. 2. For the dataset the displayed planes were labelled. Left: ankle, right: calcaneus. Red: axial, Green: sagittal, Blue: (semi-)coronal. (Color figure online)

3.2 Ablation Study

We performed an ablation study to find the best configuration of the algorithm. For all comparisons, a 5-fold cross-validation is carried out. As a performance metric we use

$$P = 0.2d + 0.6\varepsilon_n + 0.2\varepsilon_i \tag{2}$$

with d being the median error of the translation in direction of planes normal, ε_n being the error of the normal vector e_w of the plane, and ε_i marking the in-plane rotation error calculated as mean difference angle of e_u and e_v. The weights were chosen in accordance with our medical advisors and they reflect that the normal orientation is the most important metric, while in-plane rotation and correct plane translation are of subordinate importance, since the rotation does not influence the displayed information and the user scrolls through the volume during the review, actively changing the translation.

First, the influence of the description of the rotation was tested. For both body regions and voxel sizes of 1.2 mm, 2.2 mm, and 2.5 mm with respective volume sizes of 128^3, 72^3, 64^3 voxels, the performance measure for the three different representations was evaluated.

Thereafter, the performance of regressing three planes using one network compared to regressing the planes with separate networks was evaluated. For that the values of α, β, and γ were systematically varied, such that α and β were in the interval [0.1, 0.9] and γ was selected such that $\alpha + \beta + \gamma = 1$. Upon training separate networks for each plane, β was selected so that $\alpha + \beta = 1$, not enforcing any orthogonality.

For the cost-function weights optimized in this way, the comparison of using a single network and three separate networks was evaluated.

3.3 Implementation

The models are implemented in PyTorch (v.1.2) and trained on Windows 10 systems with 32 GB RAM and 24 GB NVIDIA TITAN RTX. The weights are initialized by the He et al. method [14]. The network is trained by Stochastic Gradient Descent (SGD) optimizer with momentum. The total number of epochs was set to 400. For selection of learning rate, learning rate decay, step size, momentum, and batch size, a hyper-parameter optimization using random sampling of the search space was performed using one fold and independently for the different rotation descriptions and volume input sizes. The optimization was performed for one of five folds. This method results in an offset of typically 0.1 and maximum 0.4 score points.

4 Results

Table 2 shows the results of the evaluation of the error depending on the angle representation. For both body regions, the 6D representation produces the best results for the rotation estimation. Using this method, the error of the normal and the in-plane rotation is minimal. Therefore, the next experiments were carried out using the 6D method.

Table 2. Mean results of plane regression evaluation using different rotation description parameters. Number of voxels is set to 72^3. The results reflect the mean of the single planes.

	Calcaneus				Ankle			
	D in mm	ε_n in °	ε_i in °	Score	D in mm	ε_n in °	ε_i in °	Score
Euler	14.39 ±1.64	8.93 ±1.60	11.05 ±2.10	10.45 ±1.18	7.78 ±0.36	6.99 ±0.81	8.37 ±1.23	7.42 ±0.77
Quat	9.93 ±2.25	9.96 ±1.75	9.54 ±1.59	9.87 ±1.66	5.00 ±0.09	8.16 ±0.79	8.31 ±0.71	7.56 ±0.63
6D	9.94 ±1.92	8.77 ±0.60	8.34 ±0.44	8.92 ±0.51	5.43 ±0.25	7.11 ±0.48	6.58 ±0.30	6.67 ±0.35

As can be seen in Table 3, varying the resolution has minor impact on the accuracy. While the training time increases from 20 s over 30 s to 60 s per epoch, the score does not change significantly. Utilizing the small advantage of sampling with 72 voxels, with minor impact on the training time, sampling with 72 voxels was used for the next experiments.

Using a combined network for predicting the parameters of all planes improves the outcome of the network slightly (Table 4). A small further improvement can be reached by adding a constraint on the orthogonality of the planes.

All the experiments show that the plane regression for the ankle region with orthogonal planes works better than for the calcaneus region. Typically, the score is 2 to 3 value points better, with all the error measures contributing in the same way to the improvement. For the comparison of the models, also the error of each plane is listed in Table 5, showing that the orientation of the axial MPR can be estimated best compared to the other MPR orientations. The inference time was below 0.05s.

Table 3. Mean results of plane regression evaluation using different numbers of voxels. The rotation is described using the 6D method. The results reflect the mean of the single planes.

	Calcaneus				Ankle			
	D in mm	ε_n in °	ε_i in °	Score	D in mm	ε_n in °	ε_i in °	Score
64^3	9.74 ±0.88	8.98 ±2.00	8.33 ±1.08	9.01 ±1.46	6.18 ±0.75	7.12 ±0.78	6.49 ±0.69	6.80 ±0.63
72^3	9.94 ±1.92	8.77 ±0.60	8.34 ±0.44	8.92 ±0.51	5.43 ±0.25	7.11 ±0.48	6.58 ±0.30	6.67 ±0.35
128^3	10.48 ±2.70	8.48 ±1.21	8.49 ±1.21	8.88 ±1.49	4.86 ±0.29	7.75 ±1.00	7.04 ±0.75	7.03 ±0.70

Table 4. Mean results of plane regression evaluation. Three: for each plane an independently optimized model is used (Table 6), Comb.: One model for regressing the parameters for all planes is used with $\alpha = \beta = 0.5$, Opt. Comb.: One model for regressing the parameters for all planes is used with optimized values α, β, and γ. The rotation is described using the 6D method. Number of voxels is set to 72^3. The results reflect the mean of the single planes.

	Calcaneus				Ankle			
	D in mm	ε_n in °	ε_i in °	Score	D in mm	ε_n in °	ε_i in °	Score
Three	9.46 ±1.24	9.26 ±0.84	8.94 ±0.82	9.24 ±0.70	6.52 ±0.33	7.55 ±0.28	6.77 ±0.38	7.19 ±0.09
Comb	9.94 ±1.92	8.77 ±0.60	8.34 ±0.44	8.92 ±0.51	5.43 ±0.25	7.11 ±0.48	6.58 ±0.30	6.67 ±0.35
Opt. Comb	10.38 ±1.89	8.14 ±1.21	7.91 ±0.79	8.54 ±0.83	5.20 ±0.21	6.93 ±0.58	6.86 ±0.36	6.57 ±0.44

Table 5. Regression results for each plane using the combined model with the optimized cost function. The rotation is described using the 6D method. Number of voxels is set to 72^3.

	Calcaneus				Ankle			
	D in mm	ε_n in °	ε_i in °	Score	D in mm	ε_n in °	ε_i in °	Score
Axial	10.35 ±3.62	7.38 ±1.26	7.69 ±0.77	8.04 ±1.64	6.61 ±0.40	5.20 ±0.97	7.76 ±0.67	5.89 ±0.80
Semic./coronal	13.11 ±1.21	8.71 ±1.12	7.49 ±0.89	9.35 ±1.09	4.56 ±0.54	7.57 ±0.77	6.05 ±0.37	6.67 ±0.35
Sagittal	7.77 ±0.81	8.65 ±1.68	8.34 ±1.07	8.41 ±1.39	4.73 ±0.25	9.18 ±1.22	6.89 ±0.40	7.83 ±0.86

Table 6. α, β, and γ values for the different models (Table 4).

		Calcaneus			Ankle		
		α	β	γ	α	β	γ
Three	**Axial**	0.2	0.8	0.0	0.6	0.4	0.0
	Coronal	0.2	0.8	0.0	0.2	0.8	0.0
	Sagittal	0.6	0.4	0.0	0.8	0.2	0.0
Comb		0.5	0.5	0.0	0.5	0.5	0.0
Opt. Comb		0.6	0.3	0.1	0.2	0.8	0.0

5 Discussion and Conclusion

We have presented an algorithm which allows to automatically regress the standard MPR plane parameters in neglectable time. The proposed algorithm is capable to deal with metal artifacts and strong truncation. Both kind of disturbances are common within intraoperatively acquired volumes.

In contrast to state-of-the-art algorithms, no additional volume segmentation for training or evaluation purposes is needed, consequently reducing the amount of labeling effort needed.

For describing the rotation, using the 6D method is most reliable followed by Euler angles and quaternions. In contrast to [10] and [11], the Euler angles were not regressed directly but their cosine and sine values, yielding a better result compared to quaternions.

We have shown that the algorithm is capable to regress both orthogonal and non-orthogonal planes, with a higher accuracy for the ankle region. A reason for this difference could be that the standard planes for ankle anatomy are better defined and therefore have less variability across samples. Especially the definition of the semi-coronal MPR of the calcaneus allows some variance.

The orthogonality term for the planes normal has a minor benefit for the result. This can be explained with the fact, that the labelled values already follow this constraint, so that it provides only little further information.

We also observe that the combined representation of the plane translational and rotational plane parameters is beneficial for the accuracy of the trained network. Depending on the required accuracy, also a stronger down-sampling allows for clinically sufficient results.

We see a limitation of the evaluation in the missing comparison with the results of [8]. However, due to its need of costly segmentations this was out of scope for this project.

Disclaimer. The methods and information presented here are based on research and are not commercially available.

References

1. Keil, H., et al.: Intraoperative assessment of reduction and implant placement in acetabular fractures—limitations of 3D-imaging compared to computed tomography. J. Orthop. Surg. **13**(1), 78 (2018). https://doi.org/10.1186/s13018-018-0780-7
2. Brehler, M., Görres, J., Franke, J., Barth, K., Vetter, S.Y., Grützner, P.A., Meinzer, H.P., Wolf, I., Nabers, D.: Intra-operative adjustment of standard planes in C-arm CT image data. Int. J. Comput. Assist. Radiol. Surg. **11**(3), 495–504 (2015). https://doi.org/10.1007/s11548-015-1281-3
3. Brehler, M.: Intra-Operative Visualization and Assessment of Articular Surfaces in C-Arm Computed Tomography Images (2015). In: https://archiv.ub.uni-heidelberg.de/volltextserver/19091/. Accessed 21 February 2020
4. Hirshberg, D.A., et al.: Evaluating the automated alignment of 3D human body scans. In: Proceedings of the 2nd International Conference on 3D Body Scanning Technologies, Lugano, Switzerland, pp. 25–26 October 2011, Lugano, Switzerland, pp. 76–86 (2011). https://doi.org/10.15221/11.076
5. Tan, W., et al.: An approach to extraction midsagittal plane of skull from brain CT images for oral and maxillofacial surgery. IEEE Access **7**, 118203–118217 (2019). https://doi.org/10.1109/ACCESS.2019.2920862
6. Qi, X., et al.: Ideal midline detection using automated processing of brain CT image. Open J. Med. Imaging **03**(02), 51–59 (2013). https://doi.org/10.4236/ojmi.2013.32007
7. Thomas, S., et al.: Upper ankle joint space detection on low contrast intraoperative fluoroscopic C-arm projections. In: Proceedings Medical Imaging 2017: Image-Guided Procedures, Robotic Interventions, and Modeling, vol. 10135, pp. 101351I (2017). https://doi.org/10.1117/12.2255633
8. Thomas, S.: Automatic image analysis of C-arm computed tomography images for ankle joint surgeries. Heidelberg (2020). urn:nbn:de:bsz:16-heidok-285277
9. Lu, X., Georgescu, B., Zheng, Y., Otsuki, J., Comaniciu, D.: AutoMPR: automatic detection of standard planes in 3D echocardiography. In: 2008 5th IEEE International Symposium on Biomedical Imaging: From Nano to Macro, Paris, France, pp. 1279–1282 (2008) https://doi.org/10.1109/isbi.2008.4541237
10. Li, Y., Khanal, B., Hou, B., Alansary, A., Cerrolaza, J.J., Sinclair, M., Matthew, J., Gupta, C., Knight, C., Kainz, B., Rueckert, D.: Standard plane detection in 3D fetal ultrasound using an iterative transformation network. In: Frangi, A.F., Schnabel, J.A., Davatzikos, C., Alberola-López, C., Fichtinger, G. (eds.) MICCAI 2018. LNCS, vol. 11070, pp. 392–400. Springer, Cham (2018). https://doi.org/10.1007/978-3-030-00928-1_45
11. Zhou, Y., Barnes, C., Lu, J., Yang, J., Li, H.: On the continuity of rotation representations in neural networks. In: 2019 IEEE/CVF Conference on Computer Vision and Pattern Recognition (CVPR), Long Beach, CA, USA, pp. 5738–5746 (2019). https://doi.org/10.1109/cvpr.2019.00589
12. Kausch, L., Thomas, S., Kunze, H., Privalov, M., Vetter, S., Franke, J., Mahnken, A.H., Maier-Hein, L., Maier-Hein, K.: Toward automatic C-arm positioning for standard projections in orthopedic surgery. Int. J. Comput. Assist. Radiol. Surg. **15**(7), 1095–1105 (2020). https://doi.org/10.1007/s11548-020-02204-0
13. Bui, M., Albarqouni, S., Schrapp, M., Navab, N., Ilic, S.: X-Ray PoseNet: 6 DoF pose estimation for mobile X-Ray devices. In: 2017 IEEE Winter Conference on Applications of Computer Vision (WACV), Santa Rosa, CA, USA, pp. 1036–1044 (2017). https://doi.org/10.1109/wacv.2017.120
14. He, K., Zhang, X., Ren, S., Sun, J.: Delving deep into rectifiers: surpassing human-level performance on imagenet classification. In: 2015 IEEE International Conference on Computer Vision (ICCV), Santiago, Chile, pp. 1026–1034 (2015). https://doi.org/10.1109/iccv.2015.123

Unsupervised Graph Domain Adaptation for Neurodevelopmental Disorders Diagnosis

Bomin Wang[1], Zhi Liu[1], Yujun Li[1(✉)], Xiaoyan Xiao[2],
Ranran Zhang[1], Yankun Cao[1], Lizhen Cui[3], and Pengfei Zhang[4]

[1] School of Information Science and Engineering, Shandong University,
Qingdao, China
liyujun@sdu.edu.cn
[2] Department of Nephrology, Qilu Hospital of Shandong University, Jinan,
China
[3] Joint SDU-NTU Centre for Artificial Intelligence Research (C-FAIR),
Shandong University, Jinan, China
clz@sdu.edu.cn
[4] Department of Cardiology, Qilu Hospital, Cheeloo College of Medicine,
Shandong University, Jinan, China

Abstract. Unsupervised domain adaptation (UDA) methods aim to reduce annotation efforts when generalizing deep learning models to new domains. UDA has been widely studied in medical image domains. However, UDA on graph domains has not been investigated yet. In this paper, we present the first attempt of unsupervised graph domain adaptation in medical imaging, with application to neurodevelopmental disorders (NDs) diagnosis, i.e. differentiating NDs patients from normal controls. It is of great importance to developing UDA methods for NDs because acquiring accurate diagnosis or labels of NDs can be difficult. In our work, we focus on Autism spectrum disorder and attention-deficit/hyperactivity disorder which are the two most common and frequently co-occurred NDs. We propose an unsupervised graph domain adaptation network (UGDAN) that consists of three main components including graph isomorphism encoders, progressive feature alignment, and un-supervised infomax regularizer. The progressive feature alignment module is designed to align graph representations of the source and target domains progressively and effectively, while the unsupervised infomax regularizer is introduced to further enhance the feature alignment by learning good unsupervised graph embeddings. We validate the proposed method with two experimental settings, cross-site adaptation and cross-disease adaptation, on two publicly available datasets. The experimental results reveal that the proposed UGDAN can achieve comparable performance compared to supervised methods trained on the target domain.

B. Wang and Z. Liu—Equal contributions.

Electronic supplementary material The online version of this chapter (https://doi.org/10.1007/978-3-030-59713-9_48) contains supplementary material, which is available to authorized users.

Keywords: Domain adaptation · Neurodevelopmental disorders · Graph
network

1 Introduction

Neurodevelopmental disorders comprise a group of psychiatric conditions originating
in childhood that involve some form of disruption to brain development [1]. Autism
spectrum disorder (ASD) and attention-deficit/hyperactivity disorder (ADHD) are two
common and impairing neurodevelopmental disorders that frequently co-occur [2, 3].
ASD is a disorder that impairs the ability to communicate and interact with others,
while ADHD is characterized by an ongoing pattern of inattention and/or hyperactivity-
impulsivity that interferes with functioning or development. Functional connectivity
(FC), which quantifies the operational interactions of multiple spatially-distinct brain
regions, is widely used to investigate neurodevelopmental disorders. Graphs provide a
powerful and intuitive way to analyze FC by modeling brain regions as nodes, and
functional connections between brain regions as edges. This also attracts researchers to
develop graph neural network (GNN) based methods for FC analysis and disease
prediction [4]. While yielding outstanding performance, like in the image domains,
GNN based methods still face the problems of domain shift and labor-intensive
annotation of data in graph domains.

Data collection and manual annotation for every new domain and task may be time-
consuming and expensive in medical imaging, where data can be acquired from different sites, modalities and scanners. To solve this problem, unsupervised domain
adaptation methods have been widely studied [5–8]. These methods aim to improve the
learning performance in the target domain by making use of the label information in the
source domain as well as some relation between the source and the target domains.
Previous works mainly focus on alleviating the annotation efforts of segmentation tasks
and various structures have been investigated. In addition, there are also several
methods apply domain adaptation approaches to microscopy images where cell-wise
labeling is often tedious and expensive [9, 10]. Compared to the tasks have been
considered in existing work, acquiring accurate diagnosis or labels of neurodevelopmental disorders can be much more difficult. First, the current diagnosis of neurodevelopmental disorders is behavioral-based and relies on the history and symptoms of
the disorders [11]. Clinicians have to integrate the information of past symptoms across
various sources, which heavily rely on an informant being both available and reliable.
Second, the diagnosis of neurodevelopmental disorders is a complicated process. For
instance, diagnosing ASD is complex since autism specialty clinics frequently span the
full spectrum of the disorder and often involve diverse symptom presentations, which
will be further complicated because of the interactions that occur between development
and ASD symptoms [12, 13]. Third, accurate diagnosis of neurodevelopmental disorders relies on the expertise and experience in multidisciplinary including psychology,
neurology, psychiatry, etc. This means that a neurodevelopmental disorder evaluation
is often performed by multimember diagnostic teams. However, this is not always

feasible in some clinics. In summary, there is an urgent need to alleviate the diagnosis or labeling efforts of neurodevelopmental disorders when applying machine learning methods to new domains and tasks via domain adaptation algorithms.

In this paper, we propose an unsupervised graph domain adaptation network (UGDAN) aiming to tackle two domain shift problems, i.e., cross-site domain shift and cross-disease domain shift, with application to two common neurodevelopmental disorders, ASD and ADHD. For the first problem, data from different sites can be heterogeneous because of the inter-site variation in scanners or protocols. For the second problem, since ASD and ADHD often co-occur, the functional connectivity of them may share some common patterns. This inspires us to predict ADHD without accessing the diagnosis information by utilizing the labeled ASD data. The proposed method UGDAN consists of three main components, graph isomorphism encoder, progressive feature alignment, and unsupervised infomax regularizer. The illustration of our method is shown in Fig. 1.

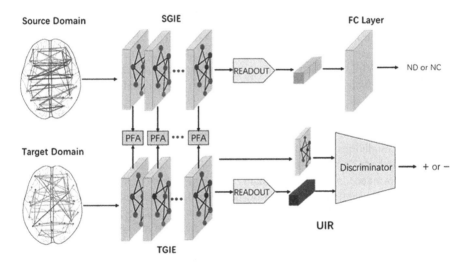

Fig. 1. Illustration of the proposed UGDAN.

2   Methodology

2.1   Problem Definition

A graph can be represented as $G = \{V, E\}$, where $V = \{v_1, v_2 \ldots, v_N\}$ is the set of N nodes and E is the set of edges. In our work, the FC data of one subject can be seen as a graph by modeling brain regions as nodes, and functional connections between brain regions as edges. For unsupervised domain adaptation in graph domain, we are given a source domain $\mathbf{D}_s = \left\{ \left(G_i^s, y_i^s\right) \right\}_{i=1}^{n_s} (G_i^s \in \mathbf{G}_s, y_i^s \in \mathbf{Y}_s)$ of n_s labeled graphs and given a target domain $\mathbf{D}_t = \left\{ \left(G_j^t\right) \right\}_{j=1}^{n_t} (G_j^t \in \mathbf{G}_t)$ of n_t unlabeled graphs. The source and target domains are drawn from the joint probability distributions $P(\mathbf{G}_s, \mathbf{Y}_s)$ and $Q(\mathbf{G}_t, \mathbf{Y}_t)$ respectively, and $P \neq Q$. We assume that by utilizing the relationship between the source and target domains, we can accurately predict the label y_j^t of each $G_j^t \in \mathbf{G}_t$ with the help of labeled graph data in $\mathbf{G}_s$.

2.2 Graph Isomorphism Encoder

To extract feature representation of graph data from both source domain and target domain, we consider GNNs which can generate node representation by repeatedly aggregating over local node neighborhood nodes. Recently, Xu et al. [14] proposed a powerful GNN called graph isomorphism network (GIN) which was shown excellent representational power compared to other GNN variants. This model generalized the Weisfeiler-Lehman test and hence achieved maximum discriminative power among GNNs. Therefore, we choose GIN as our encoders to extract discriminative graph representations. Formally, given a input graph G, the message-passing scheme in GIN is

$$H^{(k)}(G) = MLP\left((1+\gamma)^k \odot H^{(k-1)}(G) + A^T H^{(k-1)}\right) \tag{1}$$

where $H^{(k)}(G) \in \mathbb{R}^{N \times d}$ is the node representations computed after the k-th GNN layer. γ is a layer-wise learnable scalar parameter and MLP is a multi-layer perceptron with ReLU function. A^T represents the transpose matrix of the adjacency matrix. In our work, we stack several GIN layers to form the Graph Isomorphism Encoder. Next, we concatenate graph embeddings at all depths of the Graph Isomorphism Encoder and apply a READOUT function to obtain a graph level representation. That is

$$H(G) = \text{READOUT}\left(\text{CONCAT}\left(\left\{H^{(k)}\right\}_{k=1}^{K}\right)\right) \tag{2}$$

where CONCAT represents the concatenation operation and K is the number of GIN layers in the Graph Isomorphism Encoder. We use sum over mean for READOUT. In our work, we deploy two Graph Isomorphism Encoders: the encoder on the labeled source domain data (SGIE) and the encoder on the unlabeled target domain data (TGIE).

2.3 Progressive Feature Alignment

Many existing approaches for UDA in the image domain focused on learning domain-invariant representations via cross-domain translation [15]. Cross-domain image-to-image translation can be seen as a pixel-to-pixel prediction problem. However, cross-domain translation in graph domain is extremely hard because graphs are in the non-Euclidean space which has no regular grids to operate on. Here we draw inspiration from knowledge distillation techniques that enforce the predictions of the student model to be similar to the teacher model. (Here we focus on reducing) We propose a Progressive Feature Alignment (PFA) module which progressively aligns graph representations of the source and target domains via Kullback–Leibler divergence (KL-divergence) minimization. By doing so, the domain discrepancy between the source and target domains, i.e., cross-domain distribution discrepancy can be reduced by PFA. In other words, we transfer knowledge from the TGIE to the SGIE and hence endow the SGIE with the ability to make predictions in the target domain. Formally, let φ and ω denote the parameters of the SGIE and the TGIE respectively, PFA minimizes the following loss function

$$
\begin{aligned}
L_{PFA} &= -\sum_{k}^{K} D_{KL}\left(p_j^k \| q_j^k\right) \\
&= -\sum_{k}^{K} D_{KL}\left(softmax\left(H_\varphi^k\left(G_j^S\right)\right) \| softmax\left(H_\omega^k\left(G_j^T\right)\right)\right)
\end{aligned}
\tag{3}
$$

where $H_\varphi^k\left(G_j^S\right)$ and $H_\omega^k\left(G_j^T\right)$ are the graph representations of k-th layer of the SGIE and the TGIE respectively. By applying the *softmax* function, $H_\varphi^k\left(G_j^S\right)$ and $H_\omega^k\left(G_j^T\right)$ can be converted to probabilities p_j^k and q_j^k. $D_{KL}(\cdot)$ is the Kullback-Leibler divergence between two distributions which calculated as

$$
D_{KL}\left(p_j^k \| q_j^k\right) = \sum_j p_j^k \ln \frac{p_j^k}{q_j^k}
\tag{4}
$$

2.4   Unsupervised Infomax Regularizer

Inspired by the recent unsupervised representation learning method Deep Graph Infomax [16], we introduce an unsupervised infomax regularizer (UIR) term to fully utilize the unlabeled target domain data. UIR uses a discriminator which takes a (global representation, local representation) pair as input and decides whether they are from the same graph. We provide the detail of UIR in Fig. 2. The intuition behind this is that the PFA will benefit from learning a good embedding from unlabeled data while aligning graph representations between two domains. UIR maximizes the mutual information between graph-level and patch-level representations. In this way, the graph embedding can learn to encode aspects of the unlabeled target domain data that are shared across multi-scaled substructures. The graph representations of k-th layer of the TGIE $H_\omega^k\left(G_j^T\right)$ can be seen a set of local representations, where each local representation denotes the representation of node i. We concatenate $h_i^{(k)}$ at all depths of the TGIE into a summarized local representation h_ω^i centered at node i. To this end, UIR minimizes the following loss function

$$
L_{UIR} = -\sum_j \sum_{k=1}^{K} I_{\rho,w}\left(h_\omega^i; H_\omega\left(G_j^T\right)\right)
\tag{5}
$$

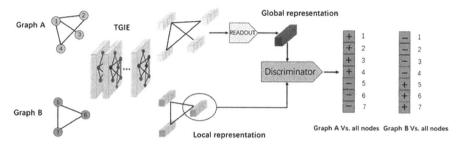

Fig. 2. The detail of UIR. We use two toy examples Graph A and Graph B for illustration.

where $H_\omega\left(G_j^T\right)$ is the output representation of the TGIE parameterized by ω, i.e., the global representation of G_j^T. $I_{\rho,w}$ is the mutual information estimator modeled by a discriminator D_ρ parameterized by ρ. Here we adopt Jensen-Shannon MI estimator

$$I_{\rho,w}\left(h_\omega; H_\omega\left(G_j^T\right)\right) :=$$

$$\mathbb{E}_{\mathbf{Q}}\left[-sp\left(-D_\rho\left(h_\omega\left(G_j^T\right), H_\omega\left(G_j^T\right)\right)\right)\right] - \mathbb{E}_{\mathbf{Q}\times Q'}\left[sp\left(D_\rho\left(h_\omega\left(G_j^T\right), H_\omega\left(G_{j'}^T\right)\right)\right)\right] \quad (6)$$

where $\mathbf{Q}'$ is a distribution that is identical to the distribution of the target domain $\mathbf{Q}$. $sp(x) = \log(1 + e^x)$ is the sigmoid function. If a local-global representation pair comes from the same graph ($h_\omega\left(G_j^T\right)$ and $H_\omega\left(G_j^T\right)$), we define the local representation as a positive pair. Otherwise, we define the pair as a negative pair ($h_\omega\left(G_{j'}^T\right)$ and $H_\omega\left(G_j^T\right)$). To generate negative pairs, we use all possible combinations of global and local representations of all graphs in a training batch.

3 Experiments

3.1 Datasets and Experimental Setups

Cross-Site Adaptation. We validated our unsupervised domain adaptation method with two tasks, cross-site adaptation, and cross-disease adaptation. For cross-site adaptation, we used data from the ABIDE database [17]. The ABIDE database aggregates fMRI data of 1112 subjects from various international acquisition sites. In our work, we chose data from NYU as the source domain and we choose UM and UCLA as two target domains. We reported the results on UM here and the results of UCLA were presented in the supplementary material. There was a total of 171 subjects in NYU, including 250 ASD patients and 218 NCs, while UM gathered 99 subjects with 45 ASD patients and 45 NCs. We use all data from NYU in our experiments and in order to fairly compare with supervised methods, we adopt a 5-fold cross-validation strategy to evaluate the performance of our method as well as the competing methods on UM data. Note that we have no access to the diagnosis information of UM data during training our method. The preprocessing of fMRI data from both source and target domains is performed according to the Configurable Pipeline for the Analysis of Connectomes (C-PAC), which includes skull striping, motion correction, slice timing correction and nuisance signal regression. Then, the mean time series for 116 regions-of-interest (ROIs). defined by AAL atlas were computed and normalized to zero mean and unit variance.

Cross-Disease Adaptation. For cross-disease adaptation, we utilized ASD data from NYU as the source domain and ADHD data from MICCAI 2019 CNI challenge as the target domain. The data used for the CNI challenge consists of fMRI data and demographic information of both ADHD patients and neurotypical controls. Data Preprocessing of CNI challenge data included slice time correction, motion correction, band-pass filtering (0.01–0.1 Hz) and normalization to the MNI template. We also used the mean time courses extracted using the standard AAL atlas.

3.2   Implementation Details

For each subject, we generated a FC matrix expressed by a 116×116 symmetrical matrix where each element in this matrix denotes the Pearson correlation coefficient between a pair of ROIs. We defined the connection between a pair of ROIs by thresholding the Pearson correlation coefficient (>0.4). For node features, since the raw mean time courses are noisy, we extracted 11 time series features which included first-order and second-order statistics, sample entropy, etc. We provide detailed definitions of these features in the supplementary material. For UGDAN, we stacked 5 GIN layers to form both SGIE and TGIE, and the discriminator in UGDAN was composed of 3 layers of fully connected networks. In the experiments, we compare our method with GCN and GIN with different experimental settings. Specifically, a model with the subscript "ST" denotes this model is trained on the source domain and test on the target domain without adaptation, while the subscript "TT" represents that both training and testing process are conducted on the target domain. We also provide an ablation study in which the performance of our model without UIR is reported.

3.3   Results

Experiment 1: Cross-Site Adaptation. Quantitative results of cross-site adaptation are illustrated in Table 1. We adopt three classification metrics including accuracy (ACC), the area under the receiver operating characteristic (ROC) curve (AUC) and precision (PRE). We can observe that directly employing a trained GNN model on source site to test on target domain yields poor results, demonstrating the difficulty of GNNs to generalize well on a new domain. UGDAN achieves significant improvement by 23.8% and 25.8% compared with the un-adapted method GCNST in terms of accuracy. It is worth noting that UGDAN consistently outperforms supervised method

GCNTT and can obtain comparable performance compared with the best supervised model GCNTT trained on the target domain under the same settings. This owes to the powerful excellent representational power of GIE encoders as well as the effectiveness of PFA and UIR. In addition, we can see from the ablation study that performance drops can be obtained by removing UIR, which also prove the effectiveness of UIR.

Table 1. Quantitative comparisons of performance on the target domain for the different models. UGDAN* denotes UGDAN without UIR.

	ACC	AUC	PRE
GCN_{ST}	0.420 ± 0.041	0.510 ± 0.044	0.493 ± 0.008
GIN_{ST}	0.475 ± 0.031	0.533 ± 0.033	0.508 ± 0.026
GCN_{TT}	0.742 ± 0.041	0.724 ± 0.064	0.817 ± 0.071
GIN_{TT}	0.784 ± 0.024	0.779 ± 0.028	0.832 ± 0.026
UGDAN*	0.712 ± 0.041	0.700 ± 0.029	0.777 ± 0.036
UGDAN	0.758 ± 0.019	0.732 ± 0.021	0.783 ± 0.046

Experiment 2: Cross-Disease Adaptation. In the experiments, we seek to predict the labels of unlabeled ADHD data using ASD data with diagnosis information. Quantitative results are shown in Fig. 3. Without adaptation, the two GNN models obtained poor performance across all metrics. Our model outperformed GCN trained on ADHD data by 0.025 in terms of ACC and AUC. UGDAN also obtained comparable performance compared to GIN trained with diagnosis information. This implies that with UGDAN, accurate diagnosis can be obtained on ADHD data without access to the true labels. To verify the effectiveness of PFA and UIR in terms of data embeddings, we also provide a t-SNE analysis in the supplementary material.

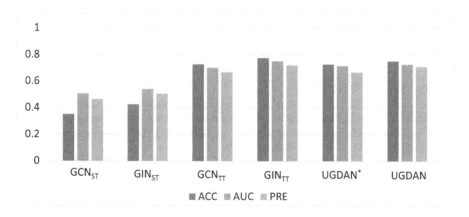

Fig. 3. Quantitative comparisons of performance on the target domain for the different models.

4 Conclusion

We have introduced an unsupervised domain adaptation method UGDAN for NDs diagnosis. In detail, we utilize two GIEs to extract graph representations, and we introduce a PFA module to align the representations source and target domains. In this way, the inter-domain discrepancy can be reduced. Besides, a UIR is designed to make full use of the unlabeled target domain data and to further enhance the feature alignment by learning good unsupervised graph embeddings. Extensive experiments demonstrate the effectiveness of UGDAN and its potential to reduce the labeling efforts when conducting graph based methods on NDs diagnosis. UGDAN can serve as a general pipeline for unsupervised domain adaptation in graph domains, and it can be used for other brain diseases like Alzheimer's Disease, bipolar disorder and Parkinson's disease, etc.

References

1. Thapar, A., Cooper, M., Rutter, M.: Neurodevelopmental disorders. Lancet Psychiatry 4(4), 339–346 (2017)
2. Jang, J., Matson, J.L., Williams, L.W., Tureck, K., Goldin, R.L., Cervantes, P.E.: Rates of comorbid symptoms in children with ASD, ADHD, and comorbid ASD and ADHD. Res. Dev. Disabil. 34(8), 2369–2378 (2013)
3. Van Steijn, D.J., et al.: The co-occurrence of autism spectrum disorder and attention-deficit/hyperactivity disorder symptoms in parents of children with ASD or ASD with ADHD. J. Child Psychol. Psychiatry 53(9), 954–963 (2012)
4. Li, X., Dvornek, N.C., Zhou, Y., Zhuang, J., Ventola, P., Duncan, J.S.: Graph neural network for interpreting task-fMRI biomarkers. In: Shen, D., et al. (eds.) MICCAI 2019. LNCS, vol. 11768, pp. 485–493. Springer, Cham (2019). https://doi.org/10.1007/978-3-030-32254-0_54
5. Yang, J., Dvornek, N.C., Zhang, F., Chapiro, J., Lin, M., Duncan, J.S.: Unsupervised domain adaptation via disentangled representations: application to cross-modality liver segmentation. In: Shen, D., et al. (eds.) MICCAI 2019. LNCS, vol. 11765, pp. 255–263. Springer, Cham (2019). https://doi.org/10.1007/978-3-030-32245-8_29
6. Ouyang, C., Kamnitsas, K., Biffi, C., Duan, J., Rueckert, D.: Data efficient unsupervised domain adaptation for cross-modality image segmentation. In: Shen, D., et al. (eds.) MICCAI 2019. LNCS, vol. 11765, pp. 669–677. Springer, Cham (2019). https://doi.org/10.1007/978-3-030-32245-8_74
7. Liu, P., Kong, B., Li, Z., Zhang, S., Fang, R.: CFEA: collaborative feature ensembling adaptation for domain adaptation in unsupervised optic disc and cup segmentation. In: Shen, D., et al. (eds.) MICCAI 2019. LNCS, vol. 11768, pp. 521–529. Springer, Cham (2019). https://doi.org/10.1007/978-3-030-32254-0_58
8. Chen, C., Dou, Q., Chen, H., Qin, J., Heng, P.A.: Synergistic image and feature adaptation: towards cross-modality domain adaptation for medical image segmentation. In: Proceedings of the AAAI Conference on Artificial Intelligence, vol. 33, pp. 865–872 (2019)
9. Pacheco, C., Vidal, R.: An unsupervised domain adaptation approach to classification of stem cell-derived cardiomyocytes. In: Shen, D., et al. (eds.) MICCAI 2019. LNCS, vol. 11764, pp. 806–814. Springer, Cham (2019). https://doi.org/10.1007/978-3-030-32239-7_89

10. Xing, F., Bennett, T., Ghosh, D.: Adversarial domain adaptation and pseudo-labeling for cross-modality microscopy image quantification. In: Shen, D., et al. (eds.) MICCAI 2019. LNCS, vol. 11764, pp. 740–749. Springer, Cham (2019). https://doi.org/10.1007/978-3-030-32239-7_82
11. Chen, C.Y., Liu, C.Y., Su, W.C., Huang, S.L., Lin, K.M.: Factors associated with the diagnosis of neurodevelopmental disorders: a population-based longitudinal study. Pediatrics **119**(2), 435–443 (2007)
12. Johnson, C.P., Myers, S.M.: Identification and evaluation of children with autism spectrum disorders. Pediatrics **120**(5), 1183–1215 (2007)
13. Lord, C., Jones, R.M.: Annual research review: re-thinking the classification of autism spectrum disorders. J. Child Psychol. Psychiatry **53**(5), 490–509 (2012)
14. Xu, K., Hu, W., Leskovec, J., Jegelka, S.: How powerful are graph neural networks?. arXiv preprint arXiv:1810.00826 (2018)
15. Kamnitsas, K., et al.: Unsupervised domain adaptation in brain lesion segmentation with adversarial networks. In: Niethammer, M., et al. (eds.) IPMI 2017. LNCS, vol. 10265, pp. 597–609. Springer, Cham (2017). https://doi.org/10.1007/978-3-319-59050-9_47
16. Veli˘ckovi´c, P., Fedus, W., Hamilton, W.L., Liò, P., Bengio, Y., Hjelm, R.D.: Deep graph infomax. arXiv preprint arXiv:1809.10341 (2018)
17. Di Martino, A., et al.: The autism brain imaging data exchange: towards a large-scale evaluation of the intrinsic brain architecture in autism. Mol. Psychiatry **19**(6), 659–667 (2014)

JBFnet - Low Dose CT Denoising by Trainable Joint Bilateral Filtering

Mayank Patwari[1,2]([⊠]) [iD], Ralf Gutjahr[2], Rainer Raupach[2], and Andreas Maier[1]

[1] Pattern Recognition Lab, Friedrich-Alexander Universität Erlangen-Nürnberg (FAU), Erlangen, Germany
mayank.patwari@fau.de
[2] Siemens Healthcare GmbH, Forchheim, Germany

Abstract. Deep neural networks have shown great success in low dose CT denoising. However, most of these deep neural networks have several hundred thousand trainable parameters. This, combined with the inherent non-linearity of the neural network, makes the deep neural network difficult to understand with low accountability. In this study we introduce JBFnet, a neural network for low dose CT denoising. The architecture of JBFnet implements iterative bilateral filtering. The filter functions of the Joint Bilateral Filter (JBF) are learned via shallow convolutional networks. The guidance image is estimated by a deep neural network. JBFnet is split into four filtering blocks, each of which performs Joint Bilateral Filtering. Each JBF block consists of 112 trainable parameters, making the noise removal process comprehendable. The Noise Map (NM) is added after filtering to preserve high level features. We train JBFnet with the data from the body scans of 10 patients, and test it on the AAPM low dose CT Grand Challenge dataset. We compare JBFnet with state-of-the-art deep learning networks. JBFnet outperforms CPCE3D, GAN and deep GFnet on the test dataset in terms of noise removal while preserving structures. We conduct several ablation studies to test the performance of our network architecture and training method. Our current setup achieves the best performance, while still maintaining behavioural accountability.

Keywords: Low dose CT denoising · Joint bilateral filtering · Precision learning · Convolutional neural networks

1 Background

Reducing the radiation dose in clinical CT is extremely desirable to improve patient safety. However, lowering the radiation dose applied results in a higher amount of quantum noise in the reconstructed image, which can obscure crucial

Electronic supplementary material The online version of this chapter (https://doi.org/10.1007/978-3-030-59713-9_49) contains supplementary material, which is available to authorized users.

© Springer Nature Switzerland AG 2020
A. L. Martel et al. (Eds.): MICCAI 2020, LNCS 12262, pp. 506–515, 2020.
https://doi.org/10.1007/978-3-030-59713-9_49

diagnostic information [17]. Therefore, it is necessary to reduce the noise content while preserving all the structural information present in the image. So far, this is achieved through non-linear filtering [13,14] and iterative reconstruction [3,5,6]. Most CT manufacturers use iterative reconstruction techniques in their latest scanners [1,18].

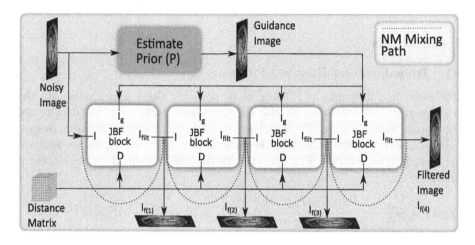

Fig. 1. The general architecture of JBFnet. There is a prior estimator and 4 JBF blocks for filtering. The NM mixing paths are shown in the diagram. The intermediate outputs of JBFnet are also shown.

In the past decade, deep learning methods [4] have shown great success in a variety of image processing tasks such as classification, segmentation, and domain translation [7]. Deep learning methods have also been applied to clinical tasks, with varying degrees of success [16,22]. One such problem where great success has been achieved is low dose CT denoising [2,8,10,20,23,25,26]. Deep learning denoising methods outperform commercially used iterative reconstruction methods [19].

Deep learning methods applied to the problem of low dose CT denoising are usually based on the concept of deep convolutional neural networks (CNN). Such CNNs have hundreds of thousands of trainable parameters. This makes the behaviour of such networks difficult to comprehend, which aggravates the difficulty of using such methods in regulated settings. This is why, despite their success, deep CNNs have only sparingly been applied in the clinic. It has been shown that including the knowledge of physical operations into a neural network can increase the performance of the neural network [11,12,21]. This also helps to comprehend the behaviour of the network.

In this work, we introduce JBFnet, a CNN designed for low dose CT denoising. JBFnet implements iterative Joint Bilateral Filtering within its architecture. A deep CNN was used to generate the guidance image for the JBF. Shallow 2 layered CNNs were used to calculate the intensity range and spatial domain functions for the JBF. JBFnet has 4 sequential filtering blocks, to implement iterative filtering. Each block learns different range and domain filters. A portion of the NM is mixed back in after filtering, which helps preserve details. We demonstrate that JBFnet significantly outperforms denoising networks with a higher number of trainable parameters, while maintaining behavioural accountability.

2   Methods

2.1   Trainable Joint Bilateral Filtering

JBF takes into account the differences in both the pixel coordinates and the pixel intensities. To achieve this, the JBF has separate functions to estimate the appropriate filtering kernels in both the spatial domain and the intensity ranges. As opposed to bilateral filtering, the JBF uses a guidance image to estimate the intensity range kernel. The operation of the JBF is defined by the following equation:

$$I_f(x) = \frac{\sum_{o \epsilon N(x)} I_n(o) B_w(I_g, x, o)}{\sum_{o \epsilon N(x)} B_w(I_g, x, o)} \; ; \; B_w(I_g, x, o) = G(x - o) F(I_g(x) - I_g(o)) \tag{1}$$

where I_n is the noisy image, I_f is the filtered image, x is the spatial coordinate, $N(x)$ is the neighborhood of x taken into account for the filtering operation, I_g is the guidance image, F is the filtering function for estimating the range kernel, and G is the filtering function for estimating the domain kernel.

We assume that G and F are defined by parameters W_g and W_f respectively. Similarly, we assume that the guidance image I_g is estimated from the noisy image I_n using a function P, whose parameters are defined by W_p. Therefore, we can rewrite Eq. 1 as:

$$I_f(x) = \frac{\sum_{o \epsilon N(x)} I_n(o) G(x - o; W_g) F[P(I_n(x); W_p) - P(I_n(o); W_p); W_f]}{\sum_{o \epsilon N(x)} G(x - o; W_g) F[P(I_n(x); W_p) - P(I_n(o); W_p); W_f]} \tag{2}$$

Since the values of W_f act on the guidance image I_g which is generated by W_p, we can rewrite W_f as a function of W_p. The chain rule can be applied to get the gradient of W_p. The least squares problem to optimize our parameters is given by:

$$L(W_g, W_f(W_p))$$
$$= \frac{1}{2} || \frac{\sum_{o \epsilon N(x)} I_n(o) G(x - o; W_g) F[P(I_n(x); W_p) - P(I_n(o); W_p); W_f]}{\sum_{o \epsilon N(x)} G(x - o; W_g) F[P(I_n(x); W_p) - P(I_n(o); W_p); W_f]} - I_f(x) ||^2 \tag{3}$$

The gradients of W_g and W_p ($\frac{\partial L}{\partial W_g}$ and $\frac{\partial L}{\partial W_f} \frac{\partial W_f}{\partial W_p}$ respectively) can be calculated using backpropogation and the parameters can by updated by standard gradient descent.

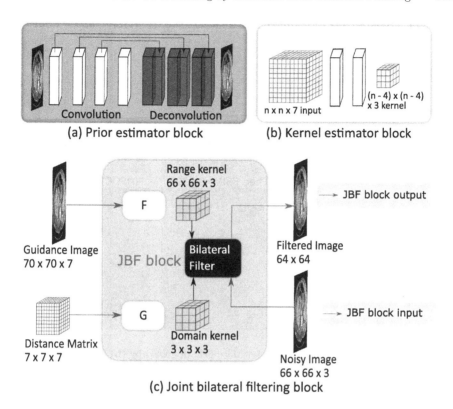

Fig. 2. The architecture of (a) the prior estimator network P, (b) the kernel estimating functions G and F and (c) the functioning of the JBF block. Figure 1 shows how to combine the components to create JBFnet.

2.2 Network Architecture

Prior Estimator Block. The guidance image is estimated from the noisy image I_n using a function P with parameters W_p. We represent this function with a deep neural network inspired by CPCE3D [20]. We have a hybrid 2D/3D network with 4 convolutional layers and 4 deconvolutional layers. Each layer has 32 filters. The filters in the convolutional layers are of size $3 \times 3 \times 3$. The filters in the deconvolutional layers are of size 3×3. Each layer is followed by a leaky ReLU activation function [9]. P takes an input of size $n \times n \times 15$, and returns an output of size $n \times n \times 7$. A structural diagram is present in Fig. 2(a).

JBF Block. We introduce a novel denoising block called the JBF block. JBFnet performs filtering through the use of JBF blocks. Each JBF block contains two filtering functions, G and F, with parameters W_g and W_f respectively. We represent each of these functions with a shallow convolutional network of 2 layers. Each layer has a single convolutional filter of size $3 \times 3 \times 3$. Each layer is followed by a ReLU activation function [9]. No padding is applied, shrinking an

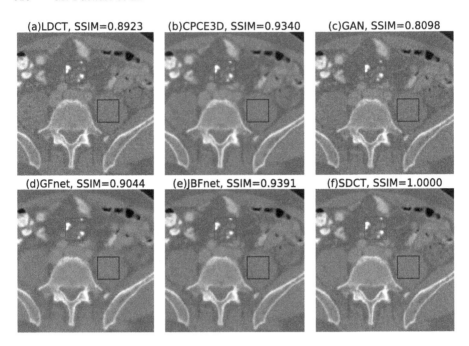

Fig. 3. We display an example of (a) low dose CT, denoised by (b) CPCE3D (c) GAN, (d) GFnet (e) JBFnet, and finally (f) the standard dose CT. The red square indicates some small low contrast features. The blue square indicates a homogenous noisy patch. SSIM scores are displayed in the captions. JBFnet achieves the best performance (SSIM = 0.9391). Images are displayed with a window of [−800, 1200]. (Color figure online)

input of $n \times n \times 7$ to $(n-4) \times (n-4) \times 3$ (Fig. 2(b)). The JBF block then executes a $3 \times 3 \times 3$ standard bilateral filtering operation with the outputs of F and G. Each JBF block is followed by a ReLU activation function. JBFnet contains four consecutive JBF blocks (Fig. 1).

Mixing in the Noise Map. After filtering with the JBF block, there is the possibilty that some important details may have been filtered out. To rectify this, we mixed an amount of the NM back into the filtered output (Fig. 1). The NM was estimated by subtracting the output of the JBF block from the input. We mix the NM by a weighted addition to the output of the JBF block. The weights are determined for each pixel by a 3×3 convolution of the NM.

Loss Functions. We optimize the values of the parameters using two loss functions. We utilize the mean squared error loss and the edge filtration loss [19]. The two loss functions are given by the following equations:

$$MSELoss(I_1, I_2) = \frac{1}{n}||I_1 - I_2||^2; EFLoss(I_1, I_2) = \frac{1}{n}||I_1' - I_2'||^2 \qquad (4)$$

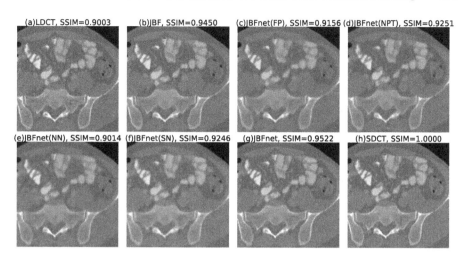

Fig. 4. We display a close-up example of (a) low dose CT, denoised by (b) JBF (c) JBFnet with frozen prior (FP), (d) JBFnet with no pre-training (NPT), (e) JBFnet with no NM (NN) added (f) JBFnet with single-weight NM (SN) added (g) JBFnet, and finally we display (e) the standard dose CT. SSIM scores are displayed in the captions. JBFnet achieves the best performance (SSIM = 0.9522). Images are displayed with a window of $[-800, 1200]$.

I_1' and I_2' were computed using $3 \times 3 \times 3$ Sobel filters on I_1 and I_2 respectively. JBFnet outputs the estimated guidance image I_g, and the outputs of the four JBF blocks, $I_{f(1)}, I_{f(2)}, I_{f(3)}$ and $I_{f(4)}$ (Fig. 1). We indicate the reference image as I. Then, our overall loss function is:

$$\lambda_1[MSELoss(I_{f(4)}, I) + 0.1 * EFLoss(I_{f(4)}, I)] + \lambda_2[MSELoss(I_g, I)]$$
$$+ \lambda_3[\sum_{i=1,2,3} MSELoss(I_{f(i)}, I) + 0.1 * EFLoss(I_{f(i)}, I)] \quad (5)$$

where the λ values are the balancing weights of the loss terms in Eq. 5.

2.3 Training and Implementation

We trained JBFnet using the body scan data from 10 different patients. The scans were reconstructed using standard clinical doses, as well as 5%, 10%, 25% and 50% of the standard dose. The standard dose volumes are our reference volumes I. The reduced dose volumes are our noisy volumes I_n.

JBFnet was fed blocks of size $64 \times 64 \times 15$. The guidance image I_g was estimated from the input, which was shrunk to $64 \times 64 \times 7$ by P. I_g was then zero-padded to $70 \times 70 \times 7$, and then shrunk to a $66 \times 66 \times 3$ range kernel by F (Fig. 2(c)). The distance matrix was constant across the whole image, and was pre-computed as a $7 \times 7 \times 7$ matrix, which was shrunk to a $3 \times 3 \times 3$ kernel by G. The noisy input I_n to the JBF blocks was a $64 \times 64 \times 3$ block, which contained the central 3 slices of the input block. I_n was zero-padded to

Table 1. Mean and standard deviations of the quality metric scores of our method and variants on the AAPM Grand Challenge dataset [15]. The number of parameters in the networks are also included. Plain JBF application yields the highest PSNR (46.79 ± 0.8685), while JBFnet yields the highest SSIM (0.9825 ± 0.0025).

	PSNR	SSIM
Low Dose CT	43.12 ± 1.021	0.9636 ± 0.0073
CPCE3D [20]	45.43 ± 0.6914	0.9817 ± 0.0029
GAN [23]	41.87 ± 0.7079	0.9398 ± 0.0079
Deep GFnet [24]	41.62 ± 0.3856	0.9709 ± 0.0029
JBF	$\mathbf{46.79 \pm 0.8685}$	0.9770 ± 0.0046
JBFnet (Frozen Prior)	42.86 ± 0.4862	0.9768 ± 0.0027
JBFnet (No pre-training)	42.75 ± 0.5076	0.9787 ± 0.0027
JBFnet (No NM)	42.64 ± 0.5001	0.9744 ± 0.0029
JBFnet (Single-weight NM)	42.07 ± 0.5421	0.9776 ± 0.0029
JBFnet	44.76 ± 0.6009	$\mathbf{0.9825 \pm 0.0025}$

$66 \times 66 \times 3$, and then shrunk to a 64×64 output by the bilateral filter (Fig. 2(c)). The NM was added in with learned pixelwise weights. After the JBF block, the output filtered image was padded by the neighbouring slices in both directions, restoring the input size to $64 \times 64 \times 3$ for future JBF blocks. This padding was not performed after the last JBF block, resulting in an overall output size of 64×64.

Training was performed for 30 epochs over the whole dataset. 32 image slabs were presented per batch. For the first ten epochs, only W_p was updated ($\lambda_1 = 0$, $\lambda_2 = 1$, $\lambda_3 = 0$). This was to ensure a good quality guidance image for the JBF blocks. From epochs 10–30, all the weights were updated ($\lambda_1 = 1$, $\lambda_2 = 0.1$, $\lambda_3 = 0.1$). JBFnet was implemented in PyTorch on a PC with an Intel Xeon E5-2640 v4 CPU and an NVIDIA Titan Xp GPU.

3   Experimental Results

3.1   Test Data and Evaluation Metrics

We tested our denoising method on the AAPM Low Dose CT Grand Challenge dataset [15]. The dataset consisted of 10 body CT scans, each reconstructed at standard doses as used in the clinic and at 25% of the standard dose. The slices were of 1 mm thickness. We aimed to map the reduced dose images onto the standard dose images. The full dose images were treated as reference images. Inference was performed using overlapping $256 \times 256 \times 15$ blocks extracted from the test data.

We used the PSNR and the SSIM to measure the performance of JBFnet. Since structural information preservation is more important in medical imaging, the SSIM is a far more important metric of CT image quality than the

PSNR. Due to the small number of patients, we used the Wilcoxon signed test to measure statistical significance. A p-value of 0.05 was used as the threshold for determining statistical significance.

3.2 Comparison to State-of-the-Art Methods

We compare the denoising performance of JBFnet to other denoising methods that use deep learning. We compare JBFnet against CPCE3D [20], GAN [23], and deep GFnet [24] (Fig. 3). All networks were trained over 30 epochs on the same dataset. JBFnet achieves significantly higher scores in both PSNR and SSIM compared to GAN and deep GFnet (w = 0.0, p = 0.005). CPCE3D achieves a significantly higher PSNR than JBFnet (w = 0.0, p = 0.005), but a significantly lower SSIM (w = 0.0, p = 0.005) (Table 1). Additionally, the JBF block consists of only 112 parameters, compared to the guided filtering block [24] which contains 1,555 parameters.

3.3 Ablation Study

Training the Filtering Functions. Usually, the filter functions F and G of the bilateral filter are assumed to be Gaussian functions. We check if representing these functions with convolutions improves the denoising performance of our network. Training the filtering functions reduces our PSNR (w = 0.0, p = 0.005) but improves our SSIM (w = 0.0, p = 0.005) (Table 1 and Fig. 4).

Pre-training the Prior Estimator. In our current training setup, we exclusively train the prior estimator P for 10 epochs, to ensure a good quality prior image. We check if avoiding this pre-training, or freezing the value of P after training improves the performance of our network. Both freezing P and not doing any pre-training reduce the PSNR and SSIM (w = 0.0, p = 0.005) (Table 1 and Fig. 4).

Pixelwise Mixing of the NM. Currently, we estimate the amount of the NM to be mixed back in by generating pixelwise coefficients from a single 3×3 convolution of the NM. We check if not adding in the NM, or adding in the NM with a fixed weight improves the denoising performance of the network. Not mixing in the NM reduces our PSNR and SSIM significantly (w = 0.0, p = 0.005). Mixing in the NM with a fixed weight reduces both PSNR and SSIM even futher (w = 0.0, p = 0.005) (Table 1 and Fig. 4).

4 Conclusion

In this study, we introduced JBFnet, a neural network which implements Joint Bilateral Filtering with learnable parameters. JBFnet significantly improves the denoising performance in low dose CT compared to standard Joint Bilateral

Filtering. JBFnet also outperforms state-of-the-art deep denoising networks in terms of structural preservation. Furthermore, most of the parameters in JBFnet are present in the prior estimator. The actual filtering operations are divided into various JBF blocks, each of which has only 112 trainable parameters. This allows JBFnet to denoise while still maintaining physical interpretability.

References

1. Angel, E.: AIDR 3D iterative reconstruction: integrated, automated and adaptive dose reduction (2012). https://us.medical.canon/download/aidr-3d-wp-aidr-3d
2. Fan, F., et al.: Quadratic autoencoder (Q-AE) for low-dose CT denoising. IEEE Trans. Med. Imaging (2019). https://doi.org/10.1109/tmi.2019.2963248
3. Gilbert, P.: Iterative methods for the three-dimensional reconstruction of an object from projections. J. Theor. Biol. **36**(1), 105–117 (1972). https://doi.org/10.1016/0022-5193(72)90180-4
4. Goodfellow, I.J., Bengio, Y., Courville, A.: Deep Learning (2014). https://doi.org/10.1016/B978-0-12-801775-3.00001-9
5. Hounsfield, G.N.: Computerized transverse axial scanning (tomography). Br. J. Radiol. **46**(552), 1016–1022 (1973). https://doi.org/10.1259/0007-1285-46-552-1016
6. Kaczmarz, S.: Angenäherte Auflösung von Systemen linearer Gleichungen. Bulletin International de l'Académie Polonaise des Sciences et des Lettres. Classe des Sciences Mathématiques et Naturelles. Série A, Sciences Mathématiques **35**, 355–357 (1937)
7. Lecun, Y., Bengio, Y., Hinton, G.: Deep learning. Nature **521**(7553), 436–444 (2015). https://doi.org/10.1038/nature14539
8. Li, M., Hsu, W., Xie, X., Cong, J., Gao, W.: SACNN: self-attention convolutional neural network for low-dose CT denoising with self-supervised perceptual loss network. IEEE Trans. Med. Imaging (2020). https://doi.org/10.1109/TMI.2020.2968472
9. Maas, A.L., Hannun, A.Y., Ng, A.Y.: Rectifier nonlinearities improve neural network acoustic models. In: International Conference on Machine Learning, vol. 30, p. 6 (2013)
10. Maier, A., Fahrig, R.: GPU denoising for computed tomography. In: Xun, J., Jiang, S. (eds.) Graphics Processing Unit-Based High Performance Computing in Radiation Therapy, 1 edn., pp. 113–128. CRC Press, Boca Raton (2015)
11. Maier, A., et al.: Precision learning: towards use of known operators in neural networks. In: Proceedings of the International Conference on Pattern Recognition, No. 2, pp. 183–188 (2018). https://doi.org/10.1109/ICPR.2018.8545553
12. Maier, A.K., et al.: Learning with known operators reduces maximum error bounds. Nature Mach. Intell. **1**(8), 373–380 (2019). https://doi.org/10.1038/s42256-019-0077-5
13. Manduca, A., et al.: Projection space denoising with bilateral filtering and CT noise modeling for dose reduction in CT. Med. Phys. **36**(11), 4911–4919 (2009). https://doi.org/10.1118/1.3232004
14. Manhart, M., Fahrig, R., Hornegger, J., Doerfler, A., Maier, A.: Guided noise reduction for spectral CT with energy-selective photon counting detectors. In: Proceedings of the Third CT Meeting, No. 1, pp. 91–94 (2014)
15. Mccollough, C.H.: Low Dose CT Grand Challenge (2016)

16. Nie, D., et al.: Medical image synthesis with context-aware generative adversarial networks. In: Medical Image Computing and Computer Assisted Intervention, vol. 3, pp. 417–425 (2017). https://doi.org/10.1007/978-3-319-66179-7
17. Oppelt, A.: Noise in computed tomography. In: Aktiengesselschaft, S. (ed.) Imaging Systems for Medical Diagnostics, chap. 13.1.4.2, 2nd edn., p. 996. Publicis Corporate Publishing (2005). https://doi.org/10.1145/2505515.2507827
18. Ramirez-Giraldo, J.C., Grant, K.L., Raupach, R.: ADMIRE: advanced modeled iterative reconstruction (2015). https://www.siemens-healthineers.com/computed-tomography/technologies-innovations/admire
19. Shan, H., et al.: Competitive performance of a modularized deep neural network compared to commercial algorithms for low-dose CT image reconstruction. Nature Mach. Intell. **1**(6), 269–276 (2019). https://doi.org/10.1038/s42256-019-0057-9
20. Shan, H., et al.: 3D convolutional encoder-decoder network for low-dose CT via transfer learning from a 2D trained network. IEEE Trans. Med. Imaging **37**(6), 1522–1534 (2018). https://doi.org/10.4172/2157-7633.1000305
21. Syben, C., et al.: Precision learning: reconstruction filter kernel discretization. In: Proceedings of the 5th International Conference on Image Formation in X-ray Computed Tomography, Salt Lake City, pp. 386–390 (2017)
22. Wolterink, J.M., Dinkla, A.M., Savenije, M.H., Seevinck, P.R., van den Berg, C.A., Isgum, I.: Deep MR to CT synthesis using unpaired data. In: Workshop on Simulation and Synthesis in Medical Imaging, pp. 14–23 (2017). https://doi.org/10.1007/978-3-319-68127-6
23. Wolterink, J.M., Leiner, T., Viergever, M.A., Isgum, I.: Generative l CT. IEEE Trans. Med. Imaging **36**(12), 2536–2545 (2017). https://doi.org/10.1109/TMI.2017.2708987
24. Wu, H., Zheng, S., Zhang, J., Huang, K.: Fast end-to-end trainable guided filter. In: Computer Vision and Pattern Recognition, pp. 1838–1847 (2018)
25. Yang, Q., et al.: Low-dose CT Image denoising using a generative adversarial network with wasserstein distance and perceptual loss. IEEE Trans. Med. Imaging **37**(6), 1348–1357 (2018). https://doi.org/10.1109/TMI.2018.2827462
26. Zhang, Y., et al.: Low-dose CT via convolutional neural network. Biomed. Opt. Express **8**(2), 679–694 (2017). https://doi.org/10.1364/boe.8.000679

MI²GAN: Generative Adversarial Network for Medical Image Domain Adaptation Using Mutual Information Constraint

Xinpeng Xie[1], Jiawei Chen[2], Yuexiang Li[2(✉)], Linlin Shen[1(✉)], Kai Ma[2], and Yefeng Zheng[2]

[1] Computer Vision Institute, Shenzhen University, Shenzhen, China
llshen@szu.edu.cn
[2] Tencent Jarvis Lab, Shenzhen, China
vicyxli@tencent.com

Abstract. Domain shift between medical images from multicentres is still an open question for the community, which degrades the generalization performance of deep learning models. Generative adversarial network (GAN), which synthesize plausible images, is one of the potential solutions to address the problem. However, the existing GAN-based approaches are prone to fail at preserving image-objects in image-to-image (I2I) translation, which reduces their practicality on domain adaptation tasks. In this paper, we propose a novel GAN (namely MI²GAN) to maintain image-contents during cross-domain I2I translation. Particularly, we disentangle the content features from domain information for both the source and translated images, and then maximize the mutual information between the disentangled content features to preserve the image-objects. The proposed MI²GAN is evaluated on two tasks—polyp segmentation using colonoscopic images and the segmentation of optic disc and cup in fundus images. The experimental results demonstrate that the proposed MI²GAN can not only generate elegant translated images, but also significantly improve the generalization performance of widely used deep learning networks (e.g., U-Net).

Keywords: Mutual information · Domain adaptation.

1 Introduction

Medical images from multicentres often have different imaging conditions, e.g., color and illumination, which make the models trained on one domain usually fail to generalize well to another. Domain adaptation is one of the effective methods

---

This work was done when Xinpeng Xie was an intern at Tencent Jarvis Lab.
J. Chen—Equal contribution.

© Springer Nature Switzerland AG 2020
A. L. Martel et al. (Eds.): MICCAI 2020, LNCS 12262, pp. 516–525, 2020.
https://doi.org/10.1007/978-3-030-59713-9_50

to boost the generalization capability of models. Witnessing the success of generative adversarial networks (GANs) [5] on image synthesis [9,20], researchers began trying to apply the GAN-based networks for image-to-image domain adaptation. For example, Chen et al. [2] used GAN to transfer the X-ray images from a new dataset to the domain of the training set before testing, which increases the test accuracy of trained models. Zhang et al. [22] proposed a task driven generative adversarial network (TD-GAN) for the cross-domain adaptation of X-ray images. Most of the existing GAN-based I2I domain adaptation methods adopted the cycle-consistency loss [10,21,25] to loose the requirement of paired cross-domain images for training. However, recent studies [8,23] proved that cycle-consistency-based frameworks easily suffer from the problem of content distortion during image translation. Let T be a bijective geometric transformation (e.g., translation, rotation, scaling, or even nonrigid transformation) with inverse transformation T^{-1}, the following generators G'_{AB} and G'_{BA} are also cycle consistent.

$$G'_{AB} = G_{AB}T, \ G'_{BA} = G_{BA}T^{-1} \tag{1}$$

where the G_{AB} and G_{BA} are the original cycle-consistent generators establishing two mappings between domains A and B. Consequently, due to lack of penalty in content disparity between source and translated images, the content of a translated image by cycle-consistency-based frameworks may be distorted by T, which is unacceptable in medical image processing.

To tackle the problem, we propose a novel **GAN** (MI^{2}GAN) to maintain the contents of **M**edical **I**mage during I2I domain adaptation by maximizing the **M**utual **I**nformation between the source and translated images. Our idea relies on two observations: 1) the content features containing the information of image-objects can be fully disentangled from the domain information; and 2) the mutual information, measuring the information that two variables share, can be used as a metric for image-object preservation. Mutual information constraint has been widely used for various medical image processing tasks, such as image registration [15]. Given two variables X and Y, the mutual information I shared by X and Y can be formulated as:

$$\mathcal{I}(X;Y) = KL(\mathbb{J}||\mathbb{M}) \tag{2}$$

where $\mathbb{J}$ and $\mathbb{M}$ are joint distribution and the product of marginals of X and Y; KL is the KL-divergence. Specifically, $\mathbb{J} = p(y|x)p(x)$ and $\mathbb{M} = p(y)p(x)$, where $x \in X$ and $y \in Y$; $p(x)$ and $p(y)$ are the distributions of X and Y, respectively; $p(y|x)$ is the conditional probability of y given x.

Since the posterior probability $p(y|x)$ is difficult to be directly estimate [4], we measure and maximize the MI between source and translated images based on the approach similar to [1,7]. Specifically, the content features of source and translated images are first extracted by the paired adversarial auto-encoders, which are then fed to a discriminator for the estimation of mutual information. Extensive experiments are conducted to validate the effectiveness of our MI^{2}GAN. The experimental results demonstrate that the proposed MI^{2}GAN

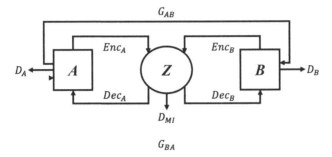

Fig. 1. The framework of our MI^{2}GAN. Similar to CycleGAN [25], our MI^{2}GAN adopts paired generators (G_{AB} and G_{BA}) and discriminators (D_B and D_A) to achieve cross-domain image translation. To preserve image-contents, X-shape dual auto-encoders ($\{Enc_A, Dec_A\}$ and $\{Enc_B, Dec_B\}$) and a mutual information discriminator (D_{MI}) are implemented.

can not only produce plausible translated images, but also significantly reduce the performance degradation caused by the domain shift.

2   MI^{2}GAN

The pipeline of our MI^{2}GAN is presented in Fig. 1. Similar to current cycle-consistency-based GAN [25], our MI^{2}GAN adopts paired generators (G_{AB} and G_{BA}) and discriminators (D_B and D_A) to achieve cross-domain image translation without paired training samples. To distill the content features from domain information, X-shape dual auto-encoders (i.e., Enc_A, Dec_A, Enc_B, and Dec_B) are implemented. The encoders (i.e., Enc_A and Enc_B) are responsible to embed the content information of source and translated images into the same latent space Z, while the decoders (i.e., Dec_A and Dec_B) aim to transform the embedded content features to their own domains using domain-related information. Therefore, to alleviate the content distortion problem during image translation, we only need to maximize the mutual information between the content features of source and translated images, which is achieved by our mutual information discriminator. In the followings, we present the modules for content feature disentanglement and mutual information maximization in details.

2.1   X-Shape Dual Auto-Encoders

We proposed the X-shape dual auto-encoders (AEs), consisting of Enc_A, Dec_A, Enc_B, and Dec_B, to disentangle the features containing content information. As the mappings between domains A and B are symmetrical, we take the content feature distillation of images from domain A as an example. The pipeline is shown in Fig. 2 (a). Given an input image (I_a), the auto-encoder (Enc_A and Dec_A) embeds it into a latent space, which can be formulated as:

$$z_a = Enc_A(I_a), \quad I'_a = Dec_A(z_a) \tag{3}$$

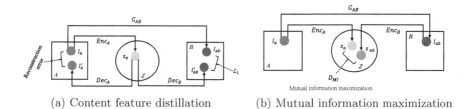

(a) Content feature distillation   (b) Mutual information maximization

Fig. 2. The pipelines of the main components contained in our framework. (a) X-shape dual auto-encoders and (b) mutual information discriminator.

where $I_{a'}$ is the reconstruction of I_a. The embedded feature z_a contains the information of content and domain A. To disentangle them, z_a is mapped to domain B via Dec_B:

$$I'_{ab} = Dec_B(z_a) \tag{4}$$

where I'_{ab} is the mapping result of z_a.

As shown in Fig. 2, apart from the X-shape dual AEs, there is another translation path between domain A and B: $I_{ab} = G_{AB}(I_a)$, where I_{ab} is the translated image yielded by G_{AB}. Through simultaneously minimizing the pixel-wise L1 norm between I_{ab} and I'_{ab}, and reconstruction error between I_a and $I_{a'}$, Dec_A and Dec_B are encouraged to recover domain-related information from the latent space (in short, the encoders remove domain information and the decoders recover it), which enable them to map the z_a to two different domains. Therefore, the information contained in z_a is highly related to the image-objects without domain bias. The content feature distillation loss ($\mathcal{L}_{dis}$), combining aforementioned two terms, can be formulated as:

$$\mathcal{L}_{dis} = ||I_{ab} - I'_{ab}||_1 + ||I_a - I_{a'}||_1. \tag{5}$$

2.2 Mutual Information Discriminator

Using our X-shape dual AEs, the content features of source I_a and translated I_{ab} images can be disentangled to z_a and z_{ab}, respectively. The content feature of translated image preserving image-objects should contain similar information to that of source image. To this end, the encoder (Enc_B) needs to implicitly impose statistical constraints onto learned representations, which thereby pushes the translated distribution of Z_{ab} to match the source Z_a (i.e., mutual information maximization between Z_a and Z_{ab}), where Z_{ab} and Z_a are two sub-spaces of Z.

Referred to adversarial training, which matches the distribution of synthesized images to that of real ones, this can be achieved by training a mutual information discriminator (D_{MI}) to distinguish between samples coming from the joint distribution, $\mathbb{J}$, and the product of marginals, $\mathbb{M}$, of the two sub-spaces Z_a and Z_{ab} [7]. We use a lower-bound to the mutual information ($\mathcal{I}$

defined in Eq. 2) based on the Donsker-Varadhan representation (DV) of the KL-divergence, which can be formulated as:

$$\mathcal{I}(Z_a; Z_{ab}) \geq \widehat{\mathcal{I}}^{(DV)}(Z_a; Z_{ab}) = \mathbb{E}_{\mathbb{J}}\left[D_{MI}(z_a, z_{ab})\right] - \log \mathbb{E}_{\mathbb{M}}\left[e^{D_{MI}(z_a, z_{ab})}\right] \quad (6)$$

where $D_{MI} : z_a \times z_{ab} \rightarrow \mathbb{R}$ is a discriminator function modeled by a neural network.

To constitute the real ($\mathbb{J}$) and fake ($\mathbb{M}$) samples for the D_{MI}, an image is randomly selected from domain B and encoded to z_b. The z_a is then concatenated to z_{ab} and z_b, respectively, which forms the samples from the joint distribution ($\mathbb{J}$) and the product of marginals ($\mathbb{M}$) for the mutual information discriminator.

Objective. With the previously defined feature distillation loss ($\mathcal{L}_{dis}$) and mutual information discriminator, the full objective $\mathcal{L}$ for the proposed MI^{2}GAN is summarized as:

$$\begin{aligned}
\mathcal{L} = {} & \mathcal{L}_{adv}(G_{BA}, D_A) + \mathcal{L}_{adv}(G_{AB}, D_B) + \alpha\mathcal{L}_{cyc}(G_{AB}, G_{BA}) \\
& + \beta\mathcal{L}_{dis}(G_{AB}, Enc_A, Dec_A, Dec_B) + \beta\mathcal{L}_{dis}(G_{BA}, Enc_B, Dec_B, Dec_A) \quad (7) \\
& + \widehat{\mathcal{I}}(G_{AB}, Enc_A, Enc_B, D_{MI}) + \widehat{\mathcal{I}}(G_{BA}, Enc_A, Enc_B, D_{MI})
\end{aligned}$$

where $\mathcal{L}_{adv}$ and $\mathcal{L}_{cyc}$ are adversarial and cycle-consistency losses, the same as that proposed in [25]. The weights α and β for $\mathcal{L}_{cyc}$ and $\mathcal{L}_{dis}$ respectively are all set to 10.

2.3   Implementation Details

Network Architecture. Consistent to the standard of CycleGAN [25], the proposed MI^{2}GAN involves paired generators (G_{AB}, G_{BA}) and discriminators (D_B, D_A). Instance normalization [18] is employed in the generators to produce elegant translation images, while PatchGAN is adopted in the discriminators [9,12] to provide patch-wise predictions. Our X-shape AEs and mutual information discriminator adopt instance normalization and leaky ReLU in their architectures, and the detailed information can be found in the *arXiv version*.

Optimization Process. The optimization of $\mathcal{L}_{dis}$ and $\widehat{\mathcal{I}}$ is performed in the same manner of $\mathcal{L}_{adv}$—fixing X-shape dual AEs, D_{MI} and D_A/D_B to optimize G_{BA}/G_{AB} first, and then optimize AEs, D_{MI} and D_A/D_B respectively, with fixed G_{BA}/G_{AB}. Therefore, similar to discriminators, our X-shape dual AEs and mutual information discriminator can directly pass the knowledge of image-objects to the generators, which helps them to improve the quality of translated results in terms of object preservation.

3   Experiments

Deep neural networks often suffer from performance degradation when applied to a new test dataset with domain shift (e.g., color and illumination) caused by different imaging instruments. Our MI^{2}GAN tries to address the problem by translating the test images to the same domain of the training set. In this section, to validate the effectiveness of the proposed MI^{2}GAN, we evaluate it on several publicly available datasets.

3.1 Datasets

Colonoscopic Datasets. The publicly available colonoscopic video datasets, i.e., CVC-Clinic [19] and ETIS-Larib [17], are selected for multicentre adaptation. The CVC-Clinic dataset is composed of 29 sequences with a total of 612 images. The ETIS-Larib consists of 196 images, which can be manually separated to 29 sequences as well. Those short videos are extracted from the colonoscopy videos captured by different centres using different endoscopic devices. All the frames of the short videos contain polyps. In this experiment, the extremely small ETIS-Larib dataset (196 frames) is used as the test set, while the relatively larger CVC-Clinic dataset (612 frames) is used for network optimization (80:20 for training and validation).

REFUGE. The REFUGE challenge dataset [14] consists of 1,200 fundus images for optic disc (OD) and optic cup (OC) segmentation, which were partitioned to training (400), validation (400) and test (400) sets by the challenge organizer. The images available in this challenge were acquired with two different fundus cameras—Zeiss Visucam 500 for the training set and Canon CR-2 for the validation and test sets, resulting in visual gap between training and validation/test samples. Since the test set is unavailable, we conduct experiment on I2I adaptation between the training and validation sets. The public training set is separated to training and validation sets according to the ratio of 80:20, and the public validation set is used as the test set.

Baselines Overview and Evaluation Criterion. Several unpaired image-to-image domain adaptation frameworks, including CycleGAN [25], UNIT [13] and DRIT [11], are taken as baselines for the performance evaluation. The direct transfer approach, which directly takes the source domain data for testing without any adaptation, is also involved for comparison. The Dice score (DSC), which measures the spatial overlap index between the segmentation results and ground truth, is adopted as the metric to evaluate the segmentation accuracy.

3.2 Ablation Study

Content Feature Distillation. We invite three experienced experts to manually tune two CVC images to the domain of ETIS (as shown in the first row of Fig. 3), i.e., tuning the image conditions such as color and saturation based on the statistical histogram of the ETIS domain. The two paired images contain the same content information but totally different domain-related knowledge. To ensure our X-shape dual auto-encoders really learn to disentangle the content features from domain information, we sent the paired images to X-shape dual AEs and visualize the content features produced by Enc_A and Enc_B using CAM

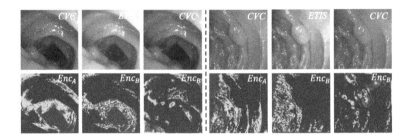

Fig. 3. Content features (the second row) produced by the encoders of our X-shape dual AEs for the input images (the first row) from different domains. The CVC images (left) are manually tuned to ETIS (middle) by experienced experts.

[24] (as illustrated in the second row of Fig. 3). For comparison, the CVC images are also sent to Enc_B for content feature distillation. It can be observed that the CVC and ETIS images respectively going through Enc_A and Enc_B result in the similar activation patterns, while the encoders yield different patterns for the CVC images. The experimental result demonstrate that the encoders of our X-shape dual AEs are domain-specific, which are able to remove the their own domain-related information from the embedding space.

Mutual Information Discriminator. To validate the contribution made by the mutual information discriminator, we evaluate the performance of MI²GAN without D_{MI}. The evaluation results are presented in Table 1. The segmentation accuracy on the test set significantly drops to 65.96%, 77.27% and 92.17% for polyp, OC and OD, respectively, with the removal of D_{MI}, which demonstrates the importance of D_{MI} for image-content preserving domain adaptation.

Table 1. DSC (%) of the polyp segmentation on colonoscopy and the segmentation of optical cup (OC) and optical disk (OD) on REFUGE fundus images, respectively.

	Colonoscopy		Fundus			
	CVC (val.)	ETIS (test)	$OC_{val.}$	$OD_{val.}$	OC_{test}	OD_{test}
Direct transfer	80.79	64.33	85.83	95.42	81.66	93.49
DRIT [11]		28.32			64.79	69.03
UNIT [13]		23.46			71.63	74.58
CycleGAN [25]		52.41			71.53	85.83
MI²GAN (Ours)		**72.86**			**83.49**	**94.87**
MI² GAN w/o D_{MI}		65.96			77.27	92.17

3.3   Comparison to State of the Art

Different I2I domain adaptation approaches are applied to the colonoscopic and fundus image datasets, respectively, which translate the test images to the domain of the training set to narrow the gap between them and improve the model generalization. The adaptation results generated by different I2I domain adaptation approaches are presented in Fig. 4. The first row of Fig. 4 shows the examplars from the training sets of colonoscopy and REFUGE datasets. Content distortions are observed in the adaptation results produced by most of the existing I2I translation approaches. In contrast, our MI^{2}GAN yields plausible adaptation results while excellently preserving the image-contents.

For quantitative analysis, we present the segmentation accuracy of deep learning networks with different adaptation approaches in Table 1. To comprehensively assess the adaptation performance of our MI^{2}GAN, we adopt two widely-used deep learning networks, i.e., ResUNet-50 [6,16] and DeepLab-V3 [3], for the polyp segmentation, and OC/OD segmentation, respectively. As shown in Table 1, due to the lack of capacity of image-content preservation, most of

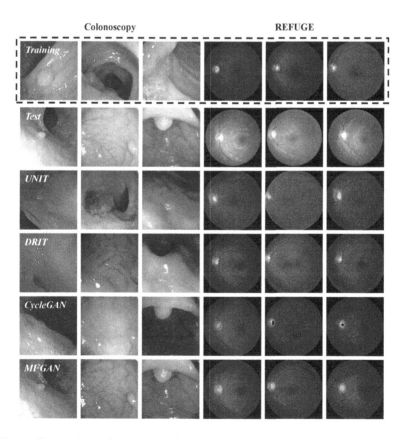

Fig. 4. Comparison of images produced by different I2I adaptation approaches.

existing I2I domain adaptation approaches degrade the segmentation accuracy for both tasks, compared to the direct transfer. The DRIT [11] yields the highest degradation of DSC, -40.87%, -16.87% and -24.46% for polyp, OC and OD, respectively. Conversely, the proposed MI^2GAN remarkably boosts the segmentation accuracy of polyp $(+8.53\%)$, OC $(+1.83\%)$, and OD $(+1.38\%)$ to the direct transfer, which are closed to the accuracy on validation set.

4   Conclusion

In this paper, we proposed a novel GAN (namely MI^2GAN) to maintain image-contents during cross-domain I2I translation. Particularly, we disentangle the content features from domain information for both the source and translated images, and then maximize the mutual information between the disentangled content features to preserve the image-objects.

Acknowledgement. This work is supported by the Natural Science Foundation of China (No. 91959108 and 61702339), the Key Area Research and Development Program of Guangdong Province, China (No. 2018B010111001), National Key Research and Development Project (2018YFC2000702) and Science and Technology Program of Shenzhen, China (No. ZDSYS201802021814180).

References

1. Belghazi, I., Rajeswar, S., Baratin, A., Hjelm, R.D., Courville, A.C.: MINE: mutual information neural estimation. In: International Conference on Machine Learning (2018)
2. Chen, C., Dou, Q., Chen, H., Heng, P.A.: Semantic-aware generative adversarial nets for unsupervised domain adaptation in chest X-ray segmentation. In: International Workshop on Machine Learning in Medical Imaging (2018)
3. Chen, L.C., Papandreou, G., Schroff, F., Adam, H.: Rethinking atrous convolution for semantic image segmentation. arXiv preprint arXiv:1706.05587 (2017)
4. Chen, X., Duan, Y., Houthooft, R., Schulman, J., Sutskever, I., Abbeel, P.: InfoGAN: interpretable representation learning by information maximizing generative adversarial nets. In: Annual Conference on Neural Information Processing Systems (2016)
5. Goodfellow, I., et al.: Generative adversarial nets. In: Annual Conference on Neural Information Processing Systems (2014)
6. He, K., Zhang, X., Ren, S., Sun, J.: Deep residual learning for image recognition. In: IEEE Conference on Computer Vision and Pattern Recognition (2016)
7. Hjelm, R.D., et al.: Learning deep representations by mutual information estimation and maximization. In: International Conference on Learning Representations (2019)
8. Huang, S., Lin, C., Chen, S., Wu, Y., Hsu, P., Lai, S.: AugGAN: cross domain adaptation with GAN-based data augmentation. In: European Conference on Computer Vision (2018)
9. Isola, P., Zhu, J.Y., Zhou, T., Efros, A.A.: Image-to-image translation with conditional adversarial networks. In: IEEE Conference on Computer Vision and Pattern Recognition (2017)

10. Kim, T., Cha, M., Kim, H., Lee, J., Kim, J.: Learning to discover cross-domain relations with generative adversarial networks. In: International Conference on Machine Learning (2017)

11. Lee, H.Y., Tseng, H.Y., Huang, J.B., Singh, M.K., Yang, M.H.: Diverse image-to-image translation via disentangled representations. In: European Conference on Computer Vision (2018)

12. Li, C., Wand, M.: Precomputed real-time texture synthesis with markovian generative adversarial networks. In: Leibe, B., Matas, J., Sebe, N., Welling, M. (eds.) ECCV 2016. LNCS, vol. 9907, pp. 702–716. Springer, Cham (2016). https://doi.org/10.1007/978-3-319-46487-9_43

13. Liu, M.Y., Breuel, T., Kautz, J.: Unsupervised image-to-image translation networks. In: Annual Conference on Neural Information Processing Systems (2017)

14. Orlando, J.I., et al.: REFUGE challenge: a unified framework for evaluating automated methods for glaucoma assessment from fundus photographs. Med. Image Anal. **59**, 101570 (2020)

15. Pluim, J.P.W., Maintz, J.B.A., Viergever, M.A.: Mutual-information-based registration of medical images: a survey. IEEE Trans. Med. Imaging **22**(8), 986–1004 (2003)

16. Ronneberger, O., Fischer, P., Brox, T.: U-net: convolutional networks for biomedical image segmentation. In: International Conference on Medical Image Computing and Computer Assisted Intervention (2015)

17. Silva, J., Histace, A., Romain, O., Dray, X., Granado, B.: Toward embedded detection of polyps in WCE images for early diagnosis of colorectal cancer. Int. J. Comput. Assist. Radiol. Surg. **9**(2), 283–293 (2013). https://doi.org/10.1007/s11548-013-0926-3

18. Ulyanov, D., Vedaldi, A., Lempitsky, V.: Instance normalization: the missing ingredient for fast stylization. arXiv preprint arXiv:1607.08022 (2016)

19. Vazquez, D., et al.: A benchmark for endoluminal scene segmentation of colonoscopy images. Journal of Healthcare Engineering, vol. 2017 (2017)

20. Wang, T., Liu, M., Zhu, J., Tao, A., Kautz, J., Catanzaro, B.: High-resolution image synthesis and semantic manipulation with conditional GANs. In: IEEE Conference on Computer Vision and Pattern Recognition (2018)

21. Yi, Z., Zhang, H., Tan, P., Gong, M.: DualGAN: unsupervised dual learning for image-to-image translation. In: IEEE International Conference on Computer Vision (2017)

22. Wolterink, J.M., Leiner, T., Išgum, I.: Graph convolutional networks for coronary artery segmentation in cardiac CT angiography. In: Zhang, D., Zhou, L., Jie, B., Liu, M. (eds.) GLMI 2019. LNCS, vol. 11849, pp. 62–69. Springer, Cham (2019). https://doi.org/10.1007/978-3-030-35817-4_8

23. Zhang, Z., Yang, L., Zheng, Y.: Translating and segmenting multimodal medical volumes with cycle- and shape-consistency generative adversarial network. In: IEEE Conference on Computer Vision and Pattern Recognition (2018)

24. Zhou, B., Khosla, A., Lapedriza, A., Oliva, A., Torralba, A.: Learning deep features for discriminative localization. In: IEEE Conference on Computer Vision and Pattern Recognition (2016)

25. Zhu, J., Park, T., Isola, P., Efros, A.A.: Unpaired image-to-image translation using cycle-consistent adversarial networks. In: IEEE International Conference on Computer Vision (2017)

Machine Learning Applications

Joint Modeling of Chest Radiographs and Radiology Reports for Pulmonary Edema Assessment

Geeticka Chauhan[1], Ruizhi Liao[1(✉)], William Wells[1,2], Jacob Andreas[1],
Xin Wang[3], Seth Berkowitz[4], Steven Horng[4], Peter Szolovits[1],
and Polina Golland[1]

[1] Massachusetts Institute of Technology, Cambridge, MA, USA
ruizhi@mit.edu
[2] Brigham and Women's Hospital, Harvard Medical School, Boston, MA, USA
[3] Philips Research North America, Cambridge, MA, USA
[4] Beth Israel Deaconess Medical Center,
Harvard Medical School, Boston, MA, USA

Abstract. We propose and demonstrate a novel machine learning algorithm that assesses pulmonary edema severity from chest radiographs. While large publicly available datasets of chest radiographs and free-text radiology reports exist, only limited numerical edema severity labels can be extracted from radiology reports. This is a significant challenge in learning such models for image classification. To take advantage of the rich information present in the radiology reports, we develop a neural network model that is trained on both images and free-text to assess pulmonary edema severity from chest radiographs at inference time. Our experimental results suggest that the joint image-text representation learning improves the performance of pulmonary edema assessment compared to a supervised model trained on images only. We also show the use of the text for explaining the image classification by the joint model. To the best of our knowledge, our approach is the first to leverage free-text radiology reports for improving the image model performance in this application. Our code is available at: https://github.com/RayRuizhiLiao/joint_chestxray.

1 Introduction

We present a novel approach to training machine learning models for assessing pulmonary edema severity from chest radiographs by jointly learning representations from the images (chest radiographs) and their associated radiology

---

G. Chauhan and R. Liao—Co-first authors.

Electronic supplementary material The online version of this chapter (https://doi.org/10.1007/978-3-030-59713-9_51) contains supplementary material, which is available to authorized users.

A. L. Martel et al. (Eds.): MICCAI 2020, LNCS 12262, pp. 529–539, 2020.
https://doi.org/10.1007/978-3-030-59713-9_51

reports. Pulmonary edema is the most common reason patients with acute congestive heart failure (CHF) seek care in hospitals [1,9,14]. The treatment success in acute CHF cases depends crucially on effective management of patient fluid status, which in turn requires pulmonary edema quantification, rather than detecting its mere absence or presence.

Chest radiographs are commonly acquired to assess pulmonary edema in routine clinical practice. Radiology reports capture radiologists' impressions of the edema severity in the form of unstructured text. While the chest radiographs possess ground-truth information about the disease, they are often time intensive (and therefore expensive) for manual labeling. Therefore, labels extracted from reports are used as a proxy for ground-truth image labels. Only limited numerical edema severity labels can be extracted from the reports, which limits the amount of labeled image data we can learn from. This presents a significant challenge for learning accurate image-based models for edema assessment. To improve the performance of the image-based model and allow leveraging larger amount of training data, we make use of free-text reports to include rich information about radiographic findings and reasoning of pathology assessment. We incorporate free-text information associated with the images by including them during our training process.

We propose a neural network model that jointly learns from images and free-text to quantify pulmonary edema severity from images (chest radiographs). At training time, the model learns from a large number of chest radiographs and their associated radiology reports, with a limited number of numerical edema severity labels. At inference time, the model computes edema severity given the input image. While the model can also make predictions from reports, our main interest is to leverage free-text information during training to improve the accuracy of image-based inference. Compared to prior work in the image-text domain that fuses image and text features [5], our goal is to decouple the two modalities during inference to construct an accurate image-based model.

Prior work in assessing pulmonary edema severity from chest radiographs has focused on using image data only [17]. To the best of our knowledge, ours is the first method to leverage the free-text radiology reports for improving the image model performance in this application. Our experimental results demonstrate that the joint representation learning framework improves the accuracy of edema severity estimates over a purely image-based model on a fully labeled subset of the data (supervised). The joint learning framework uses a ranking-based criterion [7,12], allowing for training the model on a larger dataset of unlabeled images and reports. This semi-supervised modification demonstrates a further improvement in accuracy. Additional advantages of our joint learning framework are 1) allowing for the image and text models to be decoupled at inference time, and 2) providing textual explanations for image classification in the form of saliency highlights in the radiology reports.

Related Work. The ability of neural networks to learn effective feature representations from images and text has catalyzed the recent surge of interest in joint image-text modeling. In supervised learning, tasks such as image captioning have leveraged a recurrent visual attention mechanism using recurrent neural networks (RNNs) to improve captioning performance [27]. The TieNet used this attention-based text embedding framework for pathology detection from chest radiographs [25], which was further improved by introducing a global topic vector and transfer learning [28]. A similar image-text embedding setup has been employed for chest radiograph (image) annotations [19]. In unsupervised learning, training a joint global embedding space for visual object discovery has recently been shown to capture relevant structure [11]. All of these models used RNNs for encoding text features. More recently, transformers such as the BERT model [8] have shown the ability to capture richer contextualized word representations using self-attention and have advanced the state-of-the-art in nearly every language processing task compared to variants of RNNs. Our setup, while similar to [25] and [11], uses a series of residual blocks [13] to encode the image representation and uses the BERT model to encode the text representation. We use the radiology reports during training only, to improve the image-based model's performance. This is in contrast to visual question answering [2,3,18], where inference is performed on an image-text pair, and image/video captioning [15,21,23,27], where the model generates text from the input image.

2   Data

For training and evaluating our model, we use the MIMIC-CXR dataset v2.0 [16], consisting of 377,110 chest radiographs associated with 227,835 radiology reports. The data was collected in routine clinical practice, and each report is associated with one or more images. We limited our study to 247,425 frontal-view radiographs.

Regex Labeling. We extracted pulmonary edema severity labels from the associated radiology reports using regular expressions (regex) with negation detection [6]. The keywords of each severity level ("none" = 0, "vascular congestion" = 1, "interstitial edema" = 2, and "alveolar edema" = 3) are summarized in the supplementary materials. In order to limit confounding keywords from other disease processes, we limited the label extraction to patients with congestive heart failure (CHF) based on their ED ICD-9 diagnosis code in the MIMIC dataset [10]. Cohort selection by diagnosis code for CHF was previously validated by manual chart review. This resulted in 16,108 radiology reports. Regex labeling yielded label 6,710 reports associated with 6,743 frontal-view images[1]. Hence, our dataset includes 247,425 image-text pairs, 6,743 of which are of CHF patients with edema severity labels. Note that some reports are associated with more than one image, so one report may appear in more than one image-text pair.

---

[1] The numbers of images of the four severity levels are 2883, 1511, 1709, and 640 respectively.

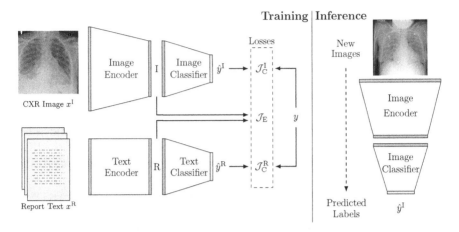

Fig. 1. The architecture of our joint model, along with an example chest radiograph x^{I} and its associated radiology report x^{R}. At training time, the model predicts the edema severity level from images and text through their respective encoders and classifiers, and compares the predictions with the labels. The joint embedding loss $\mathcal{J}_{\mathrm{E}}$ associates image embeddings I with text embeddings R in the joint embedding space. At inference time, the image stream and the text stream are decoupled and only the image stream is used. Given a new chest radiograph (image), the image encoder and classifier compute its edema severity level.

3   Methods

Let x^{I} be a 2D chest radiograph, x^{R} be the free-text in a radiology report, and $y \in \{0, 1, 2, 3\}$ be the corresponding edema severity label. Our dataset includes a set of N image-text pairs $\mathrm{X} = \{\mathrm{x}_j\}_{j=1}^{N}$, where $\mathrm{x}_j = (\mathrm{x}_j^{\mathrm{I}}, \mathrm{x}_j^{\mathrm{R}})$. The first N_{L} image-text pairs are annotated with severity labels $\mathrm{Y} = \{\mathrm{y}_j\}_{j=1}^{N_{\mathrm{L}}}$. Here we train a joint model that constructs an image-text embedding space, where an image encoder and a text encoder are used to extract image features and text features separately (Fig. 1). Two classifiers are trained to classify the severity labels independently from the image features and from the text features. This setup enables us to decouple the image classification and the text classification at inference time. Learning the two representations jointly at training time improves the performance of the image model.

Joint Representation Learning. We apply a ranking-based criterion [7,12] for training the image encoder and the text encoder parameterized by $\theta_{\mathrm{E}}^{\mathrm{I}}$ and $\theta_{\mathrm{E}}^{\mathrm{R}}$ respectively, to learn image and text feature representations $I(x^{\mathrm{I}}; \theta_{\mathrm{E}}^{\mathrm{I}})$ and $R(x^{\mathrm{R}}; \theta_{\mathrm{E}}^{\mathrm{R}})$. Specifically, given an image-text pair $(\mathrm{x}_j^{\mathrm{I}}, \mathrm{x}_j^{\mathrm{R}})$, we randomly select an impostor image $\mathrm{x}_{s(j)}^{\mathrm{I}}$ and an impostor report $\mathrm{x}_{s(j)}^{\mathrm{R}}$ from X. This selection is generated at the beginning of each training epoch. Map $s(j)$ produces a random permutation of $\{1, 2, ..., N\}$.

We encourage the feature representations between a matched pair (I_j, R_j) to be "closer" than those between mismatched pairs $(I_{s(j)}, R_j)$ and $(I_j, R_{s(j)})$ in the joint embedding space. Direct minimization of the distance between I and R could end up pushing the image and text features into a small cluster in the embedding space. Instead we encourage matched image-text features to be close while spreading out all feature representations in the embedding space for downstream classification by constructing an appropriate loss function:

$$
\begin{aligned}
\mathcal{J}_{\mathrm{E}}(\theta_{\mathrm{E}}^{\mathrm{I}}, \theta_{\mathrm{E}}^{\mathrm{R}}; x_j, x_{s(j)}) = {}& \max(0, \mathrm{Sim}(I_j, R_{s(j)}) - \mathrm{Sim}(I_j, R_j) + \eta) \\
& + \max(0, \mathrm{Sim}(I_{s(j)}, R_j) - \mathrm{Sim}(I_j, R_j) + \eta),
\end{aligned} \tag{1}
$$

where $\mathrm{Sim}(\cdot, \cdot)$ is the similarity measurement of two feature representations in the joint embedding space and η is a margin parameter that is set to $|y_j - y_{s(j)}|$ when both $j \leqslant N_{\mathrm{L}}$ and $s(j) \leqslant N_{\mathrm{L}}$; otherwise, $\eta = 0.5$. The margin is determined by the difference due to the mismatch, if both labels are known; otherwise the margin is a constant.

Classification. We employ two fully connected layers (with the same neural network architecture) on the joint embedding space to assess edema severity from the image and the report respectively. For simplicity, we treat the problem as multi-class classification, i.e. the classifiers' outputs $\hat{y}^{\mathrm{I}}(I; \theta_{\mathrm{C}}^{\mathrm{I}})$ and $\hat{y}^{\mathrm{R}}(R; \theta_{\mathrm{C}}^{\mathrm{R}})$ are encoded as one-hot 4-dimensional vectors. We use cross entropy as the loss function for training the classifiers and the encoders on the labeled data:

$$
\begin{aligned}
\mathcal{J}_{\mathrm{C}}(\theta_{\mathrm{E}}^{\mathrm{I}}, \theta_{\mathrm{E}}^{\mathrm{R}}, \theta_{\mathrm{C}}^{\mathrm{I}}, \theta_{\mathrm{C}}^{\mathrm{R}}; x_j, y_j) = {}& -\sum_{i=0}^{3} y_{ji} \log \hat{y}_i^{\mathrm{I}}(I_j(x_j^{\mathrm{I}}; \theta_{\mathrm{E}}^{\mathrm{I}}); \theta_{\mathrm{C}}^{\mathrm{I}}) \\
& - \sum_{i=0}^{3} y_{ji} \log \hat{y}_i^{\mathrm{R}}(R_j(x_j^{\mathrm{R}}; \theta_{\mathrm{E}}^{\mathrm{R}}); \theta_{\mathrm{C}}^{\mathrm{R}}),
\end{aligned} \tag{2}
$$

i.e., minimizing the cross entropy also affects the encoder parameters.

Loss Function. Combining Eq. (1) and Eq. (2), we obtain the loss function for training the joint model:

$$
\mathcal{J}(\theta_{\mathrm{E}}^{\mathrm{I}}, \theta_{\mathrm{E}}^{\mathrm{R}}, \theta_{\mathrm{C}}^{\mathrm{I}}, \theta_{\mathrm{C}}^{\mathrm{R}}; X, Y) = \sum_{j=1}^{N} \mathcal{J}_{\mathrm{E}}(\theta_{\mathrm{E}}^{\mathrm{I}}, \theta_{\mathrm{E}}^{\mathrm{R}}; x_j, x_{s(j)}) + \sum_{j=1}^{N_{\mathrm{L}}} \mathcal{J}_{\mathrm{C}}(\theta_{\mathrm{E}}^{\mathrm{I}}, \theta_{\mathrm{E}}^{\mathrm{R}}, \theta_{\mathrm{C}}^{\mathrm{I}}, \theta_{\mathrm{C}}^{\mathrm{R}}; x_j, y_j).
\tag{3}
$$

Implementation Details. The image encoder is implemented as a series of residual blocks [13], the text encoder is a BERT model that uses the beginner [CLS] token's hidden unit size of 768 and maximum sequence length of 320 [8]. The image encoder is trained from a random initialization, while the BERT model is fine-tuned during the training of the joint model. The BERT model parameters are initialized using pre-trained weights on scientific text [4]. The image features and the text features are represented as 768-dimensional vectors in the joint embedding space. The two classifiers are both 768-to-4 fully connected layers.

We employ the stochastic gradient-based optimization procedure AdamW [26] to minimize the loss in Eq. (3) and use a warm-up linear scheduler [24] for the learning rate. The model is trained on all the image-text pairs by optimizing the first term in Eq. (3) for 10 epochs and then trained on the labeled image-text pairs by optimizing Eq. (3) for 50 epochs. The mini-batch size is 4. We use dot product as the similarity metric in Eq. (1). The dataset is split into training and test sets. All the hyper-parameters are selected based on the results from 5-fold cross validation within the training set.

4   Experiments

Data Preprocessing. The size of the chest radiographs varies and is around 3000×3000 pixels. We randomly translate and rotate the images on the fly during training and crop them to 2048×2048 pixels as part of data augmentation. We maintain the original image resolution to capture the subtle differences in the images between different levels of pulmonary edema severity. For the radiology reports, we extract the *impressions*, *findings*, *conclusion* and *recommendation* sections. If none of these sections are present in the report, we use the *final report* section. We perform tokenization of the text using ScispaCy [20] before providing it to the BERT tokenizer.

Expert Labeling. For evaluating our model, we randomly selected 531 labeled image-text pairs (corresponding to 485 reports) for expert annotation. A board-certified radiologist and two domain experts reviewed and corrected the regex labels of the reports. We use the expert labels for model testing. The overall accuracy of the regex labels (positive predictive value compared against the expert labels) is 89%. The other 6,212 labeled image-text pairs and around 240K unlabeled image-text pairs were used for training. There is no patient overlap between the training set and the test set.

Model Evaluation. We evaluated variants of our model and training regimes as follows:

- **image-only**: An image-only model with the same architecture as the image stream in our joint model. We trained the image model in isolation on the 6,212 labeled images.
- A joint image-text model trained on the 6,212 labeled image-text pairs only. We compare two alternatives to the joint representation learning loss:
  - **ranking-dot, ranking-l2, ranking-cosine**: the ranking based criterion in Eq. (1) with $\mathrm{Sim}(I, R)$ defined as one of the dot product $I^\top R$, the reciprocal of euclidean distance $-\|I - R\|$, and the cosine similarity $\frac{I^\top R}{\|I\|.\|R\|}$;
  - **dot, l2, cosine**: direct minimization on the similarity metrics without the ranking based criterion.
- **ranking-dot-semi**: A joint image-text model trained on the 6,212 labeled and the 240K unlabeled image-text pairs in a semi-supervised fashion, using the ranking based criterion with dot product in Eq. (1). Dot product is

selected for the ranking-based loss based on cross-validation experiments on the supervised data comparing ranking-dot, ranking-l2, ranking-cosine, dot, l2, and cosine.

Table 1. Performance statistics for all similarity measures. Low F1 scores demonstrate the difficulty of decoupling all severity measures for multi-class classification.

Method	AUC (0 *vs* 1,2,3)	AUC (0,1 *vs* 2,3)	AUC (0,1,2 *vs* 3)	macro-F1
l2	0.78	0.76	0.83	0.42
ranking-l2	0.77	0.75	0.80	0.43
cosine	0.77	0.75	0.81	0.44
ranking-cosine	0.77	0.72	0.83	0.41
dot	0.65	0.63	0.61	0.15
ranking-dot	**0.80**	**0.78**	**0.87**	**0.45**

All reported results are compared against the expert labels in the test set. The image portion of the joint model is decoupled for testing, and the reported results are predicted from images only. To optimize the baseline performance, we performed a separate hyper-parameter search for the `image-only` model using 5-fold cross validation (while holding out the test set).

We use the area under the ROC (AUC) and macro-averaged F1-scores (macro-F1) for our model evaluation. We dichotomize the severity levels and report 3 comparisons (0 *vs* 1, 2, 3; 0, 1 *vs* 2, 3; and 0, 1, 2 *vs* 3), since these 4 classes are ordinal (e.g., $\mathbb{P}(\text{severity} = 0 \text{ or } 1) = \hat{y}_0^I + \hat{y}_1^I$, $\mathbb{P}(\text{severity} = 2 \text{ or } 3) = \hat{y}_2^I + \hat{y}_3^I$).

Results. Table 1 reports the performance statistics for all similarity measures. The findings are consistent with our cross-validation results: the ranking based criterion offers significant improvement when it is combined with the dot product as the similarity metric.

Table 2 reports the performance of the optimized baseline model (`image-only`) and two variants of the joint model (`ranking-dot` and `ranking-dot-semi`). We observe that when the joint model learns from the large number of unlabeled image-text pairs, it achieves the best performance. The unsupervised learning minimizes the ranking-based loss in Eq. (1), which does not depend on availability of labels.

It is not surprising that the model is better at differentiating the severity level 3 than other severity categories, because level 3 has the most distinctive radiographic features in the images.

Table 2. Performance statistics for the two variants of our joint model and the baseline image model.

Method	AUC (0 *vs* 1,2,3)	AUC (0,1 *vs* 2,3)	AUC (0,1,2 *vs* 3)	macro-F1
image-only	0.74	0.73	0.78	0.43
ranking-dot	0.80	0.78	0.87	0.45
ranking-dot-semi	**0.82**	**0.81**	**0.90**	**0.51**

Level 1

[CLS] frontal and lateral radiographs of the chest demonstrates slight decrease in size of the severely enlarged cardiac sil hou ette . persistent small bilateral pleural effusion s . probable small hi atal hernia . there is persistent mild pulmonary vascular congestion . clear lungs . no pneum othorax . decrease in severe enlargement of the cardiac sil hou ette likely due to decrease in peric ardial effusion with persistent small effusion s and pulmonary vascular congestion . no pneumonia [SEP]

Level 2

[CLS] surgical clips are again present in the right axi ll a . the cardiac , mediast inal and hil ar contours appear unchanged . upward tent ing of the medial right hem idia ph rag m is very similar . there is a persistent small - to - moderate pleural effusion on the right wit and a small number on the left . fiss ures are mildly thick ened . sub ple ural thickening at the right lung apex appears stable . there is a new mild interstitial abnormality including ker ley b lines and peri bro nc hi al cuff ing suggesting mild - to - moderate interstitial pulmonary edema . however , there is no definite new focal opacity . bony structures are unre mark able . findings most consistent with pulmonary edema . [SEP]

Level 3

[CLS] a trache ostomy and left - side d pic c are stable in position . widespread alveolar op aci ties have increased from are less significant in extent compared to . this likely reflects a combination of increasing edema and persistent multif ocal infection . no pleural effusion or pneum othorax is identified . the cardio media sti nal and hil ar contours are within normal limits . widespread alveolar op aci ties are increased from the most recent prior exam consistent with increasing edema in the setting of persistent multif ocal infection . [SEP]

Fig. 2. Joint model visualization. Top to bottom: (Level 1) The highlight of the Grad-CAM image is centered around the right hilar region, which is consistent with findings in pulmonary vascular congestion as shown in the report. (Level 2) The highlight of the Grad-CAM image is centered around the left hilar region which shows radiating interstitial markings as confirmed by the report heatmap. (Level 3) Grad-CAM highlights bilateral alveolar opacities radiating out from the hila and sparing the outer lungs. This pattern is classically described as "batwing" pulmonary edema mentioned in the report. The report text is presented in the form of sub-word tokenization performed by the BERT model, starting the report with a [CLS] token and ending with a [SEP].

Joint Model Visualization. As a by-product, our approach provides the possibility of interpreting model classification using text. While a method like Grad-CAM [22] can be used to localize regions in the image that are "important" to the model prediction, it does not identify the relevant characteristics of the radiographs, such as texture. By leveraging the image-text embedding association, we visualize the heatmap of text attention corresponding to the last layer of the [CLS] token in the BERT model. This heatmap indicates report tokens that are important to our model prediction. As shown in Fig. 2, we use Grad-CAM [22] to localize relevant image regions and the highlighted words (radiographic findings, anatomical structures, etc.) from the text embedding to explain the model's decision making.

5   Conclusion

In this paper, we presented a neural network model that jointly learns from images and text to assess pulmonary edema severity from chest radiographs. The joint image-text representation learning framework incorporates the rich information present in the free-text radiology reports and significantly improves the performance of edema assessment compared to learning from images alone. Moreover, our experimental results show that joint representation learning benefits from the large amount of unlabeled image-text data.

Expert labeling of the radiology reports enabled us to quickly obtain a reasonable amount of test data, but this is inferior to direct labeling of images. The joint model visualization suggests the possibility of using the text to semantically explain the image model, which represents a promising direction for the future investigation.

Acknowledgments. This work was supported in part by NIH NIBIB NAC P41EB015902, Wistron Corporation, Takeda, MIT Lincoln Lab, and Philips. We also thank Dr. Daniel Moyer for helping generate Fig. 1.

References

1. Adams Jr, K.F., et al.: Characteristics and outcomes of patients hospitalized for heart failure in the United States: rationale, design, and preliminary observations from the first 100,000 cases in the acute decompensated heart failure national registry (adhere). Am. Heart J. **149**(2), 209–216 (2005)
2. Anderson, P., et al.: Bottom-up and top-down attention for image captioning and visual question answering. In: Proceedings of the IEEE Conference on Computer Vision and Pattern Recognition, pp. 6077–6086 (2018)
3. Antol, S., et al.: VQA: visual question answering. In: Proceedings of the IEEE International Conference on Computer Vision, pp. 2425–2433 (2015)
4. Beltagy, I., Lo, K., Cohan, A.: SciBERT: a pretrained language model for scientific text. In: Proceedings of the 2019 Conference on Empirical Methods in Natural Language Processing and the 9th International Joint Conference on Natural Language Processing (EMNLP-IJCNLP), pp. 3606–3611 (2019)
5. Ben-Younes, H., Cadene, R., Cord, M., Thome, N.: MUTAN: multimodal tucker fusion for visual question answering. In: Proceedings of the IEEE International Conference on Computer Vision, pp. 2612–2620 (2017)
6. Chapman, W.W., Bridewell, W., Hanbury, P., Cooper, G.F., Buchanan, B.G.: A simple algorithm for identifying negated findings and diseases in discharge summaries. J. Biomed. Inform. **34**(5), 301–310 (2001)
7. Chechik, G., Sharma, V., Shalit, U., Bengio, S.: Large scale online learning of image similarity through ranking. J. Mach. Learn. Res. **11**, 1109–1135 (2010)

8. Devlin, J., Chang, M.W., Lee, K., Toutanova, K.: BERT: pre-training of deep bidirectional transformers for language understanding. arXiv preprint arXiv:1810.04805 (2018)
9. Gheorghiade, M., et al.: Assessing and grading congestion in acute heart failure: a scientific statement from the acute heart failure committee of the heart failure association of the European society of cardiology and endorsed by the European society of intensive care medicine. Eur. J. Heart Fail. **12**(5), 423–433 (2010)
10. Goldberger, A.L., et al.: Physiobank, physiotoolkit, and physionet: components of a new research resource for complex physiologic signals. Circulation **101**(23), e215–e220 (2000)
11. Harwath, D., Recasens, A., Surís, D., Chuang, G., Torralba, A., Glass, J.: Jointly discovering visual objects and spoken words from raw sensory input. In: Proceedings of the European Conference on Computer Vision (ECCV), pp. 649–665 (2018)
12. Harwath, D., Torralba, A., Glass, J.: Unsupervised learning of spoken language with visual context. In: Advances in Neural Information Processing Systems, pp. 1858–1866 (2016)
13. He, K., Zhang, X., Ren, S., Sun, J.: Deep residual learning for image recognition. In: Proceedings of the IEEE Conference on Computer Vision and Pattern Recognition, pp. 770–778 (2016)
14. Hunt, S.A., et al.: 2009 focused update incorporated into the ACC/AHA 2005 guidelines for the diagnosis and management of heart failure in adults: a report of the American college of cardiology foundation/American heart association task force on practice guidelines developed in collaboration with the international society for heart and lung transplantation. J. Am. Coll. Cardiol. **53**(15), e1–e90 (2009)
15. Jing, B., Xie, P., Xing, E.: On the automatic generation of medical imaging reports. arXiv preprint arXiv:1711.08195 (2017)
16. Johnson, A.E., et al.: MIMIC-CXR, a de-identified publicly available database of chest radiographs with free-text reports. Sci. Data 6 (2019)
17. Liao, R., et al.: Semi-supervised learning for quantification of pulmonary edema in chest x-ray images. arXiv preprint arXiv:1902.10785 (2019)
18. Lu, J., Yang, J., Batra, D., Parikh, D.: Hierarchical question-image co-attention for visual question answering. In: Advances in Neural Information Processing Systems, pp. 289–297 (2016)
19. Moradi, M., Madani, A., Gur, Y., Guo, Y., Syeda-Mahmood, T.: Bimodal network architectures for automatic generation of image annotation from text. In: Frangi, A.F., Schnabel, J.A., Davatzikos, C., Alberola-López, C., Fichtinger, G. (eds.) MICCAI 2018. LNCS, vol. 11070, pp. 449–456. Springer, Cham (2018). https://doi.org/10.1007/978-3-030-00928-1_51
20. Neumann, M., King, D., Beltagy, I., Ammar, W.: ScispaCy: fast and robust models for biomedical natural language processing. In: Proceedings of the 18th BioNLP Workshop and Shared Task. pp. 319–327. Association for Computational Linguistics, Florence, Italy, August 2019. https://doi.org/10.18653/v1/W19-5034. https://www.aclweb.org/anthology/W19-5034
21. Plummer, B.A., Brown, M., Lazebnik, S.: Enhancing video summarization via vision-language embedding. In: Proceedings of the IEEE Conference on Computer Vision and Pattern Recognition, pp. 5781–5789 (2017)
22. Selvaraju, R.R., Cogswell, M., Das, A., Vedantam, R., Parikh, D., Batra, D.: Grad-CAM: visual explanations from deep networks via gradient-based localization. In: Proceedings of the IEEE International Conference on Computer Vision, pp. 618–626 (2017)

23. Vasudevan, A.B., Gygli, M., Volokitin, A., Van Gool, L.: Query-adaptive video summarization via quality-aware relevance estimation. In: Proceedings of the 25th ACM International Conference on Multimedia, pp. 582–590 (2017)
24. Vaswani, A., et al.: Attention is all you need. In: Guyon, I., Luxburg, U.V., et al. (eds.) Advances in Neural Information Processing Systems, vol. 30, pp. 5998–6008. Curran Associates, Inc. (2017). http://papers.nips.cc/paper/7181-attention-is-all-you-need.pdf
25. Wang, X., Peng, Y., Lu, L., Lu, Z., Summers, R.M.: TieNet: text-image embedding network for common thorax disease classification and reporting in chest x-rays. In: Proceedings of the IEEE Conference on Computer Vision and Pattern Recognition, pp. 9049–9058 (2018)
26. Wolf, T., et al.: Huggingface's transformers: state-of-the-art natural language processing. ArXiv arXiv:1910 (2019)
27. Xu, K., et al.: Show, attend and tell: neural image caption generation with visual attention. In: International Conference on Machine Learning, pp. 2048–2057 (2015)
28. Xue, Y., Huang, X.: Improved disease classification in chest X-rays with transferred features from report generation. In: Chung, A.C.S., Gee, J.C., Yushkevich, P.A., Bao, S. (eds.) IPMI 2019. LNCS, vol. 11492, pp. 125–138. Springer, Cham (2019). https://doi.org/10.1007/978-3-030-20351-1_10

Domain-Specific Loss Design for Unsupervised Physical Training: A New Approach to Modeling Medical ML Solutions

Hendrik Burwinkel[1]([✉]), Holger Matz[4], Stefan Saur[4], Christoph Hauger[4],
Ayşe Mine Evren[1], Nino Hirnschall[5], Oliver Findl[5], Nassir Navab[1,3],
and Seyed-Ahmad Ahmadi[2]

[1] Computer Aided Medical Procedures, Technische Universität München,
Boltzmannstraße 3, 85748 Garching bei München, Germany
hendrik.burwinkel@tum.de
[2] German Center for Vertigo and Balance Disorders, Ludwig-Maximilians Universität
München, Marchioninistr. 15, 81377 München, Germany
[3] Computer Aided Medical Procedures, Johns Hopkins University,
3400 North Charles Street, Baltimore, MD 21218, USA
[4] Carl Zeiss Meditec AG, Rudolf-Eber-Str. 11, 73447 Oberkochen, Germany
[5] Vienna Hanusch Hospital, Heinrich-Collin-Str. 30, 1140 Vienna, Austria

Abstract. Today, cataract surgery is the most frequently performed ophthalmic surgery in the world. The cataract, a developing opacity of the human eye lens, constitutes the world's most frequent cause for blindness. During surgery, the lens is removed and replaced by an artificial intraocular lens (IOL). To prevent patients from needing strong visual aids after surgery, a precise prediction of the optical properties of the inserted IOL is crucial. There has been lots of activity towards developing methods to predict these properties from biometric eye data obtained by OCT devices, recently also by employing machine learning. They consider either only biometric data or physical models, but rarely both, and often neglect the IOL geometry. In this work, we propose OpticNet, a novel optical refraction network, loss function, and training scheme which is unsupervised, domain-specific, and physically motivated. We derive a precise light propagation eye model using single-ray raytracing and formulate a differentiable loss function that back-propagates physical gradients into the network. Further, we propose a new transfer learning procedure, which allows unsupervised training on the physical model and fine-tuning of the network on a cohort of real IOL patient cases. We show that our network is not only superior to systems trained with standard procedures but also that our method outperforms the current state of the art in IOL calculation when compared on two biometric data sets.

Keywords: Physical learning · Transfer learning · IOL calculation

© Springer Nature Switzerland AG 2020
A. L. Martel et al. (Eds.): MICCAI 2020, LNCS 12262, pp. 540–550, 2020.
https://doi.org/10.1007/978-3-030-59713-9_52

1   Introduction

In the field of ophthalmology, the term 'cataract' refers to an internal crystallization of the human eye lens. When untreated, this process develops an increasing and severe opacity of the lens. Due to its ubiquitous appearance especially in older generations, but also for younger populations, it constitutes the most frequent cause for blindness. Fortunately, the treatment of cataracts nowadays is a standard medical procedure and has become the world's most frequently performed ophthalmic surgery. During surgery, the capsular bag of the human eye containing the lens is opened, and the lens is destroyed with ultrasound pulses in a process called phacoemulsification. Then, an artificial intraocular lens (IOL) is inserted into the empty bag, replacing the human lens in the refractive system of the eye. To prevent the patient from the necessity of strong visual aids after the surgery (so called refractive surprises), the optical properties of the IOL have to be defined carefully already prior to surgery. The last years have seen a large body of work to predict these required properties. Today, the prediction is based on biometric data of the patient's eye measured by optical coherence tomography (OCT). Geometric distances like eye length are used to predict an IOL refractive power that satisfies the needs of the eye-specific optical system.

These IOL formulas can be clustered into 4 generations. The first and second generation are separated into physical and regression models. The physical models (e.g. Fyodorov, Binkhorst [5,9]) are based on the vergence concept, a simple wavefront-based model. The lens is approximated with no thickness and an effective lens position (ELP) [9]. The second group consists of multilinear regression formulas (SRK I, SRK II [23]) based on the measured biometric data. Only the axial eye length (AL) and the keratometry (K), the curvature of the cornea surface, are considered. Still based on the same concepts, in the third generation of IOL formulas (HofferQ, Holladay1, SRK/T [13,14,22]), these basic ideas were further developed by calculating an individual ELP. Additionally, lens constants for the used IOL types were defined. Introduced by formulas like Haigis and Barrett Universal II [3,4,27], in the fourth generation several new biometric measures including the lens thickness (LT) are considered. Still based on the vergence concept, these formulas introduced an increasing amount of parameters to fine-tune the insufficient model. Olsen and Hirnschall et al. tried to overcome these limitations by incorporating raytracing into their models [11,21]. Recently, an increasing number of formulas like the Hill-RBF or Kane also incorporate concepts of machine learning into their predictions [1,6,7,10,24,29]. Due to proprietary interests, most of the details regarding the used methods are not published, however, descriptions of e.g. the Hill-RBF formula on their website describe the model as a big data approach based on over 12,000 patients' eyes [10]. All current methods are either mainly data-driven or physically motivated models or try to combine both approaches using adjustable parameters or fine-tuning the outcome. Additionally, precise IOL geometry is often neglected.

We are overcoming these limitations by transferring precise physical model knowledge into our machine learning system. There has been recent work in the field of physically motivated learning for environment studies and mechanical simulations. In 2017, Karpatne et al. showed that a loss based on physical constraints of temperature and density as well as density and depths can maintain physical consistency and improve the accuracy of the prediction of lake temperature [17]. In 2019, the same group showed an improved RNN model for lake temperature prediction which they initialized by pre-training it on simulated data [15]. Also using pre-training, Tercan et al. stabilized their prediction of part weights manufactured by injection molding on small datasets [26]. Ba et al. [2] gave an overview of current concepts of physics-based learning, including a work to predict object falls in an image from pure physical constraints [25].

Contributions: We propose a new method for physics-based learning in the medical field of ophthalmology for improved IOL power prediction in cataract surgery. To do so, we derive a detailed single-ray raytracing model of the eye that considers the complete IOL geometry to formulate a domain-specific differentiable loss function which back-propagates physical gradients into our prediction network. This allows an entirely unsupervised training on the physics itself. Further, we propose a transfer learning procedure, which consists of unsupervised training on the physical loss and fine-tuning on real patient cases. We show that the proposed network is not only superior to systems with a standard training procedure but also significantly outperforms the current state of the art in IOL calculation on two biometric data sets. On a wider scope, our work proposes a general methodology for applying medical domain expertise within neural network approaches by designing problem-specific losses. The incorporation of physical models can drastically benefit performance, in particular when only a little amount of supervised training data is available.

2   Methodology

2.1   Mathematical Background and Unsupervised Physical Loss

General Concept: Our proposed optical network OpticNet predicts the target value $\mathbf{Y} = \mathrm{P_{IOL}}$, the refractive power of an IOL that leads to a target refraction $\mathrm{Ref_T}$ for an eye with biometric values $\mathbf{X}$. It therefore optimises the objective function $f(\mathbf{X}, \mathrm{Ref_T}) : \mathbf{X} \rightarrow \mathbf{Y}$. To significantly improve performance, OpticNet explicitly incorporates physical knowledge of the eye's optical system. We design a single-ray raytracer for the calculation of an IOL curvature radius R corresponding to a power $\mathrm{P_{IOL}}$ minimizing the refractive error for a given $\mathbf{X}$. The raytracer is used to design a physical loss $f_{phy}(R, \mathbf{X}, \mathrm{Ref_T})$ that backpropagates its physical gradients into the network during an unsupervised training. Essentially, this implements the physical properties of ray propagation in the eye within a neural network. This network is fine-tuned on real-life surgery outcomes from a cohort of patient data to account for deviations from the physical model.

Ray Transfer Matrix Analysis: In order to describe the physical optical system of the eye, we use the ray transfer matrix analysis. In this physical methodology, every light ray entering an optical system can be described by a vector with two quantities, the distance of its entry point from the optical axis r and

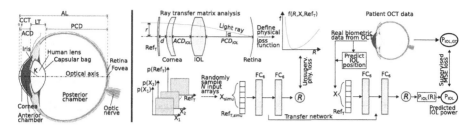

Fig. 1. Left. Human eye anatomy. **Right.** Training: Unsupervised pre-training using physical loss, fine-tuning on real data. AL: axial length, CCT: central cornea thickness, LT: lens thickness, ACD/PCD: anterior/posterior chamber depth, ACD_{IOL}/PCD_{IOL}: postoperative ACD/PCD, K: cornea curvature, FC_6: FC layer with 6 neurons.

its angle α with the axis. Every optical element inside the system is modelled by a matrix either defining the ray propagation within a medium (A) or the refraction at a surface between two media (B). For an optical path length x inside a medium and a surface with curvature radius R between two media with refractive indices n_1 and n_2, the corresponding optical elements are modelled by

$$A = \begin{bmatrix} 1 & x \\ 0 & 1 \end{bmatrix} \qquad B = \begin{bmatrix} 1 & 0 \\ \frac{(n_1-n_2)}{n_2 R} & \frac{n_1}{n_2} \end{bmatrix} \tag{1}$$

the two matrices given in Eq. 1 [8]. By successively applying these transformations to the initial ray vector, the propagation of the ray through the system can be formulated (evolution of r and α).

Unsupervised Physical Loss: As shown in Fig. 1, the eye's complex optical system can be divided into the cornea and lens surfaces, as well as their distances inside specific optical media, resulting in the biometric parameters $\mathbf{X} = [AL, ACD_{IOL}, CCT, K_{max}, K_{min}]$ and Ref_T ($K_{max/min}$ = max. and min. curvature radius of cornea for e.g. astigmatism). We want to achieve that a light ray entering the optical system parallel to the optical axis is focused onto the center of the retina (Fig. 1 right, intersection with retina has distance $0\,mm$ to axis) under an angle α, which corresponds to a refractive error of 0 diopter (D). However, often patients have specific wishes for their target refraction Ref_T to e.g. read a book easily without any visual aids (e.g. $Ref_T = -0.5$ D). To adapt this in the model, an additional very thin refractive element in front of the eye with distance d is used to effectively model this refraction within the system. Due to the thinness, we use the thin lens approximation [8] and formulate it with one matrix as a function of the compensating refraction $-Ref_T$. Therefore, the single-ray raytracer is defined as (see Fig. 1 and its caption for abbreviations):

$$\begin{bmatrix} 0 \\ \alpha \end{bmatrix} = \begin{bmatrix} 1 & PCD_{IOL} \\ 0 & 1 \end{bmatrix} \begin{bmatrix} 1 & 0 \\ \frac{(n_L - n_V)}{-n_V R} & \frac{n_L}{n_V} \end{bmatrix} \begin{bmatrix} 1 & LT(R) \\ 0 & 1 \end{bmatrix} \begin{bmatrix} 1 & 0 \\ \frac{(n_V - n_L)}{n_L R} & \frac{n_V}{n_L} \end{bmatrix} \begin{bmatrix} 1 & ACD_{IOL} \\ 0 & 1 \end{bmatrix} \cdot$$

$$\begin{bmatrix} 1 & 0 \\ \frac{(n_C - n_V)}{n_V \frac{6.8}{7.7} K} & \frac{n_C}{n_V} \end{bmatrix} \begin{bmatrix} 1 & CCT \\ 0 & 1 \end{bmatrix} \begin{bmatrix} 1 & 0 \\ \frac{(1 - n_C)}{n_C K} & \frac{1}{n_C} \end{bmatrix} \begin{bmatrix} 1 & d \\ 0 & 1 \end{bmatrix} \begin{bmatrix} 1 & 0 \\ -Ref_T & 1 \end{bmatrix} \cdot \begin{bmatrix} r \\ 0 \end{bmatrix} = \mathbf{M} \cdot \begin{bmatrix} r \\ 0 \end{bmatrix} \quad (2)$$

where the IOL thickness $LT(R)$ is a function of the curvature radius R derived from corresponding lens data, $K = (K_{max} + K_{min})/2$ is the average cornea curvature, $PCD_{IOL} = AL - CCT - ACD_{IOL} - LT$, $n_V = 1.336$ is the refractive index (RI) of the anterior/posterior chamber [20], $n_C = 1.376$ is the RI of the cornea [20], $n_L = 1.46$ is the RI of the used lens, $\frac{6.8}{7.7}$ is the Gullstrand ratio, the standard ratio of anterior and posterior cornea surface curvature [20] and matrix $\mathbf{M}$ is the result of the complete matrix multiplication. The refractive power of the IOL is directly described by the curvature radius R inside the equation. After performing the matrix multiplication to receive $\mathbf{M}$ and multiplying the ray vector, we obtain two equations. To generate our physical loss, we are only interested in the first one: $0 = \mathbf{M}[0,0](R, \mathbf{X}, Ref_T) \cdot r = \mathbf{M}[0,0](R, \mathbf{X}, Ref_T)$. This equation is linked directly to the IOL power using the optical thick lens formula, which depends on the curvature radius R and yields the corresponding P_{IOL} [8]:

$$P_{IOL}(R) = (n_L - n_V) \cdot \left(\frac{2}{R} - \frac{(n_L - n_V) \cdot LT(R)}{n_L \cdot R^2} \right) \quad (3)$$

Therefore, the calculation of R and P_{IOL} are equivalent. We can insert R into Eq. 3 to obtain the searched IOL power. Hence, the function $\mathbf{M}[0,0]$ yields zero whenever a radius R and corresponding IOL power is given which results in zero refractive error for an eye with parameters $\mathbf{X}$ and Ref_T. However, a straightforward optimization of a loss defined by the function $\mathbf{M}[0,0](R, \mathbf{X}, Ref_T)$ would not converge, because it does not have a minimum at its root, but continues into the negative number space. Instead, we use the squared function to keep the loss differentiable and at the same time enable minimization using gradient descent to reach the root. Our physical loss is therefore defined by:

$$f_{phy}(R, \mathbf{X}, Ref_T) = \mathbf{M}[0,0](R, \mathbf{X}, Ref_T)^2. \quad (4)$$

This physical loss can be fully implemented using differentiable tensors, which allow the propagation of optical refraction gradients into the network. Therefore, it allows an unsupervised training of a network which can predict the physical IOL power for inputs $\mathbf{X}$ and Ref_T. We simply need to sample an arbitrary amount of randomly distributed biometric input data, without any knowledge of the corresponding IOL power ground truth. The network weights are forced to implement the optical refraction, given the loss as a constraint.

2.2 Training Procedure Using the Unsupervised Physical Loss

Optical Network Based on Unsupervised Physical Training: For every biometric parameter X_i and target refraction Ref_T, we define a uniform distribution corresponding to a value range for this parameter that is in agreement with the range occurring in real patient eyes and fully covers its value area. We use these distributions to sample N individual parameter vectors $\mathbf{X}_{simu}$ and $\text{Ref}_{T, simu}$. We properly initialize our network (only positive outcomes to stay in the valid region of f_{phy}), train it using these N samples as input and back-propagate the physical loss $f_{phy}(R, \mathbf{X}, \text{Ref}_T)$ to update the learned weight parameters. A sufficiently large N assures that our network is not overfitting on the data. Nevertheless, we split our N samples into training and testing set to control our training process. The resulting optical network has fully incorporated the physical knowledge and is able to predict the correct physical IOL radius and power for random inputs.

Training on Real Data: Our optical network is fine-tuned on real biometric patient data. A commercial OCT device (ZEISS IOLMaster 700) was used to extract the geometric parameters $\mathbf{X}$ for all patient eyes prior and posterior to surgery by internal image processing on the obtained OCT images. Using these as input, the predicted radii R for every patient are processed by Eq. 3 to obtain P_{IOL}, then the MSE loss against the IOL power ground truth is backpropagated through Eq. 3 to update the network weights. For every surgery site, the outcome of the surgery due to technique and equipment might slightly differ. Therefore, it is preferable to fine-tune the method on site-specific data. Here, the big advantage of our method comes into play. Instead of handcrafted parameters or training on data neglecting physical knowledge, our optical network has already incorporated this prior knowledge and is able to adapt to the necessary correction as a whole. For a standard untrained network, training on such small amounts of data would inevitably lead to strong overfitting. Due to the raytracing approach, additionally, our loss is capable of incorporating the geometric IOL position ACD_{IOL} instead of a formula-depended ELP. Although unknown prior to surgery, it was shown that the biometric parameters of the eye are correlated to the IOL position [12,19]. The ACD_{IOL} prediction is a research field on its own and not topic of this work. We, therefore, train a PLS model to predict an approximated IOL position for our training using the ground truth of ACD_{IOL}, whose usability is an advantage of the method on its own.

3   Experiments

3.1   Experimental Setup

Datasets: For evaluation, we use two biometric datasets obtained during medical studies in two different surgery sites. The datasets have 76 and 130 individual patients correspondingly. The patients did not have any refractive treatment prior to surgery. For every patient, pre- and postoperative OCT images of the patient eye, as well as refractive measurements after surgery, have been collected. We, therefore, have the biometric input, the ground truth for both the position of the IOL and IOL power, as well as the refractive outcome of the surgery.

Evaluation of Methods: For performance evaluation, we follow the validation method proposed by Wang et al. [28]. For every patient, the known inserted

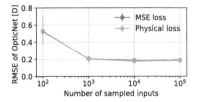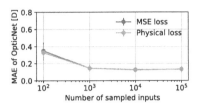

Fig. 2. Comparison of physical and MSE loss for different amounts of sampled input.

IOL power and the refractive error of the eye after surgery form a ground truth pair. We are using the measured refractive error as the new target refraction and predict a corresponding IOL power. The difference of the predicted power to the one that was actually inserted in the eye yields the IOL prediction error. We transfer this error to the resulting refractive error using the approximation proposed by Liu et al. [18]. For every dataset, we perform 10-fold cross-validation. In each run, the data is split into 60% training, 30% validation and 10% test set.

Network Setup: After performing multiple experiments with smaller and larger network structures, the neural network is designed as an MLP with 2 hidden layers and 6 hidden units each, mapping down to one output, the predicted IOL power. This setting resulted in the best and at the same time most stable performance, since it corresponds to the small input space of six parameters and the limited amount of real patient data. Settings: learning rate: 0.001, weight decay: 0.005, activation: leakyReLU, $\alpha = 0.1$, optimizer: Adam.

3.2   Model Evaluation

Physical Loss Against MSE Loss: We evaluate the training with our unsupervised physical loss against the supervised training with standard MSE loss, where full ground truth has to be present. To obtain this ground truth IOL power for the N randomly generated input samples, we have to solve the physical function $0 = \mathbf{M}[0,0](R, \mathbf{X}, \mathrm{Ref}_T)$ numerically using e.g. the Netwon-Raphson method for every input sample and insert the resulting curvature radius into Eq. 3. We compare training sizes $N \in \{10^2, 10^3, 10^4, 10^5\}$ 10 times each and use separate validation and test sets of 10000 simulated eyes. Non-converged runs (occurring for $N = 10^2$) were discarded (RMSE > 1.0 D). Shown in Fig. 2, the physical loss performs equally good or slightly better due to the explicit physical knowledge guiding the training, although it is completely unsupervised and fast to train.

Performance of OpticNet: We are comparing the performance of OpticNet against other IOL formulas using our datasets individually and combined. As explained above, unfortunately for many new formulas there is no publication available regarding detailed implementation. Some methods provide online masks for IOL calculation, but due to regulations of the clinical partner usage was not allowed on our patient data. Still, we were able to perform the calculations on several reference IOL formulas, including Barrett Universal II, which is

Table 1. Root mean sq. err. and mean abs. err. of IOL prediction and resulting eye refractive error on two biometric datasets (p-value: ≤ 0.05 *, < 0.01 **, < 0.001 ***).

Data	Method	RMSE P_{IOL}	MAE P_{IOL}	RMSE Ref.	MAE Ref.	p-val
St. 1	SRK/T	0.822 ± 0.256	0.692 ± 0.196	0.537 ± 0.159	0.449 ± 0.124	***
St. 1	Holliday1	0.619 ± 0.195	0.488 ± 0.138	0.405 ± 0.121	0.322 ± 0.092	**
St. 1	HofferQ	0.555 ± 0.149	0.416 ± 0.129	0.364 ± 0.095	0.279 ± 0.084	
St. 1	Haigis	0.538 ± 0.150	0.404 ± 0.111	0.353 ± 0.092	0.267 ± 0.072	*
St. 1	Barrett II	0.556 ± 0.119	0.449 ± 0.080	0.363 ± 0.073	0.295 ± 0.055	**
St. 1	Raytracer	0.524 ± 0.107	0.436 ± 0.087	0.338 ± 0.080	0.277 ± 0.065	**
St. 1	Solo NN	0.759 ± 0.650	0.595 ± 0.453	0.496 ± 0.414	0.390 ± 0.293	***
St. 1	**OpticNet**	$\mathbf{0.402 \pm 0.161}$	$\mathbf{0.311 \pm 0.115}$	$\mathbf{0.269 \pm 0.102}$	$\mathbf{0.206 \pm 0.073}$	/
St. 2	SRK/T	0.828 ± 0.165	0.653 ± 0.123	0.540 ± 0.106	0.428 ± 0.081	***
St. 2	Holliday1	0.746 ± 0.192	0.593 ± 0.158	0.485 ± 0.126	0.388 ± 0.105	***
St. 2	HofferQ	0.758 ± 0.187	0.610 ± 0.145	0.494 ± 0.122	0.399 ± 0.096	***
St. 2	Haigis	0.740 ± 0.172	0.594 ± 0.132	0.482 ± 0.113	0.389 ± 0.088	***
St. 2	Barrett II	0.678 ± 0.200	0.542 ± 0.162	0.441 ± 0.131	0.359 ± 0.107	**
St. 2	Raytracer	0.778 ± 0.126	0.616 ± 0.103	0.505 ± 0.080	0.400 ± 0.063	***
St. 2	Solo NN	0.916 ± 0.578	0.630 ± 0.304	0.596 ± 0.375	0.410 ± 0.200	***
St. 2	**OpticNet**	$\mathbf{0.551 \pm 0.193}$	$\mathbf{0.400 \pm 0.075}$	$\mathbf{0.360 \pm 0.120}$	$\mathbf{0.264 \pm 0.046}$	/
Both	SRK/T	0.891 ± 0.081	0.680 ± 0.080	0.581 ± 0.051	0.448 ± 0.051	***
Both	Holliday1	0.834 ± 0.138	0.673 ± 0.115	0.544 ± 0.082	0.439 ± 0.068	***
Both	HofferQ	0.820 ± 0.155	0.674 ± 0.125	0.535 ± 0.093	0.44 ± 0.076	***
Both	Haigis	0.791 ± 0.160	0.650 ± 0.130	0.516 ± 0.097	0.420 ± 0.080	***
Both	Barrett II	0.757 ± 0.156	0.604 ± 0.129	0.494 ± 0.094	0.394 ± 0.079	***
Both	Raytracer	0.846 ± 0.122	0.682 ± 0.102	0.553 ± 0.071	0.445 ± 0.063	***
Both	Solo NN	0.812 ± 0.373	0.518 ± 0.199	0.529 ± 0.238	0.340 ± 0.132	***
Both	**OpticNet**	$\mathbf{0.562 \pm 0.158}$	$\mathbf{0.376 \pm 0.099}$	$\mathbf{0.366 \pm 0.098}$	$\mathbf{0.249 \pm 0.064}$	/

currently considered state-of-the-art, even compared to recent machine learning approaches [16]. As specified, we perform 10-fold cross-validation, where each test set is unseen prior to its evaluation. For all reference IOL formulas, the latest optimized constants in literature are used. Additionally, we fine-tune every method by calculating the offsets of their averaged predictions on the training set for every fold and subtract this offset from the corresponding test set predictions. Further, we provide the result of our purely physics-based single-ray raytracer (Raytracer) and the performance of a standard network without physical pre-training (Solo NN). As expected and shown in Table 1, the Barrett formula has a good performance compared to the other state-of-the-art formulas. However, our OpticNet is significantly outperforming all methods (Wilcoxon signed-rank test). Only for the HofferQ formula on the smaller dataset significance was slightly missed (p = 0.08). Especially the NN without pre-training clearly overfits the data, the reported results were only achievable by discarding non-converged runs for the different folds.

4 Discussion and Conclusion

In this work, we introduced a new concept for optimized IOL calculation based on the explicit transfer of physical knowledge into our network, called OpticNet. By introducing a new unsupervised physical loss that allows a direct training on the optical model of the eye, a designed single-ray raytracer, physical gradients are propagated into the network. We showed that this physical loss results in equal or improved training performance compared to a standard supervised training, since explicit physical knowledge is incorporated. Additionally, we show that the transfer step on real data significantly outperforms state-of-the-art IOL formulas on two biometric datasets, even stronger when combining the datasets and therefore incorporating the distribution shift. The importance of physical knowledge becomes clear especially in the overfitting of an untrained NN.

The concept of unsupervised pre-training on a physical model is a general approach that can be transferred into various fields. It allows the transfer from a solely data-driven training to a model-based training that explicitly incorporates prior knowledge and connects the parameter-based network to the mathematics behind the domain-specific model. This direct connection is ideal for knowledge transfer, since the loss is customized for the task. The concept could be applicable in fields like Ultrasound and X-ray imaging as well, whose technology underlies physical rules and properties on its own. Potential approaches would be the incorporation of models describing the sound and electromagnetic wave propagation. Shown in this work for OCT-based IOL calculation, the concept has particular benefits when little annotated training data is available, which is a ubiquitous challenge in the medical domain.

Acknowledgements. The study was supported by the Carl Zeiss Meditec AG in Oberkochen and Jena, Germany, and the German Federal Ministry of Education and Research (BMBF) in connection with the foundation of the German Center for Vertigo and Balance Disorders (DSGZ) (grant number 01 EO 0901). Further, we thank NVIDIA Corporation for the sponsoring of a Titan V GPU.

References

1. Achiron, A., et al.: Predicting refractive surgery outcome: machine learning approach with big data. J. Refract. Surg. **33**(9), 592–597 (2017). https://doi.org/10.3928/1081597X-20170616-03
2. Ba, Y., Zhao, G., Kadambi, A.: Blending diverse physical priors with neural networks, no. 1, pp. 1–15 (2019). http://arxiv.org/abs/1910.00201
3. Barrett, G.D.: Intraocular lens calculation formulas for new intraocular lens implants. J. Cataract Refract. Surg. **13**(4), 389–396 (1987). https://doi.org/10.1016/S0886-3350(87)80037-8

4. Barrett, G.D.: An improved universal theoretical formula for intraocular lens power prediction. J. Cataract Refract. Surg. **19**(6), 713–720 (1993). https://doi.org/10.1016/S0886-3350(13)80339-2

5. Binkhorst, R.D.: Intraocular lens power calculation. Int. Ophthalmol. Clin. **19**(4), 237–254 (1979). https://doi.org/10.1097/00004397-197901940-00010

6. Clarke, G.P., Burmeister, J.: Comparison of intraocular lens computations using a neural network versus the Holladay formula. J. Cataract Refract. Surg. **23**(10), 1585–1589 (1997). https://doi.org/10.1016/S0886-3350(97)80034-X

7. Connell, B.J., Kane, J.X.: Comparison of the Kane formula with existing formulas for intraocular lens power selection. BMJ Open Ophthalmol. **4**(1), 1–6 (2019). https://doi.org/10.1136/bmjophth-2018-000251

8. Demtröder, W.: Experimentalphysik 2. SLB. Springer, Heidelberg (2013). https://doi.org/10.1007/978-3-642-29944-5

9. Fyodorov, S.N., Galin, M.A., Linksz, A.: Calculation of the optical power of intraocular lenses. Investigative Ophthalmology & Visual Science **14**(8), 625–628 (1975)

10. Hill, W.E.: Hill-RBF Method. Haag-Streit White Paper (2017)

11. Hirnschall, N., Buehren, T., Trost, M., Findl, O.: Pilot evaluation of refractive prediction errors associated with a new method for ray-tracing-based intraocular lens power calculation. J. Cartaract Refr. Surg. **45**(6), 738–744 (2019). https://doi.org/10.1016/j.jcrs.2019.01.023

12. Hirnschall, N., Farrokhi, S., Amir-Asgari, S., Hienert, J., Findl, O.: Intraoperative optical coherence tomography measurements of aphakic eyes to predict postoperative position of 2 intraocular lens designs. J. Cataract Refract. Surg. **44**(11), 1310–1316 (2018). https://doi.org/10.1016/j.jcrs.2018.07.044

13. Hoffer, K.J.: The Hoffer Q formula: a comparison of theoretic and regression formulas. J. Cataract Refract. Surg. **19**(6), 700–712 (1993). https://doi.org/10.1016/S0886-3350(13)80338-0

14. Holladay, J.T., Musgrove, K.H., Prager, T.C., Lewis, J.W., Chandler, T.Y., Ruiz, R.S.: A three-part system for refining intraocular lens power calculations. J. Cataract Refract. Surg. **14**(1), 17–24 (1988). https://doi.org/10.1016/S0886-3350(88)80059-2

15. Jia, X., et al.: Physics guided RNNs for modeling dynamical systems: a case study in simulating lake temperature profiles. In: SIAM International Conference on Data Mining, SDM 2019, pp. 558–566 (2019). https://doi.org/10.1137/1.9781611975673.63

16. Kane, J.X., Van Heerden, A., Atik, A., Petsoglou, C.: Accuracy of 3 new methods for intraocular lens power selection. J. Cataract Refract. Surg. **43**(3), 333–339 (2017). https://doi.org/10.1016/j.jcrs.2016.12.021

17. Karpatne, A., Watkins, W., Read, J., Kumar, V.: Physics-guided neural networks (PGNN): an application in lake temperature modeling (2017). http://arxiv.org/abs/1710.11431

18. Liu, Y., Wang, Z., Mu, G.: Effects of measurement errors on refractive outcomes for pseudophakic eye based on eye model. Optik **121**(15), 1347–1354 (2010). https://doi.org/10.1016/j.ijleo.2009.01.022

19. Norrby, S., Bergman, R., Hirnschall, N., Nishi, Y., Findl, O.: Prediction of the true IOL position. Br. J. Ophthalmol. **101**(10), 1440–1446 (2017). https://doi.org/10.1136/bjophthalmol-2016-309543

20. Olsen, T.: On the calculation of power from curvature of the cornea. Br. J. Ophthalmol. **70**(2), 152–154 (1986). https://doi.org/10.1136/bjo.70.2.152

21. Olsen, T., Hoffmann, P.: C constant: new concept for ray tracing-assisted intraocular lens power calculation. J. Cataract Refract. Surg. **40**(5), 764–773 (2014). https://doi.org/10.1016/j.jcrs.2013.10.037

22. Retzlaff, J.A., Sanders, D.R., Kraff, M.C.: Development of the SRK/T intraocular lens implant power calculation formula. J. Cataract Refract. Surg. **16**(3), 333–340 (1990). https://doi.org/10.1016/S0886-3350(13)80705-5

23. Sanders, D.R., Kraff, M.C.: Improvement of intraocular lens power calculation using empirical data. Am. Intraocul. Implant Soc. J. **6**(3), 263–267 (1980). https://doi.org/10.1016/S0146-2776(80)80075-9

24. Sramka, M., Vlachynska, A.: Artificial neural networks application in intraocular lens power calculation. In: Proceedings of the 9th EUROSIM Congress on Modelling and Simulation, EUROSIM 2016, The 57th SIMS Conference on Simulation and Modelling, SIMS 2016, vol. 142, pp. 25–30, December 2018. https://doi.org/10.3384/ecp1714225

25. Stewart, R., Ermon, S.: Label-free supervision of neural networks with physics and domain knowledge. In: 31st AAAI Conference on Artificial Intelligence, AAAI 2017, vol. 1, no. 1, pp. 2576–2582 (2017)

26. Tercan, H., Guajardo, A., Heinisch, J., Thiele, T., Hopmann, C., Meisen, T.: Transfer-learning: bridging the gap between real and simulation data for machine learning in injection molding. Procedia CIRP **72**, 185–190 (2018). https://doi.org/10.1016/j.procir.2018.03.087

27. Turczynowska, M., Koźlik-Nowakowska, K., Gaca-Wysocka, M., Grzybowski, A.: Effective ocular biometry and intraocular lens power calculation. Eur. Ophthalmic Rev. **10**(02), 94 (2016). https://doi.org/10.17925/EOR.2016.10.02.94

28. Wang, L., Booth, M.A., Koch, D.D.: Comparison of intraocular lens power calculation methods in eyes that have undergone laser-assisted in-situ keratomileusis. Trans. Am. Ophthalmol. Soc. **102**, 189–196 (2004). Discussion 196–7

29. Yarmahmoodi, M., Arabalibeik, H., Mokhtaran, M., Shojaei, A.: Intraocular lens power formula selection using support vector machines. Front. Biomed. Technol. **2**(1), 36–44 (2015)

Multiatlas Calibration of Biophysical Brain Tumor Growth Models with Mass Effect

Shashank Subramanian$^{(\boxtimes)}$ (iD), Klaudius Scheufele, Naveen Himthani, and George Biros

Oden Institute, University of Texas at Austin, 201 E. 24th Street, Austin, TX, USA
{shashank,naveen,biros}@oden.utexas.edu, kscheufele@austin.utexas.edu

Abstract. We present a *3D fully-automatic* method for the calibration of partial differential equation (PDE) models of glioblastoma (GBM) growth with *"mass effect"*, the deformation of brain tissue due to the tumor. We quantify the mass effect, tumor *proliferation*, tumor *migration*, and the localized *tumor initial condition* from a *single* multiparameteric Magnetic Resonance Imaging (mpMRI) patient scan. The PDE is a reaction-advection-diffusion partial differential equation coupled with linear elasticity equations to capture mass effect. The single-scan calibration model is notoriously difficult because the precancerous (healthy) brain anatomy is unknown. To solve this inherently ill-posed and ill-conditioned optimization problem, we introduce a novel inversion scheme that uses *multiple brain atlases* as proxies for the healthy precancer patient brain resulting in robust and reliable parameter estimation. We apply our method on both synthetic and clinical datasets representative of the heterogeneous spatial landscape typically observed in glioblastomas to demonstrate the validity and performance of our methods. In the synthetic data, we report calibration errors (due to the ill-posedness and our solution scheme) in the 10%–20% range. In the clinical data, we report good quantitative agreement with the observed tumor and qualitative agreement with the mass effect (for which we do not have a ground truth). Our method uses a minimal set of parameters and provides both global and local quantitative measures of tumor infiltration and mass effect.

Keywords: Glioblastoma · Mass effect · Tumor growth models · Inverse problems

1 Introduction

Gliomas are the most common primary brain tumors in adults. Glioblastomas are high grade gliomas with poor prognosis. A significant challenge in the characterization of these tumors involves mass effect (the deformation of the surrounding

Electronic supplementary material The online version of this chapter (https://doi.org/10.1007/978-3-030-59713-9_53) contains supplementary material, which is available to authorized users.

© Springer Nature Switzerland AG 2020
A. L. Martel et al. (Eds.): MICCAI 2020, LNCS 12262, pp. 551–560, 2020.
https://doi.org/10.1007/978-3-030-59713-9_53

healthy tissue due to tumor growth) along with biomarkers representing the tumor aggressiveness and growth dynamics. The integration of mathematical models with clinical imaging data holds the enormous promise of robust, minimal, and explainable models that quantify cancer growth and connect cell-scale phenomena to organ-scale, personalized, clinical observables [19,25,27]. Here, we focus on the calibration of mathematical tumor growth models with clinical imaging data from a *single* pretreatment scan in order to assist in diagnosis and treatment planning. Longitudinal pretreatment scans are rare for GBMs since most patients seek immediate treatment. For this reason, we need algorithms that rely only on one mpMRI scan.

Contributions: The single-scan calibration problem is formidable for two main reasons: the tumor initial condition (IC) and the subject's healthy precancer anatomy (the brain anatomy before the cancer begins to grow) are unknown. Using brain anatomy symmetry does not apply to all subjects; so typically an atlas is used as a proxy to the healthy subject brain. However, natural anatomical differences between the atlas and the subject interfere with tumor-related deformations; disentangling the two is hard. In light of these difficulties, our contributions are as follows: • Based on the method described in [23], we propose a novel multistage scheme for inversion: first we estimate the tumor initial conditions, then, given an atlas, we invert for the three scalar model parameters (proliferation, migration, mass effect). We repeat this step for several atlases and we compute expectations of the observables (see Sect. 2 and Sect. 2.1). These calculations are quite expensive. However, the entire method runs in parallel on GPUs so that 3D inversion in 128^3 resolution takes less than an hour and 256^3 resolution takes about six hours. • We use synthetic data (for which we know the ground truth) in order to estimate the errors associated with our numerical scheme and we validate our method on a set of clinical mpMRI scans. We report these results in Sect. 3.

Related Work: The most common mathematical models for tumor growth dynamics are based on reaction-diffusion PDEs [14,16,17,24,26], which have been coupled with mechanical models to capture mass effect [3,11,12,22]. While there have been many studies to calibrate these models using inverse problems [4–6,9,10,13,14,16,18,20], most do not invert for all unknown parameters (tumor initial condition and model parameters) or they assume the presence of multiple imaging scans (both these scenarios make the inverse problem more tractable). In [23], the authors presented a methodology to invert for tumor initial condition (IC) and cell proliferation and migration from a single scan; but their forward (growth) model does not account for mass effect. They demonstrate the importance of using a *sparse tumor IC* (see Sect. 2) to correctly reconstruct for other tumor parameters. In [7], the author considers mathematical aspects of inverting simultaneously for the tumor IC, tumor parameters, and mass effect but assumes known precancer brain anatomy. Other than those works, the current state of the art for single-scan biophysically-based tumor characterization is GLISTR [9].

The main shortcoming of GLISTR is that it requires manual seeding for the tumor IC and uses a single atlas. Further, GLISTR uses deformable registration for both anatomical variations and mass effect deformations but does not decouple them (this is extremely ill-posed), since the primary goal of GLISTR lies in image segmentation. Finally, the authors in [1] present a 2D synthetic study to quantify mass effect. However, they assume known tumor IC and precancer brain. To the best of our knowledge, we are not aware of any other framework that can fully-automatically calibrate tumor growth models with mass effect for all unknown parameters in 3D. Furthermore, our solvers employ efficient parallel algorithms and GPU acceleration.

2   Methods

Before we describe our methods, we introduce the following notations: $c = c(\boldsymbol{x}, t)$ is the tumor concentration ($\boldsymbol{x}$ is a voxel, t is time) with *observed* tumor data c_1 and *unknown* tumor IC c_0; $\boldsymbol{m}(\boldsymbol{x}, t) := (m_{\mathrm{WM}}(\boldsymbol{x}, t), m_{\mathrm{GM}}(\boldsymbol{x}, t), m_{\mathrm{CSF}}(\boldsymbol{x}, t))$ is the brain segmentation into white matter (WM), gray matter (GM), and cerebrospinal fluid (CSF) with *observed* pretreatment segmented brain $\boldsymbol{m}_1$ and *unknown* precancer healthy brain $\boldsymbol{m}_0$; Additionally, (κ, ρ, γ) are scalars that represent the *unknown* migration rate, proliferation rate, and a mass effect parameter, respectively.

Tumor Growth Mathematical Model: Following [22], we use a non-linear reaction-advection-diffusion PDE (on an Eulerian framework):

$$\partial_t c + \mathrm{div}\,(c\boldsymbol{v}) - \kappa\mathcal{D}c - \rho\mathcal{R}c = 0, \quad c(0) = c_0 \qquad \text{in } \varOmega \times [0,1] \tag{1a}$$

$$\partial_t \boldsymbol{m} + \mathrm{div}\,(\boldsymbol{m} \otimes \boldsymbol{v}) = 0, \quad \boldsymbol{m}(0) = \boldsymbol{m}_0 \qquad \text{in } \varOmega \times [0,1] \tag{1b}$$

$$\mathrm{div}\,(\lambda\nabla\boldsymbol{u} + \mu(\nabla\boldsymbol{u} + \nabla\boldsymbol{u}^{\mathsf{T}})) = \gamma\nabla c \qquad \text{in } \varOmega \times [0,1] \tag{1c}$$

$$\partial_t \boldsymbol{u} = \boldsymbol{v}, \quad \boldsymbol{u}(0) = \boldsymbol{0} \qquad \text{in } \varOmega \times [0,1], \tag{1d}$$

where $\mathcal{D} := \mathrm{div}\, m_{\mathrm{WM}}\nabla c$ is a diffusion operator; $\mathcal{R} := m_{\mathrm{WM}}c(1 - c)$ is a logistic growth operator; Eq. (1a) is coupled to a linear elasticity equation (Eq. (1c)) with forcing $\gamma\nabla c$, which is coupled back through a convective term with velocity $\boldsymbol{v}(\boldsymbol{x}, t)$ which parameterizes the displacement $\boldsymbol{u}(\boldsymbol{x}, t)$. The linear elasticity model is parameterized by Lamè coefficients $\lambda(\boldsymbol{x}, t)$ and $\mu(\boldsymbol{x}, t)$. We note here that $\gamma = 0$ implies $\boldsymbol{u} = \boldsymbol{0}$, which reduces the tumor growth model to a simple reaction-diffusion PDE with no mass effect (i.e., Eq. (1a) with $\boldsymbol{v} = \boldsymbol{0}$).

Following [23], we parameterize $c_0 = \boldsymbol{\phi}^T(\boldsymbol{x})\boldsymbol{p} = \sum_{i=1}^{m} \phi_i(\boldsymbol{x})p_i$; where $\boldsymbol{p}$ is an m-dimensional parameterization vector, $\phi_i(\boldsymbol{x}) = \phi_i(\boldsymbol{x} - \boldsymbol{x}_i, \sigma)$ is a Gaussian function centered at point $\boldsymbol{x}_i$ with standard deviation σ, and $\boldsymbol{\phi}(\boldsymbol{x}) = \{\phi_i(\boldsymbol{x})\}_{i=1}^{m}$. Here, $\boldsymbol{x}_i$ are voxels that are segmented as tumor and σ is one voxel, meaning m can be quite large ($\sim$1000). This parameterization alleviates some of the ill-posedness associated with the inverse problem [23].

Inverse Problem: The unknowns in our growth model are $(\boldsymbol{p}, \boldsymbol{m}_0, \kappa, \rho, \gamma)$. Our model is calibrated for these parameters using imaging data. We introduce an approximation to m_0 (discussed in the numerical scheme), and estimate the rest through the following inverse problem formulation:

$$\min_{p,\kappa,\rho,\gamma} \mathcal{J}(p,\kappa,\rho,\gamma) := \frac{1}{2}\|Oc(1) - c_1\|_{L_2(\Omega)}^2 + \frac{\beta}{2}\|\phi^T p\|_{L_2(\Omega)}^2 \qquad (2)$$

subject to the reaction-advection-diffusion forward (growth) model $\mathcal{F}(p,\kappa,\rho,\gamma)$ given by Eq. (1). Recall that κ, ρ, and γ are scalars, but $p \in \mathbf{R}^m$. The objective function minimizes the L_2 mismatch between the simulated tumor $c(1)$ at $t = 1$ and data c_1 and is balanced by a regularization term on the inverted initial condition (IC). O is an observation operator that defines the clearly observable tumor margin (see [21] for details). Following [23], we further introduce the following constraints to our optimization problem: $\|p\|_0 \leq s$ and $\max(\phi^T p) = 1$. The first constraint restricts the tumor initial condition to a few Gaussians while the second enforces the assumption that $t = 0$ corresponds to the first time the tumor concentration reaches one at some voxel in the domain. Note that these two constraints are *modeling assumptions*. They are introduced to alleviate the severe ill-posedness of the backward tumor growth PDE (see [23] for details).

2.1   Summary of Our Multistage Inversion Method

To solve our inverse problem, we propose the following numerical scheme:

(**S**.1) We solve Eq. (2) using the *simple* growth model with no mass effect $\mathcal{F}(p,\kappa,\rho,\gamma = 0)^1$ for (p,κ,ρ). In this step, our *precancer scan*, i.e. m_0, is approximated as the patient brain with tumor regions replaced by white matter. In order to solve the resulting non-linear inverse problem, we employ the fast *adjoint-based* algorithm outlined in [23].

(**S**.2) Next, due to mass effect, we need an estimate for m_0. For this, we use $m_0(x) = a(x)$, where $a(x)$ is an atlas or a scan from another healthy individual. We note that the *inverted* tumor initial condition p^{rec} lies in the precancer scan space from (**S**.1). In order to make sure that the p^{rec} locations do not fall in anatomical structures such as ventricles (where the tumor does not grow), we register the patient[2] and the atlas (our new precancer scan), and transfer (warp) the initial condition to the atlas with the deformation map. This registration is unnecessary if the precancer scan is known.

(**S**.3) Finally, we solve Eq. (2) constrained by $\mathcal{F}(p = p^{rec},\kappa,\rho,\gamma)$ for (κ,ρ,γ). Since we have only three unknown parameters, we use first order finite differences to approximate the gradient of the objective.

We repeat (**S**.2) and (**S**.3) for different atlases to make our inversion scheme less sensitive to the atlas selected. For the ℓ_2 regularized solve in (**S**.1) and the inverse solve in (**S**.3), we use a quasi-Newton optimization method (L-BFGS)

---

[1] $\gamma = 0$ indicates no mass effect and Eq. (1b), Eq. (1c) and Eq. (1d) are not needed.
[2] We simply mask the tumor region of the patient for this registration.

globalized by Armijo linesearch with gradient-based convergence criteria. For the registration in (**S**.2), we use the registration solver CLAIRE [15].

Solver Timings: On average, the full multistage inversion on 128^3 takes less than an hour using GPUs. The inversion in the different atlases are embarrassingly parallel. For 256^3 (the resolution of our results), the full inversion takes an average of six hours. Finally, our optimization solvers converge in an average of 20–30 quasi-Newton iterations without any failures.

3   Results

We ask the following four questions:

(Q1) Given p (tumor IC) and m_0 (precancer scan), can we reconstruct (κ, ρ, γ) using (**S**.3) from Sect. 2.1?

(Q2) Given m_0 but unknown p, can we reconstruct $(p, \kappa, \rho, \gamma)$ using (**S**.1) and (**S**.3)?

(Q3) With m_0 and p unknown, can we reconstruct $(p, \kappa, \rho, \gamma)$ using (**S**.1)–(**S**.3) taking different atlases as m_0?

(Q4) How does our scheme perform on clinical data?

We use *synthetic* data to answer (i)–(iii) and quantify our errors. For *clinical* data, since the ground truth is unknown (we do not have longitudinal data), we evaluate our scheme qualitatively.

(Q1) Known m_0, known p: In this experiment, we generate data by growing synthetic tumors resembling clinical observations in a healthy atlas using different ground truth parameter combinations. The test-cases are aimed at simulating similar tumor volumes, but with varying amount of mass effect. We consider the following variations for our parameter configurations:

(i)   *TC(a):*   no mass effect $\gamma^\star = 0$ $\rho^\star = 12$ $\kappa^\star = 0.025$
(ii) *TC(b):*   mild mass effect $\gamma^\star = 0.4$ $\rho^\star = 12$ $\kappa^\star = 0.025$
(iii) *TC(c):* moderate mass effect $\gamma^\star = 0.8$ $\rho^\star = 10$ $\kappa^\star = 0.05$
(iv) *TC(d):*   large mass effect $\gamma^\star = 1.2$ $\rho^\star = 10$ $\kappa^\star = 0.025,$

where * represents the non-dimensionalized ground truth parameters. The tumors along with the deformed atlas are visualized in Fig. 1. We report our inversion results in Table 1 ("True IC") with tumor initial condition and precancer scan taken as the ground truth. We report the relative errors in reconstructing parameters $\iota = \{\kappa, \rho, \gamma\}$ (if the ground truth is zero, then the error is absolute; *TC(a)*), relative error in the two-norm of the displacement norm u, i.e., $\|u\|_2$ (this field informs us of the extent of mass effect), relative error in the final tumor reconstruction, and the norm of the gradient to indicate convergence. We observe excellent reconstruction with relative errors less than 2%.

(Q2) Known m_0, unknown p: We use our inversion scheme outlined in Sect. 2.1, where we first invert for the tumor initial condition using (**S**.1) and then the model parameters with *known* precancer scan. We report our inversion

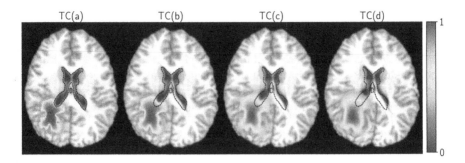

Fig. 1. Synthetic patient T1 MRIs generated with Eq. (1). The normalized tumor concentration is overlaid (color) along with the undeformed ventricles (black dashed contour) to indicate the variable extent of mass effect.

Table 1. Inversion results assuming **known** m_0, the anatomy of the subject before tumor occurrence. u^* is the ground truth displacement norm in number of voxels (1 voxel $\approx 0.9\,$mm), e_ι is the relative error for parameter ι, e_u is the relative error in the two-norm of the displacement field norm, $\|g\|_2$ is the norm of the final gradient, and μ_{T,L_2} is the relative error in final tumor concentration. We report the tumor initial condition (IC) used in "Tumor IC" – True IC indicates synthetic ground truth IC and Inverted IC indicates our reconstruction using (**S**.1) and (**S**.3)

Test-case	Tumor IC	e_γ	e_ρ	e_κ	$\|u^*\|_\infty$	e_u	$\|g\|_2$	μ_{T,L_2}
TC (a)	True IC	4.44E−6	2.26E−3	2.76E−3	0	4.63E−4	1.54E−1	3.28E−3
TC (b)	True IC	1.50E−3	2.69E−3	1.39E−2	6.1	7.27E−4	5.70E−4	3.52E−3
TC (c)	True IC	1.32E−3	2.50E−5	2.14E−4	10	1.49E−3	4.81E−5	7.18E−4
TC (d)	True IC	2.17E−4	1.00E−5	3.12E−4	14.9	4.61E−4	6.26E−5	3.35E−4
TC (a)	Inverted IC	2.95E−3	1.63E−1	2.08E−1	0	2.86E−1	1.27E−2	2.42E−1
TC (b)	Inverted IC	5.29E−2	1.29E−1	1.94E−1	6.1	4.93E−3	6.43E−4	2.12E−1
TC (c)	Inverted IC	3.54E−2	1.39E−1	2.55E−1	10	1.46E−2	3.54E−4	1.65E−1
TC (d)	Inverted IC	1.46E−2	1.59E−1	2.16E−1	14.9	3.78E−2	7.48E−4	1.88E−1

results in Table 1 under "Inverted IC". As expected, the errors increase due to the fact that we reconstructed p (IC) using the no-mass-effect model (images of the reconstructions included in supplementary Fig. S1). But we can still recover the model parameters quite well: the mass effect (indicated by the error in two-norm of the displacement norm e_u) is captured with relative errors less that 4%; the reaction and diffusion coefficient also have good estimates of around 15% and 22% average relative errors respectively. Hence, our scheme exhibits good reconstruction performance.

(Q3) Unknown m_0, unknown p: This scenario corresponds to the actual clinical problem. For this test-case, we invoke (**S**.2) and average the results using three atlases. To reiterate the scheme, the inverted tumor ICs from (**S**.1) are warped to each atlas (through registration) for the final inversion (**S**.3). We report inversion results in Table 2 and show an exemplary reconstruction of the patient using the different atlases in Fig. 2. Despite the approximation error in

Table 2. Inversion results for **unknown m_0 and p** for test-case TC (d). e_ι are the relative errors in parameter ι. Atlas (*) are the different atlases used. True atlas is the ground truth m_0. u^* is the ground truth displacement norm in number of voxels, $\|g\|_2$ is the norm of the final gradient, and μ_{T,L_2} is the relative error in the final tumor concentration. The results for other test-cases are reported in supplementary Table S2.

ID	Test-case	e_γ	e_ρ	e_κ	$\|u^*\|_\infty$	e_u	$\|g\|_2$	μ_{T,L_2}
	True atlas	2.48E−2	1.81E−1	2.54E−1	14.9	2.91E−2	5.75E−4	2.05E−1
TC (d)	Atlas (1)	1.63E−2	2.15E−1	2.65E−1	14.9	1.21E−1	2.71E−4	3.37E−1
	Atlas (2)	1.13E−1	2.35E−1	9.27E−2	14.9	2.00E−1	6.85E−4	3.04E−1
	Atlas (3)	2.50E−1	2.06E−1	5.46E−1	14.9	2.39E−1	5.43E−4	2.79E−1

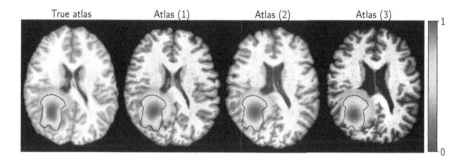

True atlas   Atlas (1)   Atlas (2)   Atlas (3)

Fig. 2. Reconstruction of the tumor concentration (color) using different atlases as the precancer scan. The tumor data segmentation is highlighted as a black dashed contour line.

m_0, we are still able to capture the parameters (average displacement relative errors of around 16% and 25% for atlases (1) and (2) respectively). Atlas (3) has poor performance because it is significantly different from the patient (for example, the ventricles are highly dissimilar). We also note that the error in tumor reconstruction is significantly higher, which is representative of the errors introduced due to the anatomical variations of each atlas from the patient.

(Q4) Clinical images: We use images from the BraTS [2] dataset. We segment the scans using a neural network based on [8] with mpMRI (T1, T2, T1-CE, and FLAIR) patient scans as input data. We use the segmented tumor core as tumor data in the reconstruction algorithm. We select four patients with visually different but, of course, unknown mass effect. For each patient, we invert in four different atlases used as m_0. For each BraTS case, we report atlas-averaged model parameters. We show each patient with the average reconstructed tumor and displacement norm in Fig. 3. We quantify mass effect using the average maximum displacement norm, $\|u\|_\infty$ and express the sensitivity to the atlas choice using the standard deviation (see supplementary Table S3 for the numbers). We observe that ABO and AAP show high mass effect ($\|u\|_\infty \sim 18$ mm ± 1). For ALU, the mass effect is moderate ($\|u\|_\infty \sim 8$ mm ± 2) and is localized at the side

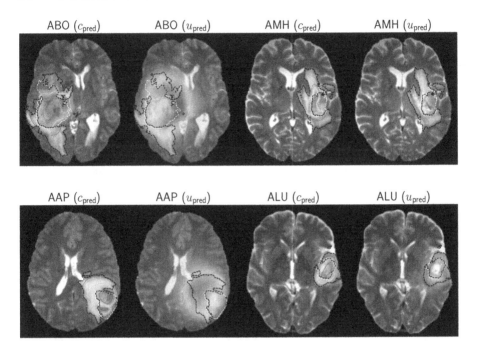

Fig. 3. Predicted normalized tumor (c_{pred}) and mass effect displacement norm (u_{pred}) for each BraTS subject. The tumor data segmentation is also overlaid: the tumor core segmentation is outlined with the white dashed contour and the edema segmentation with the black dashed contour. Higher tumor concentrations (~1) are indicated by red and lower ones by green. Higher displacement norm values ($\sim18\,\text{mm}$) are indicated by red and lower ones by green. (Color figure online)

of the brain around the tumor (see Fig. 3); the ventricles are largely undeformed. Finally, *AMH* has a mild to moderate predicted mass effect ($\|u\|_{\infty} \sim 5$ mm ± 3) but exhibits the largest variation.

4   Conclusions

Our results are very promising. First, our solver is robust (never crashes, takes excessive iterations, or needs subject-specific hyperparameter settings). Most important, our method does not require any manual preprocessing and can be run in a black box fashion. Our experiments on synthetic data with known ground truth is the first time that such a solver is verified (for example, no such verification is undertaken in GLISTR) and we demonstrated that our approximations do not introduce significant errors despite the fact that m_0 is unknown. For clinical data, the model errors are expected to dominate the errors from using an atlas and splitting the calibration procedure into two stages. Second, we tested our method on a small number of clinical scans in order to test the feasibility of our method. With a very small number of calibration parameters, our solver was able to quantitatively match the observed tumor margins and qualitatively correlate with observed mass effect. We observed significant variability across subjects and small variability with respect to the choice of atlas. Our method provides a means to quantify and localize the mass effect without relying on any assumptions on symmetry and location of the tumor. It also provides quantitative mass effect measures for comparing and stratifying subjects. The biophysical features can complement other radiogenomic features in downstream learning tasks. Our next step is a more comprehensive evaluation of our solver on a larger clinical dataset. We aim to use more atlases (with possible preselection) with the intent to characterize the effect of the atlas choice and number on parameter inversion. The eventual goal is to correlate our model parameters/predictions to useful clinical outcomes such as patient survival or regions of tumor recurrence.

References

1. Abler, D., Büchler, P., Rockne, R.C.: Towards model-based characterization of biomechanical tumor growth phenotypes. In: Bebis, G., Benos, T., Chen, K., Jahn, K., Lima, E. (eds.) ISMCO 2019. LNCS, vol. 11826, pp. 75–86. Springer, Cham (2019). https://doi.org/10.1007/978-3-030-35210-3_6
2. Bakas, S., Reyes, M., et al.: Identifying the best machine learning algorithms for brain tumor segmentation, progression assessment, and overall survival prediction in the BRATS challenge. CoRR abs/1811.02629 (2018). http://arxiv.org/abs/1811.02629
3. Clatz, O., et al.: Realistic simulation of the 3-D growth of brain tumors in MR images coupling diffusion with biomechanical deformation. IEEE Trans. Med. Imaging **24**(10), 1334–1346 (2005)
4. Colin, T., Iollo, A., Lagaert, J.B., Saut, O.: An inverse problem for the recovery of the vascularization of a tumor. J. Inverse Ill-Posed Probl. **22**(6), 759–786 (2014)
5. Ezhov, I., et al.: Neural parameters estimation for brain tumor growth modeling. In: Shen, D., et al. (eds.) MICCAI 2019. LNCS, vol. 11765, pp. 787–795. Springer, Cham (2019). https://doi.org/10.1007/978-3-030-32245-8_87
6. Gholami, A., Mang, A., Biros, G.: An inverse problem formulation for parameter estimation of a reaction–diffusion model of low grade gliomas. J. Math. Biol. **72**(1), 409–433 (2015). https://doi.org/10.1007/s00285-015-0888-x
7. Gholami, A.: Fast algorithms for biophysically-constrained inverse problems in medical imaging. Ph.D. dissertation thesis (2017)
8. Gholami, A., et al.: A novel domain adaptation framework for medical image segmentation. In: Crimi, A., Bakas, S., Kuijf, H., Keyvan, F., Reyes, M., van Walsum, T. (eds.) BrainLes 2018. LNCS, vol. 11384, pp. 289–298. Springer, Cham (2019). https://doi.org/10.1007/978-3-030-11726-9_26
9. Gooya, A., et al.: GLISTR: glioma image segmentation and registration. IEEE Trans. Med. Imaging **31**(10), 1941–1954 (2013)
10. Hogea, C., Davatzikos, C., Biros, G.: An image-driven parameter estimation problem for a reaction-diffusion glioma growth model with mass effect. J. Math. Biol. **56**, 793–825 (2008)
11. Hogea, C., Davatzikos, C., Biros, G.: Modeling glioma growth and mass effect in 3D MR images of the brain. In: Ayache, N., Ourselin, S., Maeder, A. (eds.) MICCAI 2007. LNCS, vol. 4791, pp. 642–650. Springer, Heidelberg (2007). https://doi.org/10.1007/978-3-540-75757-3_78

12. Hormuth, D.A., Eldridge, S.L., Weis, J.A., Miga, M.I., Yankeelov, T.E.: Mechanically coupled reaction-diffusion model to predict glioma growth: methodological details. In: von Stechow, L. (ed.) Cancer Systems Biology. Methods in Molecular Biology, vol. 1711, pp. 225–241. Humana Press, New York (2018). https://doi.org/10.1007/978-1-4939-7493-1_11

13. Knopoff, D.A., Fernández, D.R., Torres, G.A., Turner, C.V.: Adjoint method for a tumor growth PDE-constrained optimization problem. Comput. Math. Appl. **66**(6), 1104–1119 (2013)

14. Konukoglu, E., et al.: Image guided personalization of reaction-diffusion type tumor growth models using modified anisotropic Eikonal equations. IEEE Trans. Med. Imaging **29**(1), 77–95 (2010)

15. Mang, A., Gholami, A., Davatzikos, C., Biros, G.: CLAIRE: a distributed-memory solver for constrained large deformation diffeomorphic image registration. SIAM J. Sci. Comput. **41**(5), C548–C584 (2019). https://doi.org/10.1137/18M1207818

16. Mang, A., et al.: Biophysical modeling of brain tumor progression: from unconditionally stable explicit time integration to an inverse problem with parabolic PDE constraints for model calibration. Med. Phys. **39**(7), 4444–4459 (2012)

17. Murray, J.D.: Mathematical Biology. Biomathematics, vol. 19. Springer, New York (1989). https://doi.org/10.1007/978-3-662-08539-4

18. Petersen, J., et al.: Deep probabilistic modeling of glioma growth. In: Shen, D., et al. (eds.) MICCAI 2019. LNCS, vol. 11765, pp. 806–814. Springer, Cham (2019). https://doi.org/10.1007/978-3-030-32245-8_89

19. Rockne, R.C., et al.: The 2019 mathematical oncology roadmap. Phys. Biol. (2019). http://iopscience.iop.org/10.1088/1478-3975/ab1a09

20. Scheufele, K., Mang, A., Gholami, A., Davatzikos, C., Biros, G., Mehl, M.: Coupling brain-tumor biophysical models and diffeomorphic image registration. Comput. Methods Appl. Mech. Eng. **347**, 533–567 (2019). https://doi.org/10.1016/j.cma.2018.12.008

21. Scheufele, K., Subramanian, S., Biros, G.: Automatic MRI-driven model calibration for advanced brain tumor progression analysis. arXiv, p. arXiv-2001 (2020)

22. Subramanian, S., Gholami, A., Biros, G.: Simulation of glioblastoma growth using a 3D multispecies tumor model with mass effect. J. Math. Biol. **79**(3), 941–967 (2019). https://doi.org/10.1007/s00285-019-01383-y

23. Subramanian, S., Scheufele, K., Mehl, M., Biros, G.: Where did the tumor start? An inverse solver with sparse localization for tumor growth models. Inverse Probl. (36) (2020). https://doi.org/10.1088/1361-6420/ab649c

24. Swanson, K., Alvord, E., Murray, J.: A quantitative model for differential motility of gliomas in grey and white matter. Cell Prolif. **33**(5), 317–330 (2000)

25. Swanson, K., Rostomily, R., Alvord Jr., E.: A mathematical modelling tool for predicting survival of individual patients following resection of glioblastoma: a proof of principle. Br. J. Cancer **98**(1), 113 (2008)

26. Swanson, K.R., Alvord, E., Murray, J.: Virtual brain tumours (gliomas) enhance the reality of medical imaging and highlight inadequacies of current therapy. Br. J. Cancer **86**(1), 14–18 (2002)

27. Yankeelov, T.E., et al.: Clinically relevant modeling of tumor growth and treatment response. Sci. Transl. Med. **5**(187), 187ps9 (2013)

Chest X-Ray Report Generation Through Fine-Grained Label Learning

Tanveer Syeda-Mahmood$^{(\boxtimes)}$, Ken C. L. Wong, Yaniv Gur, Joy T. Wu,
Ashutosh Jadhav, Satyananda Kashyap, Alexandros Karargyris, Anup Pillai,
Arjun Sharma, Ali Bin Syed, Orest Boyko, and Mehdi Moradi

IBM Almaden Research Center, San Jose, CA, USA
stf@us.ibm.com

Abstract. Obtaining automated preliminary read reports for common
exams such as chest X-rays will expedite clinical workflows and improve
operational efficiencies in hospitals. However, the quality of reports generated by current automated approaches is not yet clinically acceptable as they cannot ensure the correct detection of a broad spectrum of
radiographic findings nor describe them accurately in terms of laterality,
anatomical location, severity, etc. In this work, we present a domain-
aware automatic chest X-ray radiology report generation algorithm that
learns fine-grained description of findings from images and uses their
pattern of occurrences to retrieve and customize similar reports from a
large report database. We also develop an automatic labeling algorithm
for assigning such descriptors to images and build a novel deep learning
network that recognizes both coarse and fine-grained descriptions of findings. The resulting report generation algorithm significantly outperforms
the state of the art using established metrics.

1 Introduction

Chest X-rays are the most common imaging modality read by radiologists in hospitals and tele-radiology practices today. With advances in artificial intelligence,
there is now the promise of obtaining automated preliminary reads that can
expedite clinical workflows, improve accuracy and reduce overall costs. Current
automated report generation methods are based on image captioning approaches
of computer vision [20,24] and use an encoder-decoder architecture where a Convolutional Neural Network (CNN) is used to encode images into a set of semantic topics [9] or limited findings [28] and a Recurrent Neural Network (RNN)
decoder or a hierarchical LSTM generates the most likely sentence given the
topics [6,9,11,13,18,22]. Other approaches have leveraged template sentences
to aid in paraphrasing and report generation [3,12,13]. Recent approaches have
also emphasized the role of clinical accuracy measured loosely through clinical
correlation between disease states in the objective functions [15].

Despite the progress, the quality of reports generated by current approaches
is not yet clinically acceptable as they do not ensure the correct detection of a

© Springer Nature Switzerland AG 2020
A. L. Martel et al. (Eds.): MICCAI 2020, LNCS 12262, pp. 561–571, 2020.
https://doi.org/10.1007/978-3-030-59713-9_54

comprehensive set of findings nor the description of their clinical attributes such as laterality, anatomical location, severity, etc. The emphasis is usually more on the report language generation than the visual detection of findings. In this paper we take a new approach in which we train deep learning networks on a large number of detailed finding labels that represent an in-depth and comprehensive characterization of findings in chest X-ray images. An initial set of core findings label vocabulary were derived through a multi-year chest X-ray lexicon building effort involving several radiologists and clinical experts. The detailed finding labels were then automatically derived from their associated radiology reports through a concept detection and phrasal grouping algorithm that associates detailed characterization modifiers with the initially identified core findings using natural language analysis. The resulting labels with large image support were used to train a novel deep learning network based on feature pyramids. Given a new chest X-ray image, the joint occurrence of detailed finding labels is predicted as a pattern vector from the learned model and is matched against a pre-assembled database of label patterns and their associated reports. Finally, the retrieved report is post-processed to remove mentioned findings whose evidence is absent in the predicted label pattern.

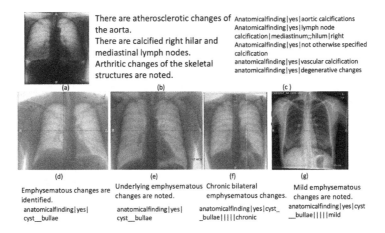

Fig. 1. Illustration of the finer description labels for capturing the essence of reports.

2   Describing Images Through Fine Finding Labels (FFL)

Consider the chest X-ray image shown in Fig. 1a. Its associated report is shown in Fig. 1b. In order to automatically produce such sentences from analyzing images, we need image labels that cover not only the core finding, such as opacity, but also its laterality, location, size, severity, appearance, etc. Specifically, a full description of the finding can be denoted by a fine finding label (FFL) as

$$F_i = < T_i | N_i | C_i | M_i^* >$$

(1)

where F_i is the FFL label, T_i is the finding type, $N_i = yes|no$ indicates a positive or negative finding (i.e. is present versus absent), C_i is the core finding itself, and M_i are one or more of the possible finding modifiers. The finding types in chest X-rays are adequately covered by six major categories namely, anatomical findings, tubes and lines and their placements, external devices, viewpoint-related issues, and implied diseases associated with findings. The vocabulary for core findings as well as possible modifiers was semi-automatically assembled through a multi-year chest X-ray lexicon development process in which a team of 4 clinicians including 3 radiologists, iteratively searched through the best practice literature such as Fleishner Society guidelines [5] and used every day use terms to expand the vocabulary by examining a large dataset of 220,000 radiology reports in a vocabulary building tool [1] addressing abbreviations, misspellings, semantical equivalence and ontological relationships. Currently, the lexicon consists of over 11000 unique terms covering the space of 78 core findings and 9 modifiers and represents the largest set of core findings assembled so far. The set of modifiers associated with each core finding also depends on the finding type and the FFL label syntax captures these for various finding types.

The FFL labels capture the essence of a report adequately as can be seen in Fig. 1c and comparing with the actual report in Fig. 1b. Further, if the FFL labels are similar, a similarity is also implied in the associated reports. Figure 1d–g show examples of similar reports all of which are characterized by similar FFL patterns. Thus if we can infer the FFL labels from the visual appearance of findings in chest X-ray images, we can expect to generate an adequate report directly from the labels.

2.1 Extraction of FFL Labels from Reports

The algorithm for extracting FFL labels from sentences in reports consists of 4 steps, namely, (a) core finding and modifier detection, (b) phrasal grouping, (c) negation sense detection, (d) pattern completion.

The vocabulary of core findings from lexicon and their synonyms were used to detect core concepts in sentences of reports using the vocabulary-driven concept extraction algorithm described in [4]. To associate modifiers with relevant core findings, we used a natural language parser called the ESG parser [16] which performed word tokenization and morpholexical analysis to create a dependency parse tree for the words in a sentence as shown in Fig. 2. The initial grouping of words is supplied directly by the parse tree such as the grouping of terms 'alveolar' and 'consolidation' into one term 'alveolar consolidation' shown in Fig. 2. Further phrasal grouping is done by clustering the lemmas using word identifiers specified in the dependency tree. For this, a connected component algorithm is used on the word positions in slots, skipping over unknowns (marked with u in tuples). This allows all modifiers present within a phrasal group containing a core finding to be automatically associated with the finding. For example, the modifier 'stable' is associated with the core finding 'alveolar consolidation' in Fig. 2. The modifiers in phrasal groups that do not contain a core finding are associated with the adjacent phrasal groups that contain a core finding.

To determine if a core finding is a positive or negative finding (e.g. "no pneumothorax"), we use a two-step approach that combines language structuring and vocabulary-based negation detection as described in [4]. The negation pattern detection algorithm iteratively identifies words within the scope of negation by iteratively expanding neighborhood of seed negation terms by traversing the dependency parse tree of a sentence. The details are described in [4].

The last step completes the FFL pattern using a priori knowledge captured in the lexicon for the associated anatomical locations of findings when these are not specified in the sentence itself as shown in Fig. 2 where the term "alveoli" is inserted from the knowledge of the location of the finding 'alveolar consolidation'. Thus the final FFL label produced may show more information than the original sentence from which it was extracted. In addition, the name of the core finding may be ontologically rolled up to the core findings as seen in Fig. 1 for 'emphysema'.

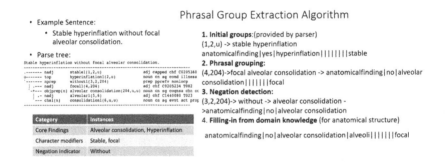

Fig. 2. Illustration of the dependency parse tree and phrasal grouping.

The FFL label extraction algorithm was applied to all sentences from a collection of 232,964 reports derived from MIMIC-4 [10] and NIH [21] datasets, to generate all possible FFL patterns from the Findings and Impression sections of reports. A total of 203,938 sentences were processed resulting in 102,135 FFL labels. By retaining only those labels with at least 100 image support, a total of 457 FFL labels were selected. As shown in the Results section, the label extraction process is highly accurate, so that spot check clinical validation is sufficient for use in image labeling. Since the FFL labels were seeded by clinically selected core findings, nearly 83% of all FFL labels extracted could be mapped into their nearest counterpart in the 457 FFL Label set. Thus the set of 457 labels were found sufficient to cover a wide variety in spoken sentences and were used as labels for building deep learning models. Of these, 78 were the original core labels (called the CFL labels) given by clinicians, and the remaining were finer description labels with modifiers extracted automatically.

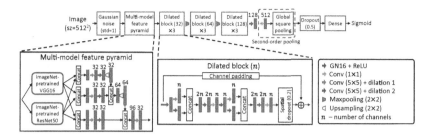

Fig. 3. Illustration of the custom deep learning network developed for large number of label recognition problem.

2.2   Learning FFL Labels from Images

The learning of FFL labels from chest X-rays is a fine-grained classification problem for which single networks used for computer vision problems may not yield the best performance, particularly since large training sets are still difficult to obtain. The work in [17] shows that concatenating different ImageNet-pretrained features from different networks can improve classification on microscopic images. Following this idea, we combine the ImageNet-pretrained features from different models through the Feature Pyramid Network in [14]. This forms the multi-model feature pyramid which combines the features in multiple scales. The VGGNet (16 layers) [19] and ResNet (50 layers) [7] are used as the feature extractors. As nature images and chest X-rays are in different domains, low-level features are used. From the VGGNet, the feature maps with 128, 256, and 512 channels are used, which are concatenated with the feature maps from the ResNet of the same spatial sizes which have 256, 512, and 1024 feature channels.

We propose dilated blocks to learn the high-level features from the extracted ImageNet features. Each dilated block is composed of dilated convolutions for multi-scale features [26], a skip connection of identity mapping to improve convergence [8], and spatial dropout to reduce overfitting. Group normalization (16 groups) [23] whose performance is independent of the training batch size is used with ReLU. Dilated blocks with different feature channels are cascaded with maxpooling to learning more abstract features. Instead of global average pooling, second-order pooling is used, which is proven to be effective for fine-grained classification [27]. Second-order pooling maps the features to a higher-dimensional space where they can be more separable. Following [27], the second-order pooling is implemented as a 1×1 convolution followed by global square pooling (Fig. 3).

Image augmentation with rigid transformations is used to avoid overfitting. As most of an image should be included, we limit the augmentation to rotation ($\pm 10°$) and shifting ($\pm 10\%$). The probability of an image to be transformed is 80%. The optimizer Nadam is used with a learning rate of 2×10^{-6}, a batch size of 48, and 20 epochs. To ensure efficient learning, we developed two instances of this network, one for the core finding labels (CFL labels) and the other for detailed FFL labels that have a support of at least 100 images for training to exploit the mutually reinforcing nature of the coarse-fine labels. Due to the variability in the size of the dataset per FFL label, the AUC per FFL label is not always a good indicator for precision on a per image level as it is dominated by the negative examples. To ensure we report as few irrelevant findings while still detecting all critical findings within an image, we select operating points on the ROC curves per label based on optimizing the F1 score, a well-known measure of accuracy, as

$$L(\theta) = -ln(\frac{1}{n}\sum_{i=1}^{n} F1_i(\theta)) \tag{2}$$

2.3 FFL Pattern-Report Database Creation

Using the FFL label detection algorithm, we can describe a report (its relevant sections) as a binary pattern vector $P = \{I_P(F_j)\}$ where $I_P(F_j) = 1$ if the FFL label $F_j \in \hat{F}$ is present in the report and zero otherwise. Here $\hat{F}$ is the set of FFL labels used in training the deep learning models. During the database creation process, we collect all reports characterized by the same binary pattern vector, and rank them based on the support provided by their constituent sentences. Let $\hat{R}_P = r_s$ be the collection of reports spanned by a pattern vector P. Then

$$Rank(r_s) = \sum_{j=1}^{M_s} h(s_j) \tag{3}$$

where M_s is the number of relevant sentences in report r_s spanned by one or more of the FFL labels in the pattern P. Here $h(s_j)$ is given by $h(s_j) = \frac{\text{Number of reports } r_i \text{ that contain } s_j}{|\hat{R}_P|}$ The highest ranked reports are then stored as associated reports with the binary pattern vectors in a database.

2.4 Report Assembly

The overall report generation workflow is illustrated in Fig. 5. An image is fed to the two deep learning networks built for CFL and FFL patterns and their predictions thresholded using the image-based precision-recall F1-score for optimization. The resulting pattern vectors are combined to result in the consolidated FFL pattern vector $Q = \{I_Q(F_j)\}$. The best matching reports are then derived from the semantically nearest pattern vectors in the database. The semantic distance between the query FFL bit pattern vector Q and a matching pattern vector from the database P is given by

$$d(Q, P) = \frac{\sqrt{\sum_{l=1}^{|\hat{F}|} w_l (I_P(F_l) - I_Q(F_l))^2}}{|\hat{F}|} \tag{4}$$

where w_l is the weight associated with the FFL label F_l. A criticality rank for each core findings on a scale of 1 to 10 was supplied by the clinicians which was normalized and used to weigh the clinical importance of a finding during

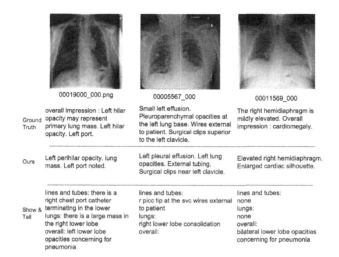

Fig. 4. Illustration of quality of reports generated by different methods.

matching. Once the matching FFL pattern is determined, the highest ranked report as given by Eq. 3 associated with the FFL pattern is retrieved as the best matching report. Finally, we drop all sentences from the retrieved report whose evidence cannot be found in the FFL label pattern of the query thus achieving the variety needed in returned reports per query. Although with 457 FFL labels, the number of possible binary patterns would be large (2^{457}), due to the sparseness of 5–7 findings per report, the actual number of distinct binary patterns in the database of over 232,000 reports was only 924 patterns corresponding to 5246 distinct sentences in the precomputed ranked list across all patterns. Thus the lookup per predicted pattern is a fairly trivial operation which is O(1) with indexing and takes less than 5 ms.

3   Results

We collected datasets from three separate sources, namely, the MIMIC [10] dataset of over 220,000 reports with associated frontal images, 2964 unique Indiana reports [2] with associated images and a set of 10,000 NIH [21] released images re-read by our team of radiologists to produce a total of 232,964 image-report pairs for our experiments.

Evaluating FFL Label Extraction Accuracy: We evaluated the accuracy of FFL label extraction by noting the number of findings missed and overcalled (which included errors in negation sense detection) as well the correctness and completeness of association of modifiers with the relevant core findings. The result of evaluation for the Indiana dataset [2] by our clinicians is shown in Table 2. As can be seen, the FFL label extraction is highly accurate in terms of the coverage of findings with around 3% error mostly due to negation sense

Table 1. Illustration of the datasets and performance of fine grained classification model for CFL and FFL labels (last column is average of AUCs weighted by the number of samples per each category).

Dataset	FFL label phases	Train	Validate	Test	Average AUC	Weighted
MIMIC-4+NIH	CFL labels	249,286	35,822	70,932	0.805	0.841
MIMIC-4+NIH	FFL labels	75,613	10,615	20,941	0.729	0.716

detection. Further, the association of modifiers to core findings given by the phrasal grouping algorithm is also accurate with over 99% precision and recall.

FFL Label Prediction from Deep Learning: The training, validation and test image datasets used for building the CFL and FFL models used MIMIC-4 and NIH datasets as shown in Table 1. The AUC averaged for all CFL labels and FFL labels is also shown in that table. In addition, using the F1-score-based optimization, the mean average image-based precision for CFL labels was 73.67% while the recall was 70.6%.

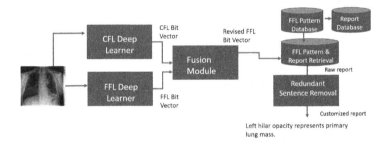

Fig. 5. Illustration of the report generation algorithm.

Table 2. The accuracy of FFL label extraction from reports.

Reports analyzed	Relevant sentences	FFL patterns extracted	Missed findings	Overcall (negated findings)	Incorrect association of modifiers	Missed modifiers
2964	3046	5245	0	168	49	11

Evaluation of Report Generation: Due to the ontological mapping used to abstract the description of findings, the match produced from our approach is at a more semantic level rather than lexical in comparison to other approaches. Figure 4 shows the reports manually and automatically produced by our approach and a comparative approach implemented from a visual attention-based captioning model [20]. We compared the performance of our algorithm with several state-of-the-art baselines from recent literature [9,12,15,20,25,28]. These included a range of approaches from visual attention-based captioning [20], knowledge-driven report generation [12], clinically accurate report generation [15], to a strawman approach using a set of template sentences manually chosen by clinicians for the FFL labels instead of the nearest report selection algorithm described earlier. Although we have tested our algorithm for very large number of images from the combined MIMIC-NIH data, for purposes of comparison, we show the results on the same Indiana test dataset that has been used most commonly by other algorithms as reported in [9]. The resulting performance using the popular scoring metrics is shown in Table 3 showing that our algorithm outperforms other approaches in all the established scoring metrics.

Conclusions: We presented an explainable AI approach to semantically correct radiology report generation. The results show superior performance both because of the detailed descriptive nature of labels, and due to a statistically informed report retrieval process that ensures a semantic match.

Table 3. Comparative performance of report generation by various methods.

Methods	BLEU-1	BLEU-2	BLEU-3	BLEU-4	METEOR	ROUGE-L
Vis-Att [20]	0.39	0.25	0.16	0.11	0.16	0.32
MM-Att [25]	0.46	0.35	0.27	0.19	0.27	0.36
KERP [12]	0.48	0.32	0.22	0.16	–	0.33
Template-based	0.28	0.29	0.32	0.27	0.35	0.34
Clinical accurate [15]	0.35	0.22	0.15	0.1	–	0.45
Co-Att [9]	0.51	0.39	0.30	0.25	0.21	0.44
Jiebo Luo [28]	0.53	0.37	0.31	0.25	0.34	0.45
CFL-only	0.49	0.39	0.36	0.32	0.48	0.52
FFL+CFL-based (ours)	**0.56**	**0.51**	**0.5**	**0.49**	**0.55**	**0.58**

References

1. Coden, A., Gruhl, D., Lewis, N., Tanenblatt, M., Terdiman, J.: SPOT the drug! An unsupervised pattern matching method to extract drug names from very large clinical corpora. In: Proceedings of the 2012 IEEE Second International Conference on Healthcare Informatics, Imaging and Systems Biology, pp. 7008–7024 (2012)
2. Demmer-Fushma, D., et al.: Preparing a collection of radiology examinations for distribution and retrieval. J. Am. Med. Inform. Assoc. (JAMIA) **23**(2), 304–310 (2014)
3. Gale, W., Oakden-Rayner, L., Carneiro, G., Bradley, A.P., Palmer, L.J.: Producing radiologist-quality reports for interpretable artificial intelligence. arXiv preprint arXiv:1806.00340 (2018)

4. Guo, Y., Kakrania, D., Baldwin, T., Syeda-Mahmood, T.: Efficient clinical concept extraction in electronic medical records. In: Thirty-First AAAI Conference on Artificial Intelligence (2017)
5. Hansell, D.M., Bankier, A.A., MacMahon, H., McLoud, T.C., Muller, N.L., Remy, J.: Fleischner society: glossary of terms for thoracic imaging. Radiology **246**(3), 697–722 (2008)
6. Harzig, P., Chen, Y.Y., Chen, F., Lienhart, R.: Addressing data bias problems for chest x-ray image report generation. arXiv preprint arXiv:1908.02123 (2019)
7. He, K., Zhang, X., Ren, S., Sun, J.: Deep residual learning for image recognition. In: IEEE Conference on Computer Vision and Pattern Recognition, pp. 770–778 (2016)
8. He, K., Zhang, X., Ren, S., Sun, J.: Identity mappings in deep residual networks. In: Leibe, B., Matas, J., Sebe, N., Welling, M. (eds.) ECCV 2016. LNCS, vol. 9908, pp. 630–645. Springer, Cham (2016). https://doi.org/10.1007/978-3-319-46493-0_38
9. Jing, B., Xie, P., Xing, E.: On the automatic generation of medical imaging reports. arXiv preprint arXiv:1711.08195 (2017)
10. Johnson, A.E.W., et al.: MIMIC-CXR: a large publicly available database of labeled chest radiographs. arXiv preprint arXiv:1901.07042 (2019)
11. Krause, J., Johnson, J., Krishna, R., Fei-Fei, L.: A hierarchical approach for generating descriptive image paragraphs. In: Proceedings of the IEEE Conference on Computer Vision and Pattern Recognition, pp. 317–325 (2017)
12. Li, C.Y., Liang, X., Hu, Z., Xing, E.P.: Knowledge-driven encode, retrieve, paraphrase for medical image report generation. arXiv preprint arXiv:1903.10122 (2019)
13. Li, Y., Liang, X., Hu, Z., Xing, E.P.: Hybrid retrieval-generation reinforced agent for medical image report generation. In: Advances in Neural Information Processing Systems, pp. 1530–1540 (2018)
14. Lin, T.Y., Dollár, P., Girshick, R., He, K., Hariharan, B., Belongie, S.: Feature pyramid networks for object detection. In: IEEE Conference on Computer Vision and Pattern Recognition, pp. 2117–2125 (2017)
15. Liu, G., et al.: Clinically accurate chest x-ray report generation. arXiv:1904.02633v (2019)
16. McCord, M.C., Murdock, J.W., Bogurae, B.K.: Deep parsing in Watson. IBM J. Res. Dev. **56**(3), 5–15 (2012)
17. Nguyen, L.D., Lin, D., Lin, Z., Cao, J.: Deep CNNs for microscopic image classification by exploiting transfer learning and feature concatenation. In: IEEE International Symposium on Circuits and Systems, pp. 1–5 (2018)
18. Rennie, S.J., Marcheret, E., Mroueh, Y., Ross, J., Goel, V.: Self-critical sequence training for image captioning. In: Proceedings of the IEEE Conference on Computer Vision and Pattern Recognition, pp. 7008–7024 (2017)
19. Simonyan, K., Zisserman, A.: Very deep convolutional networks for large-scale image recognition. arXiv:1409.1556 [cs.CV] (2014)
20. Vinyals, O., Toshev, A., Bengio, S., Erhan, D.: Show and tell: a neural image caption generator. In: Proceedings of the IEEE Conference on Computer Vision and Pattern Recognition, pp. 3156–3164 (2015)
21. Wang, X., Peng, Y., Lu, L., Lu, Z., Bagheri, M., Summers, R.M.: ChestX-ray8: hospital-scale chest x-ray database and benchmarks on weakly-supervised classification and localization of common thorax diseases. In: Proceedings of the IEEE Conference on Computer Vision and Pattern Recognition, pp. 2097–2106 (2017)

22. Wang, X., Peng, Y., Lu, L., Lu, Z., Summers, R.M.: TieNet: text-image embedding network for common thorax disease classification and reporting in chest x-rays. In: Proceedings of the IEEE Conference on Computer Vision and Pattern Recognition, pp. 9049–9058 (2018)
23. Wu, Y., He, K.: Group normalization. In: Ferrari, V., Hebert, M., Sminchisescu, C., Weiss, Y. (eds.) ECCV 2018. LNCS, vol. 11217, pp. 3–19. Springer, Cham (2018). https://doi.org/10.1007/978-3-030-01261-8_1
24. Xu, K., et al.: Show, attend and tell: neural image caption generation with visual attention. In: International Conference on Machine Learning, pp. 2048–2057 (2015)
25. Xue, Y., et al.: Multimodal recurrent model with attention for automated radiology report generation. In: Frangi, A.F., Schnabel, J.A., Davatzikos, C., Alberola-López, C., Fichtinger, G. (eds.) MICCAI 2018. LNCS, vol. 11070, pp. 457–466. Springer, Cham (2018). https://doi.org/10.1007/978-3-030-00928-1_52
26. Yu, F., Koltun, V.: Multi-scale context aggregation by dilated convolutions. arXiv:1511.07122 [cs.CV] (2015)
27. Yu, K., Salzmann, M.: Statistically-motivated second-order pooling. In: Ferrari, V., Hebert, M., Sminchisescu, C., Weiss, Y. (eds.) ECCV 2018. LNCS, vol. 11211, pp. 621–637. Springer, Cham (2018). https://doi.org/10.1007/978-3-030-01234-2_37
28. Yuan, J., Liao, H., Luo, R., Luo, J.: Automatic radiology report generation based on multi-view image fusion and medical concept enrichment. In: Shen, D., et al. (eds.) MICCAI 2019. LNCS, vol. 11769, pp. 721–729. Springer, Cham (2019). https://doi.org/10.1007/978-3-030-32226-7_80

Peri-Diagnostic Decision Support Through Cost-Efficient Feature Acquisition at Test-Time

Gerome Vivar[1,2], Kamilia Mullakaeva[1], Andreas Zwergal[2], Nassir Navab[1,3], and Seyed-Ahmad Ahmadi[1,2(✉)]

[1] Technical University of Munich (TUM), Munich, Germany
ahmadi@cs.tum.edu
[2] German Center for Vertigo and Balance Disorders (DSGZ),
Ludwig-Maximilians-Universität (LMU), Munich, Germany
[3] Whiting School of Engineering, Johns Hopkins University, Baltimore, USA

Abstract. Computer-aided diagnosis (CADx) algorithms in medicine provide patient-specific decision support for physicians. These algorithms are usually applied after full acquisition of high-dimensional multimodal examination data, and often assume feature-completeness. This, however, is rarely the case due to examination costs, invasiveness, or a lack of indication. A sub-problem in CADx, which to our knowledge has not been addressed by the MICCAI community so far, is to guide the physician during the entire peri-diagnostic workflow, including the acquisition stage. We model the following question, asked from a physician's perspective: "Given the evidence collected so far, which examination should I perform next, in order to achieve the most accurate and efficient diagnostic prediction?". In this work, we propose a novel approach which is enticingly simple: use dropout at the input layer, and integrated gradients of the trained network at test-time to attribute feature importance dynamically. We validate and explain the effectiveness of our proposed approach using two public medical and two synthetic datasets. Results show that our proposed approach is more cost- and feature-efficient than prior approaches and achieves a higher overall accuracy. This directly translates to less unnecessary examinations for patients, and a quicker, less costly and more accurate decision support for the physician.

Keywords: Computer-aided diagnosis · Peri-diagnostic decision support · Cost-sensitive feature attribution · Integrated Gradients

1 Introduction

The diagnostic workflow in medicine is "an iterative process of information gathering, information integration and interpretation" [2]. Information is first

---

G. Vivar and K. Mullakaeva contributed equally to this work.

A. L. Martel et al. (Eds.): MICCAI 2020, LNCS 12262, pp. 572–581, 2020.
https://doi.org/10.1007/978-3-030-59713-9_55

acquired through a clinical history and interview, followed by alternating examinations and working diagnoses, until sufficient information has been aggregated for a final diagnosis. The decision which examination to perform next lies in the responsibility of the physician, who has to consider its medical indication, its invasiveness towards the patient, and often also its financial cost. Machine learning (ML) and computer-aided diagnosis (CADx) have a large potential for decision support in the clinic [16]. From a ML perspective, CADx is the task to learn the mapping of a multimodal feature vector onto a diagnostic label. Most CADx algorithms studied so far, however, ignore the acquisition stage, and provide decision support only at the end of the diagnostic workflow when all examination data is acquired and the feature vector is complete. As such, current CADx approaches miss out on the opportunity to aid the physician during the entire, *peri-diagnostic workflow*, including the acquisition stage. In this work, we address this problem by i) iteratively suggesting the next most important examination/feature to acquire, while ii) considering the overall examination cost and aiming for a maximally accurate and efficient diagnostic prediction. To the best of our knowledge, the problem of *peri-diagnostic decision support* has not been addressed in the MICCAI community so far.

Related Works: In ML literature, this problem is often described as budgeted or cost-sensitive feature acquisition. Most recent approaches can be roughly categorized into reinforcement learning (RL) and non-RL approaches. Among *RL approaches*, [14] applied cost-sensitive n-step Q learning to CADx on Physionet (2012) and proprietary data. Kachuee et al. [7] classify diabetes with Deep Q-networks (DQN) and Monte-Carlo dropout, and select the feature with the maximum confidence gain of the predictor network while considering cost. [5] classify non-medical data with a DQN-variant that penalizes accumulated feature cost and incorrect predictions. RL-approaches have two important limitations: first, agent and predictor only work in tandem, neither has any utility or generalizability on its own. Second, unless agent and predictor are perfectly tuned, the network can quickly settle on a sub-optimal final classification accuracy in favor of low cost. Among the *non-RL approaches*, [3] classify fetal heartbeat patterns using Recurrent Neural Networks (RNN), which suggest the next feature through learned attention vectors as masks at every timestep. This can lead to suggesting several or repeated features at each timestep, and requires a fixed number of timesteps before its final prediction which can be inefficient. Kachuee et al. classify non-medical data [8] and detect hypothyroidism [6] using denoising autoencoders (DAE). The DAE is trained with dropout at the input layer, and learns to reconstruct complete feature vectors from incomplete inputs. Next, the encoder part is fine-tuned and trained in tandem with a predictor network towards the final prediction task. At test time, the partial derivatives of all outputs with respect to each input feature are aggregated to form the total "feature attribution". In this context, it is important to note that feature attribution needs to fulfill four axioms, which have been derived in [15]. The gradient-based attribution only with respect to the input as performed in [6,8] violates the

"Sensitivity Axiom" of feature attribution. This can lead to an acquisition of inefficient features [15] and ultimately, unnecessary patient examinations.

Contributions: 1) We propose for the first time to apply Integrated Gradients (IG), an axiomatic feature attribution method, to the problem of dynamic, budgeted feature acquisition. 2) We propose Accumulated IG (AIG), for dynamically suggesting the next most important feature to acquire at test-time, and 3) we highlight the advantages of our proposed approach on two medical datasets and two explanatory datasets for illustration of its working principles.

2 Materials and Methods

2.1 Datasets and Preprocessing

To evaluate our method, we utilized four datasets. The first two are publicly available medical datasets, to demonstrate the efficacy of our method on real-life data. The latter two datasets are non-medical, for further benchmarking as well as to illustrate the inner workings and limitations of the different methods we evaluate. For pre-processing, we perform outlier removal and scaling in NHANES and Thyroid (Winsorization of real-valued features to $[5, 95]$-percentile, normalization to range $[0, 1]$). **NHANES:** The "National Health and Examination Survey" dataset [1] contains demographic information, laboratory results, questionnaire, and physical examination data. The goal here is to predict diabetes (normal, pre-diabetes, and diabetes) based on measured fasting glucose levels. Costs for features were established in a crowd-sourcing approach [9], and represent the total 'inconvenience' of feature acquisition from a patient-perspective (including time burden, financial cost, discomfort, etc.) [7]. The cost varies from 1 to 9 on a relative, numeric scale. We use all 92062 samples and 45 features in this dataset. **Thyroid:** The UCI Thyroid disease dataset [4,13] poses a three-class classification problem (normal thyroid function vs. hyperfunction vs. or subnormal function). There are 21 features, representing demographic information, questionnaires and laboratory results that are important for thyroid disease classification. Feature costs are provided as part of the public dataset "ann-thyroid" [13], and range from 1.00 to 22.78. We use all 7200 samples and 21 features. **MNIST:** In the MNIST dataset [11], we classify handwritten digit images in vectorized form, to simulate a tabular dataset. We use all 70,000 images with 784 features. We further assume a uniform cost of 1 for every pixel, to make our results comparable to related works. **Synthesized:** We also use a synthesized dataset as in [6], to further explain and visualize the feature attribution process. The dataset consists of 16,000 samples with 64 dimensions. The first 32 dimensions contain salient information for classification, at a linearly increasing cost from 1 to 32. The second 32 dimensions contain no valuable information for classification, again at a linearly increasing cost of 1 to 32. Hence, intuitively, an efficient feature acquisition approach should choose only features from the first 32 dimensions. For a more detailed explanation, we refer the reader to [6].

2.2   Problem Setting

In this work, we consider the problem of patient-specific, dynamic feature acquisition at test time. The goal is to sequentially acquire features that can achieve the maximum prediction performance, as efficiently as possible. We aim for a model that is both cost- and feature-efficient, i.e. a model that achieves the maximum prediction performance with the least accumulated cost and smallest number of features possible. Formally, we consider the problem of predicting a target value $\hat{y} \in Y$ based on a feature vector $\bar{x} \in \mathbf{R}^d$, which initially contains incomplete information about the patient at test-time. For clarity, we denote a complete feature vector as x and an incomplete feature vector as $\bar{x}$.

2.3   Efficient Feature Acquisition at Test-Time Using Integrated Gradients

To efficiently acquire features at test-time, we propose to use feature attribution by Integrated Gradients (IG) [15]. Previous works make use of backpropagation for feature acquisition [6,8], by calculating the gradients of the network at the current input value. This, however, violates the "Sensitivity(a)" axiom of feature attribution [15], which states that if an input differs in one feature compared to a neutral baseline input, and if this leads to a different output, then that feature should be given a non-zero attribution. IG can be shown to uniquely satisfy the axiom "Sensitivity(a)", as well as the axiom "Implementation Invariance", which states that two different networks that produce the exact same outputs for the same inputs should produce the same feature attribution [15].

Where previous gradient-based approaches [6,8] only take the gradient at the current input, IG takes a path integral of the gradients while linearly blending between a baseline input $x' \in \mathbf{R}^d$ and the actual input $x \in \mathbf{R}^d$, to avoid local gradients becoming saturated [15]. The baseline input x' represents an "absence" of features and can be encoded as a zero-valued vector. Importantly, IG was originally designed for inference explanation, by computing feature attributions with respect to the known correct output class and model posterior. In our scenario, however, we do not know the output label of interest at test-time. We thus propose *Accumulated IG (AIG)*, i.e. to aggregate the attributions of all input features from all possible output classes (see Eqs. 1 and 2). In addition, since we have an input $\bar{x}$ which is initially empty at test-time, we have to use a different baseline in order to be able to calculate AIG. We thus represent missing features with a neutral baseline at the central tendency (i.e. mean), analogous to mean-imputation in regular machine learning. Here, accumulating the gradients implies combining attributions from K different functions. This follows the "Linearity Axiom" of attribution theory, keeping AIG axiomatic as in the original IG formulation [15].

To handle missing information at test-time, previous works [6,8] proposed to use denoising autoencoders (DAE). We validate a combination of DAE with AIG in our experiments, but we also propose a simplified version without the need for auto-encoding. The simplified model is a vanilla multi-layer perceptron

(MLP) trained end-to-end, while applying a Beta-distributed dropout layer to the input [6] to simulate missing information during training (see Fig. 1).

Implementation Details: We approximate the continuous IG as in [15] by a few discrete steps. We calculate the attribution along the i-th dimension with respect to one specific class (k) using:

$$\text{IG}_{i,k}^{approx}(\bar{x}_i, class_k) = (\bar{x}_i - x_i') \times \sum_{s=1}^{m} \frac{\partial F(x' + \frac{s}{m} \times (\bar{x} - x'))}{\partial \bar{x}_i} \times \frac{1}{m} \quad (1)$$

where $\bar{x}$ is the input vector and x_i' the baseline at the i-th dimension, $\frac{\partial F(.)}{\partial \bar{x}_i}$ is the partial derivative of the network's output with respect to input $\bar{x}_i$, and m is the number of approximation steps of the path integral in IG. We then sum up all the attributions for the current feature from all classes and aggregate both positive and negative gradients. To account for cost-efficiency, we scale the attribution to a feature by the inverse feature cost:

$$f^{(i)} = \frac{|\sum_{k=1}^{K} \text{IG}^{approx}(\bar{x}_i, class_k)|}{c_i} \quad (2)$$

where $f^{(i)}$ denotes the AIG feature attribute of input $\bar{x}_i$ and c_i denotes its cost. Then $f_t \in \mathbf{R}^d$ is a vector which consists of AIG attributions of all features $[f_t^{(1)}, f_t^{(2)}, ..., f_t^{(d)}]$ at timestep t. To determine which feature to acquire next, we take the index of the feature attribute with the maximum value: $a_{f_t} = \arg\max(f_t)$, where a_{f_t} denotes the feature to acquire at timestep t as illustrated in Fig. 1. Using this newly acquired feature and previously acquired features we then perform classification (a_{c_t}) on this incomplete feature vector and obtain the label y_t. We repeat this process until there are no more remaining features to acquire. Alternatively, one can set a maximum allowed cost to constrain feature acquisition to a maximum allowed budget. The network architecture and an unrolled feature acquisition process are illustrated in Fig. 1.

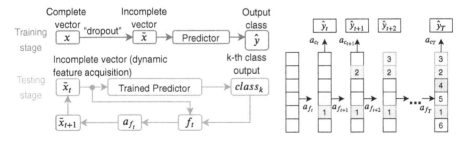

Fig. 1. Illustration of our proposed network architecture (left panel) and an unrolled feature acquisition sequence at test-time (right panel).

3 Results and Discussion

3.1 Experimental Setup

Baseline Model Comparison: We evaluate our work against several baseline and state-of-the-art approaches in budgeted feature acquisition, including a recent deep-RL [7], two non-RL [6,8], and a random feature selection based approaches. We split the datasets into 15% test set and 85% training set, where 15% of the latter is used for validation. We use Adam optimization [10] implemented in PyTorch [12] on a single-GPU workstation (Nvidia GTX 1080 Ti). For RL, we used the author implementation and parametrizations of Opportunistic Learning [7], to train an agent for 11,000 episodes. For comparison, all methods including our own are based on a two-layer multilayer perceptron (MLP) [64 and 32 units]. The non-RL approaches we compare against are Dynamic Feature Query (DPFQ) [8] and Feature Acquisition Considering Cost at Test-time (FACT) [6]. Again, we use the same MLP architecture for the encoder [64, 32], decoder [32, 64], and predictor [32, 16, K classes]. For the binary layer in FACT, we use the identical 8-bit representation as in [6]. Further, to randomly drop entries, we use a Beta-distribution with $\alpha = 1.5$ and $\beta = 1.5$, following [6].

Proposed: We use a vanilla MLP (encoder [64, 32], and predictor [32, 16, K classes]). We used the Adam optimizer in PyTorch with a low learning rate ($lr = 1e - 4$). We use $m = 50$ for the number of steps in the integral approximation in Eq. (1).

3.2 Feature Acquisition Performance

We compare our work with previous deep-RL [7] and non-RL [6,8] techniques to evaluate the effectiveness of our proposed method. We observe that our proposed AIG approach with and without DAE outperforms the SOTA methods, with a particularly large margin in the two medical datasets. Overall, our approach is the most cost- and feature-efficient (see Fig. 3) and consistently achieves the highest overall classification accuracy. The only exception is RL for Synthesized data, but otherwise RL lacks robustness. Our method's feature-efficiency is evident e.g. in Thyroid and NHANES, on average, it is able to outperform the SOTA and reach the maximum classification accuracy after just 7 ($\sim$33%) and 10 ($\sim$22%) features, respectively (see Fig. 3). Importantly, this directly translates to the avoidance of unnecessary examinations and a much faster time-to-diagnosis, without requiring patients to undergo all examinations. Apart from feature-efficiency, our approach is also cost-efficient, e.g. spending only $\sim$20 ($\sim$25%) units of cost in Thyroid, and $\sim$50 ($\sim$29%) units in NHANES to achieve maximum classification performance. Further, methods like RL or DPFQ may choose cheaper features first, despite little gain in classification accuracy (see Fig. 3, right Thyroid panel), whereas our method suggests more costly features in the beginning, at the benefit of reaching the highest classification accuracy almost instantly.

3.3   Interpretation of Patient-Specific Feature Acquisition

We also use test samples of each dataset to visualize and discuss the order of feature selection by the different methods. We show heatmaps in Fig. 2, where warmer colors denote higher priority in the feature acquisition. We plot ten test samples for datasets Thyroid, NHANES and Synthesized, and one test image from MNIST. In Fig. 2, we observe that our proposed approach initially always acquires the most informative feature, before starting to acquire features in an instance- or patient-specific manner. For example, in the Thyroid dataset, features 21, 2, and 3 are consistently selected first by the model, while at feature 1 or 19, model suggestions start to diverge which feature to acquire next. Similarly in NHANES, features 2 and 30 form an initial decision baseline, before the model diverges into patient-specific decisions at features 33 or 45. In contrast, FACT and RL may change the feature acquisition order almost instantly, already at the first or second acquisition step, which may not always be justified or effective. In MNIST, FACT heatmaps show an outlining of the digit, as FACT multiplies the output of the de-noising auto-encoder with the feature-aggregation score. This strategy prioritizes high-intensity/-amplitude features, and leads to intuitive visualizations on MNIST, but does not directly translate to an efficient feature acquisition performance, as seen e.g.. in the NHANES dataset. Further, approaches like RL may choose features in random order (MNIST), or in order of least cost instead of relevance (Synthesized). In future work, we aim at investigating such phenomena from a medical perspective.

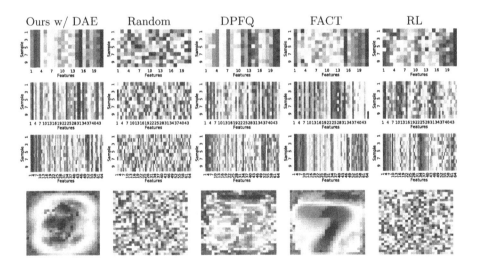

Fig. 2. Feature acquisition heatmaps for all datasets. From top to bottom: UCI-Thyroid (10 patient samples), NHANES (10 patient samples), Synthesized (10 samples), and MNIST. Warmer colors denote higher priority for feature acquisition. (Color figure online)

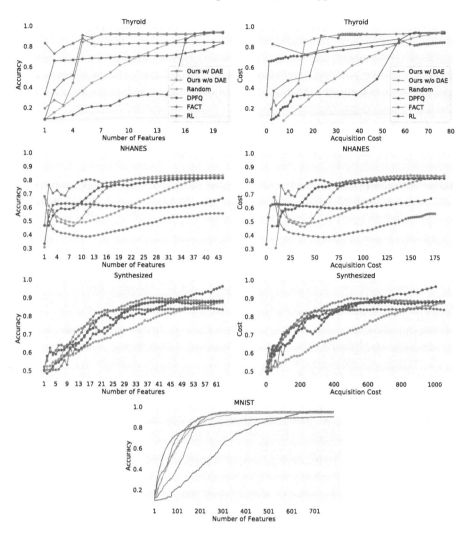

Fig. 3. Comparison of feature aggregation methods against our proposed approach, on four datasets (row 1: Thyroid; row 2: NHANES; row 3: Synthesized; row 4: MNIST with uniform feature cost). Left column: Feature count vs. accuracy curves; right column: Accumulated feature cost vs. accuracy. The compared baseline approaches denote: Random (random feature selection), DPFQ [8], FACT [6], RL [7]. Our approach is consistently most feature- and cost-efficient and achieves the highest classification final accuracy.

4   Conclusion

We propose a novel method which can efficiently acquire features at test-time, through Accumulated Integrated Gradients (AIG) and network training with dropout at the input layer. We empirically show that our approach is cost- and feature-efficient when evaluated on two medical datasets and two explanatory toy datasets. Our proposed method enables patient-specific, peri-diagnostic decision support for clinicians, which could potentially optimize spending, maximize hospital resources, and reduce examination burden for patients. Future work could address two important limitations of our work, which occur frequently in real-life clinical data, namely how to train a peri-diagnostic CADx system from data that is i) incomplete at training time and ii) made up of features from different modalities which are organized into blocks with acquisition costs that increase blockwise instead of one feature at a time.

Acknowledgments. This work was supported by the German Federal Ministry of Education and Health (BMBF) in connection with the foundation of the German Center for Vertigo and Balance Disorders (DSGZ) [grant number 01 EO 0901].

References

1. Centers for Disease Control and Prevention (CDC): National health and nutrition examination survey (NHANES). https://www.cdc.gov/nchs/nhanes/index.htm. Accessed 11 Mar 2020
2. Committee on Diagnostic Error in Health Care, Board on Health Care Services, Institute of Medicine, The National Academies of Sciences, Engineering, and Medicine: Improving Diagnosis in Health Care. National Academies Press, Washington, D.C., December 2015. https://doi.org/10.17226/21794. http://www.nap.edu/catalog/21794
3. Contardo, G., Denoyer, L., Artières, T.: Recurrent neural networks for adaptive feature acquisition. In: Hirose, A., Ozawa, S., Doya, K., Ikeda, K., Lee, M., Liu, D. (eds.) ICONIP 2016. LNCS, vol. 9949, pp. 591–599. Springer, Cham (2016). https://doi.org/10.1007/978-3-319-46675-0_65
4. Dua, D., Graff, C.: UCI machine learning repository (2017). http://archive.ics.uci.edu/ml
5. Janisch, J., Pevný, T., Lisỳ, V.: Classification with costly features using deep reinforcement learning. In: Proceedings of the AAAI Conference on Artificial Intelligence, vol. 33, pp. 3959–3966 (2019)
6. Kachuee, M., Darabi, S., Moatamed, B., Sarrafzadeh, M.: Dynamic feature acquisition using denoising autoencoders. IEEE Trans. Neural Netw. Learn. Syst. **30**(8), 2252–2262 (2018)
7. Kachuee, M., Goldstein, O., Kärkkäinen, K., Sarrafzadeh, M.: Opportunistic learning: budgeted cost-sensitive learning from data streams. In: International Conference on Learning Representations (ICLR) (2019). https://openreview.net/forum?id=S1eOHo09KX
8. Kachuee, M., Hosseini, A., Moatamed, B., Darabi, S., Sarrafzadeh, M.: Context-aware feature query to improve the prediction performance. In: 2017 IEEE Global Conference on Signal and Information Processing (GlobalSIP), pp. 838–842. IEEE (2017)
9. Kachuee, M., Karkkainen, K., Goldstein, O., Zamanzadeh, D., Sarrafzadeh, M.: Cost-sensitive diagnosis and learning leveraging public health data. Preprint https://arxiv.org/abs/1902.07102 (2019)

10. Kingma, D.P., Ba, J.: Adam: a method for stochastic optimization. In: Bengio, Y., LeCun, Y. (eds.) 3rd International Conference on Learning Representations (ICLR) (2015). http://arxiv.org/abs/1412.6980

11. Lecun, Y., Bottou, L., Bengio, Y., Haffner, P.: Gradient-based learning applied to document recognition. Proc. IEEE **86**(11), 2278–2324 (1998). https://doi.org/10.1109/5.726791. http://ieeexplore.ieee.org/document/726791/

12. Paszke, A., et al.: Pytorch: an imperative style, high-performance deep learning library. In: Wallach, H., Larochelle, H., Beygelzimer, A., d'Alché-Buc, F., Fox, E., Garnett, R. (eds.) Advances in Neural Information Processing Systems 32, pp. 8024–8035. Curran Associates, Inc., Red Hook (2019)

13. Quinlan, J.R., Compton, P.J., Horn, K.A., Lazarus, L.: Inductive knowledge acquisition: a case study. In: Proceedings of the Second Australian Conference on Applications of Expert Systems, pp. 137–156. Addison-Wesley Longman Publishing Co., Inc., Upper Saddle River (1987)

14. Shim, H., Hwang, S.J., Yang, E.: Joint active feature acquisition and classification with variable-size set encoding. In: Advances in Neural Information Processing Systems, pp. 1368–1378 (2018)

15. Sundararajan, M., Taly, A., Yan, Q.: Axiomatic attribution for deep networks. In: Proceedings of the 34th International Conference on Machine Learning - Volume 70, pp. 3319–3328. JMLR.org (2017)

16. Yanase, J., Triantaphyllou, E.: A systematic survey of computer-aided diagnosis in medicine: past and present developments. Expert Syst. Appl. **138**, 112821 (2019). https://doi.org/10.1016/j.eswa.2019.112821

A Deep Bayesian Video Analysis Framework: Towards a More Robust Estimation of Ejection Fraction

Mohammad Mahdi Kazemi Esfeh[1], Christina Luong[2], Delaram Behnami[1], Teresa Tsang[2], and Purang Abolmaesumi[1(✉)]

[1] The University of British Columbia, Vancouver, BC, Canada
purang@ece.ubc.ca
[2] Vancouver General Hospital, Vancouver, BC, Canada

Abstract. Ejection Fraction (EF) is a widely-used and critical index of cardiac health. EF measures the efficacy of the cyclic contraction of the ventricles and the outward pumpage of blood through the arteries. Timely and robust evaluation of EF is essential, as reduced EF indicates dysfunction in blood delivery during the ventricular systole, and is associated with a number of cardiac and non-cardiac risk factors and mortality-related outcomes. Automated reliable EF estimation in echocardiography (echo) has proven challenging due to low and variable image quality, and limited amounts of data for training data-driven algorithms which delays the integration of the technologies in the clinical workflow. In this paper, we introduce a Bayesian learning framework for automated EF assessment in echo videos. Our key contribution is to automatically estimate the epistemic uncertainty, *i.e.* the model uncertainty, in EF estimation. We anticipate that such information about uncertainty can be incorporated in clinical decision making. We use a ResNet18-based $(2+1)$D as the baseline architecture for video analysis and provide its side-by-side comparison of our probabilistic approach using public data from 10,031 echo exams. Our results clearly indicate the superior performance of the Bayesian model in the clinically critical lower EF population.

Keywords: Bayesian deep learning · Ejection fraction · Uncertainty · Echocardiography · Ultrasound imaging

1 Introduction

1.1 Clinical Background: Cardiac Systolic Function

The human cardiac cycle, *a.k.a* heartbeat, consists of systole (contraction) and diastole (expansion) of the heart ventricles. During ventricular systole, the filled

This work is funded in part by the Natural Sciences and Engineering Research Council of Canada (NSERC), and the Canadian Institutes of Health Research (CIHR).
M.M. Kazemi Esfeh and C. Luong—Joint first authorship.
T. Tsang and P. Abolmaesumi—Joint senior authorship.

© Springer Nature Switzerland AG 2020
A. L. Martel et al. (Eds.): MICCAI 2020, LNCS 12262, pp. 582–590, 2020.
https://doi.org/10.1007/978-3-030-59713-9_56

ventricles contract, ejecting a fraction of the blood out of the ventricles, and push the blood through the arteries to the lungs and other organs. This pumpage of blood is known as the cardiac systolic function, and is a vital process as it ensures oxygenated blood is delivered to the organs (including the brain), and deoxygenated blood is sent back to the lungs. The most commonly-used measure of assessing the functional cardiac health is left ventricular ejection fraction (LVEF or EF). EF is defined as the ratio of the blood pumped out (stroke volume) of the ventricle and the maximum amount of blood in the ventricle (end-diastolic volume). A normal EF is typically within the range $EF = 65\% \pm 10\%$. Lower EF values indicate systolic dysfunction, where blood pumpage and delivery are sub-optimal. Two-dimensional (2D) transthoracic echocardiography (echo) is widely used for a quick assessment of the systolic function. Echo relies on ultrasound technology and is popular due to its non-ionizing nature, accessibility, and ability to visualize the soft tissue of the heart in real-time. The inverse correlation between EF and mortality metrics and risks has been widely studied. EF is inversely linked with death and its associated indices for lower values, $EF \leq 45\%$ [25]. Very low EF ($EF \leq 20\%$) may inevitably lead to systolic heart failure (HF), *a.k.a.* HF with reduced EF (HFrEF), and is hence extremely dangerous. Reduced systolic function could also be due to cardiac infarction (heart attack), cardiomyopathy (weak or diseased cardiac muscle), coronary artery disease, or valve disease [2], all of which are important to be diagnosed early. One study estimated a five-year survival rate of only 50% for HF patients [12]. Low EF can have a serious impact on cardiac health as well as other systems and organs. Severely low EF readings ($EF \leq 15\%$) are correlated with chances of ischemic stroke due to a lack of blood in the brain [25]. One study showed a non-linear link between low-EF and metrics of accelerated cognitive aging, and argued that HF increases the odds of Alzheimer's and cerebrovascular disease [11]. Low-EF is also linked with outcomes of both cardiac [20] and non-cardiac [17] disease. Patients with reduced EF are at a higher risk of post-operative complications [20], and are significantly more likely to die within 90 days of surgery [17].

1.2   Related Works and Limitations

EF estimation in various imaging modalities has been explored for years by many researchers. In the past decade, advancements and ubiquity of deep learning (DL)-based algorithms have revolutionized cardiovascular image acquisition and analysis [18]. DL-based methods have proven promising in automated evaluation of systolic function in echo. Many works have attempted automated EF estimation using a variety of techniques [3,4,6,7,9,10,16,18,19]. Some of these methods rely on tracing of the endocardial borders to estimate the EF according to its formal definition [9], while others attempt to directly analyze echo videos [3,4,19]. The focus of these works is on formulating a machine learning framework, where the objective is to learn the mapping between input echo videos and corresponding EF measurements. Despite these efforts, beyond the low-risk retrospective studies done by various research groups, accurate and robust automatic assessment of EF in clinical settings has proven non-trivial and challenging. In these

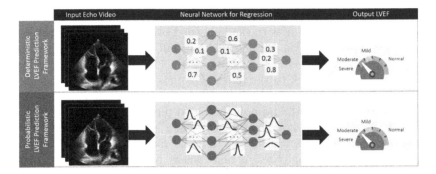

Fig. 1. A high-level overview and comparison of DL frameworks for EF assessment. The top row (blue) shows a deterministic framework [19], where each model weight graph edge is characterized by one value. The bottom row (green) shows the proposed Bayesian method with SVI, that accounts for the model's epistemic uncertainties. (Color figure online)

works, authors explore different models in terms of their architectures, searching for ways to capture spatio-temporal cardiac function-related features visible in echo videos. Nevertheless, to the best of our knowledge, none of the cited works have explored or addressed the uncertainties in the DL models that stem from our limited training data. Echo suffers from inherently low quality, as well as the variability of anatomy in terms of shape, size, and dynamics. 2D echo provides a limited amount of cross-sectional information of a three-dimensional (3D) structure and its quality is highly dependent on capturing the correct imaging plane. Furthermore, often a limited number of training data is available for such automated technologies. The impact of these factors gets more pronounced in the lower EF, and generally unhealthier population. The importance of robust EF prediction in echo was made particularly evident recently as a Food and Drug Administration (FDA) Class 2 Device recall was issued for the automated EF assessment on General Electric (GE)'s Vscan Extend model, the LVivo EF app (FDA Recall Event ID 82938 [1]). LVivo EF was released in September 2018 as one of the few commercially-available DL-based echo analysis technologies, and was recalled by FDA due to an overestimation bias in June 2019 [1].

1.3   Our Contribution

Due to its potential life-or-death implications, it is imperative to quickly and accurately identify low-EF patients. In this paper, we propose to alleviate the lack of robustness in DL frameworks for clinical assessment of EF by accounting for the epistemic uncertainty of DL models. While previous works utilize deterministic frameworks to predict the clinical EF measurement (Fig. 1 top), in this work, we use a probabilistic approach (Fig. 1 bottom), which aims to effectively take epistemic uncertainty into account. Deep deterministic supervised models, while powerful in predicting the outcome, are usually very confident in their predictions despite the lack of infinite training data. On the other hand, Bayesian

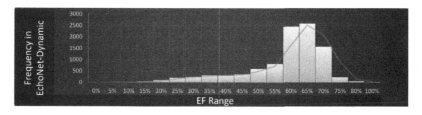

Fig. 2. A closer look at the distribution of EF labels in EchoNet-Dynamic [19] dataset. Low-EF cases are less frequent, more clinically critical and more challenging to measure.

deep methods predict the outcome by learning the weights as random variables during inference, which lets them have a measure of confidence when predicting the labels of unseen data [5]. In the case of automatic EF estimation, where the number of studies with low EF is relatively small in the training set (see e.g., Fig. 2), it is critical for the model to have a measure of confidence/uncertainty to tell "I don't know" on the out-of-distribution or shifted unseen data at the test phase. It has been shown that stochastic variational inference (SVI) methods are very promising and usually outperform other methods in terms of uncertainty quality under data shift. In particular, SVI is shown to perform better than a few other methods in terms of accuracy at high confidence(in other words, it is less confidently wrong) [23], which is desired for clinical applications. Our main contributions are: 1) formulation of a Bayesian probabilistic video analysis SVI framework; 2) validation of the proposed method for EF assessment on EchoNet-Dynamic, a public echo dataset [19]; and 3) empirical demonstration of the superiority of the proposed Bayesian method compared to its deterministic counterpart for EF assessment.

2   Materials and Methods

2.1   EF Dataset

The data used for this study come from EchoNet-Dynamic publicly available echo dataset [19], courtesy of Ouyang *et al.* . This dataset contains 10,031 echo exams, each video capturing A few cardiac cycles in the apical four-chamber (AP4) views. The corresponding left ventricular volume and EF measurements are also provided. In this study, we utilize 32 frames from each video (input) and the EF labels (output). As shown in Fig. 2, the majority of samples are associated with normal EF range and fewer low-EF cases are available. This imposes further difficulty for accurate assessment of EF for the lower end of the spectrum, which coincidentally, is more clinically critical.

2.2   Modeling Epistemic Uncertainty with a Bayesian Framework

In Bayesian DL, the goal is to find a distribution over the outputs, *i.e.* [14]

$$p(y^*|x^*, X, Y) = \int p(y^*|x^*, w)p(w|X, Y)dw; \tag{1}$$

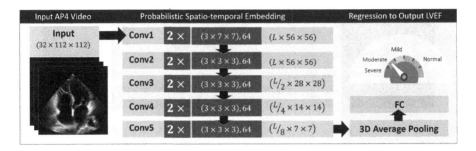

Fig. 3. The proposed architecture based on convolutional layers applied in two spatial domains and time ((2+1)D Conv) [24] for EF estimation. In the proposed probabilistic framework, all model weights are characterized as random variables.

where x^* and y^* denote the new input data point and the corresponding model prediction, respectively; X and Y are the seen data points and their labels; and w denotes the weights of the deep model. The posterior predictive $p(y^*|x^*, X, Y)$ predicts the probability that label y^* be the true label for the new input x^* at test time. To obtain the posterior predictive in Eq. (1), the integral should be computed over all of the network weights w. As the number of weights is often very large in deep neural networks, this integral is intractable even for very simple distributions. We must, therefore, approximate the integral via computing the posterior distribution $p(w|X, Y)$, which can be computed using the Bayes' rule:

$$p(w|X, Y) = \frac{p(Y|X, w)p(w)}{p(Y|X)} \qquad ; \qquad (2)$$

The distribution in the denominator could be found as

$$p(Y|X) = \int p(Y|X, w)p(w)dw \qquad ; \qquad (3)$$

where $p(w)$ is the prior distribution and is to be chosen before training the network. The integral in Eq. (3) is also intractable and should be approximated. Conventional numerical integration methods are usually too time-costly to be used in the case of DL, as the latent space is often very high-dimensional. Due to the intractability of these integrals, we usually approximate the posterior distribution with a known distribution q_θ from a tractable family, parameterized by θ. One way to find the parameters θ of the approximate posterior distribution is to minimize the Kullback-Leibler (KL)-divergence between the true and approximate posterior distributions, defined as the following:

$$\mathrm{KL}(q_\theta(w)||p(w|X, Y)) = \int q_\theta(w) \log \frac{q_\theta(w)}{p(w|X, Y)} dw. \qquad (4)$$

We assume that $q_\theta(w)$ is absolutely continuous with respect to the true posterior $p(w|X, Y)$ to make sure that the integral is defined. The integral in Eq. (4) is also intractable since it contains the posterior distribution. Hence, instead of minimizing the KL-divergence we equivalently maximize the evidence lower bound (ELBO):

$$L(\theta) = \int q_\theta(w) \log p(Y|X, w)dw - \mathrm{KL}(q_\theta(w)||p(w)). \tag{5}$$

The procedure of finding the parameters θ by optimizing the loss function in Eq. (4) is known as variational inference [13]. The total loss function that we minimize in the training phase is:

$$L^{\mathrm{total}}(\theta) = -\sum_{i=1}^{N} \int q_\theta(w) \log p(y_i|f^w(x_i))dw + \mathrm{KL}(q_\theta(w)||p(w)). \tag{6}$$

Here, N denotes the size of the training set and $f^w(x_i)$ is the output of the model for input x_i. The summation in Eq. (6) is not tractable for networks with more than one hidden layer [5]. In order to get around the computation-cost of this summation, we can use mini-batch optimization:

$$\hat{L}^{\mathrm{total}}(\theta) = -\frac{N}{M} \sum_{i \in \mathbb{I}} \int (q_\theta(w) \log p(y_i|f^w(x_i))dw + \mathrm{KL}(q_\theta(w)||p(w)), \tag{7}$$

where $\mathbb{I}$ is a random index set with $|\mathbb{I}| = M$. It can be proven that this is an unbiased stochastic estimator of the total loss function [5], i.e. $\mathbb{E}_{\mathbb{I}}[\hat{L}^{\mathrm{total}}(\theta)] = L^{\mathrm{total}}(\theta)$, and the local optima of the loss function in Eq. (7) are local optima of the loss function in Eq. (6) [22].

2.3   Neural Networks Architecture

Our network architecture is the Bayesian counterpart of ResNet18 [8] with $(2+1)$D residual convolutional layers $(\mathrm{R}(2+1)\mathrm{D}-18)$ [24]. This architecture (Fig. 3), previously used for EF estimation [19], uses convolutional layers for spatial (intra-frame) and temporal (inter-frame) feature learning. The network inputs are 32-frame 112×112 AP4 echo videos. The conv1 layer performs a spatial down-sampling by a $(1 \times 2 \times 2)$ convolutional striding, and layers conv2-5 compute features with convolutional strides of $(2 \times 2 \times 2)$. The extracted spatio-temporal embeddings are then fed through a fully-connected (FC) layer, which maps the features to the regression space $0\% \leq \mathrm{EF} \leq 100\%$.

2.4   Implementation and Training of the Bayesian Model

We assume a fully-factorized normal posterior distribution (mean-field approximation [26]) for weights with the set of parameters $\theta = \{\mu, \sigma\}$ and train the network parameters by maximizing the ELBO using the stochastic gradient methods [21,27]. The prior of each weight is assumed to be a Gaussian distribution with the mean μ equal to weights learned in the deterministic counterpart of the networks. The standard deviation of the prior distribution of all weights has been set to $\sigma = 0.7$. We use the Adam optimizer with a learning rate of 0.001 and $\beta = (0.9, 0.999)$ [15] with a batch size of 8. Training and validation were done on 7,465 and 1,288 samples, respectively.

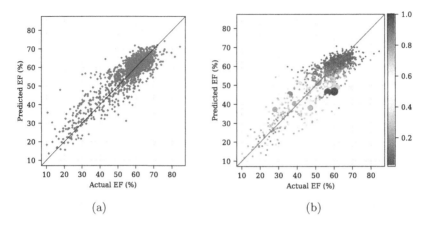

(a)   (b)

Fig. 4. Performance of the deterministic model (a) versus its proposed probabilistic counterpart (b). Colors and radii of data points in (b) illustrate the associated predictive uncertainty. Uncertainty is measured from the variance of the posterior predictive distribution, approximated from the variance of 30 drawn samples. Comparing the two models, the overestimation of EF<45% is reduced by using the probabilistic model. (Color figure online)

3 Experiments and Results

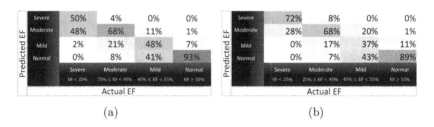

(a)   (b)

Fig. 5. Normalized confusion matrices of the deterministic model (a) versus its proposed probabilistic counterpart (b). Our proposed model performs significantly better on the high-risk severely dysfunctional cases compared to the former.

We tested the model on 1,277 unseen samples in EchoNet-Dynamic and performed a thorough comparison of the deterministic [19] and our proposed model. Figure 4 shows the scatter plot of the results, for both the Bayesian model and its deterministic counterpart. As one can see, the Bayesian network is showing much better performance in addition to giving a measure of uncertainty. It can be observed in Fig. 4(b) that the network is on average more confident on its predictions for the normal cases, $i.e.$ EF>55%. as expected while it has more uncertainty about the lower EF values, $i.e.$ EF < 30%. That is because the network has not seen as much data with lower EF as normal cases in the training phase. Furthermore, there are some points among the normal cases on which the network is very uncertain. These points correspond to dataset shift ($e.g.$ low-quality videos, zoomed videos, incorrect imaging planes, etc.). Our model achieved precision-recall area under the curve (AUC) of 0.86 and 0.74 for EF < 25% and EF < 45%, respectively. The confusion matrices in Fig. 5 show a breakdown of the model performance for four EF ranges. Compared to its deterministic counterpart, our proposed probabilistic model (Fig. 5(b)) performs significantly better for lower EF bins. The recalls for (EF < 25%, EF < 45%) are (0.72, 0.81) and (0.50, 0.77) for our model and the deterministic model, respectively. Also, our model shows more tendency to predict lower ejection fraction for the same data points in comparison with the deterministic counterpart, which is desired from the clinical point of view.

4 Conclusion

In this paper, we introduced a Bayesian deep neural network for video-based estimation of EF in echo. This framework characterizes each weight of the DL model as a random variable with Gaussian distribution, whose parameters can be learned. This probabilistic formulation is able to model the epistemic uncertainties in EF estimation. We validated our proposed method on the EchoNet-Dynamic [19] dataset and empirically showed our approach performed superior to that of the baseline deterministic approach for low EF. In order to analyze the impacts of the Bayesian formulation and highlight the advantages of the proposed framework, we performed identical experiments for the deterministic and probabilistic methods and provided a side-by-side comparison. Low-EF is a vital indicator of many serious complications, and must not be missed. Our probabilistic formulation enables robust estimation of EF, which is essential for safe and reliable clinical integration of DL-based automated technologies. Based on the presented rationale and results, we believe that the proposed approach can further be thought of as a generic method for a more robust evaluation of critical clinical indices.

References

1. Class 2 device recall Vscan Extend. https://www.accessdata.fda.gov/scripts/cdrh/cfdocs/cfRes/res.cfm?id=173162. Accessed 16 Mar 2020
2. Everything you need to know about ejection fraction. https://www.healthline.com/health/ejection-fraction#measurement. Accessed 16 Mar 2020
3. Behnami, D., et al.: Automatic detection of patients with a high risk of systolic cardiac failure in echocardiography. In: Stoyanov, D., et al. (eds.) DLMIA/ML-CDS -2018. LNCS, vol. 11045, pp. 65–73. Springer, Cham (2018). https://doi.org/10.1007/978-3-030-00889-5_8
4. Behnami, D., et al.: Dual-view joint estimation of left ventricular ejection fraction with uncertainty modelling in echocardiograms. In: Shen, D., et al. (eds.) MICCAI 2019. LNCS, vol. 11765, pp. 696–704. Springer, Cham (2019). https://doi.org/10.1007/978-3-030-32245-8_77
5. Gal, Y.: Uncertainty in deep learning. Ph.D. thesis, PhD thesis, University of Cambridge (2016)

6. Ge, R., et al.: K-net: integrate left ventricle segmentation and direct quantification of paired echo sequence. IEEE TMI **39**(5), 1690–1702 (2019)

7. Ghorbani, A., et al.: Deep learning interpretation of echocardiograms. NPJ Digit. Med. **3**(1), 1–10 (2020)

8. He, K., et al.: Deep residual learning for image recognition. In: Proceedings of the IEEE CVPR, pp. 770–778 (2016)

9. Jafari, M.H., et al.: A unified framework integrating recurrent fully-convolutional networks and optical flow for segmentation of the left ventricle in echocardiography data. In: Stoyanov, D., et al. (eds.) DLMIA/ML-CDS -2018. LNCS, vol. 11045, pp. 29–37. Springer, Cham (2018). https://doi.org/10.1007/978-3-030-00889-5_4

10. Jafari, M.H., et al.: Semi-supervised learning for cardiac left ventricle segmentation using conditional deep generative models as prior. In: 2019 IEEE 16th International Symposium on Biomedical Imaging (ISBI 2019), pp. 649–652. IEEE (2019)

11. Jefferson, A.L., et al.: Relation of left ventricular ejection fraction to cognitive aging (from the framingham heart study). Am. J. Cardiol. **108**(9), 1346–1351 (2011)

12. Jones, N., et al.: Survival of patients with chronic heart failure in the community: a systematic review and meta-analysis. Eur. J. Heart Fail. **21**(11), 1306–1325 (2019)

13. Jordan, M.I., et al.: An introduction to variational methods for graphical models. Mach. Learn. **37**(2), 183–233 (1999)

14. Kendall, A., Gal, Y.: What uncertainties do we need in Bayesian deep learning for computer vision. In: NIPS, pp. 5574–5584 (2017)

15. Kingma, D.P., Ba, J.: Adam: a method for stochastic optimization (2014). arXiv preprint arXiv:1412.6980

16. Leclerc, S., et al.: Deep learning for segmentation using an open large-scale dataset in 2D echocardiography. IEEE TMI **38**(9), 2198–2210 (2019)

17. Lerman, B.J., et al.: Association of left ventricular ejection fraction and symptoms with mortality after elective noncardiac surgery among patients with heart failure. JAMA **321**(6), 572–579 (2019)

18. Litjens, G., et al.: State-of-the-art deep learning in cardiovascular image analysis. JACC Cardiovasc. Imag. **12**(8), 1549–1565 (2019)

19. Ouyang, D., et al.: Interpretable AI for beat-to-beat cardiac function assessment. medRxiv, p. 19012419 (2019)

20. Pieri, M., et al.: Outcome of cardiac surgery in patients with low preoperative ejection fraction. BMC Anesthesiol. **16**(1), 97 (2016)

21. Ranganath, R., Gerrish, S., Blei, D.: Black box variational inference. In: Artificial Intelligence and Statistics, pp. 814–822 (2014)

22. Robbins, H., Monro, S.: A stochastic approximation method. Ann. Math. Stat. **22**(3), 400–407 (1951)

23. Snoek, J., et al.: Can you trust your model's uncertainty? evaluating predictive uncertainty under dataset shift. In: Advances in Neural Information Processing Systems, pp. 13969–13980 (2019)

24. Tran, D., et al.: A closer look at spatiotemporal convolutions for action recognition. In: Proceedings of the IEEE CVPR, pp. 6450–6459 (2018)

25. Tullio, D., et al.: Left ventricular ejection fraction and risk of stroke and cardiac events in heart failure: data from the warfarin versus aspirin in reduced ejection fraction trial. Stroke **47**(8), 2031–2037 (2016)

26. Wainwright, M.J., et al.: Graphical models, exponential families, and variational inference. Found. Trends® Mach. Learn. **1**(1–2), 1–305 (2008)

27. Wingate, D., Weber, T.: Automated variational inference in probabilistic programming (2013). arXiv preprint arXiv:1301.1299

Distractor-Aware Neuron Intrinsic Learning for Generic 2D Medical Image Classifications

Lijun Gong$^{(\boxtimes)}$, Kai Ma, and Yefeng Zheng

Tencent Jarvis Lab, Shenzhen, China
lijungong@tencent.com

Abstract. Medical image analysis benefits Computer Aided Diagnosis (CADx). A fundamental analyzing approach is the classification of medical images, which serves for skin lesion diagnosis, diabetic retinopathy grading, and cancer classification on histological images. When learning these discriminative classifiers, we observe that the convolutional neural networks (CNNs) are vulnerable to distractor interference. This is due to the similar sample appearances from different categories (i.e., small inter-class distance). Existing attempts select distractors from input images by empirically estimating their potential effects to the classifier. The essences of how these distractors affect CNN classification are not known. In this paper, we explore distractors from the CNN feature space via proposing a neuron intrinsic learning method. We formulate a novel distractor-aware loss that encourages large distance between the original image and its distractor in the feature space. The novel loss is combined with the original classification loss to update network parameters by back-propagation. Neuron intrinsic learning first explores distractors crucial to the deep classifier and then uses them to robustify CNN inherently. Extensive experiments on medical image benchmark datasets indicate that the proposed method performs favorably against the state-of-the-art approaches.

Keywords: Neuron Intrinsic Learning · Distractor-awareness · Medical Image Classification

1 Introduction

There have been continuous research investigations on medical images in Computer Aided Diagnosis (CADx) [7] as the automatic identification and analysis of diseases from medical images benefit the clinic diagnosis. Recently, convolutional neural networks (CNNs) have significantly improved the accuracy of CADx systems and reduced the workload of human screening. For 2D medical image classification tasks [19], image classification frameworks (e.g., ResNet [4] and EfficientNet [17]) are typically adopted, where a feature extraction backbone [14] pre-trained on nature images is applied to get robust low-level features [13] for

© Springer Nature Switzerland AG 2020
A. L. Martel et al. (Eds.): MICCAI 2020, LNCS 12262, pp. 591–601, 2020.
https://doi.org/10.1007/978-3-030-59713-9_57

the classification network. The collected medical image data is used to finetune the classification network for adaptation [2,7,16].

In practice, however, we observe that simple yet intuitive finetuning may not achieve favorable results because of the large appearance discrepancy between medical and natural domains. Moreover, the inter-class appearance difference in medical images is usually smaller than the one in nature images. As shown in Fig. 1, dermatoscopic, fundus and histological images share one thing in common: images in each column appear similar but belong to different categories. The visual similarity of samples from different categories deteriorates network classification accuracy. Meanwhile, noisy and blurry effects occur when generating medical images due to limitation of hardware conditions. Such effects also degrade the image quality for effective classifications. Therefore, it is desirable to properly handle these aforementioned specific cases, referred as distractors, when training CNNs for better classification accuracy.

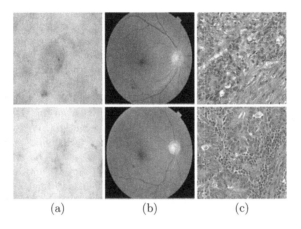

(a)   (b)   (c)

Fig. 1. 2D medical images. The three columns include (a) dermatoscopic, (b) fundus, and (c) histological images respectively. Each column consists of two similar samples from different categories.

In previous works, distractors are mainly selected from input images. For example, an online hard example selection method is proposed in [8] for image classification. The purpose is to adapt training data to fit the CNN optimizer. Similarly, an online hard example mining (OHEM) is proposed in [10] to pick up distractors from input images to improve object detection accuracy. It passes all training proposals to the CNN and selects low confidence ones (i.e., hard examples) to update the CNN parameters. More recently, a distractor-aware learning scheme is explored in [20] to select distractors from input frames for visual tracking. In sum, these methods identify distractors from input images as data augmentation to improve the classification performance. This raises a concern that whether these distractors are selected properly to improve CNNs' performance, especially from the perspective of deep feature space of medical image.

In this paper, we propose a neuron intrinsic learning method to generate distractors in the feature space and then use them to benefit CNN training. Without modifying CNN structures, we take two rounds of back-propagations during each training iteration. In the first round, we take the partial derivatives of the classification loss with respect to the input image and obtain a response map on the input layer. This response map is named as intrinsic response map (i.e., A^+) where each element reflects how much it contributes to the classification loss. Note that in this step, the loss is computed via ground truth labels and the CNN parameters are fixed. Then, we generate another pseudo label, compute the pseudo loss, and generate another intrinsic response map A^- accordingly. This pseudo label indicates the outcome of the distractor effect on the CNN. We trace this label back to the feature space via partial derivatives to generate the distractor (i.e., A^-). The elements on A^- show their contribution to produce the pseudo label, which should be kept distant from those on A^+. We organize these response maps with a distraction loss and combine it with the original classification loss. Using this combined loss, we update the CNN parameters on the second round. To this end, we explore distractors A^- which are crucial to downgrade deep classifiers (i.e., produce pseudo labels) and use them to improve classification accuracy. We validate the effectiveness of the proposed method on three 2D medical image classification tasks. The results shows that our method performs favorably against state-of-the-art approaches.

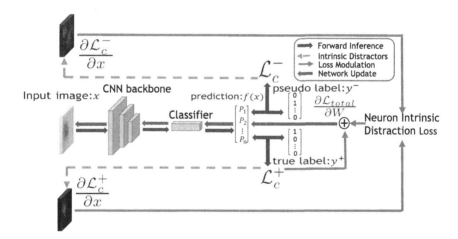

Fig. 2. Overview of neuron intrinsic learning for generic medical image classification. During each training iteration, we use two rounds of back-propagations. In the first round, we generate intrinsic response maps by taking $\frac{\partial \mathcal{L}_c^+}{\partial x}$ and $\frac{\partial \mathcal{L}_c^-}{\partial x}$. We formulate these maps as a distractor-aware loss term and integrate it into the loss objective function. In the second round, we update the network parameters by taking $\frac{\partial \mathcal{L}_{total}}{\partial W}$.

2   Proposed Method

Figure 2 shows an overview of the proposed method. We do not alter CNN structures while proposing a learning strategy to delve and update CNN parameters. During each training iteration, we use two rounds of back-propagations. In the first round, we generate neuron intrinsic response maps containing a distractor in the CNN feature space. We formulate a novel distraction loss that encourages large distance between the original image and its distractor in the feature space so as to help increase network's robustness. In the second round, we use this combined loss to update network parameters. The details are illustrated in the following.

2.1   Neuron Intrinsic Response Map

We denote an input image as x, the corresponding CNN output as $f(x)$, and loss function as $\mathcal{L}(f(x), y)$. In image classification, $f(x)$ is a vector of scores where each element represents the probability of x belonging to one predefined category, and y is the ground truth label. We can interpret $\mathcal{L}(f(x), y)$ from the Taylor expansion [11] perspective as:

$$\mathcal{L}(f(x), y) \approx A^{\mathsf{T}} x + B \tag{1}$$

where A is the derivative of the loss $\mathcal{L}(f(x), y)$ and B is a constant value. Equation 1 shows that each element in A contributes to $\mathcal{L}(f(x), y)$. Given a specific input sample x_0, we can compute A as:

$$A = \frac{\partial \mathcal{L}(f(x), y)}{\partial x}\Big|_{x=x_0} \tag{2}$$

We define A as a neuron intrinsic response map as it inherently reflects the response of the network loss. The CNN parameters are fixed when we compute A.

2.2   Distractor Synthesis in the CNN Feature Space

Distractors usually confuse CNN to have incorrect predictions. We synthesize a distractor in the CNN feature space based on intrinsic response maps. Given an input image x, we denote its ground truth label as y^+. This label is an n-dimensional vector where there are one element with a value of 1 and remaining elements of 0. The corresponding position of this non-zero element in the vector represents the ground truth category. Besides y^+, we define a pseudo label y^- where the non-zero element does not reside in the ground truth category. The classification losses (i.e., softmax and cross-entropy) computed by using y^+ and y^- can be written as follows:

$$\mathcal{L}_c^+ = -y^+ [\log(\mathrm{softmax}(f(x)))]^{\mathsf{T}} \tag{3}$$

$$\mathcal{L}_c^- = -y^- [\log(\mathrm{softmax}(f(x)))]^{\mathsf{T}} \tag{4}$$

Algorithm 1: Distractor-Aware Neuron Intrinsic Learning

Input: input image x and ground truth label y^+

1 pred $= f(x)$;

2 $d_{\text{pred}} = \text{pred.index}(f_m(x)); d = y.\text{index}(y_m^+)$;

3 $\mathcal{L}_c^+ = -y^+[\log(\text{softmax}(f(x)))]^{\mathsf{T}}$;

4 **if** $d_{pred} \neq d$ **then**

5 $\quad$ $y^- = [0, 0, ..., 0]$;

6 $\quad$ $y^-[d_{\text{pred}}] = 1$;

7 $\quad$ $\mathcal{L}_c^- = -y^-[\log(\text{softmax}(f(x)))]^{\mathsf{T}}$;

8 $\quad$ $A^+ = \frac{\partial \mathcal{L}_c^+}{\partial x}; A^- = \frac{\partial \mathcal{L}_c^-}{\partial x}$;

9 $\quad$ $\mathcal{L}_d = \frac{1}{||A^+ - A^-||_2^2 + \epsilon}$;

10 **else**

11 $\quad$ $\mathcal{L}_d = 0$;

12 **end**

13 $\mathcal{L}_{\text{total}} = \mathcal{L}_c^+ + \lambda \mathcal{L}_d$;

14 $\mathcal{L}_{\text{total}}.\text{backward}$;

15 $f.\text{update}$;

We generate two intrinsic response maps $A^+ = \frac{\partial \mathcal{L}_c^+}{\partial x}$ and $A^- = \frac{\partial \mathcal{L}_c^-}{\partial x}$ following Eq. 2. The A^- is defined as the distractor for the current input image. The elements in A^- contribute to the pseudo loss computed in Eq. 4, which induce the network to make incorrect predictions approaching to y^-. Instead specifying in the image space, we directly synthesize a distractor A^- in the CNN feature space to simulate distractions leading CNN to predict similar scores to pseudo label y^-.

Once the distractor is determined, we want to use it to enhance the network's robustness against interference from similar inter-class samples. An novel distractor-aware objective is set by increasing the distance between A^+ and A^-. The distraction loss can be written as:

$$\mathcal{L}_d = \frac{1}{||A^+ - A^-||_2^2 + \epsilon} \tag{5}$$

where $||A^+ - A^-||_2^2$ is the Euclidean distance between A^+ and A^-, and ϵ is a small constant value for stable numerical computation. By making A^+ and A^- different, the CNN is robust to overcome distractions.

2.3 Network Training

We incorporate distractors from Sect. 2.2 during network training. Algorithm 1 shows the details. During each iteration, we first perform forward propagation to compute the classification loss by using Eq. 3, and compute intrinsic response map A^+ via a back propagation. Second, we verify if CNN predicts correctly for the current input. We denote the element with maximum value of $f(x)$ as $f_{\text{m}}(x)$. When the network generates incorrect predictions, we generate y^- by setting $f_m(x)$ as 1 and the remaining elements as 0 The distraction loss in Eq. 5 can be computed together with the classification loss. The final loss function to train the CNN can be written as:

$$\mathcal{L}_{total} = \mathcal{L}_c^+ + \lambda \mathcal{L}_d \tag{6}$$

where λ is a constant value to balance these two loss terms. We take $\mathcal{L}_{total}$ to update CNN parameters.

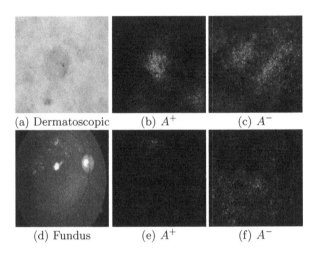

(a) Dermatoscopic   (b) A^+ (c) A^-

(d) Fundus   (e) A^+ (f) A^-

Fig. 3. Visualization of intrinsic response maps. A dermatoscopic image is shown in (a), its corresponding intrinsic response maps are shown in (b) and (c). A fundus image is shown in (d), its corresponding intrinsic response maps are shown in (e) and (f). The pixel values in these maps indicate their contribution to the loss function $\frac{\partial \mathcal{L}_c^+}{\partial x}$ and $\frac{\partial \mathcal{L}_c^-}{\partial x}$, respectively.

Visualization. We show that the learned CNN is effective to exclude distractor interference by visualizing the intrinsic response maps shown in Fig. 3. A dermatoscopic image is shown in Fig. 3(a), the corresponding A^+ and A^- are shown in Fig. 3(b) and Fig. 3(c), respectively. The high pixel values around the lesion region in (b) indicate that these pixels are extensively utilized to compute $\mathcal{L}_c^+$, while the high pixels values in the background area in Fig. 3(c) indicate these distractors mainly contribute to $\mathcal{L}_c^-$. We enlarge the distance between A^+ and A^- to empower the CNN to differentiate distractor interference in the feature space. Another example from a fundus image shows similar performance. Note that the crucial pixels determining fundus classification are around the lension area, not the optic disk area.

3   Experiments

We evaluate the proposed method on three 2D medical image classification tasks. Under each task we compare the proposed method with state-of-the-art approaches on the same benchmark. These tasks include skin lesion classification, diabetic retinopathy grading, and MSI/MSS classification on histological images. The corresponding benchmark datasets are HAM10000 [18], APTOS2019 [1], and CRC-MSI [5], respectively. In HAM10000, there are 10,015 skin lesion images with predefined 7 categories. We split images randomly to form a training set and a test set with a ratio of 7:3. In APTOS2019, there are 3,662 fundus images for grading diabetic retinopathy into five categories. We split the whole dataset randomly into training and test sets where 70% are for training and 30% are for testing. In CRC-MSI, there are 93,408 training and 98,904 testing data for binary classification of histological images. For all the benchmark datasets, we use the averaged F1 score as the evaluation metric.

There are four prevalent CNN backbones (i.e., ResNet50 [4], VGGNet11 [12], InceptionNet V4 [15], and EfficientNet-b0 [17]) utilized during evaluations. The initial weights of all the layers are from the ImageNet pretrained model [2] except for the last fully connected layer. The learning rate is set as 0.01 and the training iterations are set as 30 epochs. We set λ as $1e^{-5}$ and ϵ as $1e^{-4}$. When comparing with existing methods, we involve deep multi-task learning [3], DIL [9], and CANet [6], and report their performance on these datasets.

3.1   Ablation Study

The proposed method introduces distractor-aware neuron intrinsic learning into the original classification network. We denote this learning scheme as DANIL. On these three datasets, we show whether performance is improved by integrating DANIL into the baseline network. Table 1 and 2 show the evaluation results. The baseline performance (i.e., Base) is consistently improved by using DANIL (i.e., Base + DANIL) on all the benchmark datasets.

Table 1. The ablation study using four CNN feature backbones on three benchmarks under averaged F1 metrics.

DataSet	Configuration	ResNet50	VGGNet11	Inception V4	EfficientNet-b0
HAM10000	Base	0.658	0.686	0.671	0.700
	Base+DANIL	**0.674**	**0.700**	**0.690**	**0.710**
ATPOS2019	Base	0.617	0.634	0.659	0.642
	Base+DANIL	**0.660**	**0.666**	**0.672**	**0.671**
CRC-MSI	Base	0.643	0.654	0.650	0.641
	Base+DANIL	**0.654**	**0.683**	**0.661**	**0.652**

Table 2. The ablation study using four CNN feature backbones on three benchmarks under averaged accuracy metrics.

DataSet	Configuration	ResNet50	VGGNet11	Inception V4	EfficientNet-b0
HAM10000	Base	0.779	0.827	0.830	0.829
	Base+DANIL	**0.825**	**0.843**	**0.845**	**0.839**
ATPOS2019	Base	0.804	0.806	0.825	0.801
	Base+DANIL	**0.825**	**0.827**	**0.838**	**0.831**
CRC-MSI	Base	0.719	0.744	0.728	0.717
	Base+DANIL	**0.735**	**0.759**	**0.743**	**0.732**

3.2 Comparisons with State-of-the-Art

We compare DANIL with OHEM [8,10] on the benchmark datasets. OHEM is a state-of-the-art hard example mining method that selects distractors from images for CNN training. We denote the original CNN training configuration as Base, the CNN training with OHEM as Base + OHEM, and the CNN training with DANIL as Base + DANIL. Besides, we introduce multi-task prediction [3], DIL [9] and CANet [6] for state-of-the-art comparison.

Table 3. State-of-the-art comparison using four CNN feature backbones on three benchmarks under averaged F1 metrics.

DataSet	Method	ResNet50	VGGNet11	Inception V4	EfficientNet-b0
HAM10000	Base	0.658	0.686	0.671	0.700
	Multi-task [3]	0.667	0.690	0.679	0.702
	Base+OHEM	0.660	0.692	0.677	0.695
	Base+DANIL	**0.674**	**0.700**	**0.690**	**0.710**
ATPOS2019	Base	0.617	0.634	0.659	0.642
	DIL [9]	0.620	0.637	0.660	0.649
	CANet [6]	0.631	0.641	0.664	0.656
	Base+OHEM	0.632	0.644	0.647	0.662
	Base+DANIL	**0.660**	**0.666**	**0.672**	**0.671**
CRC-MSI	Base	0.643	0.654	0.650	0.641
	Base+OHEM	0.649	0.667	0.649	0.644
	Base+DANIL	**0.654**	**0.683**	**0.661**	**0.652**

Table 3 and 4 show the evaluation results under averaged F1 and accuracy metrics. We observe that OHEM improves the original CNN classification in the majority of backbones and benchmarks. For some cases (e.g., EfficientNet-b0 on HAM10000), OHEM deteriorates the classification performance under averaged F1 metrics. This indicates that selecting hard examples from images are not robust to improve the CNN performance. In comparison, DANIL is able to

Table 4. State-of-the-art comparison using four CNN feature backbones on three benchmarks under averaged accuracy metrics.

DataSet	Method	ResNet50	VGGNet11	Inception V4	EfficientNet-b0
HAM10000	Base	0.779	0.827	0.830	0.829
	Multi-task	0.811	0.830	0.834	0.828
	Base+OHEM	0.818	0.832	0.830	0.832
	Base+DANIL	**0.825**	**0.843**	**0.845**	**0.839**
ATPOS2019	Base	0.804	0.806	0.825	0.801
	DIL	0.810	0.806	0.825	0.802
	CANet	0.813	0.810	0.826	0.802
	Base+OHEM	0.813	0.812	0.828	0.814
	Base+DANIL	**0.825**	**0.827**	**0.838**	**0.831**
CRC-MSI	Base	0.719	0.744	0.728	0.717
	Base+OHEM	0.725	0.749	0.732	0.717
	Base+DANIL	**0.735**	**0.759**	**0.743**	**0.732**

consistently make an improvement for different backbones on different datasets. This shows the effectiveness of exploring distractors in the CNN feature space for network training. DANIL performs favorably against existing methods (i.e., multi-task learning, DIL, and CANet) with different CNN backbones as well.

4   Concluding Remarks

Medical image classification was the foundation of automatic computer aided diagnosis. Recent attempts adapted natural image classification models to address this problem. Their performance was hinged by the discrepancy between medical and natural images. In this work, we proposed DANIL to synthesize distractors in the CNN feature space for network learning. We started from the pseudo label, which was the outcome of the distractor interference, and backtraced into the CNN feature space to generate distractors via neuron intrinsic learning. The distractors were kept distant from positive samples in the CNN feature space via the proposed distraction loss. This loss was proposed to learn a more distractor-aware CNN. Extensive experiments on different medical image classification tasks and datasets demonstrated that DANIL improved the CNN classification accuracy and performed favorably against state-of-the-art approaches.

Acknowledgments. This work was funded by the Key Area Research and Development Program of Guangdong Province, China (No. 2018B010111001), National Key Research and Development Project (2018YFC2000702) and Science and Technology Program of Shenzhen, China (No. ZDSYS201802021814180).

References

1. APTOS 2019 Blindness Detection (2019). https://www.kaggle.com/c/aptos2019-blindness-detection/data
2. Deng, J., Dong, W., Socher, R., Li, L.J., Li, K., Fei-Fei, L.: Imagenet: a large-scale hierarchical image database. In: IEEE Conference on Computer Vision and Pattern Recognition, pp. 248–255. IEEE (2009)
3. Haofu, L., Luo, J.: A deep multi-task learning approach to skin lesion classification. In: Workshops at the Thirty-First AAAI Conference on Artificial Intelligence (2017)
4. He, K., Zhang, X., Ren, S., Sun, J.: Deep residual learning for image recognition. In: IEEE Conference on Computer Vision and Pattern Recognition, pp. 770–778 (2016)
5. Kather, J.N.: Histological images for MSI vs. MSS classification in gastrointestinal cancer. FFPE Samples (2019). https://doi.org/10.5281/zenodo.2530835
6. Li, X., Hu, X., Yu, L., Zhu, L., Fu, C.W., Heng, P.A.: Canet: cross-disease attention network for joint diabetic retinopathy and diabetic macular edema grading. IEEE Trans. Med. Imaging **39**, 1483–1493 (2019)
7. Litjens, G., et al.: A survey on deep learning in medical image analysis. Med. Image Anal. **42**, 60–88 (2017)
8. Loshchilov, I., Hutter, F.: Online batch selection for faster training of neural networks (2015). arXiv preprint arXiv:1511.06343
9. Rakhlin, A.: Diabetic retinopathy detection through integration of deep learning classification framework. bioRxiv, p. 225508 (2018)
10. Shrivastava, A., Gupta, A., Girshick, R.: Training region-based object detectors with online hard example mining. In: IEEE Conference on Computer Vision and Pattern Recognition, pp. 761–769 (2016)
11. Simonyan, K., Vedaldi, A., Zisserman, A.: Deep inside convolutional networks: Visualising image classification models and saliency maps (2013). arXiv preprint arXiv:1312.6034
12. Simonyan, K., Zisserman, A.: Very deep convolutional networks for large-scale image recognition (2014). arXiv preprint arXiv:1409.1556
13. Song, Y., Bao, L., He, S., Yang, Q., Yang, M.H.: Stylizing face images via multiple exemplars. Comput. Vis. Image Underst. **162**, 135–145 (2017)
14. Song, Y., et al.: Joint face hallucination and deblurring via structure generation and detail enhancement. Int. J. Comput. Vis. **127**, 785–800 (2019)
15. Szegedy, C., Vanhoucke, V., Ioffe, S., Shlens, J., Wojna, Z.: Rethinking the inception architecture for computer vision. In: IEEE Conference on Computer Vision and Pattern Recognition, pp. 2818–2826 (2016)
16. Tajbakhsh, N., et al.: Convolutional neural networks for medical image analysis: full training or fine tuning? IEEE Trans. Med. Imaging **35**(5), 1299–1312 (2016)
17. Tan, M., Le, Q.V.: Efficientnet: Rethinking model scaling for convolutional neural networks (2019). arXiv preprint arXiv:1905.11946

18. Tschandl, P., Rosendahl, C., Kittler, H.: The ham10000 dataset, a large collection of multi-source dermatoscopic images of common pigmented skin lesions. Sci. Data **5**, 180161 (2018)
19. Wang, W., et al.: Medical image classification using deep learning. In: Chen, Y.-W., Jain, L.C. (eds.) Deep Learning in Healthcare. ISRL, vol. 171, pp. 33–51. Springer, Cham (2020). https://doi.org/10.1007/978-3-030-32606-7_3
20. Zhu, Z., Wang, Q., Li, B., Wu, W., Yan, J., Hu, W.: Distractor-aware siamese networks for visual object tracking. In: European Conference on Computer Vision, pp. 101–117 (2018)

Large-Scale Inference of Liver Fat with Neural Networks on UK Biobank Body MRI

Taro Langner[1(✉)], Robin Strand[1], Håkan Ahlström[1,2], and Joel Kullberg[1,2]

[1] Uppsala University, 751 85 Uppsala, Sweden
taro.langner@surgsci.uu.se
[2] Antaros Medical, BioVenture Hub, 431 53 Mölndal, Sweden

Abstract. The UK Biobank Imaging Study has acquired medical scans of more than 40,000 volunteer participants. The resulting wealth of anatomical information has been made available for research, together with extensive metadata including measurements of liver fat. These values play an important role in metabolic disease, but are only available for a minority of imaged subjects as their collection requires the careful work of image analysts on dedicated liver MRI. Another UK Biobank protocol is neck-to-knee body MRI for analysis of body composition. The resulting volumes can also quantify fat fractions, even though they were reconstructed with a two- instead of a three-point Dixon technique. In this work, a novel framework for automated inference of liver fat from UK Biobank neck-to-knee body MRI is proposed. A ResNet50 was trained for regression on two-dimensional slices from these scans and the reference values as target, without any need for ground truth segmentations. Once trained, it performs fast, objective, and fully automated predictions that require no manual intervention. On the given data, it closely emulates the reference method, reaching a level of agreement comparable to different gold standard techniques. The network learned to rectify non-linearities in the fat fraction values and identified several outliers in the reference. It outperformed a multi-atlas segmentation baseline and inferred new estimates for all imaged subjects lacking reference values, expanding the total number of liver fat measurements by factor six.

Keywords: Magnetic resonance imaging (MRI) · Liver fat · Neural network

1 Introduction

The UK Biobank has recruited more than 500,000 volunteer participants for medical examination, 100,000 of whom are planned to undergo extensive imaging procedures [14]. Several modalities are involved, including dedicated MRI of

This research was supported by a grant from the Swedish Heart- Lung Foundation and the Swedish Research Council (2016-01040, 2019-04756), and used the UK Biobank Resource under application no. 14237.

A. L. Martel et al. (Eds.): MICCAI 2020, LNCS 12262, pp. 602–611, 2020.
https://doi.org/10.1007/978-3-030-59713-9_58

the brain, heart, pancreas, and liver, the latter of which enables measurements of accumulated liver fat [16], closely linked to type 2 diabetes and other metabolic disorders. A fat content of 5.5% has been defined as threshold for non-alcoholic fatty liver disease (NAFLD) [1], which can progress to non-alcoholic steatohepatitis (NASH) [11] and eventually to liver cirrhosis with potentially fatal outcome [19]. Both MRI and magnetic resonance spectroscopy (MRS) are non-invasive alternatives to liver biopsy for reliable quantification [13].

Due to its scale, the UK Biobank has the potential to relate liver fat as a biomarker to the wide range of metadata, such as disease outcomes, life-style factors and genetic profiles. At the time of writing, almost 40,000 participants have completed the UK Biobank imaging procedures, but only about 5,000 reference liver fat measurements are available, based on transverse slices acquired from the dedicated liver MRI [16]. Like these images, the volumes acquired by neck-to-knee body MRI for body composition analysis [15] also include the liver and encode voxel-wise proton density fat fractions (PDFF). Due to using a two- instead of a three-point Dixon technique for reconstruction, these images may encode systematically different PDFF values, and similar protocols have previously shown low agreement with other established methods for quantification of liver fat [8]. However, the UK Biobank has released more than 30,000 neck-to-knee body MRI scans, which can be evaluated with machine learning techniques.

Various biological metrics, including liver fat, can be automatically inferred on image data from these scans by convolutional neural networks for regression [9]. Similar strategies have been previously applied to a range of medical imaging modalities including MRI, with the goal of quantifying properties such as age [2, 4, 10] and structures of the heart [17], but also blood pressure and smoking status [12]. This technique is distinct from neural networks trained for segmentation, which can also be applied for liver fat measurements [7], but typically require ground truth segmentations for training.

In this work, liver fat measurements are inferred by automated analysis of the more readily available neck-to-knee body MRI with neural networks trained for regression on data from these images [15] and the UK Biobank reference values [16] as ground truth.

The following contributions are made. (1) A specialized framework for liver fat inference from UK Biobank neck-to-knee body MRI, which adapts a generalized system [9] for superior performance. (2) A three-way comparison between this method, the reference, and a simple multi-atlas segmentation baseline. (3) Inferred liver fat values for more than 30,000 UK Biobank subjects, which could be shared for medical research and as a baseline for quality control. Code samples and documentation for these implementations are publicly available.[1]

---

[1] https://github.com/tarolangner/mri-biometry.

2   Methodology

2.1   UK Biobank Data

UK Biobank participants were recruited by letter from the National Health Service and scanned at three different imaging centers in the United Kingdom. The majority of subjects reported white British ethnicity ($\sim 94\%$) with a mean age of 64 years (range 44–82, standard deviation 7.5), mean BMI of $26.6\,\mathrm{kg/m^2}$ (standard deviation 4.3) and a share of 52% males.

Reference Liver Fat Measurements. Reference liver fat measurements were available for 4,613 subjects as field 22402-2.0 of the UK Biobank [16]. These values are based on the PDFF map of a single transverse slice of the liver, acquired with a Siemens 1.5T MAGNETOM Aera and three-point Dixon technique. The reference method returns the mean PDFF of three manually placed ROIs with liver tissue, avoiding vessels and other confounding structures. To avoid confusion with relative values, liver fat percentage points are referred to as fat fractions (FF). The available values range from 0–46 FF, with a mean of 3.9 $\pm$ 4.6 FF (median 2.1 FF). Of the total 4,613 subjects, 920 (20%) have recorded liver fat values above 5.5 FF.

Neck-to-knee Body MRI. The UK Biobank neck-to-knee body MRI scans released as field 20201-2.0 were also acquired with a Siemens 1.5T MAGNETOM Aera device, using a dual-echo Dixon technique with TR=6.69 and TE=2.39/4.77 ms, and flip angle 10deg[15]. The resulting water-fat volumes cover most of the body with six overlapping stations, excluding the head and lower legs, whereas the arms and other tissues at the edges of the magnetic field are typically subject to strong image artifacts. Likewise, the borders between stations often contain motion artifacts that can affect the shape of the liver.

2.2   Experimental Setup

A neural network was trained for regression of liver fat values on the neck-to-knee body MRI of those subjects with available reference values. It was first evaluated in 10-fold cross-validation and then, after training on the full dataset, applied to the remaining subjects for inference. In both phases, independent measurements from a simple multi-atlas segmentation strategy described in one of the following sections served as a baseline.

Datasets. All 32,323 neck-to-knee body MRI scans that have been released at the time of writing were quality controlled by an operator who visually inspected two-dimensional mean intensity projections of the water and fat signals [9]. Due to water-fat swaps, noise, metal objects, unusual positioning and other artifacts, about 3,6% of the subjects were excluded, leaving 31,171 images for the experiments.

Three datasets were formed from those subjects that passed the quality control. Validation dataset A consists of those 4,418 subjects with existing reference values as ground truth for the training and validation of the network in 10-fold cross-validation. The atlas technique was also validated on this set. Inference dataset B consists of the remaining 26,753 quality-controlled subjects for whom no reference values were available. The network was applied to these subjects, but without reference measurements (and consequently no ground truth values) this dataset could not be used as a true independent test set. Instead, the atlas was applied to extract baseline measurements for comparison, but was too slow to process them all within the given time. Therefore, comparison dataset C was formed as a random subset of 1,000 subjects from dataset B and used to compare the network and atlas.

Image Formatting. The stations of the neck-to-knee body MRI were fused and resampled to a common spatial resolution of 2.23 mm × 2.23 mm × 3 mm, with 370 × 224 × 174 voxels. Next, water and fat fraction images were calculated by voxel-wise division of the water or fat signal intensity by the sum of both signals. This sum image was also used to generate body masks by applying a threshold, calculated as the mean of Otsu filter thresholds for all coronal slices of the summed signal for a given subject.

For the neural network, a highly compressed two-dimensional format with coronal and sagittal fat fraction slices was extracted, as seen in Fig. 1. Based on the body mask, the coronal slice was extracted at center of mass and the sagittal slice at a quarter of mass, typically locating the latter along the center of the right thigh. Both slices were cropped to exclude the bed and the bottom half of the body and then concatenated. The resulting image of 376 × 176 pixels was then compressed to an 8bit format, with fat fractions ranging from 0 to 50%. No body mask was applied to this format. While there is no guarantee that this strategy captures actual liver tissue it operated robustly and required only about 5 seconds per subject with a GPU implementation.

Neural Network Configuration. The convolutional neural network is based on a ResNet50 [5] modified for regression of a single value. Most hyperparameters of the generalized regression framework were retained [9], with batch size 32, online-augmentation by translations and weights pretrained on ImageNet. Accordingly, a mean squared error loss was optimized with Adam and a base learning rate of 0.0001 was reduced by factor ten in the last 1,000 iterations of 6,000 iterations in total. Each training sample consists of the two-dimensional image format as extracted from the neck-to-knee body MRI as input and the reference liver fat measurement of the same subject as ground truth value. Training on one split required about 25 min whereas prediction was almost instantaneous. All experiments ran on a Nvidia GTX 1080 Ti 11 GB GPU, a 12-core Intel Xeon W-2133, 3.60 GHz CPU and 32 GB RAM in PyTorch.

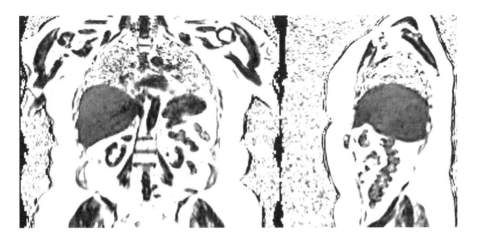

Fig. 1. Input format for the network as extracted from the neck-to-knee body MRI volumes. The upper halves of specific coronal and sagittal fat fraction slices are combined in a two-dimensional 8bit image of 376×176 pixels, encoding fractions of 0–50%. For the shown subject, the reference method lists a liver fat percentage of 15%.

Multi-atlas Baseline. A simple multi-atlas segmentation pipeline was implemented on the neck-to-knee body MRI, performing a median readout of minimal liver segmentations. Three subjects (one female, two male) with high observed variance in liver shape were manually segmented to serve as templates, using the water fraction and signal images to outline the liver and exclude vessels and adjacent tissue. When applying the atlas, these templates were transformed to each target with a graph-cut based deformable registration technique [3] on the full paired water and fat fraction volumes, after applying the body mask to remove background noise. A resolution pyramid of six levels was used together with a GPU-implementation of the normalized cross-correlation. The obtained deformation field for each template was applied to the corresponding liver segmentation. The thus aligned binary segmentations were then multiplied and subsequently eroded with a spherical kernel of seven voxels in diameter. The median fat fraction value of the remaining selected voxels was then returned. PDFF values of the neck-to-knee body MRI may not directly correspond to those reconstructed by the reference method [8]. As a final step, the raw atlas output was therefore fit to the reference values on the validation dataset A by linear regression. When applied to other datasets, the same parameters were applied.

Evaluation. The network was trained to emulate the reference method by regression. The success of this training was therefore evaluated by quantifying the quality of fit with the coefficient of determination R^2 and reporting the mean absolute error (MAE) and the 95% limits of agreement (LoA). Furthermore, Pearson's coefficient of correlation r is used to examine the randomness of errors by network and atlas relative to the reference. The measured values can also be thresholded to identify subjects above the NAFLD risk level of 5.5 FF. The area under curve (AUC) of the receiver operating characteristic (ROC) curve is reported together with sensitivity and specificity for thresholding at this level. For some outliers, manual segmentations similar to the atlas templates were created. Their extent was more conservative, and their median FF values are reported after correction with the same linear regression parameters as applied for the atlas.

3   Results and Discussion

Results are shown in Fig. 2 and Table 1. The network outperformed the atlas on validation dataset A but retained the same pattern of agreement with it on comparison dataset C, indicating robust generalization to those subjects lacking reference values. On dataset A, the network inferred outliers of up to 23 FF, with 16 subjects reaching errors above 5 FF. However, manual segmentation of the top ten outliers yielded measurements that agree for LoA of $(-0.7$ to 1.2 FF) to the atlas and $(-5.4$ to 7.3 FF) with the network, but only $(-23.5$ to 8.4 FF) with the reference. This substantial disparity shows a mismatch between the reference and liver fat as observed in the neck-to-knee body MRI in these subjects that can not be explained by the different imaging protocol alone, but might instead be the result of spurious outliers in the reference method. The errors of atlas and network relative to the reference on dataset A are highly correlated ($r = 0.715$ with LoA -1.9 to 2.0).

Table 1. Method comparison on datasets A (first three rows) and C (bottom row). The three final columns assume thresholding at 5.5 FF, with the first named method as ground truth. MAE: mean absolute error, LoA: 95% limits of agreement, ROC-AUC: area under receiver operating characteristic curve, Sens: sensitivity, Spec: specificity.

	MAE	R^2	LoA	ROC-AUC	Sens	Spec
Reference vs Network	0.77	0.940	$(-2.22$ to $2.31)$	0.992	89.3	98.2
Reference vs Atlas	1.03	0.912	$(-2.73$ to $2.73)$	0.991	78.0	99.2
Atlas vs Network	0.76	0.952	$(-1.89$ to $1.98)$	0.995	97.4	95.8
Atlas (C) vs Network (C)	0.80	0.950	$(-1.94$ to $2.11)$	0.991	93.6	95.6

The agreement between network and reference with LoA $(-2.2$ to 2.3 FF) is well within the LoA of $(-4.0$ to 3.4 FF) reported between the gold standard modalities of MRI and MRS [18]. Even within MRI, variability with LoA of $(-2.2$ to 1.8 FF) for comparable protocols and $(-2.5$ to $1.6)$ between different imaging sites has to be expected [6]. The atlas was about fifty times slower than the network and affected by systematic differences between the two- and three-point Dixon techniques [8]. It was only evaluated after correction with the parameters from linear regression ($0.9x - 0.8$ FF), which failed to resolve a non-linear structure in the range of 1–4 FF, however. Perhaps due to the outliers, the

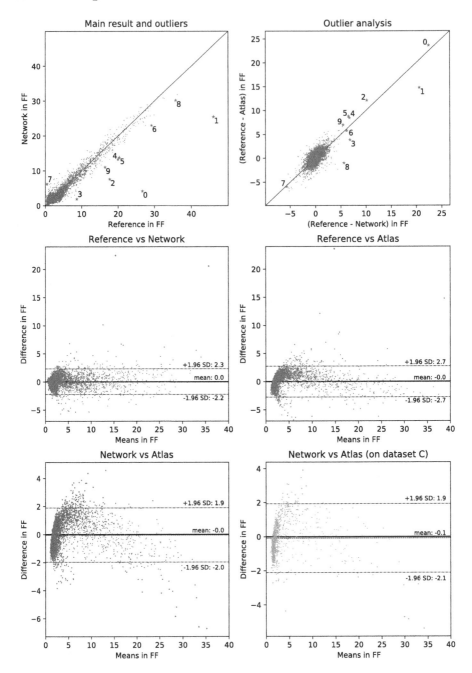

Fig. 2. Results on datasets A (blue, N = 4,418) and C (red, N = 1,000). Differences were calculated by subtracting the second named method from the first. Dashed lines denote 95% limits of agreement. Both network and atlas independently contradict the reference in the annotated top ten network prediction outliers, indicating implausible reference values. Network inference on dataset C retains its agreement with the atlas as observed on dataset A, showing no signs of deteriorated performance in generalization. (Color figure online)

corrected values still tend to overestimate low and underestimate high values. In contrast, the network learned to rectify values in the lower ranges and emulated the reference method better, possibly with information from additional image features. With LoA of (–2.7 to 2.7 FF), the atlas still surpasses the generalized regression framework, which used mean intensity projections of water and fat signals for LoA of (–4.0 to 4.2 FF) on the same UK Biobank subjects [9].

Several limitations apply to the presented results. The comparison on dataset C indicates robust generalization to UK Biobank subjects from any of the given centers with the chosen protocol. However, without an independent test set no conclusions can be drawn for other demographic groups, scanning devices or imaging protocols, which would likely at least require retraining of the network. The slice selection strategy is also not guaranteed to be optimal or even certain to capture any liver tissue, and inherently limited by encoding only values from 0–50 FF in 8bit. Conceptually, the potential of deep learning on the neck-to-knee body MRI is still not fully leveraged, as design choices regarding the compression of the volumes into two-dimensional representations still strongly affects the results, reminiscent of hand-crafted features. The proposed regression technique also generates no output segmentations, so that individual predictions can not be easily explained or corrected. In contrast, neural networks for segmentation have been proposed for automation of parts of the reference method [7] and could provide these, reducing any manual work to the initial creation of representative ground truth segmentations for training only.

However, apart from the underlying design decisions, the presented approach eliminates the need for manual intervention entirely. No human guidance, model-based assumptions or representative ground truth segmentations are required, and inference for thousands of subjects can be performed within a day. Future work will consist in expanding the training and evaluation samples, once available, which is likely to further improve performance [10]. Likewise, the planned repeat scans of up to 70,000 subjects could potentially be processed without any further changes to the presented system, enabling convenient longitudinal samples of liver fat. The reference method also provides a liver inflammation factor and iron content, and while there is no indication that these could be inferred from the neck-to-knee body MRI in the same way, the presented approach could also be evaluated on the dedicated liver MRI slices directly. Furthermore, the inferred values of dataset B could be examined more thoroughly, using known correlations to the metadata and by expanding the baseline measurements.

It is worth noting that after the completion of these experiments an alternative reference set of about 10,000 image-based liver fat measurements has been released by the UK Biobank as field 22436-2.0. A full analysis of these values is beyond the scope of this work, but preliminary results indicate that field 22436-2.0 may be of higher quality than field 22402-2.0 used here. Like the network and atlas, the new reference contradicts field 22402-2.0 on the top ten outliers found in this work. Early attempts to emulate this new reference with the proposed method yielded a closer fit. As the data collection is continuously progressing, future work will be able to examine the relationship between both references and the proposed technique in ever increasing sample sizes.

4   Conclusion

In conclusion, the proposed framework can emulate the reference measurements and infer similar liver fat values from the neck-to-knee body MRI of the UK Biobank. It outperforms the atlas and combines high speed with accurate and objective predictions, while eliminating any need for manual intervention or guidance and leaving no room to subjective variability. The inferred liver fat measurements are readily available at large scale for distribution as return data by the UK Biobank. They could be used to identify potentially erroneous outliers as observed in the reference method, but also as an approximation in medical research, enabling larger sample sizes until all subjects have been evaluated with more established gold standard techniques.

References

1. Browning, J.D., et al.: Prevalence of hepatic steatosis in an urban population in the united states: impact of ethnicity. Hepatology **40**(6), 1387–1395 (2004)
2. Cole, J.H., et al.: Predicting brain age with deep learning from raw imaging data results in a reliable and heritable biomarker. NeuroImage **163**, 115–124 (2017). https://doi.org/10.1016/j.neuroimage.2017.07.059, https://linkinghub.elsevier.com/retrieve/pii/S1053811917306407
3. Ekström, S., Malmberg, F., Ahlström, H., Kullberg, J., Strand, R.: Fast Graph-Cut Based Optimization for Practical Dense Deformable Registration of Volume Images (2018). arXiv:1810.08427 [cs]
4. Halabi, S.S., et al.: The RSNA pediatric bone age machine learning challenge. Radiology **290**(2), 498–503 (2019). https://doi.org/10.1148/radiol.2018180736, http://pubs.rsna.org/doi/10.1148/radiol.2018180736
5. He, K., Zhang, X., Ren, S., Sun, J.: Deep residual learning for image recognition. In: 2016 IEEE Conference on Computer Vision and Pattern Recognition (CVPR), pp. 770–778 (2016). https://doi.org/10.1109/CVPR.2016.90
6. Hernando, D., et al.: Multisite, multivendor validation of the accuracy and reproducibility of proton-density fat-fraction quantification at 15 t and 3t using a fat-water phantom. Magn. Res. Med. **77**(4), 1516–1524 (2017)
7. Irving, B., et al.: Deep quantitative liver segmentation and vessel exclusion to assist in liver assessment. In: Valdés Hernández, M., González-Castro, V. (eds.) MIUA 2017. CCIS, vol. 723, pp. 663–673. Springer, Cham (2017). https://doi.org/10.1007/978-3-319-60964-5_58
8. Kukuk, G.M., et al.: Comparison between modified dixon MRI techniques, MR spectroscopic relaxometry, and different histologic quantification methods in the assessment of hepatic steatosis. Eur. Radiol. **25**(10), 2869–2879 (2015)
9. Langner, T., Ahlström, H., Kullberg, J.: Large-scale biometry with interpretable neural network regression on UK biobank body MRI (2020). arXiv preprint arXiv:2002.06862

10. Langner, T., Wikström, J., Bjerner, T., Ahlström, H., Kullberg, J.: Identifying morphological indicators of aging with neural networks on large-scale whole-body MRI. IEEE Trans. Med. Imaging **39**, 1430–1437 (2019). https://doi.org/10.1109/tmi.2019.2950092
11. Pagadala, M.R., McCullough, A.J.: The relevance of liver histology to predicting clinically meaningful outcomes in nonalcoholic steatohepatitis. Clin. Liver Dis. **16**(3), 487–504 (2012)
12. Poplin, R., et al.: Prediction of cardiovascular risk factors from retinal fundus photographs via deep learning. Nat. Biomed. Eng. **2**(3), 158 (2018)
13. Reeder, S.B., Cruite, I., Hamilton, G., Sirlin, C.B.: Quantitative assessment of liver fat with magnetic resonance imaging and spectroscopy. J. Magn. Reson. Imaging **34**(4), 729–749 (2011)
14. Sudlow, C., et al.: UK Biobank: an open access resource for identifying the causes of a wide range of complex diseases of middle and old age. PLOS Med. **12**(3), e1001779 (2015). https://doi.org/10.1371/journal.pmed.1001779
15. West, J., et al.: Feasibility of MR-based body composition analysis in large scale population studies. PLoS ONE **11**(9), e0163332 (2016). https://doi.org/10.1371/journal.pone.0163332, https://www.ncbi.nlm.nih.gov/pmc/articles/PMC5035023/
16. Wilman, H.R., et al.: Characterisation of liver fat in the UK biobank cohort. PloS one **12**(2), e0172921 (2017)
17. Xue, W., Islam, A., Bhaduri, M., Li, S.: Direct multitype cardiac indices estimation via joint representation and regression learning. IEEE Trans. Med. Imaging **36**(10), 2057–2067 (2017)
18. Yokoo, T., et al.: Linearity, bias, and precision of hepatic proton density fat fraction measurements by using MR imaging: a meta-analysis. Radiology **286**(2), 486–498 (2018)
19. Younossi, Z.M., Koenig, A.B., Abdelatif, D., Fazel, Y., Henry, L., Wymer, M.: Global epidemiology of nonalcoholic fatty liver disease-meta-analytic assessment of prevalence, incidence, and outcomes. Hepatology **64**(1), 73–84 (2016)

BUNET: Blind Medical Image Segmentation Based on Secure UNET

Song Bian[1(✉)], Xiaowei Xu[2], Weiwen Jiang[3], Yiyu Shi[3], and Takashi Sato[1]

[1] Kyoto University, Kyoto, Japan
{sbian,takashi}@easter.kuee.kyoto-u.ac.jp
[2] Guangdong Provincial People's Hospital, Guangzhou, China
xiao.wei.xu@foxmail.com
[3] University of Notre Dame, Notre Dame, USA
{wjiang2,yshi4}@nd.edu

Abstract. The strict security requirements placed on medical records by various privacy regulations become major obstacles in the age of big data. To ensure efficient machine learning as a service schemes while protecting data confidentiality, in this work, we propose blind UNET (BUNET), a secure protocol that implements privacy-preserving medical image segmentation based on the UNET architecture. In BUNET, we efficiently utilize cryptographic primitives such as homomorphic encryption and garbled circuits (GC) to design a complete secure protocol for the UNET neural architecture. In addition, we perform extensive architectural search in reducing the computational bottleneck of GC-based secure activation protocols with high-dimensional input data. In the experiment, we thoroughly examine the parameter space of our protocol, and show that we can achieve up to 14x inference time reduction compared to the-state-of-the-art secure inference technique on a baseline architecture with negligible accuracy degradation.

1 Introduction

The use of neural-network (NN) based machine learning (ML) algorithms in aiding medical diagnosis, especially in the field of medical image computing, appears to be extremely successful in terms of its prediction accuracy. However, the security regulations over medical records contradicts the use of big data in the age of ML. Highly sensitive patient records are protected under the Health Insurance Portability and Accountability Act (HIPAA), where strong protection measures need to be taken over all of the electronic Protected Health Information (ePHI) possessed by a patient. In particular, access controls and client-side encryption are mandated for the distribution of all ePHI records over public networks [12,14]. In addition, while qualified professionals are allowed to handle

Electronic supplementary material The online version of this chapter (https://doi.org/10.1007/978-3-030-59713-9_59) contains supplementary material, which is available to authorized users.

ePHI, the data exposure is required to be kept minimal [10,13], i.e., just enough to accomplish the necessary professional judgements.

A central question to the real-world deployment of NN-based ML techniques in medical image processing is how the related data transfer and computations can be handled securely and efficiently. Previous security measures on medical data generally involved physical means (e.g., physically disconnected from the internet), and these techniques clearly cannot benefit from the large-scale distributed computing networks available for solving ML tasks. Recent advances in cryptography and multi-party secure computing seek alternatives to address the security concerns. In particular, the concept of secure inference (SI) is proposed, where Alice as a client wishes to inference on some of her inputs with the machine learning models provided by the server, called Bob. The security requirement is that no one, including Bob, learns anything about the inputs from Alice, while Alice also learns nothing about the models from Bob. Over the past few years, prior arts on SI flourished [6,15,17,18,22,24], where secure protocols targeted on general learning problems were proposed. In addition, we also observe protocol- and system-level optimizations [2,22] on SI. Unfortunately, most existing works mentioned above do not have a clear application in mind. Thus, the utilized network architectures and datasets (e.g., MNIST, CIFAR-10) are usually generic, without immediate practical implications.

In this work, we propose BUNET, a secure protocol for the UNET architecture [23] that enables input-hiding segmentation on medical images. In the proposed protocol, we use a combination of cryptographic building blocks to ensure that client-side encryption is enforced on all data related to the patients, and that practical inference time can also be achieved. As a result, medical institutions can take advantage of third-party machine learning service providers without violating privacy regulations. The main contributions of this work are summarized as follows.

- **Privacy-Preserving Image Segmentation:** To the best of our knowledge, we are the first to propose a secure protocol for image segmentation.
- **Architectural Exploration for Secure UNET:** We perform a search on the possible alternative UNET architectures to reduce the amount of computations (in terms of cryptographic realizations) in SI.
- **Thorough Empirical Evaluations:** By performing architectural-protocol co-design, we achieved 8x–14x inference time reduction with negligible accuracy degradation.

2 Preliminaries

2.1 Cryptographic Primitives

In this work, we mainly consider the four types of cryptographic primitives: a packed additive homomorphic encryption (PAHE) scheme based on the ring learning with errors (RLWE) problem [4,5,8,11], additive secret sharing (ASS) [9], garbled circuits (GC) [26], and multiplication triples (MT) [1,16]. In what follows, we provide a brief overview for each primitive.

PAHE: A PAHE is a cryptosystem, where the encryption (Enc) and decryption (Dec) functions act as group (additive) homomorphisms between the plaintext and ciphertext spaces. Except for the normal Enc and Dec, a PAHE scheme is equipped with the following three abstract operators. We use $[\mathbf{x}]$ to denote the encrypted ciphertext of $\mathbf{x} \in \mathbb{Z}^n$, and $n \in \mathbb{Z}$ the maximum number of plaintext integers that can be held in a single ciphertext.

- Homomorphic addition ($\boxplus$): for $\mathbf{x}, \mathbf{y} \in \mathbb{Z}^n$, $\mathsf{Dec}([\mathbf{x}] \boxplus [\mathbf{y}]) = \mathbf{x} + \mathbf{y}$. Note we can also perform homomorphic subtraction $\boxminus$, where $\mathsf{Dec}([\mathbf{x}] \boxminus [\mathbf{y}]) = \mathbf{x} - \mathbf{y}$.
- Homomorphic Hadamard product ($\boxdot$): for $\mathbf{x}, \mathbf{y} \in \mathbb{Z}^n$, $\mathsf{Dec}([\mathbf{x}] \boxdot \mathbf{y}) = \mathbf{x} \circ \mathbf{y}$, where $\circ$ is the element-wise multiplication operator.
- Homomorphic rotation (rot): for $\mathbf{x} \in \mathbb{Z}^n$, let $\mathbf{x} = (x_0, x_1, \cdots, x_{n-1})$, $\mathsf{rot}([\mathbf{x}], k) = (x_k, x_{k+1}, \cdots, x_{n-1}, x_0, \cdots, x_{k-1})$ for $k \in \{0, \cdots, n-1\}$.

ASS and Homomorphic Secret Sharing: A two-party ASS scheme consists of two operators, (Share, Rec), and some prime modulus $p_A \in \mathbb{Z}$. Each operator takes two inputs, where we have $\mathbf{s}_A = \mathsf{Share}(\mathbf{x}, \mathbf{s}_B) = (\mathbf{x} - \mathbf{s}_B) \bmod p_A$ and $\mathbf{x} = \mathsf{Rec}(\mathbf{s}_A, \mathbf{s}_B) = (\mathbf{s}_A + \mathbf{s}_B) \bmod p_A$. In [15], homomorphic secret sharing (HSS) is adopted, where ASS operates over an encrypted $\mathbf{x}$. For HSS, we have that

$$[\mathbf{s}_A] = \mathsf{Share}([\mathbf{x}], \mathbf{s}_B) = ([\mathbf{x}] \boxminus \mathbf{s}_B) \bmod p_A, \text{ and} \tag{1}$$

$$[\mathbf{x}] = \mathsf{Rec}([\mathbf{s}_A], \mathbf{s}_B) = (\mathbf{s}_A \boxplus \mathbf{s}_B) \bmod p_A. \tag{2}$$

GC: GC can be considered as a more general form of HE. In particular, the circuit garbler, Alice, "encrypts" some function f along with her input x to Bob, the circuit evaluator. Bob evaluates $f(x, y)$ using his encrypted input y that is received from Alice obliviously, and obtains the encrypted outputs. Alice and Bob jointly "decrypt" the output of the function $f(x, y)$ and one of the two parties learns the result.

MT: Beaver's MT [1] is a technique that performs multiplication on a pair of secret-shared vectors $\mathbf{x} = \mathsf{Rec}(\mathbf{s}_{A,x}, \mathbf{s}_{B,x})$ and $\mathbf{y} = \mathsf{Rec}(\mathbf{s}_{A,y} + \mathbf{s}_{B,y})$ between Alice and Bob. Here, we take computations performed by Alice as an example, and only note that the exact same procedure is also executed by Bob on his shares of secrets. To compute $\mathbf{x} \circ \mathbf{y}$, Alice and Bob first pre-share a set of respective multiplication triples $(\mathbf{a}_A, \mathbf{b}_A, \mathbf{c}_A)$ and $(\mathbf{a}_B, \mathbf{b}_B, \mathbf{c}_B)$, where we have $\mathsf{Rec}(\mathbf{a}_A, \mathbf{a}_B) \circ \mathsf{Rec}(\mathbf{b}_A, \mathbf{b}_B) = \mathsf{Rec}(\mathbf{c}_A, \mathbf{c}_B)$. Alice locally calculates $\mathbf{d}_A = \mathbf{s}_{A,x} \circ \mathbf{a}_A \bmod p_A$, $\mathbf{e}_A = \mathbf{b}_A \circ \mathbf{s}_{A,y} \bmod p_A$. Then, Alice and Bob publish their results $\mathbf{d}_A, \mathbf{d}_B$ and $\mathbf{e}_A, \mathbf{e}_B$. Finally, Alice obtains

$$\mathbf{g}_A = \left(\mathbf{c}_A + (\mathbf{e} \circ \mathbf{s}_{A,0,i}) + (\mathbf{d} \circ \mathbf{s}_{A,1,i}) - \mathbf{d} \circ \mathbf{e}\right) \bmod p_A, \tag{3}$$

where $\mathbf{d} = \mathsf{Rec}(\mathbf{d}_A, \mathbf{d}_B)$ and similarly for $\mathbf{e}$. MT guarantees that $\mathsf{Rec}(\mathbf{g}_A, \mathbf{g}_B) = \mathbf{x} \circ \mathbf{y}$, where $\mathbf{g}_B$ is the MT results computed by Bob.

2.2 Related Works on Secure Neural Network Inference

While a limited number of pioneering works have been proposed for secure inference and training with neural networks [15,17,18,24], it was not until recently that such protocols carried practical significance. For example, in [17], an inference with a single CIFAR-10 image takes more than 500 seconds to complete. Using the same neural architecture, the performance was improved to less than 8 seconds in one of the most recent arts on SI, ENSEI [15]. Unfortunately, as shown in Sect. 4, without UNET-specific optimizations and protocol designs, existing approaches carry significant performance overhead, especially on the 3D images (e.g., CT scans) used in medical applications. Hence, in this work, we establish our protocol based on the ENSEI construction, and explore UNET-specific optimizations and cryptographic protocol designs to improve the practicality of secure inference in medical segmentation.

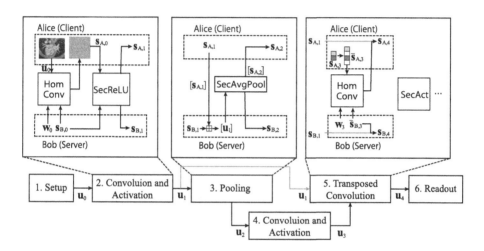

Fig. 1. The overview of the BUNET protocol.

3   Secure UNET for Blind Segmentation

3.1   BUNET: The Protocol

Figure 1 shows an example of the BUNET protocol structured as an UNET architecture, where the input image goes through four steps. The operators used in each step will be discussed in detail later in Sect. 3.2.

1. **Setup:** As our protocol takes both 2D and 3D images as inputs, the input image is of dimension $\dim(U) = c \times w \times h \times d$ ($d = 1$ for 2D inputs). Alice first raster-scans her input image into a one-dimensional vector $\mathbf{u}$ of length $\ell = c \cdot w \cdot h \cdot d$. Bob does a similar transformation on his filter weights to obtain the one-dimensional vector $\mathbf{w}$. We use $\mathbf{u}_0$ to denote the input from Alice at the 0-th layer, and $\mathbf{w}_0$ for that of Bob.

2. **Convolution and Activation:** We conduct a standard (input-hiding) convolution on the input image with activation functions followed. For each convolution layer with activation, we run

$$\mathbf{v}_0 = \mathsf{HomConv}(\mathbf{u}_0, \mathbf{w}_0), \text{ and } \mathbf{u}_1 = f_a(\mathbf{v}_0) = \mathsf{SecAct}(\mathbf{v}_0). \qquad (4)$$

for some weight vector $\mathbf{w}_0$. Here, f_a is some abstract activation function (e.g., ReLU or square activation). The output $\mathbf{u}_1 = f_a(\mathbf{v}_0)$ will become the input to the next layer. We point out that the activation function has a significant impact on both the accuracy and the inference time of UNET. Therefore, we propose a hybrid UNET architecture, where both ReLU and square activations are used. Here, the security guarantee is that Alice obtains no knowledge on $\mathbf{w}_0$, and Bob knows nothing about $\mathbf{u}_0$ and $\mathbf{v}_0$ after the protocol execution.

3. **Pooling:** While the standard UNET architecture employs max pooling [23] as the downsampling method, in the experiment, we show a large performance difference between max pooling and average pooling protocols, due to the change of underlying protocol. We also demonstrate in Table 1 that the two pooling methods result in marginal accuracy differences. Hence, we modified the UNET architecture to employ only average pooling. Consequently, the proposed protocol executes $\mathbf{u}_2 = \mathsf{SecAvgPool}(\mathbf{u}_1)$. For an input of length ℓ and pooling size $\zeta_p = z_{w,p} \times z_{h,p} \times z_{d,p}$, the pooled output have a dimension of ℓ/ζ_p.

4. **Bottom-Level Convolution and Activation:** Here, Step 2 is repeated, and we get $\mathbf{u}_3$ as output. Note that the input and output dimension is reduced by Step 3, so Step 4 is computationally lighter than Step 2.

5. **Transposed Convolution and Concatenation:** While the arithmetic procedures for transposed convolution is essentially the same as a normal convolution, protocol-level modifications are required for the image concatenation and padding operations. Concretely, after obtaining input from the previous layer, e.g., $\mathbf{u}_3$, Alice needs to zero-pad $\mathbf{u}_3$ in an interleaving manner, according to some stride size $\zeta_t = z_{w,t} \times z_{h,t} \times z_{d,t}$. The padded result, $\overline{\mathbf{u}}_3$, will have a length of $\ell \cdot \zeta_t/\zeta_p$. Subsequently, Alice uses $\overline{\mathbf{u}}_3$ as input to execute the following protocols.

$$\mathbf{v}_4 = \mathsf{HomConv}(\overline{\mathbf{u}}_3, \mathbf{w}_3), \text{ and } \mathbf{u}_4 = \mathbf{v}_4 \| \mathbf{u}_1. \qquad (5)$$

One subtlety is that, the output from the 1-st layer, $\mathbf{u}_1$, needs to be concatenated with the output from the transposed convolution layer for the rest of the normal convolutions. However, $\mathbf{u}_1$ and $\mathbf{v}_4$ will actually be encrypted under different keys. Thus, both results need to be decrypted and concatenated by Alice. The concatenated result, $\mathbf{v}_1 \| \mathbf{u}_4$, will become the inputs to later layers.

6. **Readout:** Here, we applies the $\mathsf{SecArgmax}$ function over the label dimensions. It is noted that, since the $\mathsf{Softmax}$ operator is monotonic and is only required in the learning process, we avoid using a separate protocol for $\mathsf{Softmax}$, and directly perform a secure Argmax. Since the Argmax function is a pixel-wise

comparison function across the label dimension, it can be implemented using a simple GC protocol similar to the secure ReLU protocol, and we omit a formal presentation.

Threat Model and Security. The threat model for BUNET is that both Bob and Alice are semi-honest, in the sense that both parties follow the protocol prescribed above (e.g., encrypting real data with Enc, etc.), but want to learn as much information as possible from the other party. Our protocol guarantees that Alice learns only the segmentation results, while Bob learns nothing about the inputs from Alice.

The security of the proposed protocol can be easily reduced to that of existing works [3,15], where any attack against BUNET will result in a non-negligible advantage against the ENSEI [3] and Gazelle [15] protocols.

3.2 The Cryptographic Building Blocks

Here, we discuss each of the cryptographic primitives used in the previous section in details.

HomConv: The HomConv operator obliviously convolve two vectors $\mathbf{u}$ and $\mathbf{w}$. A very recent work [3] discovered that, instead of the complex rotate-and-accumulate approach proposed by previous works [15], homomorphic convolution can be performed in the frequency domain, where the only computation needed in the homomorphic domain is the $\odot$ (homomorphic Hadamard product) operator. Therefore, the homomorphic convolution protocol proceeds as follows.

1. First, Alice performs an integer discrete Fourier transform (DFT) (i.e., number theoretic transform in [3]) on $\mathbf{u}$ and obtain its frequency-domain representation, $\hat{\mathbf{u}}$. She simply encrypts this input array into a ciphertext $[\hat{\mathbf{u}}]$ (when $\ell > n$, the vector is encrypted into multiple ciphertexts) by running $[\hat{\mathbf{u}}] = \mathsf{Enc}_{\mathcal{K}}(\hat{\mathbf{u}})$, where $\mathcal{K}$ is the encryption key. The resulting ciphertext is transferred to Bob.
2. Before receiving any input from Alice, Bob applies DFT on his filter weights to obtain $\hat{\mathbf{w}}$. In this process, the size of the filter will be padded to be ℓ. Upon receiving the inputs $[\hat{\mathbf{u}}]$ from Alice, Bob computes

$$[\hat{\mathbf{v}}] = [\hat{\mathbf{u}} \circ \hat{\mathbf{w}}] = [\hat{\mathbf{u}}] \odot \hat{\mathbf{w}} \tag{6}$$

for all ciphertexts $\hat{\mathbf{u}}$.
3. Finally, since $\hat{\mathbf{v}}$ contains information of the weights from Bob, Bob applies HSS as $[\hat{\mathbf{s}}_A] = \mathsf{Share}([\hat{\mathbf{v}}], \hat{\mathbf{s}}_B)$. Bob keeps $\hat{\mathbf{s}}_B$ and returns $[\hat{\mathbf{s}}_A]$ to Alice, where Alice decrypts and obtain $\hat{\mathbf{s}}_A$. Both Alice and Bob run inverse DFT on their shares of secrets (i.e., $\hat{\mathbf{s}}_A$ and $\hat{\mathbf{s}}_B$) and obtain $\mathbf{s}_A$ and $\mathbf{s}_B$, respectively, completing the protocol.

SecAct: The SecAct protocols are summarized as follows.

- **ReLU:** We follow the construction in [15] based on the GC protocol. Alice first garbles the circuit with her share of secret $\mathbf{s}_A$. The garbled circuit obliviously computes the following function

$$\mathbf{v} = \mathsf{Rec}(\mathbf{s}_A, \mathbf{s}_B), \text{ and } \bar{\mathbf{s}}_A = \mathsf{Share}(\mathsf{ReLU}(\mathbf{v}), \bar{\mathbf{s}}_B) \tag{7}$$

where $\bar{\mathbf{s}}_B$ is a freshly generated share of secret from Bob. After protocol execution, Alice obtains $\bar{\mathbf{s}}_A$, which contains $\mathsf{ReLU}(\mathbf{v})$ in an oblivious manner.
- **Square:** Since the square activation (i.e., $y = x^2$) is essentially evaluating a polynomial over the inputs, the computationally-light MT can be used instead of GC. To use MT, we first share the secret $\mathbf{v}$ twice among Alice and Bob. Then, the MT protocol outlined in Sect. 2.1 can be executed, where both $\mathbf{x}$ and $\mathbf{y}$ equal $\mathbf{v}$. After the protocol execution, Alice and Bob respectively obtain $\mathbf{g}_A$ and $\mathbf{g}_B$ where $\mathsf{Rec}(\mathbf{g}_A, \mathbf{g}_B) = \mathbf{v}^2$. The main observation here is that, all computations are coefficient-wise multiplications and additions over $\lg p_A$-bit integers. As a result, MT-based square activation is much faster than GC-based ReLU activation.

SecPool: As mentioned above, two types of secure pooling can be implemented, the SecMaxPool and the SecAvgPool operator. In [15], it is shown that max-pooling can be implemented using the GC protocol as in Eq. (7), where we replace the ReLU operator with the MaxPool operator. Meanwhile, for secure average pooling, we can use a simple protocol that is purely based on PAHE. Specifically, we can compute the window-wise sum of some vector $\mathbf{v}$ by calculating $\mathsf{SecAvgPool}(\mathbf{v}) = \sum_{i=0}^{\zeta_p - 1} \mathsf{rot}([\mathbf{v}], i)$, where ζ_p is the pooling window size. Since both homomorphic rotations and additions are light operations compared to GC, SecAvgPool is much faster than SecMaxPool.

4   Accuracy Experiments and Protocol Instantiation

4.1   Experiment Setup

Due to the lack of immediate existing works, we compare BUNET with the standard UNET architecture implemented by the ENSEI [3] protocol, which is the best performing protocol on secure multi-class inference. Here, the standard UNET architecture only utilizes max-pooling for pooling layers, and ReLU for activation layers. We denote this architecture as the baseline architecture. As shown in the appendix, the baseline architecture consists 19 convolution layers including three transposed convolution layers, 14 activation layers, three average pooling layers, and a readout layer implementing the SecArgmax function.

Our experiments are conducted on three datasets, GM [21], EM [7], and HVSMR [20]. Due to the space constraint, we only present the accuracy and performance results on the EM (two dimensional) and HVSMR (three dimensional) datasets (GM will be added to the appendix).

The cryptographic performance is characterized on an Intel i5-9400 2.9 GHz CPU, and the accuracy results are obtained with an NVIDIA P100 GPU. The adopted PAHE library is SEAL version 3.3.0 [25] (we set q, p to be 60-bit and 20-bit integers, respectively, and $n = 2048$, ensuring a 128-bit security) and MT/GC protocols are implemented using LowGear provided by MP-SPDZ [16,19].

4.2 Accuracy and Performance Results

In this section, we explore how the neural architecture of 3D-UNET impact on the segmentation performance. In particular, it is important to see if the proposed architectural modifications for UNET result in satisfactory prediction accuracy while accelerating the network inference time. We downsampled the

Table 1. The dice accuracy on the HVSMR dataset for different neural architectures

	ReLU+Max Float	ReLU Float	ReLU+Max 32-bit (Baseline)	ReLU 16-bit	Hybrid 20-bit	Square 16-bit
HVSMR Myo. Dice	0.74 ±0.04	0.73 ±0.04	0.73 ±0.05	0.73 ±0.05	0.71 ±0.05	0.52 ±0.14
HVSMR BP Dice	0.87 ±0.04	0.87 ±0.04	0.87 ±0.04	0.87 ±0.04	0.86 ±0.05	0.83 ±0.05
HVSMR Time (s)	-	-	42616	14205	3054.2	1118.2
EM Dice	0.9405	-	0.9411	0.9398	0.9385	0.8767
EM Time (s)	-	-	8838.1	2968.6	1077.9	227.16

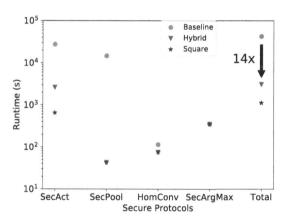

Fig. 2. Runtime distribution for different cryptographic building blocks in a single run of secure segmentation.

Table 2. Number of activation functions per layer batch

1st	2nd	3rd	4th	5th	6th	7th
33554432	8388608	2097152	524288	2097152	8388608	33554432

images in the HVSMR [20] dataset to a dimension of $64 \times 64 \times 64$ containing three class labels, i.e., background, myocardium (myo.) and blood pool (BP). The images in the EM dataset has 200×200 input for binary segmentation. Table 1 summarizes the dice scores and runtime under various architectural settings (HVSMR with 10-fold cross validation). Here, the pooling function is average pooling unless otherwise stated. Hybrid refers to the neural architecture where the first and the last layer batches use square activation, while all other layer batches adopt ReLU activation. Here, a layer batch denotes two convolution and activation layers with the same output feature dimensions and channels. We have three main observations. First, the use of average pooling instead of max pooling results in negligible accuracy degradation, on a level that can likely be compensated by parameter tuning. Second, the UNET architecture is robust in low-quantization environment, where we see little accuracy difference between floating point, 32-bit and 16-bit quantization factors, especially on the EM dataset. Lastly, replacing all ReLU activations with squares results in significant accuracy degradation for the segmentation of myocardium. However, the BP prediction can be acceptable for a quick run of cheap evaluations, and the hybrid architecture successfully achieves a good balance between accuracy and performance.

We record the total runtime for a single blind image segmentation with respect to different 3-D UNET architectures. In Table 2 and Fig. 2, we illustrate the per-layer-batch number of neuron activations and the runtime distribution for different secure protocols in BUNET. As expected, the first and last two layers contain the most amount of activations, and replacing the GC-based heavy ReLU activation with square activation results in an immediate 5x total runtime reduction. In addition, it is observed that the runtime for ReLU activation functions dominate the total runtime across architectures, while square activation is as light as a frequency-domain homomorphic convolution operation.

Compared to the baseline 32-bit ReLU architecture, we obtain 8x–14x runtime reduction with the reasonably accurate Hybrid architecture, and up to 39x reduction with the cheapest (all-square) UNET implementation on EM. Finally, we note that most NN operations can be parallelized as well as the cryptographic building blocks. Therefore, since our performance is recorded on a single-thread CPU, we expect further runtime reduction for BUNET on parallel computing architectures.

5   Conclusion

In this work, we propose BUNET to perform blind medical image segmentation on encrypted medical images. The observation we make is that protocol and network designs need to be jointly performed to achieve the best accuracy-performance trade-off. By designing UNET-specific protocols and optimizing the UNET architecture, we show that up to 8x–14x inference time reduction can be achieved with negligible accuracy degradation on several medical datasets.

Acknowledgment. This work was partially supported by JSPS KAKENHI Grant No. 20K19799, 20H04156, Edgecortix Inc, the Science and Technology Planning Project of Guangdong Province under Grant No. 2017A070701013, 2017B090904034, 2017B030314109, 2018B090944002, and 2019B020230003, Guangdong peak project under Grant No. DFJH201802, the National Key Research and Development Program under Grant No. 2018YFC1002600, the Natural Science Foundation of Guangdong Province under Grant No. 2018A030313785.

References

1. Beaver, D.: Efficient multiparty protocols using circuit randomization. In: Feigenbaum, J. (ed.) CRYPTO 1991. LNCS, vol. 576, pp. 420–432. Springer, Heidelberg (1992). https://doi.org/10.1007/3-540-46766-1_34

2. Bian, S., Jiang, W., Lu, Q., Shi, Y., Sato, T.: NASS: Optimizing secure inference via neural architecture search. arXiv preprint arXiv:2001.11854 (2020)

3. Bian, S., Wang, T., Hiromoto, M., Shi, Y., Sato, T.: ENSEI: Efficient secure inference via frequency-domain homomorphic convolution for privacy-preserving visual recognition (2020)

4. Brakerski, Z.: Fully homomorphic encryption without modulus switching from classical GapSVP. In: Safavi-Naini, R., Canetti, R. (eds.) CRYPTO 2012. LNCS, vol. 7417, pp. 868–886. Springer, Heidelberg (2012). https://doi.org/10.1007/978-3-642-32009-5_50

5. Brakerski, Z., Gentry, C., Vaikuntanathan, V.: (Leveled) fully homomorphic encryption without bootstrapping. ACM Trans. Comput. Theory (TOCT) **6**(3), 13 (2014)

6. Brutzkus, A., Gilad-Bachrach, R., Elisha, O.: Low latency privacy preserving inference. In: International Conference on Machine Learning, pp. 812–821 (2019)

7. Cardona, A., et al.: An integrated micro-and macroarchitectural analysis of the drosophila brain by computer-assisted serial section electron microscopy. PLoS Biol. **8**(10), e1000502 (2010)

8. Cheon, J.H., Kim, A., Kim, M., Song, Y.: Homomorphic encryption for arithmetic of approximate numbers. In: Takagi, T., Peyrin, T. (eds.) ASIACRYPT 2017. LNCS, vol. 10624, pp. 409–437. Springer, Cham (2017). https://doi.org/10.1007/978-3-319-70694-8_15

9. Damgård, I., Nielsen, J.B., Polychroniadou, A., Raskin, M.: On the communication required for unconditionally secure multiplication. In: Robshaw, M., Katz, J. (eds.) CRYPTO 2016. LNCS, vol. 9815, pp. 459–488. Springer, Heidelberg (2016). https://doi.org/10.1007/978-3-662-53008-5_16

10. Drolet, B.C., Marwaha, J.S., Hyatt, B., Blazar, P.E., Lifchez, S.D.: Electronic communication of protected health information: privacy, security, and hipaa compliance. J. Hand Surg. **42**(6), 411–416 (2017)

11. Fan, J., Vercauteren, F.: Somewhat practical fully homomorphic encryption. IACR Cryptology ePrint Archive 2012, 144 (2012)

12. HHS.gov (2009). https://www.hhs.gov/hipaa/for-professionals/privacy/laws-regulations/index.html. Accessed 04 Mar 2020

13. HHS.gov (2009). https://www.hhs.gov/hipaa/for-professionals/privacy/guidance/minimum-necessary-requirement/index.html. Accessed 03 Apr 2020

14. Hoffman, S., Podgurski, A.: Securing the hipaa security rule. J. Internet Law, Spring, 06–26 (2007)

15. Juvekar, C., et al.: Gazelle: a low latency framework for secure neural network inference. arXiv preprint arXiv:1801.05507 (2018)
16. Keller, M., Pastro, V., Rotaru, D.: Overdrive: making SPDZ great again. In: Nielsen, J.B., Rijmen, V. (eds.) EUROCRYPT 2018. LNCS, vol. 10822, pp. 158–189. Springer, Cham (2018). https://doi.org/10.1007/978-3-319-78372-7_6
17. Liu, J., et al.: Oblivious neural network predictions via MinioNN transformations. In: Proceedings of ACM SIGSAC Conference on Computer and Communications Security, pp. 619–631. ACM (2017)
18. Mohassel, P., et al.: Secureml: a system for scalable privacy-preserving machine learning. In: Proceedings of Security and Privacy (SP), pp. 19–38. IEEE (2017)
19. MP-SPDZ (2018). https://github.com/data61/MP-SPDZ/. Accessed 03 Oct 2020
20. Pace, D.F., Dalca, A.V., Geva, T., Powell, A.J., Moghari, M.H., Golland, P.: Interactive whole-heart segmentation in congenital heart disease. In: Navab, N., Hornegger, J., Wells, W.M., Frangi, A.F. (eds.) MICCAI 2015. LNCS, vol. 9351, pp. 80–88. Springer, Cham (2015). https://doi.org/10.1007/978-3-319-24574-4_10
21. Prados, F., Ashburner, J., Blaiotta, C., Brosch, T., Carballido-Gamio, J., Cardoso, M.J., Conrad, B.N., Datta, E., Dávid, G., De Leener, B., et al.: Spinal cord grey matter segmentation challenge. Neuroimage **152**, 312–329 (2017)
22. Riazi, M.S., Samragh, M., Chen, H., Laine, K., Lauter, K.E., Koushanfar, F.: Xonn: Xnor-based oblivious deep neural network inference. IACR Cryptology ePrint Archive 2019, 171 (2019)
23. Ronneberger, O., Fischer, P., Brox, T.: U-Net: convolutional networks for biomedical image segmentation. In: Navab, N., Hornegger, J., Wells, W.M., Frangi, A.F. (eds.) MICCAI 2015. LNCS, vol. 9351, pp. 234–241. Springer, Cham (2015). https://doi.org/10.1007/978-3-319-24574-4_28
24. Rouhani, B.D., et al.: Deepsecure: Scalable provably-secure deep learning. In: Proceedings of DAC, pp. 1–6. IEEE (2018)
25. Microsoft SEAL (release 3.3), June 2019. https://github.com/Microsoft/SEAL, microsoft Research, Redmond, WA
26. Yao, A.C.: Protocols for secure computations. In: 23rd Annual Symposium on Foundations of Computer Science, 1982. SFCS 2008, pp. 160–164. IEEE (1982)

Temporal-Consistent Segmentation of Echocardiography with Co-learning from Appearance and Shape

Hongrong Wei[1,2], Heng Cao[1,2], Yiqin Cao[1,2], Yongjin Zhou[3],
Wufeng Xue[1,2(✉)], Dong Ni[1,2], and Shuo Li[4]

[1] National-Regional Key Technology Engineering Laboratory for Medical Ultrasound, Guangdong Key Laboratory for Biomedical Measurements and Ultrasound Imaging, School of Biomedical Engineering, Health Science Center, Shenzhen University, Shenzhen, China
xuewf@szu.edu.cn

[2] Medical Ultrasound Image Computing (MUSIC) Lab, Shenzhen, China

[3] School of Biomedical Engineering, Health Science Center, Shenzhen University, Shenzhen, China

[4] Department of Medical Imaging, Western University, London, ON, Canada

Abstract. Accurate and temporal-consistent segmentation of echocardiography is important for diagnosing cardiovascular disease. Existing methods often ignore consistency among the segmentation sequences, leading to poor ejection fraction (EF) estimation. In this paper, we propose to enhance temporal consistency of the segmentation sequences with two co-learning strategies of segmentation and tracking from ultrasonic cardiac sequences where only end diastole and end systole frames are labeled. First, we design an appearance-level co-learning (CLA) strategy to make the segmentation and tracking benefit each other and provide an eligible estimation of cardiac shapes and motion fields. Second, we design another shape-level co-learning (CLS) strategy to further improve segmentation with pseudo labels propagated from the labeled frames and to enforce the temporal consistency by shape tracking across the whole sequence. Experimental results on the largest publicly-available echocardiographic dataset (CAMUS) show the proposed method, denoted as CLAS, outperforms existing methods for segmentation and EF estimation. In particular, CLAS can give segmentations of the whole sequences with high temporal consistency, thus achieves excellent estimation of EF, with Pearson correlation coefficient 0.926 and bias of 0.1%, which is even better than the intra-observer agreement.

Keywords: Echocardiography · Segmentation · Tracking

1 Introduction

In clinical practice, echocardiography is frequently-used for assessing cardiac function and diagnosing cardiovascular disease (CVD) due to its advantages of

© Springer Nature Switzerland AG 2020
A. L. Martel et al. (Eds.): MICCAI 2020, LNCS 12262, pp. 623–632, 2020.
https://doi.org/10.1007/978-3-030-59713-9_60

being real-time, radiation-free and low cost. However, it suffers from defects of large noise and low contrast, resulting in great difficulties for visual inspection. Manual annotation of the key structures is usually time-consuming and has large inter-/intra-observer variability. Concerning the ejection fraction (EF), the correlations of inter-/intra-observer agreement are only 0.801 and 0.896, respectively [9]. This indicates a demand of automatic methods for highly accurate segmentation, and precise estimation of EF from cardiac ultrasonic sequences.

Previous state-of-the-art methods [11,14] analyze each frames independently, and have achieved comparable performance with the inter-observer agreements for ventricle borders segmentation. However, eligible left ventricle segmentation does not necessary lead to accurate EF estimation, as we found in Table 3. A prerequisite for precise EF estimation is to preserve the temporal consistency of the segmentations for all frames under condition that only ED and ES frames are annotated.

Benefits of leveraging temporal information for cardiac image segmentation have been proved by previous works [1,4,5,8,10,12,13,15,16]. Recurrent neural network was employed to extract spatial-temporal features [5,10,15] for cardiac segmentation. Specially, [10] utilize hierarchically convolutional LSTMs for spatiotemporal feature embedding and a double-branch module for joint learning of LV segmentation and chamber view classification. Optical flow was utilized to extract the relative motion vector between consecutive frames, and then the learned motion vector was used as an complementary input to the original images to learn the segmentation by neural networks [4,8,16], or to propagate the labels from ED to ES [12]. Instead, [13] utilized the same encoder for joint learning of motion estimation and segmentation on cardiac magnetic resonance images, demonstrating the effectiveness of motion learning as a regularizer for the segmentation task. However, the motion estimation between frames that have large time interval in [13] may lead to large motion estimated error for echocardiographic images that have blurry (or even disappeared) boundaries. Although the temporal information in these methods could help relieve temporally inconsistent segmentation results of the whole sequence, they still suffer from two weaknesses that will be solved in our work: 1) requirements of densely labeled cardiac sequences [4,5,15,16], and 2) lack of explicit and effective enforcement of temporal consistency of the consecutive frames [4,5,8,10,12,13,15,16].

In this paper, we overcome these weaknesses and propose a novel temporally consistent segmentation method called Co-Learning of segmentation and tracking on Appearance and Shape level (CLAS) for sparsely labeled echocardiographic sequence. CLAS leverages temporal information by two co-learning strategies to improve the segmentation accuracy and enforce temporal consistency. Our contributions lies in three folds:

1) We design an appearance-level co-learning (CLA) strategy for segmentation and tracking, to collaboratively learn eligible shapes and motion fields from both labeled and unlabeled frames of the whole sequence. These results act as a warm start for further enhancement of temporal consistency.

2) We design a shape-level co-learning (CLS) strategy to enhance the temporal consistency of the segmentation results. Specifically, bi-direction tracking of cardiac shapes improve the previous motion fields learned from appearance. The segmentation is then further improved by pseudo labels deformed using these motion fields. The two tasks mutually benefit each other, leading to enhanced temporal consistency.

3) We validate the performance with a sparsely labeled echocardiography dataset, and achieve state-of-the-art performance that is comparable to the intra-observer variability for segmentation of endocardium (Endo) and epicardium (Epi). Especially, due to the temporal consistency enforcement, we achieve a correlation of 0.926 for EF estimation, which is obviously better than the inter-/intra-observer variability 0.801/0.896 and best reported method [9] 0.845.

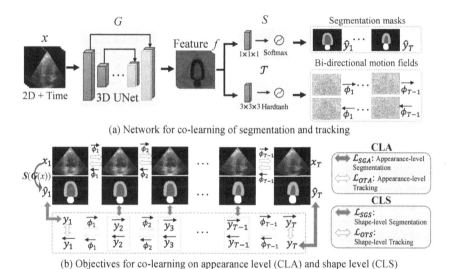

(a) Network for co-learning of segmentation and tracking

(b) Objectives for co-learning on appearance level (CLA) and shape level (CLS)

Fig. 1. Overview of the proposed CLAS, which explores the mutual benefits of segmentation and tracking with the network architecture (a) on appearance and shape levels (b) to explicitly enhance the temporal consistency of the segmentation.

2   Method

The overview of the proposed CLAS is illustrated in Fig. 1. The network architecture (Sect. 2.1) contains a 3D UNet for feature learning, on top of which two heads are used for co-learning of the segmentation and tracking. The co-learning operates on two levels. For CLA (Sect. 2.2), supervised segmentation for the ED/ES frames and unsupervised tracking for the intermediate frames are collaboratively learned to benefit the feature learning with large amount of unlabeled data, and to provide eligible estimation of the segmentation and motion field. For CLS (Sect. 2.3), pseudo labels that are deformed from the labeled ED/ES frames are used to further improve the segmentation part, while the bi-directional tracking between ED and ES are used to improve the tracking, i.e., enhance the temporal consistency of the segmented cardiac sequences. The overall objective function is described in Sect. 2.4.

2.1 Co-learning Network Architecture

We design a co-learning network architecture, which enables us to exploit the mutual benefits of the two tasks during the feature learning from cardiac appearance (CLA) and among shapes of ground truth labels, segmentation predictions, and those obtained in tracking (CLS). Our co-learning network differs from previous works that used a predetermined optical flow [4,8,16], or learned it directly from the compact representation [13].

Given a 2D ultrasonic cardiac sequence $\mathbf{x} = \{x_t\}_{t=1,2,...,T}$ with y_1, y_T as labels of ED $(t = 1)$ and ES $(t = T)$, we aim at obtaining accurate and temporally consistent segmentation masks for the whole cardiac sequence. The overall network contains three parts: 1) a basic 3D UNet G as backbone for feature extraction, 2) a segmentation head S that outputs the predicted masks $\hat{y}$ of the background, Endo, myocardium (Myo) and left atrium (LA), and 3) a tracking head $\mathcal{T}$ that outputs the bi-directional motion fields $[\overrightarrow{\phi}_t, \overleftarrow{\phi}_t](t = 1, ...T - 1)$, which can be utilized to deform images or labels to adjacent frames using spatial transformation [7].

2.2 Appearance-Level Co-learning

We design a CLA strategy to optimize the whole network with a supervised appearance-level segmentation task and an unsupervised appearance-level object tracking task. The segmentation task learns robust shape information from the appearance of the cardiac images by G and output the shape masks of cardiac structures by optimizing the segmentation objective $\mathcal{L}_{SGA}$. The tracking task learns the bi-directional motion fields of adjacent frames from the shape information in G by optimizing the tracking objective $\mathcal{L}_{OTA}$.

The two tasks benefit each other as follows: the supervised segmentation learns from ED and ES frames and makes G encode the shape information, and acts as a noise removing filter for the tracking of cardiac structures in the noisy ultrasonic images; the tracking task learns from the whole sequence, including the intermediate unlabeled frames, help feature learning of G by feeding more training images and identifying the moving objects, thus benefits the cardiac structure segmentation.

The supervised appearance-level segmentation loss $\mathcal{L}_{SGA}$ is defined by a combination of cross-entropy (CE) loss and multi-class Dice loss:

$$\mathcal{L}_{SGA} = \frac{1}{2\,|\Omega|} \sum_{t \in \{1,T\}} \sum_{c=1}^{C} \left(-y^{c,t} \cdot \log \hat{y}^{c,t} + \frac{1}{|C|} \left(1 - \frac{|y^{c,t} \cdot \hat{y}^{c,t}|}{|y^{c,t}| + |\hat{y}^{c,t}|}\right)\right) \quad (1)$$

where $c \in \{background, Endo, Myo, LA\}$, in echocardiography, y and $\hat{y}$ are ground truth and predicted probability respectively.

The unsupervised appearance-level object tracking loss $\mathcal{L}_{OTA}$ is defined by a combination of local cross-correlation (CC, [2]) and smooth loss $\mathcal{L}_{sm}$ [3]:

$$\mathcal{L}_{OTA} = \frac{-1}{2(T-1)} \sum_{t=1}^{T-1} \left(CC(x_{t+1}, x_t \circ \overrightarrow{\phi}_t) + CC(x_t, x_{t+1} \circ \overleftarrow{\phi}_t)\right) \quad (2)$$

$$+ \frac{\gamma}{2(T-1)} \sum_{t=1}^{T-1} \left(\mathcal{L}_{sm}(\overrightarrow{\phi}_t) + \mathcal{L}_{sm}(\overleftarrow{\phi}_t)\right)$$

where $\circ$ denotes spatial transformation, γ is a regularization parameter for the smooth term. Minimizing $\mathcal{L}_{OTA}$ encourages primary object tracking through the whole cardiac sequence.

In spite of the above mentioned benefits, there is still no explicit enforcement of temporal consistency for the predicted masks across the whole sequences. Therefore, we propose a shape-level co-learning scheme in next section.

2.3   Shape-Level Co-learning

We further design a CLS strategy to enforce temporal consistency of the previously estimated shape masks and motion fields. Specifically, we introduce the CLS strategy that includes: 1) unsupervised segmentation of the intermediate frames with pseudo labels that are deformed from the true masks of ED/ES, and 2) semi-supervised bi-directional tracking of the shape between ED and ES.

The shape-level tracking operates on cardiac shapes that are free of noise and background, therefore improves the motion fields learned in CLA. The enhanced motion fields can improve the segmentation performance by providing high quality pseudo labels for the intermediate frames. The improved segmentation can bring further improvement for the shape-level tracking by shape embedding in G. In the end, this mutual improvement of segmentation and tracking can effectively enforce the temporal consistency of the predicted segmentation masks.

We first generate pseudo labels $\overrightarrow{y}^{c,t}, (t = 2, ..., T)$ by forward deformation of the true ED label $y^{c,1}$ using a sequential motion fields as: $\overrightarrow{y}^{c,t} = [[y^{c,1} \circ \overrightarrow{\phi}_1] \circ \overrightarrow{\phi}_2] \circ ... \circ \overrightarrow{\phi}_{t-1}$. Similarly, $\overleftarrow{y}^{c,t}, (t = 1, ..., T-1)$ can be obtained from the true ES label $y^{c,T}$ by backward deformation. With these pseudo labels, the shape-level segmentation helps improve the segmentation network using the large amount of intermediate frames $(t = 2, ...T - 1)$:

$$\mathcal{L}_{SGS} = \frac{1}{2(T-2)|C||\Omega|} \sum_{t=2}^{T-1} \sum_{c=1}^{C} \left(2 - \frac{|\overrightarrow{y}^{c,t} \cdot \hat{y}^{c,t}|}{|\overrightarrow{y}^{c,t}| + |\hat{y}^{c,t}|} - \frac{|\overleftarrow{y}^{c,t} \cdot \hat{y}^{c,t}|}{|\overleftarrow{y}^{c,t}| + |\hat{y}^{c,t}|}\right) \quad (3)$$

The shape-level tracking helps improve the learned motion fields by matching the ED/ES labels with the pseudo labels $\overleftarrow{y}_1$, and $\overrightarrow{y}_T$:

$$\mathcal{L}_{OTS} = \frac{1}{2|C||\Omega|} \sum_{c=1}^{C} (2 - \frac{|\overrightarrow{y}^{c,T} \cdot y^{c,T}|}{|\overrightarrow{y}^{c,T}| + |y^{c,T}|} - \frac{|\overleftarrow{y}^{c,1} \cdot y^{c,1}|}{|\overleftarrow{y}^{c,1}| + |y^{c,1}|}) \qquad (4)$$

2.4 Overall Objective

We optimize the network in two stages. For stage one, CLA is used:

$$\mathcal{L}_{stage1} = \mathcal{L}_{SGA} + \mathcal{L}_{OTA} \qquad (5)$$

Minimizing $\mathcal{L}_{stage1}$ results in eligible segmentation masks and motion fields of the whole sequence that acts as a warm start for CLS. Then, for stage two, both CLA and CLS are employed:

$$\mathcal{L}_{stage2} = \mathcal{L}_{SGA} + \mathcal{L}_{OTA} + \alpha\mathcal{L}_{SGS} + \beta\mathcal{L}_{OTS} \qquad (6)$$

where α, β are trade-off parameters.

3 Experiments and Analysis

3.1 Experimental Configuration

Dataset. CAMUS [9] consists of 500 patients' 2D echocardiography from ED to ES phase with two-chamber (2CH) and four-chamber (4CH) views. Half of the patients have an EF lower than 45% (being considered at pathological risk), 19% of the images have poor quality. The ground truths of Endo, Myo, LA on ED and ES frames for both 2CH and 4CH were provided. For CAMUS challenge, 450 patients are used for model training, other 50 patients for testing, for which ground truths are not accessible. During training, ten frames starting from ED and ending with ES, were sampled with equal interval from each sequence. All the images were resized to 256×256 and the intensity was normalized to $[-1, 1]$.

Experiment Setup. We train the network for 30 epochs with batch size 4, and use Adam optimizer with initial learning rates 10^{-4} for both G and S, $0.5 * 10^{-4}$ for T, of which parameters initialization is Gaussian distribution $N(0, 10^{-5})$. Learning rate is reduce to 10^{-5} after 25 epochs. The hyperparameters α, β, and γ are set to be 0.2, 0.4 and 10, respectively. We optimize $\mathcal{L}_{stage1}$ in the first 10 epochs and $\mathcal{L}_{stage2}$ in the last 20 epochs. For testing phase, we train the model using all training data (450 patients) with same hyperparameters and experiment setups as 10-fold cross validation.

3.2   Results and Analysis

Tracking Performance. We first evaluate the effects of segmentation and shape-level tracking tasks to object tracking, which is measured by the Dice coefficients between pairs of $y_{c,1}$, $\overleftarrow{y}_{c,1}$ and $y_{c,T}$, $\overrightarrow{y}_{c,T}$ for $c \in \{endo, epi, LA\}$. Table 1 demonstrates the results of tracking performance by using appearance tracking only (OTA), two levels of tracking (OTA+OTS), and CLAS. We observed that: 1) only appearance tracking cannot lead to accurate performance given the low contrast and noisy echocardiography; 2) shape-level tracking helps eliminate the background and noisy information, thereby improving tracking of Endo, Epi and LA; 3) CLAS, with further shape-level segmentation, enforces the temporal consistency, and leads to even better tracking.

Table 1. Mean Dice of 2CH and 4CH views for object tracking on training set using 10-fold cross-validation. Three methods have significant difference with $p \ll 0.001$.

Method	ED			ES		
	Endo	Epi	LA	Endo	Epi	LA
OTA	0.816	0.900	0.802	0.798	0.889	0.807
$OTA + OTS$	0.909	0.939	0.861	0.887	0.927	0.889
CLAS	**0.923**	**0.945**	**0.876**	**0.903**	**0.935**	**0.900**

Segmentation Performance. The accuracy and consistency of segmentation across the whole cardiac sequence are examined on the test set. Figure 2 shows the visualization of segmentation results by UNet and CLAS. In the left, we observe that CLAS achieves better Dice coefficients and low EF error. The larger estimation error of UNet falsely indicates the case as a pathological one (EF below 45%). In the right of Fig. 2, from the curves of the areas of Endo, Myo, and LA, we can see that results from CLAS are smoother across the whole sequence than those of UNet, validating its better temporal consistency.

Table 2 shows segmentation performance on the test set in terms of Dice, Hausdorff distance (HD) and mean absolute distance (MAD). We concluded that the proposed CLAS achieves excellent segmentation performance that is comparable with the intra-observer variability, and outperforms the existing best methods (UNet) reported on the CAMUS challenge website, for all cardiac structures including Endo, Epi and LA.

Volume and EF Estimation. LV volumes (EDV and ESV) and EF are very important indicator of heart disease. We calculate them from segmentation masks of ED and ES frames of 2CH and 4CH views using Simpson's biplane method of discs [6], and display the results in Table 3. We concluded that CLAS gives accurate estimation of LV volumes that are better than inter-observer variability, and EF estimation that are even better than intra-observer variability. It's also clear that CLAS outperforms existing methods. Especially for EF, CLAS

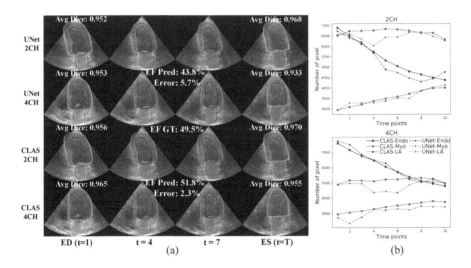

Fig. 2. Visualization of segmentation and its temporal consistency for UNet and CLAS. (a) Segmentation results of two-chamber (2CH) and four-chamber (4CH) view sequences for one patient. Dice coefficients and the predicted EF are shown. (b) Curves of predicted areas for Endo, Myo and LA in the whole cardiac sequence.

Table 2. Segmentation results on test set (50 patients). Mean performance of 2CH and 4CH views are shown. The results of UNet represent the best reported results of CAMUS challenge to date. Note that 10 poor quality patients were excluded for the inter-observer and intra-observer results [9].

ED	Endo			Epi			LA		
	Dice	HD	MAD	Dice	HD	MAD	Dice	HD	MAD
inter-observer	0.919	6.0	2.2	0.913	8.0	3.5	–	–	–
intra-observer	0.945	4.6	1.4	0.957	5.0	1.7	–	–	–
UNet[a]	0.936	5.3	1.7	0.956	5.2	1.7	0.889	5.7	2.2
ACNNs [11]	0.936	5.6	1.7	0.953	5.9	1.9	0.881	6.0	2.3
CLAS	**0.947**	**4.6**	**1.4**	**0.961**	**4.8**	**1.5**	**0.902**	**5.2**	**1.9**
ES	Endo			Epi			LA		
	Dice	HD	MAD	Dice	HD	MAD	Dice	HD	MAD
inter-observer	0.873	6.6	2.7	0.890	8.6	3.9	–	–	–
intra-observer	0.930	4.5	1.3	0.951	5.0	1.7	–	–	–
UNet[a]	0.912	5.5	1.7	0.946	5.7	1.9	0.918	5.3	2.0
ACNNs [11]	0.913	5.6	1.7	0.945	5.9	2.0	0.911	5.8	2.2
CLAS	**0.929**	**4.6**	**1.4**	**0.955**	**4.9**	**1.6**	**0.927**	**4.8**	**1.8**

[a]https://www.creatis.insa-lyon.fr/Challenge/camus/

delivers the highest correlation (0.926) with the ground truth and very low bias (0.1), compared to the best existing results (0.845 for UNet).

Table 3. LV volume and EF estimation on test set (50 patients) of CAMUS dataset.

Methods	EDV			ESV			EF		
	corr	bias(ml)	std	corr	bias(ml)	std	corr	bias (%)	std
inter-observer	0.940	18.7	12.9	0.956	18.9	9.3	0.801	−9.1	8.1
intra-observer	0.978	−2.8	7.1	0.981	−0.1	5.8	0.896	−2.3	5.7
UNet[a]	0.926	7.2	15.6	0.960	4.4	10.2	0.845	0.1	7.3
ACNNs [11]	0.928	2.8	15.5	0.954	2.0	10.1	0.807	0.3	8.3
RAL [10]	0.952	−7.5	11.0	0.960	−3.8	9.2	0.839	−0.9	6.8
CLAS	**0.958**	−0.7	15.1	**0.979**	−0.0	8.4	**0.926**	−0.1	6.7

[a]https://www.creatis.insa-lyon.fr/Challenge/camus/

4   Conclusion

We proposed a co-learning model CLAS for temporal-consistent segmentation of echocardiographic sequences with sparsely labeled data. We first used appearance-level co-learning to learn an eligible prediction for the segmentation and tracking tasks from both labeled and unlabeled frames. Then we used shape-level co-learning to enhance the temporal consistency of the predicted masks across the whole sequence. This two-level co-learning strategy iteratively improved the segmentation and tracking, and made the proposed CLAS outperforms existing methods obviously in aspects of multiple cardiac structures segmentation, as well as estimation of the clinically significant measurements (EDV, ESV, and EF).

Acknowledgement. The paper is partially supported by the National Key R&D Program of China (No. 2019YFC0118300), Shenzhen Peacock Plan (No. KQTD2016 053112051497, KQJSCX20180328095606003), Natural Science Foundation of China under Grants 61801296, the Shenzhen Basic Research JCYJ20190808115419619.

References

1. Al-Kadi, O.S.: Spatio-temporal segmentation in 3D echocardiographic sequences using fractional Brownian motion. IEEE Trans. Biomed. Eng. (2019)
2. Avants, B.B., Epstein, C.L., Grossman, M., Gee, J.C.: Symmetric diffeomorphic image registration with cross-correlation: evaluating automated labeling of elderly and neurodegenerative brain. Med. Image Anal. **12**(1), 26–41 (2008)
3. Balakrishnan, G., Zhao, A., Sabuncu, M.R., Guttag, J., Dalca, A.V.: VoxelMorph: a learning framework for deformable medical image registration. IEEE Trans. Med. Imaging **38**(8), 1788–1800 (2019)

4. Chen, S., Ma, K., Zheng, Y.: TAN: temporal affine network for real-time left ventricle anatomical structure analysis based on 2D ultrasound videos. arXiv preprint arXiv:1904.00631 (2019)

5. Du, X., Yin, S., Tang, R., Zhang, Y., Li, S.: Cardiac-DeepIED: automatic pixel-level deep segmentation for cardiac bi-ventricle using improved end-to-end encoder-decoder network. IEEE J. Transl. Eng. Health Med. **7**, 1–10 (2019)

6. Folland, E., Parisi, A., Moynihan, P., Jones, D.R., Feldman, C.L., Tow, D.: Assessment of left ventricular ejection fraction and volumes by real-time, two-dimensional echocardiography. A comparison of cineangiographic and radionuclide techniques. Circulation **60**(4), 760–766 (1979)

7. Jaderberg, M., Simonyan, K., Zisserman, A., et al.: Spatial transformer networks. In: Advances in Neural Information Processing Systems, pp. 2017–2025 (2015)

8. Jafari, M.H., et al.: A unified framework integrating recurrent fully-convolutional networks and optical flow for segmentation of the left ventricle in echocardiography data. In: Stoyanov, D., et al. (eds.) DLMIA/ML-CDS 2018. LNCS, vol. 11045, pp. 29–37. Springer, Cham (2018). https://doi.org/10.1007/978-3-030-00889-5_4

9. Leclerc, S., et al.: Deep learning for segmentation using an open large-scale dataset in 2D echocardiography. IEEE Trans. Med. Imaging **38**(9), 2198–2210 (2019)

10. Li, M., et al.: Recurrent aggregation learning for multi-view echocardiographic sequences segmentation. In: Shen, D., et al. (eds.) MICCAI 2019. LNCS, vol. 11765, pp. 678–686. Springer, Cham (2019). https://doi.org/10.1007/978-3-030-32245-8_75

11. Oktay, O., et al.: Anatomically constrained neural networks (ACNNs): application to cardiac image enhancement and segmentation. IEEE Trans. Med. Imaging **37**(2), 384–395 (2017)

12. Pedrosa, J., et al.: Fast and fully automatic left ventricular segmentation and tracking in echocardiography using shape-based b-spline explicit active surfaces. IEEE Trans. Med. Imaging **36**(11), 2287–2296 (2017)

13. Qin, C., et al.: Joint learning of motion estimation and segmentation for cardiac MR image sequences. In: Frangi, A.F., Schnabel, J.A., Davatzikos, C., Alberola-López, C., Fichtinger, G. (eds.) MICCAI 2018. LNCS, vol. 11071, pp. 472–480. Springer, Cham (2018). https://doi.org/10.1007/978-3-030-00934-2_53

14. Ronneberger, O., Fischer, P., Brox, T.: U-Net: convolutional networks for biomedical image segmentation. In: Navab, N., Hornegger, J., Wells, W.M., Frangi, A.F. (eds.) MICCAI 2015. LNCS, vol. 9351, pp. 234–241. Springer, Cham (2015). https://doi.org/10.1007/978-3-319-24574-4_28

15. Savioli, N., Vieira, M.S., Lamata, P., Montana, G.: Automated segmentation on the entire cardiac cycle using a deep learning work-flow. In: 2018 Fifth International Conference on Social Networks Analysis, Management and Security (SNAMS), pp. 153–158. IEEE (2018)

16. Yan, W., Wang, Y., Li, Z., van der Geest, R.J., Tao, Q.: Left ventricle segmentation via optical-flow-net from short-axis cine MRI: preserving the temporal coherence of cardiac motion. In: Frangi, A.F., Schnabel, J.A., Davatzikos, C., Alberola-López, C., Fichtinger, G. (eds.) MICCAI 2018. LNCS, vol. 11073, pp. 613–621. Springer, Cham (2018). https://doi.org/10.1007/978-3-030-00937-3_70

Decision Support for Intoxication Prediction Using Graph Convolutional Networks

Hendrik Burwinkel[1]([✉]), Matthias Keicher[1]([✉]), David Bani-Harouni[1],
Tobias Zellner[4], Florian Eyer[4], Nassir Navab[1,3], and Seyed-Ahmad Ahmadi[2]

[1] Computer Aided Medical Procedures, Technische Universität München,
Boltzmannstr. 3, 85748 Garching bei München, Germany
hendrik.burwinkel@tum.de
[2] German Center for Vertigo and Balance Disorders, Ludwig-Maximilians Universität
München, Marchioninistr. 15, 81377 München, Germany
[3] Computer Aided Medical Procedures, Johns Hopkins University,
3400 North Charles Street, Baltimore, MD 21218, USA
[4] Division of Clinical Toxicology and Poison Control Centre Munich, Department
of Internal Medicine II, TUM School of Medicine, Technische Universität München,
Ismaninger Str. 22, 81675 München, Germany

Abstract. Every day, poison control centers (PCC) are called for immediate classification and treatment recommendations of acute intoxication cases. Due to their time-sensitive nature, a doctor is required to propose a correct diagnosis and intervention within a minimal time frame. Usually the toxin is known and recommendations can be made accordingly. However, in challenging cases only symptoms are mentioned and doctors have to rely on clinical experience. Medical experts and our analyses of regional intoxication records provide evidence that this is challenging, since occurring symptoms may not always match textbook descriptions due to regional distinctions or institutional workflow. Computer-aided diagnosis (CADx) can provide decision support, but approaches so far do not consider additional patient data like age or gender, despite their potential value for the diagnosis. In this work, we propose a new machine learning based CADx method which fuses patient symptoms and meta data using graph convolutional networks. We further propose a novel symptom matching method that allows the effective incorporation of prior knowledge into the network and evidently stabilizes the prediction. We validate our method against 10 medical doctors with different experience diagnosing intoxications for 10 different toxins from the PCC in Munich and show our method's superiority for poison prediction.

Keywords: Graph convolutional networks · Representation learning · Disease classification

---

H. Burwinkel and M. Keicher—Both authors share first authorship.

© Springer Nature Switzerland AG 2020
A. L. Martel et al. (Eds.): MICCAI 2020, LNCS 12262, pp. 633–642, 2020.
https://doi.org/10.1007/978-3-030-59713-9_61

1   Introduction

Intoxication is undoubtedly one of the most significant factors of global suffering and death. In 2016, the abuse of alcohol alone resulted in 2.8 million deaths globally, and was accountable for 99.2 million DALYs (disability-adjusted life-years) -4.2% of all DALYs. Other drugs also summed up to 31.8 million DALYs and 451,800 deaths world-wide [7]. In case of an intoxication, fast diagnosis and treatment are essential in order to prevent permanent organ damage or even death [10]. Since not all medical practitioners are experts in the field of toxicology, specialized poison control centers (PCC) like the center in Munich were established. These institutions can be called by anyone – doctor or layman – to help in the classification and treatment of patients. Most of the time, the substance responsible for the intoxication is known. However, when this is not the case, the medical doctor (MD) working at the PCC has to reach a diagnosis solely based on the reported symptoms, without ever seeing the patient face to face and give treatment recommendations accordingly. Especially for inexperienced MDs, this is a challenging task for several reasons. First, the symptom description may not match the symptoms described in the literature that is used to diagnose the patients. This is exacerbated by inter-individual, regional, and inter-institutional differences in the description of symptoms when reaching the doctor. Secondly, not all patients react to intoxication with the same symptoms and they may have further confounding symptoms not caused by the intoxication but due to other medical conditions. Thirdly, meta information like age, gender, weight, or geographic location, are not assessed in a structured way. Current computer-aided diagnosis (CADx) systems in toxicology do not solve these problems. Most are rule-based expert systems [1,5,12] which are very sensitive towards input variations. Furthermore, they do not consider meta information or population context, despite their potential value in diagnosis. We propose a model that can solve both mentioned problems. By employing Graph Convolutional Networks (GCN) [6,9], we incorporate the meta information and population context into the diagnosis process in a natural way using graph structures. Here, each patient corresponds to a node, and patients are connected according to the similarity of their meta-information [13]. Connecting patients in this way leads to neighborhoods of similar patients. GCNs perform local filtering of graph-structured data analogous to Convolutional Neural Networks (CNN) on regular grids. This relatively novel concept [2] already led to advancements in medicine, ranging from human action prediction [16] to drug discovery [14]. It has also been used successfully in personalized disease prediction [3,8,13]. Attention mechanisms [4,11] improve filtering by weighting similarity scores between nodes based on node features, which help to compensate for locally inaccurate graph structure. We base the proposed model on inductive Graph Attention Networks (GAT) [15], one of the leading representatives of this GCN-class. Opposite to "transductive", an "inductive" GCN is able to infer on unseen data after training.

Contributions: Our approach for toxin prediction leverages structured incorporation of patient meta information to significantly boost performance.

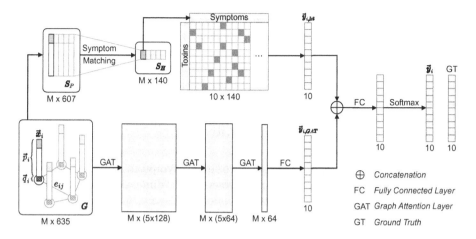

Fig. 1. Schematic architecture of ToxNet. The symptom vectors are processed in the graph-based GAT layers and the literature matching network in parallel.

We further address the issue of mismatching symptom descriptions by augmenting the GCN with a parallel network layer which learns a conceptual mapping of patient symptoms to textbook symptoms described in literature. This network branch is designed to explicitly incorporate domain prior knowledge from medical literature, and produces an alternative prediction. This stabilizes the output of the model and ensures a reasonable prediction. In a set of experiments on real PCC data, we show that our model outperforms several standard approaches. Ultimately, we compare our model to patient diagnoses made by 10 MDs on a separate real-life test set. The favorable performance of our model demonstrates its high potential for decision support in toxicology (Fig. 1).

2   Methodology

General Framework: The proposed network performs the classification of the intoxication of patients with 1D symptom vectors $\mathbf{P}$ using non-symptom meta information $\mathbf{Q}$ and literature symptom vectors $\mathbf{H}$ in an inductive graph approach. Therefore, it optimizes the objective function $f(\mathbf{P}, G(\mathbf{P}, \mathbf{Q}, \mathbf{E}), \mathbf{H}) : \mathbf{P} \rightarrow \mathbf{Y}$, where $G(\mathbf{P}, \mathbf{Q}, \mathbf{E})$ is a graph with vertices containing symptoms $\mathbf{P}$ and meta data $\mathbf{Q}$. Binary edges $\mathbf{E}$ denote connections between the vertices and $\mathbf{Y}$ is a set of poison classes. The symptom vectors contain a binary entry for every considered symptom, 1 if the symptom is present, 0 if not. Therefore, every patient has an individual symptom vector $\boldsymbol{p}_i$ with the occurring symptoms, and every poison has a vector $\boldsymbol{h}_i$ of literature symptoms that should occur for this poison, leading to the symptom sets: $\mathbf{P} = \{\boldsymbol{p}_1, \boldsymbol{p}_2, ..., \boldsymbol{p}_M\}, \boldsymbol{p}_i \in \{0, 1\}^{F_P}$, $\mathbf{H} = \{\boldsymbol{h}_1, \boldsymbol{h}_2, ..., \boldsymbol{h}_C\}, \boldsymbol{h}_i \in \{0, 1\}^{F_H}$, where M is the number of patients, C is the number of poison classes, F_P and F_H are the dimensions of the patient and literature symptom vector, respectively. Within $\mathbf{Q} = \{\boldsymbol{q}_1, \boldsymbol{q}_2, ..., \boldsymbol{q}_M\}$, every vector $\boldsymbol{q}_i$ contains the patient's meta information. For every vertex in the graph we concatenate the patient symptom vector $\boldsymbol{p}_i$ with the meta data $\boldsymbol{q}_i$ of $\mathbf{Q}$ and receive $\mathbf{X}$ with vectors $\boldsymbol{x}_i$ of dimension F. Additionally, the edges $\mathbf{E}$ are created based on the similarity of the meta information between two patients. The network processes the patient symptoms within three GAT layers and a learned explicit literature matching layer in parallel. The resulting representations are fused to predict the corresponding intoxication.

Symptom Vectors: As described above, every symptom vector corresponds to a binary encoding of all symptoms present. The dimensions F_P and F_H of the vectors refer to the total number of individual symptoms S_P and S_H that are listed within all patient cases and poison descriptions respectively. Since real patient cases also show some symptoms that are not part of the literature, $F_H < F_P$ and $S_H \subseteq S_P$. The first F_H entries of every $\boldsymbol{p}_i$ correspond to S_H.

Neighborhood Generation: The edges E represent the neighborhood of every concatenated vector or vertex $\boldsymbol{x}_i$ and define which vertices $\boldsymbol{x}_j \in N_i$ should be aggregated to update the current representation of $\boldsymbol{x}_i$ within a GAT layer. The neighborhood N_i of $\boldsymbol{x}_i$ is defined as the set of all $\boldsymbol{x}_j$ with $e_{ij} \in \mathbf{E}$. An edge e_{ij} is established when the meta information of $\boldsymbol{x}_i$ and $\boldsymbol{x}_j$ is consistent.

GAT Layer: To update the representation of the vectors $\boldsymbol{x}_i$ of $\mathbf{X}$, the GAT layer applies a shared learnable linear transformation $\mathbf{W} \in \mathbb{R}^{F' \times F}$ to all $\boldsymbol{x}_i$, resulting in a new representation with dimension F'. For every neighbor $\boldsymbol{x}_j \in N_i$, an attention coefficient α is calculated using the shared attention mechanism a. The coefficient represents the importance of $\boldsymbol{x}_j$ for the update of $\boldsymbol{x}_i$ and is calculated as $a(\mathbf{W}\boldsymbol{x}_i, \mathbf{W}\boldsymbol{x}_j) = \boldsymbol{a}^T[\mathbf{W}\boldsymbol{x}_i || \mathbf{W}\boldsymbol{x}_j]$, where $[\,||\,]$ represents the concatenation of $\mathbf{W}\boldsymbol{x}_i$ and $\mathbf{W}\boldsymbol{x}_j$, and $\boldsymbol{a} \in \mathbb{R}^{2F'}$ denotes a single feed-forward layer. To normalize every attention coefficient α and allow easy comparability between coefficients, after applying the leakyReLU activation σ, for every $\boldsymbol{x}_i$ the softmax function is applied to all coefficients corresponding to N_i.

$$\alpha_{ij} = \frac{\exp(\sigma(\boldsymbol{a}^T([\mathbf{W}\boldsymbol{x}_i || \mathbf{W}\boldsymbol{x}_j])))}{\sum_{r \in N_i} \exp(\sigma(\boldsymbol{a}^T[\mathbf{W}\boldsymbol{x}_i || \mathbf{W}\boldsymbol{x}_r]))} \tag{1}$$

To update $\boldsymbol{x}_i$, every feature representation $\mathbf{W}\boldsymbol{x}_j$ is weighted with the corresponding $\alpha_{i,j}$ and summed up to receive the new representation $\boldsymbol{x}'_i$. The GAT network repeats this step multiple times with individually learned $\mathbf{W}^k$, so-called heads, to statistically stabilize the prediction and receive individual attention coefficients α^k. The different representations $\boldsymbol{x}'_i$ are concatenated (represented as $||$) to yield the final new representation:

$$\boldsymbol{x}'_i = \|_{k=1}^{K} \sigma \left(\sum_{j \in N_i} \alpha_{ij}^k \mathbf{W}^k \boldsymbol{x}_j \right) \tag{2}$$

Here, K is the number of used heads and α_{ij}^k is the attention coefficient of head k for the vertices $\boldsymbol{x}_i$ and $\boldsymbol{x}_j$ [15].

Literature Symptom Matching: For every toxin class c_i of all toxins C, the literature provides a list of commonly occurring symptoms. These are encoded in the binary symptom vector $\boldsymbol{h}_i$ for every poison. We design a specific symptom matching layer $\mathbf{W}_{\text{symp}} \in \mathbb{R}^{F_H \times F_P}$ which learns a mapping of the patient symptom vectors $\mathbf{P}$ to the literature symptoms. This concept results in an interpretable transfer function which gives deeper insight into symptom correlations and explicitly incorporates the domain prior knowledge from literature. Due to the described setup of the symptom vectors, the first F_H entries of every $\boldsymbol{p}_i$ correspond to the literature symptoms. Since these should be preserved after the matching procedure, we initialize the first F_H learnable parameters of $\mathbf{W}_{\text{symp}}$ with the unity matrix I_{F_H} and freeze the diagonal during training. Like this, every symptom s of S_H is mapped to itself. The remaining symptoms only occurring for the patient cases are transformed into a representation of a dimension in agreement with the symptoms of the literature. As a second transformation, we create a literature layer $\mathbf{W}_{\text{lit}} \in \mathbb{R}^{C \times F_H}$ whose ith row is initialized with $\boldsymbol{h}_i$ for all classes C and that is kept constant during training. The resulting transformation $\boldsymbol{y}_{i,lit} = \mathbf{W}_{\text{lit}} \cdot \sigma(\mathbf{W}_{\text{symp}}\, \boldsymbol{p}_i)$ therefore maps the patient symptoms onto the poison classes with the explicit usage of literature information.

Representation Fusion: The output of the last GAT layer is processed by a FC layer to result in $\boldsymbol{y}_{i,GAT}$ with dimension C. The GAT and literature representations $\boldsymbol{y}_{i,GAT}$ and $\boldsymbol{y}_{i,lit}$ are concatenated, activated and transferred through a last linear transformation and a softmax function onto the class output $\boldsymbol{y}_i$.

3   Experiments and Discussion

3.1   Experimental Setup

Dataset: The dataset consists of 8995 patients and was extracted from the PCC database from the years 2001–2019. All cases were mono-intoxications, meaning only one toxin was present and the toxin was known. We chose the following toxins: ACE inhibitors (n $= 119$), acetaminophen (n $= 1376$), antidepressants (selective serotonin re-uptake-inhibitors, n $= 1073$), benzodiazepines (n $= 577$), beta blockers (n $= 288$), calcium channel antagonists (n $= 75$), cocaine (n $= 256$), ethanol (n $= 2890$), NSAIDs (excluding acetaminophen, n $= 1462$)) and opiates (n $= 879$). The ten toxin groups were chosen either because they are part of the most frequently occurring intoxications and lead to a different treatment and intervention or because they have clinically distinct features, lead to severe intoxications, have a specific treatment, and should not be missed. Accordingly, the different classes are unbalanced in their occurrence since e.g. alcohol intoxication is very common. Together with the patient symptoms, additional meta information for every case is given. From all available data, we use the parameters age group (child, adult, elder), gender, aetiology, point of entry and week day and year of intoxication to set up the graph, since these resulted in best performance.

Graph Setup: Our graph is based on the described meta information for every patient. An edge e_{ij} between patient x_i and x_j is established when the meta data is consistent for the medically relevant selection of parameters, i.e., the abovementioned meta parameters. This results in a sparse graph that at the same time has more meaningful edges (the graph increases the likelihood of patients with same poisonings to become connected). Samples in the training set are only connected to other training samples. Every validation and test sample is also only connected to training samples, but using directed connections, so they are only considered during validation and testing. Hence, the inductive GAT network can perform inference for even a single new unseen patient, since the training set provides the graph background during inference.

Network Setup: Hyperparameters: optimizer: Adam, learn. rate: 0.001, w. decay: 5e-4, loss function: cross entropy, dropout: 0.0, activation: ELU, heads: 5.

Model Evaluation: First, we evaluate our network against different benchmark approaches. Then we compare the different network components within an ablation study. By disabling different parts of the network, each individual contribution is evaluated. Here, 'GAT' refers to a setup where the GAT pipeline of ToxNet is used alone, 'LitMatch' to a setup where the literature-matching branch of ToxNet is used alone and 'MLP with meta' to a standard MLP. 'GAT' and 'MLP with meta' both receive the symptom vectors and meta data as input, 'LitMatch' uses the symptom vectors and has the literature vectors explicitly encoded in its setup (Sect. 2). Additionally, we test a sequential setting, where the literature matching is performed prior, and the learned features are transferred to the GAT (ToxNet(S)). All experiments use a 10-fold cross-validation, where each fold contains 10% of the data as test data, the remaining 90% are further split into 80% training and 20% validation. After proving the superiority of our method, we compare our network against the performance of 10 MDs, who are classifying the same unseen subset of the full test data as our method. This subset is divided into 25 individual cases for every MD, and 25 additional cases identical for all MDs, i.e., $250 + 25 = 275$ cases. In this setup, we are able to perform a statistical analysis on a larger set of cases, but also evaluate the inter-variability of the MDs to distinguish between easy and difficult cases.

Table 1. Performance comparison of different methods for poison prediction. Methods are described in detail in Sect. 3.

Method	F1 Sc. micro	F1 Sc. macro	p-val micro	p-val macro
Naive matching	0.201 ± 0.012	0.127 ± 0.007	$**$	$**$
Decision tree	0.246 ± 0.016	0.227 ± 0.016	$**$	$**$
LitMatch	0.474 ± 0.005	0.342 ± 0.023	$**$	$**$
MLP with meta	0.544 ± 0.015	0.429 ± 0.019	$**$	$**$
GAT	0.629 ± 0.010	0.458 ± 0.021	$**$	$**$
ToxNet(S)	0.637 ± 0.013	0.478 ± 0.023	$**$	$**$
ToxNet	$\mathbf{0.661} \pm 0.010$	$\mathbf{0.529} \pm 0.036$	/	/

(p-value: <0.01 $*$, <0.005 $**$)

3.2 Experimental Results

Performance Comparison Against other Methods: In Table 1, we compare the F1 micro and macro scores of different benchmark approaches against our method ToxNet. The Naive Matching provides a lower baseline by simply voting for the poison which has the most overlap between literature and patient symptoms. The decision tree was trained based on the literature symptoms and then used on the patient symptoms. Both models perform poorly, which leads to the conclusion that the available literature alone is not a good guide for poison classification. With the LitMatch neural network branch from our approach, we maintain the possibility to incorporate literature knowledge explicitly, but receive significantly better results. In the next step, a Multi-Layer Perceptron (MLP) with 3 hidden layers and $5 \cdot 128$, $5 \cdot 64$ and 64 hidden units respectively (same as GAT) was trained on the patient data to perform the prediction. In order to allow for a fair comparison, the patient's meta information was concatenated to their symptom vector, thus resulting in both the MLP and GAT using the same information. By comparing the MLP to a standard GAT network, it is observable that the usage of the meta information inside our graph method significantly boosts the classification performance, showing the value that the graph structure adds to the evaluation. Adding the literature information into the method by applying our proposed method ToxNet increases this performance even further. It needs to be stated that this enhancement is reached, although the literature data alone was shown not to be very informative for the task at hand. We therefore assume that there is a synergy effect, and an improvement of the literature might lead to an even stronger boost. To identify the individual contributions, both pipelines within ToxNet (GAT and LitMatch) are also evaluated separately as described above. Within our experiment, we found that the parallel setting of ToxNet is slightly superior to a sequential setting (ToxNet(S)). The results described above are also illustrated in the boxplot in Fig. 2 (left).

Performance Comparison Against Medical Experts: In order to evaluate the performance of our method against medical experts, we conducted a survey

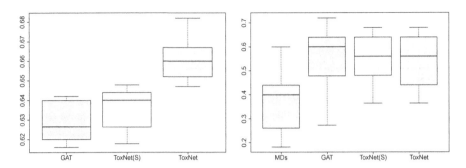

Fig. 2. Left: Comparison of ToxNet and different benchmark methods over 10-fold cross validation. **Right:** Comparison of ToxNet and benchmark methods with MDs' performance over 10 different sets evaluated by one MD each.

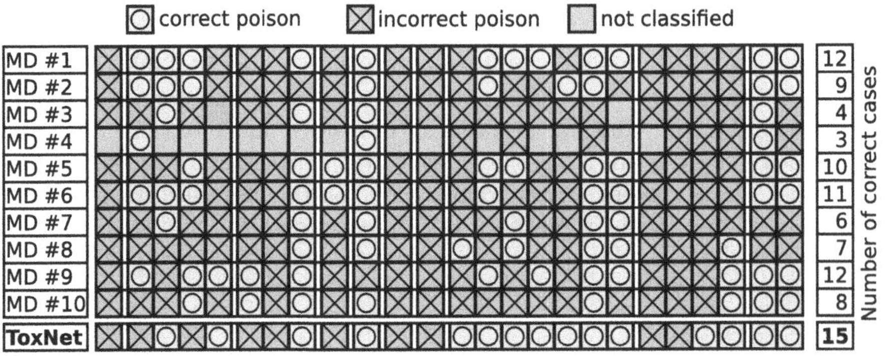

Fig. 3. Clinician inter-variability and comparison with ToxNet. Poison classes are ordered alphabetically, each group separated with a white spacing.

with 10 medical doctors (MDs) from the toxicology department of the Klinikum rechts der Isar in Munich, where each MD had to classify 50 intoxication cases that were split up as described above. Six doctors of the toxicology department were assistant doctors, the other four were specialists in pharmacology and toxicology. Figure 2 (right) shows a box plot of the performance of the 10 MDs compared with different benchmark methods as well as our method ToxNet on the ten individual sets of 25 cases each, so 250 in total. All three graph-based approaches clearly outperform the MDs due to the optimized usage of meta information. For this small subset of the full test set, the performance boost of ToxNet compared to GAT is not as severe as for the full test set. However, the overall performance is more stable (smaller margins). In Fig. 3, we performed a detailed inter-variability study on the 25 cases evaluated by all doctors. Except for one case, every intoxication case correctly classified by the majority of MDs, our method accomplished as well. Furthermore, for eight cases, where only half of the MDs or less correctly predicted the intoxication, our method still succeeded. These results demonstrate that our proposed ToxNet architecture can predict simple cases reliably and at an expert-level performance, while additionally providing a high prediction stability on cases that are challenging to a majority of doctors. Even compared to the two best MDs, who correctly classified 12 cases, our method overall resulted in 15 correct poison predictions. Six cases were wrongly classified by all doctors and our method. These are data samples with insufficient documentation quality (e.g. only a single generic symptom) which indicate intrinsic challenges from medical data in the wild.

4 Conclusion

In this work, we proposed ToxNet, a new architecture for improved intoxication prediction. The network effectively incorporates patient symptoms, meta-information like age group or residence, and domain prior knowledge from literature. We showed that the usage of meta-data within the graph structure of a graph convolutional network inside ToxNet leads to a significantly higher classification performance than all other methods investigated. In our benchmark study, we explicitly showed that a simple concatenation of the meta-data to the patient symptom vector is not sufficient – the improvement can be attributed to the patient graph. Additionally, we introduced a symptom matching method that allows the explicit usage of literature knowledge and included it into a parallel learning approach which further improved the overall network performance. Although we found that the literature information by itself was not informative enough for a satisfactory classification, we showed that a parallel integration with our graph network still led to synergy effects and an improved classification. We evaluated our network against 10 MDs with different experience levels and found a more stable prediction on both simple and highly challenging intoxication cases, given the high inter-rater variability among experts. We thus demonstrated the potential of ToxNet as a clinical decision support in this highly critical domain of medical intervention. On a wider scale, we believe that our architecture and validation provide a valuable case study: medical expertise can be regionally flavored and affect symptoms in a way that is not necessarily covered by expert literature. A proper modeling of these effects, fused with recent advances in graph-based population models, can lead to significant improvements in the field of computer-aided diagnosis and support clinical practice.

Acknowledgements. The study was supported by the Carl Zeiss Meditec AG in Oberkochen, Germany, and the German Federal Ministry of Education and Research (BMBF) in connection with the foundation of the German Center for Vertigo and Balance Disorders (DSGZ) (grant number 01 EO 0901).

References

1. Batista-Navarro, R.T.B., et al.: ESP: an expert system for poisoning diagnosis and management. Inform. Health Soc. Care **35**(2), 53–63 (2010)

2. Bronstein, M.M., Bruna, J., LeCun, Y., Szlam, A., Vandergheynst, P.: Geometric deep learning: going beyond Euclidean data. IEEE Signal Process. Mag. **34**(4), 18–42 (2016)
3. Burwinkel, H., et al.: Adaptive image-feature learning for disease classification using inductive graph networks. In: Shen, D., et al. (eds.) MICCAI 2019. LNCS, vol. 11769, pp. 640–648. Springer, Cham (2019). https://doi.org/10.1007/978-3-030-32226-7_71
4. Cheng, J., Dong, L., Lapata, M.: Long short-term memory-networks for machine reading. In: Proceedings of the 2016 Conference on Empirical Methods in Natural Language Processing, pp. 551–561. Association for Computational Linguistics, Stroudsburg (2016)
5. Darmoni, S., et al.: SETH: an expert system for the management on acute drug poisoning in adults. Comput. Methods Programs Biomed. **43**(3–4), 171–176 (1994)
6. Defferrard, M., Bresson, X., Vandergheynst, P.: Convolutional neural networks on graphs with fast localized spectral filtering. In: Advances in Neural Information Processing Systems, pp. 3844–3852 (2016)
7. Degenhardt, L., et al.: The global burden of disease attributable to alcohol and drug use in 195 countries and territories, 1990–2016: a systematic analysis for the Global Burden of Disease Study 2016. Lancet Psychiatry **5**(12), 987–1012 (2018)
8. Kazi, A., et al.: Graph convolution based attention model for personalized disease prediction. In: Shen, D., et al. (eds.) MICCAI 2019. LNCS, vol. 11767, pp. 122–130. Springer, Cham (2019). https://doi.org/10.1007/978-3-030-32251-9_14
9. Kipf, T.N., Welling, M.: Semi-supervised classification with graph convolutional networks. In: 5th International Conference on Learning Representations. (ICLR 2017) - Conference Track Proceedings, International Conference on Learning Representations ICLR (2016)
10. Kulling, P., Persson, H.: Role of the intensive care unit in the management of the poisoned patient. Med. Toxicol. **1**(5), 375–86 (1986)
11. Lin, Z., et al.: A structured self-attentive sentence embedding. In: 5th International Conference on Learning Representations. (ICLR 2017) - Conference Track Proceedings, International Conference on Learning Representations ICLR (2017)
12. Long, J.B., Zhang, Y., Brusic, V., Chitkushev, L., Zhang, G.: Antidote application. In: Proceedings of the 8th ACM International Conference on Bioinformatics, Computational Biology, and Health Informatics, pp. 442–448. ACM Press (2017)
13. Parisot, S., et al.: Spectral graph convolutions for population-based disease prediction. In: Descoteaux, M., et al. (eds.) MICCAI 2017. LNCS, vol. 10435, pp. 177–185. Springer, Cham (2017). https://doi.org/10.1007/978-3-319-66179-7_21
14. Stokes, J.M., et al.: A deep learning approach to antibiotic discovery. Cell **180**(4), 688–702 (2020)
15. Veličković, P., Casanova, A., Liò, P., Cucurull, G., Romero, A., Bengio, Y.: Graph attention networks. In: 6th International Conference on Learning Representations. (ICLR 2018) - Conference Track Proceedings, International Conference on Learning Representations ICLR (2018)
16. Yan, S., Xiong, Y., Lin, D.: Spatial temporal graph convolutional networks for skeleton-based action recognition. In: 32nd AAAI Conference on Artificial Intelligence, pp. 7444–7452. AAAI press (2018)

Latent-Graph Learning for Disease Prediction

Luca Cosmo[1,2](✉), Anees Kazi[3], Seyed-Ahmad Ahmadi[4], Nassir Navab[3,5], and Michael Bronstein[1,6,7]

[1] University of Lugano, Lugano, Switzerland
[2] Sapienza University of Rome, Rome, Italy
luca.cosmo@usi.ch
[3] Computer Aided Medical Procedures (CAMP), Technical University of Munich, Munich, Germany
[4] German Center for Vertigo and Balance Disorders, Ludwig Maximilians Universität München, München, Germany
[5] Whiting School of Engineering, Johns Hopkins University, Baltimore, USA
[6] Imperial College London, London, UK
[7] Twitter, London, UK

Abstract. Recently, Graph Convolutional Networks (GCNs) have proven to be a powerful machine learning tool for Computer Aided Diagnosis (CADx) and disease prediction. A key component in these models is to build a population graph, where the graph adjacency matrix represents pair-wise patient similarities. Until now, the similarity metrics have been defined manually, usually based on meta-features like demographics or clinical scores. The definition of the metric, however, needs careful tuning, as GCNs are very sensitive to the graph structure. In this paper, we demonstrate for the first time in the CADx domain that it is possible to learn a single, optimal graph towards the GCN's downstream task of disease classification. To this end, we propose a novel, end-to-end trainable graph learning architecture for dynamic and localized graph pruning. Unlike commonly employed spectral GCN approaches, our GCN is spatial and inductive, and can thus infer previously unseen patients as well. We demonstrate significant classification improvements with our learned graph on two CADx problems in medicine. We further explain and visualize this result using an artificial dataset, underlining the importance of graph learning for more accurate and robust inference with GCNs in medical applications.

Keywords: Graph convolution · Disease prediction · Graph learning

1 Introduction

There is a growing body of literature that recognises the potential of geometric deep learning [1] and graph convolutional networks (GCNs) in healthcare.

L. Cosmo, A. Kazi—Equal contribution.
N. Navab, M. Bronstein—Shared last authorship.

© Springer Nature Switzerland AG 2020
A. L. Martel et al. (Eds.): MICCAI 2020, LNCS 12262, pp. 643–653, 2020.
https://doi.org/10.1007/978-3-030-59713-9_62

GCNs have been applied to various problems already, including protein interaction prediction [6], metric learning on brain connectomes [15] or representation learning for medical images [2]. In this work, we focus on Computer Aided Diagnosis (CADx), i.e. disease prediction from multimodal patient data using GCNs in conjunction with population models [10,12,19,24]. The population model is realized in form of a graph, where vertices represent patients, edges represent connections between patients, and edge weights represent the patient similarity according to a defined metric [19]. Employing graph models for CADx is motivated by the success of GCNs in social network analyses and recommender systems [1]. The graph provides neighborhood information between patients, and graph signal processing [18] is used to aggregate patient features over local neighborhoods, similar to localized filters in regular convolutional neural networks (CNN). Graph deep learning is then tasked with learning a set of filters with optimal weights towards the downstream task of CADx classification. Kipf and Welling [14] proposed semi-supervised node classification using spectral GCNs. This approach was first adapted to CADx in medicine by Parisot et al. [19], who proposed to compute patient similarities from a set of meta-features such as age and sex. Importantly, the definition of the similarity metric between patients determines the graph adjacency matrix. As such, both feature selection and similarity metric definition need to be carefully tuned to avoid placing a meaningless graph structure at the core of the GCN. To alleviate this, several works have proposed using multiple graphs [10,12,22], where each graph is built from different patient features and encodes a unique latent structure about the population. Patient feature vectors are processed through each graph separately, and fused prior to the decision layer. Multi-graph approaches significantly improve the classification accuracy for CADx, and make GCNs much more robust towards the definition of individual graphs and similarity metrics [10,12]. On the other hand, they open up new challenges: i) multiple graphs limit the scalability due to the number of parameters, and ii) the multi-graph fusion layer needs to be carefully designed (e.g. self-attention [10] or LSTM-based attention [12]). An alternative approach to hand-crafting the graph adjacency matrix is to learn it end-to-end, to optimally support the downstream task. Notably, while we focus on CADx in this paper, learning a graph can benefit many other applications as well. In some cases, the learned graph might even be of higher interest than the downstream task, as it can provide important information for interpretability. There is little work so far on graph learning in general [21], and in the field of medicine in particular. Approaches proposed so far, however, are very diverse in nature. For example, Zhan et al. [26] construct multiple graph Laplacians and optimize the weighting among them during training. Franceschi et al. [5] sample graph adjacency matrices from a binomial distribution via reparameterization trick and optimize the graph structure via hypergradient descent, demonstrating its efficacy via classification accuracy e.g. in citation networks. Jang et al. [8] simultaneously learn an EEG feature representation and a brain connectivity graph based on deterministic graph sampling. Interestingly, they demonstrate that the graph makes sense from a neuroscientific perspective, setting an example that

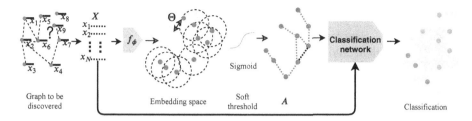

Fig. 1. Latent-graph learning architecture: Input node features are embedded into a lower dimensional space by a MLP f_ϕ. The parameter Θ is a soft-threshold applied to the distances between embedded features in order to build the adjacency matrix A. Its outcome provides the population graph model to the GCN, to further learn the node representations for the classification task.

the learned graph can have an intrinsic value regarding model interpretability and knowledge discovery.

Contribution: In this work, we propose a graph convolution-based neural network to perform patient classification, which automatically *learns to predict an underlying patients-graph* that is optimal for the downstream task, e.g. CADx. We show that using a single graph *learned end-to-end* allows both to achieve better performance and reduce the network complexity. Moreover, our method naturally *applies to inductive settings*, since we learn a function predicting the underlying graph structure which is robust to the introduction of new patients to the current population.

2   Method

In this section we describe our proposed model for graph learning and node classification. The advantages of our method are two-fold. First, despite the recent successes of multi-graph methods [10,12,22], we show that using a graph that has been learned end-to-end allows to achieve a significantly better classification performance. The second advantage is that during graph inference, a single graph is built from multiple features via Euclidean embedding, which drastically reduces the network complexity and solves the scalability problem, as an arbitrary amount of features can be embedded. Compared to CADx GCN models so far [10,11,19,24], we thus do not have to restrict the graph's adjacency context to patient meta-feature only (e.g. sex, age), but we can represent and embed a much richer patient representation into the graph structure. In the following paragraphs we detail the architecture of our graph learning module and its use in a classification network.

2.1 Latent-Graph Learning

Given an input set of nodes and associated features $\mathbf{x_i} \in \mathbb{R}^d$, the goal is to predict an optimal underlying graph structure which allows employing graph convolutional (GC) layers towards the downstream task, e.g. CADx in this work.

Predicting a discrete (e.g. binary) graph structure is a non-differentiable problem. To overcome this limitation and train the graph end-to-end, we represent it as a real-valued, i.e. weighted adjacency matrix $\mathbf{A} \in [0,1]^{N \times N}$. With this continuous relaxation we allow the output of any GC layer to be differentiable with respect to the graph structure.

In order to keep the architecture computationally and memory efficient, rather than learning on edge features [8,23], we propose to learn a function $\tilde{\mathbf{x}}_{\mathbf{i}} = f_\phi(\mathbf{x_i})$ which embeds node input features into a lower dimensional Euclidean space. The edge weight a_{ij} connecting i^{th} and j^{th} nodes is thus directly related to the embedded features distance through the following sigmoid function:

$$a_{ij} = \frac{1}{1 + e^{-t(\|\tilde{\mathbf{x}}_{\mathbf{i}} - \tilde{\mathbf{x}}_{\mathbf{j}}\|_2 + \theta)}} \tag{1}$$

where θ is a threshold parameter and t is a temperature parameter pushing values of a_{ij} to either 0 or 1. Values of t and θ are both optimized during training. We use a simple Multilayer Perceptron (MLP) as our function f_ϕ. In Fig. 1 we illustrate the graph learning pipeline.

2.2 Classification Model

We use our Latent-Graph learning to perform node classification. Given a set of N patients and associated set of multi-modal features $\mathbf{X} \in \mathbb{R}^{N \times d_1}$, the task is to predict the corresponding labels $y_i, i = 1 \ldots N$. To this end, we build a classification model composed by few graph convolutional layers followed by a fully connected layer to predict the patient label. To build the graph $\mathbf{A}$ with our latent-graph learning module, our model simply requires the full patient feature vectors X. In particular, we make use of the spatial GC layer defined as:

$$\mathbf{H_{l+1}} = \sigma(\mathbf{D}^{-1}\mathbf{A}\mathbf{H}_l\mathbf{W}) \tag{2}$$

where σ is a non-linear activation function, $\mathbf{D}$ is a normalization matrix with $d_{ii} = \sum_{ij} a_{ij}$, $\mathbf{H}_l$ is the output activation of the previous layer, and W are the model filters to be learnt. As mentioned before, both the latent-graph learning and classification MLPs are trained in an end to end manner.

3 Experiments and Results

We start this section with a proof of concept example showing the ability of our method to retrieve the ground truth graph. The remaining of the section is dedicated to comparison with the-state-of-the-art methods on two publicly available medical datasets.

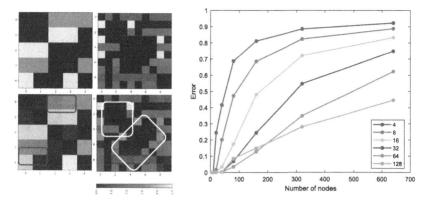

Fig. 2. Left: the ground truth graphs (top) and output of the graph optimization (bottom) for 5 and 10 nodes setting. The red boxes show errors made by our method, and white boxes show successful adjacency reconstruction. **Right:** the mean squared error between predicted and ground truth vectors as a function of number of nodes. Each curve represents different embedding dimensions.

3.1 Proof of Concept

In order to test the graph learning part of our model, we design an experimental setup on simulated data. We define a simple optimization problem in which the task is to regress a target vector y_i associated to each node.

Setting. Consider to be given a randomly generated graph $\mathcal{G}$ of N nodes with associated features $\mathbf{X}$. We define the ground-truth vector of the ith node as the sum of its neighbors' features, $y_i = \sum_{j \in \mathcal{N}_i} x_{ij}$. Further, we initialize the feature matrix as $\mathbf{X} = \mathbf{I} \in \mathbb{R}^{N \times N}$. Since our focus is to analyze the graph learning part, identity features are simple and orthogonal, which results in a minimum impact of the feature space on the task. As a result, only one possible output graph structure can minimise the categorical cross-entropy loss to zero.

Task. With this setting, we formalize the following optimization problem:

$$\arg\min_{\Phi} \sum_i (\mathbf{y}_i - \mathbf{A}(\Phi, \mathbf{X})\mathbf{X}_{:i})^2 \tag{3}$$

where $\mathbf{A}(\Phi, \mathbf{X})$ is our latent-graph learning module with input node features $\mathbf{X}$ and optimized parameters Φ. The main challenge in optimizing Eq. 3 is to choose the neighborhood between the nodes.

Results. In Fig. 2 we show results of the optimization. We analyze the behaviour of our method at increasing number of nodes. From the right plot we can see how the embedding size plays an important role on the ability of the method to retrieve the underlying graph. On the left we can appreciate how our latent-graph learning module is able to retrieve the relevant (ground-truth) affinity graph on two simulated toy graphs with 5 and 10 nodes.

Table 1. Comparison with some baseline methods on classification task on TADPOLE and UKBB datasets.

	TADPOLE		UKBB	
Baselines	Accuracy	AUC	Accuracy	AUC
Linear classifier	70,22 ± 06,32	80.26 ± 04.81	59.66 ± 1.170	80.26 ± 00.91
Spectral-GCN [19]	81.00 ± 06.40	74.70 ± 04.32	OOM	OOM
DGCNN [25]	84.59 ± 04.33	83.56 ± 04.11	58.35 ± 00.91	76.82 ± 03.03
Proposed	**92.91 ± 02.50**	94.49 ± 03.70	64.35 ± 01.11	82.35 ± 00.36

3.2 Comparisons

After our proof of concept on simulated data, we now show results of the proposed method on two publicly available medical datasets. We benchmark our classification results against three baselines and three comparative methods, which will be described in the following. We choose TADPOLE and UKBB mainly due to the difference is their size, which helps us analyze the adaptability of our proposed graph learning technique to smaller and larger datasets in medicine.

TADPOLE: This dataset [16] is a subset of the Alzheimer's Disease Neuroimaging Initiative (adni.loni.usc.edu), comprising 564 patients with 354-dimensional multi-modal features. The task is to classify each patient into Cognitively Normal (CN), Mild Cognitive Impairment (MCI), or Alzheimer's Disease (AD). Features are extracted from MR and PET imaging, cognitive tests, cerebro-spinal fluid (CSF) and clinical examinations.

UKBB: UK Biobank data provides pre-computed structural, volumetric and functional features of the brain, which are extracted from MRI and fMRI images using Freesurfer [4]. Here, we task the network to predict the age for each patient. We use a subsample of UKBB data [17], consisting of 14,503 patients with 440 features per individual. We quantize the patients' age into four decades from age 50–90 as classification targets.

Error Metrics. We use two standard error metrics to evaluate the performance of the network; accuracy and area under the ROC-curve (AUC). In multi-class classification tasks AUC is computed as the average of the AUC of each class versus all the others.

Baselines. Table 1 presents the results of some baselines in comparison with our proposed method. As a linear and non-graph baseline technique, we use a ridge regression classifier. Spectral-GCN [19] performs graph based convolution in the Fourier domain, but requires a pre-defined graph. Dynamic graph CNN (DGCNN) [25] constructs a kNN graph on the output activation during training, however the graph is not learned. All the graph-based techniques perform better than the ridge classifier (1). It can be further observed that a pre-defined graph may indeed be a sub-optimal choice, since Spectral-GCN performs worse than both DGCNN and the proposed method. Importantly, due to the polynomial filter approximation of the graph Laplacian, Spectral-GCNs can have too high memory demands, which is why UKBB could not be evaluated with this model. Further, our proposed model outperforms all the baselines with a margin of 8.32% and 6% for TADPOLE and UKBB respectively.

Comparative Methods. In Table 2 we show the performance of our model with respect to some state-of-the-art methods on the considered datasets. We choose Multi-GCN [10], which uses multiple graphs from different set of features, and reportedly performed very well on datasets similar to TADPOLE. InceptionGCN [11] is a state of the art GCN method on TADPOLE. Differentiable Graph Module (DGM) [9] is a recently proposed graph learning method. As can be seen in Table 2, the proposed model outperforms all comparative methods. For the much larger UKBB dataset, we again received out of memory (OOM) errors for the spectral GCN methods Multi-GNC and InceptionGCN. We further report the number of parameters required by each compared model in Table 2. All the experiments are performed with ten-fold stratified cross-validation. The low standard deviation of our proposed method further shows the improved robustness of our model.

Out of Sample Extension. Our method does not directly optimize a graph for a given population but it rather learn a f_ϕ that predicts the graph from input patients features. As such, it is easily extendable to previously unseen test patients, to enable inductive inference. Unseen patients can thus be added during testing time, and will be embedded into the lower dimensional space for graph representation. For a comparison to state of the art, we select DGCNN and DGM which are both inductive graph methods as well. Table 3 shows the accuracy of classification, given a data split of 90% training data vs 10% testing data. The higher accuracy and lowest standard deviation for our method confirms the superiority, robustness and precision of our model in a fully inductive setting.

Table 2. Comparison with state-of-the-art graph convolutional based methods on TADPOLE and UKBB datasets. P denotes the number of model parameters.

	TADPOLE			UKBB		
Method	Accuracy	AUC	P	Accuracy	AUC	P
Multi-GCN [10]	76.06 ± 0.72	90.32 ± 4.85	46k	OOM	OOM	46k
InceptionGCN [11]	84.11 ± 4.50	88.39 ± 4.16	58k	OOM	OOM	58k
DGM [9]	91.05 ± 5.93	**96.86 ± 1.81**	6k	61.59 ± 1.05	79.32 ±0.95	14k
Proposed	**92.91 ± 2.50**	94.49 ± 3.70	**2k**	**64.351 ± 1.11**	**82.352 ± 0.37**	**11k**

Table 3. Classification accuracy score for out of sample extension in TADPOLE and UKBB datasets.

Method	DGCNN	DGM	Proposed
TADPOLE	82.99 ± 04.91	88.12 ± 03.65	**91.85 ± 02.62**
UKBB	51.84 ± 08.16	53.37 ± 07.94	**63.91 ± 01.49**

Qualitative Results. Our quantitative results show the superiority of our proposed method, which learns a graph along with the task in an end to end fashion. However, from a qualitative point of view, it is challenging to evaluate the learnt graph, because it is not possible to compute a groundtruth graph. As an approximation, we generate a reference graph for TADPOLE from a weighted summation of the individual graphs from different modalities as is described in Multi-GCN [10]. Figure 3 shows the ground truth graph on the left and the learned graph. Interestingly, even though both the graphs are not completely identical, the overall structure shows similarities with one another. Remaining differences in the graph structure are expected, as these probably explain the improved classification performance to a degree. However, a proper interpretation from a medical perspective would require an assessment by experts, which we suggest as future work.

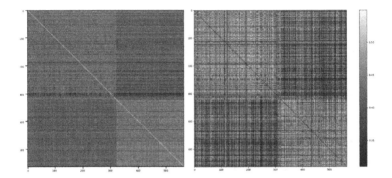

Fig. 3. The figure shows the ground truth graph (left) and learnt graph on the (right). The graph is shown for all the point in the dataset for random fold. The colorbar shows the affinity of each point with other.

Implementation Details. For both datsets we use the same architecture composed by two convolution layers ($16 \rightarrow 8$) and a final MLP ($32 \rightarrow 16 \rightarrow$ #classes) for obtaining the classification score. We use RElU non linearity and dropout (keep rate 0.9) after each layer except the last. Following [20] we reduce the input feature dimensions via *recursive feature elimination* [7] to 30 and 200 for TADPOLE and UKBB respectively and apply standard normalization. We use the Adam optimizer [13] to minimize the categorical cross-entropy loss with a learning rate of 0.01 reduced to 0.0001 at the intervals of 100 epochs in a piecewise constant fashion. The number of epochs = 600. All the experiments are implemented in TensorFlow and performed using a commercial workstation and GPU (Titan Xp, 12 GB VRAM).

4   Discussion and Conclusion

In this paper, we propose a model capable of learning the optimal relationships between the patients towards the downstream task, e.g. CADx and age prediction in medicine in this work. The entire model is trained in an end-to-end fashion, backpropagating directly through the graph adjacency. This is achieved using a soft thresholding technique with the learnable threshold Θ and the temperature parameter. Such a setting allows the update of each edge weight with respect to the loss.

As a proof of concept, we showed experiments on the simulated data where we successfully learned the given ground truth graph. Further, we showed applications to disease prediction on TADPOLE data and age prediction on UK Biobank data. In all the experiments our model performed better than baseline and state of the art approaches. Our model generalizes to inductive setting and outperforms other state of the art methods. At the same time, our model only contains up to two orders of magnitude less parameters than state of the art single- and multi-graph methods. It can be concluded from all of our experiments that a pre-defined graph might not be optimal, neither in a single - nor in a multi-graph setting. Instead, an end-to-end learning of the graph adjacency can lead to significant benefits in downstream tasks like classification. Further, the graph structure itself may have an intrinsic value, e.g. for better interpretability and knowledge discovery in the medical dataset.

In the proposed method we learned a global threshold for the entire population, but this may not be necessarily optimal. A single threshold operating globally on the Euclidean embedding might neglect the heterogeneity of the embedding structure. Therefore, learning a dedicated and patient-specific neighborhood threshold for each node might be the next step. Likewise, the Euclidean space embedding may not be optimal to learn a semantically meaningful graph either. Recent works have shown that the appropriate isometric space for embedding graphs is a negatively curved, i.e. a hyperbolic space [3]. Another direction to explore is the interpret the graph as well as the model.

In summary, we have proposed a novel graph learning method which makes it unnecessary to pre-define graph adjacencies. Until now, this was a prerequisite for employment of GCN models in medicine so far, but the definition of patient similarity metrics was not often well-motivated. The proposed graph learning method thus paves the way for further employment of graph methods in clinical decision support systems, which consistently demonstrate higher classification performances than the current state of the art in medicine.

Acknowledgement. The study was carried out with financial support of TUM-ICL incentive funding, Freunde und Förderer der Augenklinik, München, Germany and ERC Consolidator grant No. 724228 (LEMAN) and German Federal Ministry of Education and Health (BMBF) in connection with the foundation of the German Center for Vertigo and Balance Disorders (DSGZ) [grant number 01 EO0901]. The UK Biobank data is used under the application id 51541.

References

1. Bronstein, M.M., Bruna, J., LeCun, Y., Szlam, A., Vandergheynst, P.: Geometric deep learning: going beyond Euclidean data. IEEE Sig. Process. Mag. **34**(4), 18–42 (2017)
2. Burwinkel, H., et al.: Adaptive Image-Feature Learning for Disease Classification Using Inductive Graph Networks. In: Shen, D., et al. (eds.) MICCAI 2019. LNCS, vol. 11769, pp. 640–648. Springer, Cham (2019). https://doi.org/10.1007/978-3-030-32226-7_71
3. Chamberlain, B.P., Clough, J., Deisenroth, M.P.: Neural embeddings of graphs in hyperbolic space. arXiv preprint arXiv:1705.10359 (2017)
4. Fischl, B.: FreeSurfer. NeuroImage **62**(2), 774–781 (2012)
5. Franceschi, L., Niepert, M., Pontil, M., He, X.: Learning discrete structures for graph neural networks. In: Proceedings of International Conference Machine Learning (ICML). Proceedings of Machine Learning Research, vol. 97, pp. 1972–1982. PMLR (2019)
6. Gainza, P., Sverrisson, F., Monti, F., Rodola, E., Bronstein, M.M., Correia, B.E.: Deciphering interaction fingerprints from protein molecular surfaces using geometric deep learning. Nat. Methods **17**, 184–192 (2019)
7. Granitto, P.M., Furlanello, C., Biasioli, F., Gasperi, F.: Recursive feature elimination with random forest for PTR-MS analysis of agroindustrial products. Chemometr. Intell. Lab. Syst. **83**(2), 83–90 (2006)
8. Jang, S., Moon, S., Lee, J.: Brain signal classification via learning connectivity structure. arXiv abs/1905.11678 (2019)
9. Kazi, A., Cosmo, L., Navab, N., Bronstein, M.: Differentiable graph module (DGM) graph convolutional networks. arXiv preprint arXiv:2002.04999 (2020)
10. Kazi, A., Krishna, S., Shekarforoush, S., Kortuem, K., Albarqouni, S., Navab, N.: Self-attention equipped graph convolutions for disease prediction. In: 2019 IEEE 16th International Symposium on Biomedical Imaging (ISBI 2019), pp. 1896–1899 (2019)
11. Kazi, A., et al.: InceptionGCN: receptive field aware graph convolutional network for disease prediction. In: Chung, A.C.S., Gee, J.C., Yushkevich, P.A., Bao, S. (eds.) IPMI 2019. LNCS, vol. 11492, pp. 73–85. Springer, Cham (2019). https://doi.org/10.1007/978-3-030-20351-1_6
12. Kazi, A., et al.: Graph convolution based attention model for personalized disease prediction. In: Shen, D., et al. (eds.) MICCAI 2019. LNCS, vol. 11767, pp. 122–130. Springer, Cham (2019). https://doi.org/10.1007/978-3-030-32251-9_14
13. Kingma, D.P., Ba, J.: Adam: a method for stochastic optimization. arXiv preprint arXiv:1412.6980 (2014)
14. Kipf, T.N., Welling, M.: Semi-supervised classification with graph convolutional networks. arXiv preprint arXiv:1609.02907 (2016)
15. Ktena, S., et al.: Metric learning with spectral graph convolutions on brain connectivity networks. NeuroImage **169**, 431 (2018)
16. Marinescu, R., et al.: Tadpole challenge: prediction of longitudinal evolution in Alzheimer's disease. arXiv preprint arXiv:1805.03909 (2018)
17. Miller, K.L., et al.: Multimodal population brain imaging in the UK biobank prospective epidemiological study. Nat. Neurosci **19**(11), 1523 (2016)
18. Ortega, A., Frossard, P., Kovacevic, J., Moura, J.M.F., Vandergheynst, P.: Graph signal processing: overview, challenges, and applications. Proc. IEEE **106**(5), 808–828 (2018)

19. Parisot, S., Ktena, S.I., Ferrante, E., Lee, M., Guerrero, R., Glocker, B., Rueckert, D.: Disease prediction using graph convolutional networks: application to autism spectrum disorder and Alzheimer's disease. Med. Image Anal. **48**, 117–130 (2018)
20. Parisot, S., et al.: Spectral graph convolutions for population-based disease prediction. In: Descoteaux, M., Maier-Hein, L., Franz, A., Jannin, P., Collins, D.L., Duchesne, S. (eds.) MICCAI 2017. LNCS, vol. 10435, pp. 177–185. Springer, Cham (2017). https://doi.org/10.1007/978-3-319-66179-7_21
21. Qiao, L., Zhang, L., Chen, S., Shen, D.: Data-driven graph construction and graph learning: a review. Neurocomputing **312**, 336–351 (2018)
22. Valenchon, J., Coates, M.: Multiple-graph recurrent graph convolutional neural network architectures for predicting disease outcomes. In: IEEE International Conference on Acoustics, Speech and Signal Processing (ICASSP), pp. 3157–3161. IEEE (2019)
23. Veličković, P., Cucurull, G., Casanova, A., Romero, A., Lio, P., Bengio, Y.: Graph attention networks. arXiv:1710.10903 (2017)
24. Vivar, G., Zwergal, A., Navab, N., Ahmadi, S.-A.: Multi-modal disease classification in incomplete datasets using geometric matrix completion. In: Stoyanov, D., et al. (eds.) GRAIL/Beyond MIC -2018. LNCS, vol. 11044, pp. 24–31. Springer, Cham (2018). https://doi.org/10.1007/978-3-030-00689-1_3
25. Wang, Y., Sun, Y., Liu, Z., Sarma, S.E., Bronstein, M.M., Solomon, J.M.: Dynamic graph CNN for learning on point clouds. ACM TOG **38**(5), 146 (2019)
26. Zhan, K., Chang, X., Guan, J., Chen, L., Ma, Z., Yang, Y.: Adaptive structure discovery for multimedia analysis using multiple features. IEEE Trans. Cybern. **49**(5), 1826–1834 (2019)

Generative Adversarial Networks

BR-GAN: Bilateral Residual Generating Adversarial Network for Mammogram Classification

Chu-ran Wang[1,2,5], Fandong Zhang[1(✉)], Yizhou Yu[5], and Yizhou Wang[2,3,4]

[1] Center for Data Science, Peking University, Beijing, China
fd.zhang@pku.edu.cn
[2] Advanced Institute of Information Technology, Peking University, Hangzhou, China
[3] Center on Frontiers of Computing Studies, Peking University, Beijing, China
[4] Department of Computer Science, Peking University, Beijing, China
[5] Deepwise AI Lab, Beijing, China

Abstract. Mammogram malignancy classification with only image-level annotations is challenging due to a lack of lesion annotations. If we can generate the healthy version of the diseased data, we can easily explore the lesion features. An intuitive idea of such generation is to use existing Cycle-GAN based methods. They achieve the healthy generation regarding healthy images as reference domain, while maintaining the original content by cycle consistency mechanism. However, healthy mammogram patterns are diverse which may lead to uncertain generations. Moreover, the back translation from healthy to the original remains an ill-posed problem due to lack of lesion information. To address these problems, we propose a novel model called bilateral residual generating adversarial network(BR-GAN). We use the Cycle-GAN as a basic framework while regarding the contralateral as generation reference based on the bilateral symmetry prior. To address the ill-posed back translation problem, we propose a residual-preserved mechanism to try to preserve the lesion features from the original features. The generated features and the original features are aggregated for further classification. BR-GAN outperforms current state-of-the-art methods on INBreast and in-house datasets.

Keywords: Mammogram classification · Domain knowledge · Cycle consistency mechanism

1 Introduction

Breast cancer is the most commonly diagnosed cancer among women [15]. Mammography is a common examination for early breast cancer diagnosis. The mammogram malignancy classification is crucial. Most existing methods require extra annotations, such as bounding boxes for detection [1,8,12,16,17] and mask ground truth for segmentation [7]. However, the above extra annotations require

---

C. Wang—This work was done when Chu-ran Wang was an intern at Deepwise AI Lab.

ⓒ Springer Nature Switzerland AG 2020
A. L. Martel et al. (Eds.): MICCAI 2020, LNCS 12262, pp. 657–666, 2020.
https://doi.org/10.1007/978-3-030-59713-9_63

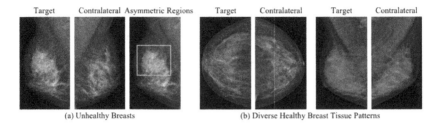

(a) Unhealthy Breasts   (b) Diverse Healthy Breast Tissue Patterns

Fig. 1. Cases to show that how the unhealthy breasts look asymmetrical, while healthy breasts are roughly symmetrical

expert domain knowledge, which is costly to obtain. Therefore, mammogram malignancy classification with the only image-level labels as the supervision is of vital significance clinically.

Exploring lesion features from a full mammogram image is the key to solve the problem. However, lesion exploration is very challenging since the lesions can be expressed as diverse appearances and the high-intensity breast tissues may partially obscure the lesions. Previous researches mainly use attention mechanism for abnormal exploration, e.g., Zhu et al. [20] and Fukui et al. [2]. However, the lack of using mammogram domain knowledge limits their performances.

Learning a healthy generation could be an effective way to exploit domain structure prior. Given a diseased image, if we know how its healthy version behaves we can localize the abnormal regions easily by the difference between the original and its healthy version. Thus such prior provides a more direct and credible way to localize abnormalities from a full mammogram. AnoGAN [13] applies such thought to anomaly detection. However, training with only healthy images restricts its effectiveness in our application. Fixed-Point GAN [14] and CycleGAN [19] can be used for the healthy generation based on the cycle consistency mechanism. An intuitive idea of applying to our application is to regard unhealthy images as a domain and healthy images as another domain. However, such approaches have two major limitations. First, we need to know which images are healthy. Moreover, healthy patterns in mammograms can be various and even similar to the lesions in some cases as shown in Fig. 1. We want to generate a healthy image that maintains all the healthy contents of the original. Regarding healthy images as generation reference may lead to a diverse generation and may conflict with our goal. Second, the cycle consistency mechanism assumes that the translated data can be translated back to the original data [5,10] and leads to the preservation of the original features, e.g., large objects and textures. However, lesions in our application can appear anywhere and have diverse appearances. It translates the healthy domain back to the original domain an ill-posed task. Thus such methods will result in undesirable lesion removal in our application.

To address the first problem, we directly regard the contralateral as generation reference by making use of the mammogram bilateral symmetry prior. To be clear, we call the image to be classified as the target, while the image of the opposite side as the contralateral. Bilateral breasts from the same person have

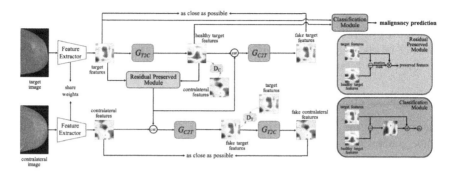

Fig. 2. The schematic overview of BR-GAN. Framework is based on CycleGAN [19] but uses contralateral features as references and adds preserved features calculated in *Residual-Preserved Module* to the back translation network. The generated features are then fed into a *Classification Module* with target features. Finally the *Classification Module* outputs labels of benign/malignant.

a roughly symmetrical glandular pattern. Most lesions only appear on one side and are invisible in the symmetrical regions of the opposite side. Therefore, the contralateral can be an effective reference for generating the healthy version of the target. Besides, a standard mammogram contains images from both sides. Thus, no extra data is required. To tackle the second problem, *i.e.*, ill-posed back translation, we try to preserve the suspicious lesion information while translating from the target data to its healthy version, and plus it when translating back. Thus, the information fed to the back translation is supposed to be sufficient.

In this paper, we propose a novel model named Bilateral Residual Generating Adversarial Network(BR-GAN) to improve mammogram malignancy classification by making use of bilateral prior and healthy generation mechanism. First, we propose a bilateral-cycle mechanism. We use contralateral images as references instead of healthy images. Due to the bilateral misalignment problem, we perform the generation in feature-level. Second, we propose a residual-preserved mechanism for better preserving lesion information during translation. While generating healthy features, we preserve the target-healthy residual features with the attention mechanism. We constrain the preserved features and the target features to share the same malignancy prediction by a residual embedding loss. In the healthy/contralateral-target translation, the preserved features are also fed into the translation network. Finally, we aggregate the generated features with the target features for further classification. Experimental results on both the public dataset and the in-house dataset demonstrate the proposed BR-GAN achieves state-of-the-art performance.

2 Bilateral Residual Generating Adversarial Network

Figure 2 outlines the overall network architecture of our framework. We first use contralateral features as references and generate the healthy version of the target features (Sect. 2.1). Then we feed both the target features and the residual between the generated features and the target features into *Classification Module* to predict labels (Sect. 2.2). We design our model in feature-level instead of pixel-level due to the bilateral misalignment.

2.1   Feature Generation

Generated features are the healthy version of the target features. To achieve feature generation, we propose a bilateral-cycle mechanism based on a Cycle-GAN framework. Due to the limitation of the cycle mechanism, we design a residual-preserved mechanism to provide lesion information for translation from healthy to unhealthy.

Bilateral-Cycle Mechanism. The GAN loss can be defined as Eq. 1.

$$\min_{G} \max_{D} \mathcal{L}_G(G, D, f_{target}, f_{reference}) := \log\left(D\left(f_{reference}\right)\right)$$
$$+ \log\left(1 - D\left(G(f_{target})\right)\right), \tag{1}$$

where $f_{reference}$ is defined as the features of the reference used in discriminator D and f_{target} is defined as the features of the target image which needs to be classified.

Most paired healthy breasts are roughly symmetrical and the abnormalities are rarely symmetrical. Thus, contralateral features are appropriate references for the healthy generation. We use contralateral features as references in our basic Cycle-GAN framework. Generator G_{T2C} tries to generate healthy features f_H^T that look similar to the contralateral features f^C from the target features f^T, while D_C aims to distinguish between translated features f_H^T and real features f^C. Generator G_{T2C} is optimized by $\min_{G_{T2C}} \max_{D_C} L_{G_{T2C}}(G_{T2C}, D_C, f^T, f^C)$, and $f_H^T = G_{T2C}(f^T)$. While generator G_{C2T} tries to translate the generated healthy features f_H^T back to the target features f^T and help f_H^T maintain the target features in lesion-free areas.

However, lesions in our application can appear anywhere and have multiple shapes. Due to the limitation of the cycle consistency mechanism mentioned in Sect. 1, the generated healthy features can not provide lesion information for back translation. Thus it will be an ill-posed problem if we feed the generated features f_H^T into the generator G_{C2T} directly. We propose a residual preserved mechanism to tackle the problem.

Residual Preserved Mechanism. While we translate the target features to its healthy version, we separate the suspicious lesion features *i.e.,* the preserved features f_P^T from the target features in *Residual Preserved Module*. The preserved features f_P^T are used as the guidance to indicate the predicted lesion information. Thus the preserved features f_P^T should contain the texture and space information of the lesions. Concatenation of the preserved features f_P^T and the generated features f_H^T will be inputs to the generator G_{C2T}, *i.e.* $G_{C2T}([f_H^T, f_P^T])$. The preserved features f_P^T will provide lesion features for a back translation and avoid the ill-posed problem.

To calculate the preserved features, we first calculate the residual between the target features and the generated features. Second, to avoid the back translation network G_{C2T} being a direct identity mapping, we do not use residual as our preserved features directly. We turn the residual features into an attention map by softmax function for normalization. If the generated features learn to be the healthy version of the target features, the locations on residual features with high values should indicate high abnormal probabilities. Third, we multiply the attention map and the target features. Finally, we get the preserved features f_P^T which is defined as:

$$f_P^T = f^T * softmax(f^T - f_H^T) \tag{2}$$

To further constrain the success of separation, we define a residual embedding loss Eq. 3 and constrain the preserved features and the target features to share the same malignancy prediction. We use the malignancy classifier to predict the malignant probabilities $p_m(\cdot)$ of the target features f^T and the preserved features f_P^T.

$$\mathcal{L}_{RE} = -p_m(f^T) * log(p_m(f_P^T)) - (1 - p_m(f^T)) * log(1 - p_m(f_P^T)). \tag{3}$$

We design the residual cycle consistency loss L_c^T measured by selected mean square error(MSE) to achieve $f^T \to G_{T2C}(f^T) \to G_{C2T}([G_{T2C}(f^T), f_P^T]) \approx f^T$.

However, with only one residual cycle consistency constrain may lead to a collapsed identical mapping from contralateral features. To avoid this problem, we design another cycle consistency loss L_c^C. L_c^C also is measured by MSE and achieves *i.e.*, $f^C \to G_{C2T}([f^C, f_P^T]) \to G_{T2C}(G_{C2T}([f^C, f_P^T])) \approx f^C$. And the generator G_{C2T} is optimized by $\min_{G_{C2T}} \max_{D_T} L_{G_{C2T}}(G_{C2T}, D_T, f^C, f^T)$, while D_T aims to distinguish between translated features $G_{C2T}([f^C, f_P^T])$ and real target features f^T.

2.2   Classification

From the feature generation procedure, we obtain the healthy features f_H^T of the target image x^T. The healthy features f_H^T and the target features f^T are fed into *Classification Module* for final classification. In the module, we first calculate the residual between the generated features f_H^T and the original target features f^T. Then concatenation of the residual and the original target features f^T which contain global semantic information is used to predict benign-malignant labels. We use the cross-entropy loss as loss function $\mathcal{L}_{CLS}$ for mammogram classification.

During training, we optimize both feature generation and classification modules jointly as in Eq. 4.

$$\mathcal{L} = \mathcal{L}_{RE} + \mathcal{L}_{CLS} + \min_{G_{T2C}} \max_{D_C} \mathcal{L}_{G_{T2C}}(G_{T2C}, D_C, f^T, f^C)$$
$$+ \min_{G_{C2T}} \max_{D_T} \mathcal{L}_{G_{C2T}}(G_{C2T}, D_T, f^C, f^T) + \mathcal{L}_c^T + \mathcal{L}_c^C \tag{4}$$

3   Experiments

3.1   Experimental Settings

Datasets. We evaluate BR-GAN on a public INBreast dataset [9] and an in-house dataset. INBreast [9] has 115 cases and 410 mammograms and provides each image a BI-RADS result as image-wise ground truth. We use the same process as Zhu *et al.* [20] (malignant if BI-RADS > 3; benign otherwise). For a fair comparison, our settings are all the same as Zhu *et al.* [20] for mass classification on the INBreast [9]. However, we discard 9 images for the lack of contralateral images and the remainings all have contralateral images. In addition, we also attempt mixed-lesion classification including mass, calcification cluster and distortion for the purpose of generalization. The in-house dataset contains 1303 images with malignancy annotations, including 589 only masses,120 only suspicious calcifications,34 only architectural distortions, 197 only asymmetries and 363 multiple lesions from 642 patients. All these 1303 images have opposite sides, i.e. 1303 pairs. We randomly divide the dataset into training, validation and testing sets as 8:1:1 in patient-wise.

Implementation Details. We use Otsus method [11] to segment the breast regions and remove backgrounds from the original images in 14-bit DICOM format. We implement all models with PyTorch and use Adam optimization. Both target and contralateral features are extracted from the last convolution layer. We use Area Under the Curve (AUC) as evaluation metrics in image-wise.

3.2   Performances

Mass Classification. The first four lines in Table 1 summarize the results of the representative methods. To be fair, we compare the results with the backbone of AlexNet [6] and ResNet50 [4] separately. Due to the slight difference of images caused by reference absence, for a fair comparison, we re-implement some representative methods for mammogram classification [20], natural image classification [2,18] and healthy generation [13,14,19] by adjusting the source codes given by the authors. We marked these methods by '*' in the table.

Mix-Lesion Classification. The performances are shown in the last two columns of Table 1 for the INBreast dataset and the in-house dataset.

Results. Attention mechanism (Zhu [20], CAM [18], ABN [2]) works but limits by the lack of mammogram domain knowledge. Only using healthy data for training highly (AnoGAN [13]) relies on the number of healthy data and is limited by the lack of reference to unhealthy data. The cycle consistency mechanism (Fixed-Point GAN [14], CycleGAN [19]) is effective to some extent but is limited by its ill-posed back translation problem in our application. However, our BR-GAN outperforms the representative methods significantly on both datasets.

To further evaluate the effectiveness of the generated features(healthy version of the target features), we calculate the mean FID [3] to measure the average of features distribution distances in the INBreast dataset. The mean FID between

Table 1. AUC evaluation on (a) INBreast for mass classification with Alexnet; (b) INBreast for mass classification with Resnet50; (c) INBreast for mixed-lesion classification with Resnet50; (d) in-house dataset for mixed-lesion classification with Alexnet.

Methodology	AUC (a)	AUC (b)	AUC (c)	AUC (d)
Pretrained CNN [1]	0.690	–	–	–
Pretrained CNN+RF [1]	0.760	–	–	–
Vanilla AlexNet, Zhu et al. [20]	0.790	–	–	–
Zhu et al. [20]	0.890	–	–	–
Vanilla*	0.820	0.827	0.780	0.697
AnoGAN [13]*	0.803	0.796	0.774	0.720
Fixed-Point GAN [14]*	0.835	0.837	0.805	0.734
CycleGAN [19]*	0.852	0.838	0.808	0.741
Zhu et al. [20]*	0.860	0.862	0.830	0.720
Vanilla*+GAP [18]*	0.857	0.827	0.780	0.718
Vanilla*+ABN [2]*	0.858	0.846	0.814	0.723
Proposed Method	**0.900**	**0.886**	**0.860**	**0.770**

Table 2. Top-1 localization error on (b) INBreast dataset for mass classification with Resnet50; (d) INBreast dataset for mixed-lesion classification with Resnet50.

Methodology	Top-1 error (b)	Top-1 error (d)
ResNet50 [4]	0.635	0.727
AnoGAN [13]*	0.684	0.789
Fixed-Point GAN [14]*	0.646	0.737
CycleGAN [19]*	0.632	0.667
ABN [2]	0.632	0.722
Zhu et al. [20]*	0.627	0.625
Proposed Method	**0.519**	**0.544**

the target and contralateral features is 63.63. The generated-contralateral mean FID is 27.54. The target-generated mean FID is 22.81 while the one after removing the lesion areas from ground truth is 0.73. Through the above comparison, we can find the generated features containing both contralateral distribution and target information in healthy areas as we want.

Localization. To verify whether the proposed model focuses on the lesion areas, we evaluate the localization error by CAM [18]. We use the top-1 localization error as ILSVRC using an inter-over-union (IOU) threshold of 0.1. As is shown in Table 2, BR-GAN largely outperforms the representative methods.

Furthermore, Fig. 3 visualizes the class activation maps of some cases. As we can see, all lesions satisfy the bilateral asymmetry prior. The proposed

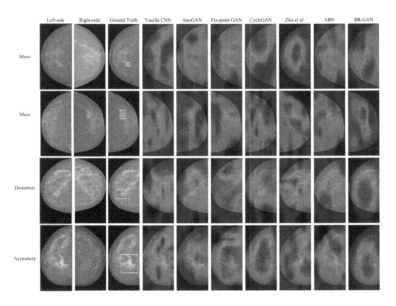

Fig. 3. Visualization of class activation maps of Vanilla CNN, AnoGAN [13], Fixed-Point GAN [14], CycleGAN [19], Zhu *et al.* [20], ABN [2] and our BR-GAN. The target containing lesions is bounded by a red rectangle. The ground truth bounding boxes are labeled by green rectangles in the third column. (Color figure online)

BR-GAN succeeds to focus on all lesions since it incorporates the bilateral asymmetry prior and modifies the cycle mechanism. The other methods show uneven results without considering bilateral information.

3.3 Ablation Experiments

To verify the effectiveness of each component, we evaluate some variant models and show results in Table 3. Here are some interpretation for the variants:

SBF: Simple Bilateral Features. The bilateral features are combined and fed into the fusion layer directly;

Single: Only use the consistency loss L_c^T;

Double: Use both consistency losses L_c^T and L_c^C;

Mask: Whether use attention mask in *Residual Preserved Module*.

Note that bilateral breasts exist misalignment, using SBF to classify is not robust enough. As shown in the above tables, the bilateral cycle mechanism, the double consistency losses, the residual preserved mechanism and attention mask for preserved features are all proved to be effective.

Table 3. Ablation experiments on (a) INBreast dataset for mass classification with AlexNet; (b) INBreast dataset for mass classification with ResNet50; (c) INBreast dataset for mixed-lesion classification with ResNet50; (d) in-house dataset for mixed-lesion classification with AlexNet.

Bilateral	L_C	L_{RE}	Mask	AUC (a)	AUC (b)	AUC (c)	AUC (d)
×	×	×	×	0.820	0.827	0.780	0.697
SBF	×	×	×	0.862	0.858	0.807	0.721
GAN	×	×	×	0.883	0.873	0.857	0.731
GAN	Single	×	✓	0.861	0.859	0.826	0.727
GAN	Double	×	✓	0.886	0.864	0.846	0.767
GAN	Double	✓	×	0.889	0.857	0.846	0.761
GAN	Double	✓	✓	**0.900**	**0.886**	**0.860**	**0.770**

4 Conclusions

In this paper, we present a novel approach called bilateral residual generating adversarial network (BR-GAN) to improve the mammogram classification performance. The approach proposes a novel way to generate the healthy version of target features to help find the abnormal features. Thus, BR-GAN enhances the interpretability of results for clinical diagnosis. Experimental results indicate that the proposed BR-GAN achieves the state-of-the-art in both the public and the in-house dataset.

Acknowledgement. This work was supported by MOST-2018AAA0102004, NSFC-61625201 and ZheJiang Province Key Research & Development Program (No. 2020C03073).

References

1. Dhungel, N., Carneiro, G., Bradley, A.P.: The automated learning of deep features for breast mass classification from mammograms. In: Ourselin, S., Joskowicz, L., Sabuncu, M.R., Unal, G., Wells, W. (eds.) MICCAI 2016. LNCS, vol. 9901, pp. 106–114. Springer, Cham (2016). https://doi.org/10.1007/978-3-319-46723-8_13
2. Fukui, H., Hirakawa, T., Yamashita, T., Fujiyoshi, H.: Attention branch network: learning of attention mechanism for visual explanation. In: Proceedings of the IEEE Conference on Computer Vision and Pattern Recognition, pp. 10705–10714 (2019)
3. Haarburger, C., et al.: Multiparametric magnetic resonance image synthesis using generative adversarial networks. In: VCBM (2019)
4. He, K., Zhang, X., Ren, S., Sun, J.: Deep residual learning for image recognition. In: Proceedings of the IEEE Conference on Computer Vision and Pattern Recognition, pp. 770–778 (2016)
5. Hu, X., Jiang, Y., Fu, C.W., Heng, P.A.: Mask-ShadowGAN: learning to remove shadows from unpaired data. In: Proceedings of the IEEE International Conference on Computer Vision, pp. 2472–2481 (2019)

6. Krizhevsky, A., Sutskever, I., Hinton, G.E.: ImageNet classification with deep convolutional neural networks. In: Pereira, F., Burges, C.J.C., Bottou, L., Weinberger, K.Q. (eds.) Advances in Neural Information Processing Systems, pp. 1097–1105 (2012)

7. Li, H., Chen, D., Nailon, W.H., Davies, M.E., Laurenson, D.: A deep dual-path network for improved mammogram image processing. In: ICASSP 2019-2019 IEEE International Conference on Acoustics, Speech and Signal Processing (ICASSP), pp. 1224–1228. IEEE (2019). https://doi.org/10.1109/ICASSP.2019.8682496

8. Lotter, W., Sorensen, G., Cox, D.: A multi-scale CNN and curriculum learning strategy for mammogram classification. In: Carneiro, G., et al. (eds.) DLMIA/ML-CDS -2017. LNCS, vol. 10553, pp. 169–177. Springer, Cham (2017). https://doi.org/10.1007/978-3-319-67558-9_20

9. Moreira, I.C., Amaral, I., Domingues, I., Cardoso, A., Cardoso, M.J., Cardoso, J.S.: Inbreast: toward a full-field digital mammographic database. Acad. Radiol. **19**(2), 236–248 (2012). https://doi.org/10.1016/j.acra.2011.09.014

10. Nizan, O., Tal, A.: Breaking the cycle-colleagues are all you need. arXiv preprint arXiv:1911.10538 (2019)

11. Otsu, N.: A threshold selection method from gray-level histograms. IEEE Trans. Syst. Man Cybern. **9**(1), 62–66 (1979). https://doi.org/10.1109/TSMC.1979.4310076

12. Ribli, D., Horváth, A., Unger, Z., Pollner, P., Csabai, I.: Detecting and classifying lesions in mammograms with deep learning. Sci. Rep. **8**(1), 4165 (2018). https://doi.org/10.1038/s41598-018-22437-z

13. Schlegl, T., Seeböck, P., Waldstein, S.M., Schmidt-Erfurth, U., Langs, G.: Unsupervised anomaly detection with generative adversarial networks to guide marker discovery. In: Niethammer, M., et al. (eds.) IPMI 2017. LNCS, vol. 10265, pp. 146–157. Springer, Cham (2017). https://doi.org/10.1007/978-3-319-59050-9_12

14. Siddiquee, M.M.R., et al.: Learning fixed points in generative adversarial networks: from image-to-image translation to disease detection and localization. In: Proceedings of the IEEE International Conference on Computer Vision, pp. 191–200 (2019)

15. Siegel, R.L., Miller, K.D., Jemal, A.: Cancer statistics. CA Cancer J. Clin. **69**(1), 7–34 (2019). https://doi.org/10.3322/caac.21551

16. Tai, S.C., Chen, Z.S., Tsai, W.T.: An automatic mass detection system in mammograms based on complex texture features. IEEE J. Biomed. Health Inf. **18**(2), 618–627 (2013). https://doi.org/10.1109/JBHI.2013.2279097

17. Wu, E., Wu, K., Cox, D., Lotter, W.: Conditional infilling GANs for data augmentation in mammogram classification. In: Stoyanov, D., et al. (eds.) RAMBO/BIA/TIA -2018. LNCS, vol. 11040, pp. 98–106. Springer, Cham (2018). https://doi.org/10.1007/978-3-030-00946-5_11

18. Zhou, B., Khosla, A., Lapedriza, A., Oliva, A., Torralba, A.: Learning deep features for discriminative localization. In: Proceedings of the IEEE Conference on Computer Vision and Pattern Recognition, pp. 2921–2929 (2016)

19. Zhu, J.Y., Park, T., Isola, P., Efros, A.A.: Unpaired image-to-image translation using cycle-consistent adversarial networks. In: Proceedings of the IEEE International Conference on Computer Vision, pp. 2223–2232 (2017)

20. Zhu, W., Lou, Q., Vang, Y.S., Xie, X.: Deep multi-instance networks with sparse label assignment for whole mammogram classification. In: Descoteaux, M., Maier-Hein, L., Franz, A., Jannin, P., Collins, D.L., Duchesne, S. (eds.) MICCAI 2017. LNCS, vol. 10435, pp. 603–611. Springer, Cham (2017). https://doi.org/10.1007/978-3-319-66179-7_69

Cycle Structure and Illumination Constrained GAN for Medical Image Enhancement

Yuhui Ma[1], Yonghuai Liu[2], Jun Cheng[1], Yalin Zheng[3], Morteza Ghahremani[4], Honghan Chen[1], Jiang Liu[5(✉)], and Yitian Zhao[1(✉)]

[1] Cixi Institute of Biomedical Engineering, Ningbo Institute of Materials Technology and Engineering, Chinese Academy of Sciences, Ningbo, China
yitian.zhao@nimte.ac.cn

[2] Department of Computer Science, Edge Hill University, Ormskirk, UK

[3] Department of Eye and Vision Science, University of Liverpool, Liverpool, UK

[4] Department of Computer Science, Aberystwyth University, Aberystwyth, UK

[5] Department of Computer Science and Engineering, Southern University of Science and Technology, Shenzhen, China
liuj@sustech.edu.cn

Abstract. The non-uniform illumination or imbalanced intensity in medical images brings challenges for automated screening, examination and diagnosis of diseases. Previously, CycleGAN was proposed to transform input images into enhanced ones without paired images. However, it did not consider many local details of the structures, which are essential for medical images. In this paper, we propose a Cycle Structure and Illumination constrained GAN (CSI-GAN), for medical image enhancement. Inspired by CycleGAN based on the global constraints of the adversarial loss and cycle consistency, the proposed CSI-GAN treats low and high quality images as those in two domains and computes local structure and illumination constraints for learning both overall characteristics and local details. To evaluate the effectiveness of CSI-GAN, we have conducted experiments over two medical image datasets: corneal confocal microscopy (CCM) and endoscopic images. The experimental results show that our method yields better performance than both conventional methods and other deep learning based methods. As a complementary output, we will release the CCM dataset to the public in the future.

Keywords: Illumination regularization · Structural loss · CycleGAN · Medical image enhancement

1 Introduction

High-quality images with adequate contrast and details are crucial for many medical imaging applications: e.g., segmentation [1] and computer-aided diagnosis [2,3]. However, medical images acquired using the same or different sensors

© Springer Nature Switzerland AG 2020
A. L. Martel et al. (Eds.): MICCAI 2020, LNCS 12262, pp. 667–677, 2020.
https://doi.org/10.1007/978-3-030-59713-9_64

usually have a large variation in quality - intensity inhomogeneity, noticeable blur and poor contrast, that are often inherited from the image acquisition process.

Figure 1(a) and (c) illustrate two examples of low-quality images captured by confocal microscopy and endoscopy - where it is difficult to observe the complete structure of corneal nerve fibers and digestive tract respectively due to the imperfect focus and poor light condition. These obstacles pose significant challenges to many subsequent image analysis tasks, such as curvilinear structure segmentation [4] and lesion detection [5]. As a consequence, fully automated and reliable image enhancement approaches have long been deemed desirable.

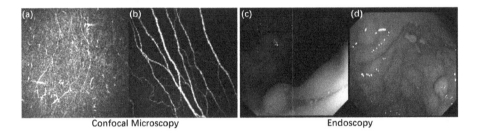

Fig. 1. Examples of low-/high-quality images captured by the same equipment. (a) and (b) corneal confocal microscopic images. (c) and (d) endoscopic images.

Many image enhancement methods have been proposed such as histogram equalization (HE) [6], dark channel prior [7], filtering-based methods [8,9] and Retinex-based methods [10,11]. However, these conventional methods usually enhance images uniformly, irrespective of whether a given region is in the foreground or background [12,13]. Deep learning has provided new avenues for image enhancement, nevertheless, a large portion of these methods [14–16] require aligned image pairs for training. In consequence, recently a few unpaired learning models have been proposed. Gatys et al. [17] proposed a neural transfer algorithm (NST) for unpaired image transformation. Zhang et al. [18] introduced a multi-style generative network (MSG-Net) to achieve real-time image style translation. Jiang et al. [19] proposed a EnlightenGAN with a global-local discriminator structure, a self-regularized perceptual loss fusion, and attention mechanism for low-light image enhancement. Nevertheless, one common limitation of these unpaired learning methods is that they often amplify noise in the dark background area and suffer from halo artifacts.

Cycle-consistent generative adversarial network (CycleGAN) [20] has an advantage of learning knowledge represented with typical images in one domain and transferring it to the other domain without paired images. However, Cycle-GAN mainly exploits global constraints on appearance and cycle-consistency, which is weak in learning local details. To address the weakness, two novel constraints including an illumination regularization and a structure loss are proposed in our new method, which we refer it to CSI-GAN for medical image enhancement. In our work, low- and high-quality images are treated as those in two different domains and high-quality images can be easily identified by clinicians, as shown in Fig. 1(b) and (d). The main contributions of this paper are summarized as follows. (1) A novel CSI-GAN is proposed to improve low-quality medical images with better illumination conditions while well-preserving structure details. (2) The proposed method has undergone rigorous quantitative and qualitative evaluation using corneal confocal microscopy and endoscopic images in an unified manner. (3) As a complementary output, we will release the CCM dataset (both poor and good quality image sets) online available to the public in the future.

2   Proposed Method

2.1   Overall Architecture

Different from conditional adversarial network (cGAN) [21], CycleGAN [20] learns a suitable translation function between the source domain A and the target domain B without the paired images in the training. In this paper, we assume that A and B are image domains with low and high quality images, respectively. CycleGAN adopts two generator/discriminator pairs ($G_{A \to B}/D_B$, $G_{B \to A}/D_A$), where $G_{A \to B}$ ($G_{B \to A}$) learns to translate an image from domain A (B) into domain B (A), and D_A (D_B) is trained to distinguish between real samples from domain A (B) and the translated images from domain B (A).

In order to prevent two generators from contradicting each other, the whole framework contains both forward and backward cycle consistency, as shown in Fig. 2. Each $a \in A$ is expected to be reconstructed as much as possible in forward cycle, which is represented as $a \to G_{A \to B}(a) \to G_{B \to A}(G_{A \to B}(a)) \approx a$. This holds for backward cycle as well: $b \to G_{B \to A}(b) \to G_{A \to B}(G_{B \to A}(b)) \approx b$. In addition, two generators are regularized as an identity mapping separately when real samples from A (B) are applied to $G_{B \to A}$ ($G_{A \to B}$), i.e., $G_{B \to A}(a) \approx a$ and $G_{A \to B}(b) \approx b$. The objective function of CycleGAN is defined as:

$$L(G_{A \to B}, G_{B \to A}, D_A, D_B)$$
$$= L_{GAN}(G_{A \to B}, D_B, A, B) + L_{GAN}(G_{B \to A}, D_A, B, A) \qquad (1)$$
$$+ \lambda_1 L_{cyc}(G_{A \to B}, G_{B \to A}, A, B) + \lambda_2 L_{identity}(G_{A \to B}, G_{B \to A}, A, B),$$

where L_{GAN} denotes the adversarial loss; L_{cyc} and $L_{identity}$ represent the cycle consistency term and the identity mapping loss with weighted coefficients λ_1 and λ_2 respectively.

In training, the generators try to minimize the objective function against the discriminators that try to maximize it, which can be formulated as:

$$G^*_{A \to B}, G^*_{B \to A} = \arg \min_{G_{A \to B}, G_{B \to A}} \max_{D_A, D_B} L(G_{A \to B}, G_{B \to A}, D_A, D_B). \qquad (2)$$

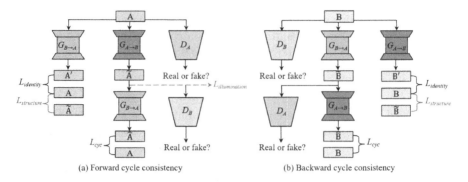

(a) Forward cycle consistency   (b) Backward cycle consistency

Fig. 2. CSI-GAN structure diagram. It comprises of two generator/discriminator pairs $(G_{A \to B}/D_B,\ G_{B \to A}/D_A)$ and two types of cycle consistency (forward/backward cycle consistency). A and B refer to low-quality and high-quality image domain, respectively. L_{cyc} and $L_{identity}$ represent the cycle consistency term and the identity mapping loss. The proposed illumination regularization and structure loss are represented as $L_{illumination}$ and $L_{structure}$, respectively.

2.2 Objective Function

Although achieving somewhat success in inter-domain image translation, Cycle-GAN has two obvious shortcomings when applied to medical images: (i) Cycle-GAN often produces unstable results due to GAN's characteristic of high freedom. In other words, the existing CycleGAN architecture lacks adequate supervision based only on the global adversarial loss and cycle consistency constraints; and (ii) For medical image enhancement, it is difficult to make sure that $G_{A \to B}$ and $G_{B \to A}$ focus on important features without extra constraints provided. On one hand, it is a challenge to remove too dark or bright regions and achieve more uniform appearance. On the other hand, subtle vital details such as curvilinear structures of corneal nerve fibers and complete morphology of digestive tract might be blurred or even lost in the translated images. To address these shortcomings, we propose to formulate and incorporate two new terms - illumination regularization and structure loss (as shown in red in Fig. 2), to guide the generator to improve illumination uniformity and structural details in the enhanced images.

• **Illumination Regularization** The proposed illumination regularization is a constraint aimed at improving overall illumination. It represents a correcting factor that reflects the non-uniformity of illumination in the enhanced image. The illumination correcting factor of the given image I is based on the following steps: (1) Calculate the average intensity of I; (2) Divide the image into $n \times m$ patches with a certain size, then calculate the average intensity of each patch and form the luminance matrix D; (3) Subtract the average intensity of I from each element of matrix D to obtain the luminance difference matrix E; (4) Resize the matrix E into the luminance distribution matrix R of the same size as I using bicubic interpolation; and (5) Calculate the average absolute value of elements in the matrix R. The illumination uniformity constraint is defined as:

$$L_{\text{illumination}}(G_{A \rightarrow B}) = E_{a \in A}\left[E_{global}\left[|upsampling\left\{E_{local}^{p \times p}[G_{A \rightarrow B}(a)]\right.\right.\right. \\ \left.\left.\left. - E_{global}[G_{A \rightarrow B}(a)]\right\}|\right]\right], \tag{3}$$

where $E_{global}[\cdot]$ denotes the global mean of the whole input image; $E_{local}^{p \times p}[\cdot]$ is intended to calculate the luminance matrix D based on each $p \times p$ patch divided in the input image; and $upsampling\{\cdot\}$ aims at rescaling the input image to the size of the original image with bicubic interpolation.

• **Structure Loss** In addition to the illumination constraint, the low-quality image and the corresponding high-quality image should share similar structural features despite vast differences in terms of intensity and contrast distribution. Structural similarity (SSIM) [22] provides relatively reasonable measurement for this task. Compared with mean squared error (MSE), SSIM effectively characterizes the similarity of image structures from three aspects: luminance, contrast and structure. Inspired from the structure comparison function in SSIM, we propose a structure loss based on the dissimilarity between the low-quality image and the corresponding high-quality image. Mathematically, it is formulated as:

$$L_{\text{structure}}(G, X) = E_{x \in X}\left[1 - \frac{1}{M}\sum_{j=1}^{M}\frac{\sigma_{x_j, G(x)_j} + c}{\sigma_{x_j}\sigma_{G(x)_j} + c}\right], \tag{4}$$

where x_j and $G(x)_j$ are the j-th local window in the image x and the corresponding generated image $G(x)$ respectively; M is the number of local windows in each of the two images; $\sigma_{x_j, G(x)_j}$ is the covariance of x_j and $G(x)_j$; σ_{x_j} and $\sigma_{G(x)_j}$ are the standard deviations of x_j and $G(x)_j$ respectively; and c is a small positive constant. Thus the objective function of the proposed CSI-GAN for image enhancement is computed as:

$$L(G_{A \rightarrow B}, G_{B \rightarrow A}, D_A, D_B) \\ = L_{GAN}(G_{A \rightarrow B}, D_B, A, B) + L_{GAN}(G_{B \rightarrow A}, D_A, B, A) \\ + \lambda_1 L_{cyc}(G_{A \rightarrow B}, G_{B \rightarrow A}, A, B) + \lambda_2 L_{identity}(G_{A \rightarrow B}, G_{B \rightarrow A}, A, B) \\ + \gamma L_{\text{illumination}}(G_{A \rightarrow B}) + \alpha L_{\text{structure}}(G_{A \rightarrow B}, A) + \beta L_{\text{structure}}(G_{B \rightarrow A}, B), \tag{5}$$

where γ, α and β are the parameters to control the weights of each part.

3   Experimental Results

To evaluate the performance of the proposed method, we applied it on two types of medical images, including the corneal confocal microscopy and the endoscopy. A clinical expert was invited to select low and high quality images of each dataset, primarily according to illumination uniformity, contrast and edge sharpness.

3.1   Experimental Settings

The proposed CSI-GAN was implemented in Python with PyTorch library. The experiments were carried out on a single NVIDIA GPU (GeForce GTX 1080,

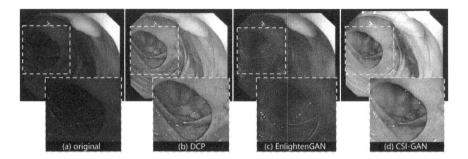

Fig. 3. Example result of different methods in endoscopic image enhancement.

8 GB). All training images were resized to 384×384. In our experiments, we selected two kinds of patch size - 48×48 and 96×96 to calculate illumination regularization and set local windows of 11×11 to calculate the structure loss. The Adam optimization with initial learning rate of 0.0002 and batch size of 1 were applied to the two adversarial pairs. The weighted parameters in the final loss function were set as: $\lambda_1 = 10$, $\lambda_2 = 5$, $\gamma = 1$, $\alpha = \beta = 5$.

3.2   Evaluation on Endoscopic Images

A total of 397 low-quality and 123 high-quality endoscopic images with the resolution of 384×288 pixels were selected. In our experiments, the low-quality images were randomly and equally divided into training and testing sets.

Five state-of-the-art approaches were selected for comparison: CLAHE [23], DCP [7], NST [17], MSG-Net [15], and EnlightenGAN [19]. An ablation study has also been conducted to justify the effectiveness of the illumination regularization and the structure loss. Due to page limitation, Fig. 3 only illustrates the enhancement results by one conventional (DCP) and one deep learning based (EnlightenGAN) enhancement method, respectively. It can be seen that DCP improves the overall brightness of the image, but it also amplifies the noise in severely dark regions and even produces some color distortion. This might be attributable to that our endoscopic images are not compatible with the haze imaging model. The result of EnlightenGAN is universally over-smoothed with many details blurred. In contrast, the proposed CSI-GAN generates a visually pleasing result with more uniform illumination and visible structural details, especially in the region with poor lighting condition. For quantitative evaluation, we adopted three no-reference assessments: Natural Image Quality Evaluator (NIQE) [24], Blind/Referenceless Image Spatial Quality Evaluator (BRISQUE) [25] and Perception based Image Quality Evaluator (PIQE) [26]. Table 1 shows the results of the enhanced endoscopic images using different methods. Significant margins have been obtained when the proposed CSI-GAN is compared with other state-of-art methods - it achieves the highest scores in terms of NIQE and PIQE, and similar score to EnlightenGAN in BRISQUE, where ours is only 0.17 lower than

Table 1. No-reference assessment results (mean ± standard deviation) of different enhancement methods.

Methods	NIQE	BRISQUE	PIQE
Original	4.40±0.67	36.70±5.42	37.27±11.50
CLAHE [23]	4.52±0.89	30.76±3.88	29.52±4.89
DCP [7]	4.13±0.64	35.45±5.59	35.09±6.37
NST [17]	9.42±1.96	30.28±3.64	25.48±6.17
MSG-Net [18]	7.26±0.55	56.72±2.29	93.43±11.96
EnlightenGAN [19]	4.38±0.88	**24.35 ± 4.18**	33.07±6.47
CycleGAN [20]	4.38±0.62	27.81±7.70	29.54±5.28
CycleGAN+I	4.33±0.61	27.70±6.71	25.25±9.35
CycleGAN+S	4.27±0.60	26.00±6.59	28.12±6.01
CSI-GAN	**3.84 ± 0.64**	24.52±5.38	**23.48 ± 6.42**

EnlightenGAN. Overall, the proposed method has demonstrated its superiority in both visual comparison and quantitative evaluation over other competing methods.

Ablation Study: Table 1 implies that our CSI-GAN is superior when compared with the results that CycleGAN enables with illumination (CycleGAN+I) or structure loss (CycleGAN+S) alone.

3.3   Evaluations on Corneal Confocal Microscopy

We further validated our image enhancement method over a publicly available CCM dataset [4]. A clinical expert was invited and selected 340 low and 288 high quality images in this dataset, respectively, for training and 60 low-quality images for test. Note, nerve fibers in all these CCM images were manually annotated at centerline pixel level. Similarly, we also compared our method with the other five methods and conduct an ablation study.

Evaluation by SNR: in order to comprehensively measure the image quality improvement of our proposed CSI-GAN, we first calculated signal-to-noise ratio (SNR) based on manually traced nerve fibers: $SNR = 10\log_{10}\left(\max(I_s)^2/\sigma_b^2\right)$, where $\max(\cdot)$ represents the maximum intensity of signal regions (manually traced nerve fibers) in the image, and σ_b is the standard deviation of the background region. In our experiments, we defined the regions after a disk-shaped dilation operation on the manually traced fibers with a radius (r) of 3, 5 and 7 pixels, respectively as the background region. The quantitative results of different enhancement approaches are shown in Table 2. The proposed method has achieved the best performance when compared with all the competing methods. It exhibits a large advantage against the original images by an increase in SNR of about 2.88 dB, 3.45 dB and 3.76 dB for $r=3$, $r=5$ and $r=7$, respectively.

Table 2. SNR and segmentation performance of the original and enhanced CCM images using different methods. (S: structure loss; I: illumination regularization).

Methods	SNR			Segmentation			
	r=3	r=5	r=7	ACC	SEN	Kappa	Dice
Original	17.472	17.611	17.650	0.969	0.421	0.528	0.541
CLAHE [23]	16.560	16.733	16.793	0.970	0.488	0.570	0.584
DCP [7]	14.587	14.879	14.986	0.964	0.708	0.615	0.633
NST [17]	16.606	16.887	17.006	0.958	0.490	0.494	0.515
MSG-Net [18]	19.122	19.915	20.217	0.964	0.441	0.495	0.512
EnlightenGAN [19]	18.407	19.257	19.699	0.960	0.671	0.580	0.601
CycleGAN [20]	19.557	20.138	20.409	0.971	0.748	0.673	0.688
CycleGAN + I	20.297	20.932	21.217	**0.977**	0.776	0.735	0.747
CycleGAN + S	20.105	20.747	21.042	0.971	0.769	0.680	0.695
CSI-GAN	**20.352**	**21.057**	**21.413**	**0.977**	**0.788**	**0.736**	**0.748**

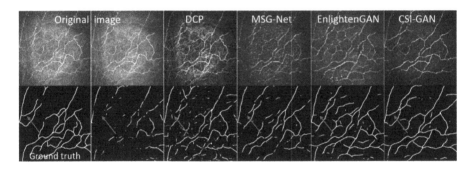

Fig. 4. An example of CCM image enhancement by different methods, and their guided nerve fiber tracing results using CS-Net. (Color figure online)

Evaluation by Nerve Fiber Segmentation: We further performed corneal nerve fiber segmentation of the enhanced images to confirm the relative benefits of the proposed method and the others. To this end, we employed a pre-trained corneal nerve fiber segmentation network, CS-Net [4], for fully automatic segmentation of corneal nerves on the low-quality images, with and without application of image enhancement methods. Then we computed *sensitivity* (SEN), *Accuracy* (ACC), Kappa score and Dice coefficient between the predicted centerlines and ground truth ones. The top row of Fig. 4 demonstrates the visual enhancement results by different methods, while its bottom row depicts the enhancement guided fiber segmentation results obtained by CS-Net. We observed that more completed fibers have been identified in our enhanced images, where indicated by the red arrows, since more uniform responses in both low and high intensity regions of the original image have been achieved by CSI-GAN. In consequence, CS-Net is able to provide relatively more sensitive segmentation on small fibers. This finding is also evidenced by the segmentation performance presented in Table 2. Our CSI-GAN method achieves the best performance and helps to improve the state-of-the-art EnlightenGAN and the baseline Cycle-GAN by 17.44% and 5.35% in SEN, 26.90% and 9.36% in Kappa, and 24.46% and 8.72% in Dice respectively, demonstrating that our method can effectively promote nerve fiber segmentation performance, especially in reducing missing rate, which is more useful for monitoring and diagnosing nerve-related diseases.

Ablation Study: Table 2 demonstrates our CSI-GAN yields significant improvements in either SNR or nerve fibre segmentation when compared with their performance using illumination regularization (CycleGAN+I) or structure loss (CycleGAN+S) alone. It further verifies that by adding illumination regularization and structure loss, our method achieves high visual image quality and well-restored structural details.

4   Conclusion

Image enhancement is helpful for improving visual quality and automatic analysis of medical images. However, it is still challenging due to different illumination conditions and diversity in quality of different medical imaging devices. This paper has proposed an unpaired learning architecture called CSI-GAN for medical image enhancement, where the low and high quality images are treated as those in two different domains. The primary advantage of this method is that it learns to migrate the features inside the high-quality images into low-quality images without paired images for training and thus has an advantage of easy implementation. Furthermore, by adding illumination regularization and structure loss, the overall illumination smoothness and well-restored structural details in the enhanced images are achieved. Compared with other traditional or deep learning-based methods, our method obtains better overall performance in enhancing images with different modalities in different metrics. In the future, we would consider further verifying our method on other medical image modalities and applying the enhanced images in the clinical settings for disease diagnosis.

Acknowledgment. This work was supported by China Postdoctoral Science Foundation (2018M640578, 2019M652156), Ningbo "2025 S & T Megaprojects" (2019B10033, 2019B10061).

References

1. Lai, M.: Deep learning for medical image segmentation. arXiv preprint arXiv:1505.02000 (2015)
2. Fu, H., et al.: Evaluation of retinal image quality assessment networks in different color-spaces. In: Shen, D., et al. (eds.) MICCAI 2019. LNCS, vol. 11764, pp. 48–56. Springer, Cham (2019). https://doi.org/10.1007/978-3-030-32239-7_6

3. Zhao, Y., et al.: Intensity and compactness enabled saliency estimation for leakage detection in diabetic and malarial retinopathy. IEEE Trans. Med. Imaging **36**(1), 51–63 (2017)

4. Mou, L., et al.: Cs-net: channel and spatial attention network for curvilinear structure segmentation. In: International Conference on Medical Image Computing and Computer-Assisted Intervention, Springer (2019)

5. Zhao, Y., et al.: Uniqueness-driven saliency analysis for automated lesion detection with applications to retinal diseases. In: Frangi, A.F., Schnabel, J.A., Davatzikos, C., Alberola-López, C., Fichtinger, G. (eds.) MICCAI 2018. LNCS, vol. 11071, pp. 109–118. Springer, Cham (2018). https://doi.org/10.1007/978-3-030-00934-2_13

6. Abdullah-Al-Wadud, M., Kabir, M., Dewan, M., Chae, O.: A dynamic histogram equalization for image contrast enhancement. IEEE Trans. Consum. Electron. **53**(2), 593–600 (2007)

7. He, K., Sun, J., Tang, X.: Single image haze removal using dark channel prior. IEEE Trans. Pattern Anal. Mach. Intell. **33**(12), 2341–2353 (2010)

8. He, K., Sun, J., Tang, X.: Guided image filtering. IEEE Trans. Pattern Anal. Mach. Intell. **35**(6), 1397–1409 (2012)

9. Zhao, Y., Zheng, Y., Liu, Y., Zhao, Y., Luo, L., Yang, S., Na, T., Yongtian, W., Liu, J.: Automatic 2d/3d vessel enhancement in multiple modality images using a weighted symmetry filter. IEEE Trans. Med. Imaging **37**(2), 438–450 (2018)

10. Jobson, D.J., Rahman, Z., Woodell, G.A.: A multiscale retinex for bridging the gap between color images and the human observation of scenes. IEEE Trans. Image Process. **6**(7), 965–976 (1997)

11. Zhao, Y., et al.: Automated tortuosity analysis of nerve fibers in corneal confocal microscopy. IEEE Transactions on Medical Imaging (2020)

12. Chen, Y.S., Wang, Y.C., Kao, M.H., Chuang, Y.Y.: Deep photo enhancer: unpaired learning for image enhancement from photographs with gans. In: Proceedings of the IEEE Conference on Computer Vision and Pattern Recognition, pp. 6306–6314 (2018)

13. Guo, X.: Lime: a method for low-light image enhancement. In: Proceedings of the 24th ACM international conference on Multimedia, pp. 87–91 (2016)

14. Lv, F., Lu, F., Wu, J., Lim, C.: Mbllen: low-light image/video enhancement using cnns. In: BMVC, pp. 220 (2018)

15. Shen, L., Yue, Z., Feng, F., Chen, Q., Liu, S., Ma, J.: Msr-net: low-light image enhancement using deep convolutional network. arXiv preprint arXiv:1711.02488 (2017)

16. Lore, K.G., Akintayo, A., Sarkar, S.: Llnet: a deep autoencoder approach to natural low-light image enhancement. Pattern Recogn. **61**, 650–662 (2017)

17. Gatys, L., Ecker, A., Bethge, M.: A neural algorithm of artistic style. J. Vis. **16**(12), 326 (2016)

18. Zhang, H., Dana, K.: Multi-style generative network for real-time transfer. arXiv preprint arXiv:1703.06953 (2017)

19. Jiang, Y., et al.: Enlightengan: deep light enhancement without paired supervision. arXiv preprint arXiv:1906.06972 (2019)

20. Zhu, J.Y., Park, T., Isola, P., Efros, A.A.: Unpaired image-to-image translation using cycle-consistent adversarial networks. In: Proceedings of the IEEE International Conference on Computer Vision, pp. 2223–2232 (2017)

21. Isola, P., Zhu, J.Y., Zhou, T., Efros, A.A.: Image-to-image translation with conditional adversarial networks. In: Proceedings of the IEEE Conference on Computer Vision and Pattern Recognition, pp. 1125–1134 (2017)

22. Wang, Z., Bovik, A.C., Sheikh, H.R., Simoncelli, E.P., et al.: Image quality assessment: from error visibility to structural similarity. IEEE Trans. Image Process. **13**(4), 600–612 (2004)
23. Zuiderveld, K.: Contrast limited adaptive histogram equalization. In: Graphics gems IV, Academic Press Professional, pp. 474–485. Inc (1994)
24. Mittal, A., Soundararajan, R., Bovik, A.C.: Making a "completely blind" image quality analyzer. IEEE Sign. Process. Lett. **20**(3), 209–212 (2012)
25. Mittal, A., Moorthy, A.K., Bovik, A.C.: No-reference image quality assessment in the spatial domain. IEEE Trans. Image Process. **21**(12), 4695–4708 (2012)
26. Venkatanath, N., Praneeth, D., Bh, M.C., Channappayya, S.S., Medasani, S.S.: Blind image quality evaluation using perception based features. In: 2015 Twenty First National Conference on Communications (NCC), pp. 1–6 IEEE (2015)

Generating Dual-Energy Subtraction Soft-Tissue Images from Chest Radiographs via Bone Edge-Guided GAN

Yunbi Liu[1,2], Mingxia Liu[2(✉)], Yuhua Xi[1], Genggeng Qin[3], Dinggang Shen[2], and Wei Yang[1(✉)]

[1] School of Biomedical Engineering, Southern Medical University, Guangzhou 510515, China
weiyanggm@gmail.com

[2] Department of Radiology and BRIC, University of North Carolina at Chapel Hill, Chapel Hill, NC 27599, USA
mxliu@med.unc.edu

[3] Nanfang Hospital, Southern Medical University, Guangzhou, China

Abstract. Generating dual-energy subtraction (DES) soft-tissue images from chest radiographs (also called bone suppression) is an important task, as it improves the detection rates for lung nodules. Previous studies focus on generating DES-like soft-tissue images from CXRs through machine/deep learning techniques. However, they usually require tedious image processing steps for bone segmentation/delineation or ignore anatomical structure information (e.g., edges of ribs and clavicles) in CXRs. In this work, we propose a bone Edge-guided Generative Adversarial Network (EGAN) to generate DES-like soft-tissue images from conventional CXRs, which does not require human intervention and can explicitly use anatomical structure information of bones in CXRs. Specifically, the edges of ribs and clavicles in an input CXR were first detected by a trained fully convolutional network. Then, the edge probability map, as well as the original CXR image, are fed into a GAN model to generate the DES-like soft-tissue image, where the detected edge information is used as the prior knowledge to directly and specifically guide the image generation process. Experimental results on 504 subjects (each equipped with a CXR, a DES bone image, and a DES soft-tissue image) demonstrate that EGAN can produce DES-like soft-tissue images with high-quality and high-resolution, compared with classic deep learning methods. We further apply the trained EGAN to CXRs acquired by different types of X-ray machines in the public JSRT and NIH ChestXray 14 datasets, and our method can also produce visually appealing DES-like soft-tissue images.

1 Introduction

A dual-energy subtraction (DES) system can produce soft-tissue-enhanced images, where bones are almost removed by using two X-ray exposures at two

© Springer Nature Switzerland AG 2020
A. L. Martel et al. (Eds.): MICCAI 2020, LNCS 12262, pp. 678–687, 2020.
https://doi.org/10.1007/978-3-030-59713-9_65

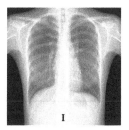 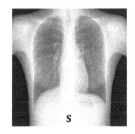 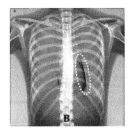

Fig. 1. Illustration of a conventional/standard chest X-ray (i.e., **I**), and its corresponding real DES soft-tissue image (i.e., **S**) and real DES bone image (i.e., **B**). The yellow dotted ellipse indicates the area with motion artifacts. (Color figure online)

different energy levels. As shown in Fig. 1, compared with conventional chest X-ray scans (CXRs), DES soft-tissue images provide improved visualization of lung areas because ribs and clavicles are almost invisible in DES images. However, it is challenging to obtain DES images clinically, as specialized equipment is required to obtain dual-energy X-ray exposures. One has to face other challenges, e.g., patients' concerns about radiation exposure and relatively high cost.

Many efforts have been made to generate DES soft-tissue images from conventional CXRs based on various image processing techniques, i.e., building the mapping/translation between these two modalities [1–4]. Due to desired soft-tissue images are expected to be bone-free, generating soft-tissue images from CXRs is also called *bone suppression*. Existing studies can be divided into *unsupervised* and *supervised* methods. Unsupervised methods usually first locate the lung areas or rib-and-clavicle edges so that they are targeted to focus on suppressing bone components [5–7]. These methods depend on the intermediate segmentation or delineation results and sometimes need human intervention. Fortunately, they have shown that prior knowledge containing location information of bones can effectively help suppress bones in CXRs. Supervised methods usually first extract features from CXRs and then learn a regressor (e.g., k-nearest neighbor, KNN) to estimate soft-tissue or bone images [4,8–11]. However, these methods simply rely on local image features and are usually time-consuming due to the high dimension (e.g., $2,000 \times 2,000$) of input images [8].

Deep learning has been recently used for bone suppression of CXRs [4,11–13] by learning image features in a data-driven manner. However, existing methods can not well handle the following challenges in translating conventional CXRs to DES soft-tissue images. *First*, the dimension of a standard CXR scan stored in a DICOM format is usually high (i.e., approximately $2,000 \times 2,000$). Therefore, it is not easy to train end-to-end deep learning models to synthesize high-resolution soft-tissue images without any prior knowledge. *Second*, previous studies usually synthesize substitutes of DES soft-tissue images, ignoring the appearance differences between substitutes and real soft-tissue images. Researchers in [4,9] regard the DES bone image (denoted by **B**) that contains more simple components than the DES soft-tissue image (denoted by **S**) as the to-be-generated target image,

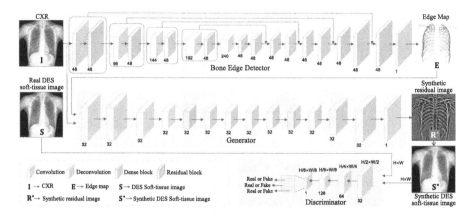

Fig. 2. Illustration of our proposed EGAN, including 1) a bone edge detector to locate edges of ribs and clavicles in CXRs, and 2) a GAN model (with a generator and a discriminator) to generate DES-like soft-tissue images with the bone edge information as guidance. The input includes a CXR $\mathbf{I}$ and a real DES soft-tissue image $\mathbf{S}$, while the output contains the synthetic residual image $\mathbf{R}^*$ and soft-tissue image $\mathbf{S}^*$. Let $\mathbf{R}$ denote the ground truth (i.e., real residual image), and those images satisfy $\mathbf{S} = \mathbf{I} - \mathbf{R}$.

and subtract the predicted bone image from the standard CXR (denoted by $\mathbf{I}$) to obtain the estimated soft-tissue image. However, $\mathbf{S} = \mathbf{I} - \mathbf{B}$ cannot be satisfied due to the sophisticated nonlinear post-processing used for raw imaging data. Another work [10] first decomposes a standard CXR into a bone image and a soft-tissue image using a DES dataset via cross projection tensor technique [14], and then regards decomposed soft-tissue images as ground truth, ignoring the differences between decomposed and real DES soft-tissue images.

To this end, we propose a bone Edge-guided GAN (EGAN) framework to generate DES soft-tissue images from conventional CXRs. As illustrated in Fig. 2, EGAN contains two major components: 1) bone edge detection, and 2) soft-tissue image generation based on CXRs and real soft-tissue images in an adversarial learning manner. The proposed EGAN does not require human intervention and can explicitly use bone edge information to guide image synthesis. And the target output (i.e., ground truth) in this work is real DES soft-tissue images rather than substitute (e.g., decomposed) soft-tissue ones.

2   Materials and Method

2.1   Data and Image Pre-processing

DES Dataset. Our DES dataset contains 504 subjects, and each subject is represented by a triplet, i.e., a CXR, a DES bone image, and a DES soft-tissue image. These images were acquired by using a digital radiography (DR) machine with a two-exposure DES unit (Discovery XR 656, GE Healthcare) at Nanfang

Hospital, Guangzhou, China. Each image was stored in DICOM format with a 14-bit depth. The size of each image ranges from $1,800 \times 1,800$ to $2,022 \times 2,022$ pixels, and the spatial resolution ranges from 0.1931×0.1931 mm^2 to 0.1943×0.1943 mm^2.

Image Pre-processing. The pre-processing steps used in this work include removing motion artifacts, normalization and generating target/ground-truth output images. *1) Removing motion artifacts.* A few motion artifacts usually exist in DES soft-tissue and bone images due to cardiac motion and breath during a two-exposure interval. To mitigate the effect of motion artifacts, we mask the regions (see yellow dotted ellipse in Fig. 1) with motion artifacts out and exclude them from the training samples as done in our previous work [10]. Specifically, we find that the correlation coefficients between the gradients of DES soft-tissue and bone images in the motion artifact regions are significantly higher than the coefficients in the other regions. Therefore, the correlation coefficients between a DES soft-tissue image and its corresponding DES bone image are computed and thresholded to generate the masks of motion artifacts. *2) Normalization.* Due to different acquisition conditions and inter-subject variance, the overall intensity and contrast differences exist in different CXR images. For normalization, we first filter each conventional CXR by a Gaussian kernel with a large sigma (i.e., 64 pixels) to get the base layer of CXR. Then, we subtract the base layer from paired $\mathbf{I}$ and $\mathbf{S}$ and normalize them to have a mean of 0 and a variance of 1. *3) Generating ground-truth images.* We subtract the normalized DES soft-tissue image from its corresponding normalized standard CXR to obtain the residual image $\mathbf{R}$ and treat this residual image as ground truth. With $\mathbf{S} = \mathbf{I} - \mathbf{R}$ satisfied, the ground-truth image is linearly related to the real DES soft-tissue image. In this way, the to-be-learned image synthesis model can generate a DES-like soft-tissue image $\mathbf{S}^*$ by subtracting the predicted residual image $\mathbf{R}^*$ from the input CXR $\mathbf{I}$.

2.2   Proposed Method

As shown in Fig. 2, our EGAN contains two components: 1) a bone edge detector to capture the edge information of ribs and clavicles in each input CXR image, and 2) a GAN model to generate DES-like soft-tissue images.

Component 1: Bone Edge Detector. In our work, we manually delineate ribs and clavicles of 82 CXRs from the used DES dataset. These 82 paired CXRs and bone edge maps are used to learn a bone edge detector, with 73 pairs for training and 9 pairs for test. We employ a classic network structure of FC-DenseNet [15] (see top panel of Fig. 2) to detect the edges of ribs and clavicles in each input CXR, where each dense block is an iterative concatenation of previous feature maps in favor of feature reuse and deep supervision. In [15], three FC-DenseNets with different layers were developed, i.e., FC-DenseNet103, FC-DenseNet63, and FC-DenseNet56, respectively. Considering that only a limited number (i.e., 82) of subjects are used for detecting bone edges, we employ the same architecture as the most compact FC-DenseNet56 model. Given a trained FC-DenseNet model for detecting the edges of ribs and clavicles, we can obtain an edge probability map for each CXR in our DES dataset.

Component 2: Edge-guided GAN. With the edge probability map as guidance, we develop an edge-guided GAN to generate DES-like soft-tissue images, consisting of a generator G and a discriminator D [16]. As shown in the bottom panel of Fig. 2, the input is the channel-wise concatenation of a conventional CXR scan and its corresponding estimated edge probability map, and the output is the synthetic residual image $\mathbf{R}^*$ (as well as the DES-like soft-tissue image $\mathbf{S}^* = \mathbf{I} - \mathbf{R}^*$), while the ground truth is the real residual image $\mathbf{R} = \mathbf{I} - \mathbf{S}$.

The generator G consists of an encoder, a decoder, and a series of residual blocks that bridge the encoder and decoder. Specifically, the encoder includes four convolution-normalization-ReLu layers (all with 32 channels) to extract features of input images (i.e., $\mathbf{I}$ and $\mathbf{E}$). The decoder consists of two deconvolution-normalization-ReLu modules with a stride of 2, a convolution-normalization-ReLu module with a stride of 1, and an output layer to produce a residual image $\mathbf{R}$ in the target domain. As we know, a residual block can be expressed in a general form: $x_{l+1} = ReLu(x_l + f(x_l, w_l))$, where x_l and x_{l+1} are input and output of the l-th block, and f is a residual function. The identity shortcut connections in each residual block can make information propagation smooth [17]. To balance the performance and convergence rate, we added a total of 7 residual blocks to bridge the encoder and decoder parts. Each residual block has 32-dimensional input and output feature maps.

Considering that the full size of each CXR is large, the image-level discriminator that classifies the whole image to be real or fake need more parameters than patch-level discriminator, and thus may be difficult to train. Here, we resort to a patch-level discriminator D to classify the real DES soft-tissue images and fake images from the generator. As shown in the bottom right of Fig. 2, this discriminator inputs a pair of real and fake soft-tissue images and outputs a probability map, where each element represents the probability that a specific $N \times N$ patch in the input image is real. We use a similar architecture as [18], including three convolution-normalization-ReLu modules with a stride of 2 and a final convolution layer with a stride of 1.

Objective Function. Both reconstruction and adversarial loss are used in EGAN. In the reconstruction loss, we minimize the l_1 distance between the synthesized residual image $G(\mathbf{x})$ and the corresponding ground-truth one $\mathbf{r}$ in both the intensity and gradient domains to encourage the consistency of $G(\mathbf{x})$ and $\mathbf{r}$. Denote $\mathbf{C}$ as the set of channel-wise concatenation of CXRs and their estimated edge probability maps. The reconstruction loss $\mathcal{L}_{rec}$ is defined as

$$
\begin{aligned}
\mathcal{L}_{rec} = E_{\mathbf{x}\sim p(\mathbf{C}),\mathbf{r}\sim p(\mathbf{R})}\|G(\mathbf{x}) - \mathbf{r}\|_1 &+ E_{\mathbf{x}\sim p(\mathbf{C}),\mathbf{r}\sim p(\mathbf{R})}\| \nabla_{1,G(\mathbf{x})} - \nabla_{1,\mathbf{r}} \|_1 \\
&+ E_{\mathbf{x}\sim p(\mathbf{C}),r\sim p(\mathbf{R})}\| \nabla_{2,G(\mathbf{x})} - \nabla_{2,\mathbf{r}} \|_1,
\end{aligned}
\tag{1}
$$

where $\mathbf{x}$ is the channel-wise concatenation of a CXR and its corresponding estimated edge probability map, $\mathbf{r}$ is the target residual image, $\nabla_{1,G(\mathbf{x})}$ and $\nabla_{2,G(\mathbf{x})}$ are the gradients of the predicted residual image in horizontal and vertical directions, respectively, and $\bigtriangledown_{1,\mathbf{r}}$ and $\bigtriangledown_{2,\mathbf{r}}$ are the gradients of the ground-truth residual image. To generate visually realistic soft-tissue images, we also introduce an adversarial loss $\mathcal{L}_{adv}$ defined as

$$\mathcal{L}_{adv} = E_{\mathbf{i}\sim p(\mathbf{I}),\mathbf{x}\sim p(\mathbf{C})} \log\left(1 - D\left(\mathbf{i} - G\left(\mathbf{x}\right)\right)\right), \qquad (2)$$

where $\mathbf{i}$ is the input CXR in the set of CXRs (denoted as $\mathbf{I}$). The adversarial loss $\mathcal{L}_{dis}$ in the discriminator is be defined as

$$\mathcal{L}_{dis} = E_{\mathbf{s}\sim p(\mathbf{S})} \log(1 - D(\mathbf{s})) + E_{\mathbf{i}\sim p(\mathbf{I}),\mathbf{x}\sim p(\mathbf{C})} \log\left(D\left(\mathbf{i} - G(\mathbf{x})\right)\right), \qquad (3)$$

where $\mathbf{s}$ is the DES soft-tissue image ($\mathbf{s} = \mathbf{i} - \mathbf{r}$) in the set of DES soft-tissue images (denoted as $\mathbf{S}$). The overall objective function of EGAN is formulated as

$$\mathbf{D}^* = \arg\min_G \mathcal{L}_{dis}, \ \mathbf{G}^* = \arg\min_G \mathcal{L}_{rec} + \lambda\mathcal{L}_{adv}, \qquad (4)$$

where λ controls the relative contributions of different losses. In this work, the adversarial loss encourages that local patches of the synthesized image should be similar to the real one, and the l_1-norm based reconstruction loss emphasizes the global shape consistency. As mentioned above, to mitigate the effect of motion artifacts on the prediction models, we mask the regions with motion artifacts out and only compute the loss in the valid regions.

Implementation. Limited by GPU memory, we first train our edge detector and then train the edge-guided GAN for image synthesis. When training the GAN, we randomly crop each training image with the size of $1,024 \times 1,024$. The generator and discriminator are trained in an iterative manner. To be specific, we first train the discriminator D with the generator G fixed for each batch. Then, we train the generator with the discriminator D fixed. The GAN is trained by mini-batches with the size of 4. We used Adam as the optimizer to train the model with a learning rate of 10^{-4}. The hyper-parameter λ in Eq. 4 that controls the relative contributions of different loss terms is empirically set as 0.1. Our EGAN is implemented in Tensorflow. It takes about 24 h to train EGAN on a workstation equipped with a GPU (TITAN X, 12G).

3 Experiments

Experimental Setup. For all 504 subjects in our DES dataset, we first treat 82 subjects (with manual delineation for ribs and clavicles) as training data, to make sure that these subjects will never be used in the test stage. We then randomly partition the remaining 422 subjects into three subsets. Finally, we partition our DES dataset into 80% for training, 10% for validation and 10% for test. Two public datasets (i.e. JSRT [19] and NIH ChestXray 14 [20]) are used as *independent* test sets to evaluate inter-dataset generalization of EGAN.

We compare our EGAN with two typical networks for image synthesis, including 1) a convolutional neural network (CNN) with only a generator, and 2) a GAN with a generator and a discriminator. For a fair comparison, CNN shares

Table 1. Performance of three different methods on the DES dataset. Metrics marked as "-R" and "-S" denote that they are used to evaluate the image quality of synthetic residual images and synthetic soft-tissue images, respectively.

Method	RMAE-R (%)	PSNR-S	SSIM-S (%)	BSR (%)
CNN	4.36 ± 0.82	37.80 ± 1.60	97.00 ± 1.00	79.40 ± 6.40
GAN	4.17 ± 0.80	38.20 ± 1.60	97.20 ± 1.10	81.40 ± 5.80
EGAN (Ours)	$\mathbf{4.00} \pm 0.77$	$\mathbf{38.50} \pm 1.60$	$\mathbf{97.50} \pm 1.00$	$\mathbf{82.70} \pm 5.70$

the same generator as our EGAN, and GAN shares the same generator and discriminator as EGAN (see Fig. 2). Note that our EGAN can employ the bone edge information (captured by an edge detector) to guide the image synthesis process, while CNN and GAN cannot take advantage of such edge information.

Two groups of experiments are performed in this work. In the *first* group, we evaluate different methods on the DES dataset and quantitatively evaluate their performance in synthesizing soft-tissue and residual images. Four metrics are used to measure the quality of those synthetic images, including 1) relative mean absolute error (RMAE), 2) peak signal-to-noise ratio (PSNR), 3) structural similarity index (SSIM), and 4) bone suppression ratio (BSR). In the *second* group, we evaluate the inter-dataset generalization capability of our EGAN for bone suppression, with the model trained on our DES dataset and test on the independent JSRT and NIH ChestXray 14 datasets. Since ground-truth DES soft-tissue images are not available for subjects in JSRT and NIH ChestXray 14, we only visually evaluate those generated images in this group of experiments.

Evaluation of Synthetic Images on DES Dataset. The quantitative results for evaluating synthetic residual and soft-tissue images generated by three methods are summarized in Table 1. We also visually show the synthetic images yielded by three methods and the corresponding target images in Fig. 3. From Table 1, we can see that our EGAN is generally superior to two competing methods regarding four evaluation metrics. For instance, our EGAN yields the BSR value of 82.70%, which is 1.3% higher than the second-best result (i.e., 81.40% of GAN). As can be seen from the enlarged area with motion artifacts (denoted as squares) in Fig. 3, the bone components in synthetic residual images generated by EGAN look visually clearer than those produced by CNN and GAN, especially at the rib edges, indicating that the bone components in the corresponding synthetic soft-tissue image are suppressed more thoroughly.

Evaluation of Cross-Dataset Generalization Capability. We further apply EGAN to synthesize soft-tissue images from CXRs acquired by using different types of X-ray machines in the JSRT and NIH ChestXray 14 datasets. Note that EGAN is trained on our DES dataset and directly applied to these two independent datasets. The visual results are shown in Fig. 4, from which one can see that the bone components in CXRs were significantly suppressed in our generated soft-tissue images and the intensity contrast of these synthetic images

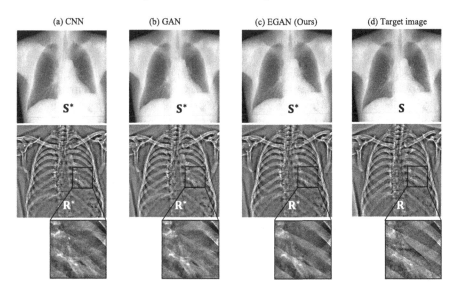

Fig. 3. Illustration of synthetic DES-like soft-tissue images (1st row) and residual images (2nd row) generated by three methods, as well as the corresponding target images in our DES dataset. Enlarged boxes locate the areas with motion artifacts.

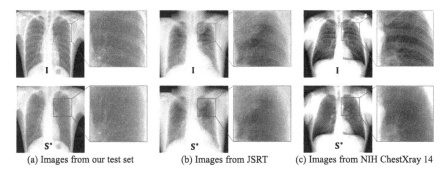

Fig. 4. Results of cross-dataset image synthesis. The 1st row shows the original CXRs from our test set, the JSRT dataset, and the NIH ChestXray 14 dataset. The 2nd row shows their corresponding DES-like soft-tissue images generated by our EGAN.

is close to their corresponding CXRs. Note that all images in Fig. 4 are recovered from their original intensity ranges. This implies that EGAN can handle different types of CXRs by providing visually appealing soft-tissue images.

4 Conclusion and Future Work

In this paper, we have presented a bone edge-guided GAN (EGAN) method to generate DES-like soft-tissue images from standard CXRs. Specifically, an edge detector is designed to detect bone edges in each CXR. With the estimated edge probability map as guidance, an edge-guided GAN model is developed to generate residual images and DES-like soft-tissue images. Experimental results on both our collected dataset and two independent datasets demonstrate that EGAN can generate reasonable soft-tissue images and also have good inter-dataset generalization. Due to the limitation of GPU memory, we currently train the edge detector and the edge-guided GAN model separately, which may lead to sub-optimal performance. It is interesting to integrate these two tasks into a unified framework by designing a more feasible strategy to effectively fuse edge information into the process of image synthesis, which will be our future work.

Acknowledgements. Y. Liu, Y. Xi and W. Yang were partially supported by the National Natural Science Foundation of China (No. 81771916) and the Guangdong Provincial Key Laboratory of Medical Image Processing (No. 2014B-030301042). A part of this work was finished when Y. Liu was visiting the University of North Carolina at Chapel Hill.

References

1. Horváth, Á., Orbán, G.G., Horváth, Á., Horváth, G.: An X-ray CAD system with ribcage suppression for improved detection of lung lesions. Periodica Polytechnica Electr. Eng. **57**(1), 19 (2013)
2. von Berg, J., et al.: A novel bone suppression method that improves lung nodule detection. Int. J. Comput. Assist. Radiol. Surg. **11**(4), 641–655 (2015). https://doi.org/10.1007/s11548-015-1278-y
3. Baltruschat, I.M., et al.: When does bone suppression and lung field segmentation improve chest X-ray disease classification? ISB **I**, 1362–1366 (2019)
4. Zarshenas, A., Liu, J., Forti, P., Suzuki, K.: Separation of bones from soft tissue in chest radiographs: anatomy-specific orientation-frequency-specific deep neural network convolution. Med. Phys. **46**(5), 2232–2242 (2019)
5. Hogeweg, L., Sánchez, C.I., van Ginneken, B.: Suppression of translucent elongated structures: applications in chest radiography. IEEE Trans. Med. Imaging **32**(11), 2099–2113 (2013)
6. Lee, J.S., Wang, J.W., Wu, H.H., Yuan, M.Z.: A nonparametric-based rib suppression method for chest radiographs. Comput. Math. Appl. **64**(5), 1390–1399 (2012)
7. Rasheed, T., Ahmed, B., Khan, M.A., Bettayeb, M., Lee, S., Kim, T.S.: RiB suppression in frontal chest radiographs: a blind source separation approach. In: International Symposium on Signal Processing and its Applications, pp. 1–4 (2007)
8. Loog, M., van Ginneken, B., Schilham, A.M.: Filter learning: application to suppression of bony structures from chest radiographs. Med. Image Anal. **10**(6), 826–840 (2006)
9. Chen, S., Suzuki, K.: Separation of bones from chest radiographs by means of anatomically specific multiple massive-training ANNs combined with total variation minimization smoothing. IEEE Trans. Med. Imaging **33**(2), 246–257 (2013)
10. Yang, W., et al.: Cascade of multi-scale convolutional neural networks for bone suppression of chest radiographs in gradient domain. Med. Image Anal. **35**, 421–433 (2017)

11. Zhou, B., Lin, X., Eck, B., Hou, J., Wilson, D.: Generation of virtual dual energy images from standard single-shot radiographs using multi-scale and conditional adversarial network. In: ACCV, pp. 298–313 (2018)
12. Oh, D.Y., Yun, I.D.: Learning bone suppression from dual energy chest X-rays using adversarial networks. arXiv preprint arXiv:1811.02628 (2018)
13. Eslami, M., Tabarestani, S., Albarqouni, S., Adeli, E., Navab, N., Adjouadi, M.: Image-to-images translation for multi-task organ segmentation and bone suppression in chest X-ray radiography. arXiv preprint arXiv:1906.10089 (2019)
14. Agrawal, A., Raskar, R., Chellappa, R.: Edge suppression by gradient field transformation using cross-projection tensors. In: CVPR, pp. 2301–2308 (2006)
15. Jégou, S., Drozdzal, M., Vazquez, D., Romero, A., Bengio, Y.: The one hundred layers tiramisu: fully convolutional DenseNets for semantic segmentation. In: CVPR Workshops, pp. 11–19 (2017)
16. Pan, Y., Liu, M., Lian, C., Zhou, T., Xia, Y., Shen, D.: Synthesizing missing PET from MRI with cycle-consistent generative adversarial networks for Alzheimer's disease diagnosis. In: Frangi, A.F., Schnabel, J.A., Davatzikos, C., Alberola-López, C., Fichtinger, G. (eds.) MICCAI 2018. Synthesizing missing PET from MRI with cycle-consistent generative adversarial networks for Alzheimer's disease diagnosis, vol. 11072, pp. 455–463. Springer, Cham (2018). https://doi.org/10.1007/978-3-030-00931-1_52
17. He, K., Zhang, X., Ren, S., Sun, J.: Identity mappings in deep residual networks. In: Leibe, B., Matas, J., Sebe, N., Welling, M. (eds.) ECCV 2016. LNCS, vol. 9908, pp. 630–645. Springer, Cham (2016). https://doi.org/10.1007/978-3-319-46493-0_38
18. Isola, P., Zhu, J.Y., Zhou, T., Efros, A.A.: Image-to-image translation with conditional adversarial networks. In: CVPR, pp. 1125–1134 (2017)
19. Shiraishi, J., et al.: Development of a digital image database for chest radiographs with and without a lung nodule: receiver operating characteristic analysis of radiologists' detection of pulmonary nodules. Am. J. Roentgenol. **174**(1), 71–74 (2000)
20. Wang, X., Peng, Y., Lu, L., Lu, Z., Bagheri, M., Summers, R.: ChestX-ray8: hospital-scale chest X-ray database and benchmarks on weakly-supervised classification and localization of common thorax diseases. In: CVPR, pp. 3462–3471 (2017)

GANDALF: Generative Adversarial Networks with Discriminator-Adaptive Loss Fine-Tuning for Alzheimer's Disease Diagnosis from MRI

Hoo-Chang Shin[1](✉), Alvin Ihsani[1], Ziyue Xu[1], Swetha Mandava[1],
Sharath Turuvekere Sreenivas[1], Christopher Forster[1], Jiook Cha[2],
and Alzheimer's Disease Neuroimaging Initiative

[1] NVIDIA Corporation, Santa Clara, USA
hshin@nvidia.com
[2] Department of Psychology, Center for REAL Intelligence,
AI Institute, Seoul National University, Seoul, South Korea
connectome@snu.ac.kr

Abstract. Positron Emission Tomography (PET) is now regarded as the gold standard for the diagnosis of Alzheimer's Disease (AD). However, PET imaging can be prohibitive in terms of cost and planning, and is also among the imaging techniques with the highest dosage of radiation. Magnetic Resonance Imaging (MRI), in contrast, is more widely available and provides more flexibility when setting the desired image resolution.

Unfortunately, the diagnosis of AD using MRI is difficult due to the very subtle physiological differences between healthy and AD subjects visible on MRI. As a result, many attempts have been made to synthesize PET images from MR images using generative adversarial networks (GANs) in the interest of enabling the diagnosis of AD from MR. Existing work on PET synthesis from MRI has largely focused on Conditional GANs, where MR images are used to generate PET images and subsequently used for AD diagnosis. There is no end-to-end training goal.

This paper proposes an alternative approach to the aforementioned, where AD diagnosis is incorporated in the GAN training objective to achieve the best AD classification performance. Different GAN losses are fine-tuned based on the discriminator performance, and the overall training is stabilized. The proposed network architecture and training regime show state-of-the-art performance for three- and four- class AD classification tasks.

Keywords: Alzheimer disease · Neuroimaging · Generative models

© Springer Nature Switzerland AG 2020
A. L. Martel et al. (Eds.): MICCAI 2020, LNCS 12262, pp. 688–697, 2020.
https://doi.org/10.1007/978-3-030-59713-9_66

1   Introduction

1.1   PET Imaging and AD Diagnosis

Alzheimer's Disease (AD) is the most common cause of dementia, affecting quality of life for many elderly people and their families. Early diagnosis and intervention of AD can improve the quality of life by significantly slowing the progression of the disease, thus it is an active area of research. Positron Emission Tomography (PET) appears to be a very promising imaging technique to assess the progression and stage of the disease by monitoring the spread of Tau-protein in the form of Neurofibrillary Tangles (NFT) and Amyloid beta ($A\beta$) [12,13,21]. As a functional imaging technique, PET uses a radioactive tracer injected into the patient, and images the distribution of the tracer over the course of minutes or hours. In AD research, PET imaging techniques measure amyloid plaque (AV45) [4,22] and tau protein aggregates (AV1451) [17,24] that are essential to understanding AD pathology and diagnosis. Compared to AV45-/AV1451-PET, FDG-PET usually helps differentiate AD from other causes of dementia, because it can characterize the patterns of glucose metabolism in the brain that are specific to AD [16]. Example T1-weighted Magnetic Resonance images, AV45-/FDG- PET brain images of CN and AD are shown in Fig. 1.

While PET plays an important role for AD diagnosis, it can be prohibitive in terms of cost and planning: *(1)* the short half life of the radioisotopes requires on-site production in remote regions; *(2)* no-show patients result in radioisotopes being wasted; *(3)* the length of imaging sessions is determined by the tracer and the use case so motion artifacts may be unavoidable, and lastly; *(4)* small variations ($\sim$5 min) in the acquisition start time may cause over- or under-estimation of quantitative parameters. It is also not as widely available as Magnetic Resonance Imaging (MRI).

1.2   Synthesizing PET Images from MR for AD Diagnosis

To address the shortcomings of PET for AD diagnosis, a number of studies have attempted AD diagnosis from T1-weighted MR images. While T1-weighted MRI is most suitable for visualizing anatomical structures in the brain, it is not optimal for AD diagnosis because it does not highlight functional or metabolic properties of brain tissues. The question arises as to whether one can leverage existing combined PET-MR image pairs (a combined imaging modality available to only large research institutions) to generate PET images from MR-only image acquisitions.

Conditional generative adversarial networks (CGAN) [11] have previously been used to generate images of a modality from a paired input image of a different modality. Frequent examples of such paired images are images and label maps, images and sketch, and pictures of the same scene from one lighting condition to another (e.g. day/night). For medical image analysis, such as AD diagnosis, PET image is generated from MRI using CGAN. The generated PET is then used to train AD classification network.

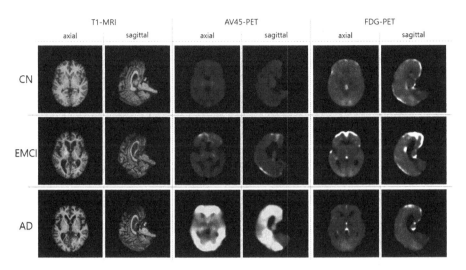

Fig. 1. Examples of T1-weighted MRI, AV45-PET, FDG-PET brain images of Cognitive Normal (CN), early mild cognitive impairment (EMCI), and Alzheimer's disease (AD). The differences are much more clearly visible on the PET images than in the MR images, especially for the EMCI case. On the EMCI case, increased accumulation of AV45 in the medial temporal, occipital, and frontal lobe (inferior frontal gyrus shown) is noticeable. AD shows reduced brain metabolism in the FDG scan, and significant uptake of AV45 compared to both CN and EMCI due to widespread accumulation of Amyloid-β. In the T1-weighted MRI images, the size increment of the ventricles is visible from CN to EMCI to AD, however it is not as clearly visible as in PET.

This work proposes an approach similar to CGAN, where CGAN is trained end-to-end with the final goal of AD classification. If trained with classification goal, then the performance of generating realistic images may be compromised. We overcome this limitation by adaptively fine-tuning the GAN losses and classification losses. Also, the overall GAN training is stabilized by the loss fine-tuning. State-of-the-art result on three- and four- class AD classification are achieved with the proposed architecture and training regime.

1.3 Dataset and Classification of Cognitive Decline

We use the publicly available ADNI (Alzheimer's Disease Neuroimaging Initiative) dataset comprised of F18-AV-45 (florbetapir) and F18-FDG (fluorodeoxyglucose) PET image pairs along with the co-registered T1-weighted MRI. The dataset contains six dementia related conditions: cognitive normal (CN), early mild cognitive impairment (EMCI), late mild cognitive impairment (LMCI), mild cognitive impairment (MCI), subjective memory complaint (SMC), and Alzheimer disease (AD). Among these conditions, SMC is difficult to subjectively distinguish from CN. Also, there may be overlaps between EMCI/LMCI and MCI. Therefore, we test binary classification of AD/CN, three-

and four- class classification of AD/MCI/CN and AD/LMCI/EMCI/CN for early AD diagnosis. Figure 1 show some examples of CN, EMCI, AD images in the dataset.

We randomly divide the dataset with 70% training, 10% validation, and 20% testing according to the patients, resulting 722/104/207 subjects for each train/validation/test set. Some subjects have multiple scans (i.e., more than one temporal scan), with the total 1,525 image triplets (AD45-/FDG-PET, T1-MRI). The images are pre-processed using FreeSurfer [1]. The T1-weighted images are skull-stripped, where non-cerebral matters such as skull and scalp are removed. Registration, re-scaling [9] and partial volume correction [8] is applied to the PET images. The T1-weigthed images are re-scaled to $1\,\mathrm{mm}^3$ with 256^3 voxels, and PET images are $2 \times 93 \times 76 \times 76$ voxels with 2 temporal resolution.

2   Related Works

Image-based AD diagnosis is regarded as a challenging task. Most of the prior works use a combination of structural and functional imaging, such as T1-weighted MRI and PET, or T1-weighted MRI and functional MRI such as DTI (diffusion tensor imaging). They also typically focus on binary classification of each state category, such as AD vs. NC, or AD vs. MCI.

A combination of T1-weighted MRI, AV45-/FDG-PET was used with multi-feature kernel supervised within-class-similar discriminative dictionary learning algorithm to demonstrate binary classification of AD/NC, MCI/NC, AD/MCI in [15]. A combination of T1-weighted MRI and FDG-PET with three-dimensional convolutional neural network (CNN) was used to demonstrate binary classification of CN/AD, CN/pMCI, sMCI/pMCI in [10]. GAN was used to generate additional PET images from T1-weighted MRI that do not have AV45-PET image pairs in [25]. MRI and real-/synthetic- PET image pairs are subsequently used to train CNN to perform binary classification of stable-MCI/progressive-MCI.

Functional MRI (fMRI) is an MRI imaging technique most similar to PET that it can measure brain activity by detecting changes associated with blood flow. A minimum spanning tree (MST) classification framework was proposed in [5] to perform binary classification of MCI/NC, AD/CN, and AD/MCI using fMRI. A combination of T1-weighted MRI and Diffusion Tensor Imaging (DTI) was used with Multiple Kernel Learning to demonstrate binary classification of CN/AD, CN/MCI, AD/MCI in [2].

More recent work demonstrates diagnosing AD from T1-weighted MRI only. Longitudinal studies with landmark-based features and support vector machines to classify CN/AD and CN/MCI in [26]. T1-weighted MRI was used with convolutional autoencoder based unsupervised learning for the CN/AD and progressive-MCI/stable-MCI classification task in [18]. Other recent works show multi-class classification using T1-weighted MRI. A variant of DenseNet CNN was used for multi-class classification of AD/MCI/NC using MRI in [23]. T1-weighted MRI was used with CNN to demonstrate binary classification of NC/AD and three-class classification of NC/AD/MCI in [6].

3   Methods

The `pix2pix` [11] CGAN architecture is widely adopted in the medical image analysis domain for synthesizing from one image modality to another. For instance, Yan et al. [25] use the CGAN to generate AV45-PET from T1-weighted MRI to supplement the training dataset with additional synthetic PET-MRI image pairs. While for generating an image of different modality may be an end-goal for computer vision domain, in medical domain we often want to diagnose a disease, such as AD, using the generated image. We hypothesize that a GAN designed and trained with this diagnosis end-goal in mind can outperform in AD diagnosis, compared to other types of CGAN application where synthesis and diagnosis are trained separately.

3.1   Conditional Generative Adversarial Networks

The `pix2pix` [11] CGAN is trained with the following objective:

$$G^* = \arg \min_G \max_D \mathcal{L}_{cGAN}(G, D) + \lambda \mathcal{L}_{L1}(G). \tag{1}$$

where $\mathcal{L}_{cGAN}(G, D)$ and and $\mathcal{L}_{L1}(G)$ are defined as

$$\mathcal{L}_{cGAN}(G, D) = \mathbb{E}_{x,y}[\log D(x, y)] + \mathbb{E}_{x,z}[\log(1 - D(x, G(x, z)))], \tag{2}$$

$$\mathcal{L}_{L1}(G) = \mathbb{E}_{x,y,z}[\|y - G(x, z)\|_1]. \tag{3}$$

where x, y and $G(x, z)$ can be regarded as MRI, PET input, and generated PET. The CGAN consists of a generator (G) that has encoder-decoder architecture, and a discriminator (D) that is a CNN classifier. The U-Net [19] architecture is usually used as the G that takes an input image and generates an output image of a same size but of different modality or characteristics. PET conventionally has lower image resolution than MRI, so we modify the U-Net architecture to take the different resolutions into account - MRI: $256 \times 256 \times 256$; PET: $2 \times 93 \times 76 \times 76$. The encoder part has eight layers while the decoder part has five. Only the middle five layers in the encoder-decoder part has the skip-connection, with the last two up-sampling (transpose convolution) layers to make the target PET resolution. The discriminator CNN has three convolutional (conv-) layers that take MRI input, and two conv-layers that take PET input. The two branches of conv-layers are merged and followed by two additional conv-layers for classification.

3.2   GAN with Discriminator-Adaptive Loss Fine-Tuning

GAN is trained with minimax objective [7] where G and D compete with each other. CGAN is trained with an additional $L1$ loss for the G, and a patch-GAN [11] classifier for the D. The D in our generative network is trained with additional AD classification losses: *(1)* based on real MRI and generated PET

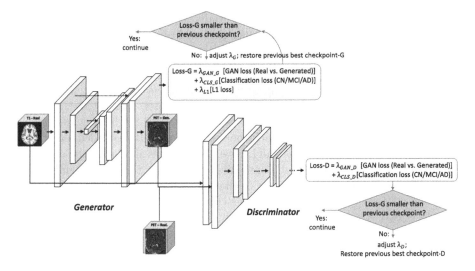

Fig. 2. Overall architecture and training pipeline. While generator and discriminator are trained independently to compete against each other, they are both trained with the additional AD classification loss that are adjusted *(1)* to generate realistic PET images, and *(2)* to perform well on AD classification. In addition, losses are monitored and weights are adjusted to stablilize the GAN training, preventing loss oscillation.

input, multiplied by a hyper-parameter λ_{GAN_D}, and *(2)* based on real MRI and PET, multiplied by λ_{CLS_D}:

$$\mathcal{L}_{\text{D}}(D, G) = \lambda_{\text{GAN}_D} \arg \min_{G} \mathcal{L}_{cGAN}(G, D) + \lambda_{\text{CLS}_D} \mathbb{E}_{x,y,\hat{y}}[\log D(\hat{y}|x, y)], \quad (4)$$

where $\hat{y}$ is the AD label.

The G is also trained with AD classification loss based on real MRI and generated PET input, in addition to the GAN loss and $L1$ loss. Each loss is multiplied with hyper-parameters to control their relative importance during the training - λ_{CLS_G}, λ_{GAN_G}, and λ_{L1}:

$$\mathcal{L}_{\text{G}}(G, D) = \lambda_{\text{GAN}_G} \arg \max_{D} \mathcal{L}_{cGAN}(G, D)$$
$$+ \lambda_{\text{CLS}_G} \mathbb{E}_{x,\hat{y}}[\log D(\hat{y}|x, G(x, z))] + \lambda_{L1}\mathcal{L}_{L1}(G). \quad (5)$$

In the earlier phase of the GAN training, generated PET likely are far from the real ones. They progressively become more realistic as the training proceeds. Therefore, D is trained initially with small λ_{GAN_D} and gradually increased during the training, while λ_{CLS_D} starts from a larger value and gradually decreased. This encourages the D to focus on AD classification when G is improving to generate more realistic PET images. The G is trained with a large λ_{GAN_G} at first so it can focus on generating realistic PET in the beginning. It is gradually decreased as λ_{CLS_G} increases from a smaller value, to emphasize AD classification using the generated PET images. We set λ_{GAN_D} and λ_{CLS_G} as 0.01 and linearly increase

10 times per epoch, λ_{CLS_D} and λ_{GAN_G} initially as 100 and decrease 1/10 times per epoch. We train for 1000 epochs using ADAM optimizer [14].

Stabilizing Training. Training D and G independently, the D and G loss can oscillate rather than being in a stable convergence state [3]. To remedy this problem we continuously monitor the D and G loss, and adjust the hyper-parameters λ for the losses if any one is lower compared to the previous epoch. Loss oscillation generally occurs when the training has well proceeded, and this is when AD classification losses get higher weights. This is similar to the approach of [20] penalizing D weights with annealing to stabilize the GAN training. For example, when the D loss starts to oscillate and becomes higher compared to the previous epoch, then *(1)* its previous checkpoint is restored, and *(2)* λ_{CLS_D} gets decreased. The overall training pipeline is shown in Fig. 2.

4 Results

We perform two- to four- class AD classification using T1-weighted MRI input. The two-class AD classification results is shown in Table 1. The CNN approach in [6] report better performance on the two-class AD/CN classification, and GANDALF show similar performance to pix2pix + CNN method. We suspect this may be because AD vs. CN is more clearly visible than AD/MCI/CN or AD/LMCI/EMCI/CN on MRI, so a deep CNN with good hyper-parameter set can provide better result and PET plays a rather limited role for the diagnosis. We did not conduct a thorough hyper-parameter search for GANDALF in this study.

Table 1. Comparison of MRI-based AD diagnosis for AD vs. CN binary classification. CNN based method [6] reports best performance which may indicate using PET and synthesized PET is more useful for early AD diagnosis.

Method	AD/CN	Acc	F_2	Prec	Rec
Esmaeilzadeh et al. [6]	200/230	**94.1**	**0.93**	**0.92**	**0.94**
pix2pix + CNN	162/428	85.2	0.83	0.84	0.83
GANDALF	162/428	85.2	0.84	0.84	0.84

Results of the three-class AD/MCI/CN classification task are shown in Table 2. We achieve state-of-the-art performance on the three-class classification compared to the prior works using T1-weighted MRI input. MCI may show more subtle difference on the MRI compared to the AD as can be seen in Fig. 1. This, and the consistent better performance of the generative methods compared to the prior works could indicate that an additional training of synthesizing PET can help achieving better performance for early AD diagnosis.

Table 2. MRI-based AD diagnosis for AD/MCI/CN three-class classification. Better performance shown by generative methods may suggest additional training to generate synthesized PET can be promising for early diagnosis of AD using MRI.

Method	AD/MCI/CN	Acc	F_2	Prec	Rec
Esmaeilzadeh et al. [6]	200/411/230	61.1	0.62	0.59	0.63
Wu et al. [23]	130/455/200	N/A	0.49	0.62	0.35
pix2pix + CNN	162/456/428	71.3	0.63	0.64	0.63
GANDALF	162/456/428	**78.7**	**0.69**	**0.83**	**0.66**

Lastly, four-class classification of AD/LMCI/EMCI/CN results are shown in Table 3. We show a meaningful first result on classifying early-MCI and late-MCI from CN and AD, a promising first step for early AD diagnosis using T1-weighted MRI. Our proposed GANDALF method also shows improved performance compared to the pix2pix + CNN method. Towards the end of the GANDALF training, the entire network acts as a classification network with T1-weighted MRI input. Finding a better/deeper classifier/discriminator architecture could improve the final classification performance. However this should be balanced with the generator architecture/depth for the GAN training with the minimax objective. A thorough hyper-parameter search could also improve the final performance.

Table 3. MRI-based AD diagnosis for AD/LMCI/EMCI/CN four-class classification. We show meaningful result that can be promising for early diagnosis on AD using T1-weighted MRI input.

Method	AD/LMCI/EMCI/CN	Acc	F_2	Prec	Rec
pix2pix + CNN	162/456/219/428	33.0	0.34	0.34	0.34
GANDALF	162/456/219/428	**37.0**	**0.40**	**0.39**	**0.40**

5   Conclusion

Early diagnosis and intervention of Alzheimer's disease (AD) can significantly slow the progression of the disease and improve patients' condition and the life quality of the patient and their caregivers. PET imaging can provide great insight for early diagnosis of AD, however, it is rarely available outside of research environments. Earlier works on MRI-based AD diagnosis use conditional generative adversarial networks (GAN) to synthesize PET from MRI, and subsequently use the generated PET for AD diagnosis.

We propose a network where AD diagnosis end-goal is incorporated into the MRI-PET synthesis and trained end-to-end, instead of first synthesizing PET and then use it for AD diagnosis. Furthermore, we suggest a training scheme to stabilize the GAN training. We achieve state-of-the-art MRI-based AD diagnosis for three-class AD classification of AD/MCI/CN. We also achieve the first meaningful result on four-class (AD/LMCI/EMCI/CN) classification that can be promising for early diagnosis of AD based on MRI, to the best of our knowledge.

Acknowledgement. The authors would like to thank Seonjoo Lee of Columbia University Medical Center for the discussion and help in data pre-processing.

References

1. FreeSurfer software suite. https://surfer.nmr.mgh.harvard.edu
2. Ahmed, O.B., Benois-Pineau, J., Allard, M., Catheline, G., Amar, C.B., Alzheimer's Disease Neuroimaging Initiative, et al.: Recognition of Alzheimer's disease and mild cognitive impairment with multimodal image-derived biomarkers and multiple kernel learning. Neurocomputing **220**, 98–110 (2017)
3. Arjovsky, M., Bottou, L.: Towards principled methods for training generative adversarial networks. In: 5th International Conference on Learning Representations. ICLR 2017, Toulon, France, 24–26 April 2017. Conference Track Proceedings (2017)
4. Berti, V., Pupi, A., Mosconi, L.: PET/CT in diagnosis of dementia. Ann. N. Y. Acad. Sci. **1228**, 81 (2011)
5. Cui, X., et al.: Classification of Alzheimer's disease, mild cognitive impairment, and normal controls with subnetwork selection and graph kernel principal component analysis based on minimum spanning tree brain functional network. Front. Comput. Neurosci. **12**, 31 (2018)
6. Esmaeilzadeh, S., Belivanis, D.I., Pohl, K.M., Adeli, E.: End-to-end Alzheimer's disease diagnosis and biomarker identification. In: Shi, Y., Suk, H.-I., Liu, M. (eds.) MLMI 2018. LNCS, vol. 11046, pp. 337–345. Springer, Cham (2018). https://doi.org/10.1007/978-3-030-00919-9_39
7. Goodfellow, I., et al.: Generative adversarial nets. In: Advances in Neural Information Processing Systems, pp. 2672–2680 (2014)
8. Greve, D.N., et al.: Different partial volume correction methods lead to different conclusions: an 18F-FDG-PET study of aging. Neuroimage **132**, 334–343 (2016)
9. Greve, D.N., et al.: Cortical surface-based analysis reduces bias and variance in kinetic modeling of brain pet data. Neuroimage **92**, 225–236 (2014)
10. Huang, Y., Xu, J., Zhou, Y., Tong, T., Zhuang, X., Alzheimer's Disease Neuroimaging Initiative (ADNI), et al.: Diagnosis of Alzheimer's disease via multi-modality 3D convolutional neural network. Front. Neurosci. **13**, 509 (2019)
11. Isola, P., Zhu, J.Y., Zhou, T., Efros, A.A.: Image-to-image translation with conditional adversarial networks. In: Proceedings of the IEEE Conference on Computer Vision and Pattern Recognition, pp. 1125–1134 (2017)
12. Johnson, K.A., et al.: Tau positron emission tomographic imaging in aging and early Alzheimer disease. Ann. Neurol. **79**, 110–119 (2016)
13. Johnson, K.A., AV45-A11 Study Group, et al.: Florbetapir (F18-AV-45) PET to assess amyloid burden in Alzheimer's disease dementia, mild cognitive impairment, and normal aging. Alzheimer's Dement. **9**(5), S72–S83 (2013)
14. Kingma, D.P., Ba, J.: Adam: a method for stochastic optimization. In: 3rd International Conference on Learning Representations (ICLR) (2015)

15. Li, Q., Wu, X., Xu, L., Chen, K., Yao, L., Alzheimer's Disease Neuroimaging Initiative, et al.: Classification of Alzheimer's disease, mild cognitive impairment, and cognitively unimpaired individuals using multi-feature kernel discriminant dictionary learning. Front. Comput. Neurosci. **11**, 117 (2018)
16. Marcus, C., Mena, E., Subramaniam, R.M.: Brain pet in the diagnosis of Alzheimer's disease. Clin. Nucl. Med. **39**(10), e413 (2014)
17. Marquié, M., et al.: Validating novel tau positron emission tomography tracer [F-18]-AV-1451 (T807) on postmortem brain tissue. Ann. Neurol. **78**(5), 787–800 (2015)
18. Oh, K., Chung, Y.C., Kim, K.W., Kim, W.S., Oh, I.S.: Classification and visualization of Alzheimer's disease using volumetric convolutional neural network and transfer learning. Sci. Rep. **9**(1), 1–16 (2019)
19. Ronneberger, O., Fischer, P., Brox, T.: U-Net: convolutional networks for biomedical image segmentation. In: Navab, N., Hornegger, J., Wells, W.M., Frangi, A.F. (eds.) MICCAI 2015. LNCS, vol. 9351, pp. 234–241. Springer, Cham (2015). https://doi.org/10.1007/978-3-319-24574-4_28
20. Roth, K., Lucchi, A., Nowozin, S., Hofmann, T.: Stabilizing training of generative adversarial networks through regularization. In: Advances in Neural Information Processing Systems, pp. 2018–2028 (2017)
21. Schwartz, A.J., et al.: Regional profiles of the candidate tau PET ligand 18F-AV-1451 recapitulate key features of Braak histopathological stages. Brain **139**, 1539–50 (2016)
22. Sevigny, J., et al.: The antibody aducanumab reduces aβ plaques in Alzheimer's disease. Nature **537**(7618), 50–56 (2016)
23. Wu, B., Sun, X., Hu, L., Wang, Y.: Learning with unsure data for medical image diagnosis. In: Proceedings of the IEEE International Conference on Computer Vision, pp. 10590–10599 (2019)
24. Xia, C.F., et al.: [18F] T807, a novel tau positron emission tomography imaging agent for Alzheimer's disease. Alzheimer's Dement. **9**(6), 666–676 (2013)
25. Yan, Yu., Lee, H., Somer, E., Grau, V.: Generation of amyloid PET images via conditional adversarial training for predicting progression to Alzheimer's disease. In: Rekik, I., Unal, G., Adeli, E., Park, S.H. (eds.) PRIME 2018. LNCS, vol. 11121, pp. 26–33. Springer, Cham (2018). https://doi.org/10.1007/978-3-030-00320-3_4
26. Zhang, J., Liu, M., An, L., Gao, Y., Shen, D.: Alzheimer's disease diagnosis using landmark-based features from longitudinal structural MR images. IEEE J. Biomed. Health Inform. **21**(6), 1607–1616 (2017)

Brain MR to PET Synthesis via Bidirectional Generative Adversarial Network

Shengye Hu[1], Yanyan Shen[1], Shuqiang Wang[1(✉)], and Baiying Lei[2]

[1] Shenzhen Institutes of Advanced Technology, Chinese Academy of Sciences,
Shenzhen, Guangdong, China
sq.wang@siat.ac.cn

[2] Shenzhen University, Shenzhen, Guangdong, China

Abstract. Fusing multi-modality medical images, such as MR and PET, can provide complementary information to improve the diagnostic performance. But compared to the substantial and available MR data, PET data is always deficient. In this paper, we propose a novel end-to-end network, called Bidirectional GAN, where image contexts and latent vector are effectively used and jointly optimized for brain MR-to-PET synthesis. Specifically, a bidirectional mapping mechanism is designed to embed the diverse brain structural details into the high-dimensional latent space. And then the superior network architecture and the modified loss functions are further utilized to enhance the quality of synthetic images. The most appealing part is that the proposed method can synthesize the plausible PET images while preserving the diverse brain structures of different subjects. The experiments demonstrate that the performance of the proposed method outperforms the state-of-the-art methods in terms of quantitative measures, qualitative evaluation and the improvement of classification accuracy.

Keywords: Medical imaging synthesis · Generative adversarial network · Bidirectional mapping mechanism

1 Introduction

Recently, medical images have become a powerful diagnostic and research tool for disease analysis and surgery planning [1]. Taking Positron emission tomography (PET) imaging as an example, it is widely used for staging and monitoring treatment in a variety of cancers [3]. In particular, complementary imaging modalities are acquired simultaneously to indicate brain disease areas and various tissue properties. Recent studies have reported that MR and PET measurements are among the best biomarkers for Alzheimer Disease (AD) progression and Mild Cognitive Impairment (MCI) conversion prediction However, the scarcity of PET data is one of the major challenges in multi-modality fusion. It occurs for various reasons, such as PET scans will increase lifetime cancer risk due to the usage of

© Springer Nature Switzerland AG 2020
A. L. Martel et al. (Eds.): MICCAI 2020, LNCS 12262, pp. 698–707, 2020.
https://doi.org/10.1007/978-3-030-59713-9_67

radioactive tracer. And it is time consuming, requiring a series of procedures and expensive facilities. In contrast, MR scan is more widely used than PET because of its low cost and unharmful radiation. Hence synthesizing PET images from MR images gradually attracts public attention.

In traditional methods, atlases-based [4] or other-based [6] methods that are based on hand-crafted features and expert knowledge have problems in synthesizing blurry PET images, and their application is very limited. In recent years, deep learning methods have made a major breakthrough in computer vision application. Due to the strong feature extraction and learning capabilities of convolutional neural networks, it has also achieved excellent performance in medical imaging synthesis tasks. For example, Li et al. [7] used a variant of convolutional neural networks (CNN) to learn a nonlinear mapping from MR to PET. And Nie et al. [8] presented a fully convolutional network (FCN) to synthesize CT images from their corresponding MR. The emergence of generative adversarial networks (GAN) [10] with outstanding generative capabilities has further encouraged the development of medical imaging synthesis. Ben et al. [2] presented a novel system combined by FCN and conditional GAN for the generation of virtual PET images using CT scans. Lei et al. [9] trained the deep GAN in a quasi-3D way and simplify the generator, which lead to faster training yet better synthesis quality. Salman et al. [5] proposed a novel pGAN for paired image synthesis, and it yielded visually and quantitatively enhanced accuracy in multi-contrast MR synthesis compared to the state-of-the-art methods.

Although these algorithms achieve remarkable results, they are confronted with the same fundamental limitation: it is difficult to generate plausible images with significantly diverse structures, because the generator learns to largely ignore the latent vectors (i.e. the noise vectors input) without any priori knowledge in the training process of GANs [11–13]. Especially for the generation of brain images that have diverse structural details (e.g. gyri and sulci) between different subjects. To copy with this challenge, in this paper we propose a novel end-to-end network, called Bidirectional GAN, where image contexts and latent vector are effectively used and jointly optimized for the brain MR to PET synthesis. To be more specific, a bidirectional mapping mechanism between the latent vector and the output image is introduced, while an advanced generator architecture is adopted to optimally extract and generate the intrinsic features of PET images. Finally, this work devises a composite loss function containing an additional pixel-wise loss and perceptual loss to encourage less blurring and yield visually more realistic results. As an attempt to bridge the gap between network generative capability and real medical images, the proposed method not only focus on synthesizing perceptually realistic images, but also concentrate on reflecting the diverse brain attributes of different subjects.

2   Method

Overview of Proposed Method. Different subjects have diverse brain structural attributes. Assuming have a dataset containing the paired brain MR images

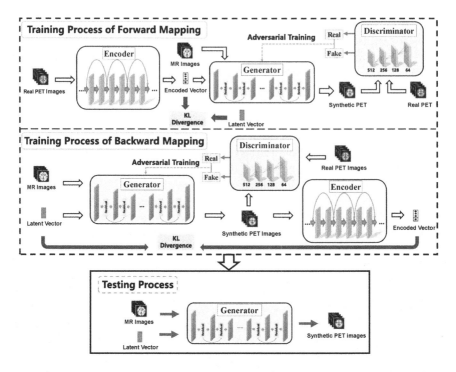

Fig. 1. An overview of the proposed Bidirectional GAN for brain MR-to-PET synthesis.

and PET images, our goal is learning a cross-modality mapping from brain MR to PET. In GAN models, in order to generate an output sample, a latent vector obtained from a known distribution (such as a standard normal distribution) is usually set as an input of the generator. In such cases, the biggest advantage of the proposed model is that it can provide the latent vector with some priori knowledges about the real images and make full use of its variability, which encourages the generator to synthesize the plausible PET images while preserving the diverse details of brain structures in different subjects. Specifically, the proposed method not only learn a mapping from the latent vector to the PET images like traditional GANs, but also learn a mapping that returns the PET images to the latent space by training an encoder at the same time. This mechanism encourages a bidirectional mapping between the latent space and the PET images, so as to the diverse brain structural features are embedded in the high-dimensional latent space.

The proposed model consists of an encoder, a generator and a discriminator. The training process of forward mapping starts with encoding the PET images and the encoded vector is trained to conform to the standard normal distribution by optimizing the KL divergence. The generator then tries to map the input MR images along with the latent vector sampled from the standard normal distribution to the synthetic PET images. But in the training process of backward mapping, the generator is first used to synthesize PET images from the MR images and the sampled latent vector. Subsequently, the synthetic PET image is fed to the encoder to reconstruct the input latent vector. Here as well, the role of discriminator in both mappings is to try to distinguish the synthetic PET from the real PET. In the testing process, only the trained generator G is used. For more details on the overall framework, please refer to Fig. 1.

Network Architecture. The generator architecture is the paramount factor to the quality of generated images. Motivated by the recent studies, the ResU-Net architecture incorporating the advantage of U-Net [14] and ResNet [15] is used. The U-Net architecture with symmetric skip connections has become an efficient way to directly shuttle the low-level underlying information between aligned MR and PET across the net. And ResNet with pure identity mappings has achieved impressive performance on many challenging computer vision tasks. The residual block is capable of boosting information exchange across different layers and alleviating the gradient vanishing issue. In the ResU-Net architecture, the residual block is deployed to instead of the traditional convolutional layers before the pooling and deconvolutional operation. Concretely, the generator consists of 6 pooling operations, 6 deconvolutional stages and 13 residual blocks. For the discriminator, 30×30 PatchGAN [11] discriminator is employed rather than that of 1×1 PixelGAN [10]. Such a patch-level architecture not only has a few parameters, but also can accurately reconstruct the local brain structure of PET images. For the encoder, the ResNet-34 is deployed to instead of plain CNN architecture to better encode the image.

Loss Function. During the adversarial training process, the generator G and the discriminator D are learned simultaneously. Upon convergence, G is capable of producing realistic counterfeit images that D cannot recognize [10]. Unfortunately, the original GAN metric cannot reasonably reflect the true difference between the distributions of real data and synthesized data. In the proposed Bidirectional GAN, the loss function of LSGAN [16] is adopted, which slightly address the mode collapse problem. To embed the diverse image features into the latent space, we enforce the KL-divergence constraint, which means the difference between the encoded PET vector and the gaussian latent vector should be minimized. Given an MR slice $x \sim P_{MR}(x)$ and its corresponding PET image $y \sim P_{PET}(y)$, the loss objective is as follows:

$$\mathcal{L}_{GAN}(D) = \mathbb{E}_{x \sim P_{MR}(x), z \sim P(z)} \left[(D(G(x,z)))^2 \right] + \mathbb{E}_{y \sim P_{PET}(y)} \left[(D(y) - 1)^2 \right] \quad (1)$$

$$\mathcal{L}_{GAN}(C)_{Forward} = \mathbb{E}_{y \sim P_{PET}(y)} \left[\emptyset_{KL}(\mathbb{E}(y) \mid \mathbb{N}(0,1)) \right] \quad (2)$$

$$\mathcal{L}_{GAN}(C)_{Backward} = \mathbb{E}_{x \sim P_{MR}(x), z \sim P(z)} \left[\emptyset_{KL}(\mathbb{E}(G(x,z)) \mid \mathbb{N}(0,1)) \right] \quad (3)$$

where $\mathbb{E}$ denotes expected value, $\emptyset_{KL}$ denotes KL divergence and $z \sim P(z)$ represents sampling a latent vector from the gaussian latent space.

Similar to many existing works, an $\mathcal{L}_1$ pixel-wise reconstruction loss is incorporated to impose an additional constraint on the generator. This means the generator is not only needed to fool the discriminator, but also needed to minimize the pixel-wise intensity difference between synthetic PET images and real

PET images. Nevertheless, these pixel-wise losses tend to produce over-smoothed outputs and fail to capture anatomical details. Thus a perceptual loss that relies on differences in higher feature representations is introduced to generate plausible images with less blurring. The aggregate loss function of generator G can be written as:

$$\mathcal{L}_{\mathcal{L}_1}(G) = \mathbb{E}_{x \sim P_{MR}(x), z \sim P(z), y \sim P_{PET}(y)} \|y - G(x, z)\|_1 \tag{4}$$

$$\mathcal{L}_{Per}(G) = \mathbb{E}_{x \sim P_{MR}(x), z \sim P(z), y \sim P_{PET}(y)} \|V(y) - V(G(x, z))\|_1 \tag{5}$$

$$\begin{aligned}\mathcal{L}_{GAN}(G) =& \mathbb{E}_{x \sim P_{MR}(x), z \sim P(z)} \left[(D(G(x, z)) - 1)^2\right] \\ &+ \lambda_1 \mathcal{L}_{\mathcal{L}_1}(G) + \lambda_2 \mathcal{L}_{Per}(G)\end{aligned} \tag{6}$$

where V is a set of feature maps right before the second max-pooling operation of pre-trained VGG-16 [17]. λ_1 and λ_2 are the hyper-parameters for controlling the relative importance of individual loss terms.

Implementation Details. The proposed Bidirectional GAN method was implemented in Python using PyTorch framework. And all experiments were carried out on a single NVIDIA GeForce GTX 2080 Ti GPU. Adam solver was used and the whole network was trained for 200 epochs. The learning rate linearly decayed the rate to 0 over the next 100 epochs after keeping the initial value 0.001 in the first 100 epochs. Parameters λ_1 and λ_2 are 10 and 5, respectively.

3   Experiments

Experimental Settings. The experiments was carried on a subset of Alzheimer Disease Neuroimaging Initiative (ADNI) database [18], including 680 subjects. And each subject had images of two modalities (i.e. MR and PET). To prevent an extremely high variability between different subjects, the redundant tissues were removed. What's more, the two modalities were also aligned with a linear registration by FSL software. For efficient use of the available data, the 10-fold cross-validation was performed to test the model performance, where 7 fold was used as the training set, 1 fold for the validation set and remaining 2 fold was used as the test set. To demonstrate the performance of the proposed method, we compared the Bidirectional GAN with the CycleGAN [20] and pGAN [5], which were the state-of-the-art and representative models for image synthesis tasks. For a fair comparison, we re-implemented these models according to the original paper. In parallel, some necessary changes were made, such as the number of network layers and the value of hyper-parameters. But the topology of models was retained. It should be stressed that all models were compared using the same training and testing data. The comparison was accomplished by using qualitative evaluation, quantitative measures, and the improvement of classification accuracy.

Qualitative Evaluation. In qualitative evaluation, some randomly selected axial slices of different subjects are shown in Fig. 2. The synthetic PET images

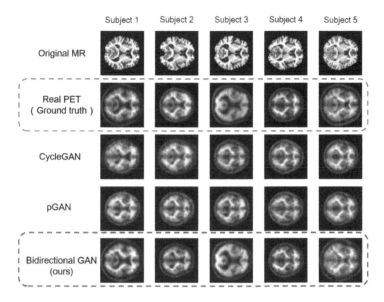

Fig. 2. Qualitative comparisons of different methods in synthesizing PET images from the corresponding MR images. The results of CycleGAN have a largely similar appearance between each other. And the pGAN also suffers from this problem. By contrast, the proposed method successfully generates the diverse brain structural details of different subjects.

derived from CycleGAN have a highly similar appearance between each other, especially for the size and style of brain structure. The diverse structural features of the brain are disappearing and only some texture details inside the brain are different. The results of pGAN suffer from the same problem, too. As mentioned in the introduction section, they have problems in generating the diverse brain structural details of different subjects. By contrast, it can be observed that the synthetic PET images derived from the proposed Bidirectional GAN have a most similar contour structure to the ground-truth. The more intuitive comparison is displayed in Fig. 3. The pseudo-color difference maps between the synthetic images and real images were calculated and the proposed method also achieved the most appearing results. But it is noticed that some details of the synthetic images are not clear enough (e.g. the image in the last row, last column of Fig. 2). We speculate that this maybe because the injecting way of the latent vectors is too simple. And this will be the future work.

Quantitative Measures. Peak signal-noise ratio (PSNR) and structural similarity index measurement (SSIM) which based on correlation, perception and pixel intensities are employed to quantitatively measure the distance between the real images and the synthetic images. Table 1 shows the quantitative results of all models using two metrics. As can be seen, Bidirectional GAN outperforms the compared models with the highest PSNR (27.36) and SSIM (0.88),

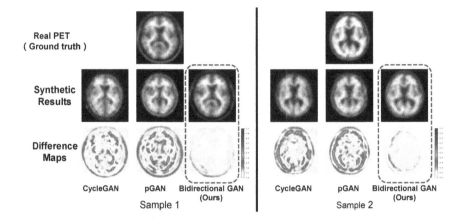

Fig. 3. The intuitive comparison of different models by calculating the pseudo-color difference maps between the synthetic PET images and real PET images.

Table 1. Quantitative measures for synthetic PET images (mean ± std).

Method	PSNR	SSIM
CycleGAN	24.68 ± 1.07	0.78 ± 0.012
pGAN	25.12 ± 1.18	0.79 ± 0.014
Bidirectional GAN(Ours)	**27.36 ± 1.22**	**0.88 ± 0.023**

indicating that the quality of the synthetic images derived from the proposed method is closest to the real PET images. It improves PSNR by at least 2.24 dB over the state-of-the-art models, which demonstrates its superiority.

Improvement of Classification Accuracy. Besides the above quantitative and qualitative comparisons, we further make a binary classification (i.e. AD vs. Normal) experiment based on the accuracy improvement after incorporating synthetic PET with original MR. As one of the most classic imaging classification networks in computer vision, VGG-16 network [17] is employed as the classifier. The same 10-fold cross-validation is used to perform the classification experiment. And the results of classification task adding synthetic PET derived from CycleGAN, pGAN and the proposed method are shown in Fig. 4. It can be observed that the classification accuracy of adding synthetic PET from any of models is higher than the MR-alone based result, due to the complementary features extracted and utilized from both modalities. In addition, the joint accuracy of adding the synthetic images from the CycleGAN (82.26%) or pGAN (83.54%) is less than that of the proposed method (87.82%). This potentially due to they cannot synthesize the PET images reflecting the diverse brain attributes, which is different from the information provided by the original MR images. According to the above experimental results, it can be concluded that the proposed

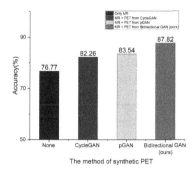

Fig. 4. Illustration of Alzheimer diseases diagnosis accuracies based on MR and synthetic PET from different methods.

Bidirectional GAN is capable of generating a substantial amount of realistic and diverse PET images with the detailed attributes of brains.

4   Conclusions

In summary, this paper proposed a Bidirectional GAN method to achieve brain MR-PET synthesis. A novel bidirectional mapping mechanism is introduced to embed the diverse brain structural features into the high-dimensional latent space. In addition, an advanced ResU-Net architecture and a modified loss function containing the adversarial loss, the reconstruction loss and the perceptual loss are employed to enhance the quality of synthetic PET images. The extensive experiments, including quantitative measures, qualitative evaluation and the improvement of classification accuracy have demonstrated that the proposed Bidirectional GAN can synthesize the plausible PET images while preserving the diverse details of brain structures in different subjects, outperforming the state-of-the-art models trained with the same dataset. In the future, the proposed framework will be extended to other medical imaging synthesis tasks. Additionally, a more efficient injecting way of the latent vector will be researched to further promote its performance.

Acknowledgements. This work was supported in part by the National Natural Science Foundations of China under Grant 61872351 and Grant 61771465, in part by the International Science and Technology Cooperation Projects of Guangdong under Grant 2019A050510030, in part by the Strategic Priority CAS Project under Grant XDB38000000, in part by the Major Projects from General Logistics Department of People's Liberation Army under Grant AWS13C008, and in part by the Shenzhen Key Basic Research Projects under Grant JCYJ2020050718250-6416.

References

1. Ernst, P., Hille, G., Hansen, C., Tönnies, K., Rak, M.: A CNN-based framework for statistical assessment of spinal shape and curvature in whole-body MRI images of large populations. In: Shen, D., et al. (eds.) MICCAI 2019. LNCS, vol. 11767, pp. 3–11. Springer, Cham (2019). https://doi.org/10.1007/978-3-030-32251-9_1
2. Ben-Cohen, A., et al.: Cross-modality synthesis from CT to PET using FCN and GAN networks for improved automated lesion detection. Eng. Appl. Artif. Intell. **78**, 186–194 (2019)
3. Li, H., et al.: A novel PET tumor delineation method based on adaptive region-growing and dual-front active contours. Med. Phys. **35**(8), 3711–3721 (2008)
4. Burgos, N., et al.: Attenuation correction synthesis for hybrid PET-MR scanners: application to brain studies. IEEE Trans. Med. Imaging **33**(12), 2332–2341 (2014)
5. Dar, S.U., Yurt, M., Karacan, L., Erdem, A., Erdem, E., Çukur, T.: Image synthesis in multi-contrast MRI with conditional generative adversarial networks. IEEE Trans. Med. Imaging **38**(10), 2375–2388 (2019)
6. Papadimitroulas, P., et al.: Investigation of realistic PET simulations incorporating tumor patient's specificity using anthropomorphic models: creation of an oncology database. Med. Phys. **40**(11), 112506 (2013)
7. Li, R., et al.: Deep learning based imaging data completion for improved brain disease diagnosis. In: Golland, P., Hata, N., Barillot, C., Hornegger, J., Howe, R. (eds.) MICCAI 2014. LNCS, vol. 8675, pp. 305–312. Springer, Cham (2014). https://doi.org/10.1007/978-3-319-10443-0_39
8. Nie, D., et al.: Medical image synthesis with context-aware generative adversarial networks. In: Descoteaux, M., Maier-Hein, L., Franz, A., Jannin, P., Collins, D.L., Duchesne, S. (eds.) MICCAI 2017. LNCS, vol. 10435, pp. 417–425. Springer, Cham (2017). https://doi.org/10.1007/978-3-319-66179-7_48
9. Xiang, L., Li, Y., Lin, W., Wang, Q., Shen, D.: Unpaired deep cross-modality synthesis with fast training. In: Stoyanov, D., et al. (eds.) DLMIA/ML-CDS - 2018. LNCS, vol. 11045, pp. 155–164. Springer, Cham (2018). https://doi.org/10.1007/978-3-030-00889-5_18
10. Goodfellow, I., et al.: Generative adversarial nets. In: Advances in Neural Information Processing Systems, pp. 2672–2680 (2014)
11. Isola, P., Zhu, J.Y., Zhou, T., Efros, A.A.: Image-to-image translation with conditional adversarial networks. In: Proceedings of the IEEE Conference on Computer Vision and Pattern Recognition, pp. 1125–1134 (2017)
12. Sangkloy, P., Lu, J., Fang, C., Yu, F., Hays, J.: Scribbler: controlling deep image synthesis with sketch and color. In: Proceedings of the IEEE Conference on Computer Vision and Pattern Recognition, pp. 5400–5409 (2017)
13. Xian, W., et al.: TextureGAN: controlling deep image synthesis with texture patches. In: Proceedings of the IEEE Conference on Computer Vision and Pattern Recognition, pp. 8456–8465 (2018)
14. Ronneberger, O., Fischer, P., Brox, T.: U-Net: convolutional networks for biomedical image segmentation. In: Navab, N., Hornegger, J., Wells, W.M., Frangi, A.F. (eds.) MICCAI 2015. LNCS, vol. 9351, pp. 234–241. Springer, Cham (2015). https://doi.org/10.1007/978-3-319-24574-4_28
15. He, K., Zhang, X., Ren, S., Sun, J.: Deep residual learning for image recognition. In: Proceedings of the IEEE Conference on Computer Vision and Pattern Recognition, pp. 770–778 (2016)

16. Mao, X., Li, Q., Xie, H., Lau, R.Y., Wang, Z., Paul Smolley, S.: Least squares generative adversarial networks. In: Proceedings of the IEEE International Conference on Computer Vision, pp. 2794–2802 (2017)
17. Simonyan, K., Zisserman, A.: Very deep convolutional networks for large-scale image recognition. arXiv preprint arXiv:1409.1556 (2014)
18. Jack Jr., C.R., et al.: The Alzheimer's disease neuroimaging initiative (ADNI): MRI methods. J. Magn. Reson. Imaging: Off. J. Int. Soc. Magn. Reson. Med. **27**(4), 685–691 (2008)
19. Li, X., Chen, H., Qi, X., Dou, Q., Fu, C.W., Heng, P.A.: H-DenseUNet: hybrid densely connected UNet for liver and tumor segmentation from CT volumes. IEEE Trans. Med. Imaging **37**(12), 2663–2674 (2018)
20. Zhu, J.Y., Park, T., Isola, P., Efros, A.A.: Unpaired image-to-image translation using cycle-consistent adversarial networks. In: Proceedings of the IEEE International Conference on Computer Vision, pp. 2223–2232 (2017)

AGAN: An Anatomy Corrector Conditional Generative Adversarial Network

Melih Engin[1][(✉)] (iD), Robin Lange[1], Andras Nemes[1](iD), Sadaf Monajemi[1],
Milad Mohammadzadeh[1], Chin Kong Goh[2], Tian Ming Tu[3],
Benjamin Y. Q. Tan[4], Prakash Paliwal[4], Leonard L. L. Yeo[4,5],
and Vijay K. Sharma[4,5]

[1] See-Mode Technologies Pty. Ltd., Melbourne, Australia
{melih,robin,andras,sadaf,milad}@see-mode.com
[2] Changi General Hospital, Singapore, Singapore
chin_kong_goh@cgh.com.sg
[3] National Neuroscience Institute, Singapore, Singapore
tu.tian.ming@singhealth.com.sg
[4] National University Health System, Singapore, Singapore
benjaminyqtan@gmail.com
[5] National University of Singapore, Singapore, Singapore
{prakash_paliwal,leonard_ll_yeo,vijay_kumar_sharma}@nuhs.edu.sg

Abstract. The accurate segmentation of medical images has important consequences in clinical applications. Noisy and artefact-heavy images can result in erroneous image segmentation and often require expert understanding of the target anatomy by clinicians to interpret and compensate for missing and obfuscated data. This is especially true in ultrasound imaging where shadowing and speckle artefacts are common. We propose a novel approach to handle such artefacts using a conditional Generative Adversarial Network called Anatomical GAN (AGAN) that can correct anatomically-invalid pixel-wise segmentation and impose shape priors in carotid artery ultrasound images by learning the underlying structure of the arteries. These anatomically accurate outputs can then be used in the clinical work flow by clinicians or be further processed by other automated methods for assistance in clinical decision making. AGAN can be chained with any pixel-wise segmentation method and is generalisable for both anatomy and artefacts. Experimental results on a longitudinal ultrasound carotid artery dataset show that AGAN can correct anatomically-invalid segmentation masks obtained with different pixel-wise segmentation methods when other state-of-the-art methods fail.

Keywords: Deep learning · Convolutional neural networks · Generative adversarial networks · Segmentation · Shape priors · Ultrasound

© Springer Nature Switzerland AG 2020
A. L. Martel et al. (Eds.): MICCAI 2020, LNCS 12262, pp. 708–717, 2020.
https://doi.org/10.1007/978-3-030-59713-9_68

a)   b)   c)   d)

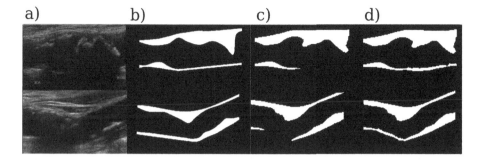

Fig. 1. a) Two ultrasound image samples of carotid artery with shadowing artefacts, b) Expert annotations of artery walls, c) Their pixel-wise segmentations by Unet++, d) the corrections made by AGAN. AGAN can make long range corrections with the accurate curvature by learning the anatomy of the arteries.

1   Introduction

Atherosclerotic carotid artery disease (CAD) is the pathogenic condition of plaque build-up on the walls of the carotid arteries. If this disease is left untreated, it can cause severe health conditions. Diagnosis of CAD is mainly done via ultrasonography. By measuring the intima-media thickness (IMT) of carotid arteries and examining the shape and texture of the artery walls, the severity of the disease is decided. This process, however, is operator dependent, error-prone, and time-consuming. Automating this analysis process would save valuable time of clinicians, and potentially lives by minimizing human errors.

In recent years, deep learning has been proven to be very effective at medical image segmentations. Most popular segmentation models (such as FCN [12], U-Net [10], U-Net++ [16]) rely on pixel-wise information. However, like all medical images, ultrasound images do not contain pixel-wise perfect information and are prone to several artefacts: i) speckle noise is present in every ultrasound image. ii) Image quality is non-uniformly distributed; ultrasound transducers focus the power mainly in the centre; edges of the images are usually noisier. iii) Ultrasound waves can be reflected or absorbed by dense structures like bones or calcified plaques causing so-called shadowing artefacts that can occlude regions of interest. These artefacts lead to erroneous pixel-wise segmentation outputs. For example, in ultrasound imaging of carotid arteries they can cause segmentations with fractured vessel walls (see Fig. 1), merged vessel walls, and false positives in the adventitia-intima border. Clinicians can deduce the correct shape of the walls by considering the anatomy of the artery, however, in automated methods, these defects can cause inaccurate IMT measurements and may result in misdiagnosis.

Generative adversarial networks are a very powerful family of neural networks that can perform tasks such as image synthesis, image-to-image transformation, style transfer, supervised and unsupervised domain adaptation. In this work, we consider mapping an anatomically-invalid segmentation mask to its anatomically correct correspondence as an image-to-image transformation problem and propose a novel generative adversarial network (GAN), called AGAN, to solve it. AGAN maps a defective pixel-wise segmentation output to its anatomically correct correspondence by learning the distribution of the defective segmentation masks as well as the distribution of the correct anatomy.

Our contributions are as follows; first, we propose a conditional GAN called AGAN that can learn from the global information to impose shape priors, as oppose the traditional methods which can only leverage local information. Furthermore, AGAN learns both the distributions of deformed samples as well as the distribution of ground truths to make pinpoint corrections. Second, in the experiments section, we demonstrate the efficacy of AGAN in two different shape imposing tasks; general denoising and pinpoint corrections. Finally, we show the efficacy of AGAN in the context of longitudinal ultrasound images of carotid arteries - a very challenging image modality.

2   Related Work

Finding and imposing shape priors to the segmentation of medical images is an active research area. Studies [2,3,11] consider the problem of learning the shape prior in an unsupervised setting. The most relevant methods to AGAN are denoising auto-encoder based ones. In [8], Oktay et al. train an autoencoder (AE) from the ground truth images. They impose the latent space of the AE as a shape prior to the training by minimizing the mean squared error between latent space representations of the ground truth and its segmentation. Similar to [6], the nature of the correction relies on the general smoothening properties of AEs. Similarly, in [9], the authors cascade an AE to the segmentation framework and take advantage of the low dimensional latent space.

In [6], Larrazabal et al. train a deep denoising AE called Post-DAE, to correct the connectivity issues in X-Ray images. Comparable to our work, Post-DAE is a deep neural network that is trained on segmentation outputs. Post-DAE encodes a given defective segmentation mask to its low-dimensional latent space, then decodes it as an anatomically plausible segmentation output.

Due to their smoothening properties, low-dimensional embeddings are effective in correcting high-frequency Gaussian or salt-and-pepper noise like artefacts, rather than performing singular pinpoint corrections. The smoothening effect also has the disadvantage of changing disease indicating information. In the case of ultrasound of carotid arteries, this could manifest itself as losing the details at the edges of a plaque. Accurately capturing the shape of a plaque is critical to diagnosis, especially the sharp outlines that determine plaque characteristics.

3   Anatomical GAN

3.1   Formulation

The training of a GAN relies on a minimax game of two modules; a discriminator and a generator. The generator tries to produce a sample that belongs to the

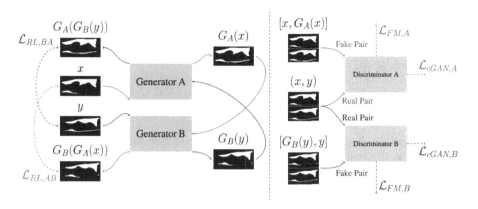

Fig. 2. Training diagram of AGAN. Red arrows indicate the flow for G_A - D_A, blue arrows indicate the flow for G_B - D_B. Dashed lines indicate the losses. (Color figure online)

target distribution and the discriminator tries to decide correctly whether a given sample was generated or not. Whereas in the case of conditional GANs, the objective is to generate samples conditioned on the input of the generator. AGAN belongs to the family of bidirectional cGANs. Its structure is based on the DualGAN [14] architecture. Consider the sets of anatomically-invalid binary segmentation masks X and their corresponding expert annotations Y. The main task is to learn a generator function $G_A : X \rightarrow Y$ that corrects the faulty masks. Similar to DualGAN, in the setting of AGAN, another generator function $G_B : Y \rightarrow X$, that is responsible of creating artefacts, is learned simultaneously. Generators G_A and G_B play the minimax game with discriminators, D_A and D_B, respectively. Discriminator function D_A tries to distinguish the real pair, (x, y), from the fake pair, $(x, G_A(x))$. Similarly, D_B tries to distinguish the real pair (x, y) from the fake one $(G_B(y), y)$.

In the literature, the joint optimization of bidirectional mappings (i.e. $X \rightarrow Y$ and $Y \rightarrow X$) are used for un-paired image-to-image translation tasks. Even though we formulate our problem as a supervised and paired image-to-image translation, we found that this setting provides more stability and consistency for our case. The joint training of G_A and G_B provides extra information and supervision as well as consistency to the generator G_A.

Alongside the adversarial losses, in order to provide extra supervision, we use feature matching loss defined as $\mathcal{L}_{FM,A}(G_A, D_A) = \sum_{i=1}^{T} \frac{1}{N_i} |D_A^i(y) - D_A^i(G_A(x))|$ and $\mathcal{L}_{FM,B}(G_B, D_B) = \sum_{i=1}^{T} \frac{1}{N_i} |D_B^i(y) - D_B^i(G_B(x))|$ as introduced in [13], where, T is the number of layers in D and i layer index. Feature matching loss is used to update the generators only; i.e. D_A and D_B are only used as feature extractors and they are not updated with this term. We also use reconstruction errors, to enforce consistency when an input is translated to the other domain and then back to its own. Reconstruction errors are defined in the following forms $\mathcal{L}_{RL,AB} = \mathcal{L}_{BCE}(G_B(G_A(x)), x)$ and $\mathcal{L}_{RL,BA} = \mathcal{L}_{BCE}(G_A(G_B(y)), y)$ where $\mathcal{L}_{BCE}$ is the binary cross-entropy loss $\mathcal{L}_{BCE} = -(y \log(x) + (1 - y) \log(1 - x))$. By combining these terms, we get the loss function of AGAN, formally defined as:

$$\mathcal{L} = \mathcal{L}_{cGAN,A}(G_A, D_A) + \mathcal{L}_{FM,A}(G_A, D_A) + \mathcal{L}_{RL,AB}(G_A, G_B) \\ + \mathcal{L}_{cGAN,B}(G_B, D_B) + \mathcal{L}_{FM,B}(G_B, D_B) + \mathcal{L}_{RL,BA}(G_B, G_A). \tag{1}$$

where $\mathcal{L}_{cGAN,A}(G_A, D_A)$ and $\mathcal{L}_{cGAN,B}(G_B, D_B)$ are the adversarial terms defined formally defined for conditional GANs as follows:

$$\mathcal{L}_{cGAN} = \mathbb{E}_{x,y}[\log D(x, y)] + \mathbb{E}_{x,z}[\log(1 - D(G(x, z)))]. \tag{2}$$

G tries the minimize this term and D tries the maximize it. Refer to Fig. 2 for the training diagram of AGAN.

Another possible design choice is to condition both generators, A and B, to the grayscale ultrasound images. This would leverage the pixel-wise information rather than just the binary segmentations. Our preliminary findings show that the generators ignore the image information and focus solely on the binary segmentations as the majority of information that is to be derived from the grayscale images is already in the binary segmentations. It is possible that with a proper architectural design and a careful parameter tuning, this way of training could potentially yield better anatomy corrections. We will regard this approach as a future research direction.

3.2 Architectural Details

GANs are powerful and versatile tools; however, their training is non-trivial. They require careful hyperparameter tuning and architecture design. To increase the stability of the training, we use spectral normalisations in the discriminators [7,15]. Attention modules [15] in the generators helps to produce more structurally coherent outputs. We use U-Net based generators with long-range connections and the discriminator architectures, both described in [4].

4   Experiments

For the experiments we use a longitudinal ultrasound carotid artery dataset that has 1323 images of 190 patients, gathered from 4 hospitals. We use annotations of an expert reader as the ground truths. We start with the pixel-wise segmentation of the dataset with two different deep learning-based models; U-Net [10] and U-Net++ [16]. Due to the difference in their robustness to the speckle noise, these two models create two different types of shape imposing and anatomy correction tasks which we investigate in depth in this section. In the first task, the distributions of artefacts is noise-like and uniformly spread in the dataset. As aforementioned, ultrasound images contain high levels of speckle noise. When such images are segmented by U-Net, the majority of segmentation masks have

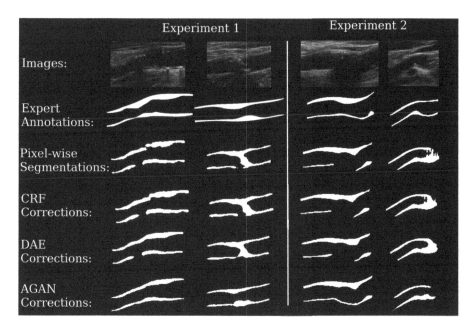

Fig. 3. Samples illustrating the different artefact types of U-Net and U-Net++. Majority of the pixel-wise segmentations in experiment 1 suffer from the winding effect at the borders due to the presence of high noise in the grayscale images. Shadowing effect (1st and 3rd columns), and merged walls (2nd and 4th columns) are also present. The winding effect can be reduced by denoising methods, however, the separation of merged walls and the connection of fragmented walls cannot be handled reliably by them, especially when the corruption area is large. In experiment 2, the winding effect is not present, however, fragmented and merged walls still appear. By learning the underlying anatomy of these arteries, AGAN can correct the deformations accurately in both cases.

false positives or false negatives at the adventitia-intima borders; which create a winding effect. Automatic evaluation of such segmentation masks can lead to under or over estimation of plaque build-ups and misdiagnosis of the severity of the disease. Refer to the first part of Fig. 3 for examples. This task can be considered as a general denoising problem.

In the second task, we aim to investigate the shape imposing task when the required anatomy correction is very specific and not noise-like, and the number of samples that require corrections is small. When the dataset is segmented by U-Net++, the segmentation masks do not show the aforementioned winding effect as this architecture is more robust to the speckle noise around the intima-adventitia border area. However, U-Net++ is still susceptible to the shadowing effect and to the presence of intense speckle noise in the lumen. These two artefacts lead to fragmented and merged walls, respectively. In the presence of these deformations, an accurate automatic analysis of the artery walls is very difficult. This is a more challenging task and require pinpoint corrections; therefore it is not possible to obtain meaningful improvements with general denoising algorithms. Refer to the second part of Fig. 3 for examples.

We compare AGAN against DAE and CRF, as both are common post-processing methods that are applied on binary masks.

For the first experiment we train all the methods in comparison with the binary segmentation masks produced by U-Net. We split the dataset 70% training, 15% validation and 15% testing and use expert annotations as the labels.

For the second experiment, we aim to fix the deformations of U-Net++. The deformations in the output of U-Net++ are very specific and small in number; 175 samples of the binary masks contain either merged walls or broken walls. Since this is an insufficient amount of data to conduct the training of deep learning methods, we separate these samples as the validation data. DAE and AGAN are trained on synthetic data generated by using the remaining sample's expert annotations and applying to them morphological operations that simulate fragmented and merged walls. We erase columns with a random width from artery walls to imitate the shadowing effect, and we insert columns of ones with a random width for the merged walls deformation.

As common in the literature, AGAN is trained using Adam [5] optimizer with the learning rate of 2×10^{-4}, $\beta_1 = 0.5$ and $\beta_2 = 0.99$. We leave all the weights of the loss terms (Eq. 1) as 1 and do not extensively tune the learning rates, β_1 and β_2 values to avoid overfitting to our limited data. Our experimental results show that these default hyper-parameters yield decent results. We implement the experiments in Tensorflow [1].

Results of Experiment 1. As presented in Table 1, AGAN yields the biggest improvement in F-1 score metric, 1.6 and 1.3% points; and in Hausdorff distance, 5.1 and 3.7 lower than CRF and DAE, respectively. Similarly, AGAN reduces the count of artefacts from 208 to 67. This number is 137 for DAE and 197 for CRF. The count of artefacts is simply the number samples that have at least one artefact. In this task, the majority of the performance increase for DAE and AGAN comes from the denoising of the aforementioned winding effect.

This mode of training, where the real data is used, is useful when the pixel-wise segmentation output shows noise-like characteristic. In such a setting, AGAN provides significant improvements without needing the creation of artificial data.

Table 1. Comparison of methods when the training data is segmentation outputs.

Archtiecture	F-1 Score	Hausdorff distance	# of artefacts
U-Net	81.7	16.1	208
U-Net & CRF	81.7	15.8	197
U-Net & DAE	82.0	14.4	137
U-Net & AGAN (Ours)	**83.3**	**10.7**	**67**

Table 2. Comparison of methods when the training data is synthetically generated.

Architecture	F-1 Score	Hausdorff distance	# of artefacts
U-Net++	87.3	8.3	175
U-Net++ & CRF	86.9	7.8	152
U-Net++ & DAE	85,2	9.7	157
U-Net++ & AGAN (Ours)	87.3	**7.7**	**49**

Results of Experiment 2. As shown in to Table 2, the number of artefacts are significantly reduced for U-Net++ & AGAN, from 175 to 49, whereas this number is 157 for DAE and 152 for CRF. In terms of F1 score and Hausdorff distance, no methods can provide significant improvements; however, in the next section we discuss the validity of these metrics in the context of anatomical correctness.

Training with synthetic data is useful when the needed correction is specific and requires pinpoint operations rather than general denoising.

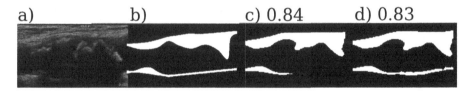

Fig. 4. a) is the grayscale image, b) is the expert annotation, c) is the segmentation, d) is the AGAN correction. The numbers above c) and d) are the corresponding F1 scores.

4.1   The Difficulty of Measuring Anatomical Correctness

We report F-1 score and Hausdorff distance for performance measurements. These metrics, as with any other metric that measures agreement between the ground truths and the predictions, do not reflect the true performance of anatomical priors. Artefacts can appear as a result of total loss of information. In such cases, it is impossible to determine the exact nature of the vessels walls. Experts deduce the missing information based on the available data. However, in many cases, there can be multiple plausible corrections which could disagree with expert annotations when measured with such metrics. If the discrepancy is greater than the deformation itself, it leads to a decrease in the score. Such examples can be seen in Fig. 4. Due to this, we include the count of artefacts.

5 Conclusion

Extracting artery walls from ultrasound images and measuring their IMTs accurately is crucial in the diagnosis of CAD. Pixel-wise segmentation models can be used to automate this process. However, due to their reliance on the local information they are not robust to noise that corrupts the locality. Given the noisy nature of the ultrasound images, relying solely on the local information results in many artefacts which lead to under or over estimation of IMTs, or to the failure of extracting the artery walls. Traditional post-processing methods like DAE and CRF fail to fix these issues. Due to this gap in the literature, we propose AGAN that can fix these issue by leveraging the global information. We consider AGAN's application to other image modalities as future work.

References

1. Abadi, M., et al.: TensorFlow: Large-scale machine learning on heterogeneous systems (2015). https://www.tensorflow.org/. Software available from tensorflow.org
2. Bouteldja, N., Merhof, D., Ehrhardt, J., Heinrich, M.P.: Deep multi-modal encoder-decoder networks for shape constrained segmentation and joint representation learning. In: Handels, H., Deserno, T.M., Maier, A., Maier-Hein, K.H., Palm, C., Tolxdorff, T. (eds.) Bildverarbeitung für die Medizin 2019, pp. 23–28. Springer Fachmedien Wiesbaden, Wiesbaden (2019)
3. Dalca, A.V., Guttag, J.V., Sabuncu, M.R.: Anatomical priors in convolutional networks for unsupervised biomedical segmentation. CoRR abs/1903.03148 (2019). http://arxiv.org/abs/1903.03148
4. Isola, P., Zhu, J.Y., Zhou, T., Efros, A.A.: Image-to-image translation with conditional adversarial networks. In: 2017 IEEE Conference on Computer Vision and Pattern Recognition (CVPR), pp. 5967–5976 (2016)
5. Kingma, D.P., Ba, J.: Adam: a method for stochastic optimization. CoRR abs/1412.6980 (2014)
6. Larrazabal, A.J., Martinez, C., Ferrante, E.: Anatomical priors for image segmentation via post-processing with denoising autoencoders. In: Shen, D., et al. (eds.) MICCAI 2019. LNCS, vol. 11769, pp. 585–593. Springer, Cham (2019). https://doi.org/10.1007/978-3-030-32226-7_65
7. Miyato, T., Kataoka, T., Koyama, M., Yoshida, Y.: Spectral normalization for generative adversarial networks. In: 6th International Conference on Learning Representations, ICLR 2018, Vancouver, BC, Canada, 30 April–3 May 2018, Conference Track Proceedings. OpenReview.net (2018). https://openreview.net/forum?id=B1QRgziT-
8. Oktay, O., et al.: Anatomically constrained neural networks (ACNN): application to cardiac image enhancement and segmentation. IEEE Trans. Med. Imaging (2017). https://doi.org/10.1109/TMI.2017.2743464
9. Ravishankar, H., Venkataramani, R., Thiruvenkadam, S., Sudhakar, P., Vaidya, V.: Learning and incorporating shape models for semantic segmentation. In: Descoteaux, M., Maier-Hein, L., Franz, A., Jannin, P., Collins, D.L., Duchesne, S. (eds.) MICCAI 2017. LNCS, vol. 10433, pp. 203–211. Springer, Cham (2017). https://doi.org/10.1007/978-3-319-66182-7_24

10. Ronneberger, O., Fischer, P., Brox, T.: U-Net: convolutional networks for biomedical image segmentation. In: Navab, N., Hornegger, J., Wells, W.M., Frangi, A.F. (eds.) MICCAI 2015. LNCS, vol. 9351, pp. 234–241. Springer, Cham (2015). https://doi.org/10.1007/978-3-319-24574-4_28
11. Sekuboyina, A., Rempfler, M., Valentinitsch, A., Kirschke, J.S., Menze, B.H.: Adversarially learning a local anatomical prior: vertebrae labelling with 2D reformations. CoRR abs/1902.02205 (2019). http://arxiv.org/abs/1902.02205
12. Shelhamer, E., Long, J., Darrell, T.: Fully convolutional networks for semantic segmentation. IEEE Trans. Pattern Anal. Mach. Intell. **39**(4), 640–651 (2017). https://doi.org/10.1109/TPAMI.2016.2572683
13. Wang, T., Liu, M., Zhu, J., Tao, A., Kautz, J., Catanzaro, B.: High-resolution image synthesis and semantic manipulation with conditional GANs. CoRR abs/1711.11585 (2017). http://arxiv.org/abs/1711.11585
14. Yi, Z., Zhang, H., Tan, P., Gong, M.: DualGAN: unsupervised dual learning for image-to-image translation. CoRR abs/1704.02510 (2017). http://arxiv.org/abs/1704.02510
15. Zhang, H., Goodfellow, I.J., Metaxas, D.N., Odena, A.: Self-attention generative adversarial networks. In: Proceedings of the 36th International Conference on Machine Learning, ICML 2019, 9–15 June 2019, Long Beach, California, USA, pp. 7354–7363 (2019). http://proceedings.mlr.press/v97/zhang19d.html
16. Zhou, Z., Rahman Siddiquee, M.M., Tajbakhsh, N., Liang, J.: UNet++: a nested U-net architecture for medical image segmentation. In: Stoyanov, D., et al. (eds.) DLMIA/ML-CDS -2018. LNCS, vol. 11045, pp. 3–11. Springer, Cham (2018). https://doi.org/10.1007/978-3-030-00889-5_1

SteGANomaly: Inhibiting CycleGAN Steganography for Unsupervised Anomaly Detection in Brain MRI

Christoph Baur[1]([✉]), Robert Graf[1], Benedikt Wiestler[4], Shadi Albarqouni[1,2], and Nassir Navab[1,3]

[1] Computer Aided Medical Procedures (CAMP), TU Munich, Munich, Germany
c.baur@tum.de
[2] Computer Vision Laboratory, ETH Zurich, Zurich, Switzerland
[3] Whiting School of Engineering, Johns Hopkins University, Baltimore, USA
[4] Department of Neuroradiology, Klinikum rechts der Isar, TU Munich, Munich, Germany

Abstract. Recently, it has been shown that CycleGANs are masters of steganography. They cannot only learn reliable mappings between two distributions without calling for paired training data, but can effectively hide information unseen during training in mapping results from which input data can be recovered almost perfectly. When preventing this during training, CycleGANs actually map samples much closer to the training distribution. Here, we propose to leverage this effect in the context of trending unsupervised anomaly detection, which primarily relies on modeling healthy anatomy with generative models. Here, we embed anomaly detection into a CycleGAN-based style-transfer framework, which is trained to translate healthy brain MR images to a simulated distribution with lower entropy and vice versa. By filtering high frequency, low amplitude signals from lower entropy samples during training, the resulting model suppresses anomalies in reconstructions of the input data at test time. Similar to Autoencoder and GAN-based anomaly detection methods, this allows us to delineate pathologies directly from residuals between input and reconstruction. Various ablative studies and comparisons to state-of-the-art methods highlight the potential of our method.

1 Introduction

Segmentation of lesions in brain MRI is a common and recurring task in the diagnosis and treatment of various neurological diseases. However, manual segmentation is time-intensive, tedious and costly. Automated methods based on

---

C. Baur and R. Graf—Contributed equally to this work.

S. Albarqouni is supported by the PRIME programme of the German Academic Exchange Service (DAAD) with funds from the German Federal Ministry of Education and Research (BMBF).

© Springer Nature Switzerland AG 2020
A. L. Martel et al. (Eds.): MICCAI 2020, LNCS 12262, pp. 718–727, 2020.
https://doi.org/10.1007/978-3-030-59713-9_69

supervised Deep Learning come close to human performance [4,14,19], but large-scale availability of manually annotated training data is paramount to their success. Autoencoders (AEs) and Generative Adversarial Networks (GANs) have recently initiated a paradigm shift towards unsupervised methods for brain lesion and anomaly segmentation. These methods do not call for pixel-precise annotations, but a set of healthy samples instead. By modeling the distribution of healthy anatomy using GANs [16,17], VAEs [3,6,12,15,21,22] and combinations of both [3], unsupervised methods have shown to be capable of segmenting arbitrary pathologies. This is essentially achieved by comparing input samples to their healthy counterparts recovered from the modeled distribution. CycleGANs, known for their ability to learn mappings between distributions at high resolution and fidelity without calling for paired training data, have been proposed in a similar, but slightly more supervised context. Andermatt et al. [1] proposed a variational CycleGAN for weakly supervised pathology segmentation by learning a mapping between the distributions of pathological and healthy brain MRI, modeling pathologies as residuals. No pixel-level annotations are required, but an additional curated set of only pathological samples needs to be provided. Instead of generating pathologous residuals, Sun et al. [18] suggest to make the pathology-to-healthy generator learn how to turn an anomaly into normal looking tissue by conditioning the translation on binary segmentation masks. Xia et al. [20] achieve a similar effect by factorizing pathologies with a separate segmentation network. Closely related is the work from Chartsias et al. [5], who generally disentangle anatomical information from imaging factors with Cycle-GANs to aid the training of domain invariant segmentation models. All latter approaches require segmentation ground-truth, though.

In contrast, here we leverage CycleGANs for unsupervised anomaly segmentation in brain MRI without requiring any pixel-level labels for anomalies. More precisely, we embed the task into a style-transfer framework which learns to translate between distributions of only healthy data in different styles. As we show, modeling style transfer among healthy data only is not sufficient to make anomalies disappear in reconstructions of input data, as the CycleGAN is a master of steganography [7]: When propagating unseen samples through a trained model, it can hide imperceptible information in the translated images, from which the input can be recovered almost perfectly. To alleviate this, we propose SteGANomaly: We display that inhibiting the steganographic capabilities of CycleGANs by removing high frequency, low amplitude signal during training forces the model to remove any anomalies present in input images during inference. As a result, anomalies can be delineated by simply comparing input samples to their reconstructions. We validate different hyperparameters and design choices of our framework in an ablative study. The best configuration of our method is compared to a variety of different State-of-the-Art methods for brain anomaly segmentation, demonstrating higher fidelity and superior performance in the tasks of unsupervised MS lesion and tumor segmentation using five different datasets.

2   Methodology

We propose to model healthy brain anatomy through an unpaired image-to-image translation framework for the task of unsupervised anomaly segmentation. To achieve this, we leverage an asymmetric CycleGAN which is trained from real and simulated healthy MR scans together with an anatomical segmentation network. The anatomical segmentation network is key to preserve anatomical coherence during style transfer.

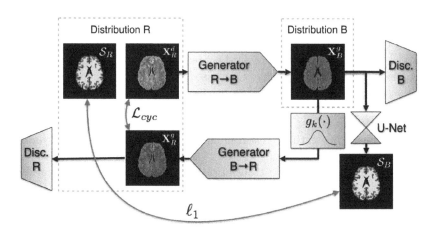

Fig. 1. The proposed CycleGAN-inspired framework for transferring real data to the simulated domain, asymmetrically enhanced by a U-Net.

2.1   Overall Concept

At the heart of our framework is a CycleGAN, composed of a generator $G_{R \to B} : \mathbf{X}_R \to \mathbf{X}_B$, which is trained to map samples coming from a domain of real, healthy MR images $\mathbf{X}_R$ to the space of simulated, healthy MRI $\mathbf{X}_B$, and another generator $G_{B \to R} : \mathbf{X}_B \to \mathbf{X}_R$ which establishes the reverse mapping. Images of both domains can either come from the respective data distribution p_d, i.e. $\mathbf{X}_R^d$ or $\mathbf{X}_B^d$ are samples from the training data, or from the generator distribution p_g, i.e. the output $\mathbf{X}_R^g$ or $\mathbf{X}_B^g$ of one the generators. In the following, we refer to the generated $\mathbf{X}_B^g$ in the style of simulated data also as the *intermediate representation*. Before passing these directly to $G_{B \to R}$ when performing a complete cycle, $\mathbf{X}_B^g$ are first convolved with an isotropic $k \times k$ Gaussian kernel $g_k(\cdot)$ to remove high frequency, low amplitude signals. As is experimentally validated, this prevents the CycleGAN from preserving anomalies during the translation process, and the degree of prevention can be controlled by the filter size. To complete the CycleGAN, there are two adversarial discriminator networks D_R and D_B, which distinguish between real and synthetic samples of each domain.

U-Net for Anatomical Segmentation—In addition, during training, we employ a pre-trained U-Net $U : \mathbf{X}_B \to \mathcal{S}_B$, whose purpose is to produce anatomical segmentations $\mathcal{S}_B$ only for the intermediate representation $\mathbf{X}_B^g$ (Fig. 1). These anatomical segmentations are matched against corresponding ground-truth anatomical segmentations $\mathcal{S}_R$ of real input samples using a pixel-wise loss function. Backpropagation through the frozen U-Net ensures that the CycleGAN preserves size and anatomical coherence of brains in intermediate translation results. Before training the CycleGAN, U is trained from data $\in \mathbf{X}_B^d$ to segment simulated, healthy axial MR slices into four different anatomical classes, i.e. background, White Matter (WM), Gray Matter (GM) and Brainstem. The model is optimized using a standard combination of Binary Cross Entropy (BCE) and DICE Loss:

$$\mathcal{L}_U = \mathcal{L}_{BCE} + \lambda \mathcal{L}_{DICE} \tag{1}$$

The U-Net not only enforces the aforementioned anatomical coherence, but also acts as a proxy to control the homogeneity of the respective anatomical regions in the simulated space and thus limits the CycleGAN in its creativity.

CycleGAN-Training—Afterwards, we fix the weights of the trained segmentation model U and attach it to the output of generator $G_{R \to B}$. The CycleGAN is trained using the standard objective to learn a style transfer between the domains of real and simulated data. However, for every generated sample $\mathbf{X}_B^g$ from the simulated domain, an anatomical segmentation is also produced by the U-Net, and the resulting segmentation is subject to a pixel-wise comparison to ground-truth anatomical segmentation $\mathcal{S}_R$ of the corresponding real sample $\mathbf{X}_R^d$ from the data distribution by means of the ℓ_1-distance. Notably, we used FreeSurfer [9] to cheaply obtain $\mathcal{S}_R$, and we assume that such $\mathcal{S}_R$ are readily available. The total loss is

$$\mathcal{L}(G_{R \to B}, G_{B \to R}, D_R, D_B) = \mathcal{L}_{GAN}(G_{R \to B}, D_B) + \mathcal{L}_{GAN}(G_{B \to R}, D_R)$$
$$+ \lambda_{cyc} L_{cyc}(G_{R \to B}, G_{B \to R})$$
$$+ \lambda_{as} \ell_1(U(G_{R \to B}), \mathcal{S}_R)$$

with

$$\mathcal{L}_{cyc}(G_{R \to B}, G_{B \to R}) = \mathbb{E}_{\mathbf{x}_R \sim p_d}[\|G_{B \to R}(g_k(G_{R \to B}(\mathbf{x}_R))) - \mathbf{x}_R\|_1]$$
$$+ \mathbb{E}_{\mathbf{x}_B \sim p_d}[\|G_{R \to B}(g_k(G_{B \to R}(\mathbf{x}_B))) - \mathbf{x}_B\|_1],$$

which is a combination of the classic CycleGAN and a one-sided ℓ_1-loss on the $\mathcal{S}_R$ of a real sample $\mathbf{x}_R$ and the U-Net segmentation of the corresponding sample $\mathbf{x}_B^g$. Note how the Gaussian filtering was incorporated into the Cycle-Consistency loss term.

2.2 Anomaly Segmentation in the Simulated Space

In order to delineate lesions in query data $\mathbf{x}^*$, we use a simple, but effective method. We compute the residuals $\mathbf{r}$ between $\mathbf{x}^*$ and its reconstruction $\hat{\mathbf{x}}^*$, which is the result of a complete cycle:

$$\hat{\mathbf{x}}^* = G_{B \to R}(g_k(G_{R \to B}(\mathbf{x}^*)))$$

$$\mathbf{r} = |\mathbf{x}^* - \hat{\mathbf{x}}^*|$$

A binary delineation can be obtained from these residuals $\mathbf{r}$ via thresholding with threshold t:

$$\mathbf{s} = \mathbf{r} \geq t$$

3   Experiments and Results

3.1   Dataset

In our experiments, we make use of five different datasets. Our clinical partners at Klinikum rechts der Isar kindly provided us with a dataset $\mathcal{D}_{healthy}$, containing brain MR scans from 100 healthy individuals, and a dataset $\mathcal{D}_{MS}$ of 49 subjects with annotated MS lesions. Both datasets consist of FLAIR and T1-weighted scans, acquired with a single Philips Achieva 3 T scanner. Further, we conduct our experiments on the publicly available $\mathcal{D}_{WMH}$ dataset comprising 60 pairs of brain MR scans (T1 and FLAIR) with expert annotations of white-matter hyper-intensities [11], and on $\mathcal{D}_{TCIA}$, which comprises MR scans of 26 subjects with brain tumors, randomly extracted from the collection described in [2]. A simulated FLAIR phantom of the healthy brain with anatomical segmentation ground-truth was created with Brainweb [8] and served as our source to generate a low-entropy intermediate distribution. A complete dataset $\mathcal{D}_B$ with $\mathcal{S}_B$ was obtained by randomly applying small elastic deformations to the single phantom and its anatomical segmentations, yielding a variety of 100 simulated FLAIR scans.

Preprocessing—The resolution and orientation of all scans has been harmonized by registering them to the SRI24 ATLAS [13], skull-stripping with ROBEX [10] and normalizing them into the range of $[-1; 1]$.

3.2   U-Net Training

Upfront, we trained an adaptation of the U-Net from axial slices in $\mathcal{D}_B$ and their ground-truth $\mathcal{S}_B$ as described in Subsect. 2.1 for 25 epochs. To rate the quality of the anatomical segmentation produced by the U-Net and to stop training, we monitored the validation loss on the original, non-deformed phantom.

3.3   Ablation Study

Here, we investigate the effect of different kernel sizes k as well as the application of the kernel during training, testing or both on the preservation of anomalies. Therefor, we train different configurations of our asymmetric CycleGAN from pathology-free data, establishing a mapping between $\mathcal{D}_{healthy}$ and $\mathcal{D}_B$, and test

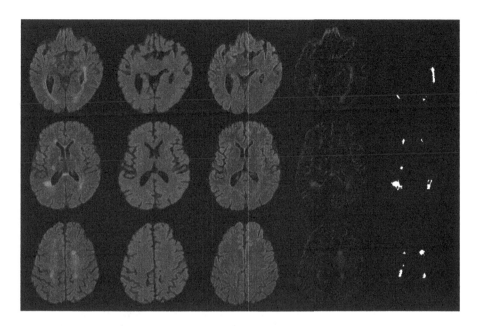

Fig. 2. A journey through our SteGANomaly. 1st column: axial input slices; 2nd column: translated, intermediate images; 3rd column: reconstructions; 4th column: residuals; 5th column: ground-truth lesion segmentations

Table 1. Ablation study on filter sizes and training/testing with/without Gaussian filtering as well as a comparison to SOTA methods on $\mathcal{D}_{MS}$.

Filter		MS		TCIA		WMH	
Training	Testing	AUPRC	⌈DICE⌉	AUPRC	⌈DICE⌉	AUPRC	⌈DICE⌉
k = 3	–	0.582	0.555	0.697	0.680	0.462	0.381
k = 3	k = 3	0.296	0.379	0.559	0.554	0.287	0.267
k = 5	–	0.563	0.546	0.724	0.696	0.461	0.369
k = 5	k = 5	0.540	0.543	0.654	0.632	0.393	0.326
k = 9	–	0.608	0.576	0.702	0.678	0.473	0.371
k = 9	k = 9	0.521	0.533	0.648	0.620	0.426	0.387
–	–	0.212	0.314	0.476	0.521	0.133	0.223
–	k = 9	0.054	0.127	0.329	0.478	0.064	0.094
SOTA method		MS		SOTA method		MS	
AE (dense) [3]		0.414	0.473	f-AnoGAN [16]		0.267	0.380
AE (spatial) [3]		0.213	0.317	Context VAE [22]		0.434	0.487
VAE (dense) [3]		0.283	0.372	GMVAE [21]		0.495	0.522

them on $\mathcal{D}_{MS}$, $\mathcal{D}_{WMH}$ as well as $\mathcal{D}_{TCIA}$. The configurations and anomaly delineation results are reported in Table 1. The performance is measured in terms of the area under the ROC and Precision-Recall curve (AUROC and AUPRC) as well as the generally best achievable DICE-score $\lceil$DICE$\rceil$ (a theoretical upper bound). Some visual examples of the style-transfer with the different configurations are provided in Fig. 3: Without any filter (2nd column, 1st row), the CycleGAN reproduces the lesions almost perfectly. Blurring the intermediate images in that model during test-time leads to a completely deformed reconstruction (2nd column, 2nd row). Training with the filter, but omitting it during testing (3rd & 4th column, 2nd row) causes the lesions to disappear as desired. Additionally using the filter during testing leads to denoised, sketchy-looking images, where hints of lesions still remain (1st row, 3rd & 4th column). Apparently, the model is able to partially recover lesions from blurry images, on which it has been trained. As can be seen from Table 1, increasing the size and strength of the filter from $k = 3$ to $k = 9$ leads to improvements in anomaly delineation.

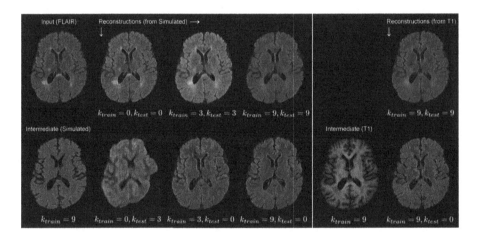

Fig. 3. Input slice, different intermediate representations and reconstructions using different SteGANomaly configurations. k_{train} and k_{test} indicate the size of the Gaussian filter used during model training and/or testing, respectively, whereby $k = 0$ means the filter has not been used.

3.4   Choice of Intermediate Distribution

In our previous experiments, we relied on a simulated dataset of healthy FLAIR MRI with low entropy to model the intermediate representation. Here, we investigate the importance of choosing the right data by replacing the simulated $\mathcal{D}_B$ with more complex, real T1 data from $\mathcal{D}_{healthy}$, i.e. we optimize for a style-transfer from healthy FLAIR to T1 and vice versa. After training a U-Net for anatomical segmentation on the T1 data, we again trained our best CycleGAN candidate configuration and compare both testing with and without Gaussian filtering. Similar to our previous experiments, visual results in Fig. 3 (column 6) reveal that pathologies are preserved when applying the filter also during testing as well (6th column, 1st row). However, when omitting the filter during testing, checkerboard patterns appear in the intermediate representation (5th column, 2nd row), and these are propagated to the reconstruction as well (6th column, 2nd row). Contrary to our expectations, suppressing anomalies still seems to work well, even though the intermediate distribution has much higher entropy.

3.5 Comparison to State-of-the-Art

Finally, we compare our best model ($k_{train} = 9$, $k_{test} = 0$) against a variety of State-of-the-Art methods on $\mathcal{D}_{MS}$ in terms of the AUPRC and $\lceil$DICE$\rceil$. All results are reported in Table 1. The overall best performing model thereby is our approach with configuration $k_{train} = 9$, $k_{test} = 0$, which shows noticeable improvements in AUPRC over all competing methods.

3.6 Discussion

With our style-transfer-based method we outperform State-of-the-Art approaches in the task of unsupervised anomaly segmentation. Our ablative studies show that i) choosing a sufficiently large Gaussian kernel for the suppression of steganographic, high frequency codes in the intermediate representation and ii) modeling an appropriate, low entropy intermediate distribution are key factors to the success of our method. Once more, we want to emphasize the importance of forcing the CycleGAN to preserve anatomical coherence when translating data to the low entropy distribution. Only when leveraging a pretrained U-Net, we were able to maintain correct geometry in reconstructions. Generally, we find the style-transfer nature of this approach to be very valuable in the context of high resolution data, on which AE-based methods are known to struggle. In contrast to those, our style-transfer preserves important details at high fidelity and works equally well on all axial slices (Fig. 2). Notably, our method is able to turn anomalies of varying size and intensity into "normal"-looking tissue, ranging from small MS lesions to large, heterogeneous tumors.

4   Conclusion

In summary, we presented a novel method towards unsupervised anomaly detection in brain MRI by embedding the modeling of healthy anatomy into a CycleGAN-based style-transfer task. Three factors are particularly important for the success of our method: i) the mapping of real healthy data to a distribution with lower entropy, ii) inhibiting the CycleGAN from hiding information in high frequency components by applying a Gaussian filter with sufficient kernel size during training and iii) enforcing the model to preserve anatomical coherence when mapping to the intermediate distribution. Our method outperforms the state-of-the-art in various measures and is able to deal with high resolution data, a current pitfall of AE-based methods. In future, we would like to embed the proposed method into a probabilistic framework to potentially replace the current residual-driven anomaly detection with likelihoods.

References

1. Andermatt, S., Horváth, A., Pezold, S., Cattin, P.: Pathology segmentation using distributional differences to images of healthy origin. In: Crimi, A., Bakas, S., Kuijf, H., Keyvan, F., Reyes, M., van Walsum, T. (eds.) BrainLes 2018. LNCS, vol. 11383, pp. 228–238. Springer, Cham (2019). https://doi.org/10.1007/978-3-030-11723-8_23
2. Bakas, S., et al.: Advancing the cancer genome atlas glioma MRI collections with expert segmentation labels and radiomic features. Sci. Data **4**, 170117 (2017)
3. Baur, C., Wiestler, B., Albarqouni, S., Navab, N.: Deep autoencoding models for unsupervised anomaly segmentation in brain MR images. arXiv preprint arXiv:1804.04488 (2018)
4. Brosch, T., Tang, L.Y., Yoo, Y., Li, D.K., Traboulsee, A., Tam, R.: Deep 3D convolutional encoder networks with shortcuts for multiscale feature integration applied to multiple sclerosis lesion segmentation. IEEE Trans. Med. Imaging **35**(5), 1229–1239 (2016)
5. Chartsias, A., et al.: Factorised spatial representation learning: application in semi-supervised myocardial segmentation. In: Frangi, A.F., Schnabel, J.A., Davatzikos, C., Alberola-López, C., Fichtinger, G. (eds.) MICCAI 2018. LNCS, vol. 11071, pp. 490–498. Springer, Cham (2018). https://doi.org/10.1007/978-3-030-00934-2_55
6. Chen, X., Konukoglu, E.: Unsupervised detection of lesions in brain MRI using constrained adversarial auto-encoders. arXiv preprint arXiv:1806.04972 (2018)
7. Chu, C., Zhmoginov, A., Sandler, M.: Cyclegan, a master of steganography. arXiv preprint arXiv:1712.02950 (2017)
8. Cocosco, C.A., Kollokian, V., Kwan, R.K.S., Pike, G.B., Evans, A.C.: Brainweb: online interface to a 3D MRI simulated brain database. In: NeuroImage, Citeseer (1997)
9. Fischl, B.: Freesurfer. Neuroimage **62**(2), 774–781 (2012)
10. Iglesias, J.E., Liu, C.Y., Thompson, P.M., Tu, Z.: Robust brain extraction across datasets and comparison with publicly available methods. IEEE Trans. Med. Imaging **30**(9), 1617–1634 (2011)
11. Kuijf, H.J., et al.: Standardized assessment of automatic segmentation of white matter hyperintensities and results of the WMH segmentation challenge. IEEE Trans. Med. Imaging **38**(11), 2556–2568 (2019)
12. Pawlowski, N., et al.: Unsupervised lesion detection in brain CT using Bayesian convolutional autoencoders (2018)
13. Rohlfing, T., Zahr, N.M., Sullivan, E.V., Pfefferbaum, A.: The SRI24 multichannel atlas of normal adult human brain structure. Hum. Brain Mapp. **31**(5), 798–819 (2009)
14. Roy, S., Butman, J.A., Reich, D.S., Calabresi, P.A., Pham, D.L.: Multiple sclerosis lesion segmentation from brain MRI via fully convolutional neural networks. arXiv preprint arXiv:1803.09172 (2018)

15. Sato, D., et al.: A primitive study on unsupervised anomaly detection with an autoencoder in emergency head CT volumes. In: Medical Imaging 2018: Computer-Aided Diagnosis, vol. 10575, p. 105751P. International Society for Optics and Photonics (2018)

16. Schlegl, T., Seeböck, P., Waldstein, S.M., Langs, G., Schmidt-Erfurth, U.: f-anogan: fast unsupervised anomaly detection with generative adversarial networks. Med. Image Anal. **54**, 30–44 (2019)

17. Schlegl, T., Seeböck, P., Waldstein, S.M., Schmidt-Erfurth, U., Langs, G.: Unsupervised anomaly detection with generative adversarial networks to guide marker discovery. In: Niethammer, M., et al. (eds.) IPMI 2017. LNCS, vol. 10265, pp. 146–157. Springer, Cham (2017). https://doi.org/10.1007/978-3-319-59050-9_12

18. Sun, L., Wang, J., Ding, X., Huang, Y., Paisley, J.: An adversarial learning approach to medical image synthesis for lesion removal. arXiv preprint arXiv:1810.10850 (2018)

19. Valverde, S., et al.: Improving automated multiple sclerosis lesion segmentation with a cascaded 3D convolutional neural network approach. NeuroImage **155**, 159–168 (2017)

20. Xia, T., Chartsias, A., Tsaftaris, S.A.: Adversarial pseudo healthy synthesis needs pathology factorization. In: MIDL (2019)

21. You, S., Tezcan, K., Chen, X., Konukoglu, E.: Unsupervised lesion detection via image restoration with a normative prior. In: International Conference on Medical Imaging with Deep Learning-Full Paper Track (2018)

22. Zimmerer, D., Kohl, S.A., Petersen, J., Isensee, F., Maier-Hein, K.H.: Context-encoding variational autoencoder for unsupervised anomaly detection. arXiv preprint arXiv:1812.05941 (2018)

Flow-Based Deformation Guidance for Unpaired Multi-contrast MRI Image-to-Image Translation

Toan Duc Bui[1(✉)], Manh Nguyen[1,2], Ngan Le[3], and Khoa Luu[3]

[1] VinAI Research, Hanoi, Vietnam
v.toanbd1@vinai.io
[2] FPT University, Hanoi, Vietnam
[3] Department of Computer Science, University of Arkansas in Fayetteville,
Fayetteville, USA

Abstract. Image synthesis from corrupted contrasts increases the diversity of diagnostic information available for many neurological diseases. Recently the image-to-image translation has experienced significant levels of interest within medical research, beginning with the successful use of the Generative Adversarial Network (GAN) to the introduction of cyclic constraint extended to multiple domains. However, in current approaches, there is no guarantee that the mapping between the two image domains would be unique or one-to-one. In this paper, we introduce a novel approach to **unpaired image-to-image translation** based on the **invertible architecture**. The invertible property of the flow-based architecture assures a cycle-consistency of image-to-image translation without additional loss functions. **We utilize the temporal information between consecutive slices to provide more constraints to the optimization for transforming one domain to another in unpaired volumetric medical images**. To capture temporal structures in the medical images, we explore the displacement between the consecutive slices using a deformation field. In our approach, the deformation field is used as a guidance to keep the translated slides realistic and consistent across the translation. The experimental results have shown that the synthesized images using our proposed approach are able to archive a competitive performance in terms of mean squared error, peak signal-to-noise ratio, and structural similarity index when compared with the existing deep learning-based methods on three standard datasets, i.e. HCP, MRBrainS13 and Brats2019.

Keywords: Flow-based generator · Image-to-image translation · cycleGAN

1 Introduction

In medical imaging, the task of obtaining diagnostic images from multiple modalities is necessary for accurate and comprehensive prediction of disease diagnosis.

© Springer Nature Switzerland AG 2020
A. L. Martel et al. (Eds.): MICCAI 2020, LNCS 12262, pp. 728–737, 2020.
https://doi.org/10.1007/978-3-030-59713-9_70

For example, T1-weighted (T1) brain images provide clear differentiate images of gray and white matter tissues, whereas T2-weighted (T2) images differentiate fluid from cortical tissue. By leveraging the information provided by both of these image modalities, we can gain a more in-depth and completed picture of the diagnosis. However, obtaining separately both images is often costly, time-consuming, and maybe corrupted by noise and artifacts. Therefore, cross-modalities synthesis is a promising application to improve the clinical feasibility and utility of multi-contrast MRI. Image-to-image translation has recently gained attention in the medical imaging community, where the task is to estimate the corresponding image in the target domain from a given source domain image of the same subject. Generally, the image-to-image translation methods can be divided into two categories including: Generative Adversarial Networks (GANs) and Flow-based Generative Networks and summarized as follows:

Generative Adversarial Networks GANs are a class of latent variable generative models that clearly identify the generator as *deterministic mapping*. The deterministic mapping represents an image as a point in the latent space without regarding its feature ambiguity. Several different GAN-based models have been used to explore image-to-image translation in a literature study [2, 3, 14, 16]. For example, Zhu et al. [16] proposed a cycleGAN method for mapping between unpaired domains by using cycle-consistency dependence to constrain the optimal solutions provided by the generative network. Balakrishnan et al. [2] proposed a RecycleGAN to explore the temporal information by learning a prediction of the next frame for video generation. Chen et al. [3] proposed a 3D cycleGAN network to learn the mapping between CT and MRI. The drawback of 3D cycleGAN is it is memory consumption and loses the global information due to working on small patch sizes.

Flow-based Generative Networks are a class of latent variable generative models that clearly identify the generator as an *invertible mapping*. The invertible mapping provides a distributional estimation of features in the latent space. Recently, many efforts making use of flow-based generative networks have been proposed to transfer between two unpaired data [4, 5, 7, 10, 12]. For example, Grover et al. [5] introduced a flow to flow (alignflow) network for unpaired image-to-image translation. Sun et al. [12] introduced a conditional dual flow-based invertible network to transfer between positron emission tomography (PET) imaging and magnetic resonance imaging (MRI) images. By using invertible properties, the flow-based methods can ensure exact cycle consistency in translation from a source domain to the target and returning to the source domain without any further loss functions.

Limitations of Existing Methods and Our Contributions. The primary drawback of the cycleGAN model is that it can not perform one-to-one mapping for accurate and unique unpaired image translation, generates biased image translations of the inverse mapping [11]. Different from the GANs-based method, the flow-based method guarantees precise cycle consistency in mapping data points from a source domain to the target and returning to the source domain. However, the flow-based methods do not take into account the temporal information between consecutive slices. To address this problem, we propose a new method by inheriting the merits of the flow-based method and exploiting temporal information between consecutive slices. Our approach provides more constraints to the optimization for transforming one domain to another domain. To capture temporal information, we employ a deformation field between consecutive slices by training a convolutional neural network. In our proposed approach, the deformation field plays a role of guidance to keep slices realistic and consistent across translation.

2   Related Work

2.1   Cycle-Consistent Adversarial Networks (cycleGAN)

Let $\{x_i\}_{i=1}^N$ and $\{y_i\}_{i=1}^M$ be unpaired data samples for two domains, i.e. the source domain X and the target domain Y, respectively. Denote D and G as a discriminator network and a generator network. The cycleGAN model [16] solves unpaired image-to-image translation between these two domains by estimating two independent mapping functions $G_{X \to Y} : X \to Y$ and $G_{Y \to X} : Y \to X$. The two mapping functions $G_{X \to Y}$ and $G_{Y \to X}$ performed by neural networks are trained to fool the discriminator D_X and D_Y respectively. The discriminator D_X, and D_Y encourage the transferred images and the real images to be similar. Hence, the cycleGAN loss is defined as:

$$
\begin{aligned}
\mathcal{L}_{cycleGAN}(G_{X \to Y}, G_{Y \to X}, D_X, D_Y) &= \mathcal{L}_{GAN}(G_{X \to Y}, D_Y) + \mathcal{L}_{GAN}(G_{Y \to X}, D_X) \\
&+ \lambda \mathcal{L}_{cycle}(G_{X \to Y}, G_{Y \to X}) + \beta \mathcal{L}_{identity}(G_{X \to Y}, G_{Y \to X})
\end{aligned}
\tag{1}
$$

where $\mathcal{L}_{GAN}$ is a GAN loss for the D network [16]. $\mathcal{L}_{cycle}$ is a cycle consistency loss that guarantees the transferred image from a time-point is able to bring back to the original image after appearance translation by the generator network G. For example, the cycle consistency loss of the data translated from $X \to Y$ via G_X and mapped back to the original domain X via G_Y is defined as:

$$
\mathcal{L}_{cycle}(G_{X \to Y}, G_{Y \to X}) = \|G_{Y \to X}(G_{X \to Y}(x)) - x\|_1
\tag{2}
$$

The identity loss $\mathcal{L}_{identity}$ is to regularize the generator to be near an identity mapping when real samples of the target domain are given as the input to the generator. The λ and β control the contribution of the two objective functions.

2.2   Flow-Based Generative Models

Flow-based Generative Models are a class of latent variable generative models that clearly identify the generator as an invertible mapping $h : Z \to X$ between a set of latent variables Z and a set of observed variables X. Let p_X and p_Z

indicate the marginal densities given by the model over X and Z, respectively. Using the change-of-variables formula, these marginal densities are defined as

$$p_X(x) = p_Z(z) \left| \det \frac{\partial h^{-1}}{\partial X} \right|_{X=x} \qquad (3)$$

where $z = h^{-1}(x)$ because of the invertibility constraints. In particular, we use a multivariate Gaussian distribution $p_Z(z) = \mathcal{N}(\mu, 0, \mathbf{I})$. Unlike adversarial training, flow models trained with maximum likelihood estimation (MLE) explicitly require a prior $p_Z(z)$ with a tractable density to evaluate model likelihoods using the change-of-variables formula in the Eq. (3).

Based on flow-based method [4], Grover et al. [5] proposed an alignflow method for unpaired image-to-image translation. In the method, the mapping between two domains $X \rightarrow Y$ can be represented through a shared feature space of latent variables Z by the composition of two invertible mapping [5]:

$$G_{X \rightarrow Y} = G_{Z \rightarrow Y} \circ G_{X \rightarrow Z}, \qquad G_{Y \rightarrow X} = G_{Z \rightarrow X} \circ G_{Y \rightarrow Z} \qquad (4)$$

where $G_{X \rightarrow Z} = G_{Z \rightarrow X}^{-1}$ and $G_{Y \rightarrow Z} = G_{Z \rightarrow Y}^{-1}$. Due to the fact that composition of invertible mappings is invertible, both $G_{X \rightarrow Y}$ and $G_{Y \rightarrow X}$ are invertible [5]. On the other hand, we can obtain $G_{X \rightarrow Y}^{-1} = G_{Y \rightarrow X}$. Thus the Eq. (2) can rewrite as

$$\begin{aligned} \mathcal{L}_{cycle}(G_{X \rightarrow Y}, G_{Y \rightarrow X}) &= \|G_{Y \rightarrow X}(G_{X \rightarrow Y}(x)) - x\|_1 \\ &= \left\|G_{X \rightarrow Y}^{-1}(G_{X \rightarrow Y}(x)) - x\right\|_1 = 0 \end{aligned} \qquad (5)$$

where $G_{X \rightarrow Y}^{-1} G_{X \rightarrow Y}$ results in an identical matrix.

Equation 5 implies that the flow-based methods can guarantee precise cycle consistency in mapping from a source domain to the target and returning to the source domain without additional loss functions. Hence, the alignflow objective loss is defined as:

$$\begin{aligned} \mathcal{L}_{flow}(G_{X \rightarrow Y}, G_{Y \rightarrow X}, D_X, D_Y) &= \mathcal{L}_{GAN}(G_{X \rightarrow Y}, D_Y) + \mathcal{L}_{GAN}(G_{Y \rightarrow X}, D_X) \\ &\quad - \lambda_X \mathcal{L}_{MLE}(G_{Z \rightarrow X}) - \lambda_Y \mathcal{L}_{MLE}(G_{Z \rightarrow Y}) \end{aligned} \qquad (6)$$

where $\lambda_Y, \lambda_Y \geq 0$ are hyperparameters that control the importance of the MLE terms for domains X and Y respectively.

Figure 1 illustrates the difference between cycleGAN and alignflow methods. Unlikes cycleGAN, the alignflow method is the full invertible architecture that guarantees the cycle-consistency translations between two unpaired domains without an additional $\mathcal{L}_{cycle}$ function.

3   Proposed Method

Our motivation is to learn a mapping between unpaired images from different domains by leveraging the temporal information between consecutive slices. We use the temporal information to constrain the mapping between two domains which should be consistent. Our method is an extension of alignflow [5] method with making use of temporal information between consecutive slides.

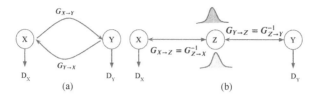

Fig. 1. A comparison between (a) cycleGAN and (b) alignflow generative model. Double-headed arrows denotes an invertible mapping

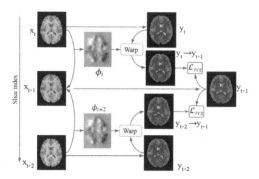

Fig. 2. Deformation Guided Temporal Constraints for domain Y

3.1   Deformation Guided Temporal Constraints

To obtain the displacement between consecutive slices, we use an unsupervised registration network [1] to learn a deformation field ϕ of a slice x_t and its consecutive slices x_k. The deformation field ϕ can be obtained using a convolutional neural network (CNN) [1] by minimizing the loss function

$$\mathcal{L}(\phi) = \|x_t - (x_k \bigcirc \phi(x_t, x_k))\|_2 + \|\nabla\phi\|_2 \tag{7}$$

where $\bigcirc$ denotes the spatial transformation operation. The first term ensures that the distance between the next slice x_t and the warped current slice $x_k \bigcirc \phi(.)$ to be close. The second term imposes regularization on $\phi(.)$.

To guarantee the consistency of the image translation, the $\mathcal{L}_1$ loss is used to measure the difference between the warping of fake images on consecutive slice t^{th} and the translation of reference slice k^{th}. We define the temporal consistency loss function for mapping $X \to Y$ and $Y \to X$ as:

$$
\begin{aligned}
\mathcal{L}_{reg}(X, G_{X \to Y}) &= \sum_{k=0, k \neq t}^{n} \|G_{X \to Y}(x_t) - G_{X \to Y}(x_k) \bigcirc \phi(x_t, x_k)\|_1 \\
\mathcal{L}_{reg}(Y, G_{Y \to X}) &= \sum_{k=0, k \neq t}^{n} \|G_{Y \to X}(y_t) - G_{Y \to X}(y_k) \bigcirc \phi(y_t, y_k)\|_1
\end{aligned}
\tag{8}
$$

Figure 2 illustrates an example for image-to-image translation from domain $X \to Y$ using temporal constraints. Let x_t, x_{t+1}, x_{t+2} be consecutive slices of

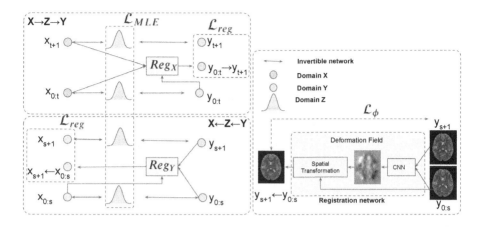

Fig. 3. Our flow-based deformation guidance approach for unpaired image-to-image translation.

real images in the source domain X. A mapping function $G_{X \to Y}$ generates the fake image y_t, y_{t+1}, y_{t+2} on target domain Y. On the source domain, we can learn displacement fields $\phi_t(.), \phi_{t+2}(.)$ between (x_t, x_{t+1}) and (x_{t+2}, x_{t+1}). To constrain the consistency of the mapping from $X \to Y$, we minimize the distance (i) between the warped fake image $y_t \bigcirc \phi_t(.)$ and y_{t+1} for mapping from t^{th} slice and $(t+1)^{th}$ slice, and (ii) between the warped fake image $y_{t+2} \bigcirc \phi_{t+2}(.)$ and y_{t+1} for mapping from $(t+2)^{th}$ slice and $(t+1)^{th}$ slice.

3.2 Network Diagram

Figure 3 illustrates the proposed network diagram for unpaired image-to-image translation. Our proposed network architecture inherits the advantages of invertible property of alignflow [5]. During training, we add two additional networks Reg_X and Reg_Y for each domain to learn the deformation field $\phi(.)$. These additional networks only use in training time, without increasing the model complexity and inference time comparison with the baseline flow-based method. The temporal constraint via $\mathcal{L}_{reg}(.)$ losses ensures the mapping of consecutive slices on the source domain should be consistent on the target domain. Finally, our objective function is defined as:

$$\begin{aligned}
\mathcal{L}_{flow\_reg}(G_{X \to Y}, G_{Y \to X}, D_X, D_Y, \phi) &= \mathcal{L}_{flow}(G_{X \to Y}, G_{Y \to X}, D_X, D_Y) \\
&+ \lambda_1 \mathcal{L}_{reg}(X, G_{X \to Y}) + \lambda_2 \mathcal{L}_{reg}(Y, G_{Y \to X}) + \beta_1 \mathcal{L}_X(\phi) + \beta_2 \mathcal{L}_Y(\phi) \\
&+ \gamma_1 \mathcal{L}_{TV}(X) + \gamma_2 \mathcal{L}_{TV}(Y)
\end{aligned}$$

$$(9)$$

where $\lambda_1, \lambda_2, \beta_1$, and β_2 control the relative importance of the temporal consistence losses and the two registration losses. $\mathcal{L}_{TV}$ denotes total variation (TV) loss to impose spatial smoothness by measuring the horizontal and vertical gradient of generated images [15]. These TV losses are weighted by γ_1, γ_2.

4   Experimental Results

4.1   Datasets and Training

We used common medical datasets to measure the robustness of our method against the existing methods: cycleGAN [16], recycleGAN [2], cycleflow [11] and alignflow [5]. cycleGAN [16] is an unpaired image-to-image translation that works on single slice level. RecycleGAN [2] built upon the cycleGAN and add a temporal predictor that is trained to predict future slice in a set of previous consecutive slices. cycleflow [11] is a flow-based method, but ignores the shared latent space Z (directly map from $X \rightarrow Y$, instead of $X \rightarrow Z \rightarrow Y$ as the alignflow method). The synthetic image from each method was quantitatively compared with the real paired image using the following performance metrics: mean squared error (MSE), peak signal-to-noise ratio (PSNR), and structural similarity index (SSIM).

Human Connectome Project (HCP) is provided by the Human Connectome project [13]. We used T1 as the source domain and T2 as the target domain. We extract the axial view of T1/T2 images into 2D images. We split the 2D images into 1150 images for training set and 500 images for testing set.

MRBrainS13: [8] contains 15 subjects for training and validation and 6 subjects for testing. For each subject, two modalities are available that include T1-weighted, and T2-FLAIR with an image size of $48 \times 240 \times 240$. We extract the dataset into 2D images with 450 images for training and 150 images for testing

Brats2019: [9] includes 210 HGG scans and 75 LGG scans. Each scan has a dimension of $240 \times 240 \times 155$. For each scan, we extract it to 2D images and use 770 images for training and 250 images for testing.

Training. All networks were implemented using the Pytorch framework and trained on the 12GB GPU. The input image is resized to 128×128 and normalized to $[-1, 1]$. We used axial slices (10 slices around the middle slice) from the each subject. The Adam optimizer with a batch size of two was used to train the network. The initialization learning rate was set as 0.0002 and was decreased ten times every 20 epochs. We trained each model for 100 epochs. The balance weights were set as $\lambda_X = \lambda_Y = 1e^{-5}, \lambda = \lambda_1 = \lambda_2 = 10, \beta_1 = \beta_2 = 1, \gamma_1 = \gamma_2 = 1$. The discriminator network is a 70×70 PatchGAN [6]. For alignflow network [5], we set the number of scale was 1, number of block was 3. We use two consecutive slices (before and later slices) to learn the temporal constraint.

4.2   Performance Evaluation

Qualitative Evaluation. Figure 4 illustrates the image translation on different datasets. The proposed methods (in the last column) provided a better synthetic image, resulting in better MSE, SSIM and PSNR scores. For example, the proposed synthetic T2 image provides a high qualitatively difference along the tumor boundary (indicated by the red arrows in the fifth row) than in existing methods using the available source T1 image as input.

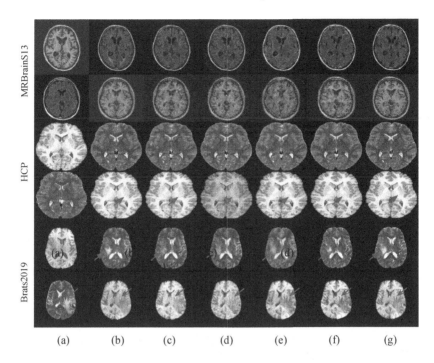

(a)   (b)   (c)   (d)   (e)   (f)   (g)

Fig. 4. A visualization of synthetic images on different datasets generated by (a) source image, (b) target image, (c) cycleGAN, (d) recycleGAN, (e) cycleflow, (f) alignflow, and (g) our method. Our method provides a good boundary on the tumor regions (red arrows in the fifth row) compared with the existing methods (Color figure online)

Quantitative Evaluation. Tables 1 reports the MSE, PSNR and SSIM values of the proposed method and existing methods. From the table, it is clear that the flow-based method (such as cycleflow [11], alignflow [5] and our method) provides competitive results with GAN-based method (such as cycleGAN, recycleGAN). By adding temporal constraints, the proposed network outperforms the baseline method (alignflow) on all performance metrics. Different from recycleGAN, that exploits temporal information via future slice prediction from consecutive slices, the proposed method measures pixel-wise temporal consistency by directly warping the synthetic slices with the deformation field of the consecutive slices from the source, and thus achieves better performances. This indicates the effectiveness of the proposed method in the unpaired image to image translation for medical image.

Table 1. Comparison between the proposed method against other image-to-image translation methods on **HCP, MRBrainS13, Brats19** datasets.

	Method	MSE ↓		PSNR ↑		SSIM ↑	
		T1 → T2	T2 → T1	T1 → T2	T2 → T1	T1 → T2	T2 → T1
HCP	cycleGAN [16]	0.0193	0.0167	23.2	24.4	0.783	0.793
	recycleGAN [2]	0.0212	0.0182	22.8	24.0	0.773	0.797
	cycleflow [11]	0.0213	0.0189	22.8	23.8	0.771	0.785
	Alignflow (baseline) [5]	0.0200	0.0158	23.1	24.6	0.785	0.811
	Our method	**0.0179**	**0.0143**	**23.5**	**25.1**	**0.80**	**0.820**
MRBrainS13	cycleGAN [16]	0.0139	**0.0235**	24.7	22.4	0.793	0.704
	recycleGAN [2]	0.0154	0.0250	24.3	22.1	0.761	0.714
	cycleflow [11]	0.0158	0.0406	24.2	20.0	0.790	0.506
	Alignflow (baseline) [5]	0.0165	0.0254	24.0	22.0	0.781	0.728
	Our method	**0.0128**	0.0236	**25.1**	**22.4**	**0.819**	**0.741**
Brats2019	cycleGAN [16]	**0.0178**	0.0281	**24.1**	22.7	0.833	0.797
	recycleGAN [2]	0.0190	0.0272	23.8	22.6	0.824	0.785
	cyclelow [11]	0.0251	0.0304	22.7	21.8	0.800	0.788
	alignflow (baseline) [5]	0.022	0.0306	23.4	21.8	0.830	0.784
	our method	0.0188	**0.0258**	23.9	**22.8**	**0.842**	**0.808**

5   Conclusion

We presented an effective method for image-to-image translation based on flow-based methods and deformation information that allows the proposed method to exploit the temporal information between consecutive slices to constrain the translation image. We show that the proposed method can provide a good translation image, yielding a better MSE, PSNR, and SSIM on various MRI datasets. Although our network is a fully invertible property, it requires more memory resource than GAN-based methods (such as cycleGAN, recycleGAN, ...).

References

1. Balakrishnan, G., Zhao, A., Sabuncu, M.R., Guttag, J., Dalca, A.V.: An unsupervised learning model for deformable medical image registration. In: Proceedings of the CVPR, pp. 9252–9260 (2018)
2. Bansal, A., Ma, S., Ramanan, D., Sheikh, Y.: Recycle-GAN: unsupervised video retargeting. In: Ferrari, V., Hebert, M., Sminchisescu, C., Weiss, Y. (eds.) ECCV 2018. LNCS, vol. 11209, pp. 122–138. Springer, Cham (2018). https://doi.org/10.1007/978-3-030-01228-1_8
3. Chen, X., et al.: One-shot generative adversarial learning for MRI segmentation of craniomaxillofacial bony structures. IEEE Trans. Med. Imaging (2019)
4. Dinh, L., Sohl-Dickstein, J., Bengio, S.: Density estimation using real nvp. arXiv preprint arXiv:1605.08803 (2016)
5. Grover, A., Chute, C., Shu, R., Cao, Z., Ermon, S.: AlignFlow: cycle consistent learning from multiple domains via normalizing flows. arXiv preprint arXiv:1905.12892 (2019)

6. Isola, P., Zhu, J.Y., Zhou, T., Efros, A.A.: Image-to-image translation with conditional adversarial networks. In: Proceedings of the IEEE Conference on Computer Vision and Pattern Recognition, pp. 1125–1134 (2017)
7. Kingma, D.P., Dhariwal, P.: Glow: generative flow with invertible 1×1 convolutions. In: Advances in Neural Information Processing Systems, pp. 10215–10224 (2018)
8. Mendrik, A.M., et al.: Mrbrains challenge: online evaluation framework for brain image segmentation in 3t MRI scans. Comput. Intell. Neurosci. **2015** (2015)
9. Menze, B.H., Jakab, A., Bauer, et al.: The multimodal brain tumor image segmentation benchmark (brats). TMI **34**(10), 1993–2024 (2015)
10. van der Ouderaa, T.F., Worrall, D.E.: Reversible GANs for memory-efficient image-to-image translation. In: Proceedings of the IEEE Conference on Computer Vision and Pattern Recognition, pp. 4720–4728 (2019)
11. Shen, Z., Zhou, S.K., Chen, Y., Georgescu, B., Liu, X., Huang, T.: One-to-one mapping for unpaired image-to-image translation. In: The IEEE Winter Conference on Applications of Computer Vision, pp. 1170–1179 (2020)
12. Sun, H., et al.: Dual-Glow: conditional flow-based generative model for modality transfer. In: Proceedings of the IEEE International Conference on Computer Vision, pp. 10611–10620 (2019)
13. Van Essen, D.C., et al.: The WU-Minn human connectome project: an overview. Neuroimage **80**, 62–79 (2013)
14. Welander, P., Karlsson, S., Eklund, A.: Generative adversarial networks for image-to-image translation on multi-contrast MR images-a comparison of cyclegan and unit. arXiv preprint arXiv:1806.07777 (2018)
15. Yuan, Y., Liu, S., Zhang, J., Zhang, Y., Dong, C., Lin, L.: Unsupervised image super-resolution using cycle-in-cycle generative adversarial networks. In: Proceedings of the IEEE Conference on Computer Vision and Pattern Recognition Workshops, pp. 701–710 (2018)
16. Zhu, J.Y., Park, T., Isola, P., Efros, A.A.: Unpaired image-to-image translation using cycle-consistent adversarial networks. In: IEEE CVPR, pp. 2223–2232 (2017)

Interpretation of Disease Evidence for Medical Images Using Adversarial Deformation Fields

Ricardo Bigolin Lanfredi[1]([✉]) [ID], Joyce D. Schroeder[2] [ID], Clement Vachet[1] [ID], and Tolga Tasdizen[1] [ID]

[1] Scientific Computing and Imaging Institute, University of Utah, Salt Lake City, UT 84112, USA
ricbl@sci.utah.edu
[2] Department of Radiology and Imaging Sciences, University of Utah, Salt Lake City, UT 84112, USA

Abstract. The high complexity of deep learning models is associated with the difficulty of explaining what evidence they recognize as correlating with specific disease labels. This information is critical for building trust in models and finding their biases. Until now, automated deep learning visualization solutions have identified regions of images used by classifiers, but these solutions are too coarse, too noisy, or have a limited representation of the way images can change. We propose a novel method for formulating and presenting spatial explanations of disease evidence, called deformation field interpretation with generative adversarial networks (DeFI-GAN). An adversarially trained generator produces deformation fields that modify images of diseased patients to resemble images of healthy patients. We validate the method studying chronic obstructive pulmonary disease (COPD) evidence in chest x-rays (CXRs) and Alzheimer's disease (AD) evidence in brain MRIs. When extracting disease evidence in longitudinal data, we show compelling results against a baseline producing difference maps. DeFI-GAN also highlights disease biomarkers not found by previous methods and potential biases that may help in investigations of the dataset and of the adopted learning methods.

Keywords: Deformation field · Deep learning interpretation · Visual attribution · Adversarial training · Disease effect · DeFI-GAN

1 Introduction

The recent surge of deep learning applications has the potential to revolutionize medical imaging in several ways, such as accessibility, efficiency, and flexibility.

Data used in preparation of this article were obtained from the Alzheimer's Disease Neuroimaging Initiative (ADNI) database (adni.loni.usc.edu).

Electronic supplementary material The online version of this chapter (https://doi.org/10.1007/978-3-030-59713-9_71) contains supplementary material, which is available to authorized users.

A. L. Martel et al. (Eds.): MICCAI 2020, LNCS 12262, pp. 738–748, 2020.
https://doi.org/10.1007/978-3-030-59713-9_71

However, decision automation leads to challenges, including the interpretability of the outcomes [21]. Understanding what kinds of evidence deep learning methods capture from an image is one approach for overcoming these challenges. The field can benefit from such understanding through improvements in user trustworthiness, patient communication, and model bias identification.

Among visual attribution methods, backpropagation through a trained classifier is commonly used to determine areas of stronger influence on the model output [2]. Several visual attribution approaches have been applied to medical imaging. The class activation map (CAM) [37] method has been employed to show low-resolution areas of focus when performing pneumonia diagnosis from a chest x-ray (CXR) [27]. CAM has also been used as part of a weakly-supervised fine segmentation of lung nodules [10]. Layer-wise relevance propagation (LRP) [4] has been used to find biases in a histopathology dataset [17]. Despite their usefulness, most visual attribution methods are too coarse or too noisy for some applications [6]. They also have limited means of representing their findings, which may obscure relevant evidence.

Spatial deformations have been used in several applications, including registration [5], generation of adversarial examples [36], and creation of atlases [9]. While deformation fields may not be adequate for producing color and texture changes, they can generate variations in position, shape, and size. Since some impacts of diseases on anatomy are linked to the latter, we hypothesize that applying deformation fields to represent and visualize model interpretations will better include such variations. We propose the deformation field interpretation with generative adversarial networks (DeFI-GAN) method, which uses adversarial training [14] to learn to produce a deformation field that alters an image to make disease signs indiscernible. Therefore, the changes caused by this field express evidence of such disease. To the best of our knowledge, no other work has used deformation fields for visual attribution in deep learning.

We perform experiments studying evidence of chronic obstructive pulmonary disease (COPD) in CXRs [13,35]. In [7], automatic assessment of COPD evidence is also performed. However, this assessment applies additive perturbations instead of deformation fields, and a different set of disease evidence is found. The method presented in [7] also models disease severity, which is not a focus of DeFI-GAN. We also employ the ADNI dataset to model the conversion from mild cognitive impairment (MCI) to Alzheimer's disease (AD) through the morphological brain changes observed in MRIs [22]. In [6], generative adversarial training is also used to assess evidence of AD in brain MRIs. This setup inspires our method, but uses additive perturbations instead of deformation fields and a different regularization loss function. We analyze the outcomes of the proposed formulation in both datasets, showing that the use of DeFI-GAN improves longitudinal prediction over a baseline [6] and highlights additional disease evidence.

2   Method

We consider binary classification problems, where class 0 is associated with healthy patients and class 1 with diseased patients. Each image in a dataset

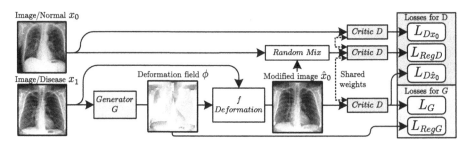

Fig. 1. Overall model architecture. The terms L_{Dx_0}, $L_{D\hat{x}_0}$, and L_G are WGAN losses, whereas L_{RegD} penalizes the gradient of D, and L_{RegG} penalizes the complexity of ϕ.

is denoted by x, and its domain by X. Images of class c are indicated by x_c. Our objective is to find a transformation f that maps from an image x_1 containing evidence for a disease to a modified image $\hat{x}_0$ where that evidence is absent. The rest of the content, e.g. patient specific anatomy, should not be modified. We propose to model f as a deformation controlled by a generated vector field, as shown in Fig. 1. Mathematically,

$$\phi = G(x_1), \quad \hat{x}_0(p) = x_1(p + \phi(p)), \ p \in X, \tag{1}$$

where G is a parameterized generator mapping from x_1 to vector field $\phi : X \to X$. Since $p + \phi(p)$ can lie between the set of coordinates on the grid for which we have defined values in x_1, we use bilinear interpolation when $X \in \mathbb{R}^2$, and trilinear interpolation when $X \in \mathbb{R}^3$. When $p + \phi(p)$ lies outside the range for which values are defined in x_1, values in $\hat{x}_0$ are set to 0.

To learn the parameters of transformation f, we follow [6] and use adversarial training. We train G jointly with a parameterized critic D tasked with discriminating modified images $\hat{x}_0$ from real images x_0. The generator G is trained to fool D, using the gradient signal from D so that the distribution of $\hat{x}_0$ approaches the distribution of x_0. For the adversarial loss, we use the WGAN [3] formulation. The critic D is trained to give high scores for modified images and low scores for real images, resulting in

$$L_D = L_{Dx_0} - L_{D\hat{x}_0} = \mathbb{E}\left[D(x_0)\right] - \mathbb{E}\left[D(\hat{x}_0)\right]. \tag{2}$$

To enforce the Lipschitz constraint required by the WGAN formulation [3], we use the penalty proposed in [15],

$$L_{RegD} = \mathbb{E}\left[(\|\nabla_{\tilde{x}} D(\tilde{x})\|_2 - 1)^2 \mid \tilde{x} \sim (1 - \alpha)x_0 + \alpha\hat{x}_0, \alpha \sim \mathcal{U}(0, 1)\right], \tag{3}$$

where $\mathcal{U}(0, 1)$ is the uniform distribution with support between 0 and 1. The expectation is approximated by sampling one α for each sampled (x_0, x_1) pair.

The generator G is trained to lower the scores for $\hat{x}_0$, resulting in

$$L_G = \mathbb{E}\left[D(\hat{x}_0)\right]. \tag{4}$$

To penalize extraneous changes and enforce smooth and realistic deformations, we use a total variation denoising [29] penalty over ϕ. This term forces nearby vectors to be aligned by penalizing $\int_X \|\nabla\phi\|_2$. After discretization,

$$L_{RegG} = \mathbb{E}\left[\frac{1}{P}\sum_{p\in X}\sum_{n\in\mathcal{N}(p)} \|\phi(p) - \phi(n)\|_2\right], \tag{5}$$

where $\mathcal{N}(p)$ is the set of neighboring pixels of pixel p (4-neighborhood for $X \in \mathbb{R}^2$ and 6-neighborhood for $X \in \mathbb{R}^3$) and P is the total number of pixels.

The complete optimization formulation is given by

$$G^* = \underset{G}{\mathrm{argmin}}(L_G + \lambda_{RegG}L_{RegG}), D^* = \underset{D}{\mathrm{argmin}}(L_D + \lambda_{RegD}L_{RegD}), \tag{6}$$

where λ_{RegG} and λ_{RegD} are hyperparameters.

3 Experiments

We tested our method[1] in two datasets, one for finding evidence of AD in brain MRIs and one for detecting signs of COPD in CXRs. We used as a baseline the visual attribution generative adversarial network (VA-GAN) method [6], where the function to obtain $\hat{x}_0$ is defined by the addition of a difference map, i.e., $\hat{x}_0 = x_1 + G(x_1)$, and the penalty on G is accordingly defined by $L_{RegG} = \|\hat{x}_0 - x_1\|_1$.

We used u-nets [28] as G's architecture, varying the number of downsamplings and channels depending on the dataset. We used the Adam optimizer [20] with a learning rate of 1e−4 and set $\lambda_{RegD} = 10$. For D's architecture, we used an ImageNet pre-trained Resnet-18 [16] for the CXR dataset and a 10-layer network as in [6] for the MRI dataset. For training the DeFI-GAN method, we performed 100 updates to D for each update to G, whereas for VA-GAN, the 100:1 update ratio was used for the first 25 updates to G, and then changed to a 5:1 ratio. It was important for the convergence of G to have X represented in pixel units.

3.1 Chronic Obstructive Pulmonary Disease in Chest X-Rays

We collected a dataset containing posterioranterior (PA) CXRs acquired at the University of Utah Hospital from 2012 to 2017. Each CXR was labeled for COPD using results from pulmonary function tests (PFTs) taken within one month of the CXR. Patients who received a lung transplant were excluded. Patients with COPD were assigned to class 1, and others were assigned to class 0. The training set was composed of patients who had only one CXR included in the dataset and contained 2,226 images from normal patients and 963 images from COPD patients. We worked under an approved Institutional Review Board (IRB)[2] and anonymized data using Orthanc[3]. Image preprocessing included center-cropping to a square, resizing to 256×256, cropping to 224×224 (randomly for training and centered for validation and testing), and equalizing histograms.

---

[1] The code is available at https://github.com/ricbl/defigan.
[2] IRB_00104019, PI: Schroeder MD.
[3] orthanc-server.com.

3.2 Alzheimer's Disease in Brain MRIs

The ADNI dataset was collected to characterize the progression of AD and contains brain MRIs of thousands of subjects followed for a few years [1]. Diagnosis of AD and MCI are provided for each exam. Following [6], we set class 0 to MCI patients, instead of healthy patients, and class 1 to AD patients, training G to provide interpretations for the conversion from MCI to AD. We also limited our study to T1-weighted MRIs from the ADNI1, ADNIGO, and ADNI2 cohorts.

Data splits and part of the preprocessing were replicated from [6][4]. For each image, we used the FMRIB Software Library (FSL) toolbox [32] v6.0 for reorientation and field of view (FOV) cropping, N4 Bias Correction [34] from Advanced Normalization Tools (ANTs) v3.0 for bias field correction, FSL to register the image to the MNI 152 space [11,12], and ROBEX [18] v1.2 for skull stripping. To correct cases where ROBEX failed, we added another step of skull stripping to the whole dataset using Brain Extraction Tool (BET) [31] with a fractional intensity threshold equal to 0.25. All volumes were rescaled to a size of $128 \times 160 \times 112$ pixels. Through visual inspection, we searched for cases where the brain volume was incorrectly oriented, or skull stripping had failed, and removed 7 volumes from 4 subjects from test and validation. The preprocessing differed from [6] in the application of BET and in the elimination of the 7 cases for which preprocessing failed. The final training set had 2,528 MCI volumes and 1,198 AD volumes, for a total of 825 subjects.

3.3 Quantitative Validation

We validated our generated disease evidence by using longitudinal scans demonstrating disease progression. For the COPD dataset, we paired all test cases where a subject was healthy in the baseline CXR with cases of the same subject after developing COPD. For each pair, we performed an affine registration of the normal case to the COPD case using SimpleITK v1.2 [24]. We then subtracted the COPD image from the registered baseline image to get a difference ground truth Δx. The final split sizes were 206 pairs of images (176 subjects) for validation and 547 pairs (354 subjects) for testing. For the ADNI dataset, we performed rigid registration instead of affine, disregarding the background for the calculations, and used MCI cases as the baseline, pairing them with AD MRIs taken with the same field strength. The final split sizes were 207 pairs (102 subjects) for validation and 259 pairs (143 subjects) for testing.

The ground truth Δx and the predicted difference $\widehat{\Delta x} = \hat{x}_0 - x_1$ were compared using normalized cross-correlation (NCC), masked to ignore regions padded during registration and skull stripping. This operation was defined by

$$\text{NCC}\left(\Delta x, \widehat{\Delta x}\right) = \frac{1}{M-1} \sum_{p \in \mathcal{M}} \frac{\left(\Delta x\left(p\right) - \mu_{\Delta x}\right)}{\sigma_{\Delta x}} \times \frac{\left(\widehat{\Delta x}\left(p\right) - \mu_{\widehat{\Delta x}}\right)}{\sigma_{\widehat{\Delta x}}}, \tag{7}$$

---

[4] VA-GAN's original preprocessing code, list of subjects, and TensorFlow implementation are available in https://github.com/baumgach/vagan-code/.

Table 1. NCC scores for the compared methods. Averages of 5 models trained with different random seeds are reported with their standard deviations.

$_{\text{Dataset}}\backslash^{\text{Method}}$	VA-GAN	DeFI-GAN
ADNI	0.332 ± 0.015	$\mathbf{0.365 \pm 0.014}$
COPD	0.174 ± 0.007	$\mathbf{0.204 \pm 0.006}$

	DeFI-GAN		VA-GAN	
Original Image	Modified Image	Modifications	Modified Image	Modifications

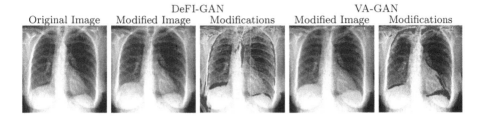

Fig. 2. Comparison of results of DeFI-GAN and VA-GAN on the COPD dataset. Green represents darkening of the image, and pink represents brightening. (Color figure online)

where M is the number of pixels in the mask $\mathcal{M}$, and μ_i and σ_i are, respectively, the average value and the unbiased standard deviation of values of pixels of image i inside mask $\mathcal{M}$. One subject may have had more than one pair of validation images. The score calculated for all pairs of each subject was averaged before calculating the final average score to avoid the over-representation of a few subjects in the final score.

Hyperparameters were chosen by considering image quality and validation NCC score. With the DeFI-GAN method, we used $\lambda_{RegG} = 25$ for the COPD dataset and $\lambda_{RegG} = 10$ for the ADNI dataset, whereas for the VA-GAN method we used, respectively, $\lambda_{RegG} = 50$ and $\lambda_{RegG} = 100$. The original VA-GAN TensorFlow implementation was used for the ADNI dataset.

Test scores are reported in Table 1, where the DeFI-GAN method is shown to outperform the baseline. The scores for the ADNI dataset when using VA-GAN are better than the ones presented in [6]. This difference is probably due to choosing the best training epoch using the NCC validation score instead of loss values. It is interesting to note that training was less noisy for the DeFI-GAN method when considering the validation scores from epoch to epoch. Relatively low NCC scores may result from the difficulty in aligning longitudinal data and from aging effects in the ground truth.

3.4   Biomarker Validation

We compared the interpretations provided by DeFI-GAN and VA-GAN and present some results in Figs. 2 and 3. The difference maps were computed using $\hat{x}_0 - x_1$. Additional results are presented in the supplementary material. For the COPD dataset, both approaches identified changes in heart silhouette and in

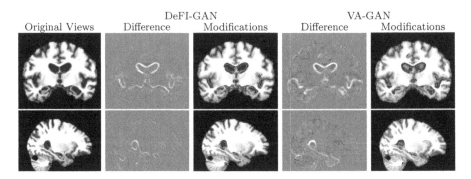

Fig. 3. Results on the ADNI dataset when removing evidence of AD from the original image. Green represents darkening of the image, and pink represents brightening. (Color figure online)

diaphragm height and curvature, which are notable characterizations of COPD in CXRs [13,35]. We found no variation in lung lucency. The lack of change in texture was expected for the DeFI-GAN method from its formulation. Interpretations from the VA-GAN method were less spatially smooth.

Saber-sheath trachea is a secondary sign of COPD, and, when it occurs, there is a narrowing of the trachea in the frontal CXR [8]. This evidence was more identifiable in models trained with DeFI-GAN when looking at quiver plot visualization, as seen in Fig. 4(a). The VA-GAN method cannot produce this kind of deformation field visualization. In the difference map produced by the VA-GAN method, the changes in the top of the image appeared as a dark fuzzy alteration. Another sign more easily identified with DeFI-GAN was the change in soft-tissue thickness, as shown in Fig. 4(b). This change is in accordance with the correlation between low body mass index (BMI) and COPD [38].

Although we focused the analysis on COPD, we also found the results in the ADNI dataset consistent with the literature. The main changes associated with AD, expansion of ventricles, hippocampus atrophy, and temporal lobe atrophy [22], are seen in Fig. 3. It was possible to find, in some cases, atrophies in the precuneus, the cerebellum, and the brainstem. These regions have been associated with AD [23,30,33]. The rest of the cortex also presented atrophies to a lesser extent. The differences highlighted by both methods were similar, except for VA-GAN having more image noise, more darkening differences, and more noticeable highlighting of the hippocampus. The highlighting of the hippocampus is related to AD [22]. A detailed representation of a deformation field in MRIs can be seen in the supplementary material.

Unexpected Changes in COPD. Some evidence pinpointed by the DeFI-GAN method in the COPD dataset is not usually listed in the literature. Figure 4(c) shows a decrease in gastric bubble size. This change may represent a confusion between the lower border of the gastric bubbles and the diaphragm, or

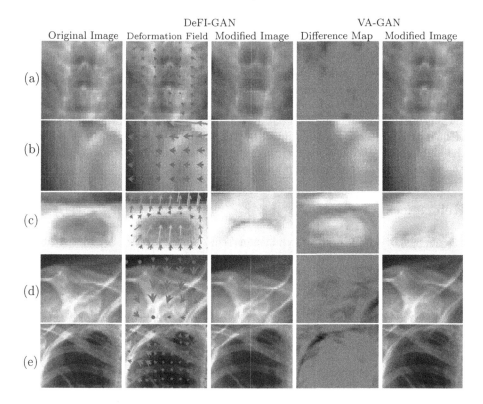

Fig. 4. Details on the generated deformation fields and comparison with VA-GAN. The color of the arrows represents their length. For rows **a–c**, their length represents the exact shift in pixels. In rows **d–e**, their length is multiplied by 2 for better visualization. The full-size images are presented in the supplementary material. From top to bottom: **(a)** trachea; **(b)** soft tissue; **(c)** gastric bubble; **(d)** shoulder; **(e)** lung apex.

be related to aerophagia due to coughing [25]. Furthermore, the shoulder position changed in a few images, as seen in Fig. 4(d). High clavicle position might be related to a decrease in shoulder mobility [26] and, consequently, difficulty in following the required body positioning for the standing CXR acquisition. It may also be related to a bias in spine position due to the prevalence of osteoporosis [19] in patients with COPD. In Fig. 4(e), the expansion of the top of the lungs contradicts larger lung volumes in COPD patients. However, when analyzed by a radiologist, lung volume still seemed larger in the original images, probably due to diaphragm height. This change may be correlated with the clavicle positioning since it increases lung volume above the clavicle. The change in gastric bubble size and the expansion of the top of the lung were, in a less comprehensible representation, also highlighted by the VA-GAN method. All these unexpected differences may also represent dataset biases. A more in-depth investigation of the causes for these changes is left to subsequent studies.

4   Conclusion

We formulated a method for producing interpretations of the disease evidence that deep learning models can capture from imaging data, using deformation fields and adversarial training. This method performed better in the proposed longitudinal quantitative validation on two medical datasets, when compared to another method with a similar goal. Furthermore, it allowed for easier discovery of qualitative disease evidence used by the model. As exemplified by some of the highlighted COPD evidence, this study has the potential to support future analyses of unexpected biases in medical imaging datasets.

References

1. About ADNI (2017). http://adni.loni.usc.edu/about/. Accessed 11 Mar 2020
2. Ancona, M., Ceolini, E., Öztireli, C., Gross, M.: Towards better understanding of gradient-based attribution methods for deep neural networks. In: ICLR (2018)
3. Arjovsky, M., Chintala, S., Bottou, L.: Wasserstein generative adversarial networks. In: ICML (2017)
4. Bach, S., Binder, A., Montavon, G., Klauschen, F., Müller, K.R., Samek, W.: On pixel-wise explanations for non-linear classifier decisions by layer-wise relevance propagation. PLoS ONE **10**(7), 1–46 (2015). https://doi.org/10.1371/journal.pone. 0130140
5. Balakrishnan, G., Zhao, A., Sabuncu, M.R., Guttag, J., Dalca, A.V.: An unsupervised learning model for deformable medical image registration. In: CVPR (2018)
6. Baumgartner, C.F., Koch, L.M., Tezcan, K.C., Ang, J.X.: Visual feature attribution using Wasserstein GANs. In: CVPR (2018). https://doi.org/10.1109/CVPR. 2018.00867
7. Bigolin Lanfredi, R., Schroeder, J.D., Vachet, C., Tasdizen, T.: Adversarial regression training for visualizing the progression of chronic obstructive pulmonary disease with chest x-rays. In: Shen, D., et al. (eds.) MICCAI 2019. LNCS, vol. 11769, pp. 685–693. Springer, Cham (2019). https://doi.org/10.1007/978-3-030-32226-7_76
8. Ciccarese, F., et al.: Saber-sheath trachea as a marker of severe airflow obstruction in chronic obstructive pulmonary disease. Radiol. Med. (Torino) **119**(2), 90–96 (2013). https://doi.org/10.1007/s11547-013-0318-3
9. Dalca, A., Rakic, M., Guttag, J., Sabuncu, M.: Learning conditional deformable templates with convolutional networks. In: NeurIPS (2019)
10. Feng, X., Yang, J., Laine, A.F., Angelini, E.D.: Discriminative localization in CNNs for weakly-supervised segmentation of pulmonary nodules. In: Descoteaux, M., Maier-Hein, L., Franz, A., Jannin, P., Collins, D.L., Duchesne, S. (eds.) MICCAI 2017. LNCS, vol. 10435, pp. 568–576. Springer, Cham (2017). https://doi.org/10. 1007/978-3-319-66179-7_65
11. Fonov, V., Evans, A.C., Botteron, K., Almli, C.R., McKinstry, R.C., Collins, D.L.: Unbiased average age-appropriate atlases for pediatric studies. NeuroImage **54**(1), 313–327 (2011). https://doi.org/10.1016/j.neuroimage.2010.07.033
12. Fonov, V., Evans, A., McKinstry, R., Almli, C., Collins, D.: Unbiased nonlinear average age-appropriate brain templates from birth to adulthood. NeuroImage **47**, S102 (2009). https://doi.org/10.1016/S1053-8119(09)70884-5

13. Foster, W.L., et al.: The emphysemas: radiologic-pathologic correlations. Radio-Graphics **13**(2), 311–328 (1993). https://doi.org/10.1148/radiographics.13.2. 8460222
14. Goodfellow, I., et al.: Generative adversarial nets. In: NeurIPS (2014)
15. Gulrajani, I., Ahmed, F., Arjovsky, M., Dumoulin, V., Courville, A.: Improved training of Wasserstein GANs. In: NeurIPS (2017)
16. He, K., Zhang, X., Ren, S., Sun, J.: Deep residual learning for image recognition. In: CVPR (2016). https://doi.org/10.1109/CVPR.2016.90
17. Hägele, M., et al.: Resolving challenges in deep learning-based analyses of histopathological images using explanation methods. CoRR abs/1908.06943 (2019)
18. Iglesias, J.E., Liu, C., Thompson, P.M., Tu, Z.: Robust brain extraction across datasets and comparison with publicly available methods. IEEE Trans. Med. Imag. **30**(9), 1617–1634 (2011). https://doi.org/10.1109/TMI.2011.2138152
19. Jaramillo, J.D., et al.: Reduced bone density and vertebral fractures in smokers. Men and COPD patients at increased risk. Ann. Am. Thorac. Soc. **12**(5), 648–656 (2015). https://doi.org/10.1513/AnnalsATS.201412-591OC
20. Kingma, D.P., Ba, J.: Adam: a method for stochastic optimization. In: ICLR (2015)
21. Langlotz, C.P., et al.: A roadmap for foundational research on artificial intelligence in medical imaging: from the 2018 NIH/RSNA/ACR/the academy workshop. Radiology **291**, 781–791 (2019)
22. Ledig, C., Schuh, A., Guerrero, R., Heckemann, R., Rueckert, D.: Structural brain imaging in Alzheimer's disease and mild cognitive impairment: biomarker analysis and shared morphometry database. Sci. Rep. **8** (2018). https://doi.org/10.1038/s41598-018-29295-9
23. Lee, J., Ryan, J., Andreescu, C., Aizenstein, H., Lim, H.K.: Brainstem morphological changes in Alzheimer's disease. NeuroReport **26**, 411–415 (2015). https://doi.org/10.1097/WNR.0000000000000362
24. Lowekamp, B., Chen, D., Ibanez, L., Blezek, D.: The design of SimpleITK. Front. Neuroinform. **7**, 45 (2013). https://doi.org/10.3389/fninf.2013.00045
25. Martin-Harris, B.: Optimal patterns of care in patients with chronic obstructive pulmonary disease. Semin. Speech Lang. **21**(04), 0311–0322 (2000). https://doi.org/10.1055/s-2000-8384
26. Morais, N., Cruz, J., Marques, A.: Posture and mobility of the upper body quadrant and pulmonary function in COPD: an exploratory study. Braz. J. Phys. Therapy **20**, 345–354 (2016). https://doi.org/10.1590/bjpt-rbf.2014.0162
27. Rajpurkar, P., Irvin, J., et al.: CheXNet: radiologist-level pneumonia detection on chest x-rays with deep learning. CoRR abs/1711.05225 (2017)
28. Ronneberger, O., Fischer, P., Brox, T.: U-Net: convolutional networks for biomedical image segmentation. In: Navab, N., Hornegger, J., Wells, W.M., Frangi, A.F. (eds.) MICCAI 2015. LNCS, vol. 9351, pp. 234–241. Springer, Cham (2015). https://doi.org/10.1007/978-3-319-24574-4_28
29. Rudin, L.I., Osher, S., Fatemi, E.: Nonlinear total variation based noise removal algorithms. Physica D **60**(1), 259–268 (1992). https://doi.org/10.1016/0167-2789(92)90242-F
30. Scahill, R.I., Schott, J.M., Stevens, J.M., Rossor, M.N., Fox, N.C.: Mapping the evolution of regional atrophy in Alzheimer's disease: unbiased analysis of fluid-registered serial MRI. PNAS **99**(7), 4703–4707 (2002). https://doi.org/10.1073/pnas.052587399
31. Smith, S.M.: Fast robust automated brain extraction. Hum. Brain Mapp. **17**(3), 143–155 (2002). https://doi.org/10.1002/hbm.10062

32. Smith, S.M., et al.: Advances in functional and structural MR image analysis and implementation as FSL. NeuroImage **23**, S208–S219 (2004). https://doi.org/10.1016/j.neuroimage.2004.07.051
33. Tabatabaei Jafari, H., Walsh, E., Shaw, M., Cherbuin, N.: The cerebellum shrinks faster than normal ageing in Alzheimer's disease but not in mild cognitive impairment. Hum. Brain Mapp. **38** (2017). https://doi.org/10.1002/hbm.23580
34. Tustison, N.J., et al.: N4ITK: improved N3 bias correction. IEEE Trans. Med. Imag. **29**(6), 1310–1320 (2010). https://doi.org/10.1109/TMI.2010.2046908
35. Washko, G.R.: Diagnostic imaging in COPD. Semin. Resp. Crit. Care **31**(3), 276–285 (2010). https://doi.org/10.1055/s-0030-1254068
36. Xiao, C., Zhu, J.Y., Li, B., He, W., Liu, M., Song, D.: Spatially transformed adversarial examples. In: ICLR (2018)
37. Zhou, B., Khosla, A., Lapedriza, A., Oliva, A., Torralba, A.: Learning deep features for discriminative localization. In: CVPR (2016). https://doi.org/10.1109/CVPR.2016.319
38. Zhou, Y., et al.: The association between BMI and COPD: the results of two population-based studies in Guangzhou, China. COPD **10**(5), 567–572 (2013). https://doi.org/10.3109/15412555.2013.781579

Spatial-Intensity Transform GANs for High Fidelity Medical Image-to-Image Translation

Clinton J. Wang[1](✉), Natalia S. Rost[2], and Polina Golland[1]

[1] Computer Science and Artificial Intelligence Lab, MIT, Cambridge, MA, USA
clintonw@csail.mit.edu
[2] Department of Neurology, Massachusetts General Hospital, Boston, MA, USA

Abstract. Despite recent progress in image-to-image translation, it remains challenging to apply such techniques to clinical quality medical images. We develop a novel parameterization of conditional generative adversarial networks that achieves high image fidelity when trained to transform MRIs conditioned on a patient's age and disease severity. The spatial-intensity transform generative adversarial network (SIT-GAN) constrains the generator to a smooth spatial transform composed with sparse intensity changes. This technique improves image quality and robustness to artifacts, and generalizes to different scanners. We demonstrate SIT-GAN on a large clinical image dataset of stroke patients, where it captures associations between ventricle expansion and aging, as well as between white matter hyperintensities and stroke severity. Additionally, SIT-GAN provides a disentangled view of the variation in shape and appearance across subjects.

Keywords: Conditional generative adversarial network ·
Image-to-image translation · Stroke

1 Introduction

Many tasks in medical image analysis require mapping images in one distribution to images in another distribution, conditioned on a set of attributes. Such mappings can be used to synthesize medical images of a specified imaging modality [12] or patient phenotype [9], while preserving most characteristics of an input image such as gross anatomy. Driven by advances in generative adversarial networks (GANs), medical image-to-image translation has been applied to tasks as diverse as data augmentation [1], super-resolution [8], MR-to-CT translation [12], and prediction of disease trajectories [9]. In such GANs, a generator is trained to map input images sampled from a source distribution to synthetic images that appear to belong to a target distribution, while an adversarial discriminator drives the generator to produce realistic images [3,17]. Medical applications

© Springer Nature Switzerland AG 2020
A. L. Martel et al. (Eds.): MICCAI 2020, LNCS 12262, pp. 749–759, 2020.
https://doi.org/10.1007/978-3-030-59713-9_72

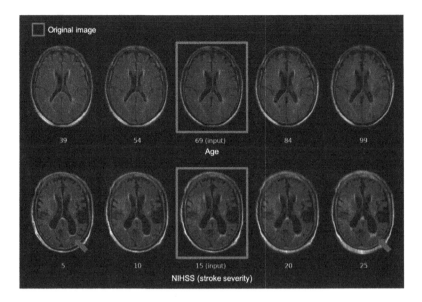

Fig. 1. Synthetic fluid-attenuated inversion recovery (FLAIR) MRIs of acute ischemic stroke patients, obtained by transforming an input MRI (center column) conditioned on changes in age (top) and stroke severity (bottom). Increasing age correlates with increasing ventricular volume, and increasing stroke severity correlates with increasing volume of periventricular white matter hyperintensities (see red arrow). (Color figure online)

of GANs have often been restricted to large datasets of high-quality research scans. When the target distribution is underrepresented in the training data or the data consists of lower quality clinical scans, GANs may introduce severe artifacts.

We address this challenge by introducing the spatial-intensity transform generative adversarial network (SIT-GAN), which constrains the generator output to transformations composed of a smooth deformation field and a sparse intensity difference map applied to the input image. This parameterization produces images with fewer artifacts and high fidelity (Fig. 1), and also yields separate visualizations of morphological and tissue intensity changes, which can be relevant to identifying and characterizing disease processes.

Previously, spatial transforms have been coupled with intensity transforms for performing medical image registration [4,16] and data augmentation in the context of semi-supervised segmentation [1]. To the best of our knowledge, this is the first work demonstrating that they are effective at regularizing the outputs of conditional generative models.

Our novel representation of image changes is complementary to prior work on conditional GANs that modify the loss function in the context of simulating aging in brain MRIs. Xia et al. [14] introduce identity-preservation and self-reconstruction losses that penalize large changes in the image for small translations in age. Ravi et al. [9] introduce biological constraints that encourage the network to follow a known hallmark of neurodegeneration, e.g., voxels should darken with age at a rate similar to neighboring voxels. The proposed representation is orthogonal to such changes in the loss function and could be combined to further improve the translation results.

We demonstrate the proposed method on a large dataset of clinical quality brain MRIs of stroke patients. Our experiments suggest that in such settings, SIT-GAN outperforms the state of the art on medical image-to-image translation.

2   Methods

2.1   Image-to-Image Translation with Partially Observed Attributes in Cross-Sectional Data

Given coordinate space Ω, a set $\mathcal{D} = \{(x_i, y_i)\}_{i=1}^{N}$ of images $x_i \in \mathcal{X} : \Omega \to \mathbb{R}$ and conditional attributes $y_i \in \mathcal{Y}$ (e.g., age and stroke severity), we want to train a generator to transform images such that their conditional attributes are shifted by a specified amount. Our network consists of a generator $G : \mathcal{X} \times \mathcal{Y} \to \mathcal{X}$,

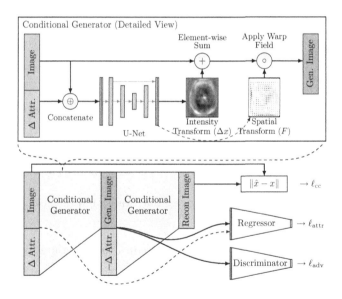

Fig. 2. The generator takes in an image and the desired change in each attribute. In our spatial-intensity transform GAN, the generated image is obtained by applying an intensity difference map and a deformation field to the input image. The parameters of the generator are updated from three loss terms: a cycle consistency loss ℓ_{cc} that discourages unnecessary changes to the input image, an attribute loss ℓ_{attr} that encourages the generated image to match the desired attribute values, and an adversarial loss ℓ_{adv} that penalizes unrealistic generated images.

discriminator $D : \mathcal{X} \to \mathbb{R}$ (logits), and regressor $R : \mathcal{X} \to \mathcal{Y}$. Here we consider continuous vector attributes $y_i = (y_{i,1}, ..., y_{i,m})$ that may have missing values. Categorical attributes can be included by expanding regressor R to produce categorical outputs (classifier).

Generator. The generator G transforms an input image such that the transformed image appears to take on different attribute values from the input image, but maintains aspects of the input image that are unrelated to the conditional attributes, such as non-pathological anatomy.

Define $z_i = (x_i, y_i)$, $z_j = (x_j, y_j)$, $\Delta y = y_j - y_i$. During training, the generator is updated using the following loss terms (Fig. 2):

$$\ell_{\mathrm{cc}} = \|G(G(x_i, \Delta y), -\Delta y) - x_i\|_1 \qquad \text{cycle consistency loss} \qquad (1)$$

$$\ell_{\mathrm{attr}} = \frac{1}{m}\|(R(G(x_i, \Delta y)) - R(x_i)) - \Delta y\|_2^2 \quad \text{relative attribute loss} \qquad (2)$$

$$\ell_{\mathrm{adv}} = -D(G(x_i, \Delta y)) \qquad \text{Wasserstein adversarial loss} \quad (3)$$

Parameterizing generator G in terms of attribute difference Δy enables evaluation of the cycle consistency loss ℓ_{cc} even when images have missing attributes [13]. To compute Δy in such cases, we introduce the convention that $y_{j,k} - y_{i,k} = 0$ if either attribute is missing. Putting the terms together,

$$\mathcal{L}_G = \mathbb{E}_{z_i, z_j}\left[\ell_{\mathrm{adv}} + \lambda_{\mathrm{attr}}\, \ell_{\mathrm{attr}} + \lambda_{\mathrm{cc}}\, \ell_{\mathrm{cc}}\right] \qquad (4)$$

where λ_{attr} and λ_{cc} are empirically determined weights.

Discriminator. We simultaneously train the discriminator D with the Wasserstein GAN losses and gradient penalty [6]:

$$\mathcal{L}_D = \mathbb{E}_{z_i, z_j}\left[D(G(x_i, \Delta y))\right] - \mathbb{E}_{z_i}\left[D(x_i)\right] - \mathbb{E}_{\hat{x}}\left[\lambda_{\mathrm{GP}}\left(\|\nabla_{\hat{x}} D(\hat{x})\|_2 - 1\right)^2\right] \quad (5)$$

where $\hat{x}$ is obtained by interpolating real and translated images as described in [6], and λ_{GP} is a weight.

Regressor. The regressor R is trained to predict the attributes of real images, using a mean squared error loss.

$$\mathcal{L}_R = \mathbb{E}_{z_i}\left[\frac{1}{m}\|R(x_i) - y_i\|_2^2\right] \qquad (6)$$

We share layers between the discriminator and regressor, so a single optimizer is assigned to both subnetworks and updated using $\mathcal{L}_D + \lambda_R \mathcal{L}_R$.

2.2   Spatial-Intensity Transform Generator

To constrain the generator to spatial-intensity transforms, we define its outputs as the deformation field $F : \Omega \to \mathbb{R}^d$ for image dimensionality d, with corresponding transform $T_F : \mathcal{X} \to \mathcal{X}$, and the intensity difference map $\Delta x : \Omega \to \mathbb{R}$.

Rather than directly producing the target image, the generator outputs the deformation field F and intensity changes Δx, then transforms the input image as $T_F(x_{\text{in}} + \Delta x)$. In addition, we added regularization terms to the generator's loss function that encourage the deformation field to be smooth and the intensity difference map to be sparse. Specifically, we used the discrete total variation norm [2] to regularize the deformation field and the L1-norm to regularize the intensity change:

$$\|F\|_{\text{TV}} = \frac{1}{|\Omega|} \sum_{\omega \in \Omega} \|\nabla F(\omega)\|_2 \tag{7}$$

$$\|\Delta x\|_1 = \frac{1}{|\Omega|} \sum_{\omega \in \Omega} |\Delta x(\omega)|, \tag{8}$$

where $\|\nabla F(\omega)\|_2$ is approximated using finite differences. The total generator loss becomes $\mathcal{L}_G = \mathbb{E}_{z_i, z_j} \left[\ell_{\text{adv}} + \lambda_{\text{attr}} \ell_{\text{attr}} + \lambda_{\text{cc}} \ell_{\text{cc}} + \lambda_{\text{TV}} \|F\|_{\text{TV}} + \lambda_{\Delta x} \|\Delta x\|_1 \right]$ for empirically determined weights λ_{TV} and $\lambda_{\Delta x}$.

2.3   Network Architecture and Implementation Details

SIT-GAN's generator was implemented as a 2D U-Net that takes in attribute difference Δy by replicating each dimension of Δy spatially and concatenating channel-wise with the input image x_{in}. The U-Net has 4 spatial resolutions, with 200 channels and 6 residual blocks at the lowest resolution. The discriminator and regressor share 5 down-sampling blocks, then split into fully connected layers of the appropriate dimension (1 output for the discriminator, m outputs for the regressor).

Batch normalization is used for all convolutional layers. Down-sampling blocks in the U-Net use convolutional layers alternating with max blur pooling [15]. Up-sampling blocks in the U-Net use bilinear upsampling between convolutional layers. The generator uses ReLU activations, the discriminator and regressor use leaky ReLU activations.

The subnetworks were trained with Adam optimizers, with one step in G's optimizer for every two steps in D/R's optimizer. D/R were trained for 50K iterations with a learning rate of 1.2×10^{-5}, and G was trained for 25K iterations with a learning rate of 1.5×10^{-4}. Both optimizers used a minibatch size of 4, and moving average parameter $\beta_1 = 0.86$. We used the following loss weights: $\lambda_R = 18$, $\lambda_{\text{attr}} = 3.5$, $\lambda_{\text{cc}} = 2.1$, $\lambda_{\text{TV}} = 16$, $\lambda_{\Delta x} = 49$, and $\lambda_{\text{GP}} = 1.1$.

2.4 Baseline Networks

To investigate the influence of different components, we compare SIT-GAN to a network whose generator does not transform the image and several networks whose generators use alternate transformations of the input image. These different parameterizations are summarized in Table 1.

Table 1. Parameterizations of the generator output.

Parameterization	G Outputs	Generated image	Regularizers
Unconstrained	x_{out}	x_{out}	N/A
Difference Transform	Δx	$x_{\mathrm{in}} + \Delta x$	$\|\Delta x\|_1$
Optical Flow	F	$T_F(x_{\mathrm{in}})$	$\|F\|_{\mathrm{TV}}$
Weighted Flow	F, w	$w \odot T_F(x_{\mathrm{in}}) + (\mathbf{1} - w) \odot x_{\mathrm{in}}$	$\|F\|_{\mathrm{TV}}$
SIT-GAN	$F, \Delta x$	$T_F(x_{\mathrm{in}} + \Delta x)$	$\|F\|_{\mathrm{TV}}, \|\Delta x\|_1$

In the unconstrained network, the generator directly synthesizes a new image. We trained two variants of this model: one that has identical hyperparameters to SIT-GAN and other baseline networks, and one in which we tuned the number of layers, types of layers, loss term weights, and type of optimizer to make it as competitive with SIT-GAN as possible. In the tuned model, the discriminator and regressor were trained with a learning rate of 8.6×10^{-5}, and the generator was trained with a learning rate of 1.1×10^{-4}. The U-Net had 3 spatial resolutions with 96 channels and 3 residual blocks at the lowest resolution. Strided convolutions were used for downsampling. The discriminator/regressor had 6 downsampling blocks using max blur pooling. The tuned network used loss weights of $\lambda_R = 21$, $\lambda_{\mathrm{attr}} = 1$, $\lambda_{\mathrm{cc}} = 4$, and $\lambda_{\mathrm{GP}} = 8$. The Adam optimizer had moving average parameter $\beta_1 = 0.46$.

In the difference transform network, the generator is constrained to a sparse intensity difference transform of the input image. It penalizes output images that differ from their inputs using the L1-norm, making it suitable for capturing image-to-image translations that only involve small regions of the image. It corresponds to SIT-GAN with $\lambda_{\mathrm{TV}} = \infty$.

The optical flow network is constrained to smooth deformations of the input image, which can capture morphological variation but not intensity changes within anatomical structures. It corresponds to SIT-GAN with $\lambda_{\Delta x} = \infty$.

The weighted flow network outputs a weighted sum of the input image and a smooth deformation of it, with pixel-wise weights computed by the generator. It is the type of model used to synthesize successive frames in video-to-video translation models [11].

2.5 Data and Evaluation

Our dataset consisted of 1821 axial brain fluid-attenuated inversion recovery (FLAIR) MRIs from 12 clinical sites in the MRI-GENIE study [5], obtained within 48 h of symptom onset in acute ischemic stroke patients. 418 images acquired from the largest site (Massachusetts General Hospital) were used for 5-fold cross validation. The models were then tested on the 1403 scans from all other clinical sites. Age was available for all patients, and stroke severity (NIHSS scale of 0–36) was available for 746 patients.

MRIs were preprocessed with resampling to isotropic 1mm resolution, N4 bias field correction, ANTS registration to a FLAIR atlas, normalisation of the white matter intensity, and cropping to 224×192. Native resolution varied, but was typically around $1\,mm \times 1\,mm \times 6\,mm$, which resulted in significant partial volume effects. The 15 middle axial slices of each subject were used, and all slices from the same subject were grouped into the same validation fold. We scaled age and stroke severity so that the empirical distribution of each attribute within the training data had a mean of 0 and a standard deviation of 1. The images were also augmented using horizontal flips and random affine transformations.

To quantify the realism of model outputs in the absence of paired data, we computed the Fréchet Inception Distance (FID) [7] between the distribution of generated images and the distribution of validation or test images. We also used Precision and Recall for Distributions (PRD) [10] to compute the $F_{1/8}$ and F_8 scores of our generator. A high $F_{1/8}$ suggests that most modes of the generated distribution belong to the true distribution, whereas a high F_8 suggests that most modes of the true distribution belong to the generated distribution. Modes are estimated by finding clusters of images in Inception v3 embedding space.

We also evaluated the effectiveness of each model in transforming the target attribute by measuring the performance of an Inception v3 regressor on our generated images. This regressor was pre-trained on ImageNet and fine-tuned on FLAIR MRIs to predict both age and stroke severity. We emphasize that this Inception v3 regressor is different from the regressor used during training of the GAN, as the generator may have learned to exploit peculiarities in the particular regressor it is trained with. Using a separately trained regressor with a different architecture eliminates any gains that the generator accrued in this manner. We measure the mean squared error (MSE) of age and stroke severity (NIHSS) respectively, normalized to the empirical standard deviation of the attribute. The MSE of the Inception regressor on held out subjects in the cross-validation set is 0.24 on age and 0.70 on NIHSS, while it is 0.34 on age and 0.62 on NIHSS in the test set.

3   Results

Our results suggest that SIT-GAN achieves better image fidelity than the unconstrained model as measured by FID, precision ($F_{1/8}$) and recall (F_8). The statistically significant improvement in both cross-validation and testing ($p<0.01$ for each metric by t-test) shows that this pattern generalizes beyond the particular clinical site it was trained on.

Even after tuning, the unconstrained model introduces artifacts in translated images such as dark streaking of the gray matter with increasing age, and partial

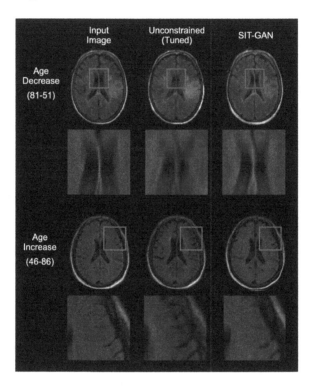

Fig. 3. Comparison of stroke MRIs translated to a different age using the baseline model and our model. While both models change the ventricle shape appropriately, the baseline model blurs the ventricles (top rows) and excessively darkens the gray matter (bottom rows).

volume-like filling of the ventricles with decreasing age. While some of these artifacts do appear in the dataset, Fig. 3 illustrates such artifacts in images that did not have them originally. SIT-GAN does not suffer from these artifacts, and still captures the growth of the ventricles correlated with aging.

The difference transform, optical flow, weighted flow, and SIT-GAN models all perform relatively well on distributional metrics but underperform the unconstrained model on target domain transfer (Table 2). This suggests that an unconstrained generator sacrifices image quality to capture more variation in the conditional attribute compared to the constrained generators.

In general, SIT-GAN attains the best image fidelity, and performs similarly to the optical and weighted flow models in target distribution matching. Often it is overly conservative in transforming input images, but when it succeeds, it is able to capture the expansion of the ventricles correlated with aging as well as the increase in white matter hyperintensities associated with stroke severity (Fig. 1), while producing less severe artifacts than the unconstrained model as seen in Fig. 3.

The learned spatial and intensity transforms also correspond to changes in morphology and tissue properties that are associated with particular patient phe-

Table 2. Performance metrics for translation of FLAIR MRIs conditioned on age and stroke severity (NIHSS), averaged over 5 runs. FID = Fréchet Inception Distance, P/R = Precision ($F_{1/8}$) and Recall (F_8) as defined in [10].

Model Type	FID	P/R	Age MSE	NIHSS MSE
Cross-validation				
Unconstrained	152.1	0.01/0.01	1.51	2.18
Unconstrained (tuned)	61.4	0.07/0.21	**0.51**	1.12
Difference transform	57.2	**0.38/0.59**	1.37	1.14
Optical flow	59.5	0.30/0.52	0.71	**1.09**
Weighted flow	60.6	0.23/0.46	0.85	1.31
SIT-GAN	**38.6**	0.35/**0.59**	0.85	1.16
Test				
Unconstrained	180.5	0.07/0.02	1.11	1.21
Unconstrained (tuned)	51.0	0.41/0.21	**0.99**	**1.01**
Difference transform	68.4	0.53/0.68	1.25	1.12
Optical flow	28.4	**0.62/0.69**	1.16	1.11
Weighted flow	35.0	0.56/0.59	1.32	1.14
SIT-GAN	**27.6**	0.53/0.66	1.28	1.12

notypes or disease processes. In FLAIR MRIs of stroke patients, the patient's age is correlated with the volume of the ventricles as well as the volume of white matter hyperintensities in the periphery of the ventricles. These effects, which would be inseparable in the unconstrained model, can be visualized separately with SIT-GAN by examining the deformation field and intensity differences individually (Fig. 4).

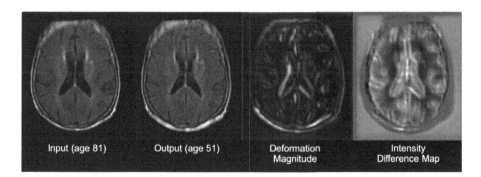

Fig. 4. The magnitude of the deformation field and intensity difference map of the spatial-intensity model for an example transformation. The shrinkage of the ventricles and sulci are well captured by the deformation field, while tissue appearance changes are reflected in the difference map.

4   Conclusion

We presented SIT-GAN, a novel parameterization of GANs for medical image-to-image translation that improves image fidelity and reduces artifacts. In many medical applications, the desired transformations can be well represented by a smooth deformation and a sparse intensity difference transform, and our method can provide robustness to artifacts. We demonstrated our model on a challenging dataset of clinical quality FLAIR MRIs of stroke patients. Our model produces high quality images that visualize correlations of the brain's shape and appearance with the patient's age and stroke severity. Additionally, our parameterization provides a disentangled view of changes in anatomical shape and tissue appearance. Such advances in image-to-image translation can help drive progress in many areas of medical image analysis, including data augmentation, data harmonization, and prediction or visualization of disease trajectories.

Because our proposed representation sacrifices the quality of target domain matching to improve image fidelity, our work leaves open questions about how to navigate or circumvent this trade-off. We suggest that promising directions include carefully relaxing the constraints (reducing regularization weights) over the course of training, or incorporating priors over anatomical structures. Future work will also extend our 2D model to 3D, using low-memory techniques to compensate for memory limitations, and more aggressive data augmentation to accommodate higher model capacity.

Acknowledgments. This work was supported by NIH NIBIB NAC P41EB015902, NIH NINDS U19NS115388, NIH NINDS R01NS086905, Wistron Corporation, Takeda Pharmaceuticals and the Siebel Foundation.

References

1. Chaitanya, K., Karani, N., Baumgartner, C.F., Donati, O., Becker, A.S., Konukoglu, E.: Semi-supervised and task-driven data augmentation. CoRR abs/1902.05396 (2019). http://arxiv.org/abs/1902.05396
2. Chambolle, A., Novaga, M., Cremers, D., Pock, T.: An introduction to total variation for image analysis. In: Theoretical Foundations and Numerical Methods for Sparse Recovery, De Gruyter (2010)
3. Choi, Y., Choi, M., Kim, M., Ha, J., Kim, S., Choo, J.: StarGAN: unified generative adversarial networks for multi-domain image-to-image translation. CoRR abs/1711.09020 (2017). http://arxiv.org/abs/1711.09020
4. Cootes, T.F., Beeston, C., Edwards, G.J., Taylor, C.J.: A unified framework for atlas matching using active appearance models. In: Kuba, A., Šáamal, M., Todd-Pokropek, A. (eds.) Information Processing in Medical Imaging, pp. 322–333. Springer, Heidelberg (1999)
5. Giese, A., et al.: Design and rationale for examining neuroimaging genetics in ischemic stroke: the MRI-genie study. Neurol. Genet. **3**(5) (2017). https://doi.org/10.1212/NXG.0000000000000180
6. Gulrajani, I., Ahmed, F., Arjovsky, M., Dumoulin, V., Courville, A.C.: Improved training of wasserstein gans. CoRR abs/1704.00028 (2017). http://arxiv.org/abs/1704.00028

7. Heusel, M., Ramsauer, H., Unterthiner, T., Nessler, B., Klambauer, G., Hochreiter, S.: Gans trained by a two time-scale update rule converge to a nash equilibrium. CoRR abs/1706.08500 (2017). http://arxiv.org/abs/1706.08500

8. Quan, T.M., Nguyen-Duc, T., Jeong, W.K.: Compressed sensing MRI reconstruction using a generative adversarial network with a cyclic loss. IEEE Trans. Med. Imaging **37**(6), 1488–1497 (2018)

9. Ravi, D., Alexander, D.C., Oxtoby, N.P.: Degenerative adversarial neuroimage nets: generating images that mimic disease progression. In: Shen, D., et al. (eds.) MICCAI 2019. LNCS, vol. 11766, pp. 164–172. Springer, Cham (2019). https:// doi.org/10.1007/978-3-030-32248-9_19

10. Sajjadi, M.S.M., Bachem, O., Lucic, M., Bousquet, O., Gelly, S.: Assessing generative models via precision and recall (2018). https://arxiv.org/abs/1806.00035

11. Wang, T., et al.: Video-to-video synthesis. CoRR abs/1808.06601 (2018). http:// arxiv.org/abs/1808.06601

12. Wolterink, J.M., Dinkla, A.M., Savenije, M.H.F., Seevinck, P.R., van den Berg, C.A.T., Išgum, I.: Deep MR to CT synthesis using unpaired data. In: Tsaftaris, S.A., Gooya, A., Frangi, A.F., Prince, J.L. (eds.) SASHIMI 2017. LNCS, vol. 10557, pp. 14–23. Springer, Cham (2017). https://doi.org/10.1007/978-3-319-68127-6_2

13. Wu, P.W., Lin, Y.J., Chang, C.H., Chang, E.Y., Liao, S.W.: RelGAN: multi-domain image-to-image translation via relative attributes. In: International Conference on Computer Vision (2019)

14. Xia, T., Chartsias, A., Wang, C., Tsaftaris, S.A.: Learning to synthesise the ageing brain without longitudinal data (2019). https://arxiv.org/abs/1912.02620

15. Zhang, R.: Making convolutional networks shift-invariant again. CoRR abs/1904.11486 (2019). http://arxiv.org/abs/1904.11486

16. Zhao, A., Balakrishnan, G., Durand, F., Guttag, J.V., Dalca, A.V.: Data augmentation using learned transforms for one-shot medical image segmentation. CoRR abs/1902.09383 (2019). http://arxiv.org/abs/1902.09383

17. Zhu, J., Park, T., Isola, P., Efros, A.A.: Unpaired image-to-image translation using cycle-consistent adversarial networks. CoRR abs/1703.10593 (2017). http://arxiv. org/abs/1703.10593

Graded Image Generation Using Stratified CycleGAN

Jianfei Liu, Joanne Li, Tao Liu, and Johnny Tam[✉]

National Eye Institute, National Institutes of Health, Bethesda, MD, USA
johnny@nih.gov

Abstract. In medical imaging, CycleGAN has been used for various image generation tasks, including image synthesis, image denoising, and data augmentation. However, when pushing the technical limits of medical imaging, there can be a substantial variation in image quality. Here, we demonstrate that images generated by CycleGAN can be improved through explicit grading of image quality, which we call stratified Cycle-GAN. In this image generation task, CycleGAN is used to upgrade the image quality and content of near-infrared fluorescent (NIRF) retinal images. After manual assignment of grading scores to a small subset of the data, semi-supervised learning is applied to propagate grades across the remainder of the data and set up the training data. These scores are embedded into the CycleGAN by adding the grading score as a conditional input to the generator and by integrating an image quality classifier into the discriminator. We validate the efficacy of the proposed stratified CycleGAN by considering pairs of NIRF images at the same retinal regions (imaged with and without correction of optical aberrations achieved using adaptive optics), with the goal being to restore image quality in aberrated images such that cellular-level detail can be obtained. Overall, stratified CycleGAN generated higher quality synthetic images than traditional CycleGAN. Evaluation of cell detection accuracy confirmed that synthetic images were faithful to ground truth images of the same cells. Across this challenging dataset, F1-score improved from $76.9 \pm 5.7\%$ when using traditional CycleGAN to $85.0 \pm 3.4\%$ when using stratified CycleGAN. These findings demonstrate the potential of stratified Cycle-GAN to improve the synthesis of medical images that exhibit a graded variation in image quality.

Keywords: CycleGAN · Semi-supervised learning · Data parsing · Image quality · Cell detection · Adaptive optics · Ophthalmology.

1 Introduction

Deep learning has catalyzed a growing interest in medical image generation [15], especially for applications in which the same anatomical structures can be imaged using different imaging instruments. A special case for image generation

This is a U.S. government work and not under copyright protection in the U.S.; foreign copyright protection may apply 2020
A. L. Martel et al. (Eds.): MICCAI 2020, LNCS 12262, pp. 760–769, 2020.
https://doi.org/10.1007/978-3-030-59713-9_73

is to start with an image that is simple or fast to acquire in a cost-effective manner (typically on a readily or commercially available device) ("low-cost"), and synthesize or generate an improved image resembling one from a sophisticated, cutting-edge device ("high-cost"). However, high-cost devices, which push the limits of technology, often exhibit substantial image quality variation. Although data curation (removing low quality images) can improve algorithm performance, it comes at a cost of shrinking the training dataset. To our knowledge, popular remedies for improving low quality data such as image denoising [14], or for augmenting training data [1,4,8,11,12], are not systematically designed to handle variations in image quality.

CycleGAN [18], originally designed for transferring image style among natural images, has become a widely-utilized deep learning method for medical image generation. In some situations, to adapt CycleGAN for medical image generation, gradient constancy loss function [4,11] or perceptual loss function [1] were incorporated to minimize the texture and perceptual discrepancies between generated and real images. The Wasserstein distance can further enhance loss functions to guide the minimization process [10]. In other situations, instead of using loss functions that highlight object boundaries, image edge maps were explicitly used as input data for CycleGAN [16,17]. Additional improvements to CycleGAN include replacing the generator network with residual network [2] or attention gates [9], or leveraging CycleGAN to generate labeled training data [5,8,12].

This paper aims to improve CycleGAN for medical image generation from the perspective of image quality variation, using near-infrared fluorescent (NIRF) retinal images as the application. Each NIRF image pair consists of one "low-cost" image acquired using a device that is commercially available and provides images with minimal image quality variation, but has poor performance for imaging cells (e.g. x_r^a, Fig. 1B) and a corresponding "high-cost" image at the same retinal region acquired using a more sophisticated device that is not commercially available [13], which has the capability to reveal cellular structures but has substantial image quality variation (e.g. x_r^b, Fig. 1B). Semi-supervised learning based on pseudo-labeling [6] is utilized to numerically grade high-cost NIRF images, for use by CycleGAN. We refer to the proposed strategy as "stratified CycleGAN" and show that synthetic images generated by stratified CycleGAN provide better and more accurate cellular details than those generated using traditional CycleGAN.

2 Methodology

The process of medical image generation through stratified CycleGAN is outlined in Fig. 1. It consists of two main steps. First, semi-supervised learning based on pseudo-labeling is used to assign numerical grades to high-cost NIRF images. Second, stratified CycleGAN is used to generate graded image sequences with different levels of image contrast which are useful for evaluating cellular details that were not readily apparent prior to image synthesis. The combined system

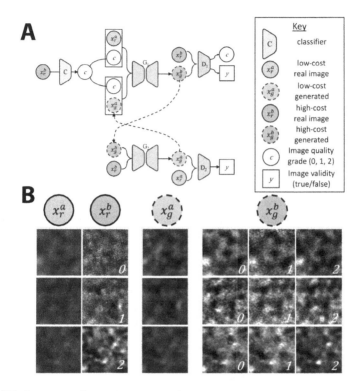

Fig. 1. (A) Overview of stratified CycleGAN to generate improved near-infrared fluorescent (NIRF) retinal medical images. After x_r^b is graded ($c \in \{0, 1, 2\}$) by the classifier C using semi-supervised learning, c is embedded with either x_r^a or x_g^a for use by the first generator G_1 to generate x_g^b. The discriminator D_1 classifies the generated image grading score c and determines the image validity y. G_2 and D_2 perform the same process, except without grading score c. (B) Examples of NIRF image pairs. Individual retinal cells can be seen (bright and dark patterns). Each NIRF image pair consists of a low-cost image with stable image quality but poor cellular quality (x_r^a) and a high-cost one with fine cellular detail but substantial image variation (x_r^b). The corresponding images generated using stratified CycleGAN are shown as (x_g^a) and (x_g^b).

enables low-cost NIRF images to be up-converted to high-cost NIRF images so that cellular-level detail can be realized. Since NIRF image pairs correspond to the same retinal region, high-cost NIRF images serve as a natural ground truth for validation.

2.1 Consistent Image Grading Using Semi-supervised Pseudo-Labeling

The purpose of this step is to consistently grade NIRF image quality using deep learning to propagate human assessment. Manual grading is a subjective process that can introduce inconsistencies into large NIRF training data.

To maximize grading consistency, manual labeling of a small amount of x_r^b training images into three grading scores (0 = poor, 1 = moderate, 2 = good) (e.g. x_r^b, Fig. 1B) was propagated using semi-supervised learning to systematically classify the remaining unlabeled images (x'^b_r).

A simple and efficient way to accomplish this is to use pseudo-labeling [6] on a NIRF image pair x_r^a and x_r^b, where x_r^b belongs to the high-cost image set that has N images. Let the user randomly pick M images of x_r^b ($M \ll N$) to assign numerical grading scores $c \in \{0, 1, 2\}$. To begin the pseudo-labeling process, the set of M labeled images are used to train a classification model C based on ResNet [3] (C, classifier, in Fig. 1A). C is then used to predict grading scores c' for unlabeled images x'^b_r. By combining labeled and unlabeled images to update C, and then by repeating the process, the accuracy of predicted c' scores is gradually improved. The essential element of pseudo-labeling is that the objective function L_C is dynamically modified to control the influence of x'^b_r on updating the classifier during the training process. L_C is given by

$$L_C = -\frac{1}{M} \sum_{i=1}^{M} \sum_{j=0}^{2} F(c_j) \cdot \log F(\hat{c}_j) - \alpha(t) \frac{1}{N-M} \sum_{i=1}^{N-M} \sum_{j=0}^{2} F(c'_j) \cdot \log F(\hat{c}'_j) \tag{1}$$

Here, $\hat{c}$ and $\hat{c}'$ are predicted grading scores of the known and unknown x_r^b images, respectively, as determined by the classifier C at the training epoch t; F is the data distribution function of the grading scores; and $\alpha(t)$ is a coefficient function to balance the relative influence of x_r^b and x'^b_r grading scores on the classifier, defined as

$$\alpha(t) = \begin{cases} 0 & t < T_1 \\ \frac{t-T_1}{T_2-T_1}\beta & T_1 \le t < T_2 \\ \beta & t \ge T_2 \end{cases} \tag{2}$$

with $\beta = 3$, $T_1 = 10$, and $T_2 = 70$ in this work. These parameters allow $\alpha(t)$ to shifts towards x'^b_r as C becomes more accurate in predicting grading scores.

2.2   Stratified CycleGAN

Each NIRF image pair (x_r^a and x_r^b) is combined with the grading score, c, associated with x_r^b. Together, this triplet comprises the input data for stratified CycleGAN. The aim is to learn the two mapping functions $G_1 : (x_r^a, c) \to x_g^b$ and $G_2 : x_r^b \to x_g^a$ (Fig. 1A), where x_g^a and x_g^b are generated images. G_1 uses c and x_r^a to generate a sequence of graded images (x_g^b), capturing the image grading variation that is contained in real images of x_r^b (x_g^b, in Fig. 1B). In contrast, G_2 can accept x_r^b with any grading score to generate x_g^a with a relatively stable image appearance. There are also two adversarial discriminators D_1 and D_2 to determine image validity, with the modification that D_1 also reports the image grading score. The objective function contains adversarial losses, cycle consistency losses, and generation losses.

The adversarial loss for G_1 and D_1 is modified to include a conditional input (grading score c for G_1) as well as classification results (grading score c from D_1), and is defined as

$$
\begin{aligned}
L_{GAN1} = E_{x_r^b \sim P(x_r^b)} &\left[\log D_1(x_r^b) + \sum_{i=0}^{2} F(c_i) \cdot \log F(\hat{c}_{r,i}) \right] \\
+ E_{x_r^a \sim P(x_r^a)} &\left[\log(1 - D_2(x_g^b)) + \sum_{i=0}^{2} F(c_i) \cdot \log F(\hat{c}_{g,i}) \right]
\end{aligned}
\tag{3}
$$

Here, P represents the data distributions of x_r^a and x_r^b, noting that $x_g^a = G_2(x_r^b)$. $x_g^b = G_1(x_r^a, c)$; $\hat{c}_r$ and $\hat{c}_g$ are the predicted grading scores of x_r^b and x_g^b, respectively, determined by D_1.

The adversarial loss for G_2 and its discriminator D_2 is similar to CycleGAN [18] and is defined as,

$$
L_{GAN2} = E_{x_r^a \sim P(x_r^a)} \left[\log D_2(x_r^a) \right] + E_{x_r^b \sim P(x_r^b)} \left[\log(1 - D_2(x_g^a)) \right]
\tag{4}
$$

The cycle consistency loss is computed based on L1 norm:

$$
\begin{aligned}
L_{cyc}(G_1, G_2, c) = E_{x_r^a \sim P(x_r^a)} &\left[\| G_2(G_1(x_r^a, c)) - x_r^a \|_1 \right] \\
+ E_{x_r^b \sim P(x_r^b)} &\left[\| G_1(G_2(x_r^b), c) - x_r^b \|_1 \right]
\end{aligned}
\tag{5}
$$

Finally, due to the similarity of NIRF image pairs (i.e. both the "low-cost" and "high-cost" instruments are based on identical contrast sources), we can also include generation losses to help stabilize the training process:

$$
L_{gen} = E_{x_r^a \sim P(x_r^a)} \left[\| G_1(x_r^a, c) - x_r^b \|_1 \right] + E_{x_r^b \sim P(x_r^b)} \left[\| G_2(x_r^b) - x_r^a \|_1 \right]
\tag{6}
$$

Note that traditional CycleGAN cannot include generation losses whenever the image modalities are changed. The overall objective function is

$$
L = L_{GAN1} + L_{GAN2} + L_{cyc} + L_{gen}
\tag{7}
$$

We implemented the network structure of CycleGAN [18] for both generator and discriminator networks in the following manner. Both G_1 and G_2 start with two stride-2 convolutions, followed with 9 residual blocks, and two fractionally-strided convolutions with the input being 256×256 images. Since the input data of G_1 also contains the grading score c, an embedding layer is created before the convolution layers to convert c into a 256×256 image, which is then concatenated as an additional channel associated with the input image x_r^a. The discriminator networks D_1 (top right, Fig. 1A) and D_2 (bottom right, Fig. 1A) also use PatchGANs with the smaller 30×30 overlapping image patches to validate whether the patches are real or fake. Such a strategy is believed to substantially reduce the number of training parameters needed [18]. D_1 also adds a fully connected layer (additional output c, Fig. 1A) to perform 3-class image grading by sharing the same input features for the PatchGAN. Minimizing Eq. 7

yields the proposed graded image generation process for synthetic NIRF images x_g^b, as illustrated in Fig. 1B. We expect the generated images to consistently incorporate additional image details as grading scores increase. All in all, the graded image generation strategy is intended to provide a strategy to utilize more training data while explicitly ensuring the quality of generated images.

2.3   Dataset and Validation Methods

A total of 1,430 NIRF pairs were collected from 10 human subjects (28.8 ± 8.6 years) for stratified CycleGAN. 1,144 image pairs (80% of them) were used for training data, with the remaining 286 pairs (20%) reserved for test data. Within the training data, 284 pairs (25%) were manually labeled with grading scores assigned to the high-cost NIRF images (large image variation), with grading scores propagated to the remaining training data using semi-supervised pseudo-labeling. The final distribution of grading scores across the complete pseudo-labeled training data was 405 image pairs (poor, 0), 415 image pairs (moderate, 1), and 324 pairs (good, 2).

Qualitative evaluation was performed by comparing the details of cellular structures generated from low-cost data to ground truth images of cells obtained using the high-cost approach. To demonstrate results in an important real-life application, quantitative evaluation was performed based on objective cell detection results on the generated images compared to detection results on ground truth images (x_r^b). Given that cellular details cannot be discerned from "low-cost" images (x_r^a), we expect that cell detection accuracy on generated images (x_g^b) should be representative of the overall image generation quality. For cell detection, we used a previously-trained LinkNet based cell detection algorithm with the same cell detection model across all images for a fair comparison [7]. Detection results on generated images (x_g^b) within the test dataset were then compared to results from paired ground truth images (x_r^b). In addition, to validate that explicit handling of image grading scores were beneficial, cell detection results from the proposed stratified CycleGAN were compared with results from (traditional, non-stratified) CycleGAN as well as from CycleGAN trained on a curated subset of only the best images (only those with high grading score, $c = 2$) ("{2} $\subset$ CycleGAN"). Precision, recall, and F1-score were used to evaluate detection accuracy.

3   Experimental Results

3.1   Image Quality Comparison

Stratified CycleGAN successfully generated synthetic image sequences with enhanced cellular detail corresponding to the increased image grading scores (Fig. 1B). From these sequences, the generated images with grade 2 were faithful to ground truth images (right two columns, Fig. 2). For evaluation purposes, only grade 2 images from the stratified CycleGAN were used for comparison.

Low-cost   CycleGAN   {2}⊂CycleGAN   Stratified CycleGAN   High-cost

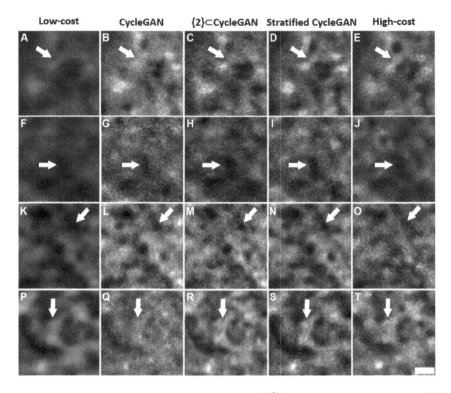

Fig. 2. Comparison of generated high-cost images (x_g^b) from low-cost NIRF images (x_r^a) using CycleGAN, $\{2\} \subset$ CycleGAN trained on a curated subset of data with only grade 2 images, and stratified CycleGAN to real ground truth NIRF images (high-cost). Each row corresponds to images from a distinct retinal region. (A–E) Two bright regions are readily distinguishable using stratified CycleGAN but are not distinguishable using either CycleGAN or $\{2\} \subset$ CycleGAN (white arrows). (F–J) Stratified CycleGAN faithfully preserved the size and shape of dark cell regions, which were either artificially decreased by CycleGAN or enlarged by $\{2\} \subset$ CycleGAN. (K–O) Image contrast was best using stratified CycleGAN. (P–T) A large amount of image noise was contained in the generated images from both CycleGAN and $\{2\} \subset$ CycleGAN, but not stratified CycleGAN. These examples illustrate the advantages of using stratified CycleGAN. Scale bar, 20 μm.

Overall, stratified CycleGAN outperformed both CycleGAN and $\{2\} \subset$ CycleGAN. CycleGAN trained without explicit handling of grading scores on a dataset that includes mixed grading scores was prone to generating degraded images (second column, Fig. 2), which is understandable since the low grade images can potentially mislead CycleGAN to generate images with low grades, when in fact, only high grade images are desired. The contrast between bright and dark regions is often poor in CycleGAN, as evidenced by loss of bright regions boundaries (white arrows, Fig. 2B), distorted dark regions (Fig. 2G), diminished image contrast (Fig. 2L), and large amounts of image noise (Fig. 2Q).

As expected, curating the data ($\{2\} \subset$ CycleGAN) improved the generated images since training data with poor image grading scores were eliminated from training (third column, Fig. 2). However, when compared to high-cost ground truth data, the image contrast at cell regions was still inferior (Fig. 2C), with examples of distorted (Fig. 2H) and blurry (Fig. 2R) image regions. In contrast, stratified CycleGAN was able to consistently generate the best images, indistinguishable from ground truth high-cost images (fourth and fifth columns, Fig. 2). Region intensities and shapes were well preserved (Figs. 2D and 2I) and image contrast were also superior in comparison with both CycleGAN and $\{2\} \subset$ CycleGAN (Figs. 2N and 2S). Taken together, these results demonstrate the efficacy of explicitly stratifying training data based on image grading scores. This strategy of progressively adding image details to generated images is a key element for generating images with consistently high grading scores.

3.2 Cell Detection Accuracy Evaluation

Table 1 compares the cell detection results compared to ground truth on generated images from CycleGAN, $\{2\} \subset$ CycleGAN, and stratified CycleGAN. The precision was highest using CycleGAN, but at the expense of recall, resulting in the worst overall F1-score. This indicates that images generated by Cycle-GAN likely misleads the detection algorithm to yield more detections (i.e. reduces false negatives but produces more false positives). Detection results on $\{2\} \subset$ CycleGAN suggest improved performance in reducing false positives to subsequently improve the overall F1-score. The best overall recall and F1-scores were achieved using stratified CycleGAN, consistent with the superior qualitative improvement in cellular detail shown in Fig. 2.

Table 1. Comparison of cell detection accuracy on three different types of generated images by CycleGAN, $\{2\} \subset$ CycleGAN trained on a subset of data with high-cost NIRF images of grade 2, and stratified CycleGAN (mean±SD).

Method	Precision (%)	Recall (%)	F1-score (%)
CycleGAN	**82.8 ± 5.7**	72.4 ± 8.7	76.9 ± 5.7
$\{2\} \subset$ CycleGAN	80.0 ± 7.2	77.1 ± 8.1	78.1 ± 4.8
Stratified CycleGAN	80.8 ± 7.6	**90.9 ± 6.5**	**85.0 ± 3.4**

4   Conclusion and Future Work

This paper introduces stratified CycleGAN, an innovative approach to explicitly integrate grading scores to better utilize training data based on image quality for CycleGAN so that the best quality images can be consistently generated. The strategy is based on enabling image details to be progressively and consistently incorporated in generated images. This was achieved by improving the generator network of CycleGAN with a conditional grading score input and by integrating a classification capability into the discriminator network. The criterion for manual grading is very simple: poor, moderate, good. It is readily performed by a naive grader after minimal instruction. More importantly, since grading scores were only used for initialization in training, as long as the general pattern of image quality was captured, stratified CycleGAN can handle the rest. The overall approach is useful for graded image generation and is inherently suitable for medical images that exhibit substantial image variations, which is a common feature in medical imaging that can be exploited. Experimental results on NIRF images demonstrate the validity of the proposed approach. We expect that stratified CycleGAN can be readily extended to enable image quality variation across both image pairs or across different medical imaging modalities, which should lead to more consistent medical image generation attuned to image quality.

Acknowledgments. This work was supported by the Intramural Research Program of the National Institutes of Health, National Eye Institute.

References

1. Armanious, K., Jiang, C., Abdulatif, S., et al.: Unsupervised medical image translation using cycle-medGAN. In: 27th European Signal Processing Conference (2018)
2. Harms, J., Lei, Y., Wang, T., et al.: Paired cycle-GAN-based image correction for quantitative cone-beam computed tomography. Med. Phys. **46**(9), 3998–4009 (2019)
3. He, K., Zhang, X., Ren, S., Sun, J.: Deep residual learning for image recognition. In: The IEEE Conference on Computer Vision and Pattern Recognition (CVPR) (2016)
4. Hiasa, Y., et al.: Cross-modality image synthesis from unpaired data using Cycle-GAN. In: Gooya, A., Goksel, O., Oguz, I., Burgos, N. (eds.) SASHIMI 2018. LNCS, vol. 11037, pp. 31–41. Springer, Cham (2018). https://doi.org/10.1007/978-3-030-00536-8_4
5. Jiang, J., et al.: Tumor-aware, adversarial domain adaptation from CT to MRI for lung cancer segmentation. In: Frangi, A.F., Schnabel, J.A., Davatzikos, C., Alberola-López, C., Fichtinger, G. (eds.) MICCAI 2018. LNCS, vol. 11071, pp. 777–785. Springer, Cham (2018). https://doi.org/10.1007/978-3-030-00934-2_86
6. Lee, D.H.: Pseudo-label : the simple and efficient semi-supervised learning method for deep neural networks. In: ICML 2013 Workshop: Challenges in Representation Learning (2013)
7. Liu, J., Han, Y.J., Liu, T., Tam, J.: Spatially aware deep learning improves identification of retinal pigment epithelial cells with heterogeneous fluorescence levels visualized using adaptive optics. In: Medical Imaging 2020: Biomedical Applications in Molecular, Structural, and Functional Imaging, vol. 11317, p. 1131719 (2020)
8. Liu, J., Shen, C., Liu, T., Aguilera, N., Tam, J.: Active appearance model induced generative adversarial network for controlled data augmentation. In: Shen, D., et al. (eds.) MICCAI 2019. LNCS, vol. 11764, pp. 201–208. Springer, Cham (2019). https://doi.org/10.1007/978-3-030-32239-7_23

9. Liu, Y., Lei, Y., Wang, T., et al.: CBCT-based synthetic CT generation using deep-attention cycleGAN for pancreatic adaptive radiotherapy. Med. Phys. (2020). https://doi.org/10.1002/mp.14121
10. McDermott, M.B.A., Yan, T., Naumann, T., et al.: Semi-supervised biomedical translation with cycle wasserstein regression GANs. In: Thirty-Second AAAI Conference on Artificial Intelligence (AAAI18) (2018)
11. Nie, D., Trullo, R., Lian, J., et al.: Medical image synthesis with deep convolutional adversarial networks. IEEE Trans. Biomed. Eng. $65(12)$, 2720–2730 (2018)
12. Sandfort, V., Yan, K., Pickhardt, P.J., Summers, R.M.: Data augmentation using generative adversarial networks (cycleGAN) to improve generalizability in CT segmentation tasks. Sci. Rep. 9, 16884 (2019)
13. Tam, J., Liu, J., Dubra, A., Fariss, R.N.: In vivo imaging of the human retinal pigment epithelial mosaic using adaptive optics enhanced indocyanine green ophthalmoscopy. Invest. Ophthalmol. Vis. Sci. $57(10)$, 4376–4384 (2016)
14. Wolterink, J.M., Leiner, T., Viergever, M.A., Išgum, I.: Generative adversarial networks for noise reduction in low-dose CT. IEEE Trans. Med. Imaging $36(12)$, 2536–2545 (2017)
15. Yi, X., Walia, E., Babyn, P.: Generative adversarial network in medical imaging: a review. Med. Image Anal. 58, 101552 (2019)
16. Yu, B., Zhou, L., Wang, L., et al.: Ea-GANs: edge-aware generative adversarial networks for cross-modality MR image synthesis. IEEE Trans. Med. Imaging $38(7)$, 1750–1762 (2019)
17. Zhang, T., et al.: SkrGAN: sketching-rendering unconditional generative adversarial networks for medical image synthesis. In: Shen, D., et al. (eds.) MICCAI 2019. LNCS, vol. 11767, pp. 777–785. Springer, Cham (2019). https://doi.org/10.1007/978-3-030-32251-9_85
18. Zhu, J.Y., Park, T., Isola, P., Efros, A.A.: Unpaired image-to-image translation using cycle-consistent adversarial networks. In: 2017 IEEE International Conference on Computer Vision (ICCV) (2017)

Prediction of Plantar Shear Stress Distribution by Conditional GAN with Attention Mechanism

Jinghui Guo[1]([✉]), Ali Ersen[2], Yang Gao[1], Yu Lin[1], Latifur Khan[1], and Metin Yavuz[2]

[1] The University of Texas at Dallas, Richardson, TX 75080, USA
jinghui.guo@utdallas.edu
[2] The University of Texas Southwestern Medical Center, Dallas, TX 75390, USA

Abstract. Diabetic foot ulcers (DFUs) are known to have multifactorial etiology. Among the biomechanical factors that lead to plantar ulcers, shear stresses have been either neglected or unmeasured due to challenges in complexity and equipment availability. The purpose of this study is to develop a software that predicts plantar shear stress using plantar pressure and temperature distributions. Thirty-one subjects, 8 of them at risk of developing DFUs were recruited, and plantar thermography, pressure and shear stress distributions were collected. We introduce the conditional generative adversarial networks (cGAN) for shear stress distribution prediction and propose an attention mechanism to improve the model's accuracy. The networks can learn the mapping from pressure to shear stress distribution. The attention mechanism can merge temperature distribution into GAN without resizing or aligning it manually. We then test on our dataset with 185 groups. The predicted anteroposterior shear stress distributions give 72.97% accuracy on peak location prediction and 14.12 kPa on global root mean square error. Our initial results are promising in terms of feasibility of our approach in predicting plantar shear stresses and this approach may benefit to address the DFU risks before ulceration.

Keywords: Conditional GAN · Positional attention · Plantar shear stress · Diabetic foot ulcers

1 Introduction

Peripheral Neuropathy is a complication of Diabetes Mellitus that leads to diabetic foot ulcers (DFU). Infected DFU usually result in lower extremity amputations. In fact, more than 100,000 amputations are performed every year in the USA due to DFU and the cost to the USA healthcare system is estimated as $30B [11,18]. Lower extremity amputations also have a significant burden on amputees' quality of life and independence. It is also known that the 5-year mortality rate of patients with DFU is comparable to colon cancer, which indicate that approximately half of those individuals will be deceased within 5 years [2].

© Springer Nature Switzerland AG 2020
A. L. Martel et al. (Eds.): MICCAI 2020, LNCS 12262, pp. 770–780, 2020.
https://doi.org/10.1007/978-3-030-59713-9_74

Although the pathology of DFU is complex, it is widely accepted that DFU mostly occur in the presence of peripheral neuropathy and the complication has a biomechanical aspect [4]. Most common causative factors are elevated plantar pressure, shear stress and temperature, besides vascular disorders, and gross foot deformities. Among the biomechanical factors that contribute to DFU, plantar pressure has been extensively studied over the last few decades. It is known that peak pressure is higher in people who are at risk for DFU; however clinical value of peak pressure in predicting ulcer location is rather low [1,11,12].

There are a number of reports on the importance of shear stresses [6,7,22–24] as another factor contributing to diabetic foot ulceration. Shear stress has been recognized as a causative factor in pressure injury etiology as the National Pressure Ulcer Advisory Panel started the shear stress initiative in 2005 [16]. Shear stress has been traditionally neglected in DFU research due to technical challenges in quantification of this variable.

Shear stresses act tangentially in anteroposterior and mediolateral directions at the foot-ground interface, which exert a complex stress pattern on the sub-layers of plantar tissue [1]. Shear stress is quite important in the formation of DFU due to several reasons: (1) Shear is significantly higher in diabetics. (2) Most DFU develop at peak shear locations. (3) Frictional shear increases the temperature of the tissue, similar to rubbing hands together. Previous research indicated that warmer tissue ulcerates much faster [3]. (4) Shear is the main causative factor behind callus formation, which often precedes DFU. Finally, the shearing action is damaging as demonstrated by the use of a chainsaw. When the engine is off, there is no "shear", it is not possible to cut a tree branch by applying only "pressure". When the engine is on, shear which is applied via the running chain, severs the branch.

The Yavuz group has been the pioneer in the measurement of plantar shear stresses and the mentioned data comes from our prior publications. Our findings on shear indicate that we need an accurate method to predict shear in order to prevent DFU. However, technology to accurately measure shear is quite costly (USD $150–200K), takes a long time to build (more than 12 months) and there are only two validated devices in the world, including ours. A few research groups developed insole-based sensing systems however they contain only 3–5 shear sensors for the entire foot sole, which is problematic as the design omits important regions of the foot. Hence, the purpose of this study was to develop a cost-effective and accurate prediction model, based on data from entire foot.

In order to do this, we have used a validated device to measure shear stress in diabetic patients [19]. To the best of our knowledge, we are the first to estimate plantar shear stress from plantar pressure and temperature images. There is one closest previous work [25] and one of the authors employed the system described in [19] which could collect data from only one half of the foot. However, previous research was limited in terms of spatial resolution (data from only forefoot) and shear prediction was based only on pressure and did not consider temperature, which depends on frictional shear.

Due to complexity of measurement and determination of shear stress under the foot [5], we have created the conditional generative adversarial networks (cGAN) [15] with attention mechanism [21] model in order to predict the amplitude of anteroposterior shear stress which may be responsible for tissue breakdown using plantar temperature and plantar pressure distribution images.

Due to advances in deep learning technology, there has been remarkable progress in image synthesis, with the emergence of Generative Adversarial Networks (GAN). More specifically, our goal is to generate shear stress images from the plantar pressure image which is an image-to-image translation task and the conditional GAN is a general-purpose solution for this kind of problem. Moreover, we consider conditional GAN as the baseline and develop novel methods by utilizing the attention mechanism and conditional GAN as our final model. The attention mechanism can merge the influence of plantar temperature images collected by another sensor into this image-to-image translation [10,27] without manually aligning images. Conditional GAN (cGAN) with attention mechanism can merge more features and generate better quality of images. We evaluate our model by the accuracy of peak shear stress location prediction, global root mean square error (RMSE) and local RMSE.

We have made the following contributions. First, we show how we can leverage GAN technologies to generate shear stress images effectively. Next, we show effectiveness of our work with other traditional approaches. Finally, our proposed solution is very cost-effective.

The paper is organized in the following ways. Section 2 presents our proposed work including adversarial Networks and attention mechanism. Section 3 presents description of our experiment and results. Section 4 presents conclusion.

2   Model Architecture

The inputs for our model are plantar pressure, temperature, real shear stress images (training cases) and the mask of plantar pressure images. The output is the generated shear stress images. Our framework is visually depicted in Fig. 1. The generator network produces samples from input distributions. In our model, it consists of encoder and decoder modules. The positional attention module is used in our final model. The discriminator network detects the difference between samples from the data distribution and the ones generated by the generator. We use gradient descent to optimize our model. After implementing the attention mechanism, our final model achieves 72.97% accuracy on peak location prediction and 14.12 kPa on global root mean square error (details can be found in Sect. 3).

2.1   Conditional GAN with Attention-Guided Loss

Generative adversarial networks (GAN) [8] are generative models that learn a mapping from random noise vector z to output image y, with $G : z \rightarrow y$. Conditional generative adversarial networks (cGAN) [15] are the conditional models

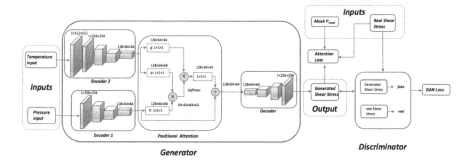

Fig. 1. Architecture of the Conditional GAN with Positional Attention PTA-GAN model. The inputs and output are circled by green boxes. The feature maps are shown as the shape of $Channel \times H \times W$. "$\otimes$" denotes matrix multiplication and "$\oplus$" denotes element-wise sum. The $1 \times 1 \times 1$ boxes in Positional Attention module denote $1 \times 1 \times 1$ convolutions. The discriminator is adversarially trained with Generator. See Sect. 2 for more details. (Color figure online)

which extends GAN. Our goal is to learn a mapping function from pressure p or temperature t data to anteroposterior shear stress data ap: $G : \{p, t\} \to ap$. The generator G, which is composed by encoder and decoder, is trained to produce anteroposterior shear stress output from pressure or temperature inputs. The generated output should not be distinguished from "real" anteroposterior shear stress data. The discriminator D, which is adversarially trained with G, should also detect the "fake" output from the generator. The discriminator D minimizes the overall distance between the real and the generated distribution.

We apply adversarial losses [15] to both mapping functions. For the mapping function $G : \{p, t\} \to ap$ and its discriminator D, we express the objective as:

$$L_{GAN}(G, D) = \mathbb{E}_{ap}[\log D(ap)] + \mathbb{E}_{p,t}[\log(1 - D(G(p, t)))] \qquad (1)$$

where G tries to minimize the objective against an adversarial D that tries to maximize it. Then we can get the objective [10,27] as:

$$G^* = \arg \max_G \min_D L_{GAN}(G, D) \qquad (2)$$

We also mix the GAN objective with the attention-guided traditional loss [10, 14], such as L1 and L2 distance [17]. The discriminator's loss function remains unchanged, but the generator's task requires not only fool the discriminator but also to be near the ground truth anteroposterior shear stress output.

Since the pressure and anteroposterior shear stress datasets are collected by the same device, the foot areas in pressure data and anteroposterior shear stress data have pixel-to-pixel correspondence. We generate the mask of pressure input to distinguish the foot and background area and then can use it to compute with anteroposterior shear stress data as an attention map. The sensor's data changes obviously from background to foot area, so we just need to set a threshold τ to get the mask.

$$p_{mask} = \begin{cases} 1 & if\,P\,(p) > \tau \\ 0 & otherwise \end{cases} \tag{3}$$

We apply p_{mask} to anteroposterior shear stress data via an element-wise product $\odot$ on each channel: $ap_{foot} = p_{mask} \odot ap$ and $ap_{background} = (1 - p_{mask}) \odot ap$.

L1 distance is used in the foot and L2 distance is used in the background area as L1 encourages less blurring. The attention-guided loss function is:

$$L_{attention}\,(G) = \mathbb{E}_{p,t,ap} \left[\| ap_{foot} - G\,(p,t)_{foot} \|_1 \right]$$
$$+ \mathbb{E}_{p,t,ap} \left[\| ap_{background} - G\,(p,t)_{background} \|_2 \right] \tag{4}$$

Finally, our objective is:

$$G^* = \arg \max_G \min_D L_{GAN}(G, D) + \lambda L_{attention}(G) \tag{5}$$

where λ balances the GAN loss and attention-guided loss.

2.2 Positional Attention in Generator

The plantar pressure data and plantar temperature data are collected by different sensors. The pressure and temperature images look quite different even though both of them are collected by the same person. On the other hand, pressure data and anteroposterior shear stress data are collected together, which means they have pixel-to-pixel correspondence. In order to combine the influence of the features from temperature data, we need to zoom and align them into the same shape and direction in traditional method. In fact, it is challenging to realize this pixel by pixel. Here, we provide the positional attention mechanism into the generator of our model to solve the problem without manually aligning them together. The positional attention mechanism can compute which positions in temperature features to pick and update the value in a particular position of pressure data so that it can generate the mapping to anteroposterior shear stress data. We define a positional attention operation [21] in our model as:

$$z_i = \frac{1}{\alpha\,(x, y)} \sum_{\forall j} f\,(x_i, y_j) g\,(y_j) \tag{6}$$

where in Fig. 1, x is the feature from encoder 1, y is the feature from encoder 2 and z is the output data which is the same size as x. i is the same index of an input and output position in latent space and j is the index that enumerates all possible positions. A pairwise function f computes a parameter which represents the relationship between i and all j. The unary function g computes a representation of the input signal at the position j. The response is normalized by a factor $\alpha\,(x, y)$.

The positional attention operation in Eq. (6) is a non-local operation since for a given position i in encoder 1, the operation can compute the parameters from all positions ($\forall j$) and return the positional attention to the encoder 2. As a comparison, the model which concatenates pressure and temperature by channel can only compute the weighted input in a local neighborhood. For example, in a one dimensional case and the kernel size is 3, j should be $i-1 \leq j \leq i+1$. It is problematical since we cannot guarantee local position i and j are from the same foot area. The fact may be even worse that i is from the foot area and j is from the background. Furthermore, the positional attention operation also supports different size of pressure and temperature inputs. On the contrary, the model concatenating pressure and temperature by channel requires pressure and temperature inputs to be same size.

Next, we discuss the choice for functions f and g [21]. We consider g as a linear embedding which is 1×1 convolution operation: $g\left(y_j\right) = W_g y_j$, where W_g is a learned weighted matrix. The pairwise function f can be defined as a dot-product similarity:

$$f\left(x_i, y_j\right) = \theta\left(x_i\right)^T \phi\left(y_j\right) \tag{7}$$

Here $\theta\left(x_i\right) = W_\theta x_i$ and $\phi\left(y_j\right) = W_\phi y_j$ are the convolution operations for input x in position i and input y in position j. For a given i, $\frac{1}{\alpha(x,y)} f\left(x_i, y_j\right)$ is the *softmax* computation [20] along the dimension j. Finally, we can get our positional attention operation as:

$$z_i = softmax\left(x_i^T W_\theta^T W_\phi y\right) g\left(y\right) \tag{8}$$

The choices for functions f and g are flexible. Here we only introduce the version we choose for our model.

3   Data and Experiments

3.1   Data Collection and Annotation

This is a part of an observational case-control study that involves analysis of temperature, shear stresses and physical activity in healthy and both neuropathic and non-neuropathic diabetic subjects. While we had three types of subjects in this pilot study, we treated all subjects as in a single cohort. Total of 31 subjects (23 healthy, 5 diabetes with neuropathy and 3 diabetes without neuropathy) visited the research lab in the morning and signed informed consent. After eligibility determination and initial check of feet-health status, we asked them to make minimum 1,400 steps and return to lab. A wearable activity tracker device (Fitbit Charge HR) was given to each subject to count the steps. In the afternoon, subjects walked on a treadmill for 10 min at self-selected speeds. At the end of walking, plantar temperature distribution of both feet was collected by using an infrared thermal camera (Flir T650C). Then, shear and pressure data during walking were collected with a semi-custom shear device [19]. The device was mounted in a $15' \times 3'$ walkway platform that the top of the shear device was flushed with the surrounding cushioning. A series of 6 barefoot steps of both sides' shear and pressure data were collected one step at a time using 2-step method [13]. Temperature, pressure and anteroposterior shear distribution data from both feet were analyzed. All procedures were approved by local Institutional Review Board.

3.2 Experimental Results and Analysis

The dataset we implement to evaluate our model contains 185 plantar pressure, 185 plantar temperature and 185 anteroposterior shear stress images. The original image size for pressure and anteroposterior shear stress is 146×199 and in order to put it into our models, we pad them into size 256×256. The temperature image is padded from 320×480 to 512×512. Since the ranges of plantar pressure, temperature and anteroposterior shear stress are quite different, we normalize them into $[0, 1]$ range. The min and max values we choose for plantar pressure is 0 kPa and 1100 kPa respectively. In case of plantar temperature data, min and max values are 5 °C and 45 °C respectively. For anteroposterior's case, shear stress's respective values are 0 kPa and 180 kPa.

We split the dataset randomly: 80% for training set and 20% for testing set, and build several models to compare the results. All the models are built with the same discriminator [10,27] which computes the conditional discriminator loss. We choose network ResNet [9] to build the generator. Model T-GAN, P-GAN, PT-GAN and PTL-GAN are using encoders and decoders as their generators. The input for T-GAN is only temperature. The input for P-GAN is only pressure. PT-GAN and PTL-GAN are using both pressure and temperature as inputs. Both of them need to concatenate pressure and temperature input by channel. The differences are as follows: for PT-GAN, the temperature input needs to resize from 512×512 to 256×256, concatenate it with pressure input, and then put them into the encoder; for PTL-GAN, the pressure and temperature inputs will pass their own encoder and merge together between encoder and decoder. Model TPA-GAN and PTA-GAN are using positional attention mechanism in their generators. For PTA-GAN, we put the pressure into encoder 1 in Fig. 1 and temperature into encoder 2, which is computed as the positional attention embedding [21]. TPA-GAN model just exchanges the inputs pressure and temperature and considers pressure as the positional attention embedding.

Table 1. The evaluation for the generated anteroposterior shear stress distribution from non-attention model T-GAN, P-GAN, PT-GAN, PTL-GAN and positional attention model TPA-GAN, PTA-GAN.

Model	GAN-loss	Attention-loss	Accuracy (%)	Global RMSE	Local RMSE
T-GAN	16.00×10^{-2}	17.30×10^{-3}	2.7	66.37	102.92
TPA-GAN	9.42×10^{-2}	20.10×10^{-3}	2.7	53.11	104.55
P-GAN	3.97×10^{-2}	$\mathbf{5.27 \times 10^{-3}}$	59.46	17.74	27.45
PT-GAN	3.42×10^{-2}	5.93×10^{-3}	56.76	25.72	$\mathbf{17.90}$
PTL-GAN	4.04×10^{-2}	5.83×10^{-3}	62.16	20.13	29.63
PTA-GAN	$\mathbf{1.72 \times 10^{-2}}$	5.33×10^{-3}	$\mathbf{72.97}$	$\mathbf{14.12}$	25.34

Table 1 shows the results for our models. GAN-loss is computed by Eq. (2). Attention-Loss is computed by Eq. (4). The learning rate is 0.0001, $\lambda = 10$ for

PTA-GAN and $\lambda = 100$ for other models, the mask threshold $\tau = 0.05$. The Accuracy presents the prediction accuracy for the peak anteroposterior shear stress location. For example, the real location is (a, b) and the generated location is (a', b'). We consider $|a - a'| \leq 1.2\,\mathrm{cm}$ *and* $|b - b'| \leq 1.2\,\mathrm{cm}$ to be correct location based on hard threshold. Multiple thresholds are utilized here. Location threshold is given by the domain expert which is fixed across all approaches. However, for mask threshold varies from approach to approach. In each approach we chose the best threshold based on cross validation. With regard to cross validation and normalized RMSE, we ran new experiments and demonstrated the superiority of our work. In particular, after a 5-fold cross validation and normalized RMSE, our method PTA-GAN (e.g., global rmse 0.089 ± 0.015, accuracy 70.27%) outperforms base lines $(0.095 \pm 0.019, 64.32\%)$.

The global RMSE (kPa) is the root-mean-square deviation which is computed by the peak anteroposterior shear stress for the whole image. The local RMSE (kPa) is computed by the values from the position where the real peak anteroposterior shear stress is.

Our final model PTA-GAN significantly improve the accuracy, reduce the global RMSE and local RMSE simultaneously which is shown in Table 1. It predicts the peak value location with 72.97% accuracy, which is higher than 59.46% in P-GAN, 56.76% in PT-GAN and 62.16% in PTL-GAN. The PTA-GAN can also decrease the root mean square error between real and generated peak anteroposterior shear stress values. The global RMSE and the local RMSE are 14.12 and 25.34 respectively, which is lower than 17.74 and 27.45 in model P-GAN, 20.13 and 29.63 in PTL-GAN.

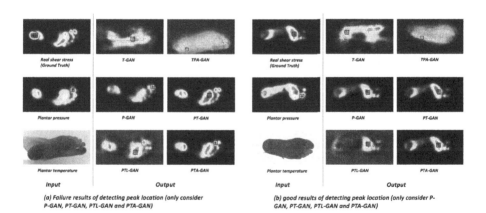

Fig. 2. Visualization of heat maps of padding plantar pressure, temperature, anteroposterior shear stress and generated ones. All the peak values in the heat maps are located in rectangular black boxes. The actual size of the box is $3\,\mathrm{cm} \times 3\,\mathrm{cm}$ in the real world. The inputs of (a) are from right foot of the same person. Inputs of (b) are from left foot of the other person. The plantar pressure and temperature images are flipped since they are collected by different sensors.

The results of the whole dataset can be viewed by heat maps as shown in Fig. 2. The reason why T-GAN and TPA-GAN cannot generate good result is because temperature data and anteroposterior shear stress data are not pixel-to-pixel correspondence. Those two models are mainly translating temperature into shear stress. Due to the small size of our data, the model cannot be well trained. For model P-GAN, PT-GAN, PTL-GAN and PTA-GAN, we can get the distribution of anteroposterior shear stress and the generated ones look quite similar with the ground truth. First, we can consider P-GAN as a baseline among these four models. It only considers translating plantar pressure to anteroposterior shear stress data. Then after merging the influence of temperature features into our model without attention mechanism, we can get PT-GAN and PTL-GAN. Without flipping or aligning temperature inputs, the generated results are unstable. PT-GAN and PTL-GAN cannot improve the performance simultaneously. Finally, we propose our final model PTA-GAN with attention mechanism. Due to the limitation of memory size, we only apply positional attention in the latent space, which is between encoder and decoder. This means the positional attention is the only difference between PTL-GAN and PTA-GAN. Compared with other models, our final model PTA-GAN generates the best result with less noise on it. Considering both evaluation and heat maps, our final model PTA-GAN gives the best performance.

4   Conclusion

Our initial results indicate the feasibility of our approach in prediction of plantar shear stresses. Quantification of plantar shear is quite challenging and there are only a couple of validated and quite expensive systems in the world that can achieve this goal. Development of a software [26] that can accurately predict shear would be very beneficial in the clinical setting. If DFU can be predicted with the help of such a system, physicians can effectively prevent ulcers before they occur. The major limitation of our study was the small sample size available to us. Our initial results are promising and warrant a larger study in order to successfully predict plantar shear stresses.

Acknowledgement. This material is based upon work supported by the National Institutes of Health: R15DK104257 & UL1TR001105; NSF awards DMS-1737978 & MRI-1828467 and FAIN award No 1906630; and an IBM faculty award (Research).

References

1. Armstrong, D.G., Peters, E.J., Athanasiou, K.A., Lavery, L.A.: Is there a critical level of plantar foot pressure to identify patients at risk for neuropathic foot ulceration? J. Foot Ankle Surg. **37**(4), 303–307 (1998)
2. Bloomgarden, Z.T.: The diabetic foot. Diabetes Care **31**(2), 372–376 (2008)
3. Boulton, A.J.: Diabetic foot-what can we learn from leprosy? Legacy of Dr Paul W. Brand. Diabetes Metab. Res. Rev. **28**(Suppl 1), 3–7 (2012)

4. Boulton, A.J.M.: The Diabetic Foot. MDText.com, Inc., South Dartmouth, MA (2000). Last Update: 26 October 2016
5. Breen, C., Khan, L., Ponnusamy, A.: Image classification using neural networks and ontologies. In: Proceedings of the 13th International Workshop on Database and Expert Systems Applications, pp. 98–102. IEEE (2002)
6. Delbridge, L., Ctercteko, G., Fowler, C., Reeve, T.S., Le Quesne, L.P.: The aetiology of diabetic neuropathic ulceration of the foot. Br. J. Surg. **72**(1), 1–6 (1985)
7. Dinsdale, S.M.: Decubitus ulcers: role of pressure and friction in causation. Arch. Phys. Med. Rehabil. **55**(4), 147–152 (1974)
8. Goodfellow, I., et al.: Generative adversarial nets. In: Advances in Neural Information Processing Systems, pp. 2672–2680 (2014)
9. He, K., Zhang, X., Ren, S., Sun, J.: Deep residual learning for image recognition. In: Proceedings of the IEEE Conference on Computer Vision and Pattern Recognition, pp. 770–778 (2016)
10. Isola, P., Zhu, J.Y., Zhou, T., Efros, A.A.: Image-to-image translation with conditional adversarial networks. In: IEEE Conference on Computer Vision and Pattern Recognition (2017)
11. Lavery, L.A., Armstrong, D.G., Wunderlich, R.P., Tredwell, J., Boulton, A.J.: Predictive value of foot pressure assessment as part of a population-based diabetes disease management program. Diabetes Care **26**(4), 1069–1073 (2003)
12. Ledoux, W.R., Shofer, J.B., Cowley, M.S., Ahroni, J.H., Cohen, V., Boyko, E.J.: Diabetic foot ulcer incidence in relation to plantar pressure magnitude and measurement location. J. Diabetes Complicat. **27**(6), 621–626 (2013)
13. McPoil, T.G., Cornwall, M.W., Dupuis, L., Cornwell, M.: Variability of plantar pressure data. A comparison of the two-step and midgait methods. J. Am. Podiatr. Med. Assoc. **89**(10), 495–501 (1999)
14. Mejjati, Y.A., Richardt, C., Tompkin, J., Cosker, D., Kim, K.I.: Unsupervised attention-guided image-to-image translation. In: Advances in Neural Information Processing Systems, pp. 3693–3703 (2018)
15. Mirza, M., Osindero, S.: Conditional generative adversarial nets. arXiv preprint arXiv:1411.1784 (2014)
16. NPIAP (2016). http://www.npuap.org/national-pressure-ulcer-advisory-panel-npuap-announces-a-change-in-terminology-from-pressure-ulcer-to-pressure-injury-and-updates-the-stages-of-pressure-injury/
17. Pathak, D., Krahenbuhl, P., Donahue, J., Darrell, T., Efros, A.A.: Context encoders: feature learning by inpainting. In: Proceedings of the IEEE Conference on Computer Vision and Pattern Recognition, pp. 2536–2544 (2016)
18. Rogers, L.C., Lavery, L.A., Armstrong, D.G.: The right to bear legs-an amendment to healthcare: how preventing amputations can save billions for the us health-care system. J. Am. Podiatr. Med. Assoc. **98**(2), 166–168 (2008)
19. Stucke, S., et al.: Spatial relationships between shearing stresses and pressure on the plantar skin surface during gait. J. Biomech. **45**(3), 619–622 (2012)
20. Vaswani, A., et al.: Attention is all you need. In: Advances in Neural Information Processing Systems, pp. 5998–6008 (2017)
21. Wang, X., Girshick, R., Gupta, A., He, K.: Non-local neural networks. In: Proceedings of the IEEE Conference on Computer Vision and Pattern Recognition, pp. 7794–7803 (2018)
22. Yavuz, M.: American society of biomechanics clinical biomechanics award 2012: plantar shear stress distributions in diabetic patients with and without neuropathy. Clin. Biomech. (Bristol, Avon) **29**(2), 223–229 (2014)

23. Yavuz, M., et al.: Plantar shear stress in individuals with a history of diabetic foot ulcer: an emerging predictive marker for foot ulceration. Diabetes Care **40**(2), e14–e15 (2017)
24. Yavuz, M., Tajaddini, A., Botek, G., Davis, B.L.: Temporal characteristics of plantar shear distribution: relevance to diabetic patients. J. Biomech. **41**(3), 556–559 (2008)
25. Yavuz, M., Ocak, H., Hetherington, V.J., Davis, B.L.: Prediction of plantar shear stress distribution by artificial intelligence methods. J. Biomech. Eng. **131**(9), 091007 (2009)
26. Yen, I.L., Goluguri, J., Bastani, F., Khan, L., Linn, J.: A component-based approach for embedded software development. In: Proceedings Fifth IEEE International Symposium on Object-Oriented Real-Time Distributed Computing, ISIRC 2002, pp. 402–410. IEEE (2002)
27. Zhu, J.Y., Park, T., Isola, P., Efros, A.A.: Unpaired image-to-image translation using cycle-consistent adversarial networks. In: 2017 IEEE International Conference on Computer Vision (ICCV) (2017)

Correction to: Acceleration of High-Resolution 3D MR Fingerprinting via a Graph Convolutional Network

Feng Cheng, Yong Chen, Xiaopeng Zong, Weili Lin, Dinggang Shen, and Pew-Thian Yap

Correction to:
Chapter "Acceleration of High-Resolution 3D MR
Fingerprinting via a Graph Convolutional Network"
in: A. L. Martel et al. (Eds.): *Medical Image Computing*
***and Computer Assisted Intervention – MICCAI 2020*,**
LNCS 12262, https://doi.org/10.1007/978-3-030-59713-9_16

The original version of this chapter was revised. The NIH grant number has been corrected to EB006733 and typographical errors have been corrected.

The updated version of this chapter can be found at
https://doi.org/10.1007/978-3-030-59713-9_16

Author Index

Printed in the United States
by Baker & Taylor Publisher Services